W9-BJS-825

AN INTRODUCTION TO THE FINITE ELEMENT METHOD

McGraw-Hill Series in Mechanical Engineering

Consulting Editors
***Jack P. Holman**, Southern Methodist University*
***John R. Lloyd**, Michigan State University*

Anderson: *Modern Compressible Flow: With Historical Perspective*
Arora: *Introduction to Optimum Design*
Bray and Stanley: *Nondestructive Evaluation: A Tool for Design, Manufacturing, and Service*
Culp: *Principles of Energy Conversion*
Dally: *Packaging of Electronic Systems: A Mechanical Engineering Approach*
Dieter: *Engineering Design: A Materials and Processing Approach*
Eckert and Drake: *Analysis of Heat and Mass Transfer*
Edwards and McKee: *Fundamentals of Mechanical Component Design*
Gebhart: *Heat Conduction and Mass Transfer*
Heywood: *Internal Combustion Engine Fundamentals*
Hinze: *Turbulence*
Howell and Buckius: *Fundamentals of Engineering Thermodynamics*
Hutton: *Applied Mechanical Vibrations*
Juvinall: *Engineering Considerations of Stress, Strain, and Strength*
Kane and Levinson: *Dynamics: Theory and Applications*
Kays and Crawford: *Convective Heat and Mass Transfer*
Kimbrell: *Kinematics Analysis and Synthesis*
Martin: *Kinematics and Dynamics of Machines*
Modest: *Radiative Heat Transfer*
Norton: *Design of Machinery*
Phelan: *Fundamentals of Mechanical Design*
Raven: *Automatic Control Engineering*
Reddy: *An Introduction to the Finite Element Method*
Rosenberg and Karnopp: *Introduction to Physics*
Schlichting: *Boundary-Layer Theory*
Shames: *Mechanics of Fluids*
Sherman: *Viscous Flow*
Shigley: *Kinematic Analysis of Mechanisms*
Shigley and Mischke: *Mechanical Engineering Design*
Shigley and Uicker: *Theory of Machines and Mechanisms*
Stiffler: *Design with Microprocessors for Mechanical Engineers*
Stoecker and Jones: *Refrigeration and Air Conditioning*
Ullman: *The Mechanical Design Process*
Vanderplaats: *Numerical Optimization: Techniques for Engineering Design, with Applications*
White: *Viscous Fluid Flow*
Zeid: *CAD/CAM Theory and Practice*

AN INTRODUCTION TO THE FINITE ELEMENT METHOD

Second Edition

J. N. Reddy

Oscar S. Wyatt Chair in Mechanical Engineering
Texas A&M University
College Station, Texas 77843

Boston, Massachusetts Burr Ridge, Illinois
Dubuque, Iowa Madison, Wisconsin New York, New York
San Francisco, California St. Louis, Missouri

This book was set in Times Roman.
The editors were John J. Corrigan and John M. Morriss;
the production supervisor was Louise Karam.
The cover was designed by Joseph Gillians.
Project supervision was done by The Universities Press.

McGraw-Hill

*A Division of The **McGraw-Hill** Companies*

This book is printed on acid-free paper.

AN INTRODUCTION TO THE FINITE ELEMENT METHOD

11 12 13 14 BKM BKM 0 9 8 7 6 5 4 3

ISBN 0-07-051355-4

Library of Congress Cataloging-in-Publication Data

Reddy, J. N. (Junuthula N.)
An introduction to the finite element method/J. N. Reddy.—2nd ed.
p. cm.—(McGraw-Hill series in mechanical engineering)
Includes bibliographical references and index.
ISBN 0-07-051355-4
1. Finite element method. I. Series.
TA347.F5R4 1993
620′.001′51535—dc20 92-29532

ABOUT THE AUTHOR

J. N. Reddy is the inaugural holder of the *Oscar S. Wyatt, Jr, Endowed Chair* in Mechanical Engineering at Texas A&M University. Prior to the current appointment, he was the *Clifton C. Garvin Chaired Professor* of Engineering Science and Mechanics at Virginia Polytechnic Institute and State University. After receiving his Ph.D. in Applied Mechanics in 1974, he joined Lockheed Missiles and Space Company in Huntsville, Alabama, as a research scientist. In 1975, he joined the School of Aerospace, Mechanical and Nuclear Engineering at the University of Oklahoma. In 1980, he was appointed as a professor of Engineering Science and Mechanics at Virginia Polytechnic Institute and State University.

Dr. Reddy has authored and coauthored over 150 journal papers and seven books on the theory and applications of the finite element method in solid and structural mechanics, fluid mechanics and heat transfer, and on the development of refined theories and finite element models of laminated composite plates and shells.

Dr. Reddy has taught many short courses on the finite element method to industry and government, and has received Certificates of Teaching Excellence from Virginia Polytechnic Institute and State University.

Dr. Reddy serves on the editorial boards of a dozen journals, including *Journal of Applied Mechanics, Meccanica, International Journal for Numerical Methods in Engineering, International Journal of Numerical Methods in Fluids, Journal of Non-Linear Mechanics,* and *Journal of Composites Technology and Research.* He has received the College of Engineering Research Award from the University of Oklahoma and the Alumni Research Award from the Virginia Polytechnic Institute and State University. Dr. Reddy was awarded the *1983 Walter L. Huber Research Prize* of the American Society of Civil Engineers and *1992 Worcester Reed Warner Medal* of the American Society of Mechanical Engineers. He is a Fellow of the American Academy of Mechanics, the American Society of Mechanical Engineers, the American Society of Civil Engineers, and the Aeronautical Society of India.

To my family
Aruna, Anita, and Anil

CONTENTS

Part 3 Finite Element Analysis of Two-Dimensional Problems

PREFACE TO THE SECOND EDITION

This second edition has the same objectives as the first, namely, an introduction to the finite element method as applied to linear, one- and two-dimensional problems of engineering and applied sciences. The revisions are mainly in the form of additional details, expansion of the topics discussed, and the addition of a few topics to make the coverage more complete.

The major organizational change from the first edition is the division of its five chapters into fourteen chapters here. These chapters are grouped into four parts. This reorganization should aid instructors in selecting suitable material for courses. Other organizational changes include putting problem sets at the ends of the chapters, providing a chapter summary for each, and reviewing pertinent equations and text in each chapter instead of referring to several chapters back. In addition, example problems in Chapters 3 and 8 are presented in separate sections on heat transfer, fluid flow, and solid mechanics.

Additional details are provided on the construction of the weak forms, time approximations (e.g., accuracy and stability of schemes, and mass lumping), alternative finite element formulations, and nonlinear finite element models. The new topics include sections on trusses and frames, the Timoshenko beam element, eigenvalue problems, and classical plate bending elements. All these changes are also reflected in the revised computer programs FEM1DV2 and FEM2DV2 (revised versions of the FEM1D, FEM2D and PLATE programs in the first edition). Therefore the sections on computer implementation and applications of FEM1DV2 and FEM2DV2 have also been modified extensively. These changes are accompanied by the addition of several figures, tables, and examples.

These extensive changes have resulted in a second edition that is 60% larger. In the interest of keeping the cost of the book within reasonable limits while retaining the basic approach and technical details, certain portions of the original manuscript have been omitted. More specifically, answers to selective problems have been included at the end of the problem statements themselves, rather than in a separate section. Interested readers and instructors can obtain a copy of the excutable programs on a diskette from the author. Fortran source programs can also be purchased from the author.

There is no doubt that this edition is more complete and thorough than the first. It can be used as a textbook for an introductory and/or intermediate level course on the finite element method at senior undergraduate as well as graduate levels. Students of engineering and applied sciences should feel comfortable with the coverage in the book.

The author gratefully acknowledges help in reading the manuscript and suggestions for constructive changes from several colleagues. These include: Hasan Akay, Purdue University at Indianapolis, Norman Knight, Jr, Clemson University; J. K. Lee, Ohio State University; William Rule, University of Alabama; Martin Sadd, University of Rhode Island; John Whitcomb, Texas A&M University, and the author's research students: Ronald Averill, Filis Kokkinos, Y. S. N. Reddy, and Donald Robbins. It is a great pleasure to acknowledge typing of the manuscript by Mrs Vanessa McCoy (Virginia Tech), without whose patience and cooperation this work would not have been completed. The author is grateful to the colleagues in the Department of Engineering Science and Mechanics of Virginia Polytechnic Institute and State University for their support and friendship throughout the author's tenure there.

J. N. Reddy

PREFACE TO THE FIRST EDITION

The motivation which led to the writing of the present book has come from my many years of teaching finite-element courses to students from various fields of engineering, meteorology, geology and geophysics, physics, and mathematics. The experience gained as a supervisor and consultant to students and colleagues in universities and industry, who have asked for explanations of the various mathematical concepts related to the finite-element method, helped me introduce the method as a variationally based technique of solving differential equations that arise in various fields of science and engineering. The many discussions I have had with students who had no background in solids and structural mechanics gave rise to my writing a book that should fill the rather unfortunate gap in the literature.

The book is designed for senior undergraduate and first-year graduate students who have had a course in linear algebra as well as in differential equations. However, additional courses (or exposure to the topics covered) in mechanics of materials, fluid flow, and heat transfer should make the student feel more comfortable with the physical examples discussed in the book.

In the present book, the finite-element method is introduced as a variationally based technique of solving differential equations. A continuous problem described by a differential equation is put into an equivalent variational form, and the approximate solution is assumed to be a linear combination, $\sum c_j\phi_j$, of approximation functions ϕ_j. The parameters c_j are determined using the associated variational form. The finite-element method provides a systematic technique for deriving the approximation functions for simple subregions by which a geometrically complex region can be represented. In the finite-element method, the approximation functions are piecewise polynomials (i.e., polynomials that are defined only on a subregion, called an element).

The approach taken in the present book falls somewhere in the middle of the approaches taken in books that are completely mathematical and those approaches that are more structural-mechanics-oriented. From my own experience as an engineer and self-taught applied mathematician, I know how unfortunate outcomes may be arrived at if one follows a "formula" without deeper insight into the problem and its approximation. Even the best theories lead ultimately to some sort of guidelines (e.g., which variational formulation is suitable, what kind of element is desirable, what is the quality of the approximation, etc.). However, without a certain theoretical knowledge of variational methods one cannot fully understand various formulations, finite-element models, and their limitations.

In the present study of variational and finite-element methods, advanced mathematics are intentionally avoided in the interest of simplicity. However, a minimum of mathematical machinery that seemed necessary is included in Chapters 1 and 2. In Chapter 2, considerable attention is devoted to the construction of variational forms since this exercise is repeatedly encountered in the finite-element formulation of differential equations. The chapter is concerned with two aspects: first, the selection of the approximation functions that meet the specified boundary condtions; second, the technique of obtaining algebraic equations in terms of the undetermined parameters. Thus, Chapter 2 not only equips readers with certain concepts and tools that are needed in Chapters 3 and 4, but it also motivates them to consider systematic methods of constructing the approximation functions, which is the main feature of the finite-element method.

In introducing the finite element method in Chapters 3 and 4, the traditional solid mechanics approach is avoided in favor of the "differential equation" approach, which has broader interpretations than a single special case. However, when specific examples are considered, the physical background of the problem is stated. Since a large number of physical problems are described by second- and fourth-order ordinary differential equations (Chapter 3), and by the Laplace operator in two dimensions (Chapter 4), considerable attention is devoted to the finite-element formulation, the derivation of the interpolation functions, and the solution of problems described by these equations. Representative examples are drawn from various fields of engineering, especially from heat transfer, fluid mechanics, and solid mechanics. Since this book is intended to serve as a textbook for a first course on the finite-element method, advanced topics such as nonlinear problems, shells, and three-dimensional analyses are omitted.

Since the practice of the finite-element method ultimately depends on one's ability to implement the technique on a digital computer, examples and exercises are designed to let the reader actually compute the solutions of various problems using computers. Ample discussion of the computer implementation of the finite-element method is given in Chapters 3 and 4. Three model programs (FEM1D, FEM2D, and PLATE) are described, and their application is illustrated via several examples. The computer programs are very easy to understand because they are designed along the same lines as the theory presented in the book. The programs are available for mainframe and IBM PC compatibles from the author for a small charge.

Numerous examples, most of which are applications of the concepts to specific problems in various fields of engineering and applied science, are provided throughout the book. The conclusion of the examples are indicated by the symbol ■. At approprate intervals in the book an extensive number of exercise problems is included to test and extend the understanding of the concepts discussed. For those who wish to gain additional knowledge of the topics covered in the book, many reference books and research papers are listed at the end of each chapter.

There are several sections that can be skipped in a first reading of the book (such sections are marked with an asterisk); these can be filled in wherever needed later. The material is intended for a quarter or a semester course, although it is better suited for a semester course.

The following schedule of topics is suggested for a first course using the present textbook:

Undergraduate		Graduate	
Chapter 1	Self-study	Chapter 1	Self-study
Chapter 2	Section 2.1 (self) Section 2.2 Sections 2.3.1–2.3.3	Chapter 2	Section 2.1 (self) Section 2.2 Section 2.3
Chapter 3	Sections 3.1–3.4 Sections 3.6–3.7	Chapter 3	Sections 3.1–3.7
Chapter 4	Sections 4.1–4.4 Section 4.7 Sections 4.8.1–4.8.4	Chapter 4 Chapter 5	Sections 4.1–4.8 Term Paper

Due to the intimate relationship between Sections 3.5 and 4.6, 3.6 and 4.7, and 3.7 and 4.8, they can be covered simultaneously. Also, it is suggested that Sections 3.6 and 3.7 (hence, 4.7 and 4.8) be covered after Section 3.2.

The author wishes to thank all those students and colleagues who have contributed by their advice and criticism to the improvement of this work. The author is also thankful to Vanessa McCoy for skillful typing of the manuscript, to Mr. N. S. Putcha and Mr. K. Chandrashekhara for proofreading the pages, and to the editors Michael Slaughter and Susan Hazlett for their help and cooperation in publishing the mauscript.

J. N. Reddy

Tejashwina vadheetamasthu
(May what we study be well studied)

PART 1

PRELIMINARIES

CHAPTER 1

INTRODUCTION

1.1 GENERAL COMMENTS

Virtually every phenomenon in nature, whether biological, geological, or mechanical, can be described with the aid of the laws of physics, in terms of algebraic, differential, or integral equations relating various quantities of interest. Determining the stress distribution in a pressure vessel with oddly shaped holes and numerous stiffeners and subjected to mechanical, thermal, and/or aerodynamic loads, finding the concentration of pollutants in seawater or in the atmosphere, and simulating weather in an attempt to understand and predict the mechanics of formation of tornadoes and thunderstorms are a few examples of many important practical problems.

Most engineers and scientists studying physical phenomena are involved with two major tasks:

1. Mathematical formulation of the physical process
2. Numerical analysis of the mathematical model

The mathematical formulation of a physical process requires background in related subjects (e.g., laws of physics) and, most often, certain mathematical tools. The formulation results in mathematical statements, often differential equations, relating quantities of interest in the understanding and/or design of the physical process. Development of the mathematical model of a process is achieved through assumptions concerning how the process works. In a numerical simulation, we use a numerical method and a computer to evaluate the mathematical model and estimate the characteristics of the process.

While the derivation of the governing equations for most problems is not unduly difficult, their solution by exact methods of analysis is a formidable task. In such cases, approximate methods of analysis provide alternative means of finding solutions. Among these, the finite difference method and the variational methods such as the Rayleigh–Ritz and Galerkin methods are most frequently used in the literature.

In the finite difference approximation of a differential equation, the derivatives in the latter are replaced by difference quotients (or the function is expanded in a Taylor series) that involve the values of the solution at discrete mesh points of the domain. The resulting algebraic equations are solved, after imposing the boundary conditions, for the values of the solution at the mesh points.

In the solution of a differential equation by a variational method, the equation is put into an equivalent weighted-integral form and then the approximate solution over the domain is assumed to be a linear combination ($\sum_j c_j\phi_j$) of appropriately chosen approximation functions ϕ_j and undetermined coefficients, c_j. The coefficients c_j are determined such that the integral statement equivalent to the original differential equation is satisfied. Various variational methods, e.g., the Rayleigh–Ritz, Galerkin, and least-squares methods, differ from each other in the choice of the integral form, weight functions, and/or approximation functions. A more complete discussion of variational methods will be given in Chapter 2. They suffer from the disadvantage that the approximation functions for problems with arbitrary domains are difficult to construct.

The finite element method overcomes the disadvantage of the traditional variational methods by providing a systematic procedure for the derivation of the approximation functions over subregions of the domain. The method is endowed with three basic features that account for its superiority over other competing methods. First, a geometrically complex domain of the problem is represented as a collection of geometrically simple subdomains, called *finite elements*. Second, over each finite element, the approximation functions are derived using the basic idea that any continuous function can be represented by a linear combination of algebraic polynomials. Third, algebraic relations among the undetermined coefficients (i.e., nodal values) are obtained by satisfying the governing equations, often in a weighted-integral sense, over each element. Thus, the finite element method can be viewed, in particular, as an element-wise application of the Rayleigh–Ritz or weighted-residual methods. In it, the approximation functions are often taken to be algebraic polynomials, and the undetermined parameters represent the values of the solution at a finite number of preselected points, called *nodes*, on the boundary and in the interior of the element. The approximation functions are derived using concepts from interpolation theory, and are therefore called *interpolation functions*. One finds that the degree of the interpolation functions depends on the number of nodes in the element and the order of the differential equation being solved.

1.2 HISTORICAL BACKGROUND

The idea of representing a given domain as a collection of discrete parts is not unique to the finite element method. It was recorded that ancient mathematicians estimated the value of π by noting that the perimeter of a polygon inscribed in a circle approximates the circumference of the latter. They predicted the value of π to accuracies of almost 40 significant digits by representing the circle as a polygon of a finitely large number of sides. In modern times, the idea found a home in aircraft structural analysis, where, for example, wings and fuselages are treated as assemblages of stringers, skins, and shear panels. In 1941, Hrenikoff introduced the so-called framework method, in which a plane elastic medium was represented as a collection of bars and beams. The use of piecewise-continuous functions defined over a subdomain to approximate an unknown function can be found in the work of Courant (1943), who used an assemblage of triangular elements and the principle of minimum total potential energy to study the St Venant torsion problem. Although certain key features of the finite element method can be found in the works of Hrenikoff (1941) and Courant (1943), its formal presentation is attributed to Argyris and Kelsey (1960) and Turner, Clough, Martin, and Topp (1956). The term "finite element" was first used by Clough in 1960. Since its inception, the literature on finite element applications has grown exponentially, and today there are numerous journals that are primarily devoted to the theory and application of the method. A review of the historical developments and the basic theory of the finite element method can be found in more than three dozen textbooks that are exclusively devoted to its introduction and application. The selective finite element books listed in References for Additional Reading at the end of this chapter are only for additional information on certain topics (e.g., three-dimensional problems, shells, structural dynamics, plasticity, and mathematics of finite elements). For the beginner, it is not necessary to consult these; the present book provides complete details of the method as applied to linear field problems, with examples from fluid mechanics, heat transfer, and solid mechanics.

1.3 THE BASIC CONCEPT OF THE FINITE ELEMENT METHOD

1.3.1 General Comments

The most distinctive feature of the finite element method that separates it from others is the division of a given domain into a set of simple subdomains, called *finite elements*. Any geometric shape that allows computation of the solution or its approximation, or provides necessary relations among the values of the solution at selected points, called *nodes*, of the subdomain, qualifies as a finite element. Other features of the method include seeking continuous, often polynomial, approximations of the solution over each element in terms of nodal values, and assembly of element equations by imposing the interelement continuity of the solution and balance of interelement forces. Here the basic ideas underlying the finite element method are introduced via two simple examples:

1. Determination of the circumference of a circle using a finite number of line segments
2. Determination of the center of mass (or gravity) of an irregular body

The first example is an expansion of an article written by the author in 1978 for a student magazine at the University of Oklahoma. Ideas expressed in the second can be found in books on statics of rigid bodies.

1.3.2 Approximation of the Circumference of a Circle

Consider the problem of determining the perimeter of a circle of radius R (see Fig. 1.1a). Ancient mathematicians estimated the value of the circumference

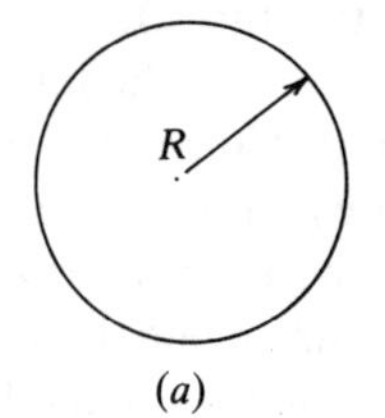

(*a*)

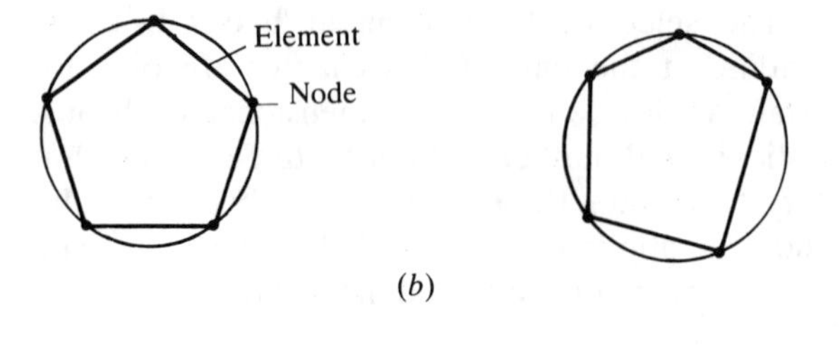

(*b*)

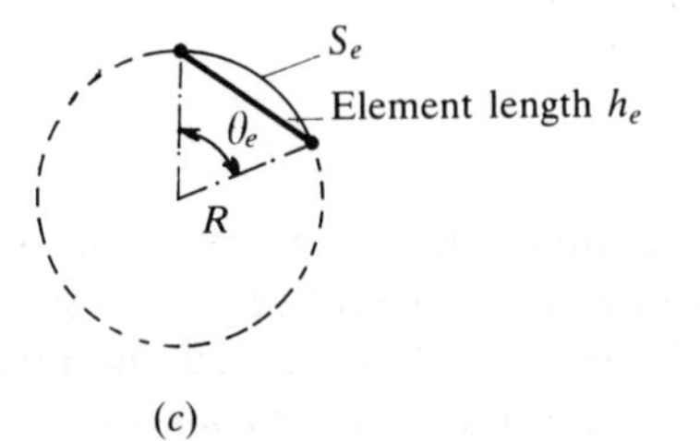

(*c*)

FIGURE 1.1
Approximation of the circumference of a circle by line elements: (a) Circle of radius R; (b) uniform and nonuniform meshes used to represent the circumference of the circle; (c) a typical element.

by approximating it by line segments, whose lengths they were able to measure. The approximate value of the circumference is obtained by summing the lengths of the line segments used to represent it. Although this is a trivial example, it illustrates several (but not all) ideas and steps involved in the finite element analysis of a problem. We outline the steps involved in computing an approximate value of the circumference of the circle. In doing so, we introduce certain terms that are used in the finite element analysis of any problem.

1. *Finite element discretization.* First, the domain (i.e., the circumference of the circle) is represented as a collection of a finite number n of subdomains, namely, line segments. This is called *discretization of the domain.* Each subdomain (i.e., line segment) is called an *element.* The collection of elements is called the *finite element mesh.* The elements are connected to each other at points called *nodes*. In the present case, we discretize the circumference into a mesh of five ($n = 5$) line segments. The line segments can be of different lengths. When all elements (i.e., line segments) are of the same length, the mesh is said to be *uniform*; otherwise, it is called a *nonuniform* mesh (see Fig. 1.1*b*).
2. *Element equations.* A typical element (i.e., line segment, Ω^e) is isolated and its required properties, i.e., length, are computed by some appropriate means. Let h_e be the length of element Ω^e in the mesh. For a typical element Ω^e, h_e is given by (see Fig. 1.1*c*)
$$h_e = 2R \sin \tfrac{1}{2}\theta_e \tag{1.1}$$
where R is the radius of the circle and $\theta_e < \pi$ is the angle subtended by the line segment. The above equations are called *element equations.* Ancient mathematicians most likely made measurements, rather than using (1.1), to find h_e.
3. *Assembly of element equations and solution.* The approximate value of the circumference (or perimeter) of the circle is obtained by putting together the element properties in a meaningful way; this process is called the *assembly* of the element equations. It is based, in the present case, on the simple idea that the total perimeter of the polygon (assembled elements) is equal to the sum of the lengths of individual elements:
$$P_n = \sum_{e=1}^{n} h_e \tag{1.2}$$
Then P_n represents an approximation to the actual perimeter, p. If the mesh is uniform, or h_e is the same for each of the elements in the mesh, then $\theta_e = 2\pi/n$, and we have
$$P_n = n\left(2R \sin \frac{\pi}{n}\right) \tag{1.3}$$
4. *Convergence and error estimate.* For this simple problem, we know the exact solution: $p = 2\pi R$. We can estimate the error in the approximation and show that the approximate solution P_n converges to the exact p in the

limit as $n \to \infty$. Consider the typical element Ω^e. The error in the approximation is equal to the difference between the length of the sector and that of the line segment (see Fig. 1.1*c*):

$$E_e = |S_e - h_e| \tag{1.4}$$

where $S_e = R\theta_e$ is the length of the sector. Thus, the error estimate for an element in the mesh is given by

$$E_e = R\left(\frac{2\pi}{n} - 2\sin\frac{\pi}{n}\right) \tag{1.5}$$

The total error (called *global* error) is given by multiplying E_e by n:

$$E = 2R\left(\pi - n\sin\frac{\pi}{n}\right) = 2\pi R - P_n \tag{1.6}$$

We now show that E goes to zero as $n \to \infty$. Letting $x = 1/n$, we have

$$P_n = 2Rn\sin\frac{\pi}{n} = 2R\frac{\sin\pi x}{x}$$

and

$$\lim_{n\to\infty} P_n = \lim_{x\to 0}\left(2R\frac{\sin\pi x}{x}\right) = \lim_{x\to 0}\left(2\pi R\frac{\cos\pi x}{1}\right) = 2\pi R \tag{1.7}$$

Hence, E_n goes to zero as $n \to \infty$. This completes the proof of convergence.

In summary, it is shown that the circumference of a circle can be approximated as closely as we wish by a finite number of piecewise-linear functions. As the number of elements is increased, the approximation improves, i.e., the error in the approximation decreases.

1.3.3 Approximate Determination of the Center of Mass

Another elementary example to illustrate the finite element concept is provided by the calculation of the center of mass of a continuous body. It should be recalled, from a first course on statics of rigid bodies, that the calculation of the center of an irregular mass or the centroid of an irregular volume makes use of the so-called method of composite bodies, in which a body is conveniently divided (mesh discretization) into several parts (elements) of simple shape for which the mass and the center of mass (element properties) can be computed readily. The center of mass of the whole body is then obtained using the moment *principle of Varignon* (a basis for the assembly of element properties):

$$(m_1 + m_2 + \ldots + m_n)\bar{X} = m_1\bar{x}_1 + m_2\bar{x}_2 + \ldots + m_n\bar{x}_n \tag{1.8}$$

where $\bar{X}$ is the x coordinate of the center of mass of the whole body, m_e is the mass of the eth part, and $\bar{x}_e$ is the x coordinate of the center of mass of the eth part. Similar expressions hold for the y and z coordinates of the center of mass of the whole body. Analogous relations hold for composite lines, areas, and volumes, wherein the masses are replaced by lengths, areas, and volumes, respectively.

When a given body is not expressible in terms of simple geometric shapes (elements) for which the mass and the center of mass can be represented mathematically, it is necessary to use a method of approximation to represent the properties of an element. As an example, consider the problem of finding the centroid $(\bar{X}, \bar{Y})$ of the irregular area (region) shown in Fig. 1.2. The region can be divided into a finite number of rectangular strips (elements), a typical element having width h_e and height b_e. The area of the eth strip is given by $A_e = h_e b_e$. The area A_e is an approximation of the true area of the element because b_e is an estimated average height of the element. The coordinates of the centroid of the region are obtained by applying the moment principle:

$$\bar{X} = \frac{\sum_e A_e \bar{x}_e}{\sum_e A_e}, \qquad \bar{Y} = \frac{\sum_e A_e \bar{y}_e}{\sum_e A_e}$$

where $\bar{x}_e$ and $\bar{y}_e$ are the coordinates of the centroid of the eth element with respect to the coordinate system used for the whole body. When the center of mass is required, A_e in the above equations is replaced by the mass $m_e = \rho_e A_e$, ρ_e being the mass density of the eth element; for a homogeneous body, ρ_e is the same for all elements.

It should be noted that the accuracy of the approximation will be improved by increasing the number of strips (decreasing their width) used. Rectangular elements are used in the present discussion for the sake of simplicity only; one may choose to use elements of any size and shape that approximate the given area to a satisfactory accuracy. For example, a trapezoidal element will require two heights to compute the area:

$$A_e = \tfrac{1}{2} h_e (b_e + b_{e+1})$$

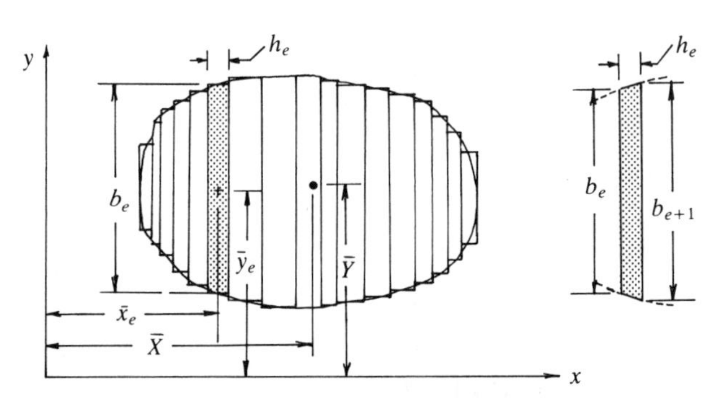

FIGURE 1.2
Approximate determination of the mass or geometric centroid of an irregular region by dividing it into a set of rectangular or trapezoidal subregions.

where b_e and b_{e+1} are the left and right heights, respectively, of the eth element.

The two examples considered above illustrate how the idea of piecewise approximation is used to approximate irregular geometries and calculate required quantities. In the first example, the circumference of a circle is approximated by a collection of line segments, whose measure is available. In the second, the geometric centroid or mass centroid of an irregular domain is located by approximating its geometry as a collection of strips that allow computation of their areas. Rectangles and trapezoids provide examples of the element geometries. Thus, subdividing a geometrically complex domain into parts that allow the evaluation of desired quantities is a very natural and practical approach. The idea can be extended to approximate functions representing physical quantities. For example, the temperature variation in a two-dimensional domain can be viewed as a curved surface, and it can be approximated over any part of the domain, i.e., over a subdomain or element, by a function of desired degree. Figure 1.3 shows a curved surface over a triangular subregion approximated by a planar surface, i.e., a linear polynomial. Such ideas form the basis of finite element approximations. The next example illustrates this idea for a one-dimensional continuous system.

1.3.4 Solution of Differential Equation

Consider the temperature variation in a composite cylinder consisting of two coaxial layers in perfect thermal contact (see Fig. 1.4). Heat dissipation from a wire (with two insulations) carrying an electric current and heat flow across a thick-walled composite circular cylindrical tube are typical examples. The temperature T is a function of the radial coordinate r. The variation of T with r is, in general, nonuniform. We wish to determine an approximation $T_e(r)$ to $T(r)$ over the thicknesses of the cylinder. The exact solution is determined by solving the dIfferential equation

$$-\frac{1}{r}\frac{d}{dr}\left(rk\frac{dT}{dr}\right) = q(r) \tag{1.9a}$$

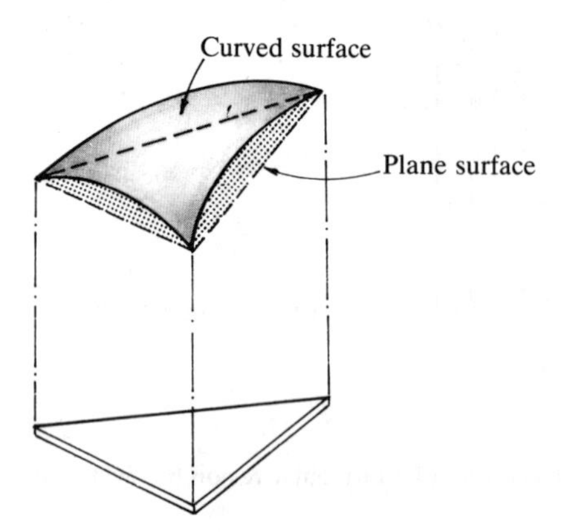

FIGURE 1.3
Approximation of curved surface (or a nonuniform function) over a triangular region by a planar surface.

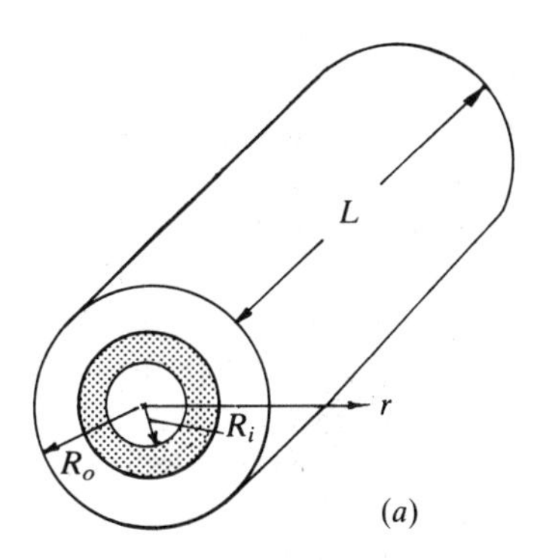

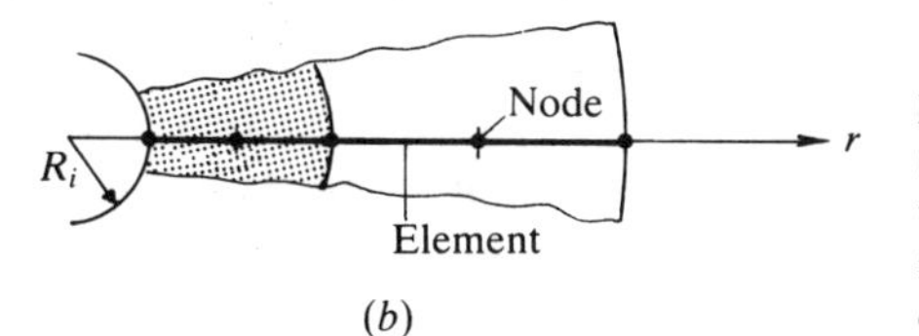

FIGURE 1.4
(*a*) Coaxial (composite) cylinder made of two different materials. (*b*) Finite element representation of a radial line of the cylinder.

subject to appropriate boundary conditions, for example, insulated at $r = R_i$ and subjected to a temperature T_o at $r = R_o$:

$$kr\frac{dT}{dr} = 0 \quad \text{at } r = R_i; \qquad T(r) = T_o \quad \text{at } r = R_o \tag{1.9b}$$

where k is the thermal conductivity, which varies from layer to layer, R_i and R_o are the inner and outer radii of the cylinder, and q is the rate of energy generation in the medium. Note that the temperature is independent of the circumferential coordinate (because of the axisymmetric geometry, boundary conditions, and loading), and it has the same variation along any radial line. When it is difficult to obtain an exact solution of the problem (1.9), either because of complex geometry and material properties or because $q(r)$ is a complicated function that does not allow exact evaluation of its integral, we seek an approximate one. In the finite element method, the domain (R_i, R_o) is divided into N subintervals, and the approximate solution is sought in the form

$$\begin{aligned}
T_1(r) &= \sum_{j=1}^{n} T_j^1 \psi_j^1(r) \quad (R_i \le r \le R_i + h_1;\ \text{first interval}) \\
T_2(r) &= \sum_{j=1}^{n} T_j^2 \psi_j^2(r) \quad (R_i + h_1 \le r \le R_i + h_1 + h_2;\ \text{second interval}) \qquad (1.10) \\
&\ \vdots \\
T_N(r) &= \sum_{j=1}^{n} T_j^N \psi_j^N(r) \quad (R_i + h_1 + \cdots + h_{N-1} \le r \le R_o;\ N\text{th interval})
\end{aligned}$$

where h_e denotes the length of the eth interval, T_j^e is the value of the temperature $T_e(r)$ at the jth geometric point of the eth interval, and ψ_j^e are polynomials on the eth interval. The continuous function $T(r)$ is approximated in each interval by a desired degree of polynomial, and the polynomial is expressed in terms of the values of the function at a selected number of points in the interval. The number of points is equal to the number of parameters in the polynomial. For example, a linear polynomial approximation of the temperature over the interval requires two values, and hence two points are identified in the interval. The endpoints of the interval are selected for this purpose because the two points also define the length of the interval (see Fig. 1.5*a*). For higher-order polynomial approximation, additional points are identified interior to the interval (see Fig. 1.5*b*). The intervals are called *finite elements,* the points used to express the polynomial approximation of the function are called *nodes,* T_j^e are called *nodal values,* and ψ_j^e are called *finite element approximation functions.* The nodal values T_j^e are determined such that $T_e(r)$ satisfies the differential equation (1.9*a*) and boundary conditions (1.9*b*) in some sense. Usually, the differential equation is satisfied in a weighted-integral sense, and boundary conditions on the function itself are satisfied exactly.

The piecewise (i.e., element-wise) approximation of the solution allows us to include any discontinuous data, such as the material properties, and to use meshes of many lower-order elements or a mesh of few higher-order elements to represent large gradients of the solution. Polynomial approximations of the form (1.10) can be derived systematically for any assumed degree of variation.

The satisfaction of the differential equation in a weighted-integral sense leads, for steady-state problems, to algebraic relations among nodal temperatures T_j^e and heats Q_j^e of the element. The algebraic equations of all elements are assembled (i.e., related to each other) such that the temperature is continuous and the heats are balanced at nodes common to elements. The

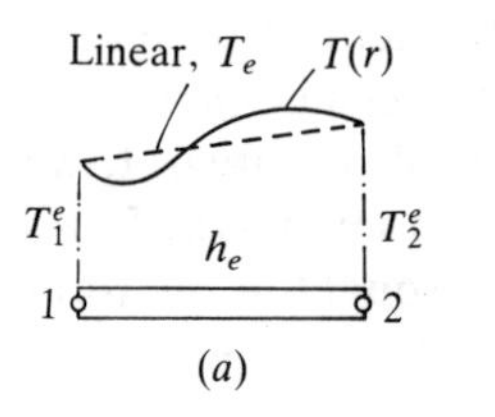

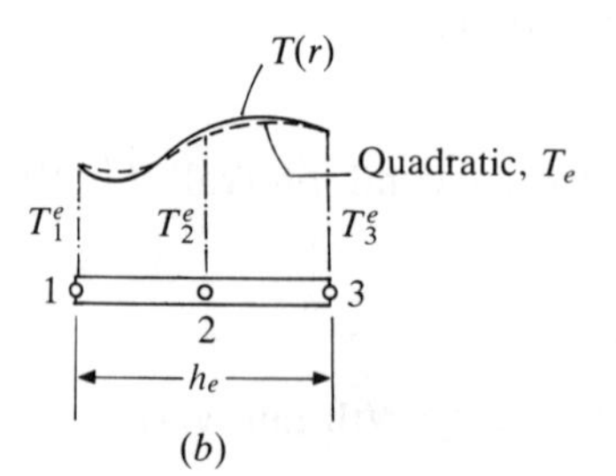

FIGURE 1.5
(*a*) Linear approximation of a function $T(r)$. (*b*) Quadratic approximation of a function $T(r)$.

assembled equations are solved for the nodal values after imposing the boundary conditions of the problem.

1.3.5 Some Remarks

In summary, in the finite element method, a given domain is divided into subdomains, called finite elements, and an approximate solution to the problem is developed over each of these. The subdivision of a whole into parts has two advantages:

1. It allows accurate representation of complex geometries and inclusion of dissimilar materials.
2. It enables accurate representation of the solution within each element, to bring out local effects (e.g., large gradients of the solution).

The three fundamental steps of the finite element method that are illustrated via the examples are:

1. Divide the whole into parts (both to represent the geometry and solution of the problem).
2. Over each part, seek an approximation to the solution as a linear combination of nodal values and approximation functions.
3. Derive the algebraic relations among the nodal values of the solution over each part, and assemble the parts to obtain the solution to the whole.

Although the above examples illustrate the basic idea of the finite element method, there are several other features that are either not present or not apparent from the discussion of the examples.

Some remarks are in order.

1. One can discretize a domain, depending on its shape, into a mesh of more than one type of element. For example, in the approximation of an irregular domain, one can use a combination of rectangles and triangles.
2. If more than one type of element is used in the representation of the domain, one of each kind should be isolated and its equations developed.
3. The governing equations are generally more complex than those considered in the first two examples. They are usually differential equations. In most cases, the equations cannot be solved over an element for two reasons. First, they do not permit the exact solution. It is here that the variational methods come into play. Second, the discrete equations obtained in the variational methods cannot be solved independent of the remaining elements, because the assemblage of the elements is subjected to certain continuity, boundary, and/or initial conditions.
4. There are two main differences in the form of the approximate solution used in the finite element method and that used in the classical variational

methods (i.e., variational methods applied to the whole domain). First, instead of representing the solution u as a linear combination ($u = \sum_j c_j \phi_j$) in terms of arbitrary parameters c_j as in the variational methods, in the finite element method the solution is often represented as a linear combination ($u = \sum_j u_j \psi_j$) in terms of the values u_j of u (and possibly its derivatives as well) at the nodal points. Second, the approximate functions in the finite element method are often polynomials that are derived using interpolation theory. However, the finite element method is *not* restricted to the use of approximations that are linear combinations of nodal values u_j and interpolation functions ψ_j that are algebraic polynomials. One can use, in addition to nodal values, nodeless variables (as in the Rayleigh–Ritz method) to represent the approximation of a function.

5. The number and the location of the nodes in an element depend on (a) the geometry of the element, (b) the degree of the polynomial approximation, and (c) the integral form of the equations. By representing the required solution in terms of its values at the nodes, one obtains directly the approximate solution at the nodes.

6. The assembly of elements, in a general case, is based on the idea that the solution (and possibly its derivatives for higher-order equations) is continuous at the interelement boundaries.

7. In general, the assemblage of finite elements is subjected to boundary and/or initial conditions. The discrete equations associated with the finite element mesh are solved only after the boundary and/or initial conditions have been imposed.

8. There are three sources of error in a finite element solution: (a) those due to the approximation of the domain (this was the only error present in the first two examples); (b) those due to the approximation of the solution; and (c) those due to numerical computation (e.g., numerical integration and round-off errors in a computer). The estimation of these errors, in general, is not a simple matter. However, under certain conditions, they can be estimated for a given element and problem (see Chapter 5).

9. The accuracy and convergence of the finite element solution depends on the differential equation, its integral form, and the element used. "Accuracy" refers to the difference between the exact solution and the finite element solution, while "convergence" refers to the accuracy as the number of elements in the mesh is increased.

10. For time-dependent problems, a two-stage formulation is usually followed. In the first stage, the differential equations are approximated by the finite element method to obtain a set of ordinary differential equations in time. In the second, the differential equations in time are solved exactly or further approximated by either variational methods or finite difference methods to obtain algebraic equations, which are then solved for the nodal values (see Chapter 6).

11. When the continuity conditions of assembly are replaced by contact conditions, the method is known as the *discrete element method* (DEM). In the discrete element method, individual elements can have finite motions (e.g., displacements and rotations). Such methods have applications in rock mechanics (mining and tunneling), ice mechanics, and other fields where a continuum is disintegrated during deformation or the original medium is a collection of individual particles (e.g., granular media and molecular biology).

1.4 THE PRESENT STUDY

This is a book on the finite element method and its applications to linear problems in engineering and applied sciences. Most introductory finite element textbooks written for use in engineering schools are intended for students of solid and structural mechanics, and these introduce the method as an offspring of matrix methods of structural analysis. A few texts that treat the method as a variationally based technique leave the variational formulations and the associated methods of approximation either to an appendix or to self-study by the student. This book is written to introduce the finite element method as a numerical technique that employs the philosophy of constructing piecewise approximations of solutions to problems described by differential equations. This viewpoint makes the student aware of the generality of the finite element concept, irrespective of the student's background. It also enables the student to see the mathematical structure common to various physical theories, and thereby to gain additional insight into various engineering problems.

1.5 SUMMARY

In a numerical simulation of a physical process, we employ a numerical method and computer to evaluate a mathematical model of the process. The finite element method is a powerful numerical technique devised to evaluate complex physical processes. The method is characterized by three features:

1. The domain of the problem is represented by a collection of simple subdomains, called finite elements. The collection of finite elements is called the finite element mesh.
2. Over each finite element, the physical process is approximated by functions of desired type (polynomials or otherwise), and algebraic equations relating physical quantities at selective points, called nodes, of the element are developed.
3. The element equations are assembled using continuity and/or "balance" of physical quantities.

In the finite element method, in general, we seek an approximate solution u to a differential equation in the form

$$u \approx \sum_{j=1}^{n} u_j \psi_j + \sum_{j=1}^{m} c_j \phi_j$$

where u_j are the values of u at the element nodes, ψ_j are the interpolation functions, c_j are the nodeless coefficients, and ϕ_j are the associated approximation functions. Direct substitution of such approximations into the governing differential equations does not always result, for an arbitrary choice of the data of the problem, in a necessary and sufficient number of equations for the undetermined coefficients u_j and c_j. Therefore, a procedure whereby a necessary and sufficient number of equations can be obtained is needed. One such procedure is provided by a weighted-integral form of the governing differential equation. Chapter 2 is devoted to the study of weighted-integral formulations of differential equations and their solution by variational methods of approximation.

There is only one *method of finite elements,* and it is characterized by the three features stated above. Of course, there can be more than one *finite element model* of the same problem. The type of model depends on the differential equations and methods used to derive the algebraic equations (i.e., the weighted-integral form used) for the undetermined coefficients over an element. Although the Rayleigh–Ritz method and polynomial approximations are used frequently to generate the finite element equations, any appropriate method or combination of methods, can be used, in principle, to generate the algebraic equations. In this spirit, the collocation method, subdomain method, boundary integral methods, and so on can be used to generate the algebraic equations among discrete values of the primary and secondary variables.

REFERENCES FOR ADDITIONAL READING

Papers

Argyris, J. H., and S. Kelsey: *Energy Theorems and Structural Analysis,* Butterworth Scientific Publications, London, 1960.

Clough, R. W.: "The Finite Element Method in Plane Stress Analysis," *Journal of Structures Division, ASCE, Proceedings of 2d Conference on Electronic Computation,* pp. 345–378, 1960.

Courant, R.: "Variational Methods for the Solution of Problems of Equilibrium and Vibration," *Bulletin of the American Mathematical Society,* vol. 49, pp. 1–43, 1943.

Hrenikoff, A.: "Solution of Problems in Elasticity by the Framework Method," *Transactions of the ASME, Journal of Applied Mechanics,* vol. 8, pp. 169–175, 1941.

Reddy, J. N.: "The Finite Element Method: A Child of the Computer Age," *Sooner Shamrock* (Engineering Student Magazine at the University of Oklahoma), pp. 23–26, Fall 1978.

Turner, M., R. W. Clough, H. H. Martin, and L. Topp: "Stiffness and Deflection Analysis of Complex Structures," *Journal of Aeronautical Science,* vol. 23, pp. 805–823, 1956.

Books

Bathe, K. J.: *Finite Element Procedures in Engineering Analysis,* Prentice-Hall, Englewood Cliffs, NJ, 1982.

Becker, E. B., G. F. Carey, and J. T. Oden: *Finite Elements, An Introduction,* vol. I, Prentice-Hall, Englewood Cliffs, NJ, 1976.

Cook, R. D., D. S. Malkus, and M. E. Plesha: *Concepts and Applications of Finite Element Analysis,* 3d ed., John Wiley, New York, 1989.

Desai, C. S., and J. F. Abel: *Introduction to the Finite Element Method,* Van Nostrand Reinhold, New York, 1972.

Gallagher, R. H.: *Finite Element Analysis Fundamentals,* Prentice-Hall, Englewood Cliffs, NJ, 1975.

Hughes, T. J. T.: *The Finite Element Method,* Prentice-Hall, Englewood Cliffs, NJ, 1987.

Irons, B. M., and S. Ahmad: *Techniques of Finite Elements,* Ellis Horwood, Chichester, U.K., 1979.

Owen, D. R. J., and E. Hinton: *Finite Elements in Plasticity: Theory and Practice,* Pineridge Press, Swansea, U.K., 1980.

Rao, S. S.: *The Finite Element Method in Engineering,* Pergamon Press, Oxford, 1982.

Reddy, J. N.: *Applied Functional Analysis and Variational Methods in Engineering,* McGraw-Hill, New York, 1986; Krieger, Melbourne, FL, 1991.

Strang, G., and G. Fix: *An Analysis of the Finite Element Method,* Prentice-Hall, Englewood Cliffs, NJ, 1973.

Zienkiewicz, O. C., and R. L. Taylor: *The Finite Element Method,* vol. 1, *Basic Formulation and Linear Problems,* McGraw-Hill, London, 1989.

—— and ——: *The Finite Element Method,* vol. 2, *Solid and Fluid Mechanics, Dynamics and Non-Linearity,* McGraw-Hill, London, 1991.

CHAPTER 2

INTEGRAL FORMULATIONS AND VARIATIONAL METHODS

2.1 NEED FOR WEIGHTED-INTEGRAL FORMS

In the finite element method, we use an integral statement to develop algebraic relations among the coefficients u_j of the approximation

$$u \approx \sum_{j=1}^{n} u_j \psi_j \tag{2.1}$$

where u represents the solution of a particular differential equation. The use of an integral statement equivalent to the governing differential equation is necessitated by the fact that substitution of (2.1) into the governing differential equation does not always result in the required number of linearly independent algebraic equations for the unknown coefficients u_j. One way to insure that there are exactly the same number n of equations as there are unknowns is to require *weighted integrals* of the error in the equation to be zero. A more detailed discussion of this idea is given in the next few paragraphs.

Suppose that we wish to determine an approximate solution of the equation

$$-\frac{d}{dx}\left(x\frac{du}{dx}\right)+u=0 \quad \text{for } 0<x<1 \tag{2.2a}$$

$$u(0)=1, \qquad \left(x\frac{du}{dx}\right)\bigg|_{x=1}=0 \tag{2.2b}$$

We seek an approximate solution, over the entire domain $\Omega=(0,1)$, in the form

$$u \approx U_N \equiv \sum_{j=1}^{N} c_j\phi_j(x)+\phi_0(x) \tag{2.3}$$

where the c_j are coefficients to be determined, and $\phi_j(x)$ and $\phi_0(x)$ are functions preselected such that the specified boundary conditions of the problem are satisfied by the N-parameter approximate solution U_N. For example, we could take $N=2$ and write the approximate solution of (2.2) in the form ($\phi_1=x^2-2x$, $\phi_2=x^3-3x$, $\phi_0=1$)

$$u \approx U_N = c_1(x^2-2x)+c_2(x^3-3x)+1$$

which satisfies the boundary conditions (2.2*b*) of the problem for any values of c_1 and c_2. The constants c_1 and c_2 are to be determined such that the approximate solution U_N in (2.3) satisfies (2.2*a*) in some sense. If we require U_N to satisfy (2.2*a*) in the exact sense, we obtain

$$\begin{aligned}-\frac{dU_N}{dx}-x\frac{d^2U_N}{dx^2}+U_N &= -2c_1(x-1)-3c_2(x^2-1)-2c_1x-6c_2x^2\\ &\quad +c_1(x^2-2x)+c_2(x^3-3x)+1=0\end{aligned}$$

Since this expression must be zero for any value of x, the coefficients of the various powers of x must be zero:

$$\begin{aligned}1+2c_1+3c_2&=0\\ -(6c_1+3c_2)&=0\\ c_1-9c_2&=0\\ c_2&=0\end{aligned}$$

The above relations are inconsistent; hence, there is *no solution* to the equations. On the other hand, we can require the approximate solution U to satisfy the differential equation (2.2*a*) in the weighted-integral sense,

$$\int_0^1 wR\,dx=0 \tag{2.4a}$$

where R is called the *residual*,

$$R \equiv -\frac{dU_N}{dx}-x\frac{d^2U_N}{dx^2}+U_N$$

and w is called a *weight function*. From (2.4a), we obtain as many linearly independent equations as there are independent functions for w. For example, if we take $w = 1$ and $w = x$, we obtain

$$0 = \int_0^1 1R\,dx = (1 + 2c_1 + 3c_2) + \tfrac{1}{2}(-6c_1 - 3c_2) + \tfrac{1}{3}(c_1 - 9c_2) + \tfrac{1}{4}c_2$$

$$0 = \int_0^1 xR\,dx = \tfrac{1}{2}(1 + 2c_1 + 3c_2) + \tfrac{1}{3}(-6c_1 - 3c_2) + \tfrac{1}{4}(c_1 - 9c_2) + \tfrac{1}{5}c_2$$

or

$$\begin{aligned} \tfrac{2}{3}c_1 + \tfrac{5}{4}c_2 &= 1 \\ \tfrac{3}{4}c_1 + \tfrac{31}{20}c_2 &= \tfrac{1}{2} \end{aligned} \tag{2.4b}$$

which provide two linearly independent equations for c_1 and c_2 (giving $c_1 = \frac{222}{23}$ and $c_2 = -\frac{100}{23}$).

Thus, integral statements of the type in (2.4a) provide means for obtaining as many algebraic equations as there are unknown coefficients in the approximation. This chapter deals with the construction of different types of integral statements used in different variational methods. A variational method is one in which approximate solutions of the type $u \approx \sum_j c_j\phi_j + \phi_0$ are sought, and the coefficients c_j are determined, as shown above, using an integral statement. The variational methods differ from each other in the choice of the weight function w and the integral statement used, which in turn dictates the choice of the approximation functions ϕ_j. In the finite element method, a given domain is viewed as an assemblage of subdomains (i.e., elements), and an approximate solution is sought over each subdomain in the same way as in variational methods. Therefore, it is informative to study variational methods before we study the finite element method.

Our goal in this chapter is to illustrate the basic steps in the integral formulations and the associated approximations of various boundary problems. Toward this goal, we first introduce necessary terminology and notation.

2.2 SOME MATHEMATICAL CONCEPTS AND FORMULAE

2.2.1 Boundary, Initial, and Eigenvalue Problems

DOMAIN AND BOUNDARY. The objective of most analyses is to determine unknown functions, called *dependent variables*, that satisfy a given set of differential equations in a given domain or region and some boundary conditions on the boundary of the domain. A domain is a collection of points in space with the property that if P is a point in the domain then all points sufficiently close to P belong to the domain. This definition implies that a domain consists only of internal points. If any two points of the domain can be joined by a line lying entirely within it then the domain is said to be *convex* and *simply connected*. The boundary of a domain is the set of points such that, in any neighborhood of each of these points, there are points that belong to the domain as well as points that do not. Note from the definition that the points on the boundary do not belong to the domain. We shall use the symbol Ω to denote an arbitrary domain and Γ to denote its boundary.

A function of several variables is said to be of class $C^m(\Omega)$ in a domain Ω if all its partial derivatives up to and including the mth order exist and are *continuous* in Ω. Thus, if f is of class C^0 in two dimensions then f is continuous (i.e., $\partial f/\partial x$ and $\partial f/\partial y$ exist but may not be continuous). The letters x and y will always be used for rectangular coordinates of a point in two dimensions.

When the dependent variables are functions of one independent variable (say, x), the domain is a line segment (i.e., one-dimensional) and the endpoints of the domain are called boundary points. When the dependent variables are functions of two independent variables (say, x and y), the (two-dimensional) domain is a surface (most often a plane) and the boundary is the closed curve enclosing it. It is not uncommon to find problems in which the dependent variable and possibly its derivatives are specified at points interior to the domain (e.g., bending of continuous beams).

A differential equation is said to describe a *boundary value problem* if the dependent variable and possibly its derivatives are required to take specified values on the boundary. An *initial value problem* is one in which the dependent variable and possibly its derivatives are specified initially (i.e., at time $t=0$). Initial value problems are generally time-dependent problems. Examples of boundary and initial value problems are given below.

BOUNDARY VALUE PROBLEM

$$-\frac{d}{dx}\left(a\frac{du}{dx}\right)=f \quad \text{for } 0<x<1 \tag{2.5}$$

$$u(0)=d_0, \qquad \left(a\frac{du}{dx}\right)\bigg|_{x=1}=g_0 \tag{2.6}$$

INITIAL VALUE PROBLEM

$$\rho\frac{d^2u}{dt^2}+au=f \quad \text{for } 0<t\leq t_0 \tag{2.7}$$

$$u(0)=u_0, \qquad \left(\frac{du}{dt}\right)\bigg|_{t=0}=v_0 \tag{2.8}$$

BOUNDARY AND INITIAL VALUE PROBLEM

$$-\frac{\partial}{\partial x}\left(a\frac{\partial u}{\partial x}\right)+\rho\frac{\partial u}{\partial t}=f(x,t) \quad \text{for } \begin{cases}0<x<1\\ 0<t\leq t_0\end{cases} \tag{2.9}$$

$$u(0,t)=d_0(t), \qquad \left(a\frac{\partial u}{\partial x}\right)\bigg|_{x=1}=g_0(t), \qquad u(x,0)=u_0(x) \tag{2.10}$$

The conditions in (2.6) are called *boundary conditions*, while those in (2.8) are called *initial conditions*. When any of the specified values (i.e., d_0, g_0, u_0, and v_0) are nonzero, the conditions are said to be *nonhomogeneous*; otherwise, they are said to be *homogeneous*. For example, $u(0) = d_0$ is a nonhomogenous boundary condition, and the associated homogeneous boundary condition is $u(0) = 0$. The set of specified quantities (e.g., a, g_0, d_0, ρ, u_0, and v_0) is called the *data* of the problem. Differential equations in which the right-hand side f is zero are called *homogeneous differential equations*.

EIGENVALUE PROBLEM. The problem of determining the values of the constant λ such that

$$\begin{gathered} -\frac{d}{dx}\left(a\frac{du}{dx}\right) = \lambda u \quad \text{for } 0 < x < 1 \\ u(0) = 0, \qquad \left(a\frac{du}{dx}\right)\bigg|_{x=1} = 0 \end{gathered} \tag{2.11}$$

is called the *eigenvalue problem* associated with the differential equation (2.5). The values of λ for which (2.11) can be satisfied are called *eigenvalues*, and the associated functions u are called *eigenfunctions*.

The *classical* (or exact) *solution* of a differential equation is the function that identically satisfies the differential equation and the specified boundary and/or initial conditions.

2.2.2 Integral Relations

Integration by parts is frequently used in the integral formulation of differential equations. In two-dimensional cases, integration by parts is better known as the gradient and divergence theorems. In this section, we derive some useful identities for future use.

INTEGRATION-BY-PARTS FORMULAE. Let u, v, and w be sufficiently differentiable functions of the coordinate x. Then the following integration-by-parts formula holds:

$$\begin{aligned} \int_a^b w\frac{dv}{dx}\,dx &= \int_a^b w\,dv = -\int_a^b v\,dw + [wv]_a^b \\ &= -\int_a^b v\frac{dw}{dx}\,dx + w(b)v(b) - w(a)v(a) \end{aligned} \tag{2.12}$$

This identity can easily be established. First, note the following identity from the product rule of differentiation:

$$\frac{d}{dx}(wv) = \frac{dw}{dx}v + w\frac{dv}{dx}$$

Therefore

$$w\frac{dv}{dx}=\frac{d}{dx}(wv)-\frac{dw}{dx}v$$

Integrating both sides over the interval (a, b), we obtain

$$\begin{aligned}\int_a^b w\frac{dv}{dx}\,dx &= \int_a^b\left[\frac{d}{dx}(wv)-\frac{dw}{dx}v\right]dx\\ &= \int_a^b \frac{d}{dx}(wv)\,dx-\int_a^b\frac{dw}{dx}v\,dx\\ &= [wv]_a^b-\int_a^b\frac{dw}{dx}v\,dx\end{aligned}$$

which is the same as (2.12).

Next, consider the expression

$$\begin{aligned}\int_a^b w\frac{d^2u}{dx^2}\,dx &= \int_a^b w\frac{d}{dx}\left(\frac{du}{dx}\right)dx\\ &= \int_a^b w\frac{dv}{dx}\,dx, \qquad v\equiv\frac{du}{dx}\end{aligned}$$

Using (2.12), we obtain

$$\begin{aligned}\int_a^b w\frac{d^2u}{dx^2}\,dx &= -\int_a^b v\frac{dw}{dx}\,dx+w(b)v(b)-w(a)v(a)\\ &= -\int_a^b\frac{du}{dx}\frac{dw}{dx}\,dx+w(b)\frac{du}{dx}(b)-w(a)\frac{du}{dx}(a)\end{aligned}\tag{2.13a}$$

or

$$-\int_a^b\frac{du}{dx}\frac{dw}{dx}\,dx=\int_a^b w\frac{d^2u}{dx^2}\,dx+w(a)\frac{du}{dx}(a)-w(b)\frac{du}{dx}(b)\tag{2.13b}$$

Similarly,

$$\begin{aligned}\int_a^b v\frac{d^4w}{dx^4}\,dx &= \int_a^b v\frac{d^2}{dx^2}\left(\frac{d^2w}{dx^2}\right)dx\\ &= \int_a^b v\frac{d^2u}{dx^2}\,dx, \qquad \text{where } u\equiv\frac{d^2w}{dx^2}\end{aligned}$$

Using (2.13a) with $w=v$, we can write the right-hand side as

$$-\int_a^b\frac{du}{dx}\frac{dv}{dx}\,dx+v(b)\frac{du}{dx}(b)-v(a)\frac{du}{dx}(a)\tag{2.14a}$$

We use (2.13b) with $w=u$ and $u=v$ to write (2.14a) as

$$\int_a^b u\frac{d^2v}{dx^2}\,dx+u(a)\frac{dv}{dx}(a)-u(b)\frac{dv}{dx}(b)+v(b)\frac{du}{dx}(b)-v(a)\frac{du}{dx}(a)\tag{2.14b}$$

and, finally, replacing u by its actual value $u = d^2w/dx^2$, we arrive at

$$\int_a^b v\frac{d^4w}{dx^4}\,dx = \int_a^b \frac{d^2w}{dx^2}\frac{d^2v}{dx^2}\,dx + \frac{d^2w}{dx^2}(a)\frac{dv}{dx}(a) - \frac{d^2w}{dx^2}(b)\frac{dv}{dx}(b)$$
$$+ v(b)\frac{d^3w}{dx^3}(b) - v(a)\frac{d^3w}{dx^3}(a) \tag{2.15}$$

Equations (2.13a) and (2.15) are useful in the weak formulation (see Section 2.3) of second- and fourth-order differential equations, respectively.

Let $\boldsymbol{\nabla}$ and ∇^2 denote, respectively, the gradient operator and the Laplacian operator in the two-dimensional cartesian rectangular coordinate system (x, y):

$$\boldsymbol{\nabla} = \hat{\mathbf{i}}\frac{\partial}{\partial x} + \hat{\mathbf{j}}\frac{\partial}{\partial y}, \qquad \nabla^2 = \boldsymbol{\nabla}\cdot\boldsymbol{\nabla} = \frac{\partial^2}{\partial x^2} + \frac{\partial^2}{\partial y^2} \tag{2.16}$$

where $\hat{\mathbf{i}}$ and $\hat{\mathbf{j}}$ denote the unit basis vectors along the x and y coordinates, respectively. The caret "^" over the vectors indicates that they are of unit length. If $F(x, y)$ and $G(x, y)$ are scalar functions of class $C^0(\Omega)$ in the two-dimensional domain Ω, the following gradient and divergence theorems hold.

GRADIENT THEOREM

$$\int_\Omega \operatorname{grad} F\,dx\,dy \equiv \int_\Omega \boldsymbol{\nabla}F\,dx\,dy = \oint_\Gamma \hat{\mathbf{n}}F\,ds$$

or (2.17a)

$$\int_\Omega \left(\hat{\mathbf{i}}\frac{\partial F}{\partial x} + \hat{\mathbf{j}}\frac{\partial F}{\partial y}\right) dx\,dy = \oint_\Gamma (n_x\hat{\mathbf{i}} + n_y\hat{\mathbf{j}})F\,ds$$

The second equation implies (because two vectors are equal if and only if their components are equal) that the following relations hold:

$$\int_\Omega \frac{\partial F}{\partial x}\,dx\,dy = \oint_\Gamma n_xF\,ds, \qquad \int_\Omega \frac{\partial F}{\partial y}\,dx\,dy = \oint_\Gamma n_yF\,ds \tag{2.17b}$$

DIVERGENCE THEOREM

$$\int_\Omega \operatorname{div}\mathbf{G}\,dx\,dy \equiv \int_\Omega \boldsymbol{\nabla}\cdot\mathbf{G}\,dx\,dy = \oint_\Gamma \hat{\mathbf{n}}\cdot\mathbf{G}\,ds$$

or (2.18)

$$\int_\Omega \left(\frac{\partial G_x}{\partial x} + \frac{\partial G_y}{\partial y}\right) dx\,dy = \oint_\Gamma (n_xG_x + n_yG_y)\,ds$$

Here the dot denotes the scalar product of vectors, $\hat{\mathbf{n}}$ denotes the unit vector normal to the surface Γ of the domain Ω, n_x and n_y (G_x and G_y) are the rectangular components of $\hat{\mathbf{n}}$ ($\mathbf{G}$), and the circle on the boundary integral

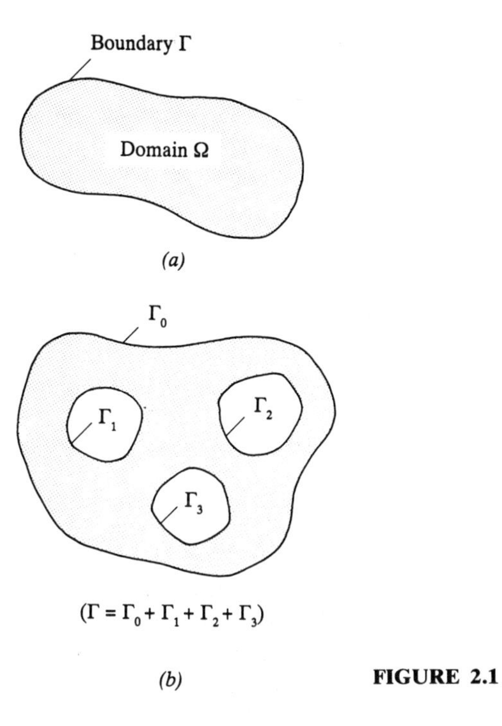

FIGURE 2.1

indicates that the integration is taken over the entire boundary (see Fig. 2.1). The direction cosines n_x and n_y of the unit vector $\hat{\mathbf{n}}$ can be written as

$$n_x = \cos(x, \hat{\mathbf{n}}), \qquad n_y = \cos(y, \hat{\mathbf{n}}) \tag{2.19}$$

where $\cos(x, \hat{\mathbf{n}})$ is the cosine of the angle between the positive x direction and the unit vector $\hat{\mathbf{n}}$.

The following identities, which can be derived using the gradient and divergence theorems, will be useful in the sequel. Let w and G be scalar functions defined in a two-dimensional domain Ω. Then

$$\int_\Omega (\boldsymbol{\nabla} G) w \, dx \, dy = -\int_\Omega (\boldsymbol{\nabla} w) G \, dx \, dy + \int_\Gamma \hat{\mathbf{n}} w G \, ds \tag{2.20a}$$

and

$$\boxed{-\int_\Omega (\nabla^2 G) w \, dx \, dy = \int_\Omega \boldsymbol{\nabla} w \cdot \boldsymbol{\nabla} G \, dx \, dy - \oint_\Gamma \frac{\partial G}{\partial n} w \, ds} \tag{2.20b}$$

where $\partial/\partial n$ denotes the normal derivative operator,

$$\frac{\partial}{\partial n} = \hat{\mathbf{n}} \cdot \boldsymbol{\nabla} = n_x \frac{\partial}{\partial x} + n_y \frac{\partial}{\partial y} \tag{2.21}$$

The following component form of (2.20*a*), with an appropriate change of variables, is useful in the sequel:

$$\int_\Omega w \frac{\partial G}{\partial x}\, dx\, dy = -\int_\Omega \frac{\partial w}{\partial x} G\, dx\, dy + \oint_\Gamma n_x w G\, ds \tag{2.22a}$$

$$\int_\Omega w \frac{\partial G}{\partial y}\, dx\, dy = -\int_\Omega \frac{\partial w}{\partial y} G\, dx\, dy + \oint_\Gamma n_y w G\, ds \tag{2.22b}$$

Equations (2.22*a, b*) can easily be established by means of (2.17*b*).

2.2.3 Functionals

An integral expression of the form

$$I(u) = \int_a^b F(x, u, u')\, dx, \qquad u = u(x), \qquad u' = \frac{du}{dx}$$

where the integrand $F(x, u, u')$ is a given function of the arguments x, u, and du/dx, is called a *functional*. The value $I(u)$ of the integral depends on u; hence the notation $I(u)$ is appropriate. However, for a given u, $I(u)$ represents a scalar value. We shall use the term *functional* to describe functions defined by integrals whose arguments themselves are functions. Loosely speaking, a functional is a "function of functions." Mathematically, a functional is an operator I mapping u into a scalar $I(u)$.

A functional $l(u)$ is said to be *linear* in u if and only if it satisfies the relation

$$l(\alpha u + \beta v) = \alpha l(u) + \beta l(v) \tag{2.23}$$

for any scalars α and β and dependent variables u and v. A functional $B(u, v)$ is said to be *bilinear* if it is linear in each of its arguments u and v:

$$\begin{aligned} &B(\alpha u_1 + \beta u_2, v) = \alpha B(u_1, v) + \beta B(u_2, v) \\ &\qquad\qquad \text{(linearity in the first argument)} \\ &B(u, \alpha v_1 + \beta v_2) = \alpha B(u, v_1) + \beta B(u, v_2) \\ &\qquad\qquad \text{(linearity in the second argument)} \end{aligned} \tag{2.24}$$

where u, u_1, u_2, v, v_1, and v_2 are dependent variables. A bilinear form $B(u, v)$ is said to be *symmetric* in its arguments u and v if

$$B(u, v) = B(v, u) \tag{2.25}$$

for all u and v.

An example of a linear functional is

$$l(v) = \int_0^L v f\, dx + \frac{dv}{dx}(L) M_0$$

where $f = f(x)$ and M_0 are known quantities. An example of a bilinear functional is

$$B(v, w) = \int_0^L a \frac{dv}{dx}\frac{dw}{dx}\,dx$$

where $a = a(x)$ is a known function.

2.2.4 The Variational Symbol

Consider the function $F = F(x, u, u')$. For an arbitrary fixed value of the independent variable x, F depends on u and u'. The change αv in u, where α is a constant and v is a function, is called the *variation* of u and is denoted by δu:

$$\delta u = \alpha v \tag{2.26}$$

The operator δ is called the *variational symbol*. The variation δu of a function u represents an admissible change in the function $u(x)$ at a *fixed* value of the independent variable x. If u is specified at a point (usually on the boundary), the variation of u is zero there because the specified value cannot be varied. thus the variation of a function u should satisfy the homogeneous form of the boundary conditions for u. The variation δu in u is a *virtual* change. Associated with this change in u (i.e., u going to $u + \alpha v$), there is a change in F. In analogy with the total differential of a function of two variables, the *first variation* of F at u is defined by

$$\delta F = \frac{\partial F}{\partial u}\,\delta u + \frac{\partial F}{\partial u'}\,\delta u' \tag{2.27}$$

Note the analogy between the first variation, (2.27), and the total differential of F,

$$dF = \frac{\partial F}{\partial x}\,dx + \frac{\partial F}{\partial u}\,du + \frac{\partial F}{\partial u'}\,du' \tag{2.28}$$

Since x is not varied during the variation of u to $u + \delta u$, $dx = 0$ and the analogy between ∂F and dF becomes apparent. That is, *δ acts as a differential operator with respect to dependent variables*. It can easily be verified that the laws of variation of sums, products, ratios, powers, and so forth are completely analogous to the corresponding laws of differentiation. For example, if $F_1 = F_1(u)$ and $F_2 = F_2(u)$ then

$$\begin{aligned}
&\textbf{1.}\quad \delta(F_1 \pm F_2) = \delta F_1 \pm \delta F_2\\
&\textbf{2.}\quad \delta(F_1 F_2) = F_2\,\delta F_1 + F_1\,\delta F_2\\
&\textbf{3.}\quad \delta\left(\frac{F_1}{F_2}\right) = \frac{F_2\,\delta F_1 - F_1\,\delta F_2}{F_2^2}\\
&\textbf{4.}\quad \delta[(F_1)^n] = n(F_1)^{n-1}\,\delta F_1
\end{aligned} \tag{2.29}$$

Furthermore, the variational operator can commute with differential and integral operators (as long as the coordinates x and y are the fixed, Lagrangian coordinates):

$$\frac{d}{dx}(\delta u) = \frac{d}{dx}(\alpha v) = \alpha\frac{dv}{dx} = \alpha v' = \delta u' = \delta\left(\frac{du}{dx}\right) \tag{2.30a}$$

$$\delta \int_a^b u(x)\,dx = \int_a^b \delta u(x)\,dx \tag{2.30b}$$

2.3 WEAK FORMULATION OF BOUNDARY VALUE PROBLEMS

2.3.1 Introduction

Recall from Section 2.1 that the motivation for integral formulations of boundary value problems comes from the fact that variational methods of approximation, e.g., the Ritz, Galerkin, least-squares, collocation, or, in general, weighted-residual methods, are based on weighted-integral statements of the governing equations. Since the finite element method is a technique for constructing approximation functions required in an element-wise application of any variational method, it is necessary to study the weighted-integral formulation and the weak formulation of differential equations. In addition to the above reason, weak formulations also facilitate, in a natural way, the classification of boundary conditions into *natural* and *essential* boundary conditions, which play a crucial role in the derivation of the approximation functions and the selection of the nodal degrees of freedom of the finite element model.

In this section, our primary objectives will be to construct the weak form of a given differential equation and to classify the boundary conditions associated with the equation. A *weak form* is a weighted-integral statement of a differential equation in which the differentiation is distributed among the dependent variable and the weight function and includes the natural boundary conditions of the problem.

2.3.2 Weighted-Integral and Weak Formulations

Consider the problem of solving the differential equation

$$-\frac{d}{dx}\left[a(x)\frac{du}{dx}\right] = q(x) \quad \text{for } 0<x<L \tag{2.31a}$$

for the solution $u(x)$, subject to the boundary conditions

$$u(0) = u_0, \qquad \left(a\frac{du}{dx}\right)\bigg|_{x=L} = Q_0 \tag{2.31b}$$

Here a and q are known functions of the coordinate x, u_0 and Q_0 are known values, and L is the length of the one-dimensional domain. The functions a and q, and constants u_0 and Q_0, along with the length L of the domain, are the data of the problem. The solution u is the *dependent variable* of the problem. When the specified values are nonzero ($u_0 \neq 0$ or $Q_0 \neq 0$), the boundary conditions are said to be nonhomogeneous; when the specified values are zero the boundary conditions are said to be homogeneous. The homogeneous form of the boundary condition $u(0) = u_0$ is $u(0) = 0$, and the homogeneous form of the boundary condition $(a\, du/dx)|_{x=L} = Q_0$ is $(a\, du/dx)|_{x=L} = 0$.

Equations of the type (2.31*a*) arise, for example, in the study of heat conduction in a heat exchanger fin or a long axisymmetric cylinder. Other examples are included in Table 3.2. In the former case, $a = kA$, with k being the thermal conductivity and A the cross-sectional area, and L being the length of the fin. For the axisymmetric case, $a = 2\pi Lkx$, x being the radial coordinate r and L the length of the cylinder (see Fig. 1.4). In both cases, q denotes the heat generation term, u_0 is the specified temperature, and Q_0 is the specified heat. Other physical problems are also described by the same equation, but with different meanings for the variables (see Table 3.2).

It should be recalled that the sole purpose of developing a weighted-integral statement of a differential equation is to have the means to obtain N linearly independent algebraic relations among the coefficients c_j of the approximation

$$u \approx U_N = \sum_{j=1}^{N} c_j \phi_j(x) + \phi_0(x) \tag{2.32}$$

This is accomplished by choosing N linearly independent weight functions in the integral statement, as will be seen shortly.

There are three steps in the development of the weak form, if it exists, of any differential equation. These steps are illustrated by means of the model differential equation and boundary conditions in (2.31).

Step 1. Move all expressions of the differential equation to one side, multiply the entire equation with a function w, called the *weight function*, and integrate over the domain $\Omega = (0, L)$ of the problem:

$$0 = \int_0^L w\left[-\frac{d}{dx}\left(a\frac{du}{dx}\right) - q\right] dx \tag{2.33}$$

We shall call the statement in (2.33) the *weighted-integral* or *weighted-residual* statement equivalent to the original equation (2.31a). The expression in the square brackets is not identically zero when u is replaced by its approximation. Mathematically, (2.33) is a statement that the error in the differential equation (due to the approximation of the solution) is zero in the weighted-integral sense. When u is the exact solution, (2.33) is trivial. The integral statement (2.33) allows us to choose N linearly independent functions for w and obtain N equations for $c_1, c_2, \ldots, c_N$ of (2.32).

Note that the weighted-integral statement of any differential equation can be developed. The weight function w in (2.33) can be any nonzero, integrable function. In general, the weight function w in the integral statement is subject to less stringent continuity requirements than the dependent variable u. The weighted-integral statement is equivalent only to the differential equation, and it does not include any boundary conditions.

Step 2. While the weighted-integral statement (2.33) allows us to obtain the necessary number N of algebraic relations among c_j for N different choices of the weight function w, it requires that the approximation functions ϕ_j be such that U_N [see (2.32)] is differentiable as many times as called for in the original differential equation and satisfies the specified boundary conditions. If this is not a concern, one can proceed with the integral statement (2.33) and obtain the necessary algebraic equations for c_j. Approximate methods based on weighted-integral statements of the form (2.33) are known as *weighted-residual* methods (see Section 2.4.3). If the differentiation is distributed between the approximate solution U_N and the weight function w, the resulting integral form will require weaker continuity conditions on ϕ_j, and hence the weighted-integral statement is called the *weak form*. As will be seen shortly, the weak formulation has two desirable characteristics. First, it requires weaker (i.e., less) continuity of the dependent variable, and often it results in a symmetric set of algebraic equations in the coefficients. Second, the *natural* boundary conditions of the problem are included in the weak form, and therefore the approximate solution U_N is required to satisfy only the *essential* boundary conditions of the problem. These two features of a weak form play an important role in the development of finite element models of a problem.

The equal distribution of differentiation among the weight function and the dependent variable is possible only if the derivatives appearing in the differential equation are of even order, as is the case with most problems studied in this book. The trading of differentiability from the dependent variable to the weight function is dictated by the need to include physically meaningful boundary terms into the weak form, regardless of the effect on the continuity requirements. On the other hand, trading of differentiation from the dependent variable to the weight function should not be performed if it leads to boundary terms that are not physically meaningful.

Returning to the integral statement (2.33), we integrate the first term of the expression by parts to obtain

$$
\begin{aligned}
0 &= \int_0^L \left\{ w\left[-\frac{d}{dx}\left(a\frac{du}{dx}\right)\right] - wq \right\} dx \\
&= \int_0^L \left(\frac{dw}{dx}a\frac{du}{dx} - wq \right) dx - \left[wa\frac{du}{dx}\right]_0^L
\end{aligned}
\tag{2.34}
$$

where the integration-by-parts formula [see (2.12) or (2.13a)]

$$
\int_0^L w\,dv = -\int_0^L v\,dw + [wv]_0^L \tag{2.35}
$$

with $v = -a\,du/dx$ is used on the first term to arrive at the second line of (2.34). The reader is asked to verify (2.34) either directly or by the use of (2.35). Note that now the weight function w is required to be differentiable at least once.

An important part of Step 2 is to identify the two types of boundary conditions associated with *any* differential equation: *natural* and *essential*. The classification is important for both the variational methods of approximation considered in this chapter and the finite element formulations presented in Chapters 3–5. The following rule is used to identify the natural boundary conditions and their form. After trading differentiation between the weight function and the variable, i.e., after completing Step 2, examine all boundary terms of the integral statement. The boundary terms will involve both the weight function and the dependent variable. Coefficients of the weight function and its derivatives in the boundary expressions are termed the *secondary variables* (SV). Specification of secondary variables on the boundary constitutes the *natural boundary conditions* (NBC). For the case at hand, the boundary term is $w(a\,du/dx)$. The coefficient of the weight function is $a\,du/dx$. Hence the secondary variable is of the form [see (2.34)] $a\,du/dx$.

The secondary variables always have physical meaning, and are often quantities of interest. In the case of heat transfer problems, the secondary variable represents heat, Q. We shall denote the secondary variable by

$$Q \equiv \left(a\frac{du}{dx}\right)n_x \tag{2.36}$$

where n_x denotes the direction cosine,

n_x = cosine of the angle between the x axis and the normal to the boundary

For one-dimensional problems, the normal at the boundary points is always along the length of the domain. Thus, $n_x = -1$ at the left end and $n_x = 1$ at the right end of the domain: $n_x(0) = -1$ and $n_x(L) = 1$.

The dependent variable of the problem, expressed in the same form as the weight function appearing in the boundary term, is called the *primary variable* (PV), and its specification on the boundary constitutes the *essential boundary condition* (EBC). For the case under consideration, the weight function appears in the boundary expression [see (2.34)] as w. Therefore, the dependent variable u is the primary variable, and the essential boundary condition involves specifying u at the boundary points.

It should be noted that the number and form of the primary and secondary variables depend on the order of the differential equation. The number of primary and secondary variables is always the same, and with each primary variable there is an associated secondary variable (e.g., displacement and force, temperature and heat, and so on). Only one of the pair, either the primary or the secondary variable, may be specified at a point of the boundary. Thus, a given problem can have its specified boundary conditions in one of three categories: (i) all specified boundary conditions are EBC; (ii) some of the specified boundary conditions are EBC and the remaining are NBC; or (iii) all specified boundary conditions are NBC. For a single second-order equation, as in the present case, there is one primary variable u and one secondary variable Q. At a boundary point, only one of the pair (u, Q) can be specified. For a fourth-order equation, such as that for the classical (i.e., Euler–Bernoulli) theory of beams, there are two of each kind (i.e., two PVs and two SVs), as will be illustrated later (see Example 2.2). In general, a $2m$th-order differential equation has m PVs and m SVs, i.e., m pairs of primary and secondary variables.

In the notation of (2.36), (2.34) takes the form

$$
\begin{aligned}
0 &= \int_0^L \left(a \frac{dw}{dx}\frac{dv}{dx} - wq \right) dx - \left[wa \frac{du}{dx} \right]_0^L \\
&= \int_0^L \left(a \frac{dw}{dx}\frac{du}{dx} - wq \right) dx - \left(wa \frac{du}{dx} n_x \right)\bigg|_{x=0} - \left(wa \frac{du}{dx} n_x \right)\bigg|_{x=L} \\
&= \int_0^L \left(a \frac{dw}{dx}\frac{du}{dx} - wq \right) dx - (wQ)_0 - (wQ)_L
\end{aligned}
\tag{2.37}
$$

Equation (2.37) is called the *weak form* of the differential equation (2.31). "Weak" refers to the reduced (i.e., weakened) continuity of u, which is required to be twice-differentiable in the weighted-integral form (2.33) but only once-differentiable in (2.37).

Step 3. The third and last step of the weak formulation is to impose the actual boundary conditions of the problem under consideration. It is here that we require the weight function w to vanish at boundary points where the essential boundary conditions are specified; i.e., w is required to satisfy the *homogeneous form* of the specified essential boundary conditions of the problem. This requirement on w might seem arbitrary for a reader not familiar with variational calculus. In weak formulations, the weight function has the meaning of a *virtual* change (or variation) of the primary variable. If a primary variable is specified at a point, the virtual change there must be zero. For more detailed discussions of this, the reader may consult books on variational methods [see Reddy (1986)]. For the problem at hand, the boundary conditions are given in (2.31*b*). By the rules of classification of the boundary conditions, $u = u_0$ is the essential boundary condition and $(a\, du/dx)|_{x=L} = Q_0$ is the natural boundary condition. Thus, the weight function w is required to satisfy

$$w(0) = 0, \quad \text{because } u(0) = u_0$$

Since $w(0) = 0$ and

$$Q(L) = \left(a \frac{du}{dx} n_x \right)\bigg|_{x=L} = \left(a \frac{du}{dx} \right)\bigg|_{x=L} = Q_0$$

(2.37) reduces to the expression

$$0 = \int_0^L \left(a \frac{dw}{dx} \frac{du}{dx} - wq \right) dx - w(L)Q_0 \tag{2.38}$$

which is the weak form equivalent to the original differentiatl equation (2.31*a*) and the natural boundary condition (2.31*b*). This completes the steps involved in the development of the weak or *variational form* of a differential equation.

The terms "variational form" and "weak form" will be used interchangeably. The weak form of a differential equation is a weighted-integral statement equivalent to the differential equation *and* the specified natural boundary conditions of the problem. Note that the weak form exists for all problems—linear or nonlinear—that are described by second- and higher-order differential equations. When the differential equation is linear and of even order, the resulting weak form will have a *symmetric* bilinear form in the dependent variable u and weight function w.

In summary, there are three steps in the development of a weak form. In the first, we put all expressions of the differential equation on one side (so that the other side is equal to zero), then multiply the entire equation by a weight function and integrate over the domain of the problem. The resulting expression is called the *weighted-integral form* of the equation. In the second step, we use integration by parts to distribute differentiation evenly between the dependent variable and the weight function, and use the boundary terms to identify the form of the primary and secondary variables. In the third step, we modify the boundary terms by restricting the weight function to satisfy the homogeneous form of the specified essential boundary conditions and replacing the secondary variables by their specified values.

It should be recalled that a weighted-integral statement or the weak form of a differential equation is needed to obtain as many algebraic equations as there are unknown coefficients in the approximation of the dependent variables of the equation. For different choices of the weight function, different algebraic equations can be obtained. Because of the restrictions placed on the weight function in Step 3 of the variational formulation, it must belong to the same space of functions as the approximation functions (i.e., $w \sim \phi_i$).

2.3.3 Linear and Bilinear Forms and Quadratic Functionals

It is informative, although not necessary for the use of variational methods or the finite element method, to see the relation between the weak form and the minimum of a quadratic functional associated with the differential equation. The weak form (2.38) contains two types of expressions: those involving both the dependent variable u and the weight function w, and those involving only the latter. We shall denote these two types of expressions by $B(w, u)$ and $l(w)$, respectively:

$$B(w, u) = \int_0^L a \frac{dw}{dx}\frac{du}{dx}\, dx, \qquad l(w) = \int_0^L wq\, dx + w(L)Q_0 \tag{2.39}$$

Hence, the weak statement (2.38) can be expressed in the form

$$0 = B(w, u) - l(w) \tag{2.40}$$

which is termed the *variational* (or weak) *problem* associated with the equations (2.31). Using the definitions of linear and bilinear forms from Section 2.2.3, it can be verified that $B(w, u)$ is bilinear and symmetric in w and u and that $l(w)$ is linear [see (2.23) and (2.24)]. The *variational problem* associated with (2.31*a, b*) can be stated as one of finding the solution u such that

$$B(w, u) = l(w) \tag{2.41}$$

holds for any w that satisfies the homogeneous form of the specified essential boundary conditions and continuity conditions implied by the weak form. The function w can be viewed as a variation (or increment) of the actual solution u^*,

$$u = u^* + w \tag{2.42}$$

and u is the variational solution, i.e., the solution of (2.41). Since both u and u^* must satisfy any specified essential boundary condition (in addition, u^* also satisfies any specified natural boundary condition), it follows that w must satisfy the homogeneous form of the specified essential boundary condition. Thus, in the notation of (2.26), w is the variation (see Section 2.2.4) of the solution:

$$w = \delta u$$

Then (2.40) can be written as

$$0 = B(\delta u, u) - l(\delta u)$$

If $B(\cdot\,, \cdot)$ is symmetric, we can write

$$\begin{aligned} &= \delta[\tfrac{1}{2}B(u, u)] - \delta[l(u)] \\ &= \delta I(u) \end{aligned} \tag{2.43a}$$

where

$$\boxed{I(u) = \tfrac{1}{2}B(u, u) - l(u)} \tag{2.43b}$$

In arriving at the second line of (2.43*a*), the following identities are used:

$$\begin{aligned} B(\delta u, u) &= \int_0^L a \frac{d\delta u}{dx}\frac{du}{dx}\, dx = \delta \int_0^L \frac{a}{2}\left[\left(\frac{du}{dx}\right)^2\right] dx \\ &= \tfrac{1}{2}\delta \int_0^L a \frac{du}{dx}\frac{du}{dx}\, dx = \tfrac{1}{2}\,\delta[B(u, u)] \end{aligned} \tag{2.44a}$$

$$l(\delta u) = \int_0^L \delta u\, q\, dx + \delta u(L)\, Q_0$$

$$= \delta\left[\int_0^L uq\, dx + u(L)Q_0\right] = \delta[l(u)] \tag{2.44b}$$

Note that the key step in the derivation of the functional $I(u)$ from the weak form is the linearity and symmetry of the bilinear form $B(w, u)$. The relation $B(\delta u, u) = \frac{1}{2}\,\delta B(u, u)$ holds only if $B(w, u)$ is bilinear and symmetric in w and u. Thus, whenever $B(w, u)$ is bilinear and symmetric, and $l(w)$ is linear, the associated quadratic functional is given by (2.43*b*). When $B(w, u)$ is not linear in w and u, but is symmetric, the functional $I(u)$ can be derived, but not from (2.43*b*). The interested reader can consult the books by Oden and Reddy (1976) and Reddy (1986).

Equation (2.43*a*) represents the necessary condition for the functional $I(u)$ to have an extremum value. For solid mechanics problems, $I(u)$ represents the total potential energy functional, and (2.43*a*) is the statement of the *total potential energy principle*:

> Of all admissible functions u, that which makes the total potential energy $I(u)$ a minimum also satisfies the differential equation and natural boundary condition in (2.31).

In other words, the weak form of a differential equation is the same as the statement of the total potential energy principle. For problems outside solid mechanics, the functional $I(u)$ may not have the meaning of energy, but it is still useful for mathematical analysis (e.g., in considering the existence and uniqueness of solutions).

As noted earlier, every differential equation admits a weighted-integral statement, and a weak form exists provided the equation is of order two or higher. However, not all equations admit the functional formulation. In order for the functional to exist, the associated bilinear form should be symmetric in its arguments. On the other hand, variational methods and the finite element method do not require a functional; an integral statement or a weak form of the equation to be solved is sufficient. If one has a functional at hand, the weak form is obtained by taking its first variation.

2.3.4 Examples

Now we consider some representative examples of differential equations in one and two dimensions, and formulate their variational equations. These examples are of primary interest in the study of the finite element method.

Example 2.1. Consider the differential equation

$$-\frac{d}{dx}\left(a\frac{du}{dx}\right) - cu + x^2 = 0 \quad \text{for } 0 < x < 1 \tag{2.45a}$$

subject to the boundary conditions

$$u(0) = 0, \quad \left(a\frac{du}{dx}\right)\bigg|_{x=1} = 1 \tag{2.45b}$$

The data are [cf. (2.31)] $q = -x^2$, $Q_0 = 1$, and $u_0 = 0$.

Following the three steps outlined above for the construction of variational statements, we obtain

$$\begin{aligned} &(1) \quad 0 = \int_0^1 w\left[-\frac{d}{dx}\left(a\frac{du}{dx}\right) - cu + x^2\right] dx \\ &(2) \quad 0 = \int_0^1 \left(a\frac{dw}{dx}\frac{du}{dx} - cwu + wx^2\right) dx - \left(wa\frac{du}{dx}\right)\bigg|_0^1 \end{aligned} \tag{2.46}$$

From the boundary term, it is clear that the specification of u is an essential boundary condition, and the specification of $a\,du/dx$ is a natural boundary condition. Since $a\,du/dx = 1$ at $x = 1$ and $w = 0$ at $x = 0$ (because u is specified there), we obtain the weak form

$$(3) \quad 0 = \int_0^1 \left(a\frac{dw}{dx}\frac{du}{dx} - cwu\right) dx + \int_0^1 wx^2\, dx - w(1) \tag{2.47a}$$

$$0 = B(w, u) - l(w) \tag{2.47b}$$

where

$$\begin{aligned} B(w, u) &= \int_0^1 \left(a\frac{dw}{dx}\frac{du}{dx} - cwu\right) dx \\ l(w) &= -\int_0^1 wx^2\, dx + w(1) \end{aligned} \tag{2.47c}$$

Since $B(\cdot\,,\cdot)$ is bilinear and symmetric, and $l(\cdot)$ is linear, we can compute the quadratic functional from (2.43):

$$I(u) = \frac{1}{2}\int_0^1 \left[a\left(\frac{du}{dx}\right)^2 - cu^2 + 2ux^2\right] dx - u(1) \tag{2.48}$$

Equations of the type of (2.45) arise in the study of the deflection of a cable or of heat transfer in a fin ($c = 0$). In the former case, u denotes the transverse deflection and a the tension in the cable. The first two terms in the quadratic functional represent the elastic strain energy, while the last term represents the work done by the distributed force in moving through the displacement u.

The next example illustrates the variational formulation of a fourth-order differential equation in one dimension.

Example 2.2. Consider the problem of finding the solution w to the differential equation

$$\frac{d^2}{dx^2}\left[b(x)\frac{d^2w}{dx^2}\right] - f(x) = 0 \quad \text{for } 0 < x < L \tag{2.49}$$

subject to the boundary conditions

$$w(0) = \left(\frac{dw}{dx}\right)\bigg|_{x=0} = 0, \quad \left(b\frac{d^2w}{dx^2}\right)\bigg|_{x=L} = M_0, \quad \left[\frac{d}{dx}\left(b\frac{d^2w}{dx^2}\right)\right]\bigg|_{x=L} = 0 \tag{2.50}$$

This equation arises, for example, in the study of the elastic bending of beams (under the Euler–Bernoulli hypothesis). In this case, w denotes the transverse deflection of the beam, L is the total length of the beam, $b(x) \geqslant 0$ is the flexural rigidity of the beam (i.e., the product of modulus of elasticity E and moment of inertia I: $b = EI$), $f(x)$ is the transverse distributed load, and M_0 is the bending moment. The solution w is the *dependent variable* of the problem, and all other quantities (L, b, f, M_0) that are known in advance are the *data* of the problem.

Since the equation contains a fourth-order derivative, we should integrate it twice by parts to distribute the derivatives equally between the dependent varaible w and the weight function v. In this case, v must be twice differentiable and satisfy the homogeneous form of EBC. Multiplying (2.49) by v, and integrating the first term by parts twice with respect to x, we obtain [see (2.15)]

$$0 = \int_0^L v\left[\frac{d^2}{dx^2}\left(b\frac{d^2w}{dx^2}\right) - f\right] dx \tag{2.51}$$

$$\begin{aligned} 0 &= \int_0^L \left[\left(-\frac{dv}{dx}\right)\frac{d}{dx}\left(b\frac{d^2w}{dx^2}\right) - vf\right] dx + \left[v\frac{d}{dx}\left(b\frac{d^2w}{dx^2}\right)\right]\Bigg|_0^L \\ &= \int_0^L \left(b\frac{d^2v}{dx^2}\frac{d^2w}{dx^2} - vf\right) dx + \left[v\frac{d}{dx}\left(b\frac{d^2w}{dx^2}\right) - \frac{dv}{dx}b\frac{d^2w}{dx^2}\right]\Bigg|_0^L \end{aligned} \tag{2.52}$$

From the last line, it follows that the specification of w and dw/dx constitutes the essential (geometric or static) boundary conditions, and the specification of

$$\frac{d}{dx}\left(b\frac{d^2w}{dx^2}\right) \equiv V \text{ (shear force)} \tag{2.53a}$$

and

$$b\left(\frac{d^2w}{dx^2}\right) \equiv M \text{ (bending moment)} \tag{2.53b}$$

constitutes the natural boundary conditions. In the present case, the specified essential boundary conditions are (because of the clamped condition)

$$w(0) = \left(\frac{dw}{dx}\right)\Bigg|_{x=0} = 0$$

Hence, the weight function v is required to satisfy the conditions

$$v(0) = \left(\frac{dv}{dx}\right)\Bigg|_{x=0} = 0 \tag{2.54}$$

The natural boundary conditions are

$$\left[\frac{d}{dx}\left(b\frac{d^2w}{dx^2}\right)\right]_{x=L} = 0, \qquad \left(b\frac{d^2w}{dx^2}\right)\Bigg|_{x=L} = M_0 \tag{2.55}$$

Using (2.54) and (2.55) in (2.52), we obtain

$$0 = \int_0^L \left(b\frac{d^2v}{dx^2}\frac{d^2w}{dx^2} - vf\right) dx - \left(\frac{dv}{dx}\right)\Bigg|_{x=L} M_0 \tag{2.56a}$$

or

$$B(v, w) = l(v) \tag{2.56b}$$

where

$$B(v, w) = \int_0^L b \frac{d^2v}{dx^2} \frac{d^2w}{dx^2}\,dx$$
$$l(v) = \int_0^L vf\,dx + \left(\frac{dv}{dx}\right)\bigg|_{x=L} M_0 \tag{2.56c}$$

The quadratic form, commonly known as the *total potential energy* of the beam, is obtained using (2.56*c*) and (2.43*b*):

$$I(w) = \int_0^L \left[\frac{b}{2}\left(\frac{d^2w}{dx^2}\right)^2 - wf\right] dx - \left(\frac{dw}{dx}\right)\bigg|_{x=L} M_0 \tag{2.57}$$

Note that for the fourth-order equation, the essential boundary conditions involve not only the dependent variable but also its first derivative. As pointed out earlier, at any boundary point, only one of the two boundary conditions (essential or natural) can be specified. For example, if the transverse deflection is specified at a boundary point then one cannot specify the shear force V at the same point, and vice versa. Similar comments apply to the slope dw/dx and the bending moment M. Note that in the present case, w and dw/dx are the primary variables, and V and M are the secondary variables.

The next example is concerned with a second-order differential equation governing conductive and convective heat transfer in two dimensions. It should be noted that the boundary condition for a convective boundary contains both primary and secondary variables.

Example 2.3. Consider steady heat conduction in a two-dimensional domain Ω, enclosed by lines AB, BC, CD, DE, EF, FG, GH, and HA (see Fig. 2.2). The governing equation is

$$-k\left(\frac{\partial^2 T}{\partial x^2} + \frac{\partial^2 T}{\partial y^2}\right) = q_0 \quad \text{in } \Omega \tag{2.58}$$

where q_0 is the uniform heat generation, k is the conductivity of the isotropic material of the domain, and T is the temperature. We wish to construct the weak form of the

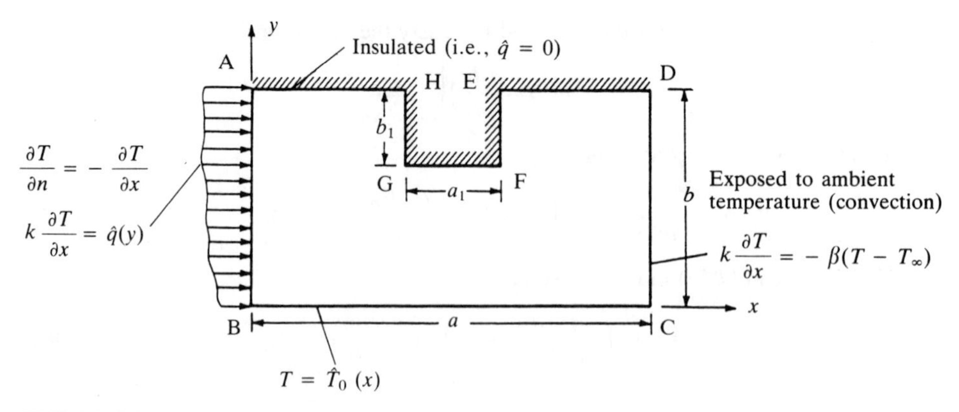

FIGURE 2.2
Conduction and convection heat transfer in two-dimensional domains.

equation. Equation (2.58), known as the Poisson equation, arises in many fields of engineering (see Table 8.1).

Proceeding as described earlier, we have

$$0 = \int_\Omega w\left[-k\left(\frac{\partial^2 T}{\partial x^2} + \frac{\partial^2 T}{\partial y^2}\right) - q_0\right] dx\, dy$$

where w denotes the weight function. Using (2.22) [with $G = \partial T/\partial x$ in (2.22*a*) and $G = \partial T/\partial y$ in (2.22*b*)], we obtain

$$0 = \int_\Omega \left[k\left(\frac{\partial w}{\partial x}\frac{\partial T}{\partial x} + \frac{\partial w}{\partial y}\frac{\partial T}{\partial y}\right) - wq_0\right] dx\, dy - \oint_\Gamma wk\left(\frac{\partial T}{\partial x} n_x + \frac{\partial T}{\partial y} n_y\right) ds \tag{2.59}$$

The reader should verify the last step [i.e. the application of (2.22)]. From the boundary expression, it follows that the secondary variable of the problem is of the form

$$k\left(\frac{\partial T}{\partial x} n_x + \frac{\partial T}{\partial y} n_y\right) = k\frac{\partial T}{\partial n} \equiv q_n$$

and the primary variable is T. The secondary variable q_n denotes the total flux across (i.e., along the normal to) the boundary. In general, q_n is composed of fluxes due to conduction, convection, and radiation.

The boundary Γ of the domain consists of several line segments, and they are subject to different types of boundary conditions (see Fig. 2.2):

on $\Gamma_1 = \text{AB}$ ($n_x = -1$, $n_y = 0$):	specified heat flux, $\hat{q}(y)$	
on $\Gamma_2 = \text{BC}$ ($n_x = 0$, $n_y = -1$):	specified temperature, $\hat{T}_0(x)$	
on $\Gamma_3 = \text{CD}$ ($n_x = 1$, $n_y = 0$):	convective boundary with ambient temperature T_∞ and film coefficient β; $k\,\partial T/\partial n + \beta(T - T_\infty) = 0$	(2.60)
on $\Gamma_4 = \text{DEFGHA}$:	insulated boundary, $\partial T/\partial n = 0$	

Using the boundary information, the boundary integral in (2.59) can be simplified as follows (note that $w = 0$ on Γ_2):

$$\begin{aligned}\oint_\Gamma w\left(k\frac{\partial T}{\partial n}\right) ds &= \int_{\Gamma_1} wq_n\, ds + \int_{\Gamma_2} 0\left(k\frac{\partial T}{\partial n}\right) ds \\ &\quad - \int_{\Gamma_3} w[\beta(T - T_\infty)]\, ds + \int_{\Gamma_4} w0\, ds \\ &= -\int_0^b w(0, y)\hat{q}(y)\, dy - \beta\int_0^b w(a, y)[T(a, y) - T_\infty]\, dy \end{aligned} \tag{2.61}$$

Substituting (2.61) into (2.59), we obtain the weak form

$$\begin{aligned}0 = \int_\Omega \left[k\left(\frac{\partial w}{\partial x}\frac{\partial T}{\partial x} + \frac{\partial w}{\partial y}\frac{\partial T}{\partial y}\right) - wq_0\right] dx\, dy + \int_0^b w(0, y)\hat{q}(y)\, dy \\ + \beta\int_0^b w(a, y)[T(a, y) - T_\infty]\, dy\end{aligned} \tag{2.62}$$

Collecting terms involving both w and T into $B(\cdot\,, \cdot)$, and those involving only w into $l(\cdot)$, we can write (2.62) in the form

$$B(w, T) = l(w) \tag{2.63a}$$

where

$$B(w, T) = \int_\Omega k\left(\frac{\partial w}{\partial x}\frac{\partial T}{\partial x} + \frac{\partial w}{\partial y}\frac{\partial T}{\partial y}\right) dx\, dy + \beta \int_0^b w(a, y)T(a, y)\, dy$$

$$l(w) = \int_\Omega w q_0\, dx\, dy - \int_0^b w(0, y)\hat{q}(y)\, dy + \beta \int_0^b w(a, y)T_\infty\, dy \tag{2.63b}$$

The quadratic functional is given by

$$I(T) = \frac{k}{2}\int_\Omega \left[\left(\frac{\partial T}{\partial x}\right)^2 + \left(\frac{\partial T}{\partial y}\right)^2\right] dx\, dy - \int_\Omega T q_0\, dx\, dy$$

$$+ \int_0^b T(0, y)\hat{q}(y)\, dy + \beta \int_0^b \tfrac{1}{2}[T^2(a, y) - 2T(a, y)T_\infty]\, dy \tag{2.63c}$$

Note that the boundary integrals in this example are defined along the y and x axes, respectively. This is because the boundaries are parallel to either the x or the y axis.

2.4 VARIATIONAL METHODS OF APPROXIMATION

2.4.1 Introduction

Our objective in this section is to study the variational methods of approximation. These include the Rayleigh–Ritz, Galerkin, Petrov–Galerkin, least-squares, and collocation methods. In all these, we seek an approximate solution in the form of a linear combination of suitable approximation functions ϕ_j and undetermined parameters c_j: $\sum_j c_j\phi_j$. The parameters c_j are determined such that the approximate solution satisfies the weighted-integral form or weak form of the governing equation or minimizes the quadratic functional associated with the equation studied. Various methods differ from each other in the choice of weight function w and approximation functions ϕ_j.

The primary objective of this section is to present a number of classical variational methods. The finite element method makes use of variational methods to formulate the discrete equations over an element. As we shall see in Chapters 3–14, the choice of the approximation functions in the finite element methods is different from that in the classical variational methods.

2.4.2 The Rayleigh–Ritz Method

In the Rayleigh–Ritz method, the coefficients c_j of the approximation are determined using the weak form of the problem, and the choice of weight functions is restricted to the approximation functions, $w = \phi_i$. Recall that the weak form contains both the governing differential equation and the natural boundary conditions of the problem, and it places less stringent continuity requirements on the approximate solution than the original differential equation or its weighted-integral form. The method is described below for a linear variational problem.

Consider the variational problem of finding the solution u such that

$$B(w, u) = l(w) \tag{2.64}$$

for all sufficiently differentiable functions w that satisfy the homogeneous form of any specified essential boundary conditions on u. When the functional B is bilinear and symmetric and l is linear, the problem in (2.64) is equivalent to minimization of the quadratic functional

$$I(u) = \tfrac{1}{2}B(u, u) - l(u) \tag{2.65}$$

In the Rayleigh–Ritz method, we seek an approximate solution to (2.64) in the form of a finite series

$$\boxed{u_N = \sum_{j=1}^{N} c_j\phi_j + \phi_0} \tag{2.66}$$

where the constants c_j, called the *Ritz coefficients*, are chosen such that (2.64) holds for $w = \phi_i$ $(i = 1, 2, \ldots, N)$; i.e., (2.64) holds for N different choices of w, so that N independent algebraic equations in c_j are obtained. The requirements on ϕ_j and ϕ_0 will be discussed shortly. The ith algebraic equation is obtained by substituting ϕ_i for w:

$$B\left(\phi_i, \sum_{j=1}^{N} c_j\phi_j + \phi_0\right) = l(\phi_i) \quad (i = 1, 2, \ldots, N)$$

If B is bilinear, the summation and constants c_j can be taken outside the operator. We have

$$\sum_{j=1}^{N} B(\phi_i, \phi_j)c_j = l(\phi_i) - B(\phi_i, \phi_0) \tag{2.67a}$$

or

$$\sum_{j=1}^{N} B_{ij}c_j = F_i, \qquad B_{ij} = B(\phi_i, \phi_j), \qquad F_i = l(\phi_i) - B(\phi_i, \phi_0) \tag{2.67b}$$

which represents the ith algebraic equation in a system of N linear algebraic equations in N constants c_j. The columns (and rows) of the matrix coefficients $B_{ij} = B(\phi_i, \phi_j)$ must be linearly independent in order that the coefficient matrix in (2.67) can be inverted.

For symmetric bilinear forms, the Rayleigh–Ritz method can also be viewed as one that seeks a solution of the form in (2.66) in which the parameters are determined by minimizing the quadratic functional corresponding to the symmetric bilinear form, that is, the functional $I(u)$ in (2.65). After substituting u_N from (2.66) for u into (2.65) and integrating, the functional $I(u)$ becomes an ordinary (quadratic) function of the parameters $c_1, c_2, \ldots$. Then the necessary condition for the minimization of $I(c_1, c_2, \ldots, c_N)$ is that its partial derivatives with respect to each of the parameters be zero:

$$\frac{\partial I}{\partial c_1} = 0, \quad \frac{\partial I}{\partial c_2} = 0, \quad \ldots, \quad \frac{\partial I}{\partial c_N} = 0 \tag{2.68}$$

Thus there are N linear algebraic equations in N unknowns, c_j ($j = 1, 2, \ldots, N$). These equations are exactly the same as those in (2.67) for all problems for which the variational problem (2.64) is equivalent to $\delta I = 0$. Of course, when $B(\cdot, \cdot)$ is not symmetric, we do not have a quadratic functional. In other words, (2.67) is more general than (2.68), and they are the same when $B(\cdot, \cdot)$ is bilinear and symmetric. In most problems of interest in the present study, we shall have a symmetric bilinear form.

Returning to the Rayleigh–Ritz approximation u_N in (2.66), we note that u_N must satisfy the specified essential boundary conditions of the problem; any specified natural boundary conditions are already included in the variational problem (2.64). The particular form of u_N in (2.66) facilitates satisfaction of specified boundary conditions. If we were to use the form

$$u_N = \sum_{j=1}^{N} c_j \phi_j(x)$$

then it would not be easy to satisfy nonhomogeneous boundary conditions. For example, suppose that u_N is required to satisfy the condition $u_N(x_0) = u_0$ at a boundary point $x = x_0$:

$$\sum_{j=1}^{N} c_j \phi_j(x_0) = u_0$$

Since c_j are unknown parameters to be determined, it is not easy to choose $\phi_j(x)$ such that this relation holds. If $u_0 = 0$ then any ϕ_j such that $\phi_j(x_0) = 0$ would meet the requirement. By writing the approximate solution u_N in the form (2.66), a sum of homogeneous and nonhomogeneous parts, the nonhomogeneous essential boundary conditions can be satisfied by ϕ_0, $\phi_0(x_0) = u_0$, and ϕ_j are required to satisfy the homogeneous form of the same boundary condition, $\phi_j(x_0) = 0$. In this way, u_N satisfies the specified boundary conditions:

$$\begin{aligned} u_N(x_0) &= \sum_{j=1}^{N} c_j \phi_j(x_0) + \phi_0(x_0) \\ &= 0 + u_0 \end{aligned}$$

If all specified essential boundary conditions are homogeneous (i.e., the specified value u_0 is zero) then ϕ_0 is taken to be zero and ϕ_j must still satisfy the same conditions, $\phi_j(x_0) = 0$. Since ϕ_i satisfy the homogeneous essential boundary conditions, the choice $w = \phi_i$ is consistent with the requirements of a weight function. The approximation functions ϕ_i satisfy the following conditions:

1. (*a*) ϕ_i should be such that $B(\phi_i, \phi_j)$ is well defined and nonzero [i.e., sufficiently differentiable as required by the bilinear form $B(\cdot, \cdot)$].
   (*b*) ϕ_i must satisfy at least the homogeneous form of the essential boundary conditions of the problem. (2.69)
2. For any N, the set $\{\phi_i\}_{i=1}^{N}$ along with the columns (and rows) of $B(\phi_i, \phi_j)$ must be linearly independent.
3. $\{\phi_i\}$ must be complete. For example, when ϕ_i are algebraic polynomials, completeness requires that the set $\{\phi_i\}$ should contain all terms of the lowest order admissible, and up to the highest order desired.

The only role that ϕ_0 plays is to satisfy the specified nonhomogeneous essential boundary conditions of the problem. Any low-order function that satisfies the specified essential boundary conditions should be used. If all specified essential boundary conditions are homogeneous then $\phi_0 = 0$ and

$$F_i = l(\phi_i) - B(\phi_i, \phi_0) = l(\phi_i) \tag{2.70}$$

Next, we consider a few examples of the application of the Rayleigh–Ritz method.

Example 2.4. Consider the differential equation [cf. Example 2.1, with $a = c = 1$]

$$-\frac{d^2u}{dx^2} - u + x^2 = 0 \quad \text{for } 0 < x < 1 \tag{2.71}$$

We consider two sets of boundary conditions:

$$\textbf{set 1:} \quad u(0) = 0, \qquad u(1) = 0 \tag{2.72a}$$

$$\textbf{set 2:} \quad u(0) = 0, \qquad \left.\left(\frac{du}{dx}\right)\right|_{x=1} = 1 \tag{2.72b}$$

Set 1. The bilinear functional and the linear functional are [see (2.47*c*)]

$$B(w, u) = \int_0^1 \left(\frac{dw}{dx}\frac{du}{dx} - wu\right) dx, \qquad l(w) = -\int_0^1 wx^2\, dx \tag{2.73}$$

Since both boundary conditions [$u(0) = u(1) = 0$] are of the essential type, we must select ϕ_i in the N-parameter Ritz approximation to satisfy the conditions $\phi_i(0) = \phi_i(1) = 0$. We choose the following functions: $\phi_0 = 0$ and

$$\phi_1 = x(1-x), \qquad \phi_2 = x^2(1-x), \qquad \ldots, \qquad \phi_N = x^N(1-x) \tag{2.74}$$

It should be pointed out that if one selects, for example, the functions $\phi_1 = x^2(1-x)$, $\phi_2 = x^3(1-x)$, etc. [not including $x(1-x)$], requirement 3 in the conditions (2.69) is violated, because the set cannot be used to generate the linear term x if the exact solution contains it. As a rule, one must start with the lowest-order admissible function and include all admissible, higher-order functions up to the desired degree.

The N-parameter Rayleigh–Ritz solution for the problem is of the form

$$u_N = c_1\phi_1 + c_2\phi_2 + \ldots + c_N\phi_N = \sum_{j=1}^{N} c_j\phi_j \tag{2.75}$$

Substituting this into the variational problem $B(w, u) = l(w)$, we obtain

$$\int_0^1 \left[\frac{d\phi_i}{dx}\left(\sum_{j=1}^{N} c_j \frac{d\phi_j}{dx}\right) - \phi_i\left(\sum_{j=1}^{N} c_j\phi_j\right)\right] dx = -\int_0^1 \phi_i x^2\, dx$$

$$\sum_{j=1}^{N} c_j \int_0^1 \left(\frac{d\phi_i}{dx}\frac{d\phi_j}{dx} - \phi_i\phi_j\right) dx = -\int_0^1 \phi_i x^2\, dx$$

or

$$\sum_{j=1}^{N} c_j B(\phi_i, \phi_j) = l(\phi_i) \tag{2.76a}$$

where the coefficients $B(\phi_i, \phi_j)$ and $l(\phi_i)$ are defined by

$$B(\phi_i, \phi_j) = \int_0^1 \left(\frac{d\phi_i}{dx}\frac{d\phi_j}{dx} - \phi_i\phi_j\right) dx, \qquad l(\phi_i) = -\int_0^1 x^2\phi_i\, dx \tag{2.76b}$$

The same result can be obtained using (2.65) [instead of (2.64)]. We have

$$I(u) = \frac{1}{2}\int_0^1 \left[\left(\frac{du}{dx}\right)^2 - u^2 + 2x^2u\right] dx$$

Substituting for $u \approx u_N$ from (2.75) into the above functional, we obtain

$$I(c_j) = \frac{1}{2}\int_0^1 \left[\left(\sum_{j=1}^{N} c_j \frac{d\phi_j}{dx}\right)^2 - \left(\sum_{j=1}^{N} c_j\phi_j\right)^2 + 2x^2\left(\sum_{j=1}^{N} c_j\phi_j\right)\right] dx \tag{2.77}$$

The necessary conditions for the minimization of I, which is a quadratic function of the variables $c_1, c_2, \ldots, c_N$, are

$$\frac{\partial I}{\partial c_i} = 0 = \int_0^1 \left[\frac{d\phi_i}{dx}\left(\sum_{j=1}^{N} c_j \frac{d\phi_j}{dx}\right) - \phi_i\left(\sum_{j=1}^{N} c_j\phi_j\right) + \phi_i x^2\right] dx$$

$$= \sum_{j=1}^{N} B_{ij}c_j - F_i$$

where

$$B_{ij} = \int_0^1 \left(\frac{d\phi_i}{dx}\frac{d\phi_j}{dx} - \phi_i\phi_j\right) dx, \qquad F_i = -\int_0^1 x^2\phi_i\, dx$$

which are the same as those in (2.76). Equations (2.76*a*, *b*) hold for any choice of admissible approximation functions ϕ_i.

For the choice of approximation functions in (2.74), the matrix coefficients $B_{ij} \equiv B(\phi_i, \phi_j)$ and vector coefficients $F_i \equiv l(\phi_i) - B(\phi_i, \phi_0) = l(\phi_i)$ can be computed using

$$\phi_i = x^i(1-x) = x^i - x^{i+1}$$

$$\frac{d\phi_i}{dx} = ix^{i-1} - (i+1)x^i$$

We have

$$B_{ij} = \int_0^1 \{[ix^{i-1} - (i+1)x^i][jx^{j-1} - (j+1)x^j] - (x^i - x^{i+1})(x^j - x^{j+1})\}\, dx$$

$$= \frac{2ij}{(i+j)[(i+j)^2 - 1]} - \frac{2}{(i+j+1)(i+j+2)(i+j+3)} \tag{2.78a}$$

$$F_i = -\int_0^1 x^2(x^i - x^{i+1})\, dx = -\frac{1}{(3+i)(4+i)} \tag{2.78b}$$

Equation (2.76) can be written in matrix form as

$$[B]\{c\} = \{F\} \tag{2.79}$$

For example, when $N = 2$, (2.79) becomes

$$\frac{1}{420}\begin{bmatrix} 126 & 63 \\ 63 & 52 \end{bmatrix}\begin{Bmatrix} c_1 \\ c_2 \end{Bmatrix} = -\frac{1}{60}\begin{Bmatrix} 3 \\ 2 \end{Bmatrix}$$

and the use of Cramer's rule to solve the equations gives

$$c_1 = -\tfrac{10}{123} = -0.0813, \qquad c_2 = -\tfrac{21}{123} = -0.1707.$$

The two-parameter Rayleigh–Ritz solution is given by

$$u_2 = c_1\phi_1 + c_2\phi_2 = (-\tfrac{10}{123})(x - x^2) + (-\tfrac{21}{123})(x^2 - x^3)$$
$$= -\tfrac{1}{123}(10x + 11x^2 - 21x^3)$$

The exact solution of (2.71) and (2.72a) is given by

$$u(x) = \frac{\sin x + 2\sin(1-x)}{\sin 1} + x^2 - 2 \tag{2.80}$$

The values of the Ritz coefficients for various values of N can be obtained by solving (2.79). A comparison of the Rayleigh–Ritz solution (2.75) with the exact solution (2.80) is presented in Table 2.1 and Fig. 2.3.

Set 2. For the second set of boundary conditions (2.72b), the bilinear form is the same as that given in (2.73) and (2.76b). The linear form is given by ($\phi_0 = 0$)

$$l(w) = -\int_0^1 wx^2\, dx + w(1) \tag{2.81a}$$

and we therefore have

$$F_i = -\int_0^1 x^2\phi_i\, dx + \phi_i(1) \tag{2.81b}$$

In this case, the ϕ_i should be selected to satisfy the condition $\phi_i(0) = 0$, because the only EBC is at $x = 0$. The following choice of ϕ_i meets the requirements:

$$\phi_i = x^i \tag{2.82}$$

The coefficients B_{ij} and F_i can be computed using (2.82) in (2.76b) and (2.81b) respectively:

$$\begin{aligned} B_{ij} &= \int_0^1 (ijx^{i+j-2} - x^{i+j})\, dx = \frac{ij}{1+j-1} - \frac{1}{i+j+1} \\ F_i &= -\int_0^1 x^{i+2}\, dx + 1 = -\frac{1}{i+3} + 1 \end{aligned} \tag{2.83}$$

TABLE 2.1
Comparison of the Rayleigh–Ritz and exact solutions of the equation

$$-\frac{d^2u}{dx^2} - u + x^2 = 0 \quad \text{for } 0 < x < 1; \qquad u(0) = u(1) = 0$$

Ritz coefficients†	x	Rayleigh–Ritz solution, $-10u$			Exact solution
		$N=1$	$N=2$	$N=3$	
$N=1$:	0.0	0.0	0.0	0.0	0.0
$c_1 = -0.1667$	0.1	0.1500	0.0885	0.0954	0.0955
$N=2$:	0.2	0.2667	0.1847	0.1890	0.1890
$c_1 = -0.0813$	0.3	0.3500	0.2783	0.2766	0.2764
$c_2 = -0.1707$	0.4	0.4000	0.3590	0.3520	0.3518
$N=3$:	0.5	0.4167	0.4167	0.4076	0.4076
$c_1 = -0.0952$	0.6	0.4000	0.4410	0.4340	0.4342
$c_2 = -0.1005$	0.7	0.3500	0.4217	0.4200	0.4203
$c_3 = -0.0702$	0.8	0.2667	0.3486	0.3529	0.3530
	0.9	0.1500	0.2115	0.2183	0.2182
	1.0	0.0	0.0	0.0	0.0

† The four-parameter Rayleigh–Ritz solution coincides with the exact solution up to four decimal places.

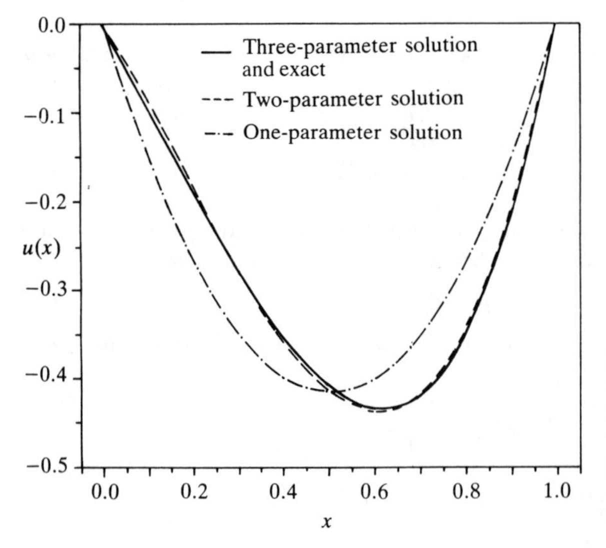

FIGURE 2.3
Comparison of the Rayleigh–Ritz solution with the exact solution of (2.71) and (2.72*a*). The three-parameter solution and the exact solution do not differ on the scale of the plot.

TABLE 2.2
Comparison of the Rayleigh–Ritz and exact solutions of the equation

$$-\frac{d^2u}{dx^2} - u + x^2 = 0 \quad \text{for } 0 < x < 1; \qquad u(0) = 0, \qquad \left(\frac{du}{dx}\right)\bigg|_{x=1} = 1$$

Ritz coefficients†	x	Rayleigh–Ritz solution, u $N=1$	$N=2$	$N=3$	Exact solution
$N=1$:	0.0	0.0	0.0	0.0	0.0
$c_1 = 1.1250$	0.1	0.1125	0.1280	0.1271	0.1262
$N=2$:	0.2	0.2250	0.2530	0.2519	0.2513
$c_1 = 1.2950$	0.3	0.3375	0.3749	0.3740	0.3742
$c_2 = -0.15108$	0.4	0.4500	0.4938	0.4934	0.4944
$N=3$:	0.5	0.5625	0.6097	0.6099	0.6112
$c_1 = 1.2831$	0.6	0.6750	0.7226	0.7234	0.7244
$c_2 = -0.11424$	0.7	0.7875	0.8325	0.8337	0.8340
$c_3 = -0.02462$	0.8	0.9000	0.9393	0.9407	0.9402
	0.9	1.0125	1.0431	1.0443	1.0433
	1.0	1.1250	1.1439	1.1442	1.1442

† The four-parameter Rayleigh–Ritz solution coincides with the exact solution up to four decimal places.

The exact solution in the present case is given by

$$u(x) = \frac{2\cos(1-x) - \sin x}{\cos 1} + x^2 - 2 \tag{2.84}$$

A comparison of the Rayleigh–Ritz solution with the exact solution is presented in Table 2.2.

Example 2.5. Consider the problem of finding the transverse deflection of a cantilever beam under a uniform transverse load of intensity f_0 per unit length and end moment M_0 using Euler–Bernoulli beam theory (see Example 2.2). The governing equations of this theory are

$$\frac{d^2}{dx^2}\left(EI\frac{d^2w}{dx^2}\right) - f_0 = 0 \quad \text{for } \begin{cases} 0 < x < L \\ EI > 0 \end{cases} \tag{2.85}$$

$$w(0) = \left(\frac{dw}{dx}\right)\bigg|_{x=0} = 0, \qquad \left(EI\frac{d^2w}{dx^2}\right)\bigg|_{x=L} = M_0, \qquad \left[\frac{d}{dx}\left(EI\frac{d^2w}{dx^2}\right)\right]\bigg|_{x=L} = 0 \tag{2.86}$$

The variational form of (2.85) (which includes the specified NBC) was derived in Example 2.2, and is given by (2.56).

We now construct an N-parameter Ritz solution using the variational form, (2.56), $B(v, w) = l(v)$, where

$$B(v, w) = \int_0^L EI\frac{d^2v}{dx^2}\frac{d^2w}{dx^2}\,dx, \qquad l(v) = \int_0^L f_0 v\,dx + \left(M_0\frac{dv}{dx}\right)\bigg|_{x=L} \tag{2.87}$$

Note that the specified EBC, $w(0)=0$ and $(dw/dx)|_{x=0}$ are homogeneous. Therefore, $\phi_0=0$. We select algebraic approximation functions ϕ_i that satisfy the continuity conditions and boundary conditions $\phi_i(0)=\phi_i'(0)=0$. The lowest-order algebraic function that meets these conditions is $\phi_1=x^2$. The next function in the sequence is $\phi_2=x^3$. Thus we have

$$\phi_1=x^2, \quad \phi_2=x^3, \quad \ldots, \quad \phi_N=x^{N+1}$$

The N-parameter Rayleigh–Ritz approximation is

$$w_N(x)=\sum_{j=1}^{N} c_j\phi_j, \quad \phi_j=x^{j+1} \tag{2.88}$$

Substituting (2.88) for w and $v=\phi_i$ into (2.87), we obtain

$$\begin{aligned} B_{ij}&=\int_0^L EI(i+1)ix^{i-1}(j+1)jx^{j-1}\,dx=\frac{EIij(i+1)(j+1)L^{i+j-1}}{i+j-1}\\ F_i&=\frac{f_0(L)^{i+2}}{i+2}+M_0(i+1)L^i \end{aligned} \tag{2.89}$$

For $N=2$ (i.e., the two-parameter solution), we have

$$\begin{aligned} EI(4Lc_1+6L^2c_2)&=\tfrac{1}{3}f_0L^3+2M_0L\\ EI(6L^2c_1+12L^3c_2)&=\tfrac{1}{4}f_0L^4+3M_0L^2 \end{aligned} \tag{2.90a}$$

or, in matrix form,

$$EI\begin{bmatrix} 4L & 6L^2\\ 6L^2 & 12L^3 \end{bmatrix}\begin{Bmatrix} c_1\\ c_2 \end{Bmatrix}=\frac{f_0L^3}{12}\begin{Bmatrix} 4\\ 3L \end{Bmatrix}+M_0L\begin{Bmatrix} 2\\ 3L \end{Bmatrix} \tag{2.90b}$$

Solving for c_1 and c_2, we obtain

$$c_1=\frac{5f_0L^2+12M_0}{24EI}, \quad c_2=\frac{-f_0L}{12EI}$$

and the solution (2.88) becomes

$$w_2(x)=\frac{5f_0L^2+12M_0}{24EI}x^2-\frac{f_0L}{12EI}x^3 \tag{2.91}$$

For the three-parameter approximation ($N=3$), we obtain the matrix equation

$$EI\begin{bmatrix} 4 & 6L & 8L^2\\ 6L & 12L^2 & 18L^3\\ 8L^2 & 18L^2 & \frac{144}{5}L^4 \end{bmatrix}\begin{Bmatrix} c_1\\ c_2\\ c_3 \end{Bmatrix}=\begin{Bmatrix} \frac{1}{3}f_0L^2+2M_0\\ \frac{1}{4}f_0L^3+3M_0L\\ \frac{1}{5}f_0L^4+4M_0L^2 \end{Bmatrix} \tag{2.92}$$

The solution of this when substituted into (2.88) for $N=3$, gives

$$w_3(x)=\frac{f_0x^2}{24EI}(6L^2-4Lx+x^2)+\frac{M_0x^2}{2EI} \tag{2.93}$$

which coincides with the exact solution of (2.85) and (2.86). If we try to compute the four-parameter solution without knowing that the three-parameter solution is exact, the parameters c_j ($j>3$) will be zero. Figure 2.4 shows a comparison of the Rayleigh–Ritz solution with the exact solution.

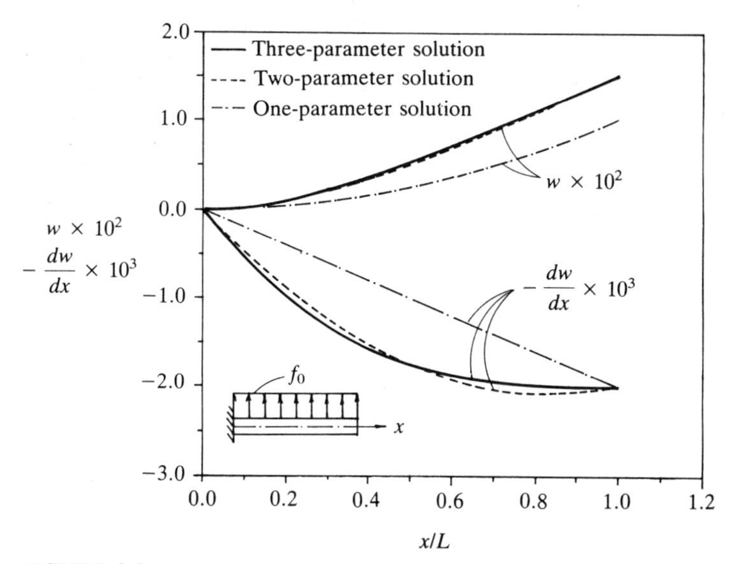

FIGURE 2.4
Comparison of the Rayleigh–Ritz solution with the exact solution of a cantilever beam under a uniform transverse load (Euler–Bernoulli beam theory).

The next example deals with two-dimensional heat conduction in a square region. Note that the dependent variable, namely the temperature, is denoted by T, consistent with the standard notation used in heat transfer books.

Example 2.6. Consider the Poisson equation in a unit square region:

$$-k\nabla^2 T = q_0 \quad \text{in } \Omega = \{(x, y): 0 < (x, y) < 1\} \tag{2.94a}$$

$$T = 0 \quad \text{on sides } x = 1 \text{ and } y = 1$$
$$\frac{\partial T}{\partial n} = 0 \quad \text{on sides } x = 0 \text{ and } y = 0 \tag{2.94b}$$

where q_0 is the rate of uniform heat generation in the region. The variational problem is of the form (see Example 2.3)

$$B(w, T) = l(w) \tag{2.95a}$$

where the bilinear and linear functionals are

$$B(w, T) = \int_0^1 \int_0^1 k\left(\frac{\partial w}{\partial x}\frac{\partial T}{\partial x} + \frac{\partial w}{\partial y}\frac{\partial T}{\partial y}\right) dx\, dy$$
$$l(v) = \int_0^1 \int_0^1 w q_0 \, dx\, dy \tag{2.95b}$$

We consider an N-parameter approximation of the form

$$T_N = \sum_{i,j=1}^{N} c_{ij} \cos \alpha_i x \cos \alpha_j y, \qquad \alpha_i = \tfrac{1}{2}(2i - 1)\pi \tag{2.96}$$

Note that (2.96) involves a double summation. Since the boundary conditions are homogeneous, we have $\phi_0 = 0$. Incidentally, ϕ_i also satisfies the natural boundary conditions of the problem. While the choice $\hat{\phi}_i = \sin i\pi x \sin i\pi y$ meets the essential boundary conditions, it is not complete, because it cannot be used to generate the solution that *does not* vanish on the sides $x = 0$ and $y = 0$. Hence, $\hat{\phi}_i$ are not admissible.

The coefficients B_{ij} and F_i can be computed by substituting (2.96) into (2.95*b*). Since the double Fourier series has two summations [see (2.96)], we introduce the notation

$$\begin{aligned} B_{(ij)(kl)} &= k \int_0^1 \int_0^1 [(\alpha_i \sin \alpha_i x \cos \alpha_j y)(\alpha_k \sin \alpha_k x \cos \alpha_l y) \\ &\quad + (\alpha_j \cos \alpha_i x \sin \alpha_j y)(\alpha_l \cos \alpha_k x \sin \alpha_l y)]\, dx\, dy \\ &= \begin{cases} 0 & \text{if } i \neq k \text{ or } j \neq l \\ \frac{1}{4}k(\alpha_i^2 + \alpha_j^2) & \text{if } i = k \text{ and } j = l \end{cases} \end{aligned} \tag{2.97a}$$

$$F_{ij} = q_0 \int_0^1 \int_0^1 \cos \alpha_i x \cos \alpha_j y\, dx\, dy = \frac{q_0}{\alpha_i \alpha_j} \sin \alpha_i \sin \alpha_j \tag{2.97b}$$

In evaluating the integrals, the following orthogonality conditions were used

$$\int_0^1 \sin \alpha_i x \sin \alpha_j x\, dx = \begin{cases} 0 & \text{if } i \neq j \\ \frac{1}{2} & \text{if } i = j \end{cases}$$

$$\int_0^1 \cos \alpha_i x \cos \alpha_j x\, dx = \begin{cases} 0 & \text{if } i \neq j \\ \frac{1}{2} & \text{if } i = j \end{cases}$$

Owing to the diagonal form of the coefficient matrix (2.97*a*), we can readily solve for the coefficients c_{ij}:

$$c_{ij} = \frac{F_{ij}}{B_{(ij)(ij)}} = \frac{4q_0}{k} \frac{\sin \alpha_i \sin \alpha_j}{(\alpha_i^2 + \alpha_j^2)\alpha_i \alpha_j} \tag{2.98}$$

The one- and two-parameter Rayleigh–Ritz solutions are

$$T_1 = \frac{32 q_0}{k\pi^4} \cos \tfrac{1}{2}\pi x \cos \tfrac{1}{2}\pi y \tag{2.99}$$

$$\begin{aligned} T_2 = \frac{q_0}{k} [&0.3285 \cos \tfrac{1}{2}\pi x \cos \tfrac{1}{2}\pi y - 0.0219(\cos \tfrac{1}{2}\pi x \cos \tfrac{3}{2}\pi y \\ &+ \cos \tfrac{3}{2}\pi x \cos \tfrac{1}{2}\pi y) + 0.0041 \cos \tfrac{3}{2}\pi x \cos \tfrac{3}{2}\pi y] \end{aligned} \tag{2.100}$$

If algebraic polynomials are to be used in the approximation of T, one can choose $\phi_1 = (1 - x)(1 - y)$ or $\phi_1 = (1 - x^2)(1 - y^2)$, both of which satisfy the (homogeneous) essential boundary conditions. However, the choice $\phi_1 = (1 - x^2)(1 - y^2)$ also meets the natural boundary conditions of the problem. The one-parameter Ritz solution for the choice $\phi_1 = (1 - x^2)(1 - y^2)$ is

$$T_1(x, y) = \frac{5q_0}{16k}(1 - x^2)(1 - y^2) \tag{2.101}$$

The exact solution of (2.94*a*, *b*) is

$$T(x, y) = \frac{q_0}{2k}\left[(1 - y^2) + 4 \sum_{n=1}^{\infty} \frac{(-1)^n \cos \alpha_n y \cosh \alpha_n x}{\alpha_n^3 \cosh \alpha_n}\right] \tag{2.102}$$

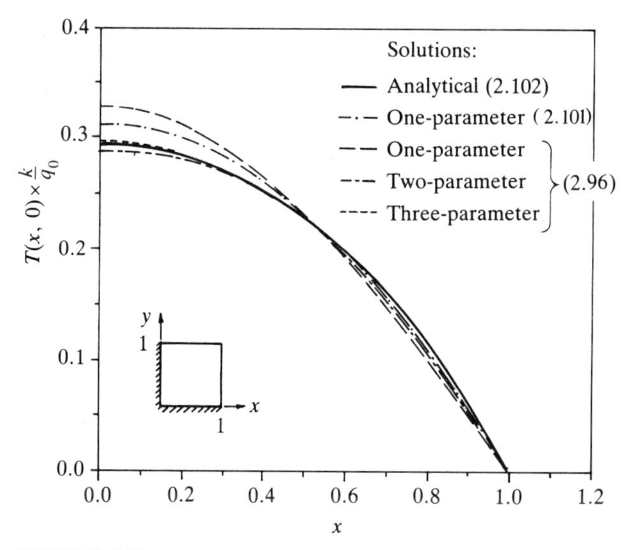

FIGURE 2.5
Comparison of the Rayleigh–Ritz solutions with the analytical solution of the Poisson equation (2.94) in two dimensions.

where $\alpha_n = \frac{1}{2}(2n-1)\pi$. The Rayleigh–Ritz solutions (2.99), (2.100), and (2.101) are compared with the exact solution (2.102) in Fig. 2.5. The analytical solution is evaluated using 20 terms of the series (2.102).

2.4.3 The Method of Weighted Residuals

As noted in Section 2.3.2, one can always write the weighted-integral form of a differential equation, whether the equation is linear or nonlinear (in the dependent variables). The weak form can be developed if the equations are second-order or higher, even if they are nonlinear. However, it is not always possible to construct a functional whose first variation is equal to the variational form. The Rayleigh–Ritz method can also be applied to all problems, including nonlinear problems, that have weak forms. In this method, the weight functions are necessarily equated to those used in the approximation. The weighted-residual method is a generalization of the Rayleigh–Ritz method in that the weight functions can be chosen from an independent set of functions, and it requires only the weighted-integral form to determine the parameters. The method of weighted residuals can be used to approximate the weighted-integral form of any equation. Since the latter form does not include any of the specified boundary conditions of the problem, the approximation functions should be selected such that the approximate solution satisfies both the natural and essential boundary conditions. In addition, the weight functions can be selected independently of the approximation functions, but are required to be linearly independent (so that the resulting algebraic equations are linearly independent). This flexibility is advantageous in certain nonlinear problems.

In this section, we discuss the general method of weighted residuals first, and then consider certain special cases that are known by specific names (e.g., the Galerkin and least-squares methods). Although a limited use of the weighted-residual method is made in this book (see Chapter 14), it is informative to have a knowledge of this method for use in the formulation of certain nonlinear problems readers might encounter in their work.

The method of weighted residuals can be described in its generality by considering the operator equation

$$A(u) = f \quad \text{in } \Omega, \tag{2.103}$$

where A is an operator (linear or nonlinear), often a differential operator, acting on the dependent variable u, and f is a known function of the independent variables. Some examples of such operators are provided by

$$\begin{aligned}
&\textbf{1. } A(u) = -\frac{d}{dx}\left(a\frac{du}{dx}\right) + cu \\
&\textbf{2. } A(u) = \frac{d^2}{dx^2}\left(b\frac{d^2u}{dx^2}\right) \\
&\textbf{3. } A(u) = -\left[\frac{\partial}{\partial x}\left(k_x\frac{\partial u}{\partial x}\right) + \frac{\partial}{\partial y}\left(k_y\frac{\partial u}{\partial y}\right)\right] \\
&\textbf{4. } A(u) = -\frac{d}{dx}\left(u\frac{du}{dx}\right) \\
&\textbf{5. } A(u, v) = u\frac{\partial u}{\partial x} + v\frac{\partial u}{\partial y} + \frac{\partial^2 u}{\partial x^2} + \frac{\partial}{\partial y}\left(\frac{\partial u}{\partial y} + \frac{\partial v}{\partial x}\right)
\end{aligned} \tag{2.104}$$

For an operator A to be *linear* in its arguments, it must satisfy the relation

$$A(\alpha u + \beta v) = \alpha A(u) + \beta A(v) \tag{2.105}$$

for any scalars α and β and dependent variables u and v. It can be easily verified that all operators in (2.104), except for 4 and 5, are linear. When an operator does not satisfy the condition (2.105), it is said to be *nonlinear*.

The function u is not only required to satisfy the operator equation (2.103), it is also required to satisfy the boundary conditions associated with the operator equation. From the examples considered so far, the boundary conditions associated with the operators defined in 1, 2, and 3 of (2.104) are obvious [see Examples 2.1–2.3].

In the weighted-residual method, the solution u is approximated, in much the same way as in the Rayleigh–Ritz method, by the expression

$$u_N = \sum_{j=1}^{N} c_j\phi_j + \phi_0 \tag{2.106}$$

except that the requirements on ϕ_0 and ϕ_j for the weighted-residual method are more stringent than those for the Rayleigh–Ritz method. Substitution of the approximate solution u_N into the left-hand side of (2.103) gives a function $f_N \equiv A(u_N)$ that, in general, is not equal to the specified function f. The difference $A(u_N) - f$, called the *residual* of the approximation, is nonzero:

$$R \equiv A(u_N) - f = A\left(\sum_{j=1}^{N} c_j\phi_j + \phi_0\right) - f \neq 0 \tag{2.107}$$

Note that the residual R is a function of position as well as of the parameters c_j. In the weighted-residual method, as the name suggests, the parameters c_j are determined by requiring the residual R to vanish in the weighted-integral sense:

$$\int_\Omega \psi_i(x, y)R(x, y, c_j)\, dx\, dy = 0 \quad (i = 1, 2, \ldots, N) \tag{2.108}$$

where Ω is a two-dimensional domain and ψ_i are *weight functions*, which, in general, are not the same as the approximation functions ϕ_i. The set $\{\psi_i\}$ must be a linearly independent set; otherwise, the equations provided by (2.108) will not be linearly independent and hence will not be solvable.

The requirements on ϕ_0 and ϕ_j for the weighted-residual method are different from those for the Rayleigh–Ritz method, which is based on the weak (integral) form of the differential equation. The differentiability requirement on ϕ_j in the weighted-residual method is dictated by the integral statement (2.108), as opposed to the weak form in the Rayleigh–Ritz method. Thus, ϕ_j must have nonzero derivatives up to the order appearing in the operator equation (2.103). Since the weighted-integral form (2.108) does not include any of the specified (either essential or natural) boundary conditions, we must also require u_N in (2.106) to satisfy all specified boundary conditions of the problem. Consequently, ϕ_0 is required to satisfy all specified boundary conditions, and ϕ_j are required to satisfy the homogeneous form of all specified boundary conditions of the problem. These requirements on ϕ_0 and ϕ_j will increase the order of the polynomial expressions used for the weighted-residual method. In general, the ϕ_j used in this method are higher-order functions than those used in the Rayleigh–Ritz method, and the functions used in the latter may not satisfy the continuity (i.e. differentiability) requirements of the weighted-residual method. Various special cases of the weighted-residual method are discussed in the following paragraphs.

THE PETROV–GALERKIN METHOD. The weighted-residual method is referred to as the *Petrov–Galerkin method* when $\psi_i \neq \phi_i$. When the operator A is linear, (2.108) can be simplified to the form

$$\sum_{j=1}^{N}\left[\int_\Omega \psi_i A(\phi_j)\, dx\, dy\right]c_j = \int_\Omega \psi_i[f - A(\phi_0)]\, dx\, dy$$

or

$$\sum_{j=1}^{N} A_{ij} c_j = F_i \tag{2.109}$$

Note that the coefficient matrix $[A]$ is not symmetric:

$$A_{ij} = \int_\Omega \psi_i A(\phi_j)\, dx\, dy \neq A_{ji} \tag{2.110}$$

THE GALERKIN METHOD. For the choice of weight function ψ_i equal to the approximation function ϕ_i, the weighted-residual method is better known as the Galerkin method. The algebraic equations of the Galerkin approximation are

$$\sum_{j=1}^{N} A_{ij} c_j = F_i \tag{2.111a}$$

where

$$A_{ij} = \int_\Omega \phi_i A(\phi_j)\, dx\, dy, \qquad F_i = \int_\Omega \phi_i [f - A(\phi_0)]\, dx\, dy \tag{2.111b}$$

Once again, we note that A_{ij} is not symmetric.

In general, the Galerkin method is *not* the same as the Rayleigh–Ritz method. This should be clear from the fact that the former uses the weighted-integral form whereas the latter uses the weak (or variational) form to determine the coefficients c_j. Consequently, the approximation functions used in the Galerkin method are required to be of higher order than those in the Rayleigh–Ritz method.

If the equation permits, and one wishes, the differentiation can be transferred from the solution u to the weight function $w = \phi_i$; and one thereby obtains the weak form to relax the continuity requirements on the approximation functions and include the specified natural boundary conditions of the problem.

The Rayleigh–Ritz and Galerkin methods yield the same solutions in two cases: (i) when the specified boundary conditions of the problem are all of the essential type, and therefore the requirements on ϕ_i in the two methods become the same and the weighted-integral form reduces to the weak form; and (ii) when the approximation functions of the Galerkin method are used in the Rayleigh–Ritz method. The reader is urged to keep the distinction between the Rayleigh–Ritz and Galerkin methods in mind.

THE LEAST-SQUARES METHOD. In this method, we determine the parameters c_j by minimizing the integral of the square of the residual (2.107):

$$\frac{\partial}{\partial c_i} \int_\Omega R^2(x, y, c_j)\, dx\, dy = 0$$

or

$$\int_\Omega \frac{\partial R}{\partial c_i} R \, dx \, dy = 0 \tag{2.112a}$$

Comparison of (2.112*a*) with (2.108) shows that $\psi_i = \partial R/\partial c_i$. If A is a linear operator, $\psi_i = A(\phi_i)$, and (2.112*a*) becomes

$$\sum_{j=1}^{N} \left[\int_\Omega A(\phi_i)A(\phi_j) \, dx \, dy \right] c_j = \int_\Omega A(\phi_i)[f - A(\phi_0)] \, dx \, dy$$

or

$$\sum_{j=1}^{N} A_{ij} c_j = F_i \tag{2.112b}$$

where

$$A_{ij} = \int_\Omega A(\phi_i)A(\phi_j) \, dx, \qquad F_i = \int_\Omega A(\phi_i)[f - A(\phi_0)] \, dx \tag{2.112c}$$

Note that the coefficient matrix A_{ij} is symmetric, but it involves the same order of differentiation as in the governing differential equation.

THE COLLOCATION METHOD. In the collocation method, we seek an approximate solution u_N to (2.103) in the form of (2.106) by requiring the residual in the equation to be identically zero at N selected points $\mathbf{x}^i \equiv (x^i, y^i)$ $(i = 1, 2, \ldots, N)$ in the domain Ω:

$$R(x^i, y^i, c_j) = 0 \quad (i = 1, 2, \ldots, N) \tag{2.113}$$

The selection of the points $\mathbf{x}^i$ is crucial in obtaining a well-conditioned system of equations and ultimately in obtaining an accurate solution. The collocation method can be shown to be a special case of (2.108) with $\psi_i = \delta(\mathbf{x} - \mathbf{x}^i)$, where $\delta(\mathbf{x})$ is the *Dirac delta function*, which is defined by

$$\int_\Omega f(\mathbf{x})\delta(\mathbf{x} - \boldsymbol{\xi}) \, dx \, dy = f(\boldsymbol{\xi}) \tag{2.114}$$

With this choice of weight functions, the weighted-residual statement becomes

$$\int_\Omega \delta(\mathbf{x} - \mathbf{x}^i) R(\mathbf{x}, c_j) \, dx \, dy = 0$$

or

$$R(\mathbf{x}^i, c_j) = 0 \tag{2.115}$$

We consider an example to illustrate the use of various cases of the weighted-residual method.

Example 2.7. Consider the differential equation [see Example 2.4 with Set 2 boundary conditions]:

$$-\frac{d^2u}{dx^2} - u + x^2 = 0, \qquad u(0) = 0, \qquad u'(1) = 1 \tag{2.116}$$

For a weighted-residual method, ϕ_0 and ϕ_i should satisfy the following conditions:

$\phi_0(0) = 0, \quad \phi_0'(1) = 1$ (satisfy actual boundary conditions)

$\phi_i(0) = 0, \quad \phi_i'(1) = 0$ (satisfy homogeneous form of the specified boundary conditions)

For a choice of algebraic polynomials, we assume $\phi_0(x) = a + bx$ and use the two conditions on ϕ_0 to determine the constants a and b. We obtain

$$\phi_0(x) = x$$

Since there are two homogeneous conditions, we must assume at least a three-parameter polynomial to obtain a nonzero function, $\phi_1 = a + bx + cx^2$. Using the conditions on ϕ_i, we obtain

$$\phi_1 = -cx(2 - x)$$

The constant c can be set equal to unity because it will be absorbed into the parameter c_1. For ϕ_2, we can assume one of the forms

$$\phi_2 = a + bx + dx^3 \quad \text{or} \quad \phi_2 = a + cx^2 + dx^3$$

with $d \neq 0$; ϕ_2 does not contain all-order terms in either case, but the approximate solution is complete because $\{\phi_1, \phi_2\}$ contains all terms up to degree three. For the first choice of ϕ_2, we obtain

$$\phi_2 = x^2(1 - \tfrac{2}{3}x)$$

The residual in the approximation of the equation is

$$\begin{aligned} R &= -\left(0 + \sum_{i=1}^{N} c_i \frac{d^2\phi_i}{dx^2}\right) - \left(\phi_0 + \sum_{i=1}^{N} c_i\phi_i\right) + x^2 \\ &= c_1(2 - 2x + x^2) + c_2(-2 + 4x - x^2 + \tfrac{2}{3}x^2) - x + x^2 \end{aligned} \tag{2.117}$$

We next consider various methods.

The Petrov–Galerkin method. Let the weight functions be

$$\psi_1 = x, \qquad \psi_2 = x^2 \tag{2.118}$$

Then

$$\int_0^1 xR\,dx = 0, \qquad \int_0^1 x^2R\,dx = 0$$

or

$$\tfrac{7}{12}c_1 + \tfrac{13}{60}c_2 - \tfrac{1}{12} = 0, \qquad \tfrac{11}{30}c_1 + \tfrac{11}{45}c_2 - \tfrac{1}{20} = 0 \tag{2.119}$$

Solving for c_i, we obtain $c_1 = \frac{103}{682}$ and $c_2 = -\frac{15}{682}$; the solution becomes

$$u_{PG} = 1.302053x - 0.173021x^2 - 0.014663x^3 \tag{2.120}$$

The Galerkin method. Taking $\psi_i = \phi_i$, we have

$$\int_0^1 x(2-x)R\,dx = 0, \qquad \int_0^1 x^2(1-\tfrac{2}{3}x)R\,dx = 0$$

or

$$\tfrac{4}{5}c_1 + \tfrac{28}{45}c_2 - \tfrac{7}{60} = 0, \qquad \tfrac{17}{90}c_1 + \tfrac{29}{315}c_2 - \tfrac{1}{36} = 0 \tag{2.121}$$

Hence, the solution becomes (with $c_1 = \frac{623}{4306}$, $c_2 = \frac{21}{4306}$),

$$u_G = 1.2894x - 0.1398x^2 - 0.00325x^3 \tag{2.122}$$

The least-squares method. Taking $\psi_i = \partial R/\partial c_i$, we have

$$\int_0^1 (2-2x+x^2)R\,dx = 0, \qquad -\int_0^1 (2-4x+x^2-\tfrac{2}{3}x^3)R\,dx = 0$$

or

$$\tfrac{28}{15}c_1 - \tfrac{47}{90}c_2 - \tfrac{13}{60} = 0, \qquad -\tfrac{47}{90}c_1 + \tfrac{253}{315}c_2 + \tfrac{1}{36} = 0 \tag{2.123}$$

The least-squares approximation of (2.116) is given by (with $c_1 = \frac{1292}{9935}$ and $c_2 = \frac{991}{19870}$)

$$u_{LS} = 1.2601x - 0.08017x^2 - 0.03325x^3 \tag{2.124}$$

The collocation method. Choosing the points $x = \frac{1}{3}$ and $x = \frac{2}{3}$ as the collocation points, we evaluate the residuals at these points and set them equal to zero:

$$\begin{aligned} R(\tfrac{1}{3}) = 0: &\quad 117c_1 - 61c_2 = 18 \\ R(\tfrac{2}{3}) = 0: &\quad 90c_1 + 34c_2 = 18 \end{aligned} \tag{2.125}$$

The solution is given by ($c_1 = \frac{1710}{9468}$ and $c_2 = \frac{486}{9468}$)

$$u_C = 1.3612x - 0.12927x^2 - 0.03422x^3 \tag{2.126}$$

The four approximate solutions are compared in Table 2.3 with the exact solution (2.84). For this problem, the Petrov–Galerkin method gives the most accurate solution.

2.5 SUMMARY

In this chapter, we have studied two major topics that are of immediate interest in the study of the finite element method in the forthcoming chapters:

1. Weighted-integral and weak formulations of differential equations
2. Solution of boundary value problems by the Rayleigh–Ritz and weighted-residual (e.g., the Galerkin, least-squares, and collocation) methods

The weighted-integral statements are required in order to generate the necessary and sufficient number of algebraic equations to solve for the parameters c_i in the approximate solution. Thus the algebraic equations are equivalent to minimizing the error introduced in the approximation of the differential equation in a weighted-integral sense.

In studying the two topics, a three-step procedure for developing the weak form of a differential equation is presented, and procedures for obtaining

TABLE 2.3
Comparison of the Rayleigh–Ritz, weighted-residual, and exact solutions of the boundary value problem in (2.116)

	Solution, $u(x)$†					
x	u_{exact}	u_{RR}	u_{PG}	u_G	u_{LS}	u_C
0.0	0.0000	0.0000	0.0000	0.0000	0.0000	0.0000
0.1	0.1262	0.1280	0.1285	0.1275	0.1252	0.1348
0.2	0.2513	0.2529	0.2536	0.2523	0.2485	0.2668
0.3	0.3742	0.3749	0.3754	0.3741	0.3699	0.3958
0.4	0.4943	0.4938	0.4941	0.4932	0.4891	0.5216
0.5	0.6112	0.6097	0.6096	0.6093	0.6058	0.6440
0.6	0.7244	0.7226	0.7221	0.7226	0.7200	0.7628
0.7	0.8340	0.8324	0.8317	0.8329	0.8314	0.8778
0.8	0.9402	0.9393	0.9384	0.9404	0.9397	0.9887
0.9	1.0433	1.0431	1.0424	1.0448	1.0449	1.0954
1.0	1.1442	1.1439	1.1437	1.1463	1.1467	1.1977

† Subscripts are as follows: *RR*, Rayleigh–Ritz; *PG*, Petrov–Galerkin; *G*, Galerkin; *LS*, least-squares; *C*, collocation.

algebraic equations in terms of the unknown parameters of the approximate solution are developed. These topics are immediately applicable in the finite element method, which is a piecewise (or element-wise) application of a variational method. Thus, the material covered in this chapter constitutes the heart of the finite element method. A few remarks are in order on the variational methods of approximation studied here.

The traditional variational methods (e.g., the Rayleigh–Ritz, Galerkin, and least-squares) presented in Section 2.4 provide a simple means of finding spatially continuous approximate solutions to physical problems. The approximate solutions obtained via these methods are continuous functions of position in the domain.

The main disadvantage, from the practical point of view, of variational methods that prevents them from being competitive with traditional finite difference methods is the difficulty encountered in selecting the approximation functions. Apart from the properties the functions are required to satisfy, there exists no unique procedure for constructing them. The selection process becomes more difficult or even impossible when the domain is geometrically complex and/or the boundary conditions are complicated.

From the preceding discussion, it is apparent that the variational methods can provide a powerful means of finding approximate solutions—provided one can find a way to systematically construct approximation functions, for almost any geometry, that depend only on the differential equation being solved and not on the boundary conditions of the problem. This property enables one to develop a computer program for a particular class of problem (each problem in the class differs from the others only in the data), that is, a *general-purpose* computer program. Since the functions must be constructed for a geometrically complex domain, it seems that (recall the discussion of the method of composites for the determination of the center of mass of an irregular shape from Chapter 1) the region must be represented (or approximated if required) as an assemblage of simple geometric shapes for which the construction of approximation functions becomes simpler. The finite element method to be discussed in the forthcoming chapters is based on these ideas.

In this method, a given domain is represented (discretized) by a collection of geometrically simple shapes (elements), and on each element of the collection, the governing equation is *formulated* using any one of the variational methods. The approximation functions are systematically generated for each (typical) element using the essential boundary conditions. The elements are *connected* together by imposing the continuity of the dependent variables across the interelement boundaries. The remaining chapters of this book are devoted to the introduction of the finite element method and its use in the analysis of several model differential equations representing mathematical models for many physical processes.

PROBLEMS

Sections 2.1–2.3

In Problems 2.1–2.9, construct the weak forms and, whenever possible, quadratic functionals.

2.1. *One-dimensional heat conduction/convection:*

$$-\frac{d}{dx}\left(a\frac{du}{dx}\right)+cu=q \quad \text{for } 0<x<1$$

$$u(0)=u_0, \qquad \left[a\frac{du}{dx}+\beta(u-u_\infty)\right]\bigg|_{x=1}=Q_0 \quad \text{at } x=1$$

where a and q are functions of x, and β, c, u_∞, and Q_0 are constants.

2.2. *Beam on elastic foundation:*

$$\frac{d^2}{dx^2}\left(b\frac{d^2w}{dx^2}\right)+kw=f \quad \text{for } 0<x<L$$

$$w=b\frac{d^2w}{dx^2}=0 \quad \text{at } x=0,\ L$$

where $b=EI$ and f are functions of x, and k is a constant (foundation modulus).

2.3. *Longitudinal deformation of a bar with an end spring:*

$$-\frac{d}{dx}\left(a\frac{du}{dx}\right)=q \quad \text{for } 0<x<L$$

$$u(0)=0, \qquad \left(a\frac{du}{dx}+ku\right)\bigg|_{x=L}=P$$

where a and q are functions of x, and k and P are constants.

2.4. *The Timoshenko (shear-deformable) beam theory:*

$$\left.\begin{aligned} -\frac{d}{dx}\left[GKA\left(\frac{dw}{dx}+\Psi\right)\right]&=f \\ -\frac{d}{dx}\left(EI\frac{d\Psi}{dx}\right)+GKA\left(\frac{dw}{dx}+\Psi\right)&=0 \end{aligned}\right\} \quad \text{for } 0<x<L$$

$$w(0)=w(L)=0, \qquad \left(EI\frac{d\Psi}{dx}\right)\bigg|_{x=0}=\left(EI\frac{d\Psi}{dx}\right)\bigg|_{x=L}=0$$

where G, K, A, E, I, and f are functions of x.

2.5. *A nonlinear equation:*

$$-\frac{d}{dx}\left(u\frac{du}{dx}\right)+f=0 \quad \text{for } 0<x<1$$

$$\left(\frac{du}{dx}\right)\bigg|_{x=0}=0, \qquad u(1)=\sqrt{2}$$

2.6. *The Euler–Bernoulli–von Kármán nonlinear theory of beams:*

$$-\frac{d}{dx}\left\{a\left[\frac{du}{dx}+\frac{1}{2}\left(\frac{dw}{dx}\right)^2\right]\right\}=q \quad \text{for } 0<x<L$$

$$\frac{d^2}{dx^2}\left(b\frac{d^2w}{dx^2}\right)-\frac{d}{dx}\left\{a\frac{dw}{dx}\left[\frac{du}{dx}+\frac{1}{2}\left(\frac{dw}{dx}\right)^2\right]\right\}=f$$

$$u=w=0 \quad \text{at } x=0, L; \qquad \left(\frac{dw}{dx}\right)\bigg|_{x=0}=0; \qquad \left(b\frac{d^2w}{dx^2}\right)\bigg|_{x=L}=M_0$$

where a, b, q, and f are functions of x, and M_0 is a constant. Here u denotes the axial displacement and w the transverse deflection of the beam.

2.7. *A second-order equation:*

$$-\frac{\partial}{\partial x}\left(a_{11}\frac{\partial u}{\partial x}+a_{12}\frac{\partial u}{\partial y}\right)-\frac{\partial}{\partial y}\left(a_{21}\frac{\partial u}{\partial x}+a_{22}\frac{\partial u}{\partial y}\right)+f=0 \quad \text{in } \Omega$$

$$u=u_0 \quad \text{on } \Gamma_1$$

$$\left(a_{11}\frac{\partial u}{\partial x}+a_{12}\frac{\partial u}{\partial y}\right)n_x+\left(a_{21}\frac{\partial u}{\partial x}+a_{22}\frac{\partial u}{\partial y}\right)n_y=t_0 \quad \text{on } \Gamma_2$$

where $a_{ij}=a_{ji}$ $(i,j=1,2)$ and f are given functions of position (x,y) in a two-dimensional domain Ω, and u_0 and t_0 are known functions on portions Γ_1 and Γ_2 of the boundary Γ: $\Gamma_1+\Gamma_2=\Gamma$.

2.8. *Navier–Stokes equations for two-dimensional flow of viscous, incompressible fluids* (primitive variables):

$$\left.\begin{aligned} u\frac{\partial u}{\partial x}+v\frac{\partial u}{\partial y}&=-\frac{1}{\rho}\frac{\partial P}{\partial x}+\nu\left(\frac{\partial^2u}{\partial x^2}+\frac{\partial^2u}{\partial y^2}\right) \\ u\frac{\partial v}{\partial x}+v\frac{\partial v}{\partial y}&=-\frac{1}{\rho}\frac{\partial P}{\partial y}+\nu\left(\frac{\partial^2v}{\partial x^2}+\frac{\partial^2v}{\partial y^2}\right) \\ \frac{\partial u}{\partial x}+\frac{\partial v}{\partial y}&=0 \end{aligned}\right\} \quad \text{in } \Omega \qquad \text{(i)}$$

$$u = u_0, \qquad v = v_0 \quad \text{on } \Gamma_1 \tag{ii}$$

$$\left.\begin{aligned} \nu\left(\frac{\partial u}{\partial x}n_x + \frac{\partial u}{\partial y}n_y\right) - \frac{1}{\rho}Pn_x = \hat{t}_x \\ \nu\left(\frac{\partial v}{\partial x}n_x + \frac{\partial v}{\partial y}n_y\right) - \frac{1}{\rho}Pn_y = \hat{t}_y \end{aligned}\right\} \quad \text{on } \Gamma_2 \tag{iii}$$

2.9. *Two-dimensional flow of viscous, incompressible fluids* (stream function–vorticity formulation):

$$\left.\begin{aligned} -\nabla^2\psi - \zeta = 0 \\ -\nabla^2\zeta + \frac{\partial\psi}{\partial x}\frac{\partial\zeta}{\partial y} - \frac{\partial\psi}{\partial y}\frac{\partial\zeta}{\partial x} = 0 \end{aligned}\right\} \quad \text{in } \Omega$$

Assume that all essential boundary conditions are specified to be zero.

Section 2.4

2.10. Compute the coefficient matrix and the right-hand side of the N-parameter Rayleigh–Ritz approximation of the equation

$$-\frac{d}{dx}\left[(1+x)\frac{du}{dx}\right] = 0 \quad \text{for } 0 < x < 1$$

$$u(0) = 0, \qquad u(1) = 1$$

Use algebraic polynomials for the approximation functions. Specialize your result for $N = 2$ and compute the Ritz coefficients.

Answer: $c_1 = \frac{55}{131}$ and $c_2 = -\frac{20}{131}$.

2.11. Use trigonometric functions for the two-parameter approximation of the equation in Problem 2.10, and obtain the Ritz coefficients.

2.12. A steel rod of diameter $D = 2$ cm, length $L = 25$ cm, and thermal conductivity $k = 50\ \text{W m}^{-1}\ {}^\circ\text{C}^{-1}$ is exposed to ambient air at $T_\infty = 20°\text{C}$ with a heat-transfer coefficient $\beta = 64\ \text{W m}^{-2}\ {}^\circ\text{C}^{-1}$. Given that the left end of the rod is maintained at a temperature of $T_0 = 120°\text{C}$ and the other end is exposed to the ambient temperature, determine the temperature distribution in the rod using a two-parameter Rayleigh–Ritz approximation with polynomial approximation functions. The equation governing the problem is given by (see Problem 2.1)

$$-\frac{d^2\theta}{dx^2} + c\theta = 0 \quad \text{for } 0 < x < 25 \text{ cm}$$

where $\theta = T - T_\infty$, T is the temperature, and c is given by

$$c = \frac{\beta P}{Ak} = \frac{\beta\pi D}{\frac{1}{4}\pi D^2 k} = \frac{4\beta}{kD} = 256\ \text{m}^{-2}$$

P being the perimeter and A the cross-sectional area of the rod. The boundary conditions are

$$\theta(0) = T(0) - T_\beta = 100°\text{C}, \qquad \left.\left(k\frac{d\theta}{dx} + \beta\theta\right)\right|_{x=L} = 0$$

Answer: For $L = 0.25$ m, $\phi_0 = 100$, $\phi_i = x^i$, the Ritz coefficients are $c_1 = -1033.385$, $c_2 = 2667.261$.

— **2.13.** Set up the equations for the N-parameter Rayleigh–Ritz approximation of the following equations associated with a simply supported beam and subjected to a uniform transverse load $f = f_0$:

$$\frac{d^2}{dx^2}\left(EI\frac{d^2w}{dx^2}\right) = f_0 \quad \text{for } 0 < x < L$$

$$w = EI\frac{d^2w}{dx^2} = 0 \quad \text{at } x = 0, L$$

(a) Use algebraic polynomials.
(b) Use trigonometric functions.

Compare the two-parameter Rayleigh–Ritz solutions with the exact solution.

Answer: (a) $c_1 = \dfrac{f_0L^2}{24EI}$, $c_2 = 0$.

— **2.14.** Repeat Problem 2.13 for $f = f_0 \sin(\pi x/L)$.

2.15. Repeat Problem 2.13 for $f = F_0\delta(x - \frac{1}{2}L)$, where $\delta(x)$ is the Dirac delta function (i.e., a point load F_0 is applied at the center of the beam).

2.16. Develop the N-parameter Rayleigh–Ritz solution for a simply supported beam under uniform transverse load using Timoshenko beam theory. The governing equations are given in Problem 2.4. Use trigonometric functions to approximate w and Ψ.

2.17. Solve the Poisson equation governing heat conduction in a square region (see Example 2.6):

$$-k\nabla^2 T = q_0$$

$$T = 0 \quad \text{on sides } x = 1 \text{ and } y = 1$$

$$\frac{\partial T}{\partial n} = 0 \text{ (insulated)} \quad \text{on sides } x = 0 \text{ and } y = 0$$

using a one-parameter Rayleigh–Ritz approximation of the form

$$T_1(x, y) = c_1(1 - x^2)(1 - y^2)$$

Answer: $c_1 = \dfrac{5q_0}{16k}$.

2.18. Solve Problem 2.12 using a two-parameter Galerkin approximation with algebraic approximation functions.

2.19. Consider the (Neumann) boundary value problem

$$-\frac{d^2u}{dx^2} = f \quad \text{for } 0 < x < L$$

$$\left(\frac{du}{dx}\right)\bigg|_{x=0} = \left(\frac{du}{dx}\right)\bigg|_{x=L} = 0$$

Find a two-parameter Galerkin approximation of the problem using trigonometric approximation functions, when (a) $f = f_0 \cos(\pi x/L)$ and (b) $f = f_0$.

Answer: (a) $\phi_i = \cos(i\pi x/L)$, $c_1 = f_0L^2/\pi^2$, $c_i = 0$ for $i \neq 1$.

— **2.20.** Find a one-parameter approximate solution of the nonlinear equation

$$-2u\frac{d^2u}{dx^2} + \left(\frac{du}{dx}\right)^2 = 4 \quad \text{for } 0 < x < 1$$

subject to the boundary conditions $u(0) = 1$ and $u(1) = 0$, and compare it with the exact solution $u_0 = 1 - x^2$. Use (*a*) the Galerkin method, (*b*) the least-squares method, and (*c*) the Petrov–Galerkin method with weight function $w = 1$.

Answer: (*a*) $(c_1)_1 = 1$, $(c_1)_2 = -3$.

2.21. Give a one-parameter Galerkin solution of the equation

$$-\nabla^2 u = 1 \quad \text{in } \Omega \ (= \text{unit square})$$

$$u = 0 \quad \text{on } \Gamma$$

Use (*a*) algebraic and (*b*) trigonometric approximation functions. What would be the one-parameter Rayleigh–Ritz solution of this problem?

Answer: (*b*) $c_{ij} = \dfrac{16}{\pi^4} \dfrac{1}{ij(i^2 + j^2)} \quad (i, j \text{ odd})$

$$\phi_{ij} = \sin i\pi x \sin j\pi y$$

2.22. Repeat Problem 2.21 for an equilateral triangular domain.

Hint: Use the product of equations of the lines representing the sides of the triangle for the approximation function.

Answer: $c_1 = -\frac{1}{2}$.

2.23. Consider the differential equation

$$-\frac{d^2u}{dx^2} = \cos \pi x \quad \text{for } 0 < x < 1$$

subject to the following three sets of boundary conditions:

(1) $u(0) = 0, \quad u(1) = 0$

(2) $u(0) = 0, \quad \left(\dfrac{du}{dx}\right)\Big|_{x=1} = 0$

(3) $\left(\dfrac{du}{dx}\right)\Big|_{x=0} = 0, \quad \left(\dfrac{du}{dx}\right)\Big|_{x=1} = 0$

Determine a three-parameter solution, with trigonometric functions, using (*a*) the Rayleigh–Ritz method, (*b*) the least-squares method, and (*c*) collocation at $x = \frac{1}{4}$, $\frac{1}{2}$, and $\frac{3}{4}$, and compare with the exact solutions:

(1) $u_0 = \pi^{-2}(\cos \pi x + 2x - 1)$

(2) $u_0 = \pi^{-2}(\cos \pi x - 1)$

(3) $u_0 = \pi^{-2} \cos \pi x$

Answer: (1*a*) $c_i = \dfrac{4}{\pi^3 i(i^2 - 1)}$.

REFERENCES FOR ADDITIONAL READING

Variational formulations and methods

Becker, M.: *The Principles and Applications of Variational Methods,* MIT Press, Cambridge, MA 1964.

Biot, M. A.: *Variational Principles in Heat Transfer,* Clarendon, London, 1972.

Finlayson, B. A.: *The Method of Weighted Residuals and Variational Principles,* Academic Press, New York, 1972.

Forray, M. J.: *Variational Calculus in Science and Engineering,* McGraw-Hill, New York, 1968.

Hildebrand, F. B.: *Methods of Applied Mathematics,* 2d ed., Prentice-Hall, New York, 1965.

Lanczos, C.: *The Variational Principles of Mechanics,* The University of Toronto Press, Toronto, 1964.

Langhaar, H. L.: *Energy Methods in Applied Mechanics,* John Wiley, New York, 1962.

Leipholz, H.: *Direct Variational Methods and Eigenvalue Problems in Engineering,* Noordhoff, Leyden, 1977.

Lippmann, H.: *Extremum and Variational Principles in Mechanics,* Springer-Verlag, New York, 1972.

Mikhlin, S. G.: *Variational Methods in Mathematical Physics,* Pergamon Press, New York, 1964.

———: *The Numerical Performance of Variational Methods,* Wolters-Noordhoff, Groningen, 1971.

Oden, J. T., and J. N. Reddy: *Variational Methods in Theoretical Mechanics,* Springer-Verlag, New York, 1976; 2d ed., 1983.

Reddy, J. N.: *Energy and Variational Methods in Applied Mechanics,* John Wiley, New York, 1984.

———: *Applied Functional Analysis and Variational Methods in Engineering,* McGraw-Hill, New York, 1986; Krieger, Melbourne, FL, 1991.

——— and M. L. Rasmussen: *Advanced Engineering Analysis,* John Wiley, New York, 1982; Krieger, Melbourne, FL, 1990.

Rektorys, K.: *Variational Methods in Mathematics, Science and Engineering,* Reidel, Boston, 1977.

Schechter, R. S.: *The Variational Methods in Engineering,* McGraw-Hill, New York, 1967.

Washizu, K.: *Variational Methods in Elasticity and Plasticity,* 2d ed., Pergamon Press, New York, 1975; 3d ed., 1982.

Weinstock, R.: *Calculus of Variations with Applications to Physics and Engineering,* McGraw-Hill, New York, 1952.

PART 2

FINITE ELEMENT ANALYSIS OF ONE-DIMENSIONAL PROBLEMS

CHAPTER 3

SECOND-ORDER BOUNDARY VALUE PROBLEMS

3.1 INTRODUCTION

The traditional variational methods (e.g., the Rayleigh–Ritz, Galerkin, and least-squares) described in Chapter 2 cease to be effective because of a serious shortcoming, namely, the difficulty in constructing the approximation functions. The approximation functions, apart from satisfying continuity, linear independence, completeness, and essential boundary conditions, are arbitrary; the selection becomes even more difficult when the given domain is geometrically complex. Since the quality of the approximation is directly affected by the choice of the approximation functions, it is discomforting to know that there exists no systematic procedure to construct them. Because of this shortcoming, despite the simplicity in obtaining approximate solutions, the traditional variational methods of approximation were never regarded as competitive computationally when compared with traditional finite difference schemes.

Ideally speaking, an effective computational method should have the following features:

1. It should have a sound mathematical as well as physical basis (i.e., yield convergent solutions and be applicable to practical problems).
2. It should not have limitations with regard to the geometry, the physical composition of the domain, or the nature of the "loading."

3. The formulative procedure should be independent of the shape of the domain and the specific form of the boundary conditions.
4. The method should be flexible enough to allow different degrees of approximation without reformulating the entire problem.
5. It should involve a systematic procedure that can be automated for use on digital computers.

The finite element method is a technique in which a given domain is represented as a collection of simple domains, called *finite elements,* so that it is possible to systematically construct the approximation functions needed in a variational or weighted-residual approximation of the solution of a problem over each element. Thus, the finite element method differs from the traditional Rayleigh–Ritz, Galerkin, least-squares, collocation, and other weighted-residual methods in the manner in which the approximation functions are constructed. But this difference is responsible for the following three basic features of the finite element method:

1. *Division of whole into parts,* which allows representation of geometrically complex domains as collections of geometrically simple domains that enable a systematic derivation of the approximation functions.
2. *Derivation of approximation functions* over each element; the approximation functions are often algebraic polynomials that are derived using interpolation theory.
3. *Assembly of elements,* which is based on continuity of the solution and balance of internal fluxes; the assemblage of elements represents a discrete analog of the original domain, and the associated system of algebraic equations represents a numerical analog of the mathematical model of the problem being analyzed.

These three features, which constitute three major steps of the fiinite element formulation, are closely related. The geometry of the elements used to represent the domain of a problem should be such that the approximation functions can be uniquely derived. The approximation functions depend not only on the geometry but also on the number and location of points, called *nodes,* in the element and the quantities to be interpolated (e.g., solution, or solution and its derivatives). Once the approximation functions have been derived, the procedure to obtain algebraic relations among the unknown coefficients (which give the values of the solution at the nodes of the finite elements) is exactly the same as that used in the Rayleigh–Ritz and weighted-residual methods. Hence, a careful reading of Chapter 2 makes the present reading easier.

The finite element method not only overcomes the shortcomings of the traditional variational methods, but it is also endowed with the features of an

TABLE 3.1
Steps involved in the finite element analysis of a typical problem

1. Discretization (or representation) of the given domain into a collection of preselected finite elements. (This step can be postponed until after the finite element formulation of the equation is completed.)
   a. Construct the finite element mesh of preselected elements.
   b. Number the nodes and elements.
   c. Generate the geometric properties (e.g., coordinates and cross-sectional areas) needed for the problem.
2. Derivation of element equations for all typical elements in the mesh.
   a. Construct the variational formulation of the given differential equation over the typical element.
   b. Assume that a typical dependent variable u is of the form

   $$u = \sum_{i=1}^{n} u_i \psi_i$$

   and substitute it into Step 2*a* to obtain element equations in the form

   $$[K^e]\{u^e\} = \{F^e\}$$

   c. Derive or select, if already available in the literature, element interpolation functions ψ_i and compute the element matrices.
3. Assembly of element equations to obtain the equations of the whole problem.
   a. Identify the interelement continuity conditions among the primary variables (relationship between the local degrees of freedom and the global degrees of freedom—connectivity of elements) by relating element nodes to global nodes.
   b. Identify the "equilibrium" conditions among the secondary variables (relationship between the local source or force components and the globally specified source components).
   c. Assemble element equations using Steps 3*a* and 3*b*.
4. Imposition of the boundary conditions of the problem.
   a. Identify the specified global primary degrees of freedom.
   b. Identify the specified global secondary degrees of freedom (if not already done in Step 3*b*).
5. Solution of the assembled equations.
6. Postprocessing of the results.
   a. Compute the gradient of the solution or other desired quantities from the primary degrees of freedom computed in Step 5.
   b. Represent the results in tabular and/or graphical form.

effective computational technique. The basic steps involved in the finite element analysis of a problem are given in Table 3.1.

In the sections that follow, our objective will be to introduce many fundamental ideas that form the basis of the finite element method. In doing so, we postpone some issues of practical and theoretical complexity to later sections of this chapter and to Chapters 4–14. The basic steps of a finite element analysis are introduced via a model second-order differential equation, which is representative of many one-dimensional systems.

3.2 BASIC STEPS OF THE FINITE ELEMENT ANALYSIS

3.2.1 Model Boundary Value Problem

Consider the problem of finding the function $u(x)$ that satisfies the differential equation

$$\boxed{-\frac{d}{dx}\left(a\frac{du}{dx}\right)+cu-q=0 \quad \text{for } 0<x<L} \tag{3.1}$$

and the boundary conditions

$$\boxed{u(0)=u_0, \qquad \left(a\frac{du}{dx}\right)\bigg|_{x=L}=Q_0} \tag{3.2}$$

where $a=a(x)$, $c=c(x)$, $q=q(x)$, u_0, and Q_0 are the data (i.e., known quantities) of the problem. Equation (3.1) arises in connection with the analytical description of many physical processes. For example, conduction and convection heat transfer in a plane wall or fin (1-D heat transfer), flow through channels and pipes, transverse deflection of cables, axial deformation of bars (see Fig. 3.1*a*), and many other physical processes are described by (3.1). Table 3.2 contains a list of several field problems described by (3.1)

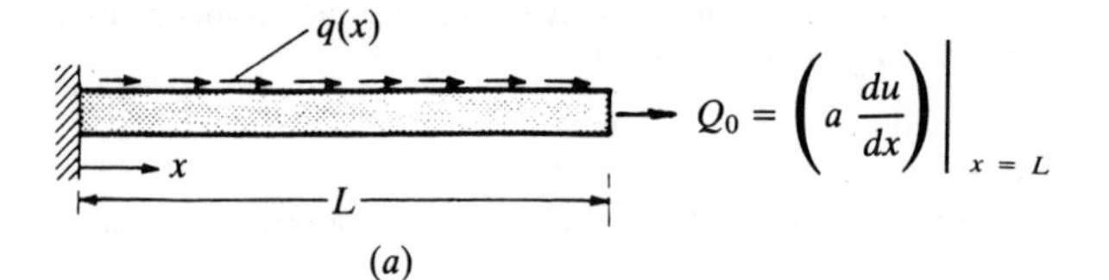

(*a*)

$q(x)$

$u=u_0\equiv 0$ $\qquad$ $a\frac{du}{dx}\equiv Q_0$

(*b*)

x

h_1 h_2 h_e h_N

1 2 2 .... e e $e+1$.. N $N+1$

Node number $\qquad$ Element number

(*c*)

FIGURE 3.1
Finite element discretization of a one-dimensional domain: (*a*) physical problem; (*b*) mathematical idealization; (*c*) finite element discretization.

TABLE 3.2
Some examples of the second-order equations in one dimension

$$-\frac{d}{dx}\left(a\frac{du}{dx}\right)=q \quad \text{for } 0<x<L$$

Essential boundary condition: $u|_{x=0}=u_0$. **Natural boundary condition:** $(a\,du/dx)|_{x=L}=Q_0$

Field	**Primary variable** u	a	**Source term** q	**Secondary variable** Q_0
1. Transverse deflection of a cable	Transverse deflection	Tension in cable	Distributed transverse load	Axial force
2. Axial deformation of a bar	Longitudinal displacement	EA (E = modulus, A = cross-sectional area)	Friction or contact force on surface of bar	Axial force
3. Heat transfer	Temperature	Thermal conductivity	Heat generation	Heat
4. Flow through pipes	Hydrostatic pressure	$\pi D^4/128\mu$ (D = diameter, μ = viscosity)	Flow source (generally zero)	Flow rate
5. Laminar incompressible flow through a channel under constant pressure gradient	Velocity	Viscosity	Pressure gradient	Axial stress
6. Flow through porous media	Fluid head	Coefficient of permeability	Fluid flux	Flow (seepage)
7. Electrostatics	Electrostatic potential	Dielectric constant	Charge density	Electric flux

when $c(x) = 0$. The mathematical structure common to apparently different fields is brought out in this table. Thus, if we can develop a numerical procedure by which (3.1) can be solved for all possible boundary conditions, the procedure can be used to solve all field problems listed in Table 3.2, as well as many others. This fact provides us with the motivation to use (3.1) as the model second-order equation in one dimension. A step-by-step procedure (see Table 3.1) for the formulation and solution of (3.1) by the finite element method is presented next.

3.2.2 Discretization of the Domain

The domain of the problem in the present case consists of all points between $x = 0$ and $x = L$: $\Omega = (0, L)$; see Fig. 3.1(*b*). The domain Ω is divided into a set of line elements, a typical element being of length h_e and located between points A and B. The collection of such elements is called the *finite element mesh* of the domain (see Fig. 3.1*c*). The reason for dividing the domain into finite elements is twofold: first, to represent the geometry of the domain; and, second, to approximate the solution over each element of the mesh in order to better represent the solution over the entire domain. Approximation of the domain in the present case is not a concern, since it is a straight line. If the domain is a curve then approximation by a set of straight or curved line elements is necessary to represent it. Approximation of the solution over each element of the mesh is simpler than its approximation over the entire domain. Recall that in the traditional variational methods, the solution is required to satisfy the boundary conditions of the problem. This places severe restrictions on the choice of approximation functions, especially when discontinuities exist in the geometry, material properties, and/or loading of the problem (see Chapter 2 for details).

To connect the elements and impose continuity of the solution at nodes common to elements, we identify the endpoints of each line element as the *element nodes*. Depending on the degree of polynomial approximation used to represent the solution, additional nodes may be identified inside the element. The nodes play the role of interpolation points, as will be seen shortly, in constructing the approximation functions over an element.

The number of elements used in a problem depends mainly on the element type and accuracy desired. Whenever a problem is solved by the finite element method for the first time, one is required to investigate the convergence characteristics of the finite element approximation by gradually *refining the mesh* (i.e., increasing the number of elements) and comparing the solution with those obtained by *higher-order* elements. The *order* of an element refers to the degree of polynomial used to represent the solution over the element.

3.2.3 Derivation of Element Equations

The derivation of finite element equations, i.e., algebraic equations that relate the primary variables to the secondary variables at the nodes of the elements, involves three steps:

1. Construct the weighted-residual or weak form of the differential equation.
2. Assume the form of the approximate solution over a typical finite element.
3. Derive the finite element equations by substituting the approximate solution into the weighted-residual or weak form.

A typical element $\Omega^e = (x_A, x_B)$, whose endpoints have the coordinates $x = x_A$ and $x = x_B$, is isolated from the mesh (see Fig. 3.2*a*). We seek an approximate solution to the governing differential equation over the element, using the Rayleigh–Ritz method discussed in Chapter 2. In principle, any method that allows the derivation of necessary algebraic relations among the nodal values of the dependent variable can be used. In this book we develop the algebraic equations using the Rayleigh–Ritz method, which is based on the weak form of the differential equation. The equations resulting from the application of a variational method are relations between the *primary* variables (i.e., those involved in the specification of the essential boundary conditions) and the *secondary* variables (i.e., those involved in the specification of the natural boundary conditions). The three steps in the derivation of finite element equations of a *typical element* of the mesh are discussed next.

STEP 1: WEAK FORM. In the finite element method, we seek an approximate solution to (3.1) over each finite element. The polynomial approximation of

x_B
x_A h_e
$\bar{x} = x - x_A$
x A $\bar{x}$ B
$\bar{x} = 0$ $\bar{x} = h_e$
(*a*)

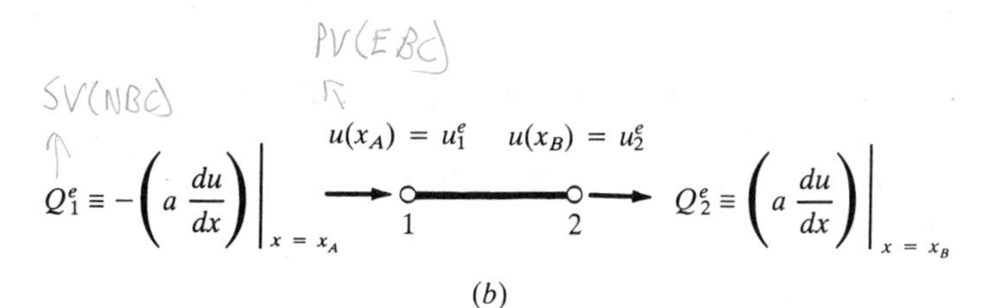

(*b*)

FIGURE 3.2.
Finite element discretization of a one-dimensional domain for the model problem in (3.1). (*a*) A typical finite element from the finite element mesh in Fig. 3.1(*c*); x = global coordinate, $\bar{x}$-local coordinate. (*b*) A typical element, with the definition of the primary (u) and secondary (Q) variables at the element nodes.

the solution within a typical finite element Ω^e is of the form

$$U^e = \sum_{j=1}^{n} u_j^e \psi_j^e(x) \tag{3.3}$$

where u_j^e are the values of the solution at the nodes of the finite element and ψ_j^e are the approximation functions over the element. The coefficients u_j^e are determined such that (3.1) is satisfied in a weighted-integral sense. As discussed in Chapter 2 (which should be consulted for additional details), the necessary and sufficient number of algebraic relations among the u_j^e can be obtained by recasting the differential equation (3.1) in a weighted-integral form:

$$0 = \int_{x_A}^{x_B} w\left[-\frac{d}{dx}\left(a\frac{du}{dx}\right) + cu - q\right] dx \tag{3.4}$$

where $w(x)$ denotes the weight function and $\Omega^e = (x_A, x_B)$ is the domain of a typical element (see Fig. 3.2*a*). For $u \approx U^e$ and each independent choice of w, we obtain an independent algebraic equation relating all u_j^e of the element. A total of n independent equations are required to solve for n values u_j^e. When w is selected to be ψ_i^e and (3.4) is used to obtain the ith equation of the required n equations, the resulting finite element model (i.e., system of algebraic equations among the nodal values) is termed the *Galerkin finite element model.* Since (3.4) contains the second derivative of U^e, the approximation functions ψ_j^e must be twice differentiable. In addition, if the secondary variables are to be included in the model, ψ_i^e must be at least cubic. Similar arguments apply for cases of the weighted-residual methods discussed in Chapter 2. For additional details of the weighted residual finite element models, see Reddy (1986) and Chapter 14.

To weaken the continuity required of the functions $\psi_j^e(x)$, we trade the differentiation in (3.4) from u to w such that both u and w are differentiated equally, once each in the present case. The resulting integral form is termed the *weak form* of (3.1). This form is not only equivalent to (3.1) but it also contains the natural boundary conditions of the problem. The three-step procedure of constructing the weak form of (3.1) was presented in Chapter 2, and is revisited in the next few paragraphs.

The first step is to multiply the governing differential equation with a weight function w and integrate over a *typical element.* The second step is to trade differentiation from u to w, using integration by parts. This is achieved as follows. Consider the identity

$$-w\left[\frac{d}{dx}\left(a\frac{du}{dx}\right)\right] = -\frac{d}{dx}\left(wa\frac{du}{dx}\right) + a\frac{dw}{dx}\frac{du}{dx} \tag{3.5a}$$

which is simply the product rule of differentiation applied to the product of two functions, $a\,du/dx$ and w. Integrating this identity over the element domain,

we obtain

$$-\int_{x_A}^{x_B} w\left[\frac{d}{dx}\left(a\frac{du}{dx}\right)\right]dx = -\int_{x_A}^{x_B} \frac{d}{dx}\left(wa\frac{du}{dx}\right)dx + \int_{x_A}^{x_B} a\frac{dw}{dx}\frac{du}{dx}\,dx$$

$$= -\left[wa\frac{du}{dx}\right]_{x_A}^{x_B} + \int_{x_A}^{x_B} a\frac{dw}{dx}\frac{du}{dx}\,dx \tag{3.5b}$$

Substituting (3.5*b*) into (3.4), we arrive at the result [cf. (2.34)]

$$0 = \int_{x_A}^{x_B}\left(a\frac{dw}{dx}\frac{du}{dx} + cwu - wq\right)dx - \left[wa\frac{du}{dx}\right]_{x_A}^{x_B} \tag{3.6}$$

The third and last step is to identify the primary and secondary variables of the variational (or weak) form. This requires us to classify the boundary conditions of each differential equation into *essential* (or geometric) and *natural* (or force) boundary conditions. The classification is made uniquely by examining the boundary term appearing in the weak form (3.6),

$$\left[wa\frac{du}{dx}\right]_{x_A}^{x_B}$$

As a rule, the coefficient of the weight function in the boundary expression is called the *secondary variable,* and its specification constitutes the *natural* (or Neumann) boundary condition. The dependent unknown *in the same form as the weight function* in the boundary expression is termed the *primary variable,* and its specification constitutes the *essential* (or Dirichlet) boundary condition. For the model equation at hand, the primary and secondary variables are

$$u \quad \text{and} \quad a\frac{du}{dx} \equiv Q$$

In writing the final form of the variational (or weak) statement, we assume that all boundary conditions at the element level are of the natural type, so that they can be included in the variational statement:

$$\boxed{-Q_A = \left(a\frac{du}{dx}\right)\bigg|_{x_A}, \qquad Q_B = \left(a\frac{du}{dx}\right)\bigg|_{x_B}.}$$

The primary and secondary variables at the nodes are shown on the typical element in Fig. 3.2(*b*). Students of engineering recognize that this figure is the *free-body diagram* of the typical element, with its internal forces (i.e., reactions) Q_1^e and Q_2^e. The quantities $Q_1^e \equiv Q_A$ and $Q_2^e \equiv Q_B$ have the meaning of forces in the axial deformation of bars; Q_1^e is a compressive force while Q_2^e is a tensile force (algebraically, both are positive, as shown in Fig. 3.2*b*). For heat conduction problems Q_1^e and Q_2^e denote the heats conducting *into* the body. The arrow on the second node should be reversed for heat transfer problems, because the Fourier heat conduction law relating the gradient of temperature to the heat flux contains a negative sign (implying heat flows from hot to cold). For additional details on heat transfer, see Section 3.3.1. With the notation in (3.7), the variational form becomes

$$0 = \int_{x_A}^{x_B} \left(a \frac{dw}{dx}\frac{du}{dx} + cwu - wq \right) dx - w(x_A)Q_A - w(x_B)Q_B \tag{3.8}$$

This completes the three-step procedure of constructing the weak form. The weak form in (3.8) contains two types of expressions: those containing both w and u; and those containing only w. We group the former type into a single expression, called the *bilinear form*:

$$B(w, u) \equiv \int_{x_A}^{x_B} \left(a \frac{dw}{dx}\frac{du}{dx} + cwu \right) dx \tag{3.9a}$$

We denote all terms containing only w (but not u) by $l(w)$, called the *linear form*:

$$l(w) = \int_{x_A}^{x_B} wq \, dx + w(x_A)Q_A + w(x_B)Q_B \tag{3.9b}$$

The variational statement (3.8) can now be expressed as

$$B(w, u) = l(w) \tag{3.10}$$

which is called the *variational problem* associated with (3.1). As will be seen later, the bilinear form results directly in the element coefficient matrix, and the linear form leads to the right-hand-side column vector of the finite element equations.

Those who have a background in applied mathematics or solid and structural mechanics will appreciate the fact that the variational problem (3.10) is nothing but the statement of the minimization of a quadratic functional or of total potential energy $I(u)$:

$$\delta I = 0$$

where δ is the variational symbol (see Section 2.3.3) and I is the quadratic functional defined by [see (2.43*b*)]

$$I(u) = \tfrac{1}{2}B(u, u) - l(u) \tag{3.11}$$

Equation (3.11) holds only when $l(u)$ is linear in u, and $B(w, u)$ is bilinear and symmetric in u and w,

$$B(w, u) = B(u, w)$$

When (3.1) describes the axial deformation of a bar, $\tfrac{1}{2}B(u, u)$ represents the elastic strain energy stored in the bar, $l(u)$ represents the work done by applied forces, and $I(u)$ represents the total potential energy of the bar element. It is important to note that finite element formulations do not require the existence of the functional $I(u)$. What is needed is a way to obtain exactly n algebraic equations among the u_j^e of (3.3) such that the governing differential equation is satisfied over the element in some meaningful way. In the present study, we use the weak form of the differential equation, i.e., (3.8) or (3.10), and the Rayleigh–Ritz method to obtain the n algebraic equations among the nodal variables u_i^e and Q_i^e.

STEP 2: APPROXIMATION OF THE SOLUTION. Recall that the weak form over an element is equivalent to the differential equation and the natural boundary conditions of the element. The essential boundary conditions of the element, say $u(x_A) = u_A$ and $u(x_B) = u_B$, are not included in the weak form. Hence, they must be included in the approximation of $u(x)$. Thus, the approximation of $u(x)$ must be an interpolant, i.e., must be equal to u_A at x_A and u_B at x_B. Since the weak form contains the first-order derivatives of u, any continuous function would be a candidate for the finite element solution. Let us denote the finite element solution over element $\Omega^e = (x_A, x_B)$ by U^e. Then we seek the approximate solution U^e in the form of algebraic polynomials. The reason for this choice is twofold: first, the interpolation theory of numerical analysis can be used to develop the approximation functions systematically over an element; second, numerical evaluation of integrals of algebraic polynomials is easy.

As in variational methods, the approximation solution U^e must fulfill certain requirements in order that it be convergent to the actual solution u as the number of elements is increased. These are:

1. The approximate solution should be continuous over the element, and differentiable, as required by the weak form.
2. It should be a *complete* polynomial, i.e., include all lower-order terms up to the highest order used.
3. It should be an interpolant of the primary variables at the nodes of the finite element.

(3.12)

The reason for the first requirement is obvious; it ensures a nonzero coefficient matrix. The second requirement is necessary in order to capture all possible states, i.e., constant, linear and so on, of the actual solution. For example, if a linear polynomial without the constant term is used to represent the temperature distribution in a one-dimensional system, the approximate solution can never be able to represent a uniform state of temperature in the element. The third requirement is necessary in order to satisfy the essential boundary conditions of the element and to enforce continuity of the primary variables at points common to several elements.

For the variational statement at hand, the minimum polynomial order is linear. A *complete* linear polynomial is of the form

$$U^e = a + bx \tag{3.13}$$

where a and b are constants. This expression meets the first two requirements in (3.12). To satisfy the third

$$U^e(x_A) = u_1^e, \qquad U^e(x_B) = u_2^e \tag{3.14}$$

we express the constants a and b in (3.13) in terms of u_1^e and u_2^e. Equations (3.14) provide two relations between (a, b) and (u_1^e, u_2^e):

$$\begin{aligned} u_1^e &= a + bx_A \\ u_2^e &= a + bx_B \end{aligned} \tag{3.15a}$$

or, in matrix form,

$$\begin{Bmatrix} u_1^e \\ u_2^e \end{Bmatrix} = \begin{bmatrix} 1 & x_A \\ 1 & x_B \end{bmatrix} \begin{Bmatrix} a \\ b \end{Bmatrix} \tag{3.15b}$$

Inverting (3.15*b*) by Cramer's rule, we obtain

$$\begin{aligned} a &= \begin{vmatrix} u_1^e & x_A \\ u_2^e & x_B \end{vmatrix} \Bigg/ \begin{vmatrix} 1 & x_A \\ 1 & x_B \end{vmatrix} = \frac{1}{h_e}(u_1^e x_B - u_2^e x_A) \equiv \frac{1}{h_e}(\alpha_1^e u_1^e + \alpha_2^e u_2^e) \\ b &= \begin{vmatrix} 1 & u_1^e \\ 1 & u_2^e \end{vmatrix} \Bigg/ \begin{vmatrix} 1 & x_A \\ 1 & x_B \end{vmatrix} = \frac{1}{h_e}(u_2^e - u_1^e) \equiv \frac{1}{h_e}(\beta_1^e u_1^e + \beta_2^e u_2^e) \end{aligned} \tag{3.15c}$$

where $h_e = x_B - x_A$ and

$$\alpha_i^e = (-1)^j x_j^e, \qquad \beta_i^e = (-1)^i; \qquad x_1^e = x_A, \qquad x_2^e = x_B \tag{3.15d}$$

In (3.15*d*), i and j permute in a natural order:

$$\text{if} \quad i = 1 \quad \text{then} \quad j = 2; \qquad \text{if} \quad i = 2 \quad \text{then} \quad j = 1$$

The α_i^e and β_i^e are introduced to show the typical form of the interpolation functions. Substitution of (3.15*c*) into (3.13) yields

$$\begin{aligned} U^e(x) &= \frac{1}{h_e}[(\alpha_1^e u_1^e + \alpha_2^e u_2^e) + (\beta_1^e u_1^e + \beta_2^e u_2^e)x] \\ &= \frac{1}{h_e}(\alpha_1^e + \beta_1^e x)u_1^e + \frac{1}{h_e}(\alpha_2^e + \beta_2^e x)u_2^e \end{aligned}$$

That is,

$$\boxed{U^e(x) = \psi_1^e(x)u_1^e + \psi_2^e(x)u_2^e = \sum_{j=1}^{2} \psi_j^e(x)u_j^e} \tag{3.16a}$$

where

$$\boxed{\psi_1^e(x) = \frac{1}{h_e}(\alpha_1^e + \beta_1^e x) = \frac{x_B - x}{x_B - x_A}, \qquad \psi_2^e(x) = \frac{1}{h_e}(\alpha_2^e + \beta_2^e x) = \frac{x - x_A}{x_B - x_A}} \tag{3.16b}$$

which are called the linear *finite element approximation functions*.

For the linear interpolation (3.16), we label the endpoints as nodes 1 and 2, and rename the secondary variables as

$$Q_A = Q_1^e, \qquad Q_B = Q_2^e \tag{3.17}$$

The global node numbers for elements connected in series can be related to the element node numbers. For linear elements, the global node numbers of element Ω^e are e and $e+1$, and the global coordinates of the element nodes are x_e and x_{e+1} (i.e., $x_A = x_e$ and $x_B = x_{e+1}$).

Note that the element interpolation functions ψ_i^e in (3.16b) are expressed in terms of the *global coordinate* x (i.e., the coordinate of the problem), but they are defined only on the element domain $\Omega^e = (x_A, x_B) = (x_e, x_{e+1})$. If we choose to express them in terms of a coordinate $\bar{x}$ with origin fixed at node 1 of the element, ψ_i^e of (3.16b) take the forms

$$\psi_1^e(\bar{x}) = 1 - \frac{\bar{x}}{h_e}, \qquad \psi_2^e(\bar{x}) = \frac{\bar{x}}{h_e} \tag{3.18}$$

The coordinate $\bar{x}$ is termed the *local* or *element coordinate*. The functions ψ_i^e are shown in Fig. 3.3(a). Note that ψ_1^e is equal to 1 at node 1 and zero at node 2, and ψ_2^e is equal to 1 at node 2 and zero at node 1. These properties of ψ_i^e are known as the interpolation properties.

The global interpolation functions Φ_I can be defined in terms of the element interpolation functions corresponding to the global node I (see Fig.

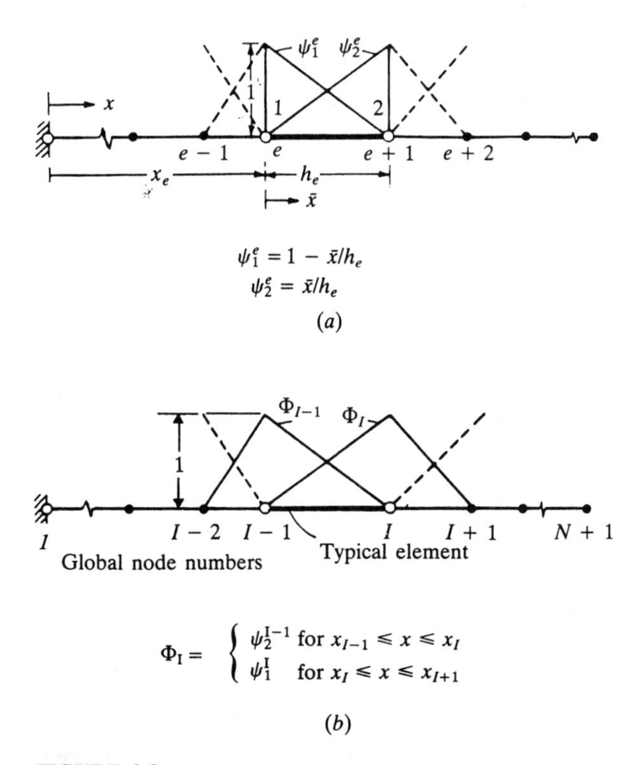

FIGURE 3.3
(a) Local and (b) global interpolation functions for the two-node (linear) element ($x_A = x_e$, $x_B = x_{e+1}$).

3.3*b*). Since $U^e(x)$ of (3.16*a*) is an interpolant of $u(x)$ over the element Ω^e, ψ_i^e are also called *interpolation functions*. Interpolation functions derived using the dependent unknown—not its derivatives—at the nodes (i.e., interpolation functions with C^0 continuity) are called the *Lagrange family of interpolation functions*. When the dependent unknown and its derivatives at the nodes are used to derive the interpolation functions, the resulting interpolation functions are known as the *Hermite family of interpolation functions* (see the classical beam element in Chapter 4).

Note that ψ_i^e are derived systematically; starting with an assumed degree of algebraic polynomial for the dependent unknown and determining the coefficients of the polynomial in terms of the primary degrees of freedom, we expressed the dependent variable as a linear combination of approximation functions and the primary nodal variables. The key in the procedure is to select the number and the location of nodes in the element so that the geometry of the latter is uniquely defined. The number of nodes must be sufficient to allow the assumed degree of interpolation of the solution in terms of the primary variables. For a linear polynomial approximation, two nodes with one primary unknown per node are sufficient to define the geometry of the element, provided the two nodes are the endpoints of the element. Since a quadratic polynomial is uniquely defined by three parameters, a total of three nodal points must be identified in the element. To define the geometry of the element, two of the nodes must be the endpoints of the element. The third can be identified inside the element.

Returning to the linear approximation (3.13), which is recast as (3.16*a*), we note that the true solution is approximated over each element by a linear polynomial $U^e(x)$ (see Fig. 3.4*a*). The error in the approximation, $E = u(x) - U^e(x)$, can be reduced by either decreasing the element size h_e or increasing the degree of the approximation (see Fig. 3.4*b*).

A quadratic approximation is of the form

$$U^e(x) = a + bx + cx^2 \tag{3.19}$$

which requires three nodes in order to rewrite U^e in terms of the values of $u(x)$ at the nodes. Two of the nodes are identified as the endpoints of the element to define the geometry, and the third node is taken interior to the element. In theory, the third node can be placed at any interior point. However, the midpoint of the element, being equidistant from the end nodes, is the best choice. Other choices (e.g., quarter-point) are dictated by special considerations (e.g., to have a certain degree of singularity in the derivative of the solution). Thus, we identify three nodes in the element of length h_e (see Fig. 3.5*a*) and rewrite $U^e(x)$ in terms of the three nodal values, (u_1^e, u_2^e, u_3^e). We have

$$\begin{aligned} u_1^e &= U^e(x_1^e) = a + bx_1^e + c(x_1^e)^2 \\ u_2^e &= U^e(x_2^e) = a + bx_2^e + c(x_2^e)^2 \\ u_3^e &= U^e(x_3^e) = a + bx_3^e + c(x_3^e)^2 \end{aligned} \tag{3.20a}$$

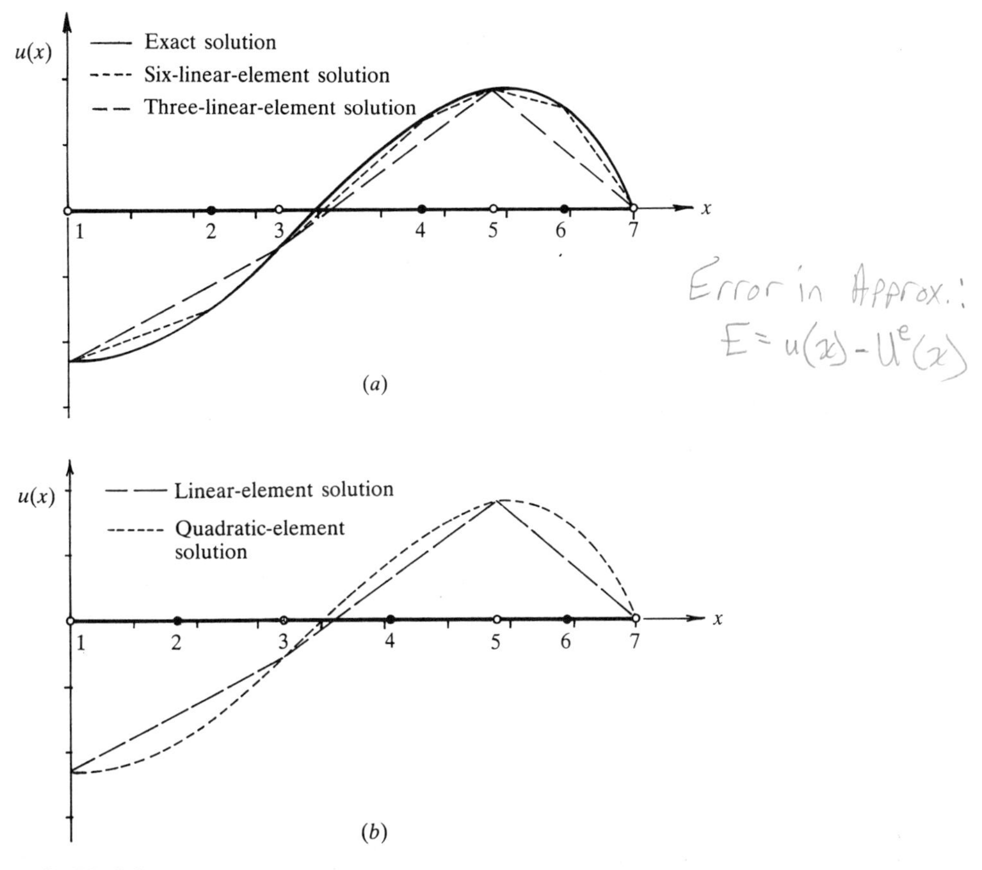

FIGURE 3.4
Refinements of finite element solutions; (*a*) mesh refinement using linear elements; (*b*) quadratic element solution using three elements.

or, in matrix form,

$$\begin{Bmatrix} u_1^e \\ u_2^e \\ u_3^e \end{Bmatrix} = \begin{bmatrix} 1 & x_1^e & (x_1^e)^2 \\ 1 & x_2^e & (x_2^e)^2 \\ 1 & x_3^e & (x_3^e)^2 \end{bmatrix} \begin{Bmatrix} a \\ b \\ c \end{Bmatrix} \tag{3.20b}$$

where x_i^e is the global coordinate of the ith node of the element Ω^e. Inverting the above relations, we obtain

$$\begin{aligned} a &= \frac{1}{D^e} \sum_{i=1}^{3} \alpha_i^e u_i^e, \qquad \alpha_i^e = x_j^e (x_k^e)^2 - x_k^e (x_j^e)^2 \\ b &= \frac{1}{D^e} \sum_{i=1}^{3} \beta_i^e u_i^e, \qquad \beta_i^e = (x_j^e)^2 - (x_k^e)^2 \\ c &= \frac{1}{D^e} \sum_{i=1}^{3} \gamma_i^e u_i^e, \qquad \gamma_i^e = -(x_j^e - x_k^e), \qquad D^e = \sum_{i=1}^{3} \alpha_i^e \end{aligned} \tag{3.21}$$

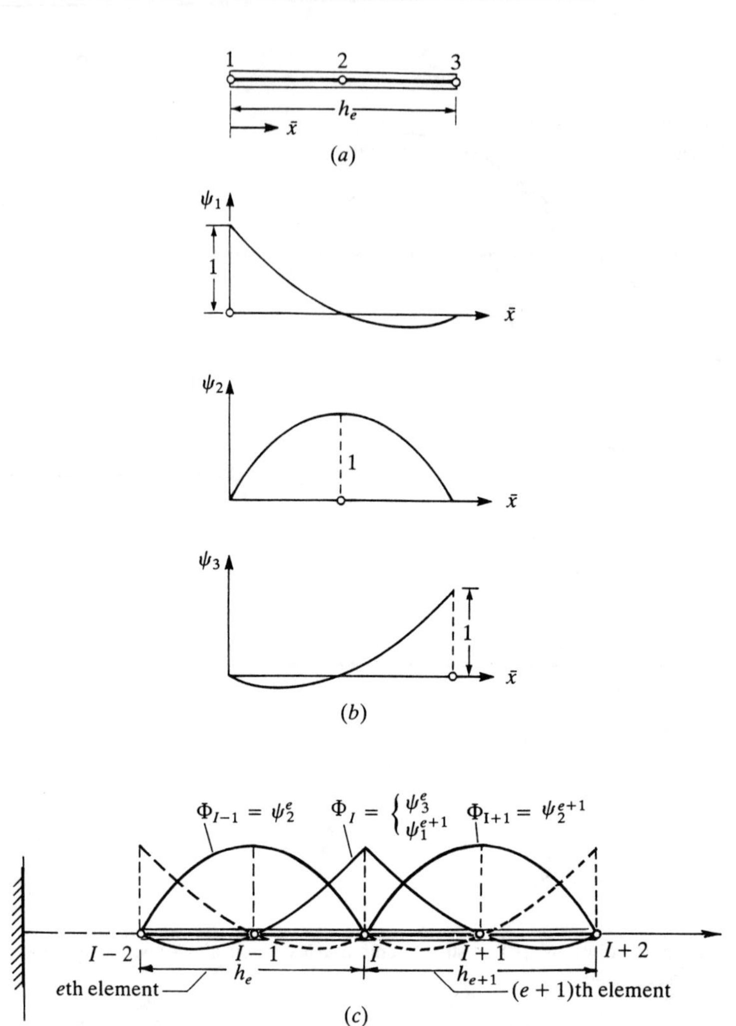

FIGURE 3.5
One-dimensional Lagrange quadratic element and its interpolation functions: (*a*) geometry of the element; (*b*) interpolation functions; (*c*) global interpolation functions corresponding to the quadratic interpolation functions. Here I denotes the global node number, e the element number, and i the element node number.

and (3.19) takes the form

$$U^e(x) = \psi_1^e(x)u_1^e + \psi_2^e(x)u_2^e + \psi_3^e(x)u_3^e = \sum_{j=1}^{3} \psi_j^e(x)u_j^e \tag{3.22}$$

where ψ_j^e are the *quadratic Lagrange interpolation functions,*

$$\psi_i^e(x) = \frac{1}{D^e}(\alpha_i^e + \beta_i^e x + \gamma_i^e x^2) \quad (i = 1, 2, 3) \tag{3.23}$$

Here D^e denotes the determinant of the matrix in (3.20b), and α_i^e, β_i^e, and γ_i^e are defined by (3.21). The subscripts used in (3.21) permute in a natural order:

$$\begin{aligned} &\text{if} \quad i=1 \quad \text{then} \quad j=2 \text{ and } k=3 \\ &\text{if} \quad i=2 \quad \text{then} \quad j=3 \text{ and } k=1 \\ &\text{if} \quad i=3 \quad \text{then} \quad j=1 \text{ and } k=2 \end{aligned} \tag{3.24}$$

For example, α_2^e, β_3^e, and γ_1^e are given by

$$\alpha_2^e = x_3^e(x_1^e)^2 - x_1^e(x_3^e)^2, \qquad \beta_3^e = (x_1^e)^2 - (x_2^e)^2, \qquad \gamma_1^e = x_3^e - x_2^e$$

The quadratic interpolation functions can be expressed in terms of a *local coordinate* $\bar{x}$, with origin fixed at node 1, the left end of the element. The global coordinate x is related to the local coordinate $\bar{x}$ by the relation

$$x = x_1^e + \bar{x} \tag{3.25}$$

where $x_1^e = x_A$ is the global coordinate of the first node of the element Ω^e. For a quadratic element with the interior node, node 2, located at $\bar{x} = \alpha h_e$, we have

$$\begin{aligned} \psi_1^e(\bar{x}) &= \left(1 - \frac{\bar{x}}{h}\right)\left(1 - \frac{1}{\alpha}\frac{\bar{x}}{h}\right) \\ \psi_2^e(\bar{x}) &= \frac{1}{\alpha(1-\alpha)}\frac{\bar{x}}{h}\left(1 - \frac{\bar{x}}{h}\right) \\ \psi_3^e(\bar{x}) &= -\frac{\alpha}{(1-\alpha)}\frac{\bar{x}}{h}\left(1 - \frac{1}{\alpha}\frac{\bar{x}}{h}\right) \end{aligned} \tag{3.26}$$

where $0 < \alpha < 1$ and $x_2^e = x_1^e + \alpha h_e$. For $\alpha = \frac{1}{2}$, i.e., when node 2 is placed at the midpoint of the element, (3.26) becomes

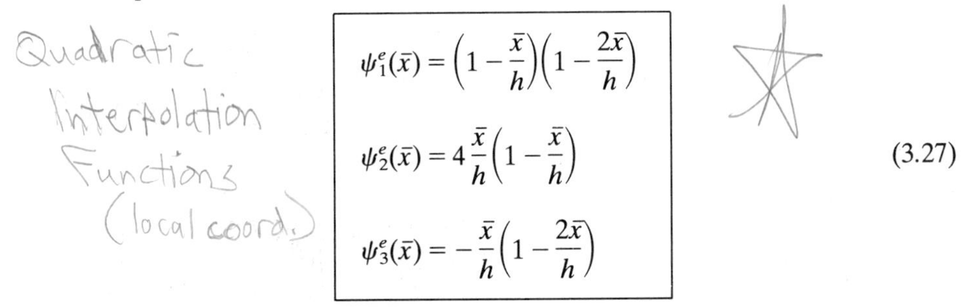

$$\boxed{\begin{aligned} \psi_1^e(\bar{x}) &= \left(1 - \frac{\bar{x}}{h}\right)\left(1 - \frac{2\bar{x}}{h}\right) \\ \psi_2^e(\bar{x}) &= 4\frac{\bar{x}}{h}\left(1 - \frac{\bar{x}}{h}\right) \\ \psi_3^e(\bar{x}) &= -\frac{\bar{x}}{h}\left(1 - \frac{2\bar{x}}{h}\right) \end{aligned}} \tag{3.27}$$

Plots of the quadratic interpolation functions are given in Fig. 3.5(*b*). The function ψ_i^e is equal to 1 at node i and zero at the other two nodes, but varies quadratically between the nodes.

All Lagrange family of interpolation functions satisfy the following properties, known as the *interpolation properties*:

$$\boxed{\begin{array}{ll} (1) & \psi_i^e(x_j^e) = \delta_{ij} \equiv \begin{cases} 0 & \text{if } i \neq j \\ 1 & \text{if } i = j \end{cases} \\[2ex] (2) & \displaystyle\sum_{j=1}^{n} \psi_j^e(x) = 1, \qquad \text{hence} \quad \sum_{j=1}^{n} \frac{d\psi_j^e}{dx} = 0 \end{array}} \tag{3.28}$$

where $n-1$ is the degree of interpolation polynomials and x_j^e is the global coordinate of node j in the element Ω^e. It can be verified that the linear interpolation functions in (3.16) and quadratic interpolation functions in (3.26) and (3.27) satisfy the two properties in (3.28). The first is a direct result of the requirement $U^e(x_j^e) = u_j^e$, and the second comes from the inclusion of a constant term in the polynomial. For example, if the approximation U^e is to represent a uniform state of solution, $U^e = U_0^e = \text{constant}$, then all $u_i^e = U_0^e$, and we have

$$U_0^e = \sum_{j=1}^{n} U_0^e \psi_j^e(x)$$

or

$$1 = \sum_{j=1}^{n} \psi_j^e(x)$$

The interpolation properties (3.28) can be used to construct the Lagrange interpolation functions of any degree. For example, the quadratic interpolation functions (3.27) can be derived using property (1) of (3.28). Since $\psi_1^e(\bar{x})$ must vanish at nodes 2 and 3, i.e., at $\bar{x} = \frac{1}{2}h_e$ and $x = h_e$, it is of the form

$$\psi_1^e(\bar{x}) = C(\bar{x} - \tfrac{1}{2}h_e)(\bar{x} - h_e)$$

The constant C is to be determined such that ψ_1^e is equal to 1 at $\bar{x} = 0$:

$$1 = C(0 - \tfrac{1}{2}h_e)(0 - h_e) \quad \text{or} \quad C = 2/h_e^2$$

This gives

$$\psi_1^e(\bar{x}) = \frac{2}{h_e^2}(\bar{x} - \tfrac{1}{2}h_e)(\bar{x} - h_e) = \left(1 - \frac{\bar{x}}{h_e}\right)\left(1 - \frac{2\bar{x}}{h_e}\right)$$

which is the same as in (3.27). The other two interpolation functions can be derived in a similar manner.

Although a detailed discussion is presented here on how to construct the Lagrange interpolation functions for one-dimensional elements, they are readily available in books on numerical analysis, and their derivation is independent of the physics of the problem to be solved. Their derivation depends only on the geometry of the element and the number and location of the nodes. The number of nodes must be equal to the number of terms in the polynomial. Thus, the interpolation functions derived above are useful not only in the finite element approximation of the problem at hand, but also in all problems that admit Lagrange interpolation of the variables, i.e., all problems for which the primary variables are the dependent unknowns of the governing equations.

STEP 3: FINITE ELEMENT MODEL. The weak form (3.8) or (3.10) is equivalent to the differential equation (3.1) over the element Ω^e and also contains the natural boundary conditions (3.7). Further, the finite element approximations (3.16*a*) or (3.22) satisfy the essential boundary conditions (3.14) of the element. The substitution of (3.16*a*) or (3.22) into (3.8) will give the necessary algebraic equations among the nodal values u_i^e and Q_i^e of the element Ω^e. In order to formulate the finite element model based on the weak form (3.8), it is *not* necessary to decide *a priori* the degree of approximation of U^e. The model can be developed for an arbitrary degree of interpolation:

$$\boxed{u \approx U^e = \sum_{j=1}^{n} u_j^e \psi_j^e(x)} \tag{3.29}$$

where ψ_j^e are the Lagrange interpolation functions of degree $n-1$. When $n>2$, the weak form in (3.8) must be modified to include nonzero secondary variables, if any, at interior nodes:

$$\boxed{0 = \int_{x_A}^{x_B} \left(a \frac{dw}{dx} \frac{du}{dx} + cwu \right) dx - \int_{x_A}^{x_B} wq \, dx - \sum_{i=1}^{n} w(x_i^e) Q_i^e} \tag{3.30}$$

where x_i^e is the global coordinate of the ith node of element Ω^e. If nodes 1 and n denote the endpoints of the element then Q_1^e and Q_n^e represent the unknown point sources, and all other Q_i^e are always known (i.e., applied point sources).

Following the Rayleigh–Ritz procedure developed in Section 2.4.2, we substitute (3.29) for u and $\psi_1^e, \psi_2^e, \ldots, \psi_n^e$ for w into the weak form (3.30) to obtain n algebraic equations:

$$0 = \int_{x_A}^{x_B} \left[a \frac{d\psi_1^e}{dx} \left(\sum_{j=1}^{n} u_j^e \frac{d\psi_j^e}{dx} \right) + c\psi_1^e \left(\sum_{j=1}^{n} u_j^e \psi_j^e(x) \right) - \psi_1^e q \right] dx - \sum_{j=1}^{n} \psi_1^e(x_j^e) Q_j^e$$

$$0 = \int_{x_A}^{x_B} \left[a \frac{d\psi_2^e}{dx} \left(\sum_{j=1}^{n} u_j^e \frac{d\psi_j^e}{dx} \right) + c\psi_2^e \left(\sum_{j=1}^{n} u_j^e \psi_j^e(x) \right) - \psi_2^e q \right] dx - \sum_{j=1}^{n} \psi_2^e(x_j^e) Q_j^e$$

$$\vdots \qquad \vdots \qquad \vdots \qquad \vdots$$

$$0 = \int_{x_A}^{x_B} \left[a \frac{d\psi_i^e}{dx} \left(\sum_{j=1}^{n} u_j^e \frac{d\psi_j^e}{dx} \right) + c\psi_i^e \left(\sum_{j=1}^{n} u_j^e \psi_j^e(x) \right) - \psi_i^e q \right] dx - \sum_{j=1}^{n} \psi_i^e(x_j^e) Q_j^e$$

(*i*th equation)

$$\vdots \qquad \vdots \qquad \vdots \qquad \vdots$$

$$0 = \int_{x_A}^{x_B} \left[a \frac{d\psi_n^e}{dx} \left(\sum_{j=1}^{n} u_j^e \frac{d\psi_j^e}{dx} \right) + c\psi_n^e \left(\sum_{j=1}^{n} u_j^e \psi_j^e(x) \right) - \psi_n^e q \right] dx - \sum_{j=1}^{n} \psi_n^e(x_j^e) Q_j^e$$

(3.31*a*)

Note that the numbering of the algebraic equations follows that of the primary variables in the element. The ith algebraic equation can be written as

$$0 = \sum_{j=1}^{n} K_{ij}^e u_j^e - f_i^e - Q_i^e \quad (i = 1, 2, \ldots, n). \tag{3.31b}$$

where

$$K_{ij}^e = \int_{x_A}^{x_B} \left(a \frac{d\psi_i^e}{dx} \frac{d\psi_j^e}{dx} + c\psi_i^e \psi_j^e \right) dx = B(\psi_i^e, \psi_j^e), \qquad f_i^e = \int_{x_A}^{x_B} q\psi_i^e \, dx = l(\psi_i^e) \tag{3.31c}$$

Note that the interpolation property (1) of (3.28) is used to write

$$\sum_{j=1}^{n} \psi_j^e(x_i^e) Q_j^e = Q_i^e \tag{3.32}$$

Equations (3.31a) can be expressed in terms of the coefficients K_{ij}^e, f_i^e, and Q_i^e as

$$\begin{aligned} K_{11}^e u_1^e + K_{12}^e u_2^e + \ldots + K_{1n}^e u_n^e &= f_1^e + Q_1^e \\ K_{21}^e u_1^e + K_{22}^e u_2^e + \ldots + K_{2n}^e u_n^e &= f_2^e + Q_2^e \\ &\vdots \\ K_{n1}^e u_1^e + K_{n2}^e u_2^e + \ldots + K_{nn}^e u_n^e &= f_n^e + Q_n^e \end{aligned} \tag{3.33a}$$

In matrix notation, the linear algebraic equations (3.33a) can be written as

$$[K^e]\{u^e\} = \{f^e\} + \{Q^e\} \tag{3.33b}$$

The matrix $[K^e]$ is called the *coefficient matrix*, or *stiffness matrix* in structural mechanics applications. The column vector $\{f^e\}$ is the *source vector*, or *force vector* in structural mechanics problems. Note that (3.33) contains $2n$ unknowns: $(u_1^e, u_2^e, \ldots, u_n^e)$ and $(Q_1^e, Q_2^e, \ldots, Q_n^e)$, called primary and secondary element *nodal degrees of freedom*; hence, it cannot be solved without having an additional n conditions. Some of these are provided by the boundary conditions and the remainder by balance of the secondary variables Q_i^e at nodes common to several elements. This balance can be implemented by putting the elements together (i.e., assembling the element equations). Upon assembly and imposition of boundary conditions, we shall obtain exactly the same number of algebraic equations as the number of unknown primary and secondary degrees of freedom. The ideas underlying the assembly procedure are discussed in the next section.

The coefficient matrix $[K^e]$, which is symmetric, and source vector $\{f^e\}$ can be evaluated for a given element and data (a, c, and q). For element-wise-constant values of a, c, and q (say, a_e, c_e, and q_e) the coefficients K_{ij}^e and f_i^e can easily be evaluated for a typical element.

Linear element. For a mesh of linear elements, the element Ω^e is located between the global nodes $x_A = x_e$ and $x_B = x_{e+1}$ (see Fig. 3.2). Hence,

$$K_{ij}^e = \int_{x_e}^{x_{e+1}} \left(a_e \frac{d\psi_i^e}{dx} \frac{d\psi_j^e}{dx} + c_e \psi_i^e \psi_j^e \right) dx, \qquad f_i^e = \int_{x_e}^{x_{e+1}} q_e \psi_i^e \, dx$$

or, in the local coordinate system $\bar{x}$,

$$K_{ij}^e = \int_0^{h_e} \left(a_e \frac{d\psi_i^e}{d\bar{x}} \frac{d\psi_j^e}{d\bar{x}} + c_e \psi_i^e \psi_j^e \right) d\bar{x}, \qquad f_i^e = \int_0^{h_e} q_e \psi_i^e \, d\bar{x}$$

where $x = x_1^e + \bar{x}$ and

$$dx = d\bar{x}, \qquad \frac{d\psi_i^e}{dx} = \frac{d\psi_i^e}{d\bar{x}}$$

The ψ_i^e can be expressed in terms of $\bar{x}$ as [see (3.18)]

$$\psi_1^e(\bar{x}) = 1 - \bar{x}/h_e, \qquad \psi_2^e(\bar{x}) = \bar{x}/h_e$$

We can compute K_{ij}^e and f_i^e by evaluating the integrals. We have

$$K_{11}^e = \int_0^{h_e} \left[a_e \left(-\frac{1}{h_e}\right)\left(-\frac{1}{h_e}\right) + c_e \left(1 - \frac{\bar{x}}{h_e}\right)\left(1 - \frac{\bar{x}}{h_e}\right) \right] d\bar{x}$$

$$= \frac{a_e}{h_e} + \tfrac{1}{3} c_e h_e$$

$$K_{12}^e = \int_0^{h_e} \left[a_e \left(-\frac{1}{h_e}\right)\frac{1}{h_e} + c_e \left(1 - \frac{\bar{x}}{h_e}\right)\frac{\bar{x}}{h_e} \right] d\bar{x}$$

$$= -\frac{a_e}{h_e} + \tfrac{1}{6} c_e h_e = K_{21}^e \quad \text{(by symmetry)}$$

$$K_{22}^e = \int_0^{h_e} \left(a_e \frac{1}{h_e}\frac{1}{h_e} + c_e \frac{\bar{x}}{h_e}\frac{\bar{x}}{h_e} \right) d\bar{x} = \frac{a_e}{h_e} + \tfrac{1}{3} c_e h_e$$

Similarly,

$$f_1^e = \int_0^{h_e} q_e \left(1 - \frac{\bar{x}}{h_e}\right) d\bar{x} = \tfrac{1}{2} q_e h_e, \qquad f_2^e = \int_0^{h_e} q_e \frac{\bar{x}}{h_e} \, d\bar{x} = \tfrac{1}{2} q_e h_e$$

Thus, for constant q_e, the total source $q_e h_e$ is equally distributed to the two nodes. The coefficient matrix and column vector are

$$[K^e] = \frac{a_e}{h_e} \begin{bmatrix} 1 & -1 \\ -1 & 1 \end{bmatrix} + \frac{c_e h_e}{6} \begin{bmatrix} 2 & 1 \\ 1 & 2 \end{bmatrix} \tag{3.34a}$$

$$\{f^e\} = \frac{q_e h_e}{2} \begin{Bmatrix} 1 \\ 1 \end{Bmatrix} \tag{3.34b}$$

If $a = a_e x$ and $c = c_e$, the coefficient matrix $[K^e]$ can be evaluated as

$$[K^e] = \frac{a_e}{h_e} \left(\frac{x_e + x_{e+1}}{2} \right) \begin{bmatrix} 1 & -1 \\ -1 & 1 \end{bmatrix} + \frac{c_e h_e}{6} \begin{bmatrix} 2 & 1 \\ 1 & 2 \end{bmatrix} \tag{3.35}$$

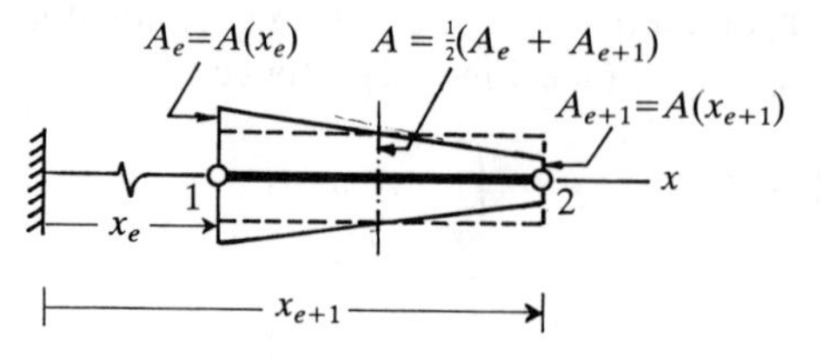

FIGURE 3.6
Approximation of an element with linearly varying cross-section by an equivalent element, with constant cross-section.

The reader should verify this. Note that when a is a linear function of x, this is equivalent to replacing a in the coefficient matrix with its average value [compare (3.34a) with (3.35)]:

$$a_{\text{avg}} = \tfrac{1}{2}(x_e + x_{e+1})a_e \tag{3.36}$$

For example, in the study of bars with linearly varying cross-section

$$a = EA(x) = E\left(A_e + \frac{A_{e+1} - A_e}{h_e}\bar{x}\right)$$

this amounts to replacing the varying cross-section with a constant cross-section within each element, the cross-sectional area of the constant section being the average area of cross section of the linearly varying element (see Fig. 3.6). Here A_e denotes the cross-sectional area at x_e and A_{e+1} is that at $x = x_{e+1}$.

When a, c, and q are algebraic polynomials in x, the evaluation of K_{ij}^e and f_j^e is straightforward. When they are complicated functions of x, numerical evaluation of the integrals in $[K^e]$ and $\{f^e\}$ will be sought. A complete discussion of the numerical evaluation of integrals is presented in Chapter 7.

When a and q are element-wise-constant and $c = 0$, the finite element equations corresponding to the linear element are

$$\frac{a_e}{h_e}\begin{bmatrix} 1 & -1 \\ -1 & 1 \end{bmatrix}\begin{Bmatrix} u_1^e \\ u_2^e \end{Bmatrix} = \frac{q_e h_e}{2}\begin{Bmatrix} 1 \\ 1 \end{Bmatrix} + \begin{Bmatrix} Q_1^e \\ Q_2^e \end{Bmatrix} \tag{3.37a}$$

$$\begin{aligned} \frac{a_e}{h_e}u_1^e - \frac{a_e}{h_e}u_2^e &= \tfrac{1}{2}q_e h_e + Q_1^e \\ -\frac{a_e}{h_e}u_1^e + \frac{a_e}{h_e}u_2^e &= \tfrac{1}{2}q_e h_e + Q_2^e \end{aligned} \tag{3.37b}$$

Quadratic element. For a quadratic-element mesh, the element Ω^e is located between global nodes $x_A = x_{2e-1}$ and $x_B = x_{2e+1}$. Hence,

$$\begin{aligned} K_{ij}^e &= \int_{x_{2e-1}}^{x_{2e+1}} \left(a_e \frac{d\psi_i^e}{dx}\frac{d\psi_j^e}{dx} + c_e\psi_i^e\psi_j^e\right)dx = \int_0^{h_e}\left(a_e\frac{d\psi_i^e}{d\bar{x}}\frac{d\psi_j^e}{d\bar{x}} + c_e\psi_i^e\psi_j^e\right)d\bar{x} \\ f_i^e &= \int_{x_{2e-1}}^{x_{2e+1}} \psi_i^e q_e\,dx = \int_0^{h_e}\psi_i^e q_e\,d\bar{x} \end{aligned} \tag{3.38}$$

where the Lagrange quadratic interpolation functions $\psi_i^e(\bar{x})$ $(i = 1, 2, 3)$ are given in (3.27). Evaluating the integrals in (3.38), we obtain

$$K_{11}^e = \int_0^{h_e} \left\{ a_e \left(-\frac{3}{h_e} + \frac{4\bar{x}}{h_e^2} \right) \left(-\frac{3}{h_e} + \frac{4\bar{x}}{h_e^2} \right) + c_e \left[1 - \frac{3\bar{x}}{h_e} + 2\left(\frac{\bar{x}}{h_e}\right)^2 \right] \left[1 - \frac{3\bar{x}}{h_e} + 2\left(\frac{\bar{x}}{h_e}\right)^2 \right] \right\} d\bar{x}$$

$$= \frac{7}{3}\frac{a_e}{h_e} + \frac{2}{15} c_e h_e$$

$$K_{12}^e = K_{21}^e = \int_0^{h_e} \left\{ a_e \left(-\frac{3}{h_e} + \frac{4\bar{x}}{h_e^2} \right) \left(\frac{4}{h_e} - \frac{8\bar{x}}{h_e} \right) + c_e \left[\frac{1 - 3\bar{x}}{h_e} + 2\left(\frac{\bar{x}}{h_e}\right)^2 \right] \left[4\frac{\bar{x}}{h_e}\left(1 - \frac{\bar{x}}{h_e}\right) \right] \right\} d\bar{x}$$

$$= -\frac{8}{3}\frac{a_e}{h_e} + \frac{2}{30} c_e h_e$$

and so on. Similarly,

$$f_1^e = \int_0^{h_e} q_e \left[1 - \frac{3\bar{x}}{h_e} + 2\left(\frac{\bar{x}}{h_e}\right)^2 \right] d\bar{x} = \tfrac{1}{6} q_e h_e = f_3^e \quad \text{(by symmetry)}$$

$$f_2^e = \int_0^{h_e} q_e \left[4\frac{\bar{x}}{h_e}\left(1 - \frac{\bar{x}}{h_e}\right) \right] d\bar{x} = \tfrac{4}{6} q_e h_e$$

Note that, for quadratic elements, the total source $q_e h_e$ is not distributed equally between the nodes. The distribution is *not* equivalent to that of two linear elements of lengths $\frac{1}{2}h_e$. Therefore, the computation of f_i^e should be based on the interpolation functions of that element. The sum of f_i^e for *any element* should always be equal to the integral of $q(x)$ over the element:

$$\sum_{i=1}^{n} f_i^e = \int_{x_A}^{x_B} q(x)\, dx \tag{3.39}$$

In summary, for element-wise-constant values of a, c, and q, the element matrices of a quadratic element are

$$[K^e] = \frac{a_e}{3h_e} \begin{bmatrix} 7 & -8 & 1 \\ -8 & 16 & -8 \\ 1 & -8 & 7 \end{bmatrix} + \frac{c_e h_e}{30} \begin{bmatrix} 4 & 2 & -1 \\ 2 & 16 & 2 \\ -1 & 2 & 4 \end{bmatrix} \tag{3.40a}$$

$$\{f^e\} = \frac{q_e h_e}{6} \begin{Bmatrix} 1 \\ 4 \\ 1 \end{Bmatrix} \tag{3.40b}$$

3.2.4 Connectivity of Elements

In deriving the element equations, we isolated a typical element (the eth) from the mesh and formulated the variational problem (or weak form) and developed its finite element model. To solve the total problem, we must put the elements back into their original positions. In doing this before discretization, we impose the continuity of the primary variables and balance of the secondary variables at the connecting nodes between elements. Continuity of the primary variables refers here to the single-valued nature of the solution; balance of secondary variables refers to the equilibrium of point sources at the junction of several elements. Thus, the assembly of elements is carried out by imposing the following two conditions:

1. Continuity of primary variables at connecting nodes:

$$u_n^e = u_1^{e+1} \tag{3.41a}$$

i.e., the last nodal value of the element Ω^e is the same as the first nodal value of the adjacent element Ω^{e+1}.

2. Balance of secondary variables at connecting nodes:

$$Q_n^e + Q_1^{e+1} = \begin{cases} 0 & \text{if no external point source is applied} \\ Q_0 & \text{if an external point source of magnitude } Q_0 \text{ is applied} \end{cases} \tag{3.41b}$$

In writing (3.41), it is assumed that elements are connected in a sequence. The continuity of primary variables $u_2^e = u_1^{e+1}$ and balance of secondary variables $Q_2^e + Q_1^{e+1}$ for a mesh of linear elements is illustrated in Fig. 3.7. The balance of secondary variables can be interpreted as the continuity of $a\,du/dx$ (*not* $a\,dU^e/dx$) at the point common to elements Ω^e and Ω^{e+1} (when no change in

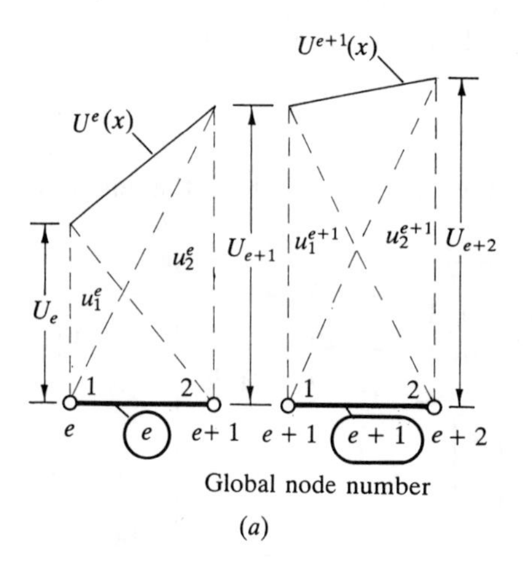

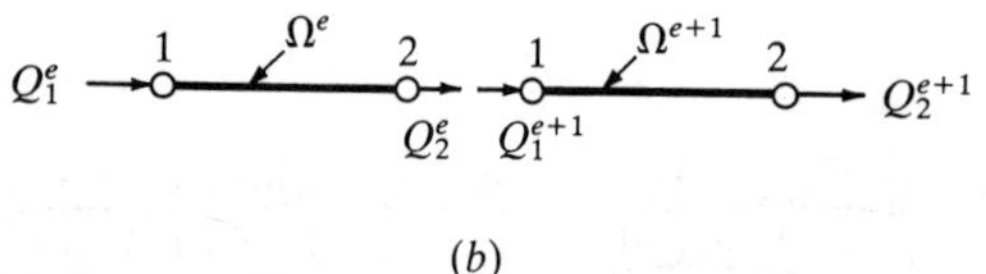

FIGURE 3.7
Assembly of two linear Lagrange elements: (*a*) continuity of the primary variable; (*b*) balance of the secondary variables.

$a\,du/dx$ is imposed externally):

$$\left(a\frac{du}{dx}\right)^e = \left(a\frac{du}{dx}\right)^{e+1}$$

or

$$\left(a\frac{du}{dx}\right)^e + \left(-a\frac{du}{dx}\right)^{e+1} = 0$$
$$Q_2^e + Q_1^{e+1} = 0 \tag{3.42}$$

The interelement continuity of the primary variables is imposed by renaming the two variables u_n^e and u_1^{e+1} at $x = x_N$ as one and the same, namely the value of u at the global node N:

$$u_n^e = u_1^{e+1} \equiv U_N \tag{3.43}$$

where $N = (n-1)e + 1$ is the global node number corresponding to node n of the element Ω^e and node 1 of the element Ω^{e+1}. For example, for a mesh of E linear finite elements ($n = 2$), we have

$$\begin{aligned} u_1^1 &= U_1 \\ u_2^1 = u_1^2 &= U_2 \\ u_2^2 = u_1^3 &= U_3 \\ &\vdots \\ u_2^{E-1} = u_1^E &= U_E \\ u_2^E &= U_{E+1} \end{aligned} \tag{3.44}$$

To enforce balance of the secondary variables Q_i^e, (3.41b), it is clear that we can set $Q_n^e + Q_1^{e+1}$ equal to zero or a specified value only if we have such expressions in our equations. To obtain such expressions, we must add the nth equation of the element Ω^e to the first equation of the element Ω^{e+1}; that is, we add

$$\sum_{j=1}^{n} K_{nj}^e u_j^e = f_n^e + Q_n^e$$

and

$$\sum_{j=1}^{n} K_{1j}^{e+1} u_j^{e+1} = f_1^{e+1} + Q_1^{e+1}$$

to give

$$\begin{aligned} \sum_{j=1}^{n} (K_{nj}^e u_j^e + K_{1j}^{e+1} u_j^{e+1}) &= f_n^e + f_1^{e+1} + (Q_n^e + Q_1^{e+1}) \\ &= f_n^e + f_1^{e+1} + Q_0 \end{aligned} \tag{3.45}$$

This process reduces the number of equations from $2E$ to $E + 1$. The first equation of the first element and the last equation of the last element will remain unchanged, except for renaming of the primary variables. The left-hand side of (3.45) can be written in terms of the global nodal values as

$$
\begin{aligned}
&(K_{n1}^e u_1^e + K_{n2}^e u_2^e + \ldots + K_{nn}^e u_n^e) + (K_{11}^{e+1} u_1^{e+1} + K_{12}^{e+1} u_2^{e+1} + \ldots + K_{1n}^{e+1} u_n^{e+1}) \\
&\quad = (K_{n1}^e U_N + K_{n2}^e U_{N+1} + \ldots + K_{nn}^e U_{N+n-1}) \\
&\qquad + (K_{11}^{e+1} U_{N+n-1} + K_{12}^{e+1} U_{N+n} + \ldots + K_{1n}^{e+1} U_{N+2n-2}) \\
&\quad = K_{n1}^e U_N + K_{n2}^e U_{N+1} + \ldots + K_{n(n-1)}^e U_{N+n-2} \\
&\qquad + (K_{nn}^e + K_{11}^{e+1}) U_{N+n-1} + K_{12}^{e+1} U_{N+n} + \ldots + K_{1n}^{e+1} U_{N+2n-2}
\end{aligned} \tag{3.46}
$$

where $N = (n-1)e + 1$.

For a mesh of E linear elements ($n = 2$), we have

$$
\begin{aligned}
K_{11}^1 U_1 + K_{12}^1 U_2 &= f_1^1 + Q_1^1 \quad \text{(unchanged)} \\
K_{21}^1 U_1 + (K_{22}^1 + K_{11}^2) U_2 + K_{12}^2 U_3 &= f_2^1 + f_1^2 + Q_2^1 + Q_1^2 \\
K_{21}^2 U_2 + (K_{22}^2 + K_{11}^3) U_3 + K_{12}^3 U_4 &= f_2^2 + f_1^3 + Q_2^2 + Q_1^3 \\
&\vdots \\
K_{21}^{E-1} U_{E-1} + (K_{22}^{E-1} + K_{11}^E) U_E + K_{12}^E U_{E+1} &= f_2^{E-1} + f_1^E + Q_2^{E-1} + Q_1^E \\
K_{21}^E U_E + K_{22}^E U_{E+1} &= f_2^E + Q_2^E \quad \text{(unchanged)}
\end{aligned} \tag{3.47a}
$$

These are called the *assembled equations*. They contain the sum of coefficients and source terms at nodes common to two elements. Note that the numbering of the global equations corresponds to the numbering of the global primary degrees of freedom, U_I. This correspondence carries the symmetry of element matrices to the global matrix. Equations (3.47a) can be expressed in matrix form as

$$
\begin{bmatrix}
K_{11}^1 & K_{12}^1 & & & \\
K_{21}^1 & K_{22}^1 + K_{11}^2 & K_{12}^2 & & 0 \\
 & K_{21}^2 & K_{22}^2 + K_{11}^3 & & \\
\cdots & \cdots & \cdots & \cdots & \cdots \\
0 & & & K_{22}^{E-1} + K_{11}^E & K_{12}^E \\
 & & & K_{21}^E & K_{22}^E
\end{bmatrix}
\begin{Bmatrix} U_1 \\ U_2 \\ U_3 \\ \vdots \\ U_E \\ U_{E+1} \end{Bmatrix}
= \begin{Bmatrix} f_1^1 \\ f_2^1 + f_1^2 \\ f_2^2 + f_1^3 \\ \vdots \\ f_2^{E-1} + f_1^E \\ f_2^E \end{Bmatrix}
+ \begin{Bmatrix} Q_1^1 \\ Q_2^1 + Q_1^2 \\ Q_2^2 + Q_1^3 \\ \vdots \\ Q_2^{E-1} + Q_1^E \\ Q_2^E \end{Bmatrix} \tag{3.47b}
$$

Recall that all the above discussion of assembly is based on the assumption that elements are connected in series. In general, several elements can be connected at a node, and the elements do not have to be consecutively numbered. In that case, the above idea still holds, with the change that coefficients coming from all elements connected at one node will add up. For

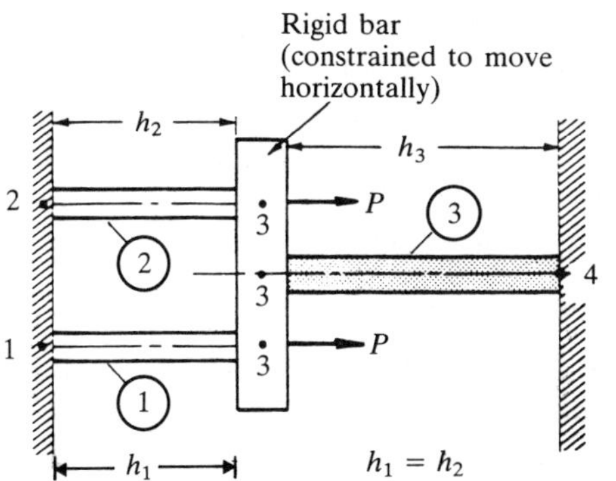

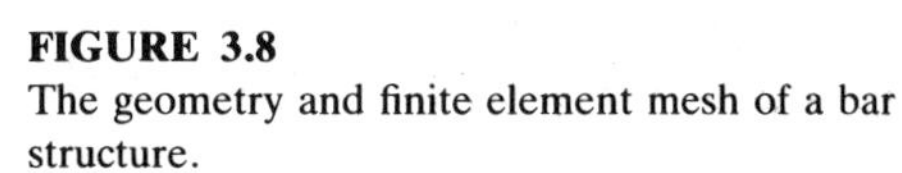

FIGURE 3.8
The geometry and finite element mesh of a bar structure.

example, consider the structure consisting of three bar elements shown in Fig. 3.8. Suppose that the connecting bar is rigid (i.e., not deformable) and is constrained to remain horizontal at all times. Then the continuity and force balance conditions for the structure are

$$u_2^1 = u_1^3 = u_2^2 \equiv U_3, \qquad Q_2^1 + Q_1^3 + Q_2^2 = 2P \tag{3.48}$$

To enforce these conditions, we must add the second equation of element 1, the first equation of element 3, and the second equation of element 2:

$$\begin{aligned}(K_{21}^1 u_1^1 + K_{22}^1 u_2^1) + (K_{11}^3 u_1^3 + K_{12}^3 u_2^3) + (K_{21}^2 u_1^2 + K_{22}^2 u_2^2)&\\ = f_2^1 + f_1^3 + f_2^2 + Q_2^1 + Q_1^3 + Q_2^2&\end{aligned} \tag{3.49}$$

We note the following correspondence of local and global nodal values (see Fig. 3.8):

$$u_1^1 = U_1, \qquad u_1^2 = U_2, \qquad u_2^1 = u_1^3 = u_2^2 = U_3, \qquad u_2^3 = U_4$$

Hence, (3.49) becomes

$$\begin{aligned}K_{21}^1 U_1 + K_{21}^2 U_2 + (K_{22}^1 + K_{11}^3 + K_{22}^2) U_3 + K_{12}^3 U_4 &= f_2^1 + f_1^3 + f_2^2 + Q_2^1 + Q_1^3 + Q_2^2\\ &= f_2^1 + f_1^3 + f_2^2 + 2P\end{aligned}$$

The other equations remain unchanged, except for renaming of the primary variables. The assembled equations are

$$\begin{bmatrix} K_{11}^1 & 0 & K_{12}^1 & 0 \\ 0 & K_{11}^2 & K_{12}^2 & 0 \\ K_{21}^1 & K_{21}^2 & \hat{K} & K_{12}^3 \\ 0 & 0 & K_{21}^3 & K_{22}^3 \end{bmatrix} \begin{Bmatrix} U_1 \\ U_2 \\ U_3 \\ U_4 \end{Bmatrix} = \begin{Bmatrix} f_1^1 \\ f_1^2 \\ f_2^1 + f_1^3 + f_2^2 \\ f_2^3 \end{Bmatrix} + \begin{Bmatrix} Q_1^1 \\ Q_1^2 \\ Q_2^1 + Q_1^3 + Q_2^2 \\ Q_2^3 \end{Bmatrix} \tag{3.50}$$

where $\hat{K} = K_{22}^1 + K_{11}^3 + K_{22}^2$.

The coefficients of the assembled matrix can be obtained directly. We note that the global coefficient K_{IJ} is a physical property of the system, relating global node I to global node J. For axial deformation of bars, K_{IJ} denotes the force required at node I to induce a unit displacement at node J, while the displacements at all other nodes are zero. Therefore, K_{IJ} is equal to the sum of all K_{ij}^e for which i corresponds to I and j corresponds to J, and i and j are the local nodes of the element Ω^e. Thus, if we have a correspondence between element node numbers and global node numbers then the assembled global coefficients can readily be written in terms of the element coefficients. The correspondence can be expressed through a matrix $[B]$, called the *connectivity matrix*, whose coefficient b_{ij} has the following meaning:

b_{ij} is the global node number corresponding to the jth node of element i

For example, for the structure shown in Fig. 3.8, the matrix $[B]$ is of order 3×2 (3 elements and 2 nodes per element):

$$[B] = \begin{bmatrix} 1 & 3 \\ 2 & 3 \\ 3 & 4 \end{bmatrix}$$

This array can be used in a variety of ways—not only for assembly, but also in the computer implementation of finite element computations. The matrix $[B]$ is used to assemble coefficient matrices as follow:

$K_{11}^1 = K_{11}$, because local node 1 of element 1 corresponds to global node 1

$K_{12}^1 = K_{13}$, because local nodes 1 and 2 of element 1 correspond to global nodes 1 and 3, respectively

and so on. When more than one element is connected at a global node, the element coefficients are to be added. For example, global node 3 appears in all three rows (i.e., elements) of the matrix $[B]$, implying that all three elements are connected at global node 3. More specifically, it indicates that node 2 of element 1, node 2 of element 2, and node 1 of element 3 are the same as global node 3. Hence

$$K_{22}^1 + K_{22}^2 + K_{11}^3 = K_{33}$$

Assembly on paper can be carried out by examining the finite element mesh of the problem. For the mesh shown in Fig. 3.8, we have

$K_{23} = K_{12}^2$, because global node 2 is the same as node 1 and global node 3 is the same as node 2 of element 2

$K_{24} = 0$, because global nodes 2 and 4 do not belong to the same element

$K_{33} = K_{22}^1 + K_{22}^2 + K_{11}^3$

and so on.

In summary, assembly of finite elements is carried out by imposing interelement continuity of primary variables and balance of secondary variables [see (3.41)]. Renaming the elemental primary variables in terms of the global primary variables and using the correspondence between the local and global nodes allows the assembly. When certain primary nodal values are not required to be continuous (by the variational formulation) across elements, such variables may be *condensed* out at the element level before assembling elements.

3.2.5 Imposition of Boundary Conditions

Up to this point, the specific nature of the problem has not been used in the development of the finite element model or in the assembly of finite elements. In other words, the discussion in Sections 3.2.1–3.2.4 is valid for any differential equation that is a special case of the model equation (3.1). One particular problem differs from others in the specification of the data and boundary conditions. Here we discuss how to impose the boundary conditions of a problem on the assembled set of algebraic equations. To this end, we use the problem in Fig. 3.8. Its boundary conditions are evident from the structure. The known primary degrees of freedom (i.e., displacements) are

$$u_1^1 = U_1 = 0, \quad u_1^2 = U_2 = 0, \quad u_2^3 = U_4 = 0 \tag{3.51a}$$

The known secondary degrees of freedom (i.e., forces) are

$$Q_2^1 + Q_2^2 + Q_1^3 = 2P \tag{3.51b}$$

The forces Q_1^1, Q_1^2, and Q_2^3 are unknown (reaction forces), and they can be determined in the post-computation, i.e., after the primary degrees of freedom are determined.

Imposing the boundary conditions (3.51) on the assembled system of equations (3.50) and for f_i^e, we obtain

$$\begin{bmatrix} K_{11}^1 & 0 & K_{12}^1 & 0 \\ 0 & K_{11}^2 & K_{12}^2 & 0 \\ K_{21}^1 & K_{21}^2 & K_{22}^1 + K_{22}^2 + K_{11}^3 & K_{12}^3 \\ 0 & 0 & K_{21}^3 & K_{22}^3 \end{bmatrix} \begin{Bmatrix} U_1 = 0 \\ U_2 = 0 \\ U_3 \\ U_4 = 0 \end{Bmatrix} = \begin{Bmatrix} Q_1^1 \\ Q_1^2 \\ 2P \\ Q_2^3 \end{Bmatrix} \tag{3.52}$$

This contains four equations in four unknowns: U_3, Q_1^1, Q_1^2, and Q_2^3.

3.2.6 Solution of Equations

As a standard procedure in finite element analysis, the unknown primary degrees of freedom are determined first by *considering the algebraic equations corresponding to the unknown primary variables.* Thus, in the present case, we consider the third equation in (3.52) to solve for U_3:

$$K_{21}^1 U_1 + K_{21}^2 U_2 + (K_{22}^1 + K_{22}^2 + K_{11}^3) U_3 + K_{12}^3 U_4 = 2P$$

or

$$(K_{22}^1 + K_{22}^2 + K_{11}^3) U_3 = 2P - (K_{21}^1 U_1 + K_{21}^2 U_2 + K_{12}^3 U_4) \tag{3.53}$$

Equation (3.53) is called the *condensed equation* for the unknown U_3. The term in parentheses on the right-hand side is zero because all specified displacements are zero in the present problem. Hence, the solution is given by

$$U_3 = 2P/(K_{22}^1 + K_{22}^2 + K_{11}^3) \tag{3.54a}$$

The unknown secondary variables are determined by considering the remaining equations of (3.52), i.e., those that contain the unknown secondary variables:

$$\begin{Bmatrix} Q_1^1 \\ Q_1^2 \\ Q_2^3 \end{Bmatrix} = \begin{bmatrix} K_{11}^1 & 0 & K_{12}^1 & 0 \\ 0 & K_{11}^2 & K_{12}^2 & 0 \\ 0 & 0 & K_{21}^3 & K_{22}^3 \end{bmatrix} \begin{Bmatrix} U_1 \\ U_2 \\ U_3 \\ U_4 \end{Bmatrix}$$

$$= \begin{Bmatrix} K_{12}^1 U_3 \\ K_{12}^2 U_3 \\ K_{21}^3 U_3 \end{Bmatrix}, \quad \text{because } U_1,\ U_2 \text{ and } U_4 \text{ are zero} \tag{3.54b}$$

It is possible, although not common with computer programs, to move all the unknowns to the left-hand side in (3.52) and solve for them all at once. But this process requires more computational time in practical problems.

In general, the assembled finite element equations can be partitioned conveniently into the following form:

$$\begin{bmatrix} [K^{11}] & [K^{12}] \\ [K^{21}] & [K^{22}] \end{bmatrix} \begin{Bmatrix} \{U^1\} \\ \{U^2\} \end{Bmatrix} = \begin{Bmatrix} \{F^1\} \\ \{F^2\} \end{Bmatrix} \tag{3.55}$$

where $\{U^1\}$ is the column of known primary variables, $\{U^2\}$ is the column of unknown primary variables, $\{F^1\}$ is the column of unknown secondary variables, and $\{F^2\}$ is the column of known secondary variables. Writing (3.55) as two matrix equations, we obtain

$$[K^{11}]\{U^1\} + [K^{12}]\{U^2\} = \{F^1\} \tag{3.56a}$$

$$[K^{21}]\{U^1\} + [K^{22}]\{U^2\} = \{F^2\} \tag{3.56b}$$

From (3.56*b*), we have

$$\boxed{\{U^2\} = [K^{22}]^{-1}(\{F^2\} - [K^{21}]\{U^1\})} \tag{3.56c}$$

Once $\{U^2\}$ is known, $\{F^1\}$ can be computed from (3.56*a*).

3.2.7 Postprocessing of the Solution

The solution of the finite element equations gives the nodal values of the primary unknown (e.g., displacement, velocity, or temperature). Postprocessing of the results includes one or more of the following:

1. Computation of any secondary variables (e.g., the gradient of the solution).
2. Interpretation of the results to check whether the solution makes sense (an understanding of the physical process and experience are the guides when other solutions are not available for comparison).
3. Tabular and/or graphical presentation of the results.

To determine the solution u as a continuous function of position x, we return to the approximation (3.29) over each element:

$$\boxed{u(x) \approx \begin{cases} U^1(x) = \sum_{j=1}^{n} u_j^1 \psi_j^1(x) \\ U^2(x) \approx \sum_{j=1}^{n} u_j^2 \psi_j^2(x) \\ \vdots \\ U^N(x) = \sum_{j=1}^{n} u_j^N \psi_j^N(x) \end{cases}} \tag{3.57}$$

where N is the number of elements in the mesh. Depending on the value of x, the corresponding element equation from (3.57) is used. The derivative of the solution is obtained by differentiating (3.57):

$$\frac{du}{dx} \approx \begin{cases} \sum_{j=1}^{n} u_j^1 \dfrac{d\psi_j^1}{dx} \\ \sum_{j=1}^{n} u_j^2 \dfrac{d\psi_j^2}{dx} \\ \vdots \\ \sum_{j=1}^{n} u_j^N \dfrac{d\psi_j^N}{dx} \end{cases} \tag{3.58}$$

Note that the derivative dU^e/dx of the linear finite element solution U^e is constant within each element, and it is discontinuous at the nodes because the continuity of the derivative of the finite element solution at the connecting nodes is not imposed:

$$\frac{dU^e}{dx} \neq \frac{dU^{e+1}}{dx}$$

The derivative calculated from different elements meeting at a node is always discontinuous in all C^0 approximations (i.e., approximations in which only the function values are interpolated), unless the approximate solution coincides with the actual solution.

The secondary variables Q_j^e can be computed in two different ways. In (3.54*b*), we determined the unknown secondary variables Q_1^1, Q_1^2, and Q_2^3 from the assembled equations of the problem in Fig. 3.8. Since the assembled equations often represent the equilibrium relations of a system, the Q_i^e computed from them will be denoted by $(Q_i^e)_{\text{equil}}$. The Q_i^e can also be determined using the definitions in (3.7), replacing u with U. We shall denote Q_i^e computed in this way by $(Q_i^e)_{\text{def}}$. Since $(Q_i^e)_{\text{def}}$ are calculated using the approximate U^e, they are not as accurate as $(Q_i^e)_{\text{equil}}$. However, in finite element computer codes, $(Q_i^e)_{\text{def}}$ are calculated instead of $(Q_i^e)_{\text{equil}}$. This is primarily because of computational aspects. Recall that, in arriving at the result (3.54*b*), we used part of the assembled coefficient matrix. In the numerical solution of simultaneous algebraic equations in a computer, the original assembled coefficient matrix is often modified, and therefore the coefficients needed for the determination of the secondary variables are not available, unless they are saved in an additional array. For the problem in Fig. 3.8, we have

$$(Q_1^1)_{\text{def}} = -\left(EA\frac{dU}{dx}\right)\bigg|_{x=0} = -EA\frac{U_3 - U_1}{h_1} = -\frac{EA}{h_1}U_3 = K_{12}^1 U_3$$

$$(Q_1^2)_{\text{def}} = -\left(EA\frac{dU}{dx}\right)\bigg|_{x=0} = K_{12}^2 U_3$$

$$(Q_2^3)_{\text{def}} = \left(EA\frac{dU}{dx}\right)\bigg|_{x=h_1+h_3} = EA\frac{U_4 - U_3}{h_3}$$

$$= -\frac{EAU_3}{h_3} = K_{21}^3 U_3 \tag{3.59}$$

where h_1 and h_3 are the lengths of elements 1 and 3, respectively.

The Qs computed using the definitions (3.7) are the same as those derived from the assembled equations for the problem in Fig. 3.8. This equality is *not to be expected in general.* In fact, when the source vector q is not zero, the secondary variables computed from the definitions (3.7) will be in error compared with those computed from the assembled equations. The error decreases as the number of elements or the degree of interpolation is increased.

This completes the basic steps involved in the finite element analysis of the model equation (3.1). A few remarks are in order on the steps described above for the model equation.

Remark 1. Although the Rayleigh–Ritz method was used to set up the element equations, any other method, such as a weighted-residual (e.g., the least-squares or Galerkin) method, could be used.

Remark 2. Steps 1–6 (see Table 3.1) are common for any problem. The derivation of interpolation functions depends only on the element geometry, and on the number and position of nodes in the element. The number of nodes in the element and the degree of approximation used are related.

Remark 3. The finite element equations (3.31) are derived for the linear operator equation

$$A(u) = q, \qquad \text{where} \quad A = -\frac{d}{dx}\left(a\frac{d}{dx}\right) + c$$

Hence, they are valid for any physical problem that is described by the operator equation $A(u) = q$ or its special cases. One need only interpret the quantities appropriately. Examples of problems described by this operator are listed in Table 3.2. Thus, a computer program written for the finite element analysis of (3.1) can be used to analyze any of the problems in this table. Also, note that the data $a = a(x)$, $c = c(x)$, and $q = q(x)$ can be different in each of the elements.

Remark 4. Integration of the element matrices in (3.31c) can be implemented on a computer using numerical integration (see Chapter 7). When these integrals are algebraically complicated, one has no other choice but numerical integration.

Remark 5. As noted in (3.48) and (3.51b), the point sources at the nodes are included in the finite element model via the balance of sources at the nodes. Thus, in constructing finite element meshes, one should include nodes at the locations of point sources. If a point source does not occur at a node, it is possible to "distribute" it to the element nodes. Let Q_0 denote a point source at point x_0, $x_A \leqslant x_0 \leqslant x_B$. The point source Q_0 can be represented as a "function" by

$$q(x) = Q_0 \delta(x - x_0)$$

where the Dirac delta function $\delta(\cdot)$ is defined by

$$\int_{-\infty}^{\infty} F(x)\delta(x - x_0)\, dx = F(x_0)$$

The contribution of the function $q(x)$ to the nodes of the element $\Omega^e = (x_A, x_B)$ is computed from [see (3.31c)]

$$f_i^e = \int_{x_A}^{x_B} q(x)\psi_i^e(x)\, dx = \int_{x_A}^{x_B} Q_0 \delta(x - x_0)\psi_i^e(x)\, dx = Q_0 \psi_i^e(x_0) \tag{3.60}$$

where ψ_i^e are the interpolation functions of the element Ω^e. Thus, the point source Q_0 is distributed to the element node i by the value $Q_0\psi_i^e(x_0)$. Equation (3.60) holds for any element, irrespective of the degree of the interpolation, the nature of the interpolation (i.e., Lagrange or Hermite polynomials), or the dimension (i.e., 1-D, 2-D, or 3-D) of the element. For linear Lagrange interpolation functions in 1-D, (3.60) yields

$$f_1^e = Q_0 \frac{x_B - x_0}{h_e} = \alpha Q_0, \qquad f_2^e = Q_0 \frac{x_0 - x_A}{h_e} = (1 - \alpha) Q_0$$

where $\alpha = (x_B - x_0)/h_e$ is the ratio of the distance between node 2 and the source to the length of the element.

Remark 6. There are three sources of error that may contribute to the inaccuracy of the finite element solution of a problem:

1. *Domain approximation error,* which is due to the approximation of the domain.

2. *Computational errors,* which are due to inexact evaluation of the coefficients K_{ij}^e and f_i^e, or are introduced owing to the finite arithmetic in a computer.
3. *Approximation error,* which is due to approximation of the solution by piecewise polynomials.

Since the geometry of the problem is exactly represented, and the linear approximation is able to represent the exact solution at the nodes (for $a = EA = \text{constant}$, $c = 0$, and $f = 0$), the first and third type of errors are zero in the problem of Fig. 3.8. The only error that can be introduced into the final numerical results is possibly due to the computer evaluation of the coefficients K_{ij}^e and f_i^e and the solution of algebraic equations. Additional discussion of the errors in the finite element approximation is presented in Chapter 5.

Remark 7. The approach used in matrix methods of structural analysis to solve the problem in Fig. 3.8 is not much different than that presented here. The difference lies only in the derivation of the element equations (3.37a). In matrix methods of structural analysis, the element equations are obtained directly from the definitions of stress and strain and their relationship. For example, consider the free-body diagram of a bar element (see Fig. 3.2*b*). From a course on deformable bodies, we have

$$\begin{aligned} \text{force} &= \text{stress} \times \text{area of cross section} \\ \text{stress} &= \text{Young's modulus} \times \text{strain} \\ \text{strain} &= \text{elongation/original length} \end{aligned}$$

The strain defined above is the average (or engineering) strain. Mathematically, strain is defined as $\varepsilon = du/dx$, u being the displacement, which includes rigid body motion as well as elongation of the bar. Hence, the force at the left end of the bar element is

$$P_1^e = A^e\sigma_1^e = A^eE^e\varepsilon_1^e = A^eE^e\frac{u_1^e - u_2^e}{h_e} = \frac{a_e}{h_e}(u_1^e - u_2^e)$$

where σ is the stress and E is Young's modulus. Similarly, the force at the right end is

$$P_2^e = \frac{a_e}{h_e}(u_2^e - u_1^e)$$

In matrix form, these relations can be expressed as

$$\frac{a_e}{h_e}\begin{bmatrix} 1 & -1 \\ -1 & 1 \end{bmatrix}\begin{Bmatrix} u_1^e \\ u_2^e \end{Bmatrix} = \begin{Bmatrix} P_1^e \\ P_2^e \end{Bmatrix} \tag{3.61}$$

which is the same as (3.37*a*) with $P_i^e = Q_i^e + f_i^e$. Note that in deriving the element equations, we have used knowledge of the mechanics of materials and the assumption that the strain is constant (or the displacement is linear) over the length of the element. Equations of the type (3.61) can also be derived for a spring element, a pipe-flow element, an electrical resistor element, and so on (see the problems at the end of this chapter). If a higher-order representation of the strain (or displacement) is required, we cannot write the force–displacement relations (3.61) directly. We must use the principle of virtual displacements, which is equivalent to the weak form of the governing equation. For more details, see Reddy (1984).

Remark 8. Another interpretation of (3.37) can be given in terms of the finite difference approximation. The axial force at any point x is given by $P(x) = EA\, du/dx$. Using the forward difference approximation, we approximate the derivative du/dx and write

$$-P_1^e \equiv P(x)\big|_{x_e} = E^e A^e[u(x_{e+1}) - u(x_e)]/h_e \tag{3.62a}$$

$$P_2^e \equiv P(x)\big|_{x_{e+1}} = E^e A^e[u(x_{e+1}) - u(x_e)]/h_e \tag{3.62b}$$

which are the same as (3.61), with $u_1^e = u(x_e)$ and $u_2^e = u(x_{e+1})$. Note that no approximation of $u(x)$ itself is assumed in writing (3.62). To obtain the value of u at a point other than the nodes (or mesh points), linear interpolation is often used.

Remark 9. For the model problem considered, the element matrices $[K^e]$ in (3.31b) are symmetric: $K_{ij}^e = K_{ji}^e$. This enables one to compute K_{ij}^e $(i = 1, 2, \ldots, n)$ for $j \le i$ only. In other words, one need compute only the diagonal terms and the upper or lower diagonal terms. Because of the symmetry of the element matrices, the assembled global matrix will also be symmetric. Thus, one need store only the upper triangle, including the diagonal, of the assembled matrix in a finite element program. Another property characteristic of the finite element method is the *sparseness* of the assembled matrix. Since $K_{IJ} = 0$ if global nodes I and J do not belong to the same element, the global coefficient matrix is *banded*, i.e., all coefficients beyond a certain distance from the diagonal are zero. The maximum of the distances between the diagonal element, including the latter, of a row and the last nonzero coefficient in that row is called the *half-bandwidth*, and can be computed from the equation

$$\text{half-bandwidth} = \max_{1 \le i \le E} \left(|b_{i1} - b_{in}| + 1\right) \times \text{NDF}$$

where E is the number of elements in the mesh, NDF is the number of degrees of freedom per node, n is the number of nodes per element, and b_{ij} are the coefficients of the connectivity matrix. When a matrix is banded and symmetric, one need store only the entries in the upper or lower band of the matrix. Equation solvers written for the solution of banded symmetric equations are available for use in such cases. The symmetry of the coefficient matrix depends on the type of the differential equation, its variational form, and the numbering of the finite element equations. The sparseness of the matrix is a result of the finite element interpolation functions, which have nonzero values only over an element of the domain.

Remark 10. The balance (or "equilibrium") of the secondary variables (or "forces") Q_i^e at the interelement boundaries is expressed by (3.41b). This amounts to imposing the condition that the secondary variable $a\,du/dx$ at the node, where u is the actual solution, be continuous. However, this does not imply continuity of $a\,dU^e/dx$, where U^e is the finite element solution. Thus, in general, we have

$$Q_2^e + Q_1^{e+1} = 0 \text{ or } Q_0 \tag{3.63a}$$

but

$$\left(a\frac{dU^e}{dx}\right)\Big|_{x_e} + \left(-a\frac{dU^{e+1}}{dx}\right)\Big|_{x_e} \neq 0 \text{ or } Q_0 \tag{3.63b}$$

Note. In most books on the finite element method, this point is not made clear to the reader. These books consider the quadratic form (3.11) of the total problem and omit the sum of the interelement contributions (for linear elements),

$$\sum_{e=1}^{N}\left(\sum_{i=1}^{2} Q_i^e u_i^e\right) \tag{i}$$

in the quadratic form of the problem. However, this amounts to imposing equilibrium conditions of the form (3.63a). When the secondary variable is specified to be nonzero (say, Q_0) at an interelement boundary (say, at global node 2), we have

$$Q_2^1 + Q_1^2 = Q_0 \tag{ii}$$

In other books, Q_0 is included in the functional as $Q_0 U_2$, where U_2 is the value of u at global node 2.

To fix ideas, consider (3.1). The variational form of this equation over the entire domain is given by (when $c = 0$)

$$0 = \int_0^L \left(a\frac{dv}{dx}\frac{du}{dx} - vq\right) dx - v(x_2)Q_0 \tag{iii}$$

When u is approximated by functions that are defined only on a local interval (which is the case in the finite element method), use of the above variational form implies the omission of the sum of the interelement contributions of (i).

Since ψ_i^e $(e = 1, 2, 3)$ is zero in any element Ω^f for $e \neq f$ (see Fig. 3.3b), the (global) finite element solution for the entire domain is given by

$$U_e(x) = \sum_{e=1}^{3}\left(\sum_{i=1}^{2} u_i^e \psi_i^e\right) \equiv \sum_{I=1}^{4} U_I \Phi_I(x) \tag{iv}$$

where $\Phi_I(x)$ $(I = 1, 2, 3, 4)$ are the piecewise-continuous *global interpolation functions*,

$$\Phi_I(x) = \begin{cases} \psi_2^{(I-1)}(x) & \text{for } x_{I-1} \leq x \leq x_I \\ \psi_1^{(I)}(x) & \text{for } x_I \leq x \leq x_{I+1} \end{cases} \tag{v}$$

Substituting (iv) for u and $v = \Phi_I$ into (iii), we obtain

$$0 = \int_0^L \left[a\frac{d\Phi_I}{dx}\left(\sum_{J=1}^{4} U_J \frac{d\Phi_J}{dx}\right) - \Phi_I q\right] dx - \Phi_I(x_2)Q_0 \tag{vi}$$

Since Φ_I is nonzero only between x_{I-1} and x_{I+1}, the integral becomes

$$0 = \int_{x_{I-1}}^{x_{I+1}} \left[a \frac{d\Phi_I}{dx} \left(U_{I-1} \frac{d\Phi_{I-1}}{dx} + U_I \frac{d\Phi_I}{dx} + U_{I+1} \frac{d\Phi_{I+1}}{dx} \right) - \Phi_I q \right] dx - \Phi_I(x_2) Q_0 \quad \text{(vii)}$$

and we have (for a three-element mesh)

$$\begin{aligned}
I = 1: \quad & 0 = \int_{x_1=0}^{x_2} \left[a \frac{d\Phi_1}{dx} \left(U_1 \frac{d\Phi_1}{dx} + U_2 \frac{d\Phi_2}{dx} \right) - \Phi_1 q \right] dx - \Phi_1(x_2) Q_0 \\
I = 2: \quad & 0 = \int_{x_1=0}^{x_3} \left[a \frac{d\Phi_2}{dx} \left(U_1 \frac{d\Phi_1}{dx} + U_2 \frac{d\Phi_2}{dx} + U_3 \frac{d\Phi_3}{dx} \right) - \Phi_2 q \right] dx - \Phi_2(x_2) Q_0 \\
I = 3: \quad & 0 = \int_{x_2}^{x_4=L} \left[a \frac{d\Phi_3}{dx} \left(U_2 \frac{d\Phi_2}{dx} + U_3 \frac{d\Phi_3}{dx} + U_4 \frac{d\Phi_4}{dx} \right) - \Phi_3 q \right] dx - \Phi_2(x_2) Q_0 \\
I = 4: \quad & 0 = \int_{x_3}^{x_4=L} \left[a \frac{d\Phi_4}{dx} \left(U_3 \frac{d\Phi_3}{dx} + U_4 \frac{d\Phi_4}{dx} \right) - \Phi_4 q \right] dx - \Phi_4(x_2) Q_0
\end{aligned} \quad \text{(viii)}$$

These equations, upon performing the integrations, yield (3.47), with the last column (containing Qs) in the latter replaced by

$$\begin{Bmatrix} 0 \\ Q_0 \\ 0 \\ 0 \end{Bmatrix} \quad \text{(ix)}$$

Although this procedure gives the assembled equations directly, it is algebraically complicated (especially for two-dimensional problems) and not amenable to simple computer implementation.

3.2.8 Radially Symmetric Problems

The equations governing physical processes in cylindrical or spherical geometries are described analytically in terms of cylindrical or spherical coordinates. When the geometry, loading, and boundary conditions are dependent only on the radial direction and independent of the other two coordinates, the governing equations are one-dimensional. The equations governing radially symmetric problems in cylindrical geometries are of the form [an analogue of (3.1)]

$$-\frac{1}{r}\frac{d}{dr}\left[a(r) \frac{du}{dr} \right] = q(r) \quad \text{for } R_i < r < R_o \quad (3.64)$$

where r is the radial coordinate, a and q are known functions of r, and u is the dependent variable. Such equations arise, for example, in connection with radial heat flow in a long circular cylinder of inner radius R_i and outer radius R_o. The radially symmetric conditions require that both a and q be functions only of r. Since the cylinder is long, the temperature distribution at any section along its length (except perhaps at the ends) is the same, and it is sufficient to consider any cross-section away from the ends, i.e., the problem is reduced from a 3-D problem to a 2-D one. Since a and q are independent of the circumferential direction θ, the temperature distribution along any radial line is the same, reducing the 2-D problem to a 1-D one, as described by (3.64).

In developing the weak form of (3.64), we multiply (3.64) with a weight function $w(r)$ and integrate over the volume of the cylinder of unit length.

$$
\begin{aligned}
(1)\quad 0 &= \int_V w\left[-\frac{1}{r}\frac{d}{dr}\left(a\frac{du}{dr}\right) - q\right] r\,dr\,d\theta\,dz \\
&= \int_0^1 \int_0^{2\pi} \int_{r_A}^{r_B} w\left[-\frac{1}{r}\frac{d}{dr}\left(a\frac{du}{dr}\right) - q\right] r\,dr\,d\theta\,dz \\
&= 2\pi \int_{r_A}^{r_B} w\left[-\frac{1}{r}\frac{d}{dr}\left(a\frac{du}{dr}\right) - q\right] r\,dr
\end{aligned}
$$

where (r_A, r_B) is the domain of the element along the radial direction. Next, we carry out the remaining two steps of the variational formulation:

$$(2)\quad 0 = 2\pi \int_{r_A}^{r_B} \left(a\frac{dw}{dr}\frac{du}{dr} - rwq\right) dr - \left[w 2\pi a \frac{du}{dr}\right]_{r_A}^{r_B}$$

$$(3)\quad 0 = 2\pi \int_{r_A}^{r_B} \left(a\frac{dw}{dr}\frac{du}{dr} - rwq\right) dr - w(r_A)Q_1^e - w(r_B)Q_2^e \tag{3.65a}$$

where

$$Q_1^e \equiv -2\pi\left(a\frac{du}{dr}\right)\bigg|_{r_A}, \qquad Q_2^e \equiv 2\pi\left(a\frac{du}{dr}\right)\bigg|_{r_B} \tag{3.65b}$$

The finite element model is obtained by substituting the approximation

$$u(r) \approx \sum_{j=1}^{n} u_j \psi_j^e(r)$$

into (3.65a):

$$[K^e]\{u^e\} = \{f^e\} + \{Q^e\} \tag{3.66a}$$

where

$$K_{ij}^e = 2\pi \int_{r_A}^{r_B} a\frac{d\psi_i^e}{dr}\frac{d\psi_j^e}{dr}\,dr, \qquad f_i^e = 2\pi \int_{r_A}^{r_B} \psi_i^e q r\,dr \tag{3.66b}$$

and ψ_i^e are the interpolation functions expressed in terms of the radial coordinate r. For example, the linear interpolation functions are of the forms

$$\psi_1^e(r) = (r_B - r)/h_e, \qquad \psi_2^e(r) = (r - r_A)/h_e \tag{3.66c}$$

The explicit forms of the coefficients K_{ij}^e and f_i^e for $a = a_e r$ and $q = q_e$ are given below.

Linear element

$$[K^e] = \frac{2\pi a_e}{h_e}\left(r_A + \tfrac{1}{2}h_e\right)\begin{bmatrix} 1 & -1 \\ -1 & 1 \end{bmatrix}, \qquad \{f^e\} = \frac{2\pi q_e h_e}{6}\begin{Bmatrix} 3r_A + h_e \\ 3r_A + 2h_e \end{Bmatrix}$$

Quadratic element

$$[K^e] = \frac{2\pi a_e}{6h_e}\begin{bmatrix} 3h_e + 14r_A & -(4h_e + 16r_A) & h_e + 2r_A \\ -(4h_e + 16r_A) & 16h_e + 32r_A & -(12h_e + 16r_A) \\ h_e + 2r_A & -(12h_e + 16r_A) & 11h_e + 14r_A \end{bmatrix}$$

$$\{f^e\} = \frac{2\pi q_e h_e}{6}\begin{Bmatrix} r_A \\ 4r_A + 2h_e \\ r_A + h_e \end{Bmatrix}$$

In the next section, we consider several examples to illustrate the steps involved in the finite element analysis of one-dimensional second-order equations arising in heat transfer, fluid mechanics, and solid mechanics. While the notation used for the dependent variables, independent coordinates, and data of problems from various fields is different, the reader should keep the common mathematical structure in mind and not get confused with the change of notation from field to field or problem to problem.

3.3 APPLICATIONS

3.3.1 Heat Transfer

Heat flows from high-temperature regions to low-temperature regions. This transfer of heat within the medium is called *conduction heat transfer*. The Fourier heat conduction law for one-dimensional systems states that the heat flow Q is related to the temperature gradient $\partial T/\partial x$ by the relation (with heat flow in the positive direction of x),

$$Q = -kA\frac{\partial T}{\partial x} \tag{3.67}$$

where k is the thermal conductivity of the material, A the cross-sectional area, and T the temperature. The negative sign in (3.67) indicates that heat flows downhill on the temperature scale. The balance of energy in an element of length dx requires that

energy into the element + energy generated within the element

= change in internal energy + energy out of the element

$$-kA\frac{\partial T}{\partial x} + qA\,dx = \rho cA\frac{\partial T}{\partial t}dx - \left[kA\frac{\partial T}{\partial x} + \frac{\partial}{\partial x}\left(kA\frac{\partial T}{\partial x}\right)dx\right]$$

or

$$\frac{\partial}{\partial x}\left(kA\frac{\partial T}{\partial x}\right) + Aq = \rho cA\frac{\partial T}{\partial t} \tag{3.68}$$

where q is the heat energy generated per unit volume, ρ is the density, c is the specific heat of the material, and t is time. Equation (3.68) governs the transient heat conduction in a slab or fin (i.e., a one-dimensional system) when the heat flow in the normal direction is zero. The following metric units will be used:

$$
\begin{array}{ll}
T & ^\circ\mathrm{C}\ \text{(celsius)} \\
k & \mathrm{W\,m^{-1}\,{}^\circ C^{-1}}\ \text{(watts per meter per degree Celsius)} \\
q & \mathrm{W\,m^{-3}} \\
\rho & \mathrm{kg\,m^{-3}} \\
c & \mathrm{J\,kg^{-1}\,{}^\circ C^{-1}}\ \text{(joules per kilogram per degree Celsius or } \mathrm{m^2\,s^{-2}\,{}^\circ C^{-1}})
\end{array}
\tag{3.69}
$$

In the case of radially symmetric problems with cylindrical geometries, (3.68) takes a different form (see Section 3.2.8). Consider a long cylinder of inner radius R_i, outer radius R_o, and length L. When L is very large compared with the diameter, it is assumed that heat flows in the radial direction r. Thus the surface area for heat flow in the cylindrical system is (see Fig. 3.9)

$$A = 2\pi r L \tag{3.70}$$

Hence, the transient radially symmetric heat flow in the cylinder is governed by

$$\frac{\partial}{\partial r}\left(kA\frac{\partial T}{\partial r}\right) + Aq = \rho c A\frac{\partial T}{\partial t} \tag{3.71a}$$

or

$$\boxed{\frac{1}{r}\frac{\partial}{\partial r}\left(kr\frac{\partial T}{\partial r}\right) + q = \rho c\frac{\partial T}{\partial t}} \tag{3.71b}$$

A cylindrical fuel element of a nuclear reactor, a current-carrying electrical wire, and a thick-walled circular tube provide examples of one-dimensional radial systems.

For the radial flow in a sphere, the cross-sectional area is

$$A = 4\pi^2 r^2$$

and the governing equation takes the form

$$\frac{1}{r^2}\frac{\partial}{\partial r}\left(kr^2\frac{\partial T}{\partial r}\right) + q = \rho c\frac{\partial T}{\partial t} \tag{3.72}$$

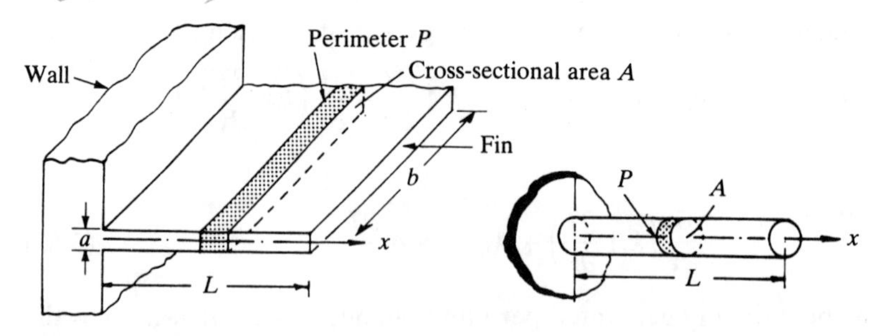

FIGURE 3.9
Convective heat transfer in a fin.

The boundary conditions for heat conduction involve specifying either temperature T or the heat flow Q at a point:

$$T = T_0 \quad \text{or} \quad Q \equiv -kA\frac{\partial T}{\partial x} = Q_0 \tag{3.73}$$

We know that when a heated surface is exposed to a cooling medium, such as air or liquid, the surface will cool faster. We say that the heat is convected away. The *convection heat transfer* between the surface and the medium in contact is given by *Newton's law of cooling*:

$$Q = \beta A(T_s - T_\infty) \tag{3.74}$$

where T_s is the surface temperature, T_∞ is the temperature of the surrounding medium (the *ambient temperature*), A is the surface area, and β is the *convection heat transfer coefficient* or *film conductance* (or film coefficient). The units of β are $\text{W m}^{-2}\,^\circ\text{C}^{-1}$. The heat flow due to conduction and convection at a boundary point must be in balance with the applied flow Q_0:

$$\boxed{\pm kA\frac{\partial T}{\partial x} + \beta A(T - T_\infty) + Q_0 = 0} \tag{3.75}$$

The sign of the first term in (3.75) is negative when the heat flow is from the fluid at T_∞ to the surface at the left end of the element, and it is positive when the heat flow is from the fluid at T_∞ to the surface at the right end.

Convection of heat from a surface to the surrounding fluid can be increased by attaching thin strips of conducting metal to the surface. The metal strips are called *fins*. For a fin with heat flow along its axis, heat can convect across the lateral surface of the fin unless it is insulated (see Fig. 3.9). To account for the convection of heat through the surface, we must add the rate of heat loss by convection to the right-hand side of (3.68):

$$\frac{\partial}{\partial x}\left(Ak\frac{\partial T}{\partial x}\right) + Aq = \rho cA\frac{\partial T}{\partial t} + P\beta(T - T_\infty) \tag{3.76a}$$

where P is the perimeter and β is the film coefficient. Equation (3.76*a*) can be expressed in the alternative form

$$\rho cA\frac{\partial T}{\partial t} - \frac{\partial}{\partial x}\left(kA\frac{\partial T}{\partial x}\right) + P\beta T = Aq + P\beta T_\infty \tag{3.76b}$$

For a steady state, we set the time derivatives in (3.68), (3.71), (3.72), and (3.76) equal to zero. The steady-state equations for various one-dimensional systems are summarized below:

Plane wall and fin

$$-\frac{d}{dx}\left(kA\frac{dT}{dx}\right) + \hat{c}T = Aq + \hat{c}T_\infty, \qquad \hat{c} = P\beta \tag{3.77}$$

Cylindrical system

$$-\frac{1}{r}\frac{d}{dr}\left(2\pi kr\frac{dT}{dr}\right) = 2\pi q \tag{3.78}$$

Spherical system

$$-\frac{1}{r^2}\frac{d}{dr}\left(4\pi^2 kr^2\frac{dT}{dr}\right) = 4\pi^2 q \tag{3.79}$$

For a plane wall and insulated lateral surfaces of a bar, we set $\hat{c} = 0$ in (3.77). The essential and natural boundary conditions associated with these equations are

$$T = T_0, \qquad Q + \beta A(T - T_\infty) + Q_0 = 0 \tag{3.80}$$

The weak form and finite element model of (3.77) can be developed using the ideas presented in Section 3.2.3. Since (3.77) is a special case of the model boundary value problem with $a = kA$, $c = P\beta$, and $q \to Aq + P\beta T_\infty$, we can immediately write the finite element model of (3.77) from (3.31):

$$[K^e]\{T^e\} = \{f^e\} + \{Q^e\} \tag{3.81a}$$

where

$$K_{ij}^e = \int_{x_A}^{x_B}\left(kA\frac{d\psi_i^e}{dx}\frac{d\psi_j^e}{dx} + P\beta\psi_i^e\psi_j^e\right)dx, \qquad f_i^e = \int_{x_A}^{x_B}\psi_i^e(Aq + P\beta T_\infty)\,dx$$
$$Q_1^e = \left(-kA\frac{dT}{dx}\right)\bigg|_{x_A}, \qquad Q_2^e = \left(kA\frac{dT}{dx}\right)\bigg|_{x_B} \tag{3.81b}$$

where Q_1^e and Q_2^e denote heat flow *into* the element at the nodes.

Equations (3.78) and (3.79) are also special cases of the model boundary value problem. However, in developing the weak forms of (3.78) and (3.79), the integration must be carried over a typical volume element of each system, as illustrated in Section 3.2.8 for a radially symmetric cylindrical problem. The weak form of (3.78) is given in (3.65), and the finite element model is given by (3.66), with $\{u^e\} = \{T^e\}$. Similarly, the weak form of (3.79) can be developed using a volume element of a sphere:

$$dV = r^2\,dr\,d\theta\,d\phi \quad \text{for } 0 \leqslant \theta < 2\pi,\ 0 \leqslant \phi < 2\pi$$

The weak form of (3.79) is

$$0 = (2\pi)^2\int_{r_A}^{r_B}\left(k\frac{dw}{dr}\frac{dT}{dr} - qw\right)r^2\,dr - Q_1^e w(r_A) - Q_2^e w(r_B) \tag{3.82}$$

and the finite element model of (3.79) is

$$[K^e]\{T^e\} = \{f^e\} + \{Q^e\} \tag{3.83a}$$

where

$$K_{ij}^e = (2\pi)^2 \int_{r_A}^{r_B} k \frac{d\psi_i^e}{dr}\frac{d\psi_j^e}{dr} r^2\, dr, \qquad f_i^e = (2\pi)^2 \int_{r_A}^{r_B} q\psi_i^e r^2\, dr$$
$$Q_1^e = -(2\pi)^2\left(r^2 k \frac{dT}{dr}\right)\bigg|_{r_A}, \qquad Q_2^e = (2\pi)^2\left(r^2 k \frac{dT}{dr}\right)\bigg|_{r_B} \tag{3.83b}$$

In the following examples, we consider some typical applications of the finite element models (3.81) and (3.66).

Example 3.1. Consider a slab of thickness L and constant thermal conductivity k ($\text{W m}^{-1}\,{}^\circ\text{C}^{-1}$). Suppose that energy at a uniform rate of q_0 (W m^{-3}) is generated in the wall. We wish to determine the temperature distribution in the wall when the boundary surfaces of the wall are subject to the following three different sets of boundary conditions:

Set 1 $\quad T(0) = T_1, \quad T(L) = T_2 \qquad (3.84a)$

Set 2 $\quad \left(-k\dfrac{dT}{dx}\right)\bigg|_{x=0} = g_0\ (\text{W m}^{-2}), \quad \left[k\dfrac{dT}{dx} + \beta(T - T_\infty)\right]\bigg|_{x=L} = 0 \qquad (3.84b)$

Set 3 $\quad T(0) = T_1, \quad \left(k\dfrac{dT}{dx}\right)\bigg|_{x=L} = g_0\ (\text{W m}^{-2}) \qquad (3.84c)$

The governing differential equation for this problem is given by (3.77) with $\hat{c} = 0$. Hence, the finite element model in (3.81) is applicable here. We must select the order of approximation (or type of element) to evaluate the coefficients K_{ij}^e and f_i^e in (3.81b). For the choice of linear elements and the data $a = kA = \text{constant}$ and $q = Aq_0 = \text{constant}$, (3.81$a$) takes the form [see (3.37a)]

$$\frac{kA}{h_e}\begin{bmatrix} 1 & -1 \\ -1 & 1 \end{bmatrix}\begin{Bmatrix} T_1^e \\ T_2^e \end{Bmatrix} = \frac{Aq_0 h_e}{2}\begin{Bmatrix} 1 \\ 1 \end{Bmatrix} + \begin{Bmatrix} Q_1^e \\ Q_2^e \end{Bmatrix}$$

For a uniform mesh of N elements, i.e., a mesh of same-size elements, $h_1 = h_2 = \cdots = L/N \equiv h$, the assembled equations are

$$\frac{kA}{h}\begin{bmatrix} 1 & -1 & & & & \\ -1 & 1+1 & -1 & & 0 & \\ & -1 & 1+1 & & & \\ \cdots & \cdots & \cdots & \cdots & \cdots & \cdots \\ & 0 & & & 1+1 & -1 \\ & & & & -1 & 1 \end{bmatrix}\begin{Bmatrix} U_1 \\ U_2 \\ U_3 \\ \vdots \\ U_N \\ U_{N+1} \end{Bmatrix} = \frac{Aq_0 h}{2}\begin{Bmatrix} 1 \\ 1+1 \\ 1+1 \\ \vdots \\ 1+1 \\ 1 \end{Bmatrix} + \begin{Bmatrix} Q_1^1 \\ Q_2^1 + Q_1^2 \\ Q_2^2 + Q_1^3 \\ \vdots \\ Q_2^{N-1} + Q_1^N \\ Q_2^N \end{Bmatrix} \tag{3.85}$$

where U_I $(I = 1, 2, \ldots, N+1)$ denotes the temperature at global node I. The assembled equations (3.85) are valid for all three sets of boundary conditions in (3.84). We now specialize the finite element equations (3.85) for each set of boundary conditions.

Set 1. The boundary conditions (3.84*a*) imply that

$$U_1 = T_1, \qquad U_{N+1} = T_2 \tag{3.86a}$$

and the balance of the heats at global nodes 2, 3, ..., N requires

$$Q_2^{e-1} + Q_1^e = 0 \quad \text{for } e = 2, 3, \ldots, N \tag{3.86b}$$

The condensed equations for the unknown primary variables $U_2, U_3, \ldots, U_N$ are obtained by considering the 2nd, 3rd, ..., Nth equations of (3.85) (and omitting those corresponding to the known temperatures):

$$\frac{kA}{h}\begin{bmatrix} 2 & -1 & & 0 \\ -1 & 2 & & \\ \cdots & \cdots & \cdots & \cdots \\ & & 2 & -1 \\ 0 & & -1 & 2 \end{bmatrix}\begin{Bmatrix} U_2 \\ U_3 \\ \vdots \\ U_{N-1} \\ U_N \end{Bmatrix} = \frac{Aq_0h}{2}\begin{Bmatrix} 2 \\ 2 \\ \vdots \\ 2 \\ 2 \end{Bmatrix} - \frac{kA}{h}\begin{Bmatrix} -T_1 \\ 0 \\ \vdots \\ 0 \\ -T_2 \end{Bmatrix} \tag{3.87}$$

The unknown secondary variables Q_1^1 and Q_2^N are computed from the first and last of the assembled equations (3.85):

$$(Q_1^1)_{\text{equil}} = -\tfrac{1}{2}Aq_0h + \frac{kA}{h}(U_1 - U_2), \qquad (Q_2^N)_{\text{equil}} = -\tfrac{1}{2}Aq_0h + \frac{kA}{h}(U_{N+1} - U_N) \tag{3.88a}$$

They can also be computed from the definition (3.81*b*):

$$\begin{aligned} (Q_1^1)_{\text{def}} &\equiv -\left(kA\frac{dT}{dx}\right)\bigg|_{x=0} = kA\sum_{j=1}^{2} T_j^1\left(\frac{d\psi_j^1}{dx}\right)\bigg|_{x=0} = -\frac{kA}{h}(U_2 - U_1) \\ (Q_2^N)_{\text{def}} &\equiv +\left(kA\frac{dT}{dx}\right)\bigg|_{x=L} = kA\sum_{j=1}^{2} T_j^N\left(\frac{d\psi_j^N}{dx}\right)\bigg|_{x=L} = \frac{kA}{h}(U_{N+1} - U_N) \end{aligned} \tag{3.88b}$$

The values computed in (3.88*b*) are in error compared with those in (3.88*a*) by the nodal sources, $\frac{1}{2}q_0hA$, for $h = L/N$.

For any number of elements, the solution can be computed from (3.87). For this set of boundary conditions and a linear-element mesh, the minimum number of elements is 2.

For $N = 2$ $(h = \frac{1}{2}L)$,

$$\frac{kA}{h}\begin{bmatrix} 1 & -1 & 0 \\ -1 & 2 & -1 \\ 0 & -1 & 1 \end{bmatrix}\begin{Bmatrix} T_1 \\ U_2 \\ T_2 \end{Bmatrix} = \frac{Aq_0h}{2}\begin{Bmatrix} 1 \\ 2 \\ 1 \end{Bmatrix} + \begin{Bmatrix} Q_1^1 \\ 0 \\ Q_2^2 \end{Bmatrix}$$

and the solution is

$$U_2 = \frac{q_0h^2}{2k} + \tfrac{1}{2}(T_1 + T_2)$$

$$(Q_1^1)_{\text{equil}} = -q_0hA + \frac{kA}{2h}(T_1 - T_2), \qquad (Q_2^2)_{\text{equil}} = -q_0hA + \frac{kA}{2h}(T_2 - T_1)$$

For $N = 3$ $(h = \frac{1}{3}L)$,

$$\frac{kA}{h}\begin{bmatrix} 1 & -1 & 0 & 0 \\ -1 & 2 & -1 & 0 \\ 0 & -1 & 2 & -1 \\ 0 & 0 & -1 & 1 \end{bmatrix}\begin{Bmatrix} T_1 \\ U_2 \\ U_3 \\ T_2 \end{Bmatrix} = \frac{Aq_0h}{2}\begin{Bmatrix} 1 \\ 2 \\ 2 \\ 1 \end{Bmatrix} + \begin{Bmatrix} Q_1^1 \\ 0 \\ 0 \\ Q_2^3 \end{Bmatrix}$$

or

$$\frac{kA}{h}\begin{bmatrix} 2 & -1 \\ -1 & 2 \end{bmatrix}\begin{Bmatrix} U_2 \\ U_3 \end{Bmatrix} = Aq_0h\begin{Bmatrix} 1 \\ 1 \end{Bmatrix} + \frac{kA}{h}\begin{Bmatrix} T_1 \\ T_2 \end{Bmatrix}$$

$$\begin{Bmatrix} Q_1^1 \\ Q_2^3 \end{Bmatrix}_{\text{equil}} = \frac{kA}{h}\begin{bmatrix} 1 & -1 & 0 & 0 \\ 0 & 0 & -1 & 1 \end{bmatrix}\begin{Bmatrix} T_1 \\ U_2 \\ U_3 \\ T_2 \end{Bmatrix} - \frac{Aq_0h}{2}\begin{Bmatrix} 1 \\ 1 \end{Bmatrix}$$

The solution is given by

$$U_2 = \frac{q_0h^2}{k} + \tfrac{1}{3}(2T_1 + T_2), \qquad U_3 = \frac{q_0h^2}{k} + \tfrac{1}{3}(T_1 + 2T_2)$$

$$(Q_1^1)_{\text{equil}} = -\tfrac{3}{2}q_0hA + \frac{kA}{3h}(T_1 - T_2), \qquad (Q_2^3)_{\text{equil}} = -\tfrac{3}{2}q_0hA + \frac{kA}{3h}(T_2 - T_1)$$

The exact solution of (3.77) (with $\hat{c} = 0$), subject to the boundary conditions (3.84*a*), is

$$T(x) = \frac{q_0L^2}{2k}\left[\frac{x}{L} - \left(\frac{x}{L}\right)^2\right] + (T_2 - T_1)\frac{x}{L} + T_1$$

$$Q(x) = -kA\frac{dT}{dx} = -\frac{q_0LA}{2}\left(1 - 2\frac{x}{L}\right) - \frac{kA}{L}(T_2 - T_1)$$

Note that the finite element solution at the nodes coincides with the exact solution for *any* number of elements. In fact, for constant $a = kA$, it can be shown that the finite element solution at the nodes is exact [see Reddy (1986), pp. 403, 404]; the only error would be that due to the round-off error introduced in electronic computations. The finite element solution, being linear, will not be exact at points other than the nodes.

If we use quadratic elements, we can improve the solution at points between the nodes. Let us consider a mesh of one quadratic element. The element coefficient matrix and source vector are given in (3.40). We have $(c_e = 0)$

$$\frac{kA}{3h}\begin{bmatrix} 7 & -8 & 1 \\ -8 & 16 & -8 \\ 1 & -8 & 7 \end{bmatrix}\begin{Bmatrix} U_1 \\ U_2 \\ U_3 \end{Bmatrix} = \frac{Aq_0h}{6}\begin{Bmatrix} 1 \\ 4 \\ 1 \end{Bmatrix} + \begin{Bmatrix} Q_1^1 \\ 0 \\ Q_3^1 \end{Bmatrix}$$

where $h = L$. The solution is

$$U_2 = \tfrac{4}{16}\frac{Aq_0h}{6}\frac{3h}{Ak} + \tfrac{8}{16}(T_1 + T_2) = \tfrac{1}{8}\frac{q_0h^2}{k} + \tfrac{1}{2}(T_1 + T_2)$$

$$(Q_1^1)_{\text{equil}} = -\tfrac{1}{2}Aq_0h + \frac{kA}{h}(T_1 - T_2), \qquad (Q_3^1)_{\text{equil}} = -\tfrac{1}{2}Aq_0h + \frac{kA}{h}(T_2 - T_1)$$

If we compute the Q_1^1 using the definition (3.81*b*), we obtain

$$(Q_1^1)_{\text{def}} \equiv -\left(kA\frac{dT_e}{dx}\right)\bigg|_{x=0} = -kA\sum_{j=1}^{3} T_j^1\left(\frac{d\psi_j^1}{dx}\right)\bigg|_{x=0}$$

$$= -kA\left[U_1\left(-\frac{3}{h}+\frac{4x}{h^2}\right) + U_2\left(\frac{4}{h}-\frac{8x}{h^2}\right) + U_3\left(-\frac{1}{h}+\frac{4x}{h^2}\right)\right]_{x=0}$$

$$= -\tfrac{1}{2}Aq_0h + \frac{kA}{h}(T_1 - T_2)$$

which coincides with the exact value. Recall that $(Q_1^1)_{\text{def}}$ for a linear-element mesh does not coincide with the exact value. Also, the values of $T(x)$ at $x = \frac{1}{4}L$, for example, from the two linear- and one quadratic-element meshes can be computed using the interpolations

$$T(\tfrac{1}{4}L) = \tfrac{1}{16}\frac{q_0L^2}{k} + \tfrac{1}{4}(3T_1 + T_2) \qquad \text{(linear element)}$$

$$T(\tfrac{1}{4}L) = \tfrac{3}{32}\frac{qL^2}{k} + \tfrac{1}{4}(3T_1 + T_2) \qquad \text{(quadratic element)}$$

The quadratic interpolation gives the exact value. Thus, the finite element solution given by the quadratic element is exact at all points, because the exact solution varies quadratically.

Set 2. In this case, the surface at $x = 0$ is subjected to a uniform heat flux g_0 (W m^{-2}) (if it is insulated, $g_0 = 0$), and heat is dissipated by convection into a fluid of temperature T_∞ at the boundary surface at $x = L$. These boundary conditions imply that

$$Q_1^1 \equiv \left(-kA\frac{dT}{dx}\right)\bigg|_{x=0} = Ag_0$$

$$Q_2^N \equiv \left(kA\frac{dT}{dx}\right)\bigg|_{x=L} = -A\beta(T - T_\infty)|_{x=L} = -A\beta(U_{N+1} - T_\infty)$$

where A is the cross-sectional area normal to heat flow and β is the heat transfer coefficient. Equations (3.86*b*) are also valid for the present case. For a one-element mesh ($N = 1$, $h = L$), the finite element equations are

$$\frac{kA}{h}\begin{bmatrix} 1 & -1 \\ -1 & 1 \end{bmatrix}\begin{Bmatrix} U_1 \\ U_2 \end{Bmatrix} = \frac{Aq_0h}{2}\begin{Bmatrix} 1 \\ 1 \end{Bmatrix} + \begin{Bmatrix} Ag_0 \\ -A\beta U_2 + A\beta T_\infty \end{Bmatrix}$$

or

$$\frac{kA}{h}\begin{bmatrix} 1 & -1 \\ -1 & 1 + \beta h/k \end{bmatrix}\begin{Bmatrix} U_1 \\ U_2 \end{Bmatrix} = \begin{Bmatrix} \frac{1}{2}Aq_0h + Ag_0 \\ \frac{1}{2}Aq_0h + A\beta T_\infty \end{Bmatrix}$$

Their solution is

$$U_1 = -\left(\frac{1+\alpha}{1-\alpha}\right)\frac{q_0h^2}{2k} - \left(\frac{\alpha}{1-\alpha}\right)\frac{g_0h}{k} + T_\infty$$

$$U_2 = -\left(\frac{1}{1-\alpha}\right)\frac{q_0h^2}{k} - \left(\frac{1}{1-\alpha}\right)\frac{g_0h}{k} + T_\infty$$

where $\alpha = 1 + \beta h/k$. The exact solution is

$$T(x) = \frac{q_0 L^2}{2k}\left(1 + \frac{2k}{\beta L} - \frac{x^2}{L^2}\right) + \frac{g_0 L}{k}\left(1 + \frac{k}{\beta L} - \frac{x}{L}\right) + T_\infty$$

The finite element solution agrees with this at the nodes.

Set 3. In this case, the boundary surface at $x = 0$ is maintained at a temperature T_1, while at the other boundary surface, $x = L$, a heat flux at the rate of g_0 is removed. These boundary conditions can be expressed in terms of the primary and secondary variables as

$$U_1 = T_1, \qquad Q_2^N = Ag_0$$

and (3.86*b*) are still valid. For a two-element mesh with $h_1 = h_2 = \frac{1}{2}L \equiv h$, we have

$$\frac{kA}{h}\begin{bmatrix} 1 & -1 & 0 \\ -1 & 2 & -1 \\ 0 & -1 & 1 \end{bmatrix}\begin{Bmatrix} U_1 \\ U_2 \\ U_3 \end{Bmatrix} = \frac{Aq_0 h}{2}\begin{Bmatrix} 1 \\ 2 \\ 1 \end{Bmatrix} + \begin{Bmatrix} Q_1^1 \\ 0 \\ Ag_0 \end{Bmatrix}$$

or

$$\frac{kA}{h}\begin{bmatrix} 2 & -1 \\ -1 & 1 \end{bmatrix}\begin{Bmatrix} U_2 \\ U_3 \end{Bmatrix} = \frac{Aq_0 h}{2}\begin{Bmatrix} 2 \\ 1 \end{Bmatrix} + \begin{Bmatrix} \frac{kA}{h}T_1 \\ Ag_0 \end{Bmatrix}$$

$$(Q_1^1)_{\text{equil}} = -\tfrac{1}{2}Aq_0 h + \frac{kA}{h}(U_1 - U_2)$$

We have

$$U_2 = \frac{3}{2}\frac{q_0 h^2}{k} + \frac{g_0 h}{k} + T_1, \qquad U_3 = 2\frac{q_0 h^2}{k} + \frac{2g_0 h}{h} + T_1$$

$$(Q_1^1)_{\text{equil}} = -(2AQ_0 h + Ag_0) \quad [= Q(0)]$$

For a three-element mesh ($h = \frac{1}{3}L$), we have

$$\frac{kA}{h}\begin{bmatrix} 2 & -1 & 0 \\ -1 & 2 & -1 \\ 0 & -1 & 1 \end{bmatrix}\begin{Bmatrix} U_2 \\ U_3 \\ U_4 \end{Bmatrix} = \frac{Aq_0 h}{2}\begin{Bmatrix} 2 \\ 2 \\ 1 \end{Bmatrix} + \begin{Bmatrix} 0 \\ 0 \\ Ag_0 \end{Bmatrix} + \frac{kA}{h}\begin{Bmatrix} T_1 \\ 0 \\ 0 \end{Bmatrix}$$

Using *Cramer's rule*, we can solve for U_2, U_3, and U_4:

$$U_2 = \frac{h}{kAD}\begin{vmatrix} F_1 & -1 & 0 \\ F_2 & 2 & -1 \\ F_3 & -1 & 1 \end{vmatrix}, \quad U_3 = \frac{h}{kAD}\begin{vmatrix} 2 & F_1 & 0 \\ -1 & F_2 & -1 \\ 0 & F_3 & 1 \end{vmatrix}, \quad U_4 = \frac{h}{kAD}\begin{vmatrix} 2 & -1 & F_1 \\ -1 & 2 & F_2 \\ 0 & -1 & F_3 \end{vmatrix}$$

where

$$D = \begin{vmatrix} 2 & -1 & 0 \\ -1 & 2 & -1 \\ 0 & -1 & 1 \end{vmatrix} = 2 + (-1) = 1$$

and $\{F\} = \{F_1 \quad F_2 \quad F_3\}^T$ is the right-hand-side vector:

$$\{F\} = \begin{Bmatrix} Aq_0h + \dfrac{kAT_1}{h} \\ Aq_0h \\ \tfrac{1}{2}Aq_0h + Ag_0 \end{Bmatrix}$$

Evaluating the determinants, we obtain

$$U_2 = \frac{h}{kA}(F_1 + F_2 + F_3) = \frac{h}{k}(\tfrac{5}{2}q_0h + g_0) + T_1$$

$$U_3 = \frac{h}{kA}(F_1 + 2F_2 + 2F_3) = \frac{h}{k}(4q_0h + 2g_0) + T_1$$

$$U_4 = \frac{h}{kA}(F_1 + 2F_2 + 3F_3) = \frac{h}{k}(\tfrac{9}{2}q_0h + 3g_0) + T_1$$

$$(Q_1^1)_{\text{equil}} = -\tfrac{1}{2}Aq_0h + \frac{kA}{h}(U_1 - U_2) = -A(g_0 + 3q_0h)$$

$$(Q_1^1)_{\text{def}} = -A(g_0 + \tfrac{5}{2}q_0h)$$

The exact solution of (3.77) with $\hat{c} = 0$ and subject to the boundary conditions (3.84c) is given by

$$\begin{aligned} T(x) &= \frac{L^2}{k}\left[\frac{q_0}{2}\left(2\frac{x}{L} - \frac{x^2}{L^2}\right) + \frac{g_0x}{L^2}\right] + T_1 \\ Q(x) &= -kA\frac{dT}{dx} = -A\left[q_0L\left(1 - \frac{x}{L}\right) + g_0\right] \end{aligned} \tag{3.89}$$

Note that the finite element solution at the nodes, for any number of elements, coincides with the exact solution. The exact value of Q_1^1 is

$$(Q_1^1)_{\text{exact}} = -\left(kA\frac{du}{dx}\right)\bigg|_{x=0} = Q(0) = -A(q_0L + g_0)$$

The value of Q_1^1 computed from the assembled equations is the same as the exact one; however, when computed using the definition $Q_1^1 = -kA(dU_e/dx)|_{x=0}$, it is in error by an amount $f_1^1 = \frac{1}{2}Aq_0h$. As the number of elements is increased, the value of $(Q_1^1)_{\text{def}}$ approaches the exact value. Of course, when quadratic elements are used, we shall obtain a more accurate (or even exact) value of $(Q_1^1)_{\text{def}}$.

The next example deals with radially symmetric heat transfer in a cylinder.

Example 3.2. Consider a long solid cylinder of radius R_0 in which energy is generated at a constant rate q_0 (W m^{-3}). The boundary surface at $r = R_0$ is maintained at a constant temperature T_0. We wish to calculate the temperature distribution $T(r)$ and heat flux $q(r) = -k\,dT/dr$ (or heat $Q = -Ak\,dT/dr$).

The governing equation for this problem is given by (3.78) with $q = q_0$. The boundary conditions are

$$T(R_0) = T_0, \qquad \left(2\pi kr\frac{dT}{dr}\right)\bigg|_{r=0} = 0 \tag{3.90}$$

The zero-flux boundary condition at $r = 0$ is a result of the radial symmetry at $r = 0$. If the cylinder is hollow with inner radius R_i then the boundary condition at $r = R_i$ can be specified temperature, specified heat flux, or convection boundary condition, depending on the situation.

The finite element model of the governing equation is given in (3.66) (for unit height of the cylinder and $a = kr$):

$$[K^e]\{T^e\} = \{f^e\} + \{Q^e\} \tag{3.91a}$$

where

$$\begin{gathered} K^e_{ij} = 2\pi\int_{r_A}^{r_B} kr\frac{d\psi^e_i}{dr}\frac{d\psi^e_j}{dr}\,dr, \qquad f^e_i = 2\pi\int_{r_A}^{r_B}\psi^e_i q_0 r\,dr \\ Q^e_1 = -2\pi k\left(r\frac{dT}{dr}\right)\bigg|_{r_A}, \qquad Q^e_2 = 2\pi k\left(r\frac{dT}{dr}\right)\bigg|_{r_B} \end{gathered} \tag{3.91b}$$

and (r_A, r_B) are the coordinates of the element $\Omega^e = (r_A, r_B)$.

For the choice of linear interpolation functions ψ^e_i as [see (3.16b) and (3.66c)]

$$\psi^e_1 = (r_B - r)/h_e, \qquad \psi^e_2 = (r - r_A)/h_e$$

the element equations for a typical linear element are

$$\frac{2\pi k}{h_e}\frac{r_{e+1} + r_e}{2}\begin{bmatrix} 1 & -1 \\ -1 & 1 \end{bmatrix}\begin{Bmatrix} T^e_1 \\ T^e_2 \end{Bmatrix} = 2\pi q_0\frac{h_e}{6}\begin{Bmatrix} 2r_e + r_{e+1} \\ r_e + 2r_{e+1} \end{Bmatrix} + \begin{Bmatrix} Q^e_1 \\ Q^e_2 \end{Bmatrix} \tag{3.92}$$

The element equations for individual elements are obtained from these by giving the element length h_e and the global coordinates of the element nodes, $r_e = r_A$ and $r_{e+1} = r_B$.

For the mesh of one linear element, we have $r_1 = 0$, $r_2 = h_e = R_0$, and

$$\pi k\begin{bmatrix} 1 & -1 \\ -1 & 1 \end{bmatrix}\begin{Bmatrix} U_1 \\ U_2 \end{Bmatrix} = \frac{\pi q_0 R_0}{3}\begin{Bmatrix} R_0 \\ 2R_0 \end{Bmatrix} + \begin{Bmatrix} Q^1_1 \\ Q^1_2 \end{Bmatrix}$$

The boundary conditions in (3.90) imply $U_2 = T_0$ and $Q^1_1 = 0$. Hence the temperature at (global) node 1 is

$$U_1 = q_0R_0^2/3k + T_0$$

and the heat at $r = R_0$ is

$$Q^1_2 = \pi k(U_2 - U_1) - \tfrac{2}{3}\pi q_0 R_0^2 = -\pi q_0 R_0^2$$

The negative sign indicates that heat is removed from the body (because $dT/dr < 0$). The one-element solution as a function of the radial coordinate r is

$$T^1(r) = U_1\psi^1_1(r) + U_2\psi^1_2(r) = \frac{q_0R_0^2}{3k}\left(1 - \frac{r}{R_0}\right) + T_0 \tag{3.93}$$

and the heat flux is

$$q(r) \equiv -k\frac{dT^1}{dr} = \tfrac{1}{3}q_0R_0 \tag{3.94}$$

The exact solution of the problem can be obtained by integrating (3.78) and evaluating the constants of integration with the help of the boundary conditions in (3.90):

$$T(r) = \frac{q_0 R_0^2}{4k}\left[1 - \left(\frac{r}{R_0}\right)^2\right] + T_0$$
$$q(r) = \tfrac{1}{2} q_0 r \ (\text{in W m}^{-2}), \qquad Q(R_0) = -\left(2\pi k r \frac{dT}{dr}\right)\Big|_{R_0} = \pi q_0 R_0^2 \tag{3.95}$$

For a mesh of two linear elements, we take $h_1 = h_2 = \tfrac{1}{2}R_0$, $r_1 = 0$, $r_2 = h_1 = \tfrac{1}{2}R_0$, and $r_3 = h_1 + h_2 = R_0$. The two-element assembly gives

$$\pi k \begin{bmatrix} 1 & -1 & 0 \\ -1 & 1+3 & -3 \\ 0 & -3 & 3 \end{bmatrix} \begin{Bmatrix} U_1 \\ U_2 \\ U_3 \end{Bmatrix} = \frac{\pi q_0 R_0}{6} \begin{Bmatrix} \tfrac{1}{2}R_0 \\ R_0 + 2R_0 \\ \tfrac{1}{2}R_0 + 2R_0 \end{Bmatrix} + \begin{Bmatrix} Q_1^1 \\ Q_2^1 + Q_1^2 \\ Q_2^2 \end{Bmatrix} \tag{3.96}$$

Imposing the boundary conditions $U_3 = T_o$ and $Q_1^1 = 0$, the condensed equations are

$$\pi k \begin{bmatrix} 1 & -1 \\ -1 & 4 \end{bmatrix} \begin{Bmatrix} U_1 \\ U_2 \end{Bmatrix} = \frac{\pi q_0 R_0^2}{12} \begin{Bmatrix} 1 \\ 6 \end{Bmatrix} + \pi k \begin{Bmatrix} 0 \\ 3T_0 \end{Bmatrix} \tag{3.97}$$

Their solution is

$$U_1 = \tfrac{5}{18} \frac{q_0 R_0^2}{k} + T_0, \qquad U_2 = \tfrac{7}{36} \frac{q_0 R_0^2}{k} + T_0 \tag{3.98a}$$

From equilibrium, Q_2^2 is computed as

$$Q_2^2 = -\tfrac{5}{12}\pi q_0 R_0^2 + 3\pi k (U_3 - U_2) = -\pi q_0 R_0^2 \tag{3.98b}$$

The finite element solution becomes

$$T_{\text{fem}}(r) = \begin{cases} U_1 \psi_1^1 + U_2 \psi_2^1 = \left(\tfrac{5}{18} \dfrac{q_0 R_0^2}{k} + T_0\right) \dfrac{R_0 - 2r}{R_0} + \left(\tfrac{7}{36} \dfrac{q_0 R_0^2}{k} + T_0\right) \dfrac{2r}{R_0} \\ U_2 \psi_1^2 + U_3 \psi_2^2 = \left(\tfrac{7}{36} \dfrac{q_0 R_0^2}{k} + T_0\right) \dfrac{2(R_0 - r)}{R_0} + T_0 \dfrac{2r - R_0}{R_0} \end{cases}$$
$$= \begin{cases} \dfrac{q_0 R_0^2}{18k}\left(r - \dfrac{3r}{R_0}\right) + T_0 & \text{for } 0 \le r \le \tfrac{1}{2}R_0 \\ \dfrac{7}{18} \dfrac{q_0 R_0^2}{k}\left(1 - \dfrac{r}{R_0}\right) + T_0 & \text{for } \tfrac{1}{2}R_0 \le r \le R_0 \end{cases} \tag{3.99}$$

Note that the heat flow at $r = R_0$ is predicted accurately by both one- and two-element models. The temperature at the center of the cylinder according to the exact solution is $T(0) = q_0 R_0^2/4k + T_0$, whereas it is $q_0 R_0^2/3k + T_0$ and $q_0 R_0^2/18k + T_0$ according to the one- and two-element models, respectively.

The finite element solutions obtained using one-, two-, four-, and eight-element meshes of linear elements are compared with the exact solution in Table 3.3. Convergence of the finite element solutions to the exact solution with an increasing number of elements is clear (see Fig. 3.10). Figure 3.11 shows plots of $q(r) = -dT/dr$, as computed by the finite element interpolation and the exact formula. The figure also shows the plot of the exact $Q = -2\pi k r\, dT/dr$ versus r.

TABLE 3.3
Comparison of the finite element and exact solutions for heat transfer in a radially symmetric cylinder (Example 3.2) ($R_0 = 0.01$ m, $q_0 = 2 \times 10^8$ W m^{-3}, $k = 20$ W m^{-1} °C^{-1}, $T_0 = 100$ °C)

	Temperature $u(r)$†				
r/R_0	One element‡	Two elements	Four elements	Eight elements	Exact
0.00	433.33	377.78	358.73	352.63	350.00
0.125	391.67	356.24	348.31	347.42	346.09
0.250	350.00	335.11	337.90	335.27	334.38
0.375	308.33	315.28	313.59	315.48	314.84
0.500	266.67	294.44	289.29	287.95	287.50
0.625	225.00	245.83	249.70	252.65	252.34
0.750	183.33	197.22	210.12	209.56	209.38
0.875	141.67	148.61	155.06	158.68	158.59
1.000	100.00	100.00	100.00	100.00	100.00

† The underlined terms are nodal values and others are interpolated values.
‡ Uniform meshes (i.e., equal-length elements) of linear elements.

3.3.2 Fluid Mechanics

All bulk matter in nature exists in one of two forms: solid or fluid. A solid body is characterized by relative immobility of its molecules, whereas a fluid state is characterized by their relative mobility. Fluids can exist either as gases or liquids. The field of fluid mechanics is concerned with the motion of fluids

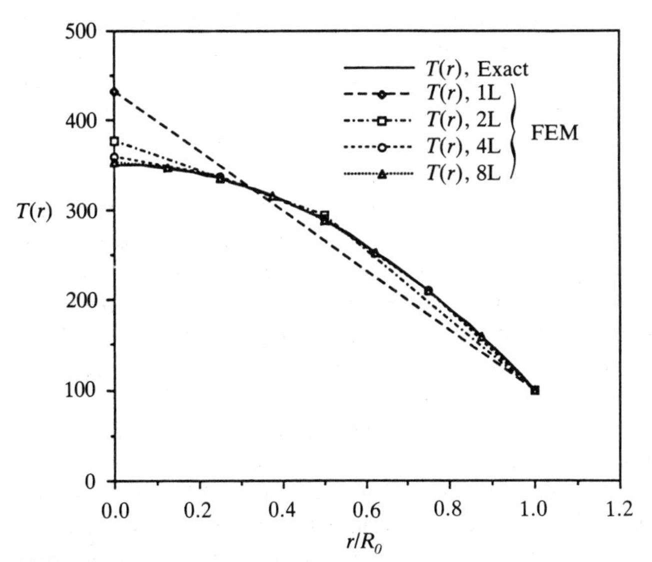

FIGURE 3.10
Comparison of the finite element solutions with the exact solution for heat transfer in a radially symmetric problem with cylindrical geometry (Example 3.2).

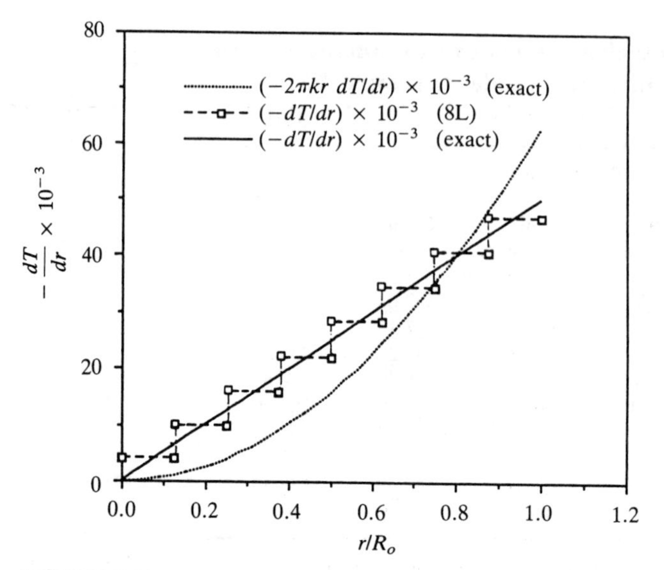

FIGURE 3.11
Comparison of the finite element solution with the exact solution for the temperature gradient in a radially symmetric problem with cylindrical geometry (Example 3.2).

and the conditions affecting the motion. In this section, we review the basic equations of fluid mechanics and develop finite element models of certain one-dimensional fluid systems. Several numerical examples will be discussed.

The basic equations of fluid mechanics are derived from the global laws of conservation of mass, momentum, and energy. Conservation of mass gives the continuity equation, while the conservation of momentum results in the equations of motion. The conservation of energy, considered in the last section, is the first law of thermodynamics and results in (3.68), and (3.77)–(3.79) for various one-dimensional systems when thermal–fluid coupling is omitted. Before we review the basic equations of fluid mechanics, it is informative to consider various types of fluids.

An *ideal* (or *perfect*) *fluid* is one that has zero viscosity and is incompressible. A *real fluid* or *viscous fluid* is one with finite viscosity, and may or may not be incompressible. *Nonviscous fluids* are those with zero viscosity, and again may or may not be incompressible. A viscous fluid is said to be a *Newtonian fluid* if its viscosity coefficient is independent of the velocity gradient (i.e., the viscosity is constant). *Non-Newtonian* fluids are those for which the viscosity is a function of the velocity gradient.

Two different viewpoints are used in the analytical description of the equations of a continuous medium. In the first, one considers the motion of all matter passing through a *fixed spatial location*. Here one is concerned with changes (e.g., in velocity field, pressure, and density) that are taking place in the matter that happens to occupy the fixed spatial location. This viewpoint, known as the *Eulerian description,* is the one that is most commonly used in fluid mechanics. In the study of the motion of fluids, one is interested in the flow characteristics of fluids occuping a fixed region rather than a fixed set of fluid particles. In the other viewpoint, one considers the motion of a *fixed set of material particles* (i.e., fixed matter), irrespective of its spatial location. This viewpoint is known as the *Lagrangian description,* and it is used in the study of the motion of solid bodies.

The conservation of mass in the Eulerian description is expressed as

$$\frac{\partial \rho}{\partial t} + \boldsymbol{\nabla} \cdot (\rho \mathbf{u}) = 0 \tag{3.100a}$$

or, for two-dimensional flow,

$$\frac{\partial \rho}{\partial t} + \frac{\partial}{\partial x}(\rho u) + \frac{\partial}{\partial y}(\rho v) = 0 \tag{3.100b}$$

where ρ is the density (in kg m^{-3}) and (u, v) are the velocity components (in m s^{-1}) in the x and y directions. All problems of fluid mechanics require that the continuity equation (3.100) be satisfied (because of the Eulerian description). In the steady-state case, we have $\partial/\partial t = 0$, and (3.100*a*,*b*) become

$$\boldsymbol{\nabla} \cdot (\rho \mathbf{u}) = 0, \qquad \frac{\partial}{\partial x}(\rho u) + \frac{\partial}{\partial y}(\rho v) = 0 \tag{3.101a,b}$$

If a fluid is incompressible, the density is constant, and the continuity equations (3.101*a*,*b*) take the forms

$$\boxed{\boldsymbol{\nabla} \cdot \mathbf{u} = 0, \qquad \frac{\partial u}{\partial x} + \frac{\partial v}{\partial y} = 0} \tag{3.102a,b}$$

For flows of incompressible fluids, the conservation of linear momentum results in the following equations of motion:

$$\boldsymbol{\nabla} \cdot \boldsymbol{\sigma} + \mathbf{f} = \rho \left(\frac{\partial \mathbf{u}}{\partial t} + \mathbf{u} \cdot \boldsymbol{\nabla} \mathbf{u} \right) \tag{3.103a}$$

or, for two-dimensional systems,

$$\boxed{\begin{aligned} \frac{\partial \sigma_x}{\partial x} + \frac{\partial \sigma_{xy}}{\partial y} + f_x &= \rho \left(\frac{\partial u}{\partial t} + u \frac{\partial u}{\partial x} + v \frac{\partial u}{\partial y} \right) \\ \frac{\partial \sigma_{xy}}{\partial x} + \frac{\partial \sigma_y}{\partial y} + f_y &= \rho \left(\frac{\partial v}{\partial t} + u \frac{\partial v}{\partial x} + v \frac{\partial v}{\partial y} \right) \end{aligned}} \tag{3.103b}$$

where $(\sigma_x, \sigma_y, \sigma_{xy})$ are the total stress components (in N m^{-2}) and (f_x, f_y) are the x and y components of the body force vector (measured per unit volume). The total stress components can be expressed in terms of the viscous stress components $(\tau_x, \tau_y, \tau_{xy})$ and the hydrostatic pressure P (in N m^{-2}):

$$\sigma_x = \tau_x - P, \qquad \sigma_y = \tau_y - P, \qquad \sigma_{xy} = \tau_{xy} \tag{3.104}$$

The viscous components of stress are related to the velocity gradients by Newton's law of viscosity. For isotropic, Newtonian fluids, these are

$$\tau_x = 2\mu \frac{\partial u}{\partial x}, \qquad \tau_y = 2\mu \frac{\partial v}{\partial y}, \qquad \tau_{xy} = \mu\left(\frac{\partial u}{\partial y} + \frac{\partial v}{\partial x}\right) \tag{3.105}$$

where μ is the viscosity (in $\mathrm{kg\,m^{-1}\,s^{-1}}$) of the fluid. Combining (3.103)–(3.105), we obtain

$$\begin{aligned}
\frac{\partial}{\partial x}\left(2\mu \frac{\partial u}{\partial x} - P\right) + \frac{\partial}{\partial y}\left[\mu\left(\frac{\partial u}{\partial y} + \frac{\partial v}{\partial x}\right)\right] + f_x &= \rho\left(\frac{\partial u}{\partial t} + u\frac{\partial u}{\partial x} + v\frac{\partial u}{\partial y}\right) \\
\frac{\partial}{\partial x}\left[\mu\left(\frac{\partial u}{\partial y} + \frac{\partial v}{\partial x}\right)\right] + \frac{\partial}{\partial y}\left(2\mu \frac{\partial v}{\partial y} - P\right) + f_y &= \rho\left(\frac{\partial v}{\partial t} + u\frac{\partial v}{\partial x} + v\frac{\partial v}{\partial y}\right)
\end{aligned} \tag{3.106}$$

These are known as the *Navier–Stokes equations*. When the viscosity is zero, we have

$$\begin{aligned}
-\frac{\partial P}{\partial x} + f_x &= \rho\left(\frac{\partial u}{\partial t} + u\frac{\partial u}{\partial x} + v\frac{\partial u}{\partial y}\right) \\
-\frac{\partial P}{\partial y} + f_y &= \rho\left(\frac{\partial v}{\partial t} + u\frac{\partial v}{\partial x} + v\frac{\partial v}{\partial y}\right)
\end{aligned} \tag{3.107}$$

which are known as the *Euler equations*.

Conservation of energy for incompressible fluids is expressed by

$$\boxed{\rho c\left(\frac{\partial T}{\partial t} + u\frac{\partial T}{\partial x} + v\frac{\partial T}{\partial y}\right) = k\left(\frac{\partial^2 T}{\partial x^2} + \frac{\partial^2 T}{\partial y^2}\right) + q + \Phi} \tag{3.108}$$

where c is the mean heat capacity at constant volume, q is the internal heat generation, k is the thermal conductivity of the (isotropic) fluid, and Φ is the viscous dissipation,

$$\Phi = \mu\left[\left(\frac{\partial u}{\partial x}\right)^2 + \left(\frac{\partial v}{\partial y}\right)^2\right] + \mu\left(\frac{\partial u}{\partial y} + \frac{\partial v}{\partial x}\right)^2 \tag{3.109}$$

which is zero for nonviscous fluids. For fluids of low viscosity and for velocities less than the sonic velocity, Φ has a magnitude that is small compared with the other terms in the equation.

In summary, the two-dimensional flows of viscous incompressible fluids are governed by (3.102), (3.106), and (3.108). There are four unknowns (u, v, P, T) and four equations. The two-dimensional flows of nonviscous fluids are governed by (3.100), (3.107), and (3.108) (with $\Phi = 0$). There are five unknowns (u, v, T, P, ρ) in four equations. The fifth equation is provided by the equation of state, $F(\rho, T, P) = 0$, which relates the density ρ, temperature T, and pressure P.

For certain flows, the equations can be further simplified to 1-D equations. These are discussed below.

Consider the flow of a viscous, incompressible fluid between two stationary parallel flat walls separated by a distance $2L$. We assume that there are no body forces. The velocity of the fluid at the wall is zero, because of adhesion, and reaches a maximum at the center of the channel. The velocity is constant on a plane parallel to the center plane, with its magnitude being proportional to the distance from the latter; the individual planes slide over each other, the velocity field being purely axial everywhere. Such fluid flows are called *laminar*. At a sufficiently large distance from the entrance, where the velocity is uniform, the velocity distribution becomes independent of the coordinate along the flow direction. The portion of the flow region beyond this distance is called the *hydrodynamically developed* or *fully developed* region, and the flow is called *fully developed flow*. For this case, the governing field equations (3.102), (3.106), and (3.108) can be simplified because of the specific nature of the velocity field:

$$u = u(x, y), \qquad v = 0$$

where the x and y coordinates are chosen along and normal to the flow direction, respectively. The continuity equation (3.012) reduces to

$$\frac{\partial u}{\partial x} = 0, \qquad \text{which implies that} \quad u = u(y)$$

The y momentum equation in (3.106), for $f_y = 0$, simplifies to

$$\frac{\partial P}{\partial y} = 0, \qquad \text{which implies that} \quad P = P(x)$$

The x momentum equation in (3.106), subject to the requirements $v = 0$, $u = u(y)$, $P = P(x)$, and $f_x = 0$, simplifies to

$$\boxed{\mu \frac{d^2u}{dy^2} = \frac{dP}{dx}} \tag{3.110}$$

The energy equation (3.108) for this problem reduces to

$$\boxed{\rho c u \frac{\partial T}{\partial x} = k\left(\frac{\partial^2 T}{\partial x^2} + \frac{\partial^2 T}{\partial y^2}\right) + \mu\left(\frac{du}{dy}\right)^2} \tag{3.111}$$

In the case of the steady flow of an incompressible, Newtonian fluid inside a circular tube of radius r, (3.110) and (3.111) take the forms

$$\frac{\mu}{r}\frac{d}{dr}\left(r\frac{du}{dr}\right) = \frac{dP}{dz} \tag{3.112}$$

$$\rho c u \frac{\partial T}{\partial z} = k\left[\frac{1}{r}\frac{\partial}{\partial r}\left(r\frac{\partial T}{\partial r}\right) + \frac{\partial^2 T}{\partial z^2}\right] + \mu\left(\frac{du}{dr}\right)^2 \tag{3.113}$$

where $u = u(r)$ is the axial velocity, $T = T(r, z)$ is the temperature, $P = P(z)$ is the hydrostatic pressure, and the r and z coordinates are chosen along and normal to the direction of main flow. This flow is also known as *Poiseuille flow*.

This completes the review of the pertinent equations of fluid mechanics that are considered in this chapter. Next, we consider the finite element analysis of one-dimensional problems governed by (3.110)–(3.113).

Example 3.3. Consider parallel flow between two long flat walls separated by a distance $2L$. We wish to determine the velocity distribution $u(y)$ for a given constant pressure gradient dP/dx, using the finite element method. The governing equation (3.110) is a special case of the model equation (3.1) with the following correspondence:

$$q = -\frac{dP}{dx} = \text{constant} \equiv q_0, \qquad a = \mu = \text{constant}, \qquad c = 0, \qquad x = y$$

Hence the finite element model in (3.31) is valid for this problem:

$$[K^e]\{u^e\} = \{f^e\} + \{Q^e\} \tag{3.114a}$$

where

$$K_{ij}^e = \int_{y_A}^{y_B} \mu \frac{d\psi_i^e}{dy}\frac{d\psi_j^e}{dy}\,dy, \qquad f_i^e = \int_{y_A}^{y_B}\left(-\frac{dP}{dx}\right)\psi_i^e\,dy$$
$$Q_1^e = -\left(\mu\frac{du}{dy}\right)\bigg|_A, \qquad Q_2^e = \left(\mu\frac{du}{dy}\right)\bigg|_B \tag{3.114b}$$

For the choice of linear finite elements, (3.114a) is the same as (3.37), with $a_e = \mu$ and $q_e = q_0$. For quadratic finite finite elements, $[K^e]$ is given by (3.40a) with $c_e = 0$, and $\{f^e\}$ is given by (3.40b).

For a two-element mesh of linear elements ($h = L$), we have

$$\frac{\mu}{h}\begin{bmatrix} 1 & -1 & 0 \\ -1 & 2 & -1 \\ 0 & -1 & 1 \end{bmatrix}\begin{Bmatrix} U_1 \\ U_2 \\ U_3 \end{Bmatrix} = \frac{q_0 h}{2}\begin{Bmatrix} 1 \\ 2 \\ 1 \end{Bmatrix} + \begin{Bmatrix} Q_1^1 \\ Q_2^1 + Q_1^2 \\ Q_2^2 \end{Bmatrix}$$

We consider two sets of boundary conditions

$$\begin{aligned} &\textbf{Set 1} \quad u(-L) = u(L) = 0 \qquad \text{or} \quad U_1 = U_3 = 0 \\ &\textbf{Set 2} \quad u(-L) = 0, \quad u(L) = U_0 \qquad \text{or} \quad U_1 = 0, \quad U_3 = U_0 \end{aligned} \tag{3.115}$$

The finite element solutions for these two cases are

$$\begin{aligned} U_2 &= \frac{q_0 L^2}{2\mu} && \text{for Set 1} \\ U_2 &= \frac{q_0 L^2}{2\mu} + \tfrac{1}{2}U_0 && \text{for Set 2} \end{aligned} \tag{3.116}$$

For a one-element mesh of the quadratic element ($h = 2L$), we have

$$\frac{\mu}{6L}\begin{bmatrix} 7 & -8 & 1 \\ -8 & 16 & -8 \\ 1 & -8 & 7 \end{bmatrix}\begin{Bmatrix} U_1 \\ U_2 \\ U_3 \end{Bmatrix} = \frac{q_0 L}{3}\begin{Bmatrix} 1 \\ 4 \\ 1 \end{Bmatrix} + \begin{Bmatrix} Q_1^1 \\ 0 \\ Q_3^1 \end{Bmatrix}$$

and the solution is

$$\begin{aligned} U_2 &= \frac{q_0 L^2}{2\mu} && \text{for Set 1} \\ U_2 &= \frac{q_0 L^2}{2\mu} + \tfrac{1}{2} U_0 && \text{for Set 2} \end{aligned} \qquad (3.117)$$

Although the nodal values predicted in the linear- and quadratic-element meshes are the same, they vary linearly and quadratically, respectively, between nodes.

The exact solutions of (3.110) for the two sets of boundary conditions in (3.115) are $(-L \leq y \leq L)$

$$\begin{aligned} u(y) &= \frac{q_0 L^2}{2\mu}\left(1 - \frac{y^2}{L^2}\right) && \text{for Set 1} \\ u(y) &= U_0 \frac{1}{2}\left(1 + \frac{y}{L}\right) + \frac{q_0 L^2}{2\mu}\left(1 - \frac{y^2}{L^2}\right) && \text{for Set 2} \end{aligned} \qquad (3.118)$$

Note that the finite element solutions at the nodes are exact, as expected (see the comments made in Example 3.1). The quadratic-element solution agrees with the exact solutions (3.118) for every value of y.

3.3.3 Solid Mechanics

Solid mechanics is that branch of mechanics dealing with the motion and deformation of solid bodies. The Lagrangian description of motion is used to express the global conservation laws. The conservation of mass for solid bodies is trivially satisfied because of the fixed material viewpoint used in the Lagrangian description. The conservation of momentum is nothing but Newton's second law of motion. Under isothermal conditions, the energy equation uncouples from the momentum equations, and we need only consider the equations of motion.

Unlike in fluid mechanics, here the equations governing solid bodies undergoing different forms of deformations are derived directly, without specializing the three-dimensional elasticity equations. Various types of load-carrying members are called by different names, e.g., bars, beams, and plates. A bar is a structural member that is subjected to only axial loads, while a beam is a member that is subjected to bending loads. The equations governing the motion of such structural elements are not obtained directly from (3.103). They are derived either by considering an element of the member with all its proper forces and using Newton's second law, or by using an energy principle. Examples of the governing equations of the bending of beams and bars were discussed in Example 2.2 in Chapter 2. Energy principles provide an alternative to Newton's laws, and they are more suitable for finite element modeling because energy principles are nothing but the weak forms used in the development of the finite element models. Here we illustrate an application of Newton's second law and the energy principle, namely, the *principle of virtual displacements,* to a bar element. Other examples will be presented in connection with the finite element modeling of beams (Chapter 4) and plates (Chapter 12) later in this book.

Consider a material body of length L and cross-sectional area A. Suppose that the material of the body is homogeneous (i.e., material properties are independent of position), the cross-section is either constant or gradually varying in the axial (i.e., lengthwise) direction, and the applied loads are axial and symmetrically positioned with respect to the geometric centroid of the cross-section. Then the axial stress in the member, except near the points of load application, will be uniform. The only nonzero stress component is $\sigma_x = \sigma_x(x, t)$. Such members are called *bars*. For this case, $u = u(x, t)$, $v = 0$, and (3.103) reduce to the single equation (here u denotes the displacement)

$$\frac{\partial \sigma_x}{\partial x} + f_x = \rho \frac{\partial^2 u}{\partial t^2} \tag{3.119}$$

However, the above equation does not include cross-sectional properties of the member, and it is not useful in its present form. The governing equations of bars of axially varying cross-sections can be derived from Newton's second law (e.g., summation of forces along the x direction) applied to an element of the body (see Fig. 3.12); i.e.,

$$\sum F_x = ma$$

giving

$$-\sigma_x A + (\sigma_x + d\sigma_x)(A + dA) + f\,dx = \rho[A + (A + dA)]\frac{dx}{2}\frac{\partial^2 u}{\partial t^2}$$

or

$$\sigma_x\, dA + d\sigma_x\, A + d\sigma_x\, dA + f\,dx = \rho(A + \tfrac{1}{2}\, dA)\, dx \frac{\partial^2 u}{\partial t^2}$$

Dividing throughout by dx and taking the limit $dx \to 0$, we obtain

$$\frac{\partial}{\partial x}(\sigma_x A) + f = \rho A \frac{\partial^2 u}{\partial t^2} \tag{3.120}$$

where f is the body force per unit length.

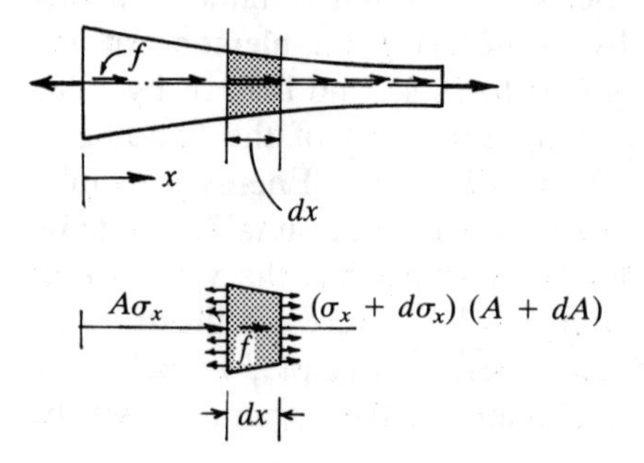

FIGURE 3.12
Axially loaded member and a typical portion of a member of length dx with its axial forces.

From the strain–stress relation $\varepsilon_x = (\sigma_x - \nu\sigma_y)/E$, we have (because $\sigma_y = 0$)

$$\sigma_x = E\varepsilon_x = E\frac{\partial u}{\partial x}$$

Substituting for σ_x from here into (3.120), we obtain

$$\frac{\partial}{\partial x}\left(EA\frac{\partial u}{\partial x}\right) + f = \rho A\frac{\partial^2 u}{\partial t^2}$$

or

$$\boxed{\rho A\frac{\partial^2 u}{\partial t^2} - \frac{\partial}{\partial x}\left(EA\frac{\partial u}{\partial x}\right) = f(x, t)} \tag{3.121}$$

For static problems, this reduces to

$$-\frac{d}{dx}\left(EA\frac{du}{dx}\right) = f(x) \tag{3.122a}$$

Equation (3.122*a*) can be used to determine the displacement $u(x)$ of a material point located at a distance x along the axis of a uniaxially loaded member, called a *bar*. It should be recalled that (3.122*a*) is derived under the assumption that all material points on the line $x =$ constant move by the same distance $u(x)$ (i.e., the stress at any cross-section is uniform). Equation (3.122*a*) is the same as the model equation (3.1), with $a = EA$ and $q = f(x)$.

The average transverse deflection $u(x)$ of a cable made of elastic material is also governed by an equation of the form (3.122*a*):

$$-\frac{d}{dx}\left(T\frac{du}{dx}\right) = f(x) \tag{3.122b}$$

where T is the uniform tension in the cable and f is the distributed transverse force.

Example 3.4. A bridge is supported by several concrete piers, and the geometry and loads of a typical pier are shown in Fig. 3.13. The load 20 kN m^{-2} represents the weight of the bridge and an assumed distribution of the traffic on the bridge. The concrete weighs approximately 25 kN m^{-3} and its modulus is $E = 28 \times 10^6$ kN m^{-2}. We wish to analyze the pier for displacements and stresses using the finite element method.

The pier is indeed a three-dimensional structure. However, we wish to approximate the deformation and stress fields in the pier as one-dimensional. To this end, we represent the distributed force at the top of the pier as a point force

$$F = (0.5 \times 0.5)20 = 5 \text{ kN}$$

The weight of the concrete is represented as the body force per unit length. The total force at any distance x is equal to the weight of the concrete above that point. The weight at a distance x is equal to the product of the volume of the body above x and the

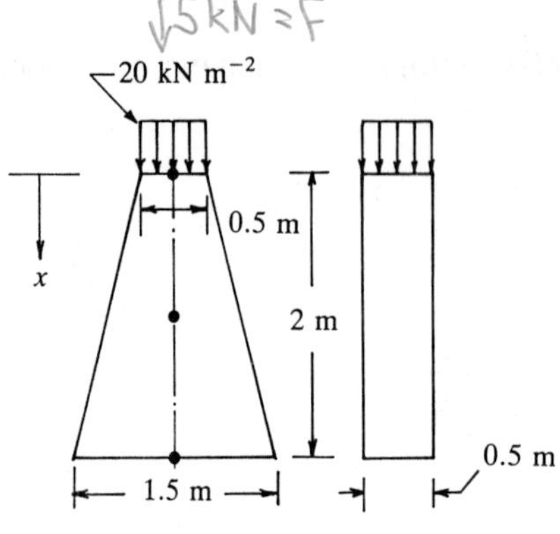

FIGURE 3.13
The geometry and loading in the concrete pier problem of Example 3.4.

specific weight of the concrete:

$$W(x) = 0.5\frac{0.5 + (0.5 + 0.5x)}{2}x \times 25.0 = 6.25(1 + 0.5x)x$$

The body force per unit length is computed from

$$f = \frac{dW}{dx} = 6.25(1 + x)$$

This completes the load representation of the problem.

The governing differential equation for the problem is given by (3.122a), with $E = 28 \times 10^6$ kN m^{-2} and cross-sectional area $A(x)$:

$$A(x) = (0.5 + 0.5x)0.5 = \tfrac{1}{4}(1 + x)$$

Thus

$$-\frac{d}{dx}\left[\tfrac{1}{4}E(1 + x)\frac{du}{dx}\right] = 6.25(1 + x) \tag{3.123a}$$

subject to the boundary conditions

$$\left[\tfrac{1}{4}E(1 + x)\frac{du}{dx}\right]\Bigg|_{x=0} = -5, \qquad u(2) = 0 \tag{3.123b}$$

The finite element model is

$$[K^e]\{u^e\} = \{f^e\} + \{Q^e\} \tag{3.124a}$$

where

$$K_{ij}^e = \int_{x_A}^{x_B} EA\frac{d\psi_i^e}{dx}\frac{d\psi_j^e}{dx}\,dx, \qquad f_i^e = \int_{x_A}^{x_B} f(x)\psi_i^e\,dx$$

$$Q_A^e = \left(-EA\frac{du}{dx}\right)\bigg|_A, \qquad Q_B^e = \left(EA\frac{du}{dx}\right)\bigg|_B \tag{3.124b}$$

For the choice of linear interpolation functions (3.16b), we have

$$K_{11}^e = \int_{x_e}^{x_{e+1}} E\tfrac{1}{4}(1 + x)\left(-\frac{1}{h}\right)^2 dx = \frac{E}{4h_e}[1 + \tfrac{1}{2}(x_e + x_{e+1})]$$

$$f_1^e = \int_{x_e}^{x_{e+1}} 6.25(1 + x)\psi_1^e\,dx = 6.25h_e[\tfrac{1}{2} + \tfrac{1}{6}(x_{e+1} + 2x_e)]$$

Similarly, other components can be evaluated:

$$[K^e] = \frac{E}{4h_e}[1 + \tfrac{1}{2}(x_e + x_{e+1})]\begin{bmatrix} 1 & -1 \\ -1 & 1 \end{bmatrix}$$
$$\{f^e\} = 6.25\frac{h_e}{2}\left(\begin{Bmatrix} 1 \\ 1 \end{Bmatrix} + \frac{1}{3}\begin{Bmatrix} x_{e+1} + 2x_e \\ 2x_{e+1} + x_e \end{Bmatrix}\right) \tag{3.125}$$

Let us consider a two-element mesh with $h_1 = h_2 = 1\,m$. We have

$$[K^1] = \frac{E}{4}\begin{bmatrix} 1.5 & -1.5 \\ -1.5 & 1.5 \end{bmatrix}, \quad \{f^1\} = \frac{6.25}{6}\begin{Bmatrix} 3+1 \\ 3+2 \end{Bmatrix} = \begin{Bmatrix} 4.167 \\ 5.208 \end{Bmatrix}$$
$$[K^2] = \frac{E}{4}\begin{bmatrix} 2.5 & -2.5 \\ -2.5 & 2.5 \end{bmatrix}, \quad \{f^2\} = \frac{6.25}{6}\begin{Bmatrix} 3+4 \\ 3+5 \end{Bmatrix} = \begin{Bmatrix} 7.292 \\ 8.333 \end{Bmatrix}$$

The assembled equations are

$$E\begin{bmatrix} 0.375 & -0.375 & 0.000 \\ -0.375 & 1.000 & -0.625 \\ 0.000 & -0.625 & 0.625 \end{bmatrix}\begin{Bmatrix} U_1 \\ U_2 \\ U_3 \end{Bmatrix} = \begin{Bmatrix} 4.167 \\ 12.500 \\ 8.333 \end{Bmatrix} + \begin{Bmatrix} Q_1^1 \\ Q_2^1 + Q_1^2 \\ Q_2^2 \end{Bmatrix}$$

The boundary and equilibrium conditions require

$$U_3 = 0, \quad Q_2^1 + Q_1^2 = 0, \quad Q_1^1 = 5\text{ kN}$$

The condensed equations are

$$E\begin{bmatrix} 0.375 & -0.375 \\ -0.375 & 1.000 \end{bmatrix}\begin{Bmatrix} U_1 \\ U_2 \end{Bmatrix} = \begin{Bmatrix} 9.167 \\ 12.500 \end{Bmatrix}, \quad Q_2^2 = -0.6255U_2 - 8.333$$

and the solution is given by

$$U_1 = 2.111 \times 10^{-6}\text{ m}, \quad U_2 = 1.238 \times 10^{-6}\text{ m}, \quad Q_2^2 = -30\text{ kN} \tag{3.126}$$

Hence the stress at the fixed end is given by

$$\sigma_x = Q_2^2/A = -30/0.75 = -40\text{ kN m}^{-2}$$

The exact solution of (3.123) is

$$u(x) = \frac{1}{E}\left[56.25 - 6.25(1+x)^2 - 7.5\ln\left(\frac{1+x}{3}\right)\right] \tag{3.127}$$

The exact values of u at nodes 1 and 2 are

$$u(0) = 2.08 \times 10^{-6}\text{ m}, \quad u(1) = 1.225 \times 10^{-6}\text{ m}$$

The four-element model gives 2.008×10^{-6} m and 1.228×10^{-6} m, respectively. The finite element solution at the nodes is not exact because $a = EA$ is not a constant in the problem.

3.4 SUMMARY

In this chapter, the finite element formulation of a second-order differential equation in a single variable has been presented systematically, following a step-by-step procedure. The basic steps of the formulation and analysis of a typical equation are described in Table 3.1. The model equation is representative of the equations arising in various fields of engineering (see Table 3.2). The finite element model is developed following three steps:

1. Weak formulation of the differential equation over an element.
2. Finite element interpolation of the primary variables of the weak fomulation.
3. Finite element formulation over a typical element.

The weak formulation itself involves a three-step procedure, which enables identification of primary variables (i.e., variables that are required to be continuous throughout the domain, including the nodes at which elements are connected). The finite element interpolation functions have been developed here on the basis of continuity, completeness, and linear independence. The finite element model has been developed by substituting appropriate interpolation of the primary variable into the weak form of the differential equation.

Applications of the model to the solution of problems of heat transfer, fluid mechanics, and solid mechanics have been presented. To aid the reader, a brief review of the basic terminology and governing equations of each of the three fields has also been given. The numerical examples should aid the reader in deeper understanding of the steps involved in the finite element analysis of one-dimensional second-order differential equations.

It has been shown that the secondary variables of a problem can be computed using either the global algebraic equations of the finite element mesh (i.e., condensed equations for the secondary variables) or by their original definition through finite element interpolation. The former method gives more accurate results, which will satisfy the equilibrium at interelement nodes, whereas the latter gives less accurate results, which are discontinuous at such nodes. The secondary variables computed using the Lagrange linear elements are element-wise-constant, while they are element-wise-linear for the Lagrange quadratic elements.

PROBLEMS

Many of the following problems are designed for hand calculation while some are intended specifically for computer calculations using the program FEM1DV2. This should give the student deeper understanding of what is involved in the formulation and solution of a problem by the finite element method. The hand calculations can be verified, in most cases, by solving the same problems using FEM1DV2, which is described in Chapter 7.

Section 3.2

3.1. Develop the weak form and the finite element model of the following differential equation over an element:

$$-\frac{d}{dx}\left(a\frac{du}{dx}\right)+\frac{d^2}{dx^2}\left(b\frac{d^2u}{dx^2}\right)+cu=f \quad \text{for } x_A<x<x_B$$

where a, b, c, and f are known functions of position x. Ensure that the element coefficient matrix $[K^e]$ is symmetric. What is the nature of the interpolation functions for the problem?

3.2. Construct the weak form and the finite element model of the differential equation

$$-\frac{d}{dx}\left(a\frac{du}{dx}\right)-b\frac{du}{dx}=f \quad \text{for } 0<x<L$$

over a typical element $\Omega^e=(x_A, x_B)$. Here a, b, and f are known functions of x, and u is the dependent variable. The natural boundary condition should *not* involve the function $b(x)$. What type of interpolation functions can be used for u?

3.3. Construct the weak form and associated finite element model of the equation

$$-\frac{d}{dx}\left(a\frac{du}{dx}\right)+cu=f \quad \text{for } 0<x<L$$

such that the natural boundary condition of the type

$$a\frac{du}{dx}+k(u-u_0)=Q$$

is included in the weak form. Here a, c, and f are known functions of x, while k, u_0, and Q are constants.

3.4. Derive the Lagrange cubic interpolation functions for a four-node (one-dimensional) element (with equally spaced nodes) using the alternative procedure based on interpolation properties (3.28). Use the local coordinate $\bar{x}$ for simplicity.

3.5. Verify (3.34*a*,*b*) by actual evaluation of $K^e_{ij}=K^e_{ji}$ and f^e_i.

3.6. Evaluate the following coefficient matrices and source vector using the linear Lagrange interpolation functions:

$$K^e_{ij}=\int_{x_A}^{x_B}(a_0+a_1x)\frac{d\psi^e_i}{dx}\frac{d\psi^e_j}{dx}\,dx$$

$$m^e_{ij}=\int_{x_A}^{x_B}(c_0+c_1x)\psi^e_i\psi^e_j\,dx, \qquad f^e_i=\int_{x_A}^{x_B}(q_0+q_1x)\psi^e_i\,dx$$

where a_0, a_1, c_0, c_1, q_0, and q_1 are constants.

3.7. Verify the coefficients in (3.40*a*,*b*) by actual evaluation of $K^e_{ij}=K^e_{ji}$ and f^e_i.

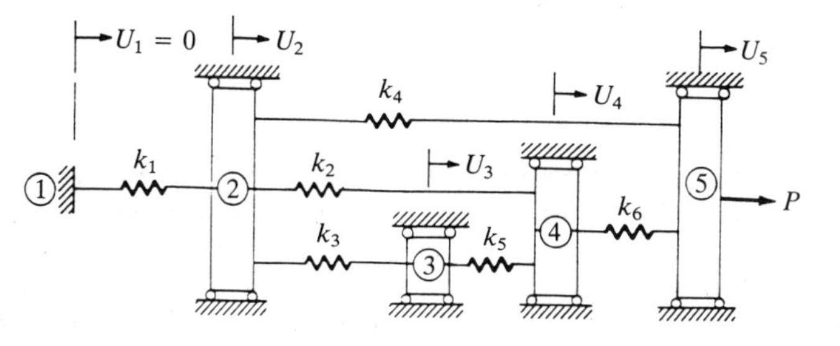

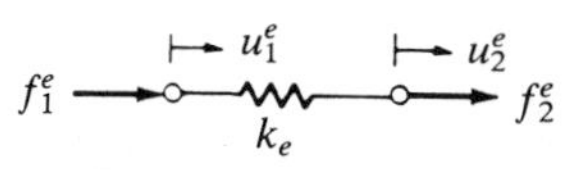

Typical element

FIGURE P3.8

3.8. Consider the system of linear elastic springs shown in Fig. P3.8. Write the force–displacement relationship for a typical (single) spring element, and assemble the element equations to obtain the force–displacement relations for the entire system. Use the boundary conditions to write the condensed equations for the unknown displacements and forces.

3.9. Consider the hydraulic network (the flow is assumed to be laminar) shown in Fig. P3.9. A typical element (which is a circular pipe of constant diameter) with two nodes is also shown in the figure. The unknown primary degree of freedom at each node is the pressure P, and the secondary degree of freedom is the flow (or discharge) Q. The element equations relating the primary variables to the secondary variables are given by

$$\frac{c(d_e)^4}{h_e}\begin{bmatrix} 1 & -1 \\ -1 & 1 \end{bmatrix}\begin{Bmatrix} P_1^e \\ P_2^e \end{Bmatrix} = \begin{Bmatrix} Q_1^e \\ Q_2^e \end{Bmatrix}, \qquad c = \frac{\pi}{128\mu}$$

where d_e is the diameter of the pipe, h_e is its length, and μ is the viscosity of the fluid. Write the condensed equations for the unknown pressures and flows (use the minimum number of elements.)

Answer: $P_1 = \frac{39}{14}Qa$, $P_2 = \frac{12}{7}Qa$, $P_3 = \frac{15}{14}Qa$.

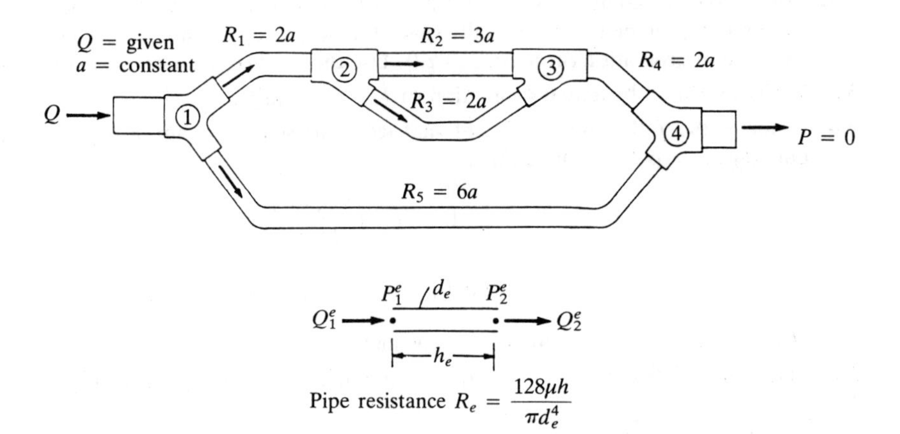

FIGURE P3.9

3.10. Consider the direct current electric network shown in Fig. P3.10. We wish to determine the voltages V and currents I in the network using the finite element method. A typical finite element in this case consists of a resistor R, with the primary degree of freedom being the voltage and the secondary degree of freedom being the current. The element equations are provided by Ohm's law:

$$\frac{1}{R_e}\begin{bmatrix} 1 & -1 \\ -1 & 1 \end{bmatrix}\begin{Bmatrix} V_1^e \\ V_2^e \end{Bmatrix} = \begin{Bmatrix} I_1^e \\ I_2^e \end{Bmatrix}$$

The continuity conditions at the interelement nodes require that the net current flow into any junction (node) always be zero in a closed loop. Set up the algebraic equations (i.e., condensed equations) for the unknown voltages and currents.

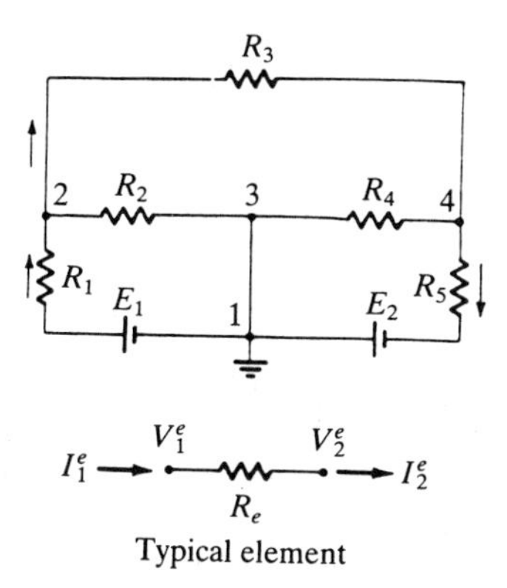

FIGURE P3.10

3.11. Consider the composite structure of axially loaded members shown in Fig. P3.11. Write the continuity conditions (i.e., the correspondence of element nodal values to global nodal values) and the equilibrium conditions (i.e., the relationships between Q_i^e at the interelement nodes) for the structure. Derive the assembled coefficient (stiffness) matrix for the structure, and set up the condensed equations for the unknown displacements and forces.

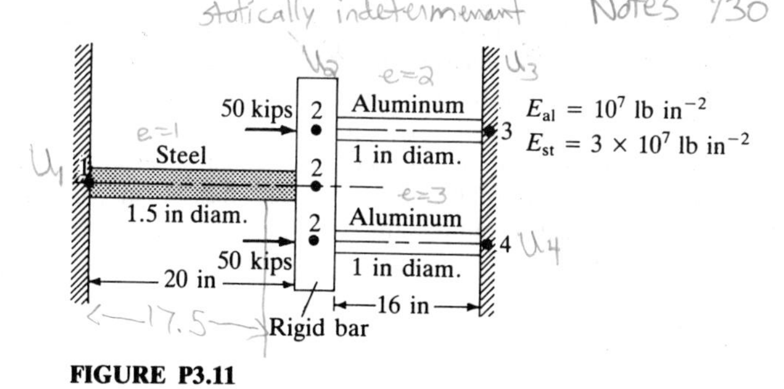

FIGURE P3.11

3.12. Use the finite element method to solve the differential equation

$$-\frac{d^2u}{dx^2} - cu + x^2 = 0 \quad \text{for } 0 < x < 1$$

for the essential or (Dirichlet) boundary conditions $u(0) = 0$ and $u(1) = 0$. Use a uniform mesh of three linear elements, and compare the finite element solution with the exact solution for $c = 1$.

Answer: $U_2 = -0.02999$, $U_3 = -0.04257$, $(Q_2^3)_{\text{def}} = 0.12771$.

3.13. Solve the differential equation in Problem 3.12 for the mixed boundary conditions

$$u(0) = 0, \qquad \left(\frac{du}{dx}\right)\bigg|_{x=1} = 0$$

Use three linear elements.

Answer: $U_2 = -0.18805$, $U_3 = -0.34144$, $U_4 = -0.40708$, $(Q_1^1)_{\text{def}} = 5.6414$.

3.14. Solve the differential equation in Problem 3.12 for the *natural* (or Neumann) boundary conditions

$$\left(\frac{du}{dx}\right)\bigg|_{x=0} = 1, \qquad \left(\frac{du}{dx}\right)\bigg|_{x=1} = \frac{4}{3}$$

Use a uniform mesh of three linear finite elements to solve the problem. Verify your solution with the analytical solution.

Note that for Neumann boundary conditions, none of the primary dependent variables is specified, and therefore the solution can be determined within a constant; the coefficient matrix of the finite element equations for this case remains unaltered. When $c = 0$, the coefficient matrix is singular and cannot be inverted. In such cases, one of the U_i should be set equal to a constant (e.g., zero) to remove the "rigid-body" mode (i.e., to determine the arbitrary constant in the solution).

Answer: $U_1 = -0.56862$, $U_2 = -0.20932$, $U_4 = 0.18696$, $U_5 = 0.61334$.

3.15. The governing equation for an unconfined aquifer with flow in the radial direction is given by the differential equation

$$-\frac{1}{r}\frac{d}{dr}\left(rk\frac{du}{dr}\right) = f$$

where k is the coefficient of permeability, f the recharge, and u the piezometric head. Pumping is considered to be a negative recharge. Consider the following problem. A well penetrates an aquifer and pumping is performed at $r = 0$ at a rate $Q = 150\, m^2\, h^{-1}$. The permeability of the aquifer is $k = 25\, \text{m}^3\, \text{h}^{-1}\, \text{m}^{-2}$. A constant head $u_0 = 50$ m exists at a radial distance $L = 200$ m. Determine the piezometric head at radial distances of 0, 10, 20, 40, 80, and 140 m (see Fig. P3.15). You are required to set up the finite element equations for the unknowns using a nonuniform mesh of five linear elements.

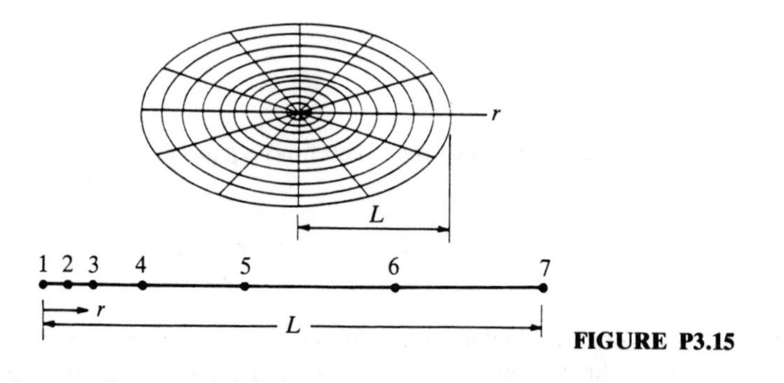

FIGURE P3.15

Section 3.3
Heat transfer

3.16. The following differential equation arises in connection with heat transfer in an insulated rod:

$$-\frac{d}{dx}\left(k\frac{dT}{dx}\right) = q \quad \text{for } 0 < x < L$$

$$T(0) = T_0, \qquad \left[k\frac{dT}{dx} + \beta(T - T_\infty) + \hat{q}\right]\bigg|_{x=L} = 0$$

where T is the temperature, k the thermal conductivity, and q the heat generation. Take the following values for the data: $q = 0$, $\hat{q} = 0$, $L = 0.1$ m, $k = 0.01$ W m^{-1} °C^{-1}, $\beta = 25$ W m^{-2} °C^{-1}. $T_0 = 50$ °C, and $T_\infty = 5$ °C. Solve the problem using two linear finite elements for temperature values at $x = \frac{1}{2}L$ and L.

Answer: $U_2 = 27.59$ °C, $U_3 = 5.179$ °C, $Q_1^1 = 4.482$ W m^{-2} $= -Q_2^2$.

3.17. An insulating wall is constructed of three homogeneous layers with conductivities k_1, k_2, and k_3 in intimate contact (see Fig. P3.17). Under steady-state conditions, the temperatures at the boundaries of the layers are characterized by the external surface temperatures T_1 and T_4 and the interface temperatures T_2 and T_3. Formulate the problem to determine the temperatures T_i $(i = 1, \ldots, 4)$ when the ambient temperatures T_0 and T_5 and the (surface) film coefficients β_0 and β_5 are known. Assume that there is no internal heat generation and that the heat flow is one-dimensional $(\partial T/\partial y = 0)$.

Answer: $U_1 = 84.489$ °C, $U_2 = 68.977$ °C, $U_3 = 50.881$ °C, $U_4 = 45.341$ °C

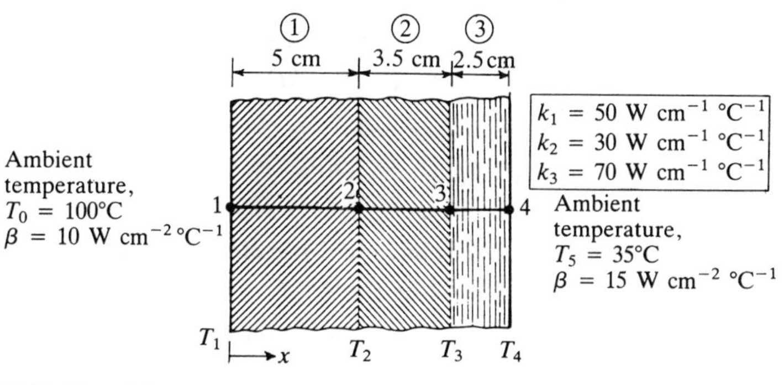

FIGURE P3.17

3.18. Consider the rectangular cooling fin shown in Fig. 3.9. The governing equation (i.e., balance of energy) is

$$-\frac{d^2T}{dx^2} + \frac{\beta}{ka}(T - T_\infty) = 0$$

where T is the temperature, k the thermal conductivity, β the film coefficient, a the thickness, and T_∞ the temperature of the surrounding fluid (i.e., ambient temperature). The boundary conditions of the problem are

$$T(0) = T_w \quad \text{(wall temperature)}, \qquad \left(kA\frac{dT}{dx}\right)\bigg|_{x=L} = 0$$

The equations can be recast by introducing the following nondimensional quantities:

$$\theta = \frac{T - T_\infty}{T_w - T_\infty}, \qquad \xi = \frac{x}{L}, \qquad N = \left(\frac{\beta L^2}{ka}\right)^{1/2}$$

They then take the forms

$$-\frac{d^2\theta}{d\xi^2} + N^2\theta = 0, \qquad \theta(0) = 1, \qquad \left(\frac{d\theta}{d\xi}\right)\bigg|_{\xi=1} = 0$$

Solve the problem using (*a*) two linear elements and (*b*) one quadratic element, and compare the finite solutions at four equal distances with the analytical solution

$$\theta(\xi) = \theta(0)\frac{\cosh N(L-\xi)}{\cosh NL}$$

3.19. A steel rod of diameter $D = 2$ cm, length $L = 5$ cm, and thermal conductivity $k = 50\ \mathrm{W\,m^{-1}\,°C^{-1}}$ is exposed to ambient air at $T_\infty = 20$ °C with a heat transfer coefficient $\beta = 100\ \mathrm{W\,m^{-2}\,°C^{-1}}$. If the left end of the rod is maintained at temperature $T_0 = 320$ °C (see Fig. P3.19), determine the temperatures at distances 25 mm and 50 mm from the left end, and the heat at the left end. The governing equation of the problem is

$$-\frac{d^2\theta}{dx^2} + N^2\theta = 0 \quad \text{for } 0 < x < L$$

where $\theta = T - T_\infty$, T is the temperature, and N^2 is given by

$$N^2 = \frac{\beta P}{Ak} = \frac{\beta\pi D}{\frac{1}{4}\pi D^2 k} = \frac{4\beta}{kD}$$

The boundary conditions are

$$\theta(0) = T(0) - T_\infty = 300\ °\mathrm{C}, \qquad \left(\frac{d\theta}{dx}\right)\bigg|_{x=L} + \frac{\beta}{k}\theta(L) = 0$$

Use (*a*) two linear elements and (*b*) one quadratic element to solve the problem by the finite element method. The exact solution is

$$\theta(x) = \theta(0)\frac{[\cosh N(L-x) + (\beta/Nk)\sinh N(L-x)]}{\cosh NL + (\beta/Nk)\sinh NL}$$

Answer: (*a*) $U_1 = 300$ °C, $U_2 = 211.97$ °C, $U_3 = 179.24$ °C, $(Q_1^1)_{\text{def}} = 3521.1\ \mathrm{W/m^2}$.
(*b*) $U_2 = 300$ °C, $U_2 = 213.07$ °C, $U_3 = 180.77$ °C, $(Q_1^1)_{\text{def}} = 4564.9\ \mathrm{W/m^2}$.

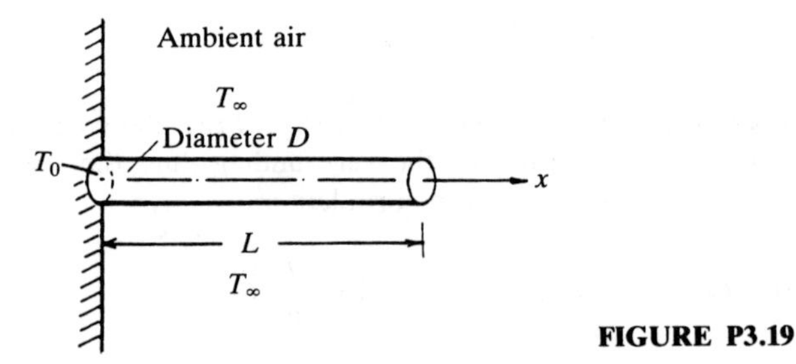

FIGURE P3.19

3.20. Find the temperature distribution in the tapered fin shown in Fig. P3.20. Assume that the temperature at the root of the fin is 250 °F, the conductivity $k = 120\ \mathrm{BTu\,h^{-1}\,ft^{-1}\,°F^{-1}}$, and the film coefficient $\beta = 15\ \mathrm{Btu\,h^{-1}\,ft^{-2}\,°F^{-1}}$, and use three linear elements. The ambient temperature at the top and bottom of the fin is $T_\infty = 75$ °F.

Answer: $T_1(\text{tip}) = 166.23$ °F, $T_2 = 191.1$ °F, $T_3 = 218.89$ °F.

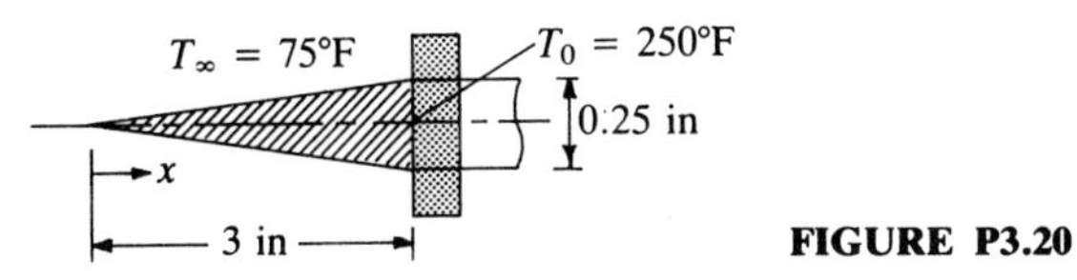

FIGURE P3.20

3.21. Consider steady heat conduction in a wire of circular cross-section with an electrical heat source. Suppose that the radius of the wire is R_0, its electrical conductivity is K_e (Ω^{-1} cm^{-1}), and it is carrying an electric current density of I (A cm^{-2}). During the transmission of an electric current, some of the electrical energy is converted into thermal energy. The rate of heat production per unit volume is given by $q_e = I^2/K_e$. Assume that the temperature rise in the wire is sufficiently small that the dependence of the thermal or electric conductivity on temperature can be neglected. The governing equations of the problem are

$$-\frac{1}{r}\frac{d}{dr}\left(rk\frac{dT}{dr}\right) = q_e \quad \text{for } 0 \leqslant r \leqslant R_0, \qquad \left(kr\frac{dT}{dr}\right)\bigg|_{r=0} = 0, \qquad T(R_0) = T_0$$

Determine the distribution of temperature in the wire using (*a*) two linear elements and (*b*) one quadratic element, and compare the finite element solution at eight equal intervals with the exact solution

$$T(r) = T_0 + \frac{q_e R_0^2}{4k}\left[1 - \left(\frac{r}{R_0}\right)^2\right]$$

Also, determine the heat flow $Q = -2\pi R_0 k(dT/dr)|_{R_0}$ at the surface using (i) the temperature field and (ii) the balance equations.

3.22. Consider a nuclear fuel element of spherical form, consisting of a sphere of "fissionable" material surrounded by a spherical shell of aluminum "cladding" as shown in Fig. P3.22. Nuclear fission is a source of thermal energy, which varies nonuniformly from the center of the sphere to the interface of the fuel element and the cladding. We wish to determine the temperature distribution in the nuclear fuel element and the aluminum cladding.

The governing equations for the two regions are the same, with the exception that there is no heat source term for the aluminum cladding. We have

$$-\frac{1}{r^2}\frac{d}{dr}\left(r^2 k_1 \frac{dT_1}{dr}\right) = q \quad \text{for } 0 \leqslant r \leqslant R_F$$

$$-\frac{1}{r^2}\frac{d}{dr}\left(r^2 k_2 \frac{dT_2}{dr}\right) = 0 \quad \text{for } R_F \leqslant r \leqslant R_C$$

where subscripts 1 and 2 refer to the nuclear fuel element and cladding, respectively. The heat generation in the nuclear fuel element is assumed to be of the form

$$q_1 = q_0\left[1 + c\left(\frac{r}{R_F}\right)^2\right]$$

where q_0 and c are constants depending on the nuclear material. The boundary conditions are

$$kr^2 \frac{dT_1}{dr} = 0 \quad \text{at } r = 0$$

$$T_1 = T_2 \quad \text{at } r = R_F$$

$$T_2 = T_0 \quad \text{ar } r = R_C$$

Use two linear elements to determine the finite element solution for the temperature distribution, and compare the nodal temperatures with the exact solution

$$T_1 - T_0 = \frac{q_0 R_F^2}{6k_1}\left\{\left[1 - \left(\frac{r}{R_F}\right)^2\right] + \tfrac{3}{10}c\left[1 - \left(\frac{r}{R_F}\right)^4\right]\right\} + \frac{q_0 R_F^2}{3k_2}(1 + \tfrac{3}{5}c)\left(1 - \frac{R_F}{R_c}\right)$$

$$T_2 - T_0 = \frac{q_0 R_F^2}{3k_2}(1 + \tfrac{3}{5}c)\left(\frac{R_F}{r} - \frac{R_F}{R_C}\right)$$

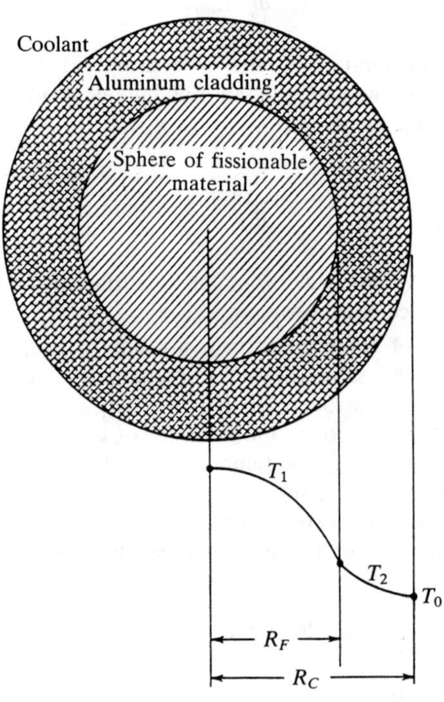

FIGURE P3.22

Fluid mechanics

3.23. Consider the flow of a Newtonian viscous fluid on an inclined flat surface, as shown in Fig. P3.23. Examples of such flow can be found in wetted-wall towers and the application of coatings to wallpaper rolls. The momentum equation, for a fully developed steady laminar flow along the z coordinate, is given by

$$-\mu \frac{d^2 w}{dx^2} = \rho g \cos \beta$$

where w is the z component of the velocity, μ is the viscosity of the fluid, ρ is the density, g is the acceleration due to gravity, and β is the angle between the inclined surface and the vertical. The boundary conditions associated with the problem are that the shear stress is zero at $x = 0$ and the velocity is zero at $x = L$:

$$\left(\frac{dw}{dx}\right)\bigg|_{x=0} = 0, \qquad w(L) = 0$$

Use (*a*) two linear finite elements of equal length and (*b*) one quadratic finite element in the domain (0, L) to solve the problem and compare the two finite element solutions at four points $x = 0$, $\frac{1}{4}L$, $\frac{1}{2}L$, and $\frac{3}{4}L$ of the domain with the exact solution

$$w_e = \frac{\rho g L^2 \cos\beta}{2\mu}\left[1 - \left(\frac{x}{L}\right)^2\right]$$

Evaluate the shear stress ($\tau_{xz} = -\mu\, dw/dx$) at the wall using (i) the velocity fields and (ii) the equilibrium equations, and compare with the exact value.

Answer: (*a*) $U_1 = \frac{1}{2}f_0$, $U_2 = \frac{3}{8}f_0$, $f_0 = (\rho g \cos\beta)L^2/\mu$.

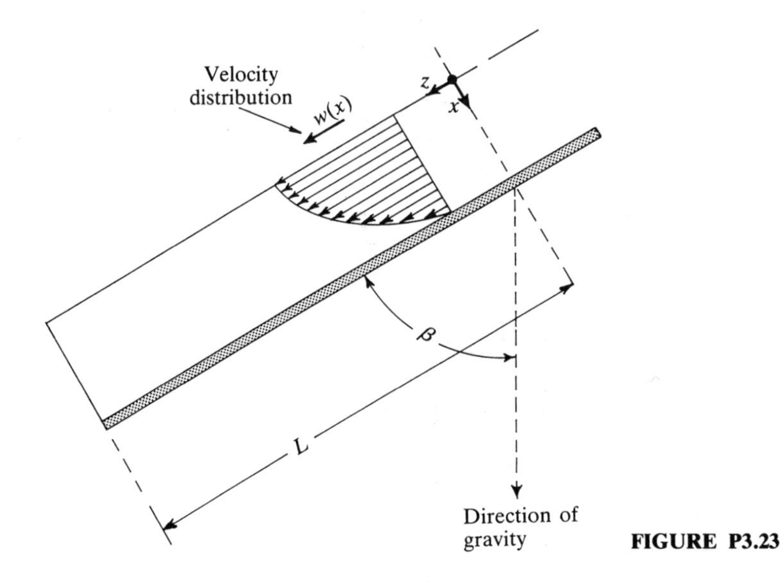

FIGURE P3.23

3.24. Consider the steady laminar flow of a viscous fluid through a long circular cylindrical tube. The governing equation is

$$-\frac{1}{r}\frac{d}{dr}\left(r\mu\frac{dw}{dr}\right) = \frac{P_0 - P_L}{L} \equiv f_0$$

where w is the axial (i.e., z) component of velocity, μ is the viscosity, and f_0 is the gradient of pressure (which includes the combined effect of static pressure and gravitational force). The boundary conditions are

$$\left(r\frac{dw}{dr}\right)\bigg|_{r=0} = 0, \qquad w(R_0) = 0$$

Using the symmetry and (*a*) two linear elements, (*b*) one quadratic element, determine the velocity field and compare with the exact solution at the nodes:

$$w_e(r) = \frac{f_0 R_0^2}{4\mu}\left[1 - \left(\frac{r}{R_0}\right)^2\right]$$

3.25. In the problem of the flow of a viscous fluid through a circular cylinder (Problem 3.24), assume that the fluid slips at the cylinder wall; i.e., instead of assuming that $w = 0$ at $r = R_0$, use the boundary condition that

$$kw = -\mu \frac{dw}{dr} \quad \text{at } r = R_0$$

in which k is the "coefficient of sliding friction." Solve the problem with two linear elements.

3.26. Consider the steady laminar flow of a Newtonian fluid with constant density in a long annular region between two coaxial cylinders of radii R_i and R_o (see Fig. P3.26). The differential equation for this case is given by

$$-\frac{1}{r}\frac{d}{dr}\left(r\mu\frac{dw}{dr}\right) = \frac{P_1 - P_2}{L} \equiv f_0$$

where w is the velocity along the cylinders (i.e. the z component of velocity), μ is the viscosity, L is the length of the region along the cylinders in which the flow is

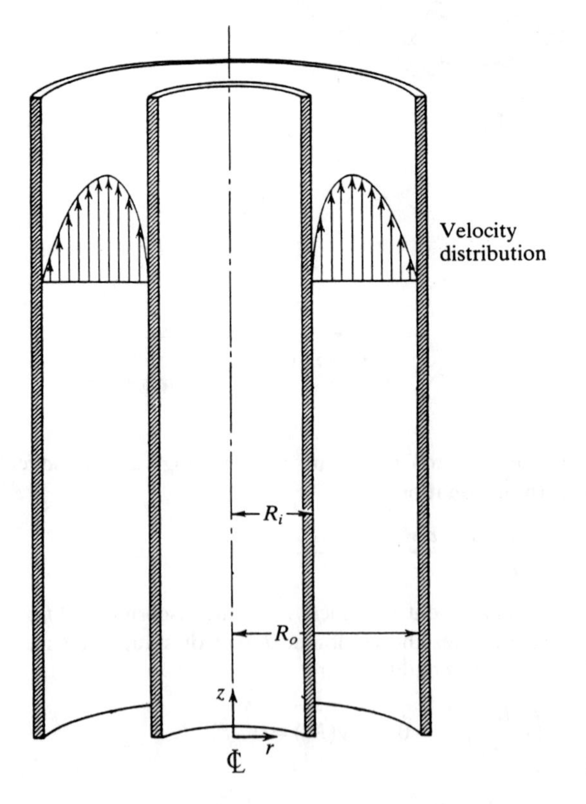

FIGURE P3.26

fully developed, and P_1 and P_2 are the pressures at $z = 0$ and $z = L$, respectively (P_1 and P_2 represent the combined effect of static pressure and gravitational force). The boundary conditions are

$$w = 0 \quad \text{at } r = R_o \text{ and } R_i$$

Solve the problem using (a) two linear elements and (b) one quadratic element, and compare the finite element solutions with the exact solution at the nodes:

$$w_e(r) = \frac{f_0 R_0^2}{4\mu}\left[1 - \left(\frac{r}{R_o}\right)^2 + \frac{1-k^2}{\ln(1/k)}\ln\left(\frac{r}{R_o}\right)\right]$$

where $k = R_i/R_o$. Determine the shear stress $\tau_{rz} = -\mu\, dw/dr$ at the walls using (i) the velocity field and (ii) the equilibrium equations, and compare with the exact values. (Note that the steady laminar flow of a viscous fluid through a long cylinder or a circular tube can be obtained as a limiting case of $k \to 0$.)

3.27. Consider the steady laminar flow of two immiscible incompressible fluids in a region between two parallel stationary plates under the influence of a pressure gradient. The fluid rates are adjusted such that the lower half of the region is filled with Fluid I (the denser and more viscous fluid) and the upper half is filled with Fluid II (the less dense and less viscous fluid), as shown in Fig. P3.27. We wish to determine the velocity distributions in each region using the finite element method.

The governing equations for each fluid are

$$-\mu_1 \frac{d^2u_1}{dx^2} = f_0, \qquad -\mu_2 \frac{d^2u_2}{dx^2} = f_0$$

where $f_0 = (P_0 - P_L)/L$ is the pressure gradient. The boundary conditions are

$$u_1(-b) = 0, \qquad u_2(b) = 0, \qquad u_1(0) = u_2(0)$$

Solve the problem using (a) four linear elements and (b) two quadratic elements, and compare the finite element solutions with the exact solution at the nodes

$$u_i = \frac{f_0 b^2}{2\mu_i}\left[\frac{2\mu_i}{\mu_1 + \mu_2} + \frac{\mu_1 - \mu_2}{\mu_1 + \mu_2}\frac{y}{b} - \left(\frac{y}{b}\right)^2\right] \quad (i = 1, 2)$$

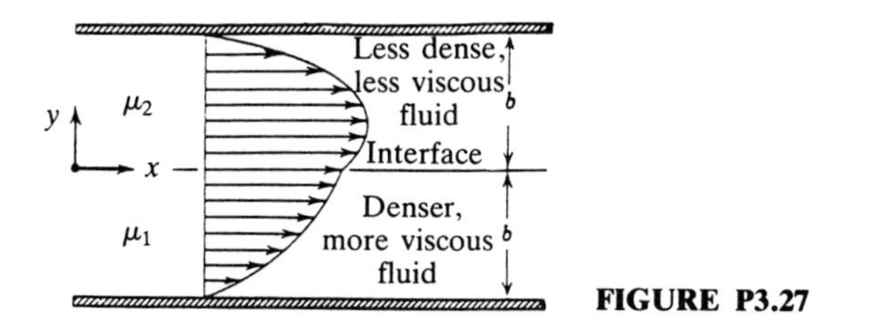

FIGURE P3.27

Solid mechanics

3.28. Find the three-element finite element solution to the stepped-bar problem (axial deformation of a bar). See Fig. P3.28 for the geometry and data.

Answer: $U_2 = 6P/\pi E_a$, $U_3 = 38P/\pi E_a$, $U_4 = 130\, P/3\pi E_a$.

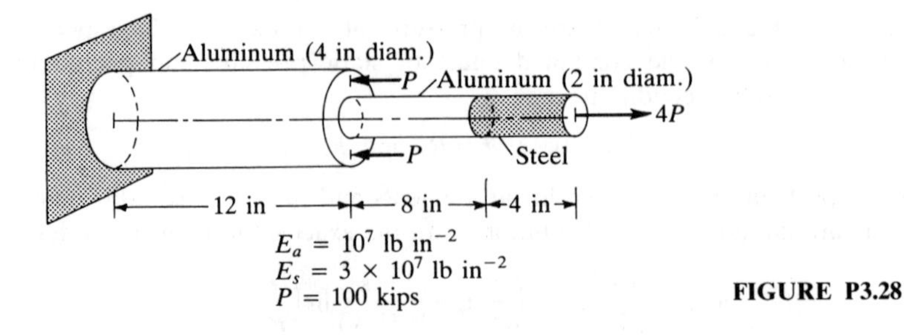

FIGURE P3.28

3.29. The equation governing the axial deformation of an elastic bar in the presence of applied mechanical loads f and P and a temperature change T is

$$-\frac{d}{dx}\left[EA\left(\frac{du}{dx} - \alpha T\right)\right] = f \quad \text{for } 0 < x < L$$

where α is the thermal expansion coefficient, E the modulus of elasticity, and A the cross-sectional area. Using three linear finite elements, determine the axial displacements in a nonuniform rod of length 30 in, fixed at the left end and subjected to an axial force $P = 400$ lb and a temperature change of 60 °F. Take $A(x) = 6 - \frac{1}{10}x$ in^2, $E = 30 \times 10^6$ lb in^{-2}, and $\alpha = 12 \times 10^{-6}$ in in^{-1} °F^{-1}.

3.30. Analyze the stepped bar with its right end supported by a linear axial spring (see Fig. P3.30). The boundary condition at $x = 24$ in is

$$EA\frac{du}{dx} + ku = 0$$

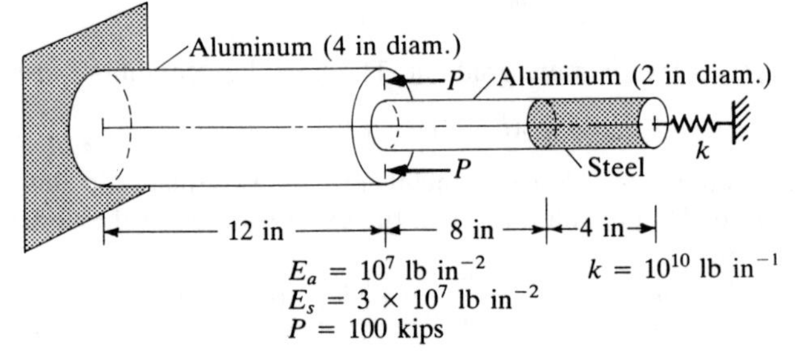

FIGURE P3.30

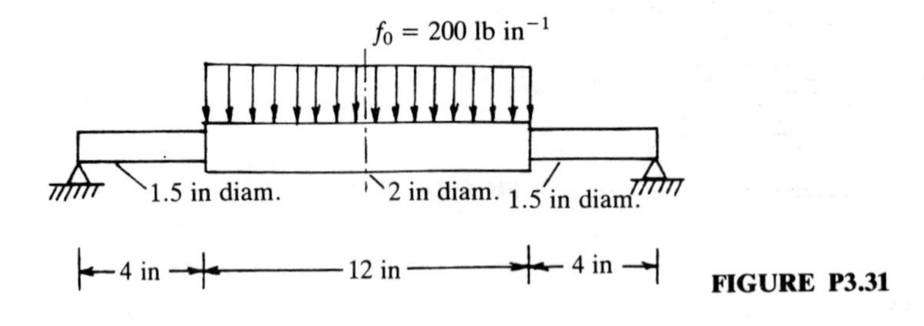

FIGURE P3.31

3.31. Consider the steel ($E = 30 \times 10^6$ psi) beam shown in Fig. P3.31. Determine the transverse deflection using two linear elements. Exploit the symmetry of the beam.

3.32. Determine the axial deformation of a nonuniform bar, $A = A_0 + A_1 x$, under its own weight (f_0 per unit length). Use two linear elements. The bar is fixed at $x = 0$.

3.33. Turbine disks are often thick near their hub and taper down to a smaller thickness at the periphery. The equation governing a variable-thickness $t = t(r)$ disk is

$$\frac{d}{dr}(tr\sigma_r) - t\sigma_\theta + t\rho\omega^2 r^2 = 0$$

where ω^2 is the angular speed of the disk and

$$\sigma_r = c\left(\frac{du}{dr} + \nu\frac{u}{r}\right), \qquad \sigma_\theta = c\left(\frac{u}{r} + \nu\frac{du}{dr}\right), \qquad c = \frac{E}{1-\nu^2}$$

(*a*) Construct the weak integral form of the governing equation such that the bilinear form is symmetric and the natural boundary condition involves specifying the quantity $tr\sigma_r$.

(*b*) Develop the finite element model associated with the weak form derived in part (*a*).

3.34. Determine the axial deformation of a varying cross-section member (see Fig. P3.34) under its own weight. Use one quadratic element.

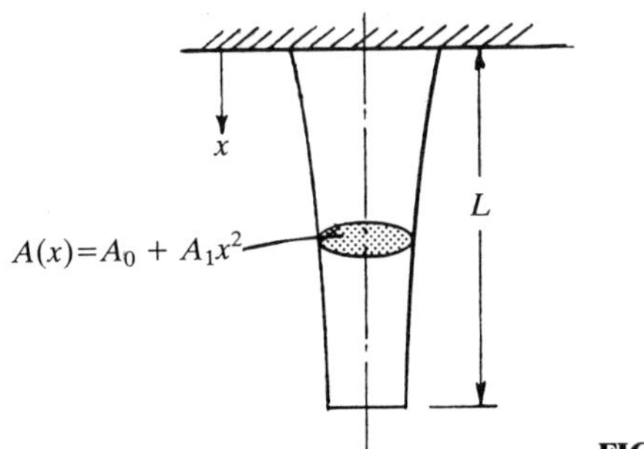

FIGURE P3.34

REFERENCES FOR ADDITIONAL READING

Fluid mechanics

Bird, R. B., W. E. Stewart, and E. N. Lightfoot: *Transport Phenomena,* John Wiley, New York, 1960.

Duncan, W. J., A. S. Thom, and A. D. Young: *Mechanics of Fluids,* 2d ed., Elsevier, New York, 1970.

Harr, M. E.: *Ground Water and Seepage,* McGraw-Hill, New York, 1962.

Shames, I. H.: *Mechanics of Fluids,* McGraw-Hill, New York, 1962.

Vallentine, H. R.: *Applied Hydrodynamics,* Butterworths, London 1959.

Heat transfer

Kreith, F.: *Principles of Heat Transfer,* 3d ed., Harper & Row, New York, 1973.

Myers, G. E.: *Analytical Methods in Conduction Heat Transfer,* McGraw-Hill, New York, 1972.

Nagotov, E. P.: *Applications of Numerical Methods to Heat Transfer,* McGraw-Hill, New York, 1978.

Solid mechanics

Boresi, A. P., and P. P. Lynn: *Elasticity in Engineering Mechanics,* Prentice-Hall, Englewood Cliffs, NJ, 1974.
Dym, C. L., and I. H. Shames: *Solid Mechanics: A Variational Approach,* McGraw-Hill, New York, 1973.
Reddy, J. N.: *Energy and Variational Methods in Applied Mechanics,* John Wiley, New York, 1984.
Timenshenko, S. P., and J. N. Goodier: *Theory of Elasticity,* McGraw-Hill, New York, 1970.

Variational methods

(See also References for Additional Reading in Chapter 2)

Crandall, S. H.: *Engineering Analysis,* McGraw-Hill, New York, 1956.
Mikhlin, S. G.: *Variational Methods in Mathematical Physics,* Pergamon Press, New York, 1964.
———: *The Numerical Performance of Variational Methods,* Wolters-Noordhoff, Groningen, 1971.
Oden, J. T., and J. N. Reddy: *Variational Methods in Theoretical Mechanics,* Springer-Verlag, New York, 1976; 2d ed., 1983.
Reddy, J. N.: *Energy and Variational Methods in Applied Mechanics,* John Wiley, New York, 1984.
———: *Applied Functional Analysis and Variational Methods in Engineering,* McGraw-Hill, New York, 1986; Krieger, Melbourne, FL, 1991.
——— and M. L. Rasmussen: *Advanced Engineering Analysis,* John Wiley, New York, 1982; Krieger, Melbourne, FL, 1990.
Rektorys, K.: *Variational Methods in Mathematics, Science and Engineering,* Reidel, Boston, 1977.
Schechter, R. S.: *The Variational Methods in Engineering,* McGraw-Hill, New York, 1967.
Washizu, K.: *Variational Methods in Elasticity and Plasticity,* 2d ed., Pergamon Press, New York, 1975; 3d ed., 1982.

CHAPTER 4

BENDING OF BEAMS

4.1 INTRODUCTION

Here we consider the finite element formulation of the one-dimensional fourth-order differential equation that arises in the Euler–Bernoulli beam theory and the pair of one-dimensional second-order equations associated with Timoshenko beam theory. The formulations of a fourth-order equation and two coupled second-order equations involve the same steps as described in Section 3.2 for a second-order equation, but the mathematical details are somewhat different, especially in the finite element formulation of the equations.

4.2 THE EULER–BERNOULLI BEAM ELEMENT

4.2.1 Governing Equation

In the Euler–Bernoulli beam theory, it is assumed that plane cross-sections perpendicular to the axis of the beam remain plane and perpendicular to the axis after deformation. In this theory, the transverse deflection w of the beam

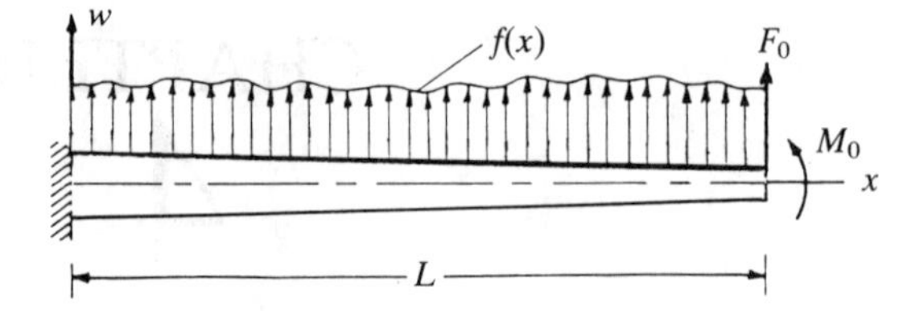

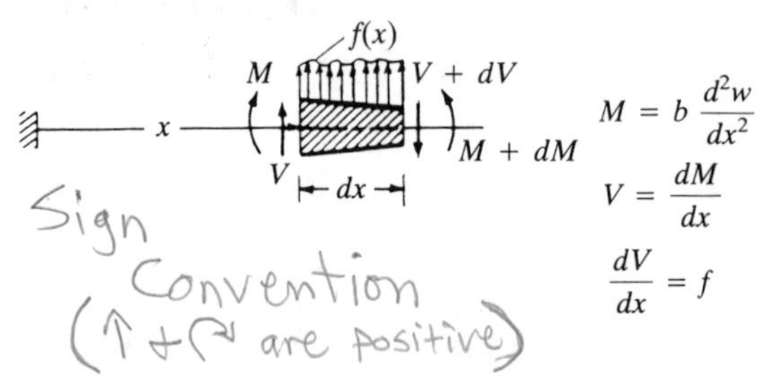

FIGURE 4.1
Bending of beams. The shear force–bending moment–deflection relations and the sign convention.

is governed by the fourth-order differential equation

$$\frac{d^2}{dx^2}\left(b\frac{d^2w}{dx^2}\right)=f(x) \quad \text{for } 0<x<L \tag{4.1}$$

where $b=b(x)$ and $f=f(x)$ are given functions of x (i.e., data), and w is the dependent variable. The sign convention used in the derivation of (4.1) is shown in Fig. 4.1. The function $b=EI$ is the product of the modulus of elasticity E and the moment of inertia I of the beam, f is the transversely distributed load, and w is the transverse deflection of the beam. In addition to satisfying the differential equation (4.1), w must also satisfy appropriate boundary conditions; since the equation is of fourth order, four boundary conditions are needed to solve it. The weak formulation of the equation will provide the form of these four boundary conditions. A step-by-step procedure for the finite-element analysis of (4.1) is presented next.

4.2.2 Discretization of the Domain

The domain of the structure (i.e., length of the beam) is divided into a set (say, N) of line elements, each element having at least the two end nodes (see Fig. 4.2*a*). Although the element is geometrically the same as that used for bars, the number and form of the primary and secondary unknowns at each node are dictated by the variational formulation of the differential equation (4.1). In most practical problems, the discretization of a given structure into a minimum number of elements is often dictated by the geometry, loading, and material properties.

4.2.3 Derivation of Element Equations

In this step we isolate a typical element $\Omega^e=(x_e, x_{e+1})$ (see Fig. 4.2*b*) and construct the weak form of (4.1) over the element. The variational formulation

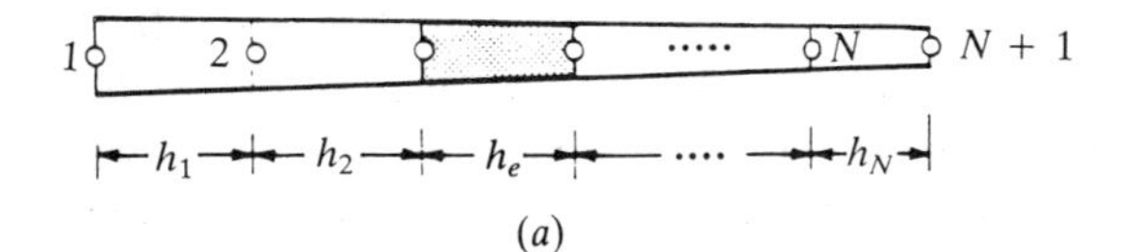

(a)

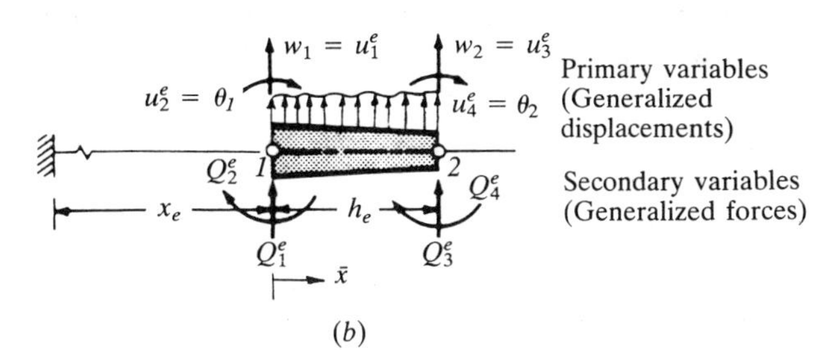

(b)

FIGURE 4.2
Discretization of a beam using Euler–Bernoulli beam elements. The generalized displacements and generalized forces are shown on a typical beam element.

provides the primary and secondary variables of the problem. Then suitable approximations for the primary variables are selected, interpolation functions are developed, and the element equations are derived.

WEAK FORM. The weak forms of problems in solid mechanics can be developed either from the principle of virtual work (i.e., the principle of virtual displacements or virtual forces) or from the governing differential equations. Here we start with a given differential equation and using the three-step procedure to obtain the weak form. Following the three-step procedure developed in Chapter 2 and revisited in Section 3.2.3, we write

$$0 = \int_{x_e}^{x_{e+1}} v\left[\frac{d^2}{dx^2}\left(b\frac{d^2w}{dx^2}\right) - f\right] dx$$

$$= \int_{x_e}^{x_{e+1}} \left[-\frac{dv}{dx}\frac{d}{dx}\left(b\frac{d^2w}{dx^2}\right) - vf\right] dx + \left[v\frac{d}{dx}\left(b\frac{d^2w}{dx^2}\right)\right]_{x_e}^{x_{e+1}}$$

$$= \int_{x_e}^{x_{e+1}} \left(b\frac{d^2v}{dx^2}\frac{d^2w}{dx^2} - vf\right) dx + \left[v\frac{d}{dx}\left(b\frac{d^2w}{dx^2}\right) - \frac{dv}{dx}b\frac{d^2w}{dx^2}\right]_{x_e}^{x_{e+1}} \tag{4.2}$$

where $v(x)$ is a weight function that is twice differentiable with respect to x. Note that, in the present case, the first term of the equation is integrated twice by parts to trade two differentiations to the weight function v, while retaining two derivatives of the dependent variable w; i.e., the differentiation is distributed equally between the weight function v and the dependent variable w. Because of the two integrations by parts, there appear two boundary expressions (see Example 2.2), which are to be evaluated at the two boundary points $x = x_e$ and $x = x_{e+1}$. Examination of the boundary terms indicates that the essential boundary conditions involve the specification of the deflection w and slope dw/dx, and the natural boundary conditions involve the specification of the bending moment $b\,d^2w/dx^2$ and shear force $(d/dx)(b\,d^2w/dx^2)$ at the endpoints of the element. Thus, there are two essential boundary conditions and two natural boundary conditions; therefore, we must identify w and dw/dx as the primary varaibles at each node (so that the essential boundary conditions are included in the interpolation). The natural boundary conditions always remain in the weak form and end up on the right-hand side (i.e., the source vector) of the matrix equation. For the sake of mathematical convenience, we introduce the following notation: $\theta = -dw/dx$ and

$$\boxed{\begin{aligned} Q_1^e &\equiv \left[\frac{d}{dx}\left(b\frac{d^2w}{dx^2}\right)\right]\bigg|_{x_e}, & Q_2^e &\equiv \left(b\frac{d^2w}{dx^2}\right)\bigg|_{x_e} \\ Q_3^e &\equiv -\left[\frac{d}{dx}\left(b\frac{d^2w}{dx^2}\right)\right]\bigg|_{x_{e+1}}, & Q_4^e &\equiv -\left(b\frac{d^2w}{dx^2}\right)\bigg|_{x_{e+1}} \end{aligned}} \tag{4.3}$$

where Q_1^e and Q_3^e denote the shear forces, and Q_2^e and Q_4^e denote the bending moments (see Fig. 4.2*b*). Since the quantities Q_i^e contain bending moments, which can also be viewed as "bending forces," the set $\{Q_1^e, Q_2^e, Q_3^e, Q_4^e\}$ is often referred to as the *generalized forces*. The corresponding displacements and rotations are called the *generalized displacements*.

With the notation in (4.3), the weak form (4.2) can be expressed as

$$\begin{aligned} 0 &= \int_{x_e}^{x_{e+1}} \left(b\frac{d^2v}{dx^2}\frac{d^2w}{dx^2} - vf\right) dx - v(x_e)Q_1^e - \left(-\frac{dv}{dx}\right)\bigg|_{x_e} Q_2^e \\ &\quad - v(x_{e+1})Q_3^e - \left(-\frac{dv}{dx}\right)\bigg|_{x_{e+1}} Q_4^e \\ &\equiv B(v, w) - l(v) \end{aligned} \tag{4.4a}$$

We can identify the bilinear and linear forms of the problem as

$$\begin{aligned} B(v, w) &= \int_{x_e}^{x_{e+1}} b\frac{d^2v}{dx^2}\frac{d^2w}{dx^2}\,dx \\ l(v) &= \int_{x_e}^{x_{e+1}} vf\,dx + v(x_e)Q_1^e + \left(-\frac{dv}{dx}\right)\bigg|_{x_e} Q_2^e \\ &\quad + v(x_{e+1})Q_3^e + \left(-\frac{dv}{dx}\right)\bigg|_{x_{e+1}} Q_4^e \end{aligned} \tag{4.4b}$$

Equation (4.4*a*) is a statement of the principle of virtual displacements for the Euler–Bernoulli beam theory. The quadratic functional, known as the *total potential energy* of the beam element, is given by [from (2.43*b*)]

$$\begin{aligned} I_e(w) &= \int_{x_e}^{x_{e+1}} \left[\frac{b}{2}\left(\frac{d^2w}{dx^2}\right)^2 - wf\right] dx - w(x_e)Q_1^e - \left(-\frac{dw}{dx}\right)\bigg|_{x_e} Q_2^e \\ &\quad - w(x_{e+1})Q_3^e - \left(-\frac{dw}{dx}\right)\bigg|_{x_{e+1}} Q_4^e \end{aligned} \tag{4.5}$$

The first term in the square brackets represents the elastic strain energy due to bending, while the second is the work done by the distributed load; the remaining terms account for the work done by the generalized forces Q_i^e in moving through the generalized displacements of the element.

INTERPOLATION FUNCTIONS. The variational form (4.4a) requires that the interpolation functions of an element be continuous with nonzero derivatives up to order two. The approximation of the primary variables over a finite element should be such that it satisfies the interpolation properties (i.e., that it satisfies the essential boundary conditions of the element):

$$w(x_e) = w_1, \qquad w(x_{e+1}) = w_2; \qquad \theta(x_e) = \theta_1, \qquad \theta(x_{e+1}) = \theta_2 \tag{4.6}$$

In satisfying the essential boundary conditions (4.6), the approximation automatically satisfies the continuity conditions. Hence, we pay attention to the satisfaction of (4.6), which form the basis for the interpolation procedure.

Since there is a total of four conditions in an element (two per node), a four-parameter polynomial must be selected for w:

$$w(x) = c_1 + c_2 x + c_3 x^2 + c_4 x^3 \tag{4.7}$$

Note that the continuity conditions (i.e., the existence of a nonzero second derivative of w in the element) are automatically met. The next step involves expressing c_i in terms of the primary nodal variables (i.e., generalized displacements)

$$u_1^e \equiv w(x_e), \qquad u_2^e \equiv \left(-\frac{dw}{dx}\right)\bigg|_{x=x_e}, \qquad u_3^e \equiv w(x_{e+1}), \qquad u_4^e \equiv \left(-\frac{dw}{dx}\right)\bigg|_{x=x_{e+1}}$$

such that the conditions (4.6) are satisfied:

$$\begin{aligned}
u_1^e &= w(x_e) = c_1 + c_2 x_e + c_3 x_e^2 + c_4 x_e^3 \\
u_2^e &= \left(-\frac{dw}{dx}\right)\bigg|_{x=x_e} = -c_2 - 2c_3 x_e - 3c_4 x_e^2 \\
u_3^e &= w(x_{e+1}) = c_1 + c_2 x_{e+1} + c_3 x_{e+1}^2 + c_4 x_{e+1}^3 \\
u_4^e &= \left(-\frac{dw}{dx}\right)\bigg|_{x_{e+1}} = -c_2 - 2c_3 x_{e+1} - 3c_4 x_{e+1}^2
\end{aligned} \tag{4.8a}$$

or

$$\begin{Bmatrix} u_1^e \\ u_2^e \\ u_3^e \\ u_4^e \end{Bmatrix} = \begin{bmatrix} 1 & x_e & x_e^2 & x_e^3 \\ 0 & -1 & -2x_e & -3x_e^2 \\ 1 & x_{e+1} & x_{e+1}^2 & x_{e+1}^3 \\ 0 & -1 & -2x_{e+1} & -3x_{e+1}^2 \end{bmatrix} \begin{Bmatrix} c_1 \\ c_2 \\ c_3 \\ c_4 \end{Bmatrix} \tag{4.8b}$$

Inverting this matrix equation to express c_i in terms of u_1^e, u_2^e, u_3^e, and u_4^e, and substituting the result into (4.7), we obtain

$$w^e(x) = u_1^e\phi_1^e + u_2^e\phi_2^e + u_3^e\phi_3^e + u_4^e\phi_4^e = \sum_{j=1}^{4} u_j^e\phi_j^e \tag{4.9a}$$

where (with $x_{e+1} = x_e + h_e$)

$$\begin{aligned}
\phi_1^e &= 1 - 3\left(\frac{x - x_e}{h_e}\right)^2 + 2\left(\frac{x - x_e}{h_e}\right)^3, \qquad \phi_2^e = -(x - x_e)\left(1 - \frac{x - x_e}{h_e}\right)^2 \\
\phi_3^e &= 3\left(\frac{x - x_e}{h_e}\right)^2 - 2\left(\frac{x - x_e}{h_e}\right)^3, \qquad \phi_4^e = -(x - x_e)\left[\left(\frac{x - x_e}{h_e}\right)^2 - \frac{x - x_e}{h_e}\right]
\end{aligned} \tag{4.9b}$$

Note that the cubic interpolation functions in (4.9) are derived by interpolating w *and* its derivative at the nodes. Such polynomials are known as the *Hermite family of interpolation functions*, and ϕ_i^e in (4.9*b*) are called the *Hermite cubic interpolation* (or *cubic spline*) *functions*. Recall that the Lagrange cubic interpolation functions are derived to interpolate a function, but not its derivatives, at the nodes. Hence, a Lagrange cubic element will have four nodes, with the dependent variable, not its derivative, as the nodal degree of freedom. Since the slope (or derivative) of the dependent variable is also required by the weak form to be continuous at the nodes for Euler–Bernoulli beam theory, the Lagrange cubic interpolation of w, although it meets the continuity requirement for w, is *not admissible* in the finite element approximation of Euler–Bernoulli beam theory.

The interpolation functions ϕ_i^e can be expressed in terms of the local coordinate $\bar{x}$:

$$\begin{aligned}
\phi_1^e &= 1 - 3\left(\frac{\bar{x}}{h_e}\right)^2 + 2\left(\frac{\bar{x}}{h_e}\right)^3, \qquad \phi_2^e = -\bar{x}\left(1 - \frac{\bar{x}}{h_e}\right)^2 \\
\phi_3^e &= 3\left(\frac{\bar{x}}{h_e}\right)^2 - 2\left(\frac{\bar{x}}{h_e}\right)^3, \qquad \phi_4^e = -\bar{x}\left[\left(\frac{\bar{x}}{h_e}\right)^2 - \frac{\bar{x}}{h_e}\right]
\end{aligned} \tag{4.10a}$$

The first, second, and third derivatives of ϕ_i^e with respect to $\bar{x}$ are

$$\begin{aligned}
\frac{d\phi_1^e}{d\bar{x}} &= -\frac{6}{h_e}\frac{\bar{x}}{h_e}\left(1 - \frac{\bar{x}}{h_e}\right), \qquad \frac{d\phi_2^e}{d\bar{x}} = -\left[1 + 3\left(\frac{\bar{x}}{h_e}\right)^2 - 4\frac{\bar{x}}{h_e}\right] \\
\frac{d\phi_3^e}{d\bar{x}} &= -\frac{d\phi_1^e}{d\bar{x}}, \qquad \frac{d\phi_4^e}{d\bar{x}} = -\frac{\bar{x}}{h_e}\left(3\frac{\bar{x}}{h_e} - 2\right) \\
\frac{d^2\phi_1^e}{d\bar{x}^2} &= -\frac{6}{h_e^2}\left(1 - 2\frac{\bar{x}}{h_e}\right), \qquad \frac{d^2\phi_2^e}{d\bar{x}^2} = -\frac{2}{h_e}\left(3\frac{\bar{x}}{h_e} - 2\right) \\
\frac{d^2\phi_3^e}{d\bar{x}^2} &= -\frac{d^2\phi_1^e}{d\bar{x}^2}, \qquad \frac{d^2\phi_4^e}{d\bar{x}^2} = -\frac{2}{h_e}\left(3\frac{\bar{x}}{h_e} - 1\right) \\
\frac{d^3\phi_1^e}{d\bar{x}^3} &= \frac{12}{h_e^3}, \quad \frac{d^3\phi_2^e}{d\bar{x}^3} = -\frac{6}{h_e^2}, \quad \frac{d^3\phi_3^e}{d\bar{x}^3} = -\frac{12}{h_e^3}, \quad \frac{d^3\phi_4^e}{d\bar{x}^3} = -\frac{6}{h_e^2}
\end{aligned} \tag{4.10b}$$

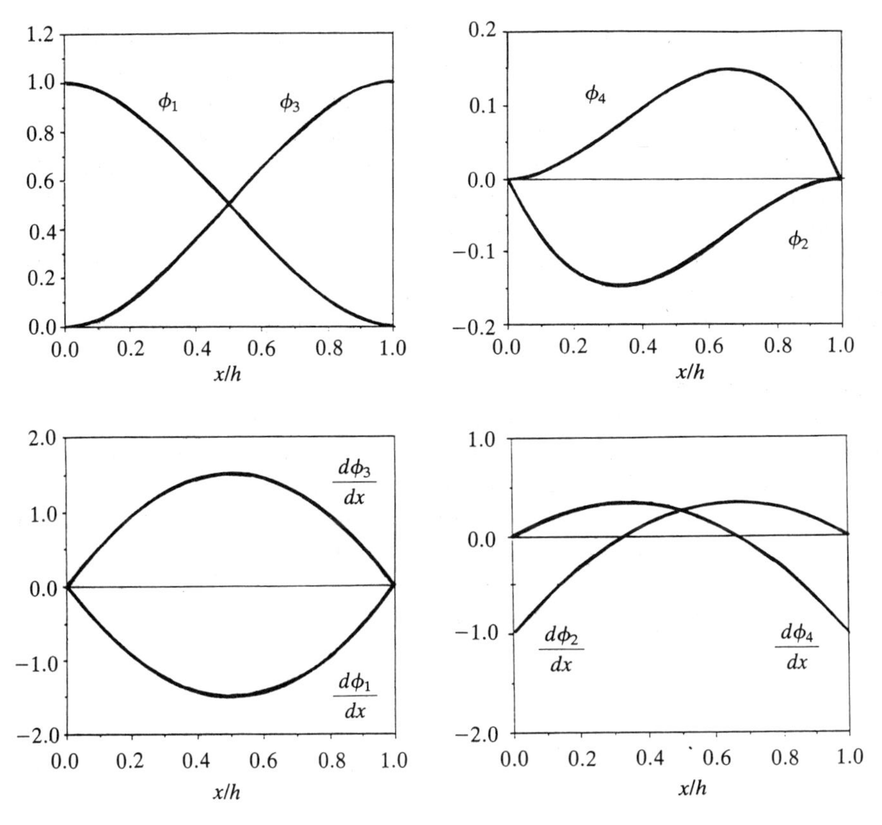

FIGURE 4.3
Hermite cubic interpolation functions and their first derivatives used in the finite element model of Euler–Bernoulli beam theory.

The Hermite cubic interpolation functions (4.9) satisfy the following interpolation properties (see Fig. 4.3):

$$
\begin{aligned}
&\phi_1^e(x_e) = 1, && \phi_i^e(x_e) = 0 \quad (i \neq 1) \\
&\phi_3^e(x_{e+1}) = 1, && \phi_i^e(x_{e+1}) = 0 \quad (i \neq 3) \\
&\left(-\frac{d\phi_2^e}{dx}\right)\bigg|_{x_e} = 1, && \left(\frac{d\phi_i^e}{dx}\right)\bigg|_{x_e} = 0 \quad (i \neq 2) \\
&\left(-\frac{d\phi_4^e}{dx}\right)\bigg|_{x_{e+1}} = 1, && \left(\frac{d\phi_i^e}{dx}\right)\bigg|_{x_{e+1}} = 0 \quad (i \neq 4)
\end{aligned}
\tag{4.11a}
$$

These can be stated in compact form as $(i, j = 1, 2)$

$$
\boxed{
\begin{aligned}
&\phi_{2i-1}^e(\bar{x}_j) = \delta_{ij}, \qquad \phi_{2i}^e(\bar{x}_j) = 0, \qquad \sum_{i=1}^{2} \phi_{2i-1}^e = 1 \\
&\left(\frac{d\phi_{2i-1}^e}{dx}\right)\bigg|_{\bar{x}_j} = 0, \qquad \left(-\frac{d\phi_{2i}^e}{dx}\right)\bigg|_{\bar{x}_j} = \delta_{ij}
\end{aligned}
}
\tag{4.11b}
$$

where $\bar{x}_1 = 0$ and $\bar{x}_2 = h_e$ are the local coordinates of nodes 1 and 2 of the element $\Omega^e = (x_e, x_{e+1})$.

It should be noted that the order of the interpolation functions derived above is the minimum required for the variational formulation (4.4). If a higher-order (i.e., higher than cubic) approximation of w is desired, one must either identify additional primary unknowns at each of the two nodes or add additional nodes with the two degrees of freedom $(w, -dw/dx)$. For example, if we add d^2w/dx^2 as the primary unknown at each of the two nodes, or add a third node, there will be a total of six conditions, and a fifth-order polynomial is required to interpolate the end conditions (see Problems 4.27 and 4.28). However, continuity of dw^2/dx^2 is not required, in general, and such elements should be used only in problems where d^2w/dx^2 is continuous everywhere.

FINITE ELEMENT MODEL. The finite element model of the Euler–Bernoulli beam is obtained by substituting the finite element interpolation (4.9a) for w and the ϕ_j for the weight function v into the weak form (4.4a). Since there are four nodal variables u_i^e, four different choices are used for v, $v = \phi_1^e$, $v = \phi_2^e$, $v = \phi_3^e$, and $v = \phi_4^e$, to obtain a set of four algebraic equations. The ith algebraic equation of the finite element model is (for $v = \phi_i^e$)

$$0 = \sum_{j=1}^{4} \left(\int_{x_e}^{x_{e+1}} b \frac{d^2\phi_i^e}{dx^2} \frac{d^2\phi_j^e}{dx^2} \, dx \right) u_j^e - \int_{x_e}^{x_{e+1}} \phi_i^e f \, dx - Q_i^e \tag{4.12a}$$

or

$$\boxed{\sum_{j=1}^{4} K_{ij}^e u_j^e - F_i^e = 0} \tag{4.12b}$$

where

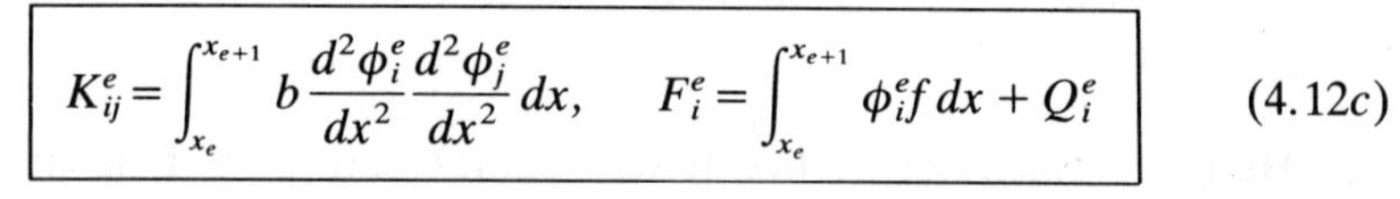

$$\boxed{K_{ij}^e = \int_{x_e}^{x_{e+1}} b \frac{d^2\phi_i^e}{dx^2} \frac{d^2\phi_j^e}{dx^2} \, dx, \qquad F_i^e = \int_{x_e}^{x_{e+1}} \phi_i^e f \, dx + Q_i^e} \tag{4.12c}$$

Note that the coefficients K_{ij}^e are symmetric: $K_{ij}^e = K_{ji}^e$. In matrix notation, (4.12b) can be written as

$$\begin{bmatrix} K_{11}^e & K_{12}^e & K_{13}^e & K_{14}^e \\ K_{21}^e & K_{22}^e & K_{23}^e & K_{24}^e \\ K_{31}^e & K_{32}^e & K_{33}^e & K_{34}^e \\ K_{41}^e & K_{42}^e & K_{43}^e & K_{44}^e \end{bmatrix} \begin{Bmatrix} u_1^e \\ u_2^e \\ u_3^e \\ u_4^e \end{Bmatrix} = \begin{Bmatrix} f_1^e \\ f_2^e \\ f_3^e \\ f_4^e \end{Bmatrix} + \begin{Bmatrix} Q_1^e \\ Q_2^e \\ Q_3^e \\ Q_4^e \end{Bmatrix} \tag{4.13}$$

This represents the finite element model of (4.1). Here $[K^e]$ is the *stiffness matrix* and $\{F^e\}$ is the *force vector* of a beam element. When a transverse point force F_0^e is applied at a point x_0 inside the element, it is distributed to the element nodes by the relation [see Remark 5 in Chapter 3: (3.60)]:

$$F_i^e = F_0^e \phi_i^e(x_0) \tag{4.14}$$

which contains both transverse forces (F_1^e and F_3^e) and bending moments (F_2^e and F_4^e).

For the case in which b $(= EI)$ and f are constant over an element, the element stiffness matrix $[K^e]$ and force vector $\{F^e\}$ have the specific forms (see Fig. 4.2 for the element displacement and force degrees of freedom)

$$[K^e] = \frac{2b}{h^3}\begin{bmatrix} 6 & -3h & -6 & -3h \\ -3h & 2h^2 & 3h & h^2 \\ -6 & 3h & 6 & 3h \\ -3h & h^2 & 3h & 2h^2 \end{bmatrix} \quad (b = EI = \text{constant})$$

$$\{F^e\} = \frac{fh}{12}\begin{Bmatrix} 6 \\ -h \\ 6 \\ h \end{Bmatrix} + \begin{Bmatrix} Q_1 \\ Q_2 \\ Q_3 \\ Q_4 \end{Bmatrix} \quad (f = \text{constant}) \tag{4.15}$$

It can be verified that the generalized force vector in (4.15) represents the "statically equivalent" forces and moments at nodes 1 and 2 due to the uniformly distributed load over the element. For any given function f, (4.12c) provides a straightforward way of computing the components of the generalized force vector $\{f^e\}$.

4.2.4 Assembly of Element Equations

The assembly procedure for beam elements is the same as that used for bar elements, except that we must take into account the two degrees of freedom at each node. Recall that the assembly of elements is based on: (a) interelement continuity of the primary variables (deflection and slope) and (b) interelement equilibrium of the secondary variables (shear force and bending moment) at the nodes common to elements. To demonstrate the assembly procedure, we select a two-element model (see Fig. 4.4). There are three global nodes and a total of six global generalized displacements and six generalized forces in the problem. The continuity of the primary variables implies the following relation between the element degrees of freedom u_i^e and the global degrees of freedom U_i (see Fig. 4.4):

$$\begin{aligned} &u_1^1 = U_1, \quad u_2^1 = U_2, \quad u_3^1 = u_1^2 = U_3 \\ &u_4^1 = u_2^2 = U_4, \quad u_3^2 = U_5, \quad u_4^2 = U_6 \end{aligned} \tag{4.16}$$

In general, the equilibrium of the generalized forces at a node between two connecting elements Ω^e and Ω^f requires that

$$\begin{aligned} Q_3^e + Q_1^f &= \text{applied external point force} \\ Q_4^e + Q_2^f &= \text{applied external bending moment} \end{aligned} \tag{4.17}$$

If no external applied forces are given, the sum should be equated to zero. In equating the sums to the applied generalized forces (i.e., force or moment), the sign convention for the element force degrees of freedom (see Fig. 4.2)

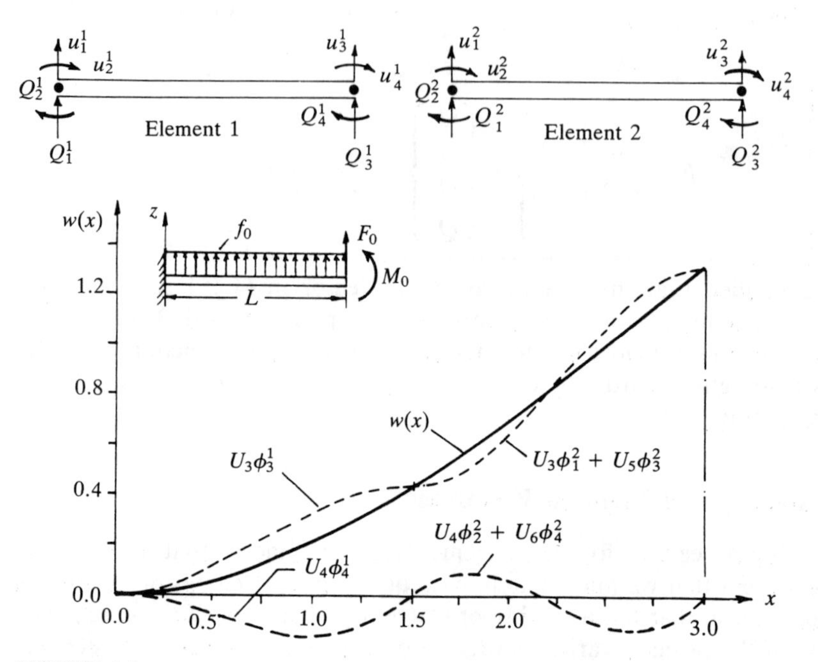

FIGURE 4.4
Assembly of two Euler–Bernoulli (or classical) beam finite elements, and the finite element solution as a linear combination of the nodal values and interpolation functions.

should be followed. Forces are taken positive acting upward and moments are taken positive acting clockwise.

To impose the equilibrium of forces in (4.17), it is necessary to add the third and fourth equations (corresponding to the second node) of Ω^e to the first and second equations (corresponding to the first node) of Ω^f. Consequently, the global stiffnesses K_{33}, K_{34}, K_{43}, and K_{44} associated with global node 2 are the superpositions of the element stiffnesses:

$$K_{33} = K_{33}^1 + K_{11}^2, \qquad K_{34} = K_{34}^1 + K_{12}^2, \qquad K_{43} = K_{43}^1 + K_{21}^2, \qquad K_{44} = K_{44}^1 + K_{22}^2 \tag{4.18}$$

In general, the assembled stiffness matrix and force vector for beam elements connected in series have the following forms:

$$
[K] = \begin{array}{c}
\text{Global nodes} \\
\end{array}
\left[\begin{array}{cc:cc:cc}
\multicolumn{2}{c}{\overbrace{\hspace{3em}}^{1}} & \multicolumn{2}{c}{\overbrace{\hspace{8em}}^{2}} & \multicolumn{2}{c}{\overbrace{\hspace{3em}}^{3}} \\
K_{11}^1 & K_{12}^1 & K_{13}^1 & K_{14}^1 & & \\
K_{21}^1 & K_{22}^1 & K_{23}^1 & K_{24}^1 & & \\
\hdashline
K_{31}^1 & K_{32}^1 & K_{33}^1 + K_{11}^2 & K_{34}^1 + K_{12}^2 & K_{13}^2 & K_{14}^2 \\
K_{41}^1 & K_{42}^1 & K_{43}^1 + K_{21}^2 & K_{44}^1 + K_{22}^2 & K_{23}^2 & K_{24}^2 \\
\hdashline
 & & K_{31}^2 & K_{32}^2 & K_{33}^2 & K_{34}^2 \\
 & & K_{41}^2 & K_{42}^2 & K_{43}^2 & K_{44}^2
\end{array}\right]
\begin{array}{l}
\\
\left.\begin{array}{c} \\ \\ \end{array}\right\} 1 \\
\left.\begin{array}{c} \\ \\ \end{array}\right\} 2 \\
\left.\begin{array}{c} \\ \\ \end{array}\right\} 3
\end{array}
\tag{4.19a}
$$

$$
\{F\} = \begin{Bmatrix} F_1^1 \\ F_2^1 \\ F_3^1 + F_1^2 \\ F_4^1 + F_2^2 \\ F_3^2 \\ F_4^2 \end{Bmatrix} \tag{4.19b}
$$

The connectivity matrix $[B]$ (which will be used in computer implementation) for the two-element mesh is

$$
[B] = \begin{bmatrix} 1 & 2 \\ 2 & 3 \end{bmatrix}
$$

Since there are two primary degrees of freedom per node, repetition of a number in $[B]$ indicates that the coefficients associated with both degrees of freedom will add up. For example, the repetition of the global node number 2 (which corresponds to global degrees of freedom 3 and 4) in rows 1 and 2 indicates that the global K_{33}, K_{34}, K_{43}, and K_{44} have contributions from both elements 1 and 2.

The assembled system of equations for a mesh of two elements with $h_1 = h_2 = \frac{1}{2}L = h$, and constant EI and f [hence, $[K^e]$ and $\{f^e\}$ are given by

(4.15)] is

$$\frac{2EI}{h^3}\begin{bmatrix} 6 & -3h & -6 & -3h & 0 & 0 \\ -3h & 2h^2 & 3h & h^2 & 0 & 0 \\ -6 & 3h & 6+6 & 3h-3h & -6 & -3h \\ -3h & h^2 & 3h-3h & 2h^2+2h^2 & 3h & h^2 \\ 0 & 0 & -6 & 3h & 6 & 3h \\ 0 & 0 & -3h & h^2 & 3h & 2h^2 \end{bmatrix}\begin{Bmatrix} U_1 \\ U_2 \\ U_3 \\ U_4 \\ U_5 \\ U_6 \end{Bmatrix}$$

$$= \frac{fh}{12}\begin{Bmatrix} 6 \\ -h \\ 6+6 \\ h-h \\ 6 \\ h \end{Bmatrix} + \begin{Bmatrix} Q_1^1 \\ Q_2^1 \\ Q_3^1+Q_1^2 \\ Q_4^1+Q_2^2 \\ Q_3^2 \\ Q_4^2 \end{Bmatrix} \tag{4.20}$$

The reader is cautioned that (4.20) does not represent the assembled equations of any two beam elements; it is based on the assumption that $h_1 = h_2$, $(EI)_1 = (EI)_2$, and $(f)_1 = (f)_2$. Equations (4.19) are more general.

4.2.5 Imposition of Boundary Conditions

At this step of the analysis, we must specify the particular boundary conditions, i.e., geometric constraints and forces applied, of the particular problem to be analyzed. The type of essential (also known as *geometric*) boundary conditions for a specific beam problem depend on the nature of the geometric support. Table 4.1 contains a list of commonly used geometric supports for beams. For the sake of completeness, the boundary conditions on the axial displacement u are also included. The natural (also called *force*) boundary conditions involve the specification of generalized forces when the corresponding primary variables are not constrained. Here we consider a cantilever beam (i.e., a beam fixed at one end and free at the other) of length L, flexural rigidity EI (= constant), and subjected to a uniformly distributed force f_0, end force F_0, and end moment M_0 (see Fig. 4.4).

First, we write the equilibrium conditions for the generalized forces. At global node 1, Q_1^1 and Q_2^1 (the shear force and the bending moment, respectively; i.e., the reactions at the fixed end) are not known. At global node 2, there are no externally applied shear forces and bending moment. Hence,

$$Q_3^1 + Q_1^2 = 0, \qquad Q_4^1 + Q_2^2 = 0 \tag{4.21a}$$

At global node 3, the shear force is given as F_0, and the bending moment as M_0 (note the sign convention for F_0 and M_0 from Fig. 4.2):

$$Q_3^2 \equiv -\left[\frac{d}{dx}\left(EI\frac{d^2w}{dx^2}\right)\right]\bigg|_{x=L} = F_0, \qquad Q_4^2 \equiv -\left(b\frac{d^2w}{dx^2}\right)\bigg|_{x=L} = -M_0 \tag{4.21b}$$

TABLE 4.1
Types of commonly used support conditions for beams and frames

Type of support	Displacement boundary conditions	Force boundary conditions
FREE	None	All, as specified
PINNED	$u=0$ $w=0$	Moment is specified
ROLLER (vertical)	$u=0$	Transverse force and moment are specified
ROLLER (horizontal)	$w=0$	Horizontal force and bending moment are specified
FIXED or CLAMPED	$u=0$ $w=0$ $dw/dx=0$	None specified

Next, we identify and impose the specified generalized displacements. Since the beam is clamped at global node 1, it follows that the deflection w and the slope dw/dx are zero there:

$$u_1^1 \equiv w_1^1 = U_1 = 0, \qquad u_2^1 \equiv \theta_1^1 = U_2 = 0 \tag{4.22}$$

Using (4.21) and (4.22) in (4.20), we obtain

$$\frac{2EI}{h^3}\begin{bmatrix} 6 & -3h & -6 & -3h & 0 & 0 \\ -3h & 2h^2 & 3h & h^2 & 0 & 0 \\ -6 & 3h & 12 & 0 & -6 & -3h \\ -3h & h^2 & 0 & 4h^2 & 3h & h^2 \\ 0 & 0 & -6 & 3h & 6 & 3h \\ 0 & 0 & -3h & h^2 & 3h & 2h^2 \end{bmatrix}\begin{Bmatrix} U_1=0 \\ U_2=0 \\ U_3 \\ U_4 \\ U_5 \\ U_6 \end{Bmatrix} = \frac{f_0 h}{12}\begin{Bmatrix} 6 \\ -h \\ 12 \\ 0 \\ 6 \\ h \end{Bmatrix} + \begin{Bmatrix} Q_1^1 \\ Q_2^1 \\ 0 \\ 0 \\ F_0 \\ -M_0 \end{Bmatrix} \tag{4.23}$$

4.2.6 Solution

Equation (4.23) contains six algebraic equations in six unknowns: $(Q_1^1, Q_2^1, U_3, U_4, U_5, U_6)$. Because the algebraic equations for the unknown generalized displacements (U_3, U_4, U_5, U_6) do not contain the unknown generalized forces (Q_1^1, Q_2^1), equations 3, 4, 5, and 6 of the system (4.23) can be solved independently; the known values of the displacements U_1 and U_2 are used in equations 3 and 4. This provides us with the motivation to partition (shown by dashed lines) the matrix equation (4.23), which can be recast in the form

$$\begin{bmatrix} [K^{11}] & [K^{12}] \\ [K^{21}] & [K^{22}] \end{bmatrix} \begin{Bmatrix} \{U^1\} \\ \{U^2\} \end{Bmatrix} = \begin{Bmatrix} \{F^1\} \\ \{F^2\} \end{Bmatrix} \tag{4.24}$$

where $\{U^1\}$ denotes the column of known generalized displacements, $\{U^2\}$ the column of unknown generalized displacements, $\{F^1\}$ the column of unknown forces, and $\{F^2\}$ the column of known forces. Equation (4.24) can be written, after carrying out the matrix multiplication, in the form [cf. (3.56)]

$$[K^{11}]\{U^1\} + [K^{12}]\{U^2\} = \{F^1\}$$

$$[K^{21}]\{U^1\} + [K^{22}]\{U^2\} = \{F^2\}$$

or

$$[K^{22}]\{U^2\} = \{F^2\} - [K^{21}]\{U^1\} \tag{4.25a}$$

$$\{F^1\} = [K^{11}]\{U^1\} + [K^{12}]\{U^2\} \tag{4.25b}$$

These are the condensed equations for the generalized displacements and forces, respectively. Since $\{U^1\}$ and $\{F^2\}$ are known $(\{U^1\} = \{0\})$, we can use (4.25a) to solve for $\{U^2\}$, and then use (4.25b) to compute the unknown reactions $\{F^1\}$:

$$\{U^2\} \equiv \begin{Bmatrix} U_3 \\ U_4 \\ U_5 \\ U_6 \end{Bmatrix} = \frac{h^3}{2EI} \begin{bmatrix} 12 & 0 & -6 & -3h \\ 0 & 4h^2 & 3h & h^2 \\ -6 & 3h & 6 & 3h \\ -3h & h^2 & 3h & 2h^2 \end{bmatrix}^{-1} \begin{Bmatrix} f_0 h \\ 0 \\ F_0 + \frac{1}{2} f_0 h \\ -M_0 + \frac{1}{12} f_0 h^2 \end{Bmatrix} \tag{4.26}$$

Inverting the matrix (say, by Gaussian elimination) and performing the matrix multiplication (i.e., solving the equations), we obtain

$$\{U^2\} = \frac{h}{6EI}\begin{bmatrix} 2h^2 & -3h & 5h^2 & -3h \\ -3h & 6 & -9h & 6 \\ 5h^2 & -9h & 16h^2 & -12h \\ -3h & 6 & -12h & 12 \end{bmatrix}\begin{Bmatrix} f_0 h \\ 0 \\ F_0 + \frac{1}{2}f_0 h \\ -M_0 + \frac{1}{12}f_0 h^2 \end{Bmatrix}$$

$$= \frac{h}{6EI}\begin{Bmatrix} 5h^2F_0 + 3hM_0 + \frac{17}{4}f_0h^3 \\ -9hF_0 - 6M_0 - 7f_0h^2 \\ 16h^2F_0 + 12hM_0 + 12f_0h^3 \\ -12hF_0 - 12M_0 - 8f_0h^2 \end{Bmatrix} \tag{4.27}$$

The reactions Q_1^1 and Q_2^1 can be obtained by substituting (4.27) into (4.25*b*). The Q_i^e obtained from the element (equilibrium) equations are more accurate than those obtained from the definitions (4.3), wherein the derivatives of w are obtained by differentiation of its finite element interpolation. The reactions from equilibrium $(Q_i^e)_{\text{equil}}$ are

$$\{F^1\} = \begin{Bmatrix} Q_1^1 \\ Q_2^1 \end{Bmatrix} = \frac{2EI}{h^3}\begin{bmatrix} -6 & -3h & 0 & 0 \\ 3h & h^2 & 0 & 0 \end{bmatrix}\begin{Bmatrix} U_3 \\ U_4 \\ U_5 \\ U_6 \end{Bmatrix} - \frac{f_0 h}{12}\begin{Bmatrix} 6 \\ -h \end{Bmatrix}$$

$$= \begin{Bmatrix} -(F_0 + 2f_0h) \\ 2h(F_0 + f_0h) + M_0 \end{Bmatrix} \tag{4.28}$$

It can be verified that the reactions Q_1^1 and Q_2^1 in (4.28) satisfy the static equilibrium equations of the beam:

$$Q_1^1 + (F_0 + 2f_0h) = 0, \qquad Q_2^1 - (2F_0h + 2f_0h^2 + M_0) = 0$$

The reactions Q_1^1 and Q_2^1 can also be computed using the definitions (4.3):

$$Q_1^1 \equiv \frac{d}{dx}\left(EI\frac{d^2w}{dx^2}\right)\bigg|_{x=0}, \qquad Q_2^1 \equiv \left(EI\frac{d^2w}{dx^2}\right)\bigg|_{x=0} \tag{4.29a}$$

From (4.10*b*), we note that the second derivative of the Hermite cubic interpolation functions is linear over the element and the third derivative is constant over the element. Therefore, the reactions, i.e., bending moment and shear force, computed using the definition (4.3) are element-wise linear and constant, respectively. Further, at nodes connecting two elements, they yield discontinuous values because the second and third derivatives of w are not made continuous across the interelement nodes. Substituting (4.10b) into (4.29a), we obtain the values

$$Q_1^1 = EI\left(U_3\frac{d^3\phi_3^1}{dx^3} + U_4\frac{d^3\phi_4^1}{dx^3}\right)\bigg|_{x=0} = EI\left[U_3\left(-\frac{12}{h_e^3}\right) + U_4\left(-\frac{6}{h_e^2}\right)\right] = -(F_0 + \tfrac{3}{2}f_0h) \tag{4.29b}$$

$$Q_2^1 = EI\left(U_3\frac{6}{h_e^2} + U_4\frac{2}{h_e}\right) = (M_0 + 2F_0h + \tfrac{23}{12}f_0h^2)$$

which are in error by $f_1^e = \frac{1}{2}f_0h$ and $f_2^e = \frac{1}{12}f_0h^2$ compared with those in (4.28).

4.2.7 Postprocessing of the Solution

The finite element solution as a function of position x is given by

$$w^e(x) = \begin{cases} U_3\phi_3^1 + U_4\phi_4^1 & \text{for } 0 \leqslant x \leqslant h \\ U_3\phi_1^2 + U_4\phi_2^2 + U_5\phi_3^2 + U_6\phi_4^2 & \text{for } h \leqslant x \leqslant 2h \end{cases} \tag{4.30a}$$

where [see (4.10a)]

$$\begin{aligned} \phi_3^1 &= 3\left(\frac{x}{h}\right)^2 - 2\left(\frac{x}{h}\right)^3, \quad \phi_4^1 = h\left[\left(\frac{x}{h}\right)^2 - \left(\frac{x}{h}\right)^3\right] \\ \phi_1^2 &= 1 - 3\left(1-\frac{x}{h}\right)^2 - 2\left(1-\frac{x}{h}\right)^3, \quad \phi_2^2 = h\left(1-\frac{x}{h}\right)\left(2-\frac{x}{h}\right)^2 \\ \phi_3^2 &= 3\left(1-\frac{x}{h}\right)^2 + 2\left(1-\frac{x}{h}\right)^3, \quad \phi_4^2 = h\left[\left(1-\frac{x}{h}\right)^3 + \left(1-\frac{x}{h}\right)^2\right] \end{aligned} \tag{4.30b}$$

The exact solution of (4.1) subject to the boundary condition (4.21) can be obtained by direct integration, and is given by

$$\left.\begin{aligned} EIw(x) &= \tfrac{1}{24}f_0x^4 - \tfrac{1}{6}(F_0 + f_0L)x^3 + \tfrac{1}{2}(M_0 + F_0L + \tfrac{1}{2}f_0L^2)x^2 \\ EI\theta(x) &= -\tfrac{1}{6}f_0x^3 + \tfrac{1}{2}(F_0 + f_0L)x^2 - (M_0 + F_0L + \tfrac{1}{2}f_0L^2) \\ M(x) &= \tfrac{1}{2}f_0x^2 - (F_0 + f_0L)x + M_0 + F_0L + \tfrac{1}{2}f_0L^2 \end{aligned}\right\} \text{for } 0 \leqslant x \leqslant L \tag{4.31}$$

The finite element solution (4.30) and the exact solution (4.31) are compared in Table 4.2 for the data

$$f_0 = 24\ \text{kN m}^{-1}, \quad F_0 = 60\ \text{kN}, \quad L = 3\ \text{m}, \quad M_0 = 0\ \text{kN m}$$

$$E = 200 \times 10^6\ \text{kN m}^{-2}, \quad I = 29 \times 10^6\ \text{mm}^4 \quad (EI = 5800\ \text{kN m}^2)$$

As expected, the finite element solution for w and θ coincides with the exact solution at the nodes. At other points, the difference between the finite element and exact solutions is less than 2%.

This completes the finite element formulation and analysis of the fourth-order differential equation (4.1) governing the Euler–Bernoulli beam theory. Whenever the flexural rigidity $b \equiv EI$ is a constant in each element, the

TABLE 4.2
Comparison of the finite element solution with the exact solution of the cantilever beam of Fig. 4.4 (2 elements; $L = 3$ m, $EI = 5800$ kN m^2, $f_0 = 24$ kN m^{-1}, $F_0 = 60$ kN, $M_0 = 0$ kN m)

	w (m)		$-\theta = dw/dx$		M/EI (kN m)	
x (m)	**FEM**	**Exact**	**FEM**	**Exact**	**FEM**	**Exact**
0.00	0.0000†	0.0000	0.0000†	0.0000	0.0489†	0.0497
0.1875	0.0008	0.0008	0.0088	0.0089	0.0452	0.0455
0.3750	0.0033	0.0033	0.0169	0.0171	0.0415	0.0414
0.5625	0.0071	0.0072	0.0244	0.0245	0.0378	0.0375
0.7500	0.0124	0.0124	0.0311	0.0311	0.0341	0.0338
0.9375	0.0188	0.0188	0.0372	0.0371	0.0305	0.0301
1.1250	0.0263	0.0263	0.0426	0.0425	0.0268	0.0266
1.3125	0.0347	0.0347	0.0472	0.0471	0.0231	0.0234
1.5000	0.0439†	0.0439	0.0512†	0.0512	0.0194†	0.0202
1.6875	0.0539	0.0539	0.0546	0.0547	0.0169	0.0171
1.8750	0.0644	0.0644	0.0575	0.0576	0.0144	0.0143
2.0625	0.0754	0.0755	0.0600	0.0600	0.0118	0.0115
2.2500	0.0868	0.0869	0.0620	0.0620	0.0093	0.0089
2.4375	0.0986	0.0987	0.0635	0.0634	0.0068	0.0065
2.6250	0.1106	0.1107	0.0645	0.0644	0.0043	0.0042
2.8025	0.1228	0.1228	0.0651	0.0650	0.0017	0.0020
3.0000	0.1350†	0.1350	0.0652†	0.0652	−0.0008†	0.0000

† Nodal values; all others are computed by interpolation.

element stiffness matrix (4.15) can be used directly. The finite element solution for the generalized displacements at the nodes is exact for all problems for which EI is constant and the applied transverse load f is a polynomial expression. Further, the solution is exact at all points if the distributed load is such that the exact solution is a cubic.

The bending moment at any point in the element Ω^e of the beam can be computed from the formula

$$M = EI \frac{d^2 w}{dx^2} = EI \sum_{j=1}^{4} u_j^e \frac{d^2 \phi_j^e}{dx^2} \tag{4.32a}$$

For beams of rectangular cross-section with height H (and width B) the maximum bending stress is

$$\sigma_x = \mp \frac{MH}{2I} = \mp \frac{EH}{2} \frac{d^2 w}{dx^2} = \mp \frac{EH}{2} \sum_{j=1}^{4} u_j^e \frac{d^2 \phi_j^e}{dx^2} \tag{4.32b}$$

The minus sign is for the top and the plus sign for the bottom of the beam.

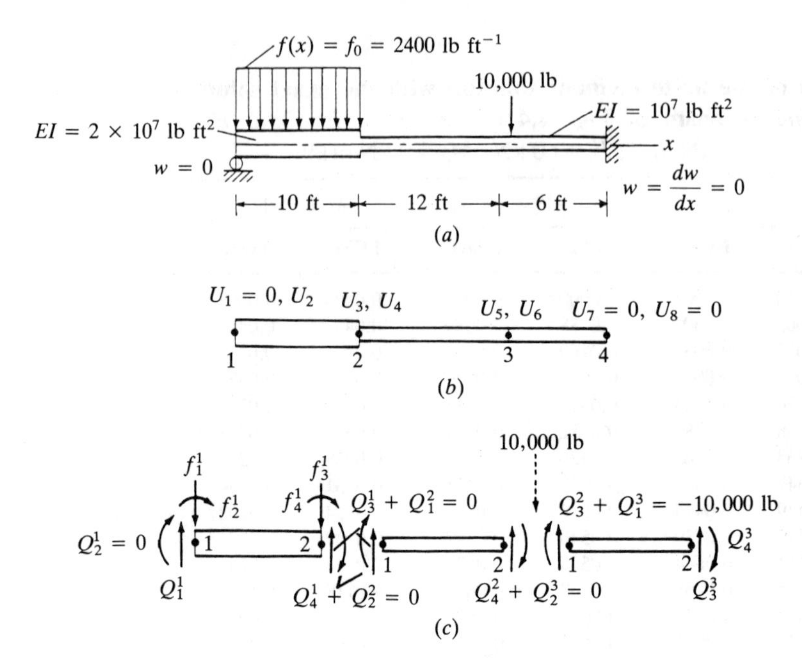

FIGURE 4.5
Finite element mesh and equilibrium conditions for the beam bending problem considered in Example 4.1: (*a*) physical problem; (*b*) finite element mesh of three elements; (*c*) equilibrium conditions among the generalized forces (i.e., secondary variables).

4.2.8 Examples

Example 4.1. Consider the beam shown in Fig. 4.5(*a*). The differential equation (4.1) is valid with the following discontinuous data:

$$EI = \begin{cases} 2\times 10^7 \text{ lb ft}^2 & \text{for } 0 \leq x \leq 10 \text{ ft} \\ 10^7 \text{ lb ft}^2 & \text{for } 10 \text{ ft} \leq x \leq 28\, ft \end{cases} \tag{4.33a}$$

$$f(x) = \begin{cases} -2400 \text{ lb ft}^{-1} & \text{for } 0 \leq x \leq 10 \text{ ft} \\ 0 & \text{for } 10 \text{ ft} \leq x \leq 28 \text{ ft} \end{cases} \tag{4.33b}$$

The geometry and loading in the present case require us to use at least three elements (see Fig. 4.5*b*) to represent the domain $\Omega = (0, L)$, $L = 28$ ft. It is possible to use two elements if we choose to distribute the point load $-10{,}000$ lb by (4.14). We shall use three elements to analyze the problem. There are four nodes and eight global degrees of freedom in the mesh of three beam elements.

Since EI and f are element-wise-constant, the element stiffness matrix and the force vector are given by (4.15), with $f_0 = -2400 \text{ lb ft}^{-1}$ in element 1 and $f = 0$ in elements 2 and 3. The point load will be directly included in the global force vector through equilibrium of forces. The assembled equations are

$$
10^7\begin{bmatrix}
0.024 & -0.12 & -0.024 & -0.12 & & & & \\
 & 0.80 & 0.12 & 0.40 & & & & \\
 & & 0.0309 & 0.0783 & -0.00694 & -0.04167 & & \\
 & & & 1.133 & 0.0417 & 0.167 & & \\
 & & & & 0.0625 & -0.125 & -0.0556 & -0.167 \\
 & & & & & 1.0 & 0.1667 & 0.333 \\
\text{symmetric} & & & & & & 0.0556 & 0.1667 \\
 & & & & & & & 0.6667
\end{bmatrix}
\times\begin{Bmatrix} U_1 \\ U_2 \\ U_3 \\ U_4 \\ U_5 \\ U_6 \\ U_7 \\ U_8 \end{Bmatrix}
=\begin{Bmatrix} -12{,}000 \\ 20{,}000 \\ -12{,}000 \\ -20{,}000 \\ 0 \\ 0 \\ 0 \\ 0 \end{Bmatrix}
+\begin{Bmatrix} Q_1^1 \\ Q_2^1 \\ Q_3^1+Q_1^2 \\ Q_4^1+Q_2^2 \\ Q_3^2+Q_1^3 \\ Q_4^2+Q_2^3 \\ Q_3^3 \\ Q_4^3 \end{Bmatrix}
\tag{4.34}
$$

where the interelement continuity conditions on the primary variables (i.e., displacements and rotations) have already been used. The equilibrium of the secondary variables (i.e., internal forces and moments) is given by

$$
Q_2^1=0, \quad Q_3^1+Q_1^2=0, \quad Q_4^1+Q_2^2=0, \quad Q_3^2+Q_1^3=-10{,}000, \quad Q_4^2+Q_2^3=0 \tag{4.35a}
$$

Note that the forces Q_1^1 and Q_3^3 and the moment Q_4^3 (the reactions at the supports) are not known. The boundary conditions on the primary degrees of freedom are

$$
w(0)=0 \Rightarrow U_1=0; \quad w(28)=0 \Rightarrow U_7=0; \quad \left(\frac{dw}{dx}\right)\bigg|_{x=28}=0 \Rightarrow U_8=0 \tag{4.35b}
$$

Using the known forces and displacements in (4.35), we can partition the global system of equations to obtain the condensed equations for the unknown generalized displacements and forces. Since the specified generalized displacements are all zero, the condensed stiffness matrix associated with the unknown generalized displacements can be obtained by deleting the columns and rows corresponding to the known U_i [see the submatrix enclosed by the dashed lines in (4.34)]:

$$
10^6\begin{bmatrix}
8.000 & 1.200 & 4.000 & 0.000 & 0.000 \\
 & 0.309 & 0.783 & -0.069 & -0.417 \\
 & & 11.333 & 0.417 & 1.667 \\
\text{symmetric} & & & 0.625 & -1.250 \\
 & & & & 10.000
\end{bmatrix}
\begin{Bmatrix} U_2 \\ U_3 \\ U_4 \\ U_5 \\ U_6 \end{Bmatrix}
=\begin{Bmatrix} 20{,}000 \\ -12{,}000 \\ -20{,}000 \\ -10{,}000 \\ 0 \end{Bmatrix}
\tag{4.36}
$$

The unknown reactions can be computed from the remaining equations, i.e., equations 1, 7, and 8 of (4.34):

$$\begin{Bmatrix} Q_1^1 \\ Q_3^3 \\ Q_4^3 \end{Bmatrix} = \begin{Bmatrix} 12{,}000 \\ 0 \\ 0 \end{Bmatrix} + 10^7 \begin{bmatrix} -0.12 & -0.024 & -0.12 & 0.0 & 0.0 \\ 0.0 & 0.0 & 0.0 & -0.0556 & 0.1667 \\ 0.0 & 0.0 & 0.0 & -0.1667 & 0.3333 \end{bmatrix} \begin{Bmatrix} U_2 \\ U_3 \\ U_4 \\ U_5 \\ U_6 \end{Bmatrix} \tag{4.37}$$

The algebraic equations (4.36) are solved first for the generalized displacements, and then (4.37) are used to obtain the unknown generalized forces of the problem.

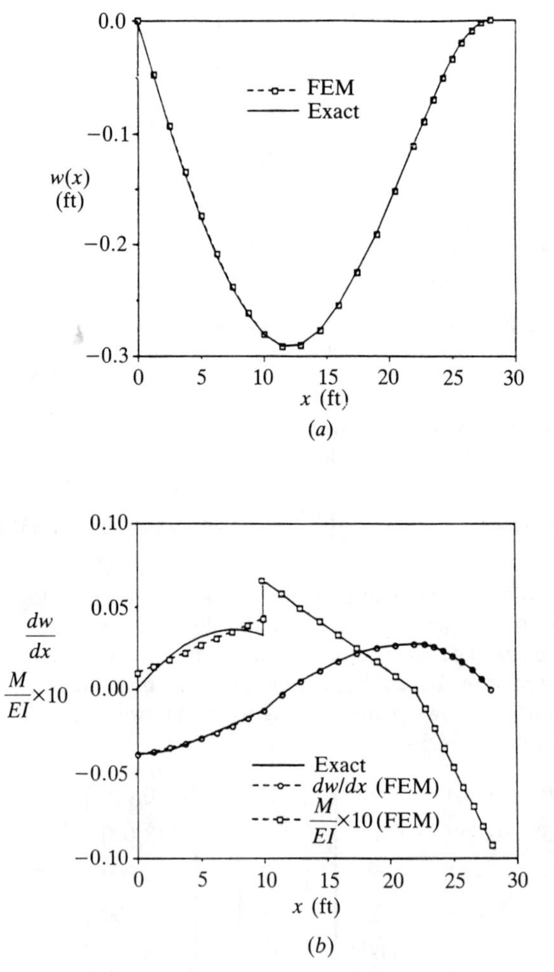

FIGURE 4.6
Comparison of the finite element solutions for deflection, slope, and bending moment of a cantilever beam with the exact solutions (Example 4.2).

Equations (4.36) are solved with the help of a computer; the solution is

$$U_2 = 0.03856, \quad U_3 = -0.2808 \text{ ft}, \quad U_4 = 0.01214$$
$$U_5 = -0.1103 \text{ ft}, \quad U_6 = -0.02752 \tag{4.38a}$$

The reaction forces, from the element equilibrium equations, are

$$Q_1^1 = 18{,}565.54 \text{ lb}, \quad Q_3^3 = 15{,}434.46 \text{ lb}, \quad Q_4^3 = 92{,}165 \text{ ft lb} \tag{4.38b}$$

Plots of the finite element solution for the transverse deflection w, rotation $\theta = -dw/dx$, and bending moment $M = EI\, d^2w/dx^2$ are shown in Fig. 4.6. Because of the discontinuity in the loading and the flexural rigidity, the exact solution $w(x)$ is also defined by three separate expressions for the three regions:

$$w(x) = \begin{cases} \frac{1}{12}R_1x^2 - \frac{1}{48}A_0x^4 - A_1x & \text{for } 0 \leq x \leq 10 \text{ ft} \\ \frac{1}{6}R_1x^3 - \frac{1}{6}A_0h_1(x - \frac{1}{2}h_1)^3 - A_2x + A_3 & \text{for } 10 \text{ ft} \leq x \leq 22 \text{ ft} \\ \frac{1}{6}R_1x^3 - \frac{1}{6}A_0h_1(x - \frac{1}{2}h_1)^3 - \frac{1}{6}F_0(x - h_1 - h_2)^3 - A_2x + A_3 & \text{for } 22 \text{ ft} \leq x \leq 28 \text{ ft} \end{cases} \tag{4.39a}$$

where

$$R_1 = 0.001856, \quad A_0 = 2400 \times 10^{-7}$$
$$A_1 = 0.03856, \quad A_2 = 0.07497, \quad A_3 = 0.20943 \tag{4.39b}$$
$$h_1 = 10 \text{ ft}, \quad h_2 = 12 \text{ ft}$$

In the next example we consider a beam with linearly varying distributed transverse load and with clamped and spring-supported ends. This requires us to evaluate the integral for f_i^e [see (4.12c)], since it is not available when $f(x)$ is a linear function of x.

Example 4.2. Consider a beam that has a clamped support at one end and is spring-supported at the other (see Fig. 4.7). A linearly varying transverse load is applied in the span, 4 ft $\leq x \leq$ 10 ft. We must use at least two elements, as dictated by the loading. A one-element model could be used when the load is distributed as discussed in Remark 5 in Section 3.2.7. The force vector $\{f^2\}$ due to the distributed load on element 2 can be computed according to (4.12c):

$$f_i^2 = \int_0^6 f(\bar{x})\phi_i^2(\bar{x})\, d\bar{x}, \quad f(\bar{x}) = -\frac{100}{6}\bar{x} \tag{4.40}$$

and ϕ_i^2 are the Hermite interpolation functions given in (4.10a). Carrying out the integration, we obtain

$$\{f^2\} = \begin{Bmatrix} -90 \text{ lb} \\ 120 \text{ lb ft} \\ -210 \text{ lb} \\ -180 \text{ lb ft} \end{Bmatrix} \tag{4.41}$$

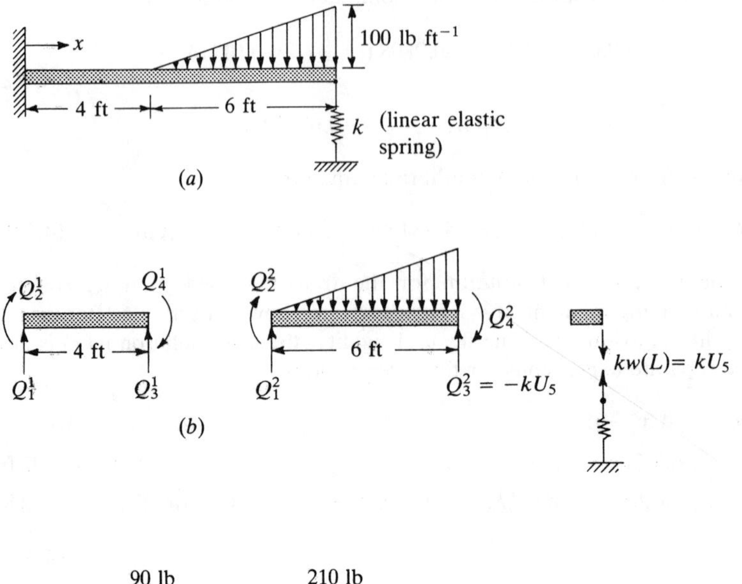

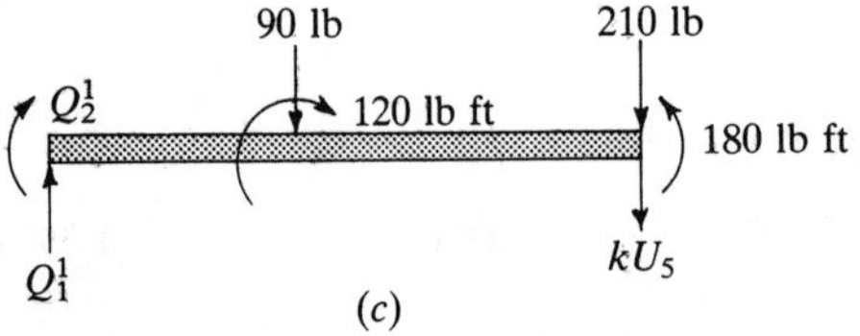

FIGURE 4.7
Finite element analysis of a cantilevered beam with discontinuous loading: (*a*) physical problem; (*b*) finite element discretization and generalized element forces; (*c*) assembly of finite element and generalized global forces (Example 4.3).

The assembled equations are

$$\frac{EI}{144}\begin{bmatrix} 27 & -54 & -27 & -54 & 0 & 0 \\ -54 & 144 & 54 & 72 & 0 & 0 \\ -27 & 54 & 27+8 & 54-24 & -8 & -24 \\ -54 & 72 & 54-24 & 144+96 & 24 & 48 \\ 0 & 0 & -8 & 24 & 8 & 24 \\ 0 & 0 & -24 & 48 & 24 & 96 \end{bmatrix}\begin{Bmatrix} U_1 \\ U_2 \\ U_3 \\ U_4 \\ U_5 \\ U_6 \end{Bmatrix} = \begin{Bmatrix} 0 \\ 0 \\ -90 \\ 120 \\ -210 \\ -180 \end{Bmatrix} + \begin{Bmatrix} Q_1^1 \\ Q_2^1 \\ Q_3^1 + Q_1^2 \\ Q_4^1 + Q_2^2 \\ Q_3^2 \\ Q_4^2 \end{Bmatrix} \tag{4.42}$$

The displacement boundary conditions at $x = 0$ and force boundary conditions at $x = L$ are

$$w(0) = 0, \quad \left(\frac{dw}{dx}\right)\bigg|_{x=0} = 0$$
$$\left[-\frac{d}{dx}\left(EI\frac{d^2w}{dx^2}\right)\right]\bigg|_{x=L} = -kw(L), \quad \left(EI\frac{d^2w}{dx^2}\right)\bigg|_{x=0} = 0 \tag{4.43a}$$

which can be expressed in terms of the nodal degrees of freedom as

$$U_1 = 0, \quad U_2 = 0, \quad Q_3^2 = -kU_5, \quad Q_4^2 = 0 \tag{4.43b}$$

The condensed equations for the generalized displacements U_3, U_4, U_5, and U_6 are given by the last four equations of (4.42):

$$\frac{EI}{144}\begin{bmatrix} 35 & 30 & -8 & -24 \\ 30 & 240 & 24 & 48 \\ -8 & 24 & 8+c & 24 \\ -24 & 48 & 24 & 96 \end{bmatrix}\begin{Bmatrix} U_3 \\ U_4 \\ U_5 \\ U_6 \end{Bmatrix} = \begin{Bmatrix} -90 \\ 120 \\ -210 \\ -180 \end{Bmatrix} \tag{4.44}$$

where $c = 144k/EI$. The solution of these equations for different values of k/EI can be computed as follows:

For $k/EI = 0$ (no spring),

$$U_3 = \frac{-0.16 \times 10^5}{EI} \text{ (ft)}, \quad U_4 = \frac{0.72 \times 10^4}{EI} \text{ (rad)}$$
$$U_5 = \frac{-0.7108 \times 10^5}{EI} \text{ (ft)}, \quad U_6 = \frac{0.99 \times 10^4}{EI} \text{ (rad)} \tag{4.45}$$

for $k/EI = 10^{-2}$ (a soft spring),

$$U_3 = \frac{-0.4627 \times 10^4}{EI} \text{ (ft)}, \quad U_4 = \frac{0.1951 \times 10^4}{EI} \text{ (rad)}$$
$$U_5 = \frac{-0.164 \times 10^5}{EI} \text{ (ft)}, \quad U_6 = \frac{0.1699 \times 10^4}{EI} \text{ (rad)} \tag{4.46}$$

for $k/EI = 10^2$ (a hard spring),

$$U_3 = \frac{-0.1216 \times 10^4}{EI} \text{ (ft)}, \quad U_4 = \frac{0.3765 \times 10^3}{EI} \text{ (rad)}$$
$$U_5 = \frac{-2.132}{EI} \text{ (ft)}, \quad U_6 = \frac{-0.7617 \times 10^3}{EI} \text{ (rad)} \tag{4.47}$$

The solutions coincide with the exact solution at the nodes:

$$w(x) = \begin{cases} \dfrac{1}{EI}\left(\frac{1}{6}R_1x^3 + \frac{1}{2}M_1x^2\right) & \text{for } 0 \leqslant x \leqslant a \\ \dfrac{1}{EI}\left[\frac{1}{6}R_1x^3 + \frac{1}{2}M_1x^2 - \dfrac{f_0}{120b}(x-a)^5\right] & \text{for } a \leqslant x \leqslant L \end{cases} \tag{4.48a}$$

where

$$R_1 = \left(\frac{f_0 b}{40}\right)\frac{60EI + 10kL^2b - kb^3}{2EI + kL^3}, \quad M_1 = \tfrac{1}{6}f_0b^2 - R_1L$$
$$f_0 = 100 \text{ lb ft}^{-1}, \quad a = 4 \text{ ft}, \quad b = 6 \text{ ft}, \quad L = 10 \text{ ft} \tag{4.48b}$$

Figure 4.8 shows plots of the transverse deflection, slope, and bending moment for the case $k/EI = 100$. The finite element solution obtained using two elements is in good agreement with the exact solution. Improvement can be expected if the region $4\text{ ft} \leqslant x \leqslant 10\text{ ft}$ is subdivided into two or more elements (note that the particular scale

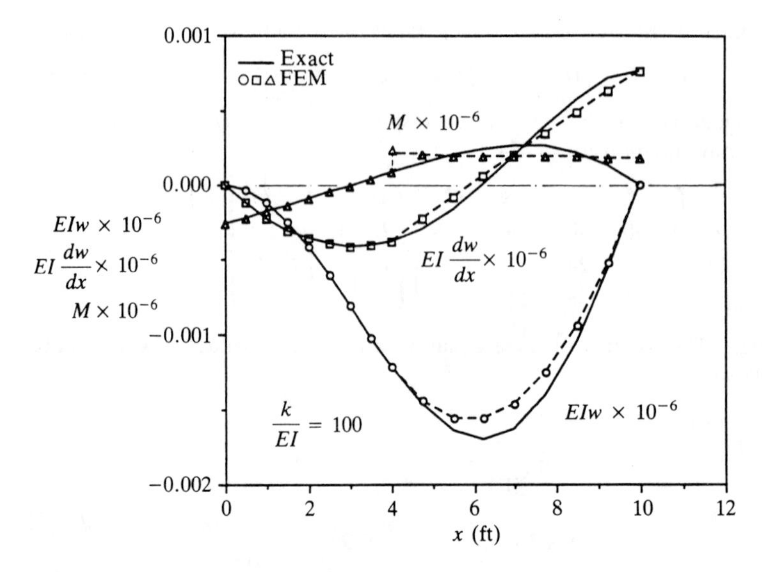

FIGURE 4.8
Comparison of the finite element solutions with the exact solutions for a clamped spring-supported beam (see Fig. 4.7 for the geometry and loading).

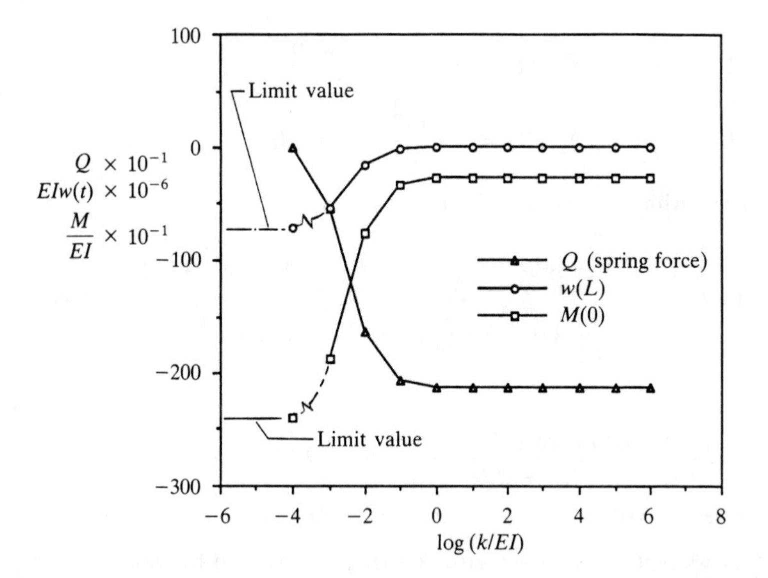

FIGURE 4.9
The effect of spring stiffness on the deflection, bending moment, and spring force for the problem in Example 4.8.

used in Fig. 4.8 brings out the difference between the finite element and exact solutions). The effect of the spring on the end deflection $w(L)$, the bending moment $M(0)$ at the fixed end, and the spring force $Q \equiv Q_3^2 = -kw(L)$ can be seen from the results presented in Fig. 4.9. For $k/EI > 1$, the spring acts essentially like a rigid support.

This completes the development and applications of the Euler–Bernoulli beam element.

4.3 PLANE TRUSS AND EULER–BERNOULLI FRAME ELEMENTS

Structures composed of bar elements and beam elements are classified as truss and frame structures, respectively. By definition, bars can only carry axial loads and deform axially, whereas beams can take transverse loads and bending moments about an axis perpendicular to the plane of the member. All members of a truss are subjected to only axial loads, and no transverse shearing forces and bending moments are experienced by any member. All members are connected to each other through pins that allow free rotation about the pin axis. On the other hand, members of a frame are connected by a rigid connection (e.g., welded or rivetted), so that axial and transverse forces and bending moments can be developed in the members. A truss can be looked upon as a special case of a frame structure. Thus, a typical truss member can be modeled using the bar finite element developed in Section 3.3.3. A member of a frame structure can be modeled by a superposition of the bar element with the beam element of Section 4.2.

The objective of this section is to formulate, with the help of the information from Sections 3.3.3 and 4.2, the truss finite element and frame finite element. The formulation will be based on matrix notation. Since a truss finite element is a special case of the frame element, the derivation is presented for the frame element only.

In many truss and frame structures, the bar and beam structural elements are found in many different orientations (see Fig. 4.10). Analysis of such structures for displacements and stresses requires the setting up of a global coordinate system and referencing of all quantities (i.e. displacements, forces, and stiffnesses) of individual elements to the common (global) coordinate system in order to assemble the elements and impose boundary conditions on the whole structure. When a truss element is oriented at an angle from the global axis, its axial displacements at the nodes have components along the global axes. Thus, every node of a truss will have two displacements in the global coordinates: one along the global x axis and another transverse to the x axis. Therefore, the element will have two degrees of freedom per node in the global coordinate system. In order to facilitate the transformation from element equations to global equations, we append the element equations for the axial displacements to those associated with the transverse displacements. Since there are no stiffnesses associated with the transverse displacements

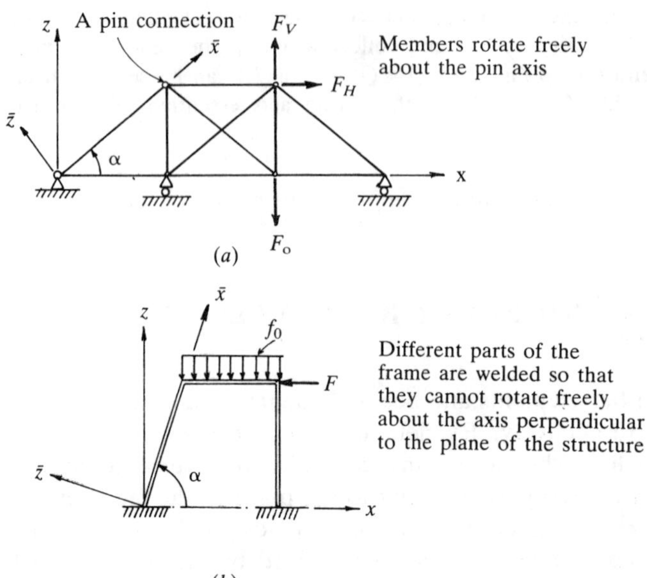

FIGURE 4.10
Typical examples of the plane truss and plane frame: (*a*) a plane truss structure (all elements carry only axial loads); (*b*) a typical plane frame structure (all members may carry axial and transverse loads and bending moments).

(because the element experiences axial deformation only), the entries in the stiffness matrix corresponding to the transverse displacements are set equal to zero. For example, when the linear Lagrange interpolation is used for the axial displacement, the element stiffness matrix in the element coordinates for a truss element with two displacements (axial and transverse) per node can be written as

$$[K^e] = \frac{A_e E_e}{h_e} \begin{bmatrix} 1 & 0 & -1 & 0 \\ 0 & 0 & 0 & 0 \\ -1 & 0 & 1 & 0 \\ 0 & 0 & 0 & 0 \end{bmatrix} \tag{4.49}$$

Similarly, a superposition of the bar element of Section 3.3.3 with the beam element of Section 4.2 gives three primary degrees of freedom per node (see Fig. 4.11*a*) of the frame element: $(u, w, -dw/dx)$. When the axial stiffness AE, bending stiffness EI, and axial distributed force q and transverse distributed force f are element-wise-constant, the superposition of the linear bar element with the Hermite cubic Euler–Bernoulli beam element gives the following element equations:

$$[K^e]\{u^e\} = \{F^e\} \tag{4.50a}$$

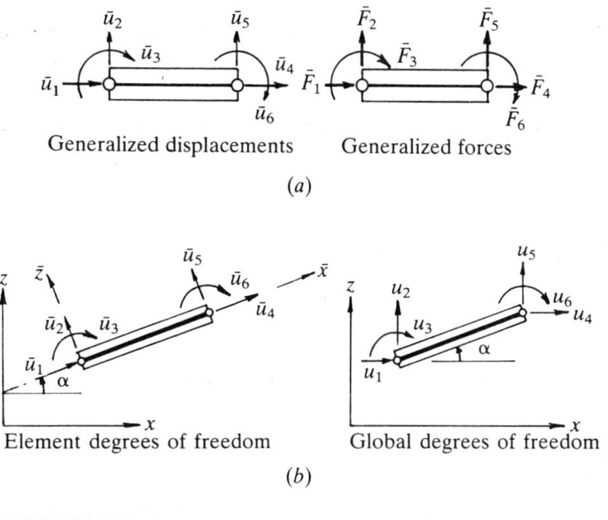

FIGURE 4.11
The frame element with primary and secondary variables (or degree of freedom) in the local and global coordinate systems. (*a*) The generalized displacements and forces in the element coordinate system ($\bar{x}, \bar{y}, \bar{z}$). (*b*) The generalized displacements in the element coordinate system and global coordinate system (x, y, z); the y axis is into the plane of the paper. The angle of orientation is measured counter-clockwise from the global x axis to the element $\bar{x}$ axis. The truss element is obtained from the frame element by omitting the rotation and moment degrees of freedom at the nodes (also, $\bar{u}_2 = \bar{u}_4 = 0$ for the truss element).

where

$$\{u^e\} = \begin{Bmatrix} u_1 \\ w_1 \\ \theta_1 \\ u_2 \\ w_2 \\ \theta_2 \end{Bmatrix}^e, \qquad \{F^e\} = \begin{Bmatrix} \frac{1}{2}qh \\ \frac{1}{2}fh \\ -\frac{1}{12}fh^2 \\ \frac{1}{2}qh \\ \frac{1}{2}fh \\ \frac{1}{12}fh^2 \end{Bmatrix} + \begin{Bmatrix} Q_1^e \\ Q_2^e \\ Q_3^e \\ Q_4^e \\ Q_5^e \\ Q_6^e \end{Bmatrix} \tag{4.50b}$$

and $q = q_0$ and $f = f_0$ are constants over an element.

The element stiffness matrix $[K^e]$ is of order 6×6:

$$[K] = \frac{2EI}{h^3} \begin{bmatrix} \mu & 0 & 0 & -\mu & 0 & 0 \\ 0 & 6 & -3h & 0 & -6 & -3h \\ 0 & -3h & 2h^2 & 0 & 3h & h^2 \\ -\mu & 0 & 0 & \mu & 0 & 0 \\ 0 & -6 & 3h & 0 & 6 & 3h \\ 0 & -3h & h^2 & 0 & 3h & 2h^2 \end{bmatrix}, \qquad \mu = \frac{Ah^2}{2I} \tag{4.50c}$$

In the following paragraphs, we develop transformation relations to express the element equations (4.50*c*)—valid in the element coordinate system—to the global coordinate system.

Let $\bar{u}$ and $\bar{w}$ denote the axial and transverse displacements referred to the local coordinate system $(\bar{x}, \bar{z})$. The local coordinates $(\bar{x}, \bar{y}, \bar{z})$ are related to the global coordinates (x, y, z) by (see Fig. 4.11*b*)

$$\begin{Bmatrix} \bar{x} \\ \bar{y} \\ \bar{z} \end{Bmatrix} = \begin{bmatrix} \cos\alpha & 0 & \sin\alpha \\ 0 & 1 & 0 \\ -\sin\alpha & 0 & \cos\alpha \end{bmatrix} \begin{Bmatrix} x \\ y \\ z \end{Bmatrix} \tag{4.51}$$

where the angle α is measured counter-clockwise from x axis to $\bar{x}$ axis. Note that the y and $\bar{y}$ coordinates are parallel to each other, and they are *into* the plane of the paper. The same transformation relations hold for displacements (u, v, w) along the global coordinates (x, y, z) and displacements $(\bar{u}, \bar{v}, \bar{w})$ in the local coordinates $(\bar{x}, \bar{y}, \bar{z})$. Since we are considering 2-D structures, we have $v = \bar{v} = 0$. The rotation $\theta \equiv -dw/dx$ about the y axis remains unchanged: $\theta = \bar{\theta}$. Hence, the relationship between (u, w, θ) and $(\bar{u}, \bar{w}, \bar{\theta})$ can be written as

$$\begin{Bmatrix} \bar{u} \\ \bar{w} \\ \bar{\theta} \end{Bmatrix} = \begin{bmatrix} \cos\alpha & \sin\alpha & 0 \\ -\sin\alpha & \cos\alpha & 0 \\ 0 & 0 & 1 \end{bmatrix} \begin{Bmatrix} u \\ w \\ \theta \end{Bmatrix} \tag{4.52}$$

Therefore, the six element nodal degrees of freedom $\bar{u}_i^e$ in the $(\bar{x}, \bar{y}, \bar{z})$ system are related to the six degrees of freedom u_i^e in the (x, y, z) system by $(\alpha = \alpha^e)$

$$\begin{Bmatrix} \bar{u}_1 \\ \bar{u}_2 \\ \bar{u}_3 \\ \bar{u}_4 \\ \bar{u}_5 \\ \bar{u}_6 \end{Bmatrix}^e = \left[\begin{array}{ccc|ccc} \cos\alpha & \sin\alpha & 0 & & & \\ -\sin\alpha & \cos\alpha & 0 & & 0 & \\ 0 & 0 & 1 & & & \\ \hline & & & \cos\alpha & \sin\alpha & 0 \\ & 0 & & -\sin\alpha & \cos\alpha & 0 \\ & & & 0 & 0 & 1 \end{array}\right] \begin{Bmatrix} u_1 \\ u_2 \\ u_3 \\ u_4 \\ u_5 \\ u_6 \end{Bmatrix}^e \tag{4.53a}$$

or

$$\{\bar{u}^e\} = [T^e]\{u^e\} \tag{4.53b}$$

The inverse of (4.53) is

$$\{u^e\} = [T^e]^{-1}\{\bar{u}^e\} \equiv [R^e]\{\bar{u}^e\}$$

and it can be shown that

$$[R^e] = [T^e]^T \tag{4.54}$$

i.e., the inverse of $[T^e]$ is equal to its transpose.

Analogously to (4.53*b*), the element force vectors in the local and global coordinate systems are related according to

$$\{\bar{F}^e\} = [T^e]\{F^e\} \tag{4.55}$$

To obtain expressions for the element stiffness matrix and the force vector referred to the global coordinates in terms of the element stiffness matrix and force vector in the local coordinates, we use (4.53)–(4.55) in the element equations

$$[\bar{K}^e]\{\bar{u}^e\} = \{\bar{F}^e\} \tag{4.56}$$

Substituting the transformation equations into (4.56), we obtain

$$[\bar{K}^e][T^e]\{u^e\} = [T^e]\{F^e\}$$

Premultiplying both sides with $[T^e]^{-1} \equiv [T^e]^T$, we obtain

$$[T^e]^T[\bar{K}^e][T^e]\{u^e\} = \{F^e\} \tag{4.57}$$

which gives

$$\boxed{[K^e] = [T^e]^T[\bar{K}^e][T^e], \qquad \{F^e\} = [T^e]^T\{\bar{F}^e\}} \tag{4.58a, b}$$

Thus, if we know the element matrices $[\bar{K}^e]$ and $\{\bar{F}^e\}$ of an element Ω^e in the local coordinate system $(\bar{x}, \bar{y}, \bar{z})$ then the element matrices of the element in the global coordinate system, which is obtained by rotating the element coordinate system through an angle α in the counter-clockwise direction about the y axis, are given by (4.58). Note that the angle α is measured in the counter-clockwise direction from the positive global x axis.

Inserting the element stiffness from (4.50*c*) for $[\bar{K}^e]$ into (4.58*a*) and carrying out the indicated matrix multiplications, we arrive at the element stiffness matrix $[K^e]$ referred to the global coordinates:

$$[K^e] = \frac{2EI}{h^3} \times \begin{bmatrix} \mu\cos^2\alpha + 6\sin^2\alpha & & & & & \\ (\mu-6)\cos\alpha\sin\alpha & \mu\sin^2\alpha + 6\cos^2\alpha & & \text{symmetric} & & \\ 3h\sin\alpha & -3h\cos\alpha & 2h^2 & & & \\ -(\mu\cos^2\alpha + 6\sin^2\alpha) & -(\mu-6)\sin\alpha\cos\alpha & -3h\sin\alpha & \mu\cos^2\alpha + 6\sin^2\alpha & & \\ -(\mu-6)\cos\alpha\sin\alpha & -(\mu\sin^2\alpha + 6\cos^2\alpha) & 3h\cos\alpha & (\mu-6)\cos\alpha\sin\alpha & \mu\sin^2\alpha + 6\cos^2\alpha & \\ 3h\sin\alpha & -3h\cos\alpha & h^2 & -3h\sin\alpha & 3h\cos\alpha & 2h^2 \end{bmatrix} \tag{4.59a}$$

where

$$\mu = Ah^2/2I \tag{4.59b}$$

Equation (4.58*b*), after multiplication, yields

$$\begin{Bmatrix} F_1 \\ F_2 \\ F_3 \\ F_4 \\ F_5 \\ F_6 \end{Bmatrix}^e = \begin{Bmatrix} \bar{F}_1 \cos\alpha - \bar{F}_2 \sin\alpha \\ \bar{F}_1 \sin\alpha + \bar{F}_2 \cos\alpha \\ \bar{F}_3 \\ \bar{F}_4 \cos\alpha - \bar{F}_5 \sin\alpha \\ \bar{F}_4 \sin\alpha + \bar{F}_5 \cos\alpha \\ \bar{F}_6 \end{Bmatrix}^e \tag{4.60}$$

which is the element force vector referred to the global coordinates.

We next consider an example each of a truss and frame structure.

Example 4.3. Consider the three-member truss shown in Fig. 4.12(*a*). All members of the truss have identical cross-sectional area A and modulus E. The hinged supports at points 1, 2, and 3 allow free rotation of the members about the y axis (taken as positive into the plane of the paper). We wish to determine the horizontal and vertical displacements at the joint 3 and the forces in each member of the structure.

Since all joints are hinged, and the applied forces are acting at the nodes, the members are subjected to only axial forces. Hence, the structure is a truss. We use three finite elements to model the structure. Any further subdivision of the members

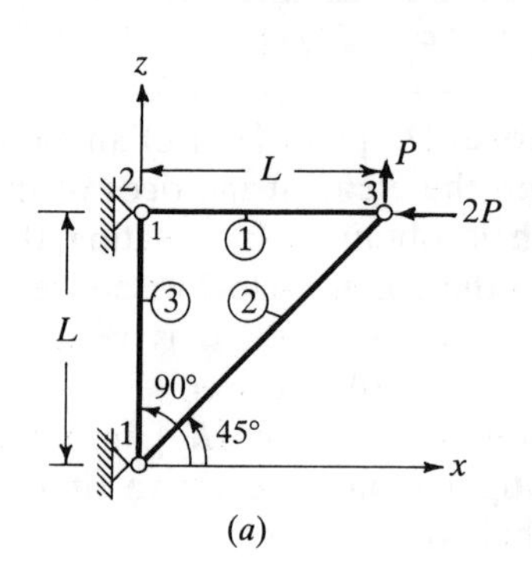

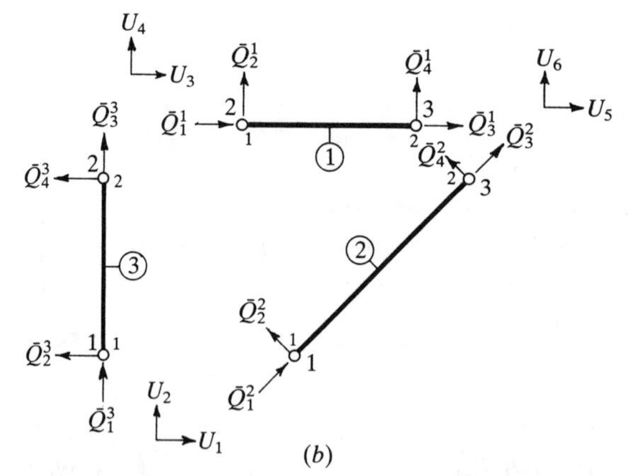

FIGURE 4.12
Geometry and finite element discretization of a plane truss: (*a*) geometry and loading; (*b*) element numbers, global node numbers of element nodes, and element nodal forces in the element coordinates.

does not add to the accuracy, because for all truss problems the finite element solutions are exact. This is a consequence of the fact that all truss members with constant cross-section are governed by the homogeneous differential equation

$$EA\frac{d^2u}{dx^2} = 0 \tag{4.61}$$

whose solution is of the form $u(x) = c_1x + c_2$. Thus, linear interpolation of the displacements should give the exact result.

The global node numbers and element numbers are shown in Fig. 4.12(b). There are two degrees of freedom, horizontal and vertical displacements, at each node of the element. The element stiffness matrix in the local coordinate system is

$$[\bar{K}^e] = \frac{A_eE_e}{h_e}\begin{bmatrix} 1 & 0 & -1 & 0 \\ 0 & 0 & 0 & 0 \\ -1 & 0 & 1 & 0 \\ 0 & 0 & 0 & 0 \end{bmatrix} \tag{4.62}$$

The (transformed) stiffness matrix of the element Ω_e in the global coordinate system is given by

$$[K^e] = \frac{EA}{h_e}\left[\begin{array}{cc|cc} \cos^2\alpha & \cos\alpha\sin\alpha & -\cos^2\alpha & -\cos\alpha\sin\alpha \\ \cos\alpha\sin\alpha & \sin^2\alpha & -\cos\alpha\sin\alpha & -\sin^2\alpha \\ \hline -\cos^2\alpha & -\cos\alpha\sin\alpha & \cos^2\alpha & \cos\alpha\sin\alpha \\ -\cos\alpha\sin\alpha & -\sin^2\alpha & \cos\alpha\sin\alpha & \sin^2\alpha \end{array}\right] \tag{4.63}$$

which is obtained from (4.59) by deleting the rows and columns corresponding to the bending degrees of freedom and setting all bending stiffnesses to zero. The element stiffness matrix is 4×4 in the global coordinate system because of the horizontal and vertical displacement degrees of freedom at each node. The element data for the problem are as follows.

Element number	Global nodes of the element		Geometric properties	Material property	Orientation
1	2	3	$A, h_1 = L$	E	$\alpha = 0°$
2	1	3	$A, h_2 = \sqrt{2}\,L$	E	$\alpha = 45°$
3	1	2	$A, h_2 = L$	E	$\alpha = 90°$

The assembled stiffness and force coefficients are given by

$$[K] = \left[\begin{array}{cccccc|c} 1 & 2 & 3 & 4 & 5 & 6 & \\ K_{11}^2 + K_{11}^3 & K_{12}^2 + K_{12}^3 & K_{13}^3 & K_{14}^3 & K_{13}^2 & K_{14}^2 & 1 \\ & K_{22}^2 + K_{22}^3 & K_{23}^3 & K_{24}^3 & K_{23}^2 & K_{24}^2 & 2 \\ & & K_{33}^3 + K_{11}^1 & K_{34}^3 + K_{12}^1 & K_{13}^1 & K_{14}^1 & 3 \\ & & & K_{44}^3 + K_{22}^1 & K_{23}^1 & K_{24}^1 & 4 \\ & \text{symmetric} & & & K_{33}^2 + K_{33}^1 & K_{34}^2 + K_{34}^1 & 5 \\ & & & & & K_{44}^2 + K_{44}^1 & 6 \end{array}\right] \tag{4.64}$$

The force vector can be written directly by including the known applied forces:

$$\{F\} = \begin{Bmatrix} Q_1^3 + Q_1^2 \\ Q_2^3 + Q_2^2 \\ Q_3^3 + Q_1^1 \\ Q_4^3 + Q_2^1 \\ Q_3^1 + Q_3^2 \\ Q_4^1 + Q_4^2 \end{Bmatrix} \tag{4.65}$$

Substituting the element data into (4.63) and the result into the assembled equations, we obtain

$$[K] = \frac{EA}{L}\begin{bmatrix} 0.3536 & 0.3536 & 0.0 & 0.0 & -0.3536 & -0.3536 \\ & 1.3536 & 0.0 & -1.0 & -0.3536 & -0.3536 \\ & & 1.0 & 0.0 & -1.0 & 0.0 \\ & & & 1.0 & 0.0 & 0.0 \\ & \text{symmetric} & & & 1.3536 & 0.3536 \\ & & & & & 0.3536 \end{bmatrix}$$

The specified displacement degrees of freedom are

$$U_1 = U_2 = U_3 = U_4 = 0 \tag{4.66}$$

The first two correspond to the horizontal and vertical displacements at node 1, and the last two to the horizontal and vertical displacements at node 2. The condensed equations for the unknown displacements and forces are

$$\frac{EA}{L}\begin{bmatrix} 1.3536 & 0.3536 \\ 0.3536 & 0.3536 \end{bmatrix}\begin{Bmatrix} U_5 \\ U_6 \end{Bmatrix} = \begin{Bmatrix} -2P \\ P \end{Bmatrix} \tag{4.67}$$

$$\begin{Bmatrix} Q_1^3 + Q_1^2 \\ Q_2^3 + Q_2^2 \\ Q_3^3 + Q_1^1 \\ Q_4^3 + Q_2^1 \end{Bmatrix} = \frac{EA}{L}\begin{bmatrix} -0.3536 & -0.3536 \\ -0.3536 & -0.3536 \\ -1.0 & 0.0 \\ 0.0 & 0.0 \end{bmatrix}\begin{Bmatrix} U_5 \\ U_6 \end{Bmatrix} \tag{4.68}$$

where, for example, $Q_1^2 + Q_1^3$ is the horizontal force and $Q_2^2 + Q_2^3$ the vertical force at node 1. Solving (4.67) and computing the reaction forces (4.68), we obtain

$$U_5 = -\frac{3PL}{EA}, \qquad U_6 = (3 + 2\sqrt{2})\frac{PL}{EA} = 5.828\frac{PL}{EA}$$
$$F_1 \equiv Q_1^2 + Q_1^3 = -P, \qquad F_2 = -P, \qquad F_3 = 3P, \qquad F_4 = 0.0 \tag{4.69}$$

The stress in each element can be computed from the relation

$$\sigma^e = \frac{\bar{Q}_3^e}{A_e}$$

where $\bar{Q}_3^e$ is the axial force at node 2 of the element Ω^e. Note that, for a truss element with constant cross-section, the following relations hold:

$$\bar{Q}_2^e = 0, \qquad \bar{Q}_4^e = 0, \qquad \bar{Q}_1^e = -\bar{Q}_3^e$$

The $\bar{Q}_3^e$ can be determined from the element equilibrium equations

$$\begin{Bmatrix} \bar{Q}_1^e \\ \bar{Q}_2^e \\ \bar{Q}_3^e \\ \bar{Q}_4^e \end{Bmatrix} = \frac{E_e A_e}{h_e} \begin{bmatrix} 1 & 0 & -1 & 0 \\ 0 & 0 & 0 & 0 \\ -1 & 0 & 1 & 0 \\ 0 & 0 & 0 & 0 \end{bmatrix} \begin{Bmatrix} \bar{u}_1^e \\ \bar{u}_2^e \\ \bar{u}_3^e \\ \bar{u}_4^e \end{Bmatrix} \tag{4.70}$$

Hence,

$$\sigma^e = \frac{E_e}{h_e}(\bar{u}_3^e - \bar{u}_1^e), \qquad \bar{u}_1^e = 0$$

From the global displacements U_5 and U_6, we have

$$\bar{u}_3^1 = -\frac{3PL}{AE}, \qquad \bar{u}_3^2 = U_5 \cos\alpha + U_6 \sin\alpha = 2\frac{PL}{AE}, \qquad \bar{u}_3^3 = 0 \tag{4.71}$$

Therefore, we have

$$\sigma^1 = -\frac{3P}{A}, \qquad \sigma^2 = \sqrt{2}\frac{P}{A}, \qquad \sigma^3 = 0 \tag{4.72}$$

The next example deals with a frame structure. It also illustrates how to incorporate a point load between nodes.

Example 4.4. The frame structure shown in Fig. 4.13 is to be analyzed for displacements and forces. Both members of the structure are made of the same material (E) and have the same geometric properties (A, I). The element stiffness matrices and force vectors in the global coordinate system (x, y, z) can be computed from (4.59) and (4.60). The geometric and material properties of each element are as follows.

Element 1

$$L = 144 \text{ in}, \quad A = 10 \text{ in}^2, \quad I = 10 \text{ in}^2, \quad \cos\alpha = 0.0, \quad \sin\alpha = 1.0$$
$$E = 10^6 \text{ psi}, \quad f_1 = \tfrac{1}{72}P \text{ lb in}^{-1} \tag{4.73}$$

Element 2

$$L = 180 \text{ in}, \quad A = 10 \text{ in}^2, \quad I = 10 \text{ in}^2, \quad \cos\alpha = 0.8, \quad \sin\alpha = 0.6$$
$$E = 10^6 \text{ psi}, \quad f_2 = 0 \tag{4.74}$$

The load $F_0 = 4P$ at the center of the element is distributed to the nodes according to (4.14).

The assembled stiffness matrix and force vectors are obtained by superposing the last three rows and columns of element 1 on the first three rows and columns of element 2; i.e., the 3×3 submatrix associated with rows and columns 4, 5, and 6 of element 1, and the 3×3 submatrix associated with rows and columns 1, 2, and 3 of element 2 overlap in the global stiffness matrix.

The known geometric boundary conditions are

$$U_1 = 0, \quad U_2 = 0, \quad U_3 = 0, \quad U_7 = 0, \quad U_8 = 0, \quad U_9 = 0 \tag{4.75a}$$

The force boundary conditions are

$$Q_4^1 + Q_1^2 = 0, \quad Q_5^1 + Q_2^2 = -2P, \quad Q_6^1 + Q_3^2 = 0 \tag{4.75b}$$

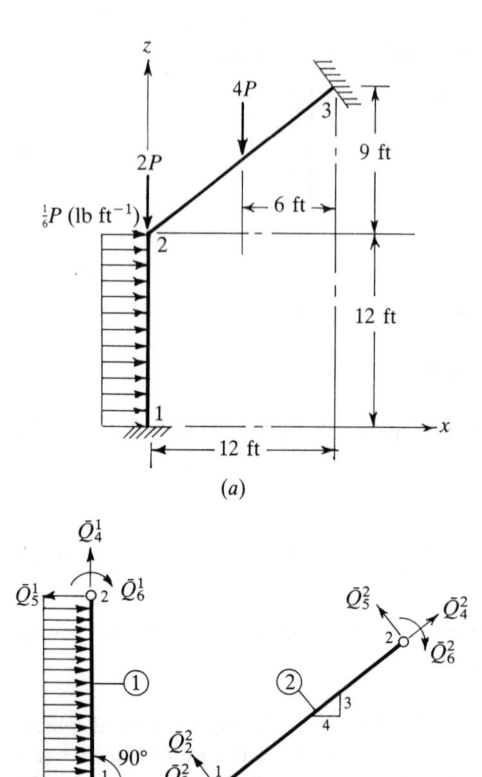

FIGURE 4.13
Geometry, loading and finite element discretization of a plane frame structure: (*a*) geometry and loading; (*b*) finite element discretization, element numbers, and element forces in the element coordinates.

Since all specified values of the known boundary conditions on the primary variables are zero, the condensed equations for the unknown global displacement degrees of freedom are

$$10^5\begin{bmatrix} 0.3560 & 0.2666 & -0.0178 \\ 0.2666 & 0.8846 & -0.0148 \\ -0.0178 & -0.0148 & 5.0000 \end{bmatrix}\begin{Bmatrix} U_4 \\ U_5 \\ U_6 \end{Bmatrix} = \begin{Bmatrix} 1 \\ -4 \\ 48 \end{Bmatrix} P \tag{4.76}$$

whose solution is

$$U_4 = 0.839 \times 10^{-4}P \text{ (in)}, \quad U_5 = -0.681 \times 10^{-4}P \text{ (in)}, \quad U_6 = 0.961 \times 10^{-4}P \text{ (rad)} \tag{4.77}$$

The reactions and forces in each member in the global coordinates can be computed from the element equations

$$\{Q^e\} = [K^e]\{u^e\} - \{f^e\} \tag{4.78a}$$

The forces $\{Q^e\}$ can be transformed to those in the element coordinate system by means of (4.55):

$$\{\bar{Q}^e\} = [T^e]\{Q^e\} \tag{4.78b}$$

We obtain

$$\{\bar{Q}^1\} = \begin{Bmatrix} 4.731 \\ 0.725 \\ -10.900 \\ -4.731 \\ 1.275 \\ 50.450 \end{Bmatrix} P, \quad \{\bar{Q}^2\} = \begin{Bmatrix} 1.458 \\ -0.180 \\ 21.550 \\ -1.458 \\ 0.180 \\ 10.870 \end{Bmatrix} P \tag{4.79}$$

4.4 THE TIMOSHENKO BEAM AND FRAME ELEMENTS

4.4.1 Governing Equations

Recall that the Euler–Bernoulli beam theory is based on the assumption that plane cross-sections remain plane *and perpendicular* to the longitudinal axis after bending. This assumption implies that all transverse shear strains are zero. When the normality assumption is not used, i.e., plane sections remain plane but not necessarily normal to the longitudinal axis after deformation, the transverse shear strain ε_{xz} is not zero. Therefore, the rotation of a transverse normal plane about the y axis is not equal to $-dw/dx$. Beam theory based on these relaxed assumptions is called a *shear deformation beam theory,* most commonly known as the *Timoshenko beam theory.* We denote the rotation about the y axis by an independent function $\Psi(x)$. The governing equilibrium equations of the Timoshenko beam theory are

$$\frac{d}{dx}\left[GAK_s\left(\Psi + \frac{dw}{dx}\right)\right] + f = 0 \tag{4.80a}$$

$$\frac{d}{dx}\left(EI\frac{d\Psi}{dx}\right) - GAK_s\left(\Psi + \frac{dw}{dx}\right) = 0 \tag{4.80b}$$

where G is the shear modulus and K_s is the *shear correction coefficient,* which is introduced to account for the difference in the constant state of shear stress in this theory and the parabolic variation of the actual (i.e., as predicted by equilibrium equations) shear stress through the thickness. When the second equation is substituted into the first for $GAK_s(\Psi + dw/dx)$, and Ψ is replaced with $-dw/dx$, we obtain governing equation (4.1) of the Euler–Bernoulli beam theory.

4.4.2 Weak Form

The weak form of (4.80) over an element $\Omega^e = (x_A, x_B)$ is developed using the usual procedure, now applied to each equation. We multiply (4.80a) wtih a weight function $-w_1$ and (4.80b) with a weight function $-w_2$, and integrate over the element length:

$$0 = \int_{x_A}^{x_B} -w_1\left\{\frac{d}{dx}\left[GAK_s\left(\Psi + \frac{dw}{dx}\right)\right] + f\right\} dx$$

$$0 = \int_{x_A}^{x_B} -w_2\left[\frac{d}{dx}\left(EI\frac{d\Psi}{dx}\right) - GAK_s\left(\Psi + \frac{dw}{dx}\right)\right] dx$$

Integrating the first term of each integral once by parts, we obtain

$$0 = \int_{x_A}^{x_B} \left[\frac{dw_1}{dx} GAK_s\left(\Psi + \frac{dw}{dx}\right) - w_1 f\right] dx - \left[w_1 GAK_s\left(\Psi + \frac{dw}{dx}\right)\right]_{x_A}^{x_B} \tag{4.81a}$$

$$0 = \int_{x_A}^{x_B} \left[\frac{dw_2}{dx} EI\frac{d\Psi}{dx} + w_2 GAK_s\left(\Psi + \frac{dw}{dx}\right)\right] dx - \left[w_2 EI\frac{d\Psi}{dx}\right]_{x_A}^{x_B} \tag{4.81b}$$

The coefficients of the weight functions w_1 and w_2 in the boundary integrals are

$$GAK_s\left(\Psi + \frac{dw}{dx}\right) \equiv V \quad \text{and} \quad EI\frac{d\Psi}{dx} \equiv M \tag{4.82}$$

where V is the shear force and M is the bending moment; these coefficients constitute the secondary variables of the weak form. The weight functions w_1 and w_2 must have the physical interpretations that give $w_1 V$ and $w_2 M$ units of work. Clearly, w_1 must be equivalent to (the variation of) the transverse deflection w, and w_2 must be equivalent to (the variation of) the rotation function Ψ:

$$w_1 \sim w, \qquad w_2 \sim \Psi$$

Hence, the primary variables of the formulation are w and Ψ. Denoting the shear forces and bending moments at the endpoints of the element by the expressions

$$\begin{aligned} Q_1^e &\equiv -\left[GAK_s\left(\Psi + \frac{dw}{dx}\right)\right]\Big|_{x_A}, \qquad Q_2^e \equiv -\left(EI\frac{d\Psi}{dx}\right)\Big|_{x_A} \\ Q_3^e &\equiv \left[GAK_s\left(\Psi + \frac{dw}{dx}\right)\right]\Big|_{x_B}, \qquad Q_4^e \equiv \left(EI\frac{d\Psi}{dx}\right)\Big|_{x_B} \end{aligned} \tag{4.83}$$

the weak statements in (4.81) can be written in the final form

$$\begin{aligned} 0 &= \int_{x_A}^{x_B} \left[GAK_s \frac{dw_1}{dx}\left(\Psi + \frac{dw}{dx}\right) - w_1 f\right] dx - w_1(x_A)Q_1^e - w_1(x_B)Q_3^e \\ 0 &= \int_{x_A}^{x_B} \left[EI\frac{dw_2}{dx}\frac{d\Psi}{dx} + GAK_s w_2\left(\Psi + \frac{dw}{dx}\right)\right] dx - w_2(x_A)Q_2^e - w_2(x_B)Q_4^e \end{aligned} \tag{4.84}$$

4.4.3 Finite Element Model

A close examination of the terms in (4.84) shows that both w and Ψ are differentiated only once with respect to x. Since the primary variables are the dependent unknowns themselves (and do not include their derivatives), the Lagrange interpolation of w and Ψ is appropriate here. The minimum admissible degree of interpolation is linear, so that $dw/dx \neq 0$ and $d\Psi/dx \neq 0$. The variables w and Ψ do not have the same physical units; they can be interpolated, in general, with different degrees of interpolation.

Let us consider Lagrange interpolation of w and Ψ in the form

$$w = \sum_{j=1}^{m} w_j \psi_j^{(1)}, \qquad \Psi = \sum_{j=1}^{n} s_j \psi_j^{(2)} \tag{4.85}$$

where $\psi_j^{(1)}$ and $\psi_j^{(2)}$ are the Lagrange interpolation functions of degree $m-1$ and $n-1$, respectively. In general, m and n are independent of each other, although $m = n$ is most common. However, when $m = n = 2$ (i.e., linear interpolation is used), the derivative of w is

$$\left(\frac{dw}{dx}\right)^e = \frac{w_2^e - w_1^e}{h_e}$$

which is element-wise-constant. The rotation function Ψ, being linear, is not consistent with that predicted by $w(x)$. For thin beams, the transverse shear deformation is negligible, and we have $\Psi = -dw/dx$, which requires

$$s_1^e \frac{x_B - x}{h_e} + s_2^e \frac{x - x_A}{h_e} = -\frac{w_2^e - w_1^e}{h_e} \tag{4.86a}$$

or, equivalently (by equating like coefficients on both sides),

$$s_1^e x_B - s_2^e x_A = -(w_2^e - w_1^e), \qquad s_2^e - s_1^e = 0 \tag{4.86b}$$

which in turn requires

$$s_1^e = s_2^e = -\frac{w_2^e - w_1^e}{h_e} \tag{4.87}$$

This implies that $\Psi(x)$ is a constant:

$$\Psi(x) = s_1^e \frac{x_B - x}{h_e} + s_2^e \frac{x - x_A}{h_e} = s_1^e \ (= s_2^e) \tag{4.88}$$

However, a constant state of $\Psi(x)$ is not admissible, because the bending energy of the element,

$$\int_{x_A}^{x_B} \frac{EI}{2} \left(\frac{d\Psi}{dx}\right)^2 dx \tag{4.89a}$$

would be zero. This numerical problem is known as *shear locking*.

To circumvent this, two alternative procedures have been developed in the literature:

1. Use a *consistent interpolation* for w and Ψ such that dw/dx and Ψ are polynomials of the same order (i.e., $m = n + 1$).

2. Use equal interpolation (i.e. $m = n$) for w and Ψ, but evaluate the bending energy with actual interpolation of Ψ and the shear energy

$$\int_{x_A}^{x_B} \frac{GAK_s}{2}\left(\frac{dw}{dx} + \Psi\right)^2 dx \tag{4.89b}$$

with a polynomial that is one order lower.

The latter can be achieved computationally without using different interpolations of Ψ in (4.89*a*, *b*). In the numerical evaluation of the integral (4.89*b*), we use a quadrature rule (see Section 7.1 for details) that is necessary to calculate the integral

$$\int_{x_A}^{x_B} GAK_s\left(\frac{dw}{dx}\right)^2 dx \tag{4.89c}$$

exactly. For example, when the linear interpolation of w and Ψ is used, we use a one-point quadrature rule to evaluate (4.89*b*) because a one-point quadrature would give the exact value of the integral (4.89*c*) when GA = constant. A two-point quadrature is needed to evaluate the integral in (4.89*b*) exactly. The use of *reduced integration,* i.e., one-point quadrature on the integral (4.89*b*), would result in the linear term in the approximation of Ψ not contributing to the shear energy. For illustrative purposes, we take a detailed look at the expression

$$\frac{GAK_s}{2}\int_{x_A}^{x_B}\left(\frac{dw}{dx} + \Psi\right)^2 dx = \frac{GAK_s}{2}\left[\left(\frac{dw}{dx} + \Psi\right)^2\bigg|_{x=x_A+h_e/2}\right]h_e$$

where $x = x_A + \frac{1}{2}h_e$ is the midpoint of the element and h_e is its length. Substituting (4.85) into this expression (with $m = n = 2$), we have

$$\frac{GAK_sh_e}{2}\left(\frac{w_2^e - w_1^e}{h_e} + s_1^e\frac{x_B - x}{h_e} + s_2^e\frac{x - x_A}{h_e}\right)^2\bigg|_{x=x_A+h_e/2}$$
$$= \frac{GAK_sh_e}{2}\left(\frac{w_2^e - w_1^e}{h_e} + \frac{s_1^e + s_2^e}{2}\right)^2 \tag{4.90}$$

which is a weaker requirement than (4.87), i.e., if (4.87) holds then (4.90) also holds, but (4.90) does not imply (4.87). We note that (4.87) must hold only for problems for which the transverse shear energy (4.89) is negligible.

In summary, we use either consistent interpolation ($m = n + 1$) or equal interpolation with reduced integration in the evaluation of the transverse shear stiffness coefficients in the Timoshenko beam element. We consider both forms of elements here.

Substitution of (4.85) for w and Ψ, and $w_1 = \psi_i^{(1)}$ and $w_2 = \psi_i^{(2)}$ into the weak forms (4.84), we obtain the finite element equations:

$$\boxed{\begin{aligned} 0 &= \sum_{j=1}^{m} K_{ij}^{11}w_j + \sum_{j=1}^{n} K_{ij}^{12}s_j - F_i^1 \quad (i = 1, 2, \ldots, m)\\ 0 &= \sum_{j=1}^{m} K_{ij}^{21}w_j + \sum_{j=1}^{n} K_{ij}^{22}s_j - F_i^2 \quad (i = 1, 2, \ldots, n)\end{aligned}} \tag{4.91a}$$

where

$$\boxed{\begin{aligned} K_{ij}^{11} &= \int_{x_A}^{x_B} GAK_s \frac{d\psi_i^{(1)}}{dx}\frac{d\psi_j^{(1)}}{dx}\,dx \\ K_{ij}^{12} &= \int_{x_A}^{x_B} GAK_s \frac{d\psi_i^{(1)}}{dx}\psi_j^{(2)}\,dx = K_{ji}^{21} \\ K_{ij}^{22} &= \int_{x_A}^{x_B} \left(EI\frac{d\psi_i^{(2)}}{dx}\frac{d\psi_j^{(2)}}{dx} + GAK_s\psi_i^{(2)}\psi_j^{(2)}\right)dx \\ F_j^1 &= \int_{x_A}^{x_B} f\psi_i^{(1)}\,dx + Q_{2i-1}, \qquad F_i^2 = Q_{2i} \end{aligned}} \tag{4.91b}$$

Equations (4.91a) can be written in matrix form as

$$\begin{bmatrix} [K^{11}] & [K^{12}] \\ [K^{21}] & [K^{22}] \end{bmatrix} \begin{Bmatrix} \{w\} \\ \{s\} \end{Bmatrix} = \begin{Bmatrix} \{F^1\} \\ \{F^2\} \end{Bmatrix} \tag{4.92}$$

A CONSISTENT INTERPOLATION ELEMENT (CIE). To illustrate the use of consistent interpolation, we select $\psi_i^{(1)}$ to be Lagrange quadratic polynomials and $\psi_i^{(2)}$ Lagrange linear polynomials. For this choice of interpolation, $[K^{11}]$ is 3×3, $[K^{12}]$ is 3×2, and $[K^{22}]$ is 2×2. The explicit forms of the matrices, when EI and GAK_s are constant, are

$$[K^{11}] = \frac{GAK_s}{3h_e}\begin{bmatrix} 7 & -8 & 1 \\ -8 & 16 & -8 \\ 1 & -8 & 7 \end{bmatrix}, \qquad [K^{12}] = \frac{GAK_s}{6}\begin{bmatrix} -5 & -1 \\ 4 & -4 \\ 1 & 5 \end{bmatrix}$$
$$[K^{22}] = \frac{EI}{h_e}\begin{bmatrix} 1 & -1 \\ -1 & 1 \end{bmatrix} + \frac{GAK_s h_e}{6}\begin{bmatrix} 2 & 1 \\ 1 & 2 \end{bmatrix} \tag{4.93}$$

The element equations become

$$GAK_s \begin{bmatrix} \frac{7}{3h} & -\frac{8}{3h} & \frac{1}{3h} & -\frac{5}{6} & -\frac{1}{6} \\ -\frac{8}{3h} & \frac{16}{3h} & -\frac{8}{3h} & \frac{4}{6} & -\frac{4}{6} \\ \frac{1}{3h} & -\frac{8}{3h} & \frac{7}{3h} & \frac{1}{6} & \frac{5}{6} \\ -\frac{5}{6} & \frac{4}{6} & \frac{1}{6} & \frac{\alpha}{h}+\frac{h}{3} & -\frac{\alpha}{h}+\frac{h}{6} \\ -\frac{1}{6} & -\frac{4}{6} & \frac{5}{6} & -\frac{\alpha}{h}+\frac{h}{6} & \frac{\alpha}{h}+\frac{h}{3} \end{bmatrix} \begin{Bmatrix} w_1 \\ w_3 \\ w_2 \\ s_1 \\ s_2 \end{Bmatrix}^e = \begin{Bmatrix} f_1 \\ f_3 \\ f_2 \\ 0 \\ 0 \end{Bmatrix}^e + \begin{Bmatrix} Q_1 \\ \hat{Q} \\ Q_3 \\ Q_2 \\ Q_4 \end{Bmatrix}^e \tag{4.94}$$

where $\alpha = EI/GAK_s$, $\hat{Q}$ is any specified transverse force at node 3, and

Q_1, Q_3) are shear forces and (Q_2, Q_4) bending moments at the end nodes 1 and 2 of the element.

Note that node 3, which is the middle node of the element, is not connected to other elements, and the only degree of freedom there is the transverse deflection. Thus, there are different number of degrees of freedom at different nodes of the element, and this therefore complicates the assembly of elements and its implementation on a computer. Hence, we eliminate the node 3 dependence in the system of element equations by condensing out w_3. We obtain

$$\begin{bmatrix} \dfrac{GAK_s}{h} & -\dfrac{GAK_s}{h} & -\tfrac{1}{2}GAK_s & -\tfrac{1}{2}GAK_s \\ -\dfrac{GAK_s}{h} & \dfrac{GAK_s}{h} & \tfrac{1}{2}GAK_s & \tfrac{1}{2}GAK_s \\ -\tfrac{1}{2}GAK_s & \tfrac{1}{2}GAK_s & \tfrac{1}{4}GAK_s h + \dfrac{EI}{h} & \tfrac{1}{4}GAK_s h - \dfrac{EI}{h} \\ -\tfrac{1}{2}GAK_s & \tfrac{1}{2}GAK_s & \tfrac{1}{4}GAK_s h - \dfrac{EI}{h} & \tfrac{1}{4}GAK_s h + \dfrac{EI}{h} \end{bmatrix} \begin{Bmatrix} w_1 \\ w_2 \\ s_1 \\ s_2 \end{Bmatrix}^e = \begin{Bmatrix} f_1 + \tfrac{1}{2}\hat{f}_3 \\ f_2 + \tfrac{1}{2}\hat{f}_3 \\ -\tfrac{1}{8}\hat{f}_3 h \\ \tfrac{1}{8}\hat{f}_3 h \end{Bmatrix}^e + \begin{Bmatrix} Q_1 \\ Q_3 \\ Q_2 \\ Q_4 \end{Bmatrix}^e \tag{4.95}$$

where $\hat{f}_3 = f_3 + \hat{Q}$. This is obtained by solving the second equation of (4.94) for w_3 and substituting for w_3 in the first four equations of (4.94). The forces f_i are obtained using

$$f_i^e = \int_{x_A}^{x_B} f(x)\psi_i^e\,dx, \qquad \psi_i^e = [\psi_i^{(1)}]^e = \text{quadratic interpolation functions} \tag{4.96}$$

The element equations (4.95) can be assembled in the same manner as the classical beam equations, with two degrees of freedom (w, s) per node. However, it should be noted that w is interpolated with quadratic polynomials.

For constant $f = f_0$, the specified load vector in (4.95) takes the form [because $f_1^e = \frac{1}{6}f_0 h$, $f_2^e = \frac{1}{6}f_0 h$, $f_3^e = \frac{4}{6}f_0 h$, $\hat{Q}_0 = 0$; see (3.40*b*)]

$$\{f^e\} = \begin{Bmatrix} \tfrac{1}{2}f_0 h \\ \tfrac{1}{2}f_0 h \\ -\tfrac{1}{12}f_0 h^2 \\ \tfrac{1}{12}f_0 h^2 \end{Bmatrix} \tag{4.97}$$

which is precisely the same as that obtained in the classical beam element [see (4.15)].

AN EQUAL INTERPOLATION, REDUCED INTEGRATION ELEMENT (RIE). When equal interpolation is used ($m = n$), all submatrices in (4.92) are of the same order: $n \times n$. The element coefficient matrices K_{ij}^{11} and K_{ij}^{12} are evaluated exactly, as is the first part of K_{ij}^{22}. The second part of K_{ij}^{22} is to be evaluated using reduced integration. For the choice of linear interpolation functions, and for constant values of GAK_s and EI, the matrices have the following explicit values:

$$[K^{11}] = \frac{GAK_s}{h_e}\begin{bmatrix} 1 & -1 \\ -1 & 1 \end{bmatrix}, \quad [K^{12}] = \frac{GAK_s}{2}\begin{bmatrix} -1 & -1 \\ 1 & 1 \end{bmatrix}$$
$$[K^{22}] = \frac{EI}{h_e}\begin{bmatrix} 1 & -1 \\ -1 & 1 \end{bmatrix} + \frac{GAK_s h_e}{4}\begin{bmatrix} 1 & 1 \\ 1 & 1 \end{bmatrix} \tag{4.98}$$

where one-point integration is used to evaluate the second part of $[K^{22}]$. Note that $[K^{11}]$, $[K^{12}]$, and the first part of $[K^{22}]$ are also evaluated exactly with one-point quadrature when EI and GAK_s are constant. Hence the uniform one-point integration for $[K^{\alpha\beta}]$ satisfies all requirements.

The element equations are

$$\begin{bmatrix} \frac{GAK_s}{h} & -\frac{1}{2}GAK_s & -\frac{GAK_s}{h} & -\frac{1}{2}GAK_s \\ -\frac{1}{2}GAK_s & \frac{1}{4}GAK_s h + \frac{EI}{h} & \frac{1}{2}GAK_s & \frac{1}{4}GAK_s h - \frac{EI}{h} \\ -\frac{GAK_s}{h} & \frac{1}{2}GAK_s & \frac{GAK_s}{h} & \frac{1}{2}GAK_s \\ -\frac{1}{2}GAK_s & \frac{1}{4}GAK_s h - \frac{EI}{h} & \frac{1}{2}GAK_s & \frac{1}{4}GAK_s h + \frac{EI}{h} \end{bmatrix} \begin{Bmatrix} w_1 \\ s^1 \\ w_2 \\ s_2 \end{Bmatrix}^e = \begin{Bmatrix} f_1 \\ 0 \\ f_2 \\ 0 \end{Bmatrix}^e + \begin{Bmatrix} Q_1 \\ Q_2 \\ Q_3 \\ Q_4 \end{Bmatrix}^e \tag{4.99}$$

It is interesting to note that the element stiffness matrix in (4.99) is the same as that in (4.95) obtained from the consistent interpolation with quadratic approximation of w and linear approximation of Ψ, except that the nodal variables are listed in a different order. The only difference is the load representation. In the consistent interpolation, the load vector is equivalent to that of the Euler–Bernoulli beam theory, whereas, in the equal interpolation with reduced integration element, the load vector does not contain any moment components due to the distributed load.

The quadratic interpolation of both w and Ψ with full integration of the element coefficient matrices also suffers slightly from the shear-locking phenomenon. A uniform two-point quadrature rule has the desired effect on $[K^{11}]$, $[K^{12}]$, and $[K^{22}]$; i.e., $[K^{11}]$, $[K^{12}]$, and the first term of $[K^{22}]$ will be evaluated exactly and the second term of $[K^{22}]$ approximately. As the degree of approximation and/or the number of elements in the mesh is increased, the shear locking will disappear and reduced integration is not necessary.

Example 4.5. Here we reconsider the indeterminate beam problem of Example 4.2 (see Fig. 4.7), and analyze it with the reduced integration element (RIE) and consistent interpolation element (CIE). Unlike the Euler–Bernoulli beam element, the Timoshenko beam element does not yield exact values at the nodes, even when EI and GAK_s are constant. This is because of the coupled nature of the equations relating w and Ψ.

The beam is modeled using two ($h_1 = 4$, $h_2 = 6$) and four ($h_1 = 2$, $h_2 = 2$, $h_3 = 3$, $h_4 = 3$) linear elements with reduced integration. The element equations for a typical element are the same as those in (4.99), where f_i^e for the elements with distributed force are given by

$$f_i^e = \int_{x_e-4}^{x_{e+1}-4} [-\tfrac{100}{6}\bar{x}\psi_i^e(\bar{x})]\, d\bar{x} \quad \text{for } x_e \geq 4 \tag{4.100}$$

where $\bar{x}$ is a coordinate such that $x = \bar{x} + 4$. We obtain

$$\{f^2\} = -\frac{h_2 f_0}{6}\begin{Bmatrix}1\\2\end{Bmatrix}$$

for the two-element case and

$$\{f^3\} = -\frac{h_3 f_0}{12}\begin{Bmatrix}1\\2\end{Bmatrix}, \qquad \{f^4\} = -\frac{h_3 f_0}{12}\begin{Bmatrix}1\\2\end{Bmatrix} - \frac{h_4 f_0}{4}\begin{Bmatrix}1\\1\end{Bmatrix}$$

for the four-element case ($h_3 = h_4$).

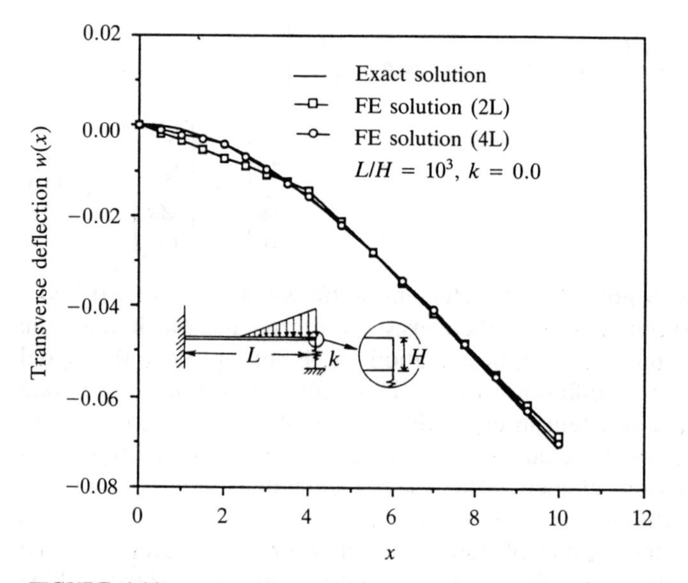

FIGURE 4.14
Comparison of the transverse deflection obtained using Timoshenko beam elements with the exact deflection of a cantilever beam (see Fig. 4.7).

Figure 4.14 shows plots of transverse deflection versus x for the case with $k = 0$ (i.e., a centilever beam), beam length-to-height ratio $L/H = 10^3$, and $EI = 10^6$ ft^2 lb. The ratio L/H is a measure of the thinness of the beam. When it is large (i.e., for very thin beams), the transverse shear deformation is negligible and the Euler–Bernoulli and Timoshenko beam theories give the same results. For small values of L/H, say 10 (i.e., $GAK_s = 4 \times 10^6$ lb), Timoshenko beam theory predicts larger deflections, and for the problem at hand the difference is not noticeable enough to be seen if the deflections from both theories were included in Fig. 4.14. The finite element solutions shown in the figure are obtained with the reduced integration elements.

Figure 4.15 contains results of a convergence study, which includes two and four linear and quadratic element meshes for the case of $k/EI = 1.0$ and $L/H = 10$. The reduced integration elements are used. The scale used for plotting the deflections is such that the difference between the finite element solutions and the exact solution can be seen clearly. The mesh of four quadratic elements gives the converged solution. To show the effect of the transverse shear deformation, the exact solution for the thin beam ($L/H = 10^3$) is also shown in Fig. 4.15. It is clear from the results that the convergence is rapid and the quadratic elements yield faster convergence.

The accuracy of the reduced integration element (RIE) relative to the full integration element (FIE) and the consistent interpolation element (CIE) can be seen from the results shown in Fig. 4.16 (for $L/H = 10$ and $k/EI = 1$). Clearly, the CIE element gives more accurate results, followed by the reduced integration element. Recall that the stiffness matrices of the two elements are identical, the only difference being in their force vectors [see (4.95) and (4.99)]. Of course, in the CIE element, the transverse deflection is interpolated using quadratic functions. The plots shown in Fig. 4.17 indicate that the reduced integration is necessry even for quadratic elements, although they are not as sensitive to locking.

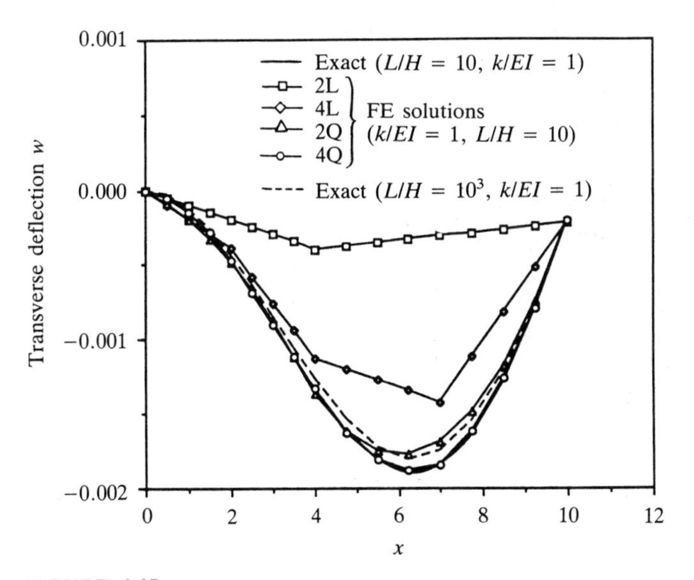

FIGURE 4.15
Convergence of the finite element solution to the exact solution of the beam in Fig. 4.7 ($k/EI = 1$); the Timoshenko beam element is used.

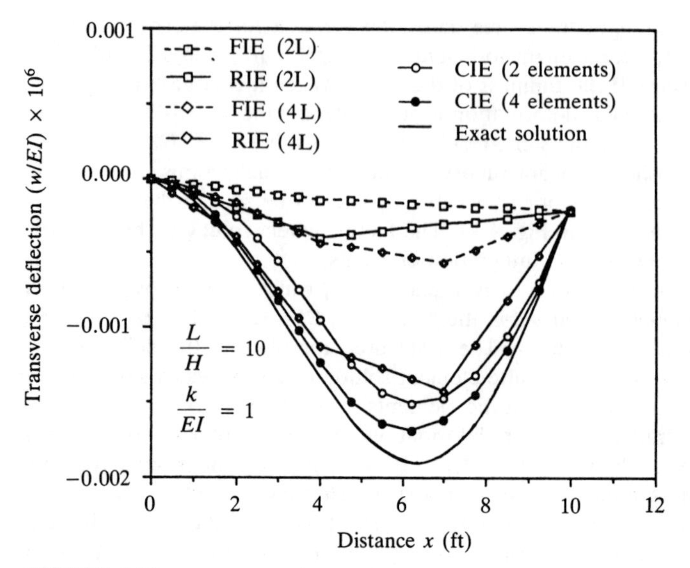

FIGURE 4.16
Comparison of the transverse deflection $w(x)$ obtained with reduced integration elements (RIE), full integration elements (FIE), and consistent interpolation elements (CIE) with the exact solution: 2L, two linear elements; 4L, four linear elements.

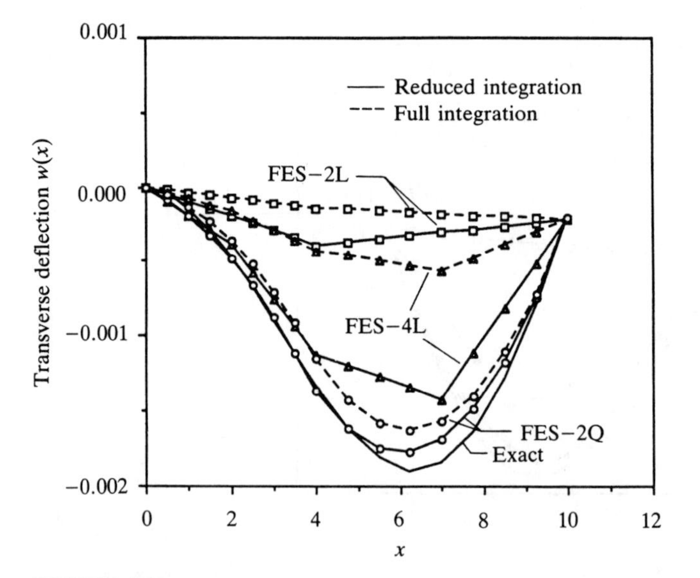

FIGURE 4.17
The effect of full and reduced integration on the accuracy of the finite element solution (FES) obtained using the Timoshenko beam element.

The following general observations can be made about various finite element models based on the Timoshenko beam theory:

1. The reduced integration element (RIE) exhibits less locking compared with the full integration element (FIE).
2. As the number of elements in the mesh is increased or the degree of approximation is increased (i.e., higher-order elements are used), the finite element solutions obtained by both RIE and FIE elements improve; i.e., the effect of locking is reduced with mesh refinements and higher-order elements.
3. The consistent interpolation element (i.e., quadratic approximation of w and linear approximation of Ψ) with full integration yields a more accurate solution than that predicted by the RIE element. This is due to a better representation of the distributed load.
4. The element with quadratic approximation of both w and Ψ and reduced integration of the coefficients yields more accurate results than the consistent interpolation element with quadratic approximation of w and linear approximation of Ψ and with full integration of the coefficients.

The frame element based on shear deformation beam theory can be obtained by superposing the bar and beam stiffnesses.

4.5 INCLUSION OF CONSTRAINT EQUATIONS

When the support plane of a roller support is at an angle to the global coordinate system (see Fig. 4.18), the boundary conditions on the displacements and forces at the roller are

$$u_n^e = 0, \qquad Q_t^e = Q_0 \tag{4.101}$$

where u_n^e is the normal component of the displacement, Q_t^e is the tangential component of the force at node 1 of the element Ω^e, and Q_0 is any specified tangential force. These conditons must be expressed in terms of the global

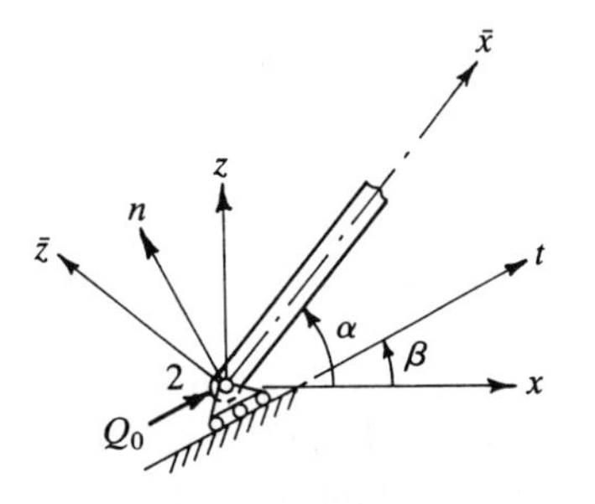

$$u_t = u\cos\beta + w\sin\beta$$
$$u_n = w\cos\beta - u\sin\beta$$

FIGURE 4.18
Transformation of specified boundary conditions from a local coordinate system to the global coordinate system.

components of displacements and forces by means of the transformation (4.53) and (4.55):

$$u_n^e = -u_1^e \sin\beta + u_2^e \cos\beta = 0 \tag{4.102a}$$

$$Q_t^e = Q_1^e \cos\beta + Q_2^e \sin\beta = Q_0 \tag{4.102b}$$

where (u_1^e, u_2^e) and (Q_1^e, Q_2^e) are the x and z components of the displacements and forces at the support.

Equations (4.102) can be incorporated into the global system of equations as follows. Consider the assembled system of equations,

$$\begin{aligned} K_{11}U_1 + K_{12}U_2 + \cdots + K_{1n}U_n &= F_1 \\ K_{21}U_1 + K_{22}U_2 + \cdots + K_{2n}U_n &= F_2 \\ &\vdots \\ K_{n1}U_1 + K_{n2}U_2 + \cdots + K_{nn}U_n &= F_n \end{aligned} \tag{4.103}$$

where U_i are the global generalized displacement degrees of freedom and F_i are the sums of the applied (f_i^e) and internal (Q_i^e) generalized force degrees of freedom. Suppose that the roller support is at the first node of the element Ω^e, and that the corresponding displacement degrees of freedom are U_I, U_J, and U_K. The forces Q_1^e, Q_2^e, and Q_3^e corresponding to these displacements end up in the forces F_I, F_J, and F_K, respectively. Hence, to include the force boundary condition of (4.102b), we must add $\cos\beta$ times the Ith global equation to $\sin\beta$ times the Jth equation:

$$(\cos\beta)(K_{I1}U_1 + K_{I2}U_2 + \ldots + K_{II}U_I + K_{IJ}U_J + \ldots + K_{IN}U_N) = (\cos\beta)F_I$$
$$(\sin\beta)(K_{J1}U_1 + K_{J2}U_2 + \ldots + K_{JI}U_I + K_{JJ}U_J + \ldots + K_{JN}U_N) = (\sin\beta)F_J$$

and obtain

$$\begin{aligned} (K_{I1}\cos\beta + K_{J1}\sin\beta)U_1 + \ldots &+ (K_{IN}\cos\beta + K_{JN}\sin\beta)U_N \\ &= (F_I\cos\beta + F_J\sin\beta) \\ &= (f_1^e\cos\beta + f_2^e\sin\beta) + (Q_1^e\cos\beta + Q_2^e\sin\beta) \\ &= (f_1^e\cos\beta + f_2^e\sin\beta) + Q_0 \end{aligned} \tag{4.104}$$

We replace the Jth equation with this and replace the Ith equation with (4.102a):

$$(\sin\beta)U_I + (-\cos\beta)U_J = 0 \tag{4.105}$$

These modifications of the global set of algebraic equations violate the symmetry of the resulting global stiffness matrix.

To retain the symmetry of the global stiffness matrix, an alternative method of incorporating the conditions (4.102) is presented here. Equations (4.102a) or (4.105) can be viewed as the constraint equations among the global displacements, which have a companion relation among the associated forces, i.e., (4.102b). In the following paragraphs, we present a general procedure by which constraint equations of the type

$$\begin{Bmatrix} \{U^1\} \\ \{U^2\} \end{Bmatrix} = [A]\{U^1\} \tag{4.106a}$$

among the primary variables

$$\{U\} = \begin{Bmatrix} \{U^1\} \\ \{U^2\} \end{Bmatrix} \tag{4.106b}$$

can be implemented in the assembled system of equations (4.103). In (4.106), $\{U^1\}$ denotes the $n \times 1$ vector of generalized nodal displacements that are selected as the independent nodal variables, and $\{U^2\}$ denotes the $m \times 1$ vector of generalized nodal displacements that are selected as the dependent nodal variables. The matrix operator $[A]$ is of order $(n+m) \times n$.

For example, consider a frame problem with N nodes and a total of $3N$ $(=m+n)$ degrees of freedom in the problem. If one of the supports is a roller support, there will be one constraint equation of the type (4.102a). If the global node number corresponding to the roller is I then $p = 3(I-1)+1$ is the first degree of freedom at the support. Hence, the constraint condition (4.102a) can be expressed in terms of the generalized global displacement degrees of freedom U_p and U_{p+1} as

$$-U_p \sin\beta + U_{p+1} \cos\beta = 0 \tag{4.107}$$

Suppose that U_p is selected as the independent variable and U_{p+1} as the dependent variable. Then the number of independent variables is $n = 3N+2$, and there is only one dependent variable (i.e., $m=1$). For this case, the constraint (4.106a) has the form

$$\begin{Bmatrix} U_1 \\ U_2 \\ \vdots \\ U_p \\ U_{p+2} \\ \vdots \\ U_{m+n} \\ U_{p+1} \end{Bmatrix} = \begin{bmatrix} 1 & 0 & \dots & 0 & 0 & 0 & \dots & 0 \\ 0 & 1 & \dots & 0 & 0 & 0 & \dots & 0 \\ \vdots & \vdots & & \vdots & \vdots & \vdots & & \vdots \\ 0 & 0 & \dots & 0 & 1 & 0 & \dots & 0 \\ 0 & 0 & \dots & & & \ddots & & 0 \\ \vdots & \vdots & & & & & \ddots & \vdots \\ 0 & 0 & \dots & 0 & \underset{\uparrow\ p\text{th column}}{\tan\beta} & 0 & \dots & 0 \end{bmatrix} \begin{Bmatrix} U_1 \\ U_2 \\ \vdots \\ U_p \\ U_{p+2} \\ \vdots \\ U_{m+n} \end{Bmatrix} \tag{4.108}$$

Equation (4.106a) represents a transformation similar to (4.55a), between two sets of global generalized nodal displacements, with $[A]$ being the transformation matrix. Therefore, the discussion presented in the paragraph following (4.58) can be used to transform the equations in $m+n$ variables to those in only n variables.

Consider the assembled equations

$$[K]\{U\} = \{F\} \tag{4.109}$$

which can be rearranged as

$$[\bar{K}]\begin{Bmatrix}\{U^1\}\\ \{U^2\}\end{Bmatrix} = \{\bar{F}\} \tag{4.110}$$

where $\{U^1\}$ and $\{U^2\}$ denote the vectors defined in (4.106*a*). Using the transformation (4.106*a*), we obtain

$$[\bar{K}][A]\{U^1\} = \{\bar{F}\} \tag{4.111}$$

To obtain a symmetric coefficient matrix, we premultiply both sides with $[A]^T$, and arrive at

$$[A]^T[\bar{K}][A]\{U^1\} = [A]^T\{\bar{F}\}$$

or

$$[\hat{K}]\{U^1\} = \{\hat{F}\} \tag{4.112a}$$

where

$$[\hat{K}] = [A]^T[\bar{K}][A], \qquad \{\hat{F}\} = [A]^T\{\bar{F}\} \tag{4.112b}$$

Equations (4.112) are now ready for the implementation of boundary conditions and solution.

As an example, consider the frame element in Fig. 4.18. We have one constraint equation,

$$U_5 = U_4 \tan\beta$$

The transformation equation (4.108) has the form

$$\begin{Bmatrix} U_1\\ U_2\\ U_3\\ U_4\\ U_6\\ \hdashline U_5 \end{Bmatrix} = \begin{bmatrix} 1 & 0 & 0 & 0 & 0\\ 0 & 1 & 0 & 0 & 0\\ 0 & 0 & 1 & 0 & 0\\ 0 & 0 & 0 & 1 & 0\\ 0 & 0 & 0 & 0 & 1\\ \hdashline 0 & 0 & 0 & \tan\beta & 0 \end{bmatrix} \begin{Bmatrix} U_1\\ U_2\\ U_3\\ U_4\\ U_6 \end{Bmatrix}$$

and the stiffness matrix $[\bar{K}]$ is the same as $[K]$, with the fifth and sixth rows and columns interchanged:

$$[\bar{K}] = \begin{bmatrix} K_{11} \cdots K_{14} & K_{16} & K_{15}\\ \vdots \quad\;\; \vdots & \vdots & \vdots\\ K_{41} \cdots K_{44} & K_{46} & K_{45}\\ K_{61} \cdots K_{64} & K_{66} & K_{65}\\ K_{51} \cdots K_{54} & K_{56} & K_{55} \end{bmatrix}$$

The transformed stiffness matrix and force vector are, from (4.112*b*), of order 5×5 and 5×1, respectively. The elements of $[\hat{K}]$ and $\{\hat{F}\}$ are given by

$$\hat{K}_{ij} = \bar{K}_{ij} \quad (i, j = 1, 2, 3, 5)$$

$$\left.\begin{aligned} \hat{K}_{4i} &= \bar{K}_{4i} + \bar{K}_{6i} \tan\beta \\ \hat{K}_{i4} &= \bar{K}_{i4} + \bar{K}_{i6} \tan\beta \end{aligned}\right\} \quad (i = 1, 2, \ldots, 5; i \neq 4)$$

$$\hat{K}_{44} = \bar{K}_{44} + (\bar{K}_{46} + \bar{K}_{64}) \tan\beta + \bar{K}_{66} \tan^2\beta$$

$$\hat{F}_i = F_i \quad (i = 1, 2, 3, 5), \qquad \hat{F}_4 = \bar{F}_4 + \bar{F}_6 \tan\beta$$

During the imposition of the boundary conditions of the problem, the force $\hat{F}_4$ will be replaced by Q_0.

4.6 SUMMARY

In this chapter, finite element models of the classical (i.e., Euler–Bernoulli) and Timoshenko beam theories have been developed. The classical beam theory is governed by a fourth-order differential equation, and therefore results in a weak form whose primary variables contain the transverse deflection and its first derivative. Therefore, Hermite interpolation of the transverse deflection is required in order to impose the continuity of the deflection and its derivative at the nodes between elements. In the case of the Timoshenko beam theory, there are two second-order equations governing the transverse deflection and the rotation. The weak forms of the equations require Lagrange interpolation of the transverse deflection and rotation. Since the rotation function is like the (negative of the) derivative of the transverse deflection, the degree of the interpolation used for the rotation should be one less than that used for the transverse deflection. Such selective interpolation of the variables is called *consistent interpolation.* When the same interpolation functions are used to approximate the transverse deflection and the rotation, the resulting stiffness matrix is often too stiff—especially when the number of elements used is small—to yield good solutions. This is due to the inconsistency of interpolation of the variables, and the phenomenon is known as *shear locking.* It is overcome by the use of reduced integration to evaluate the stiffness coefficients associated with transverse shear strains. Both reduced integration elements (RIE) and consistent interpolation elements (CIE) have been developed here. It has been shown that consistent interpolation with linear approximation of the rotation and quadratic interpolation of the transverse deflection yields the same stiffness matrix as that obtained with a reduced integration element with linear approximation of the rotation as well as the deflection. However, the load vector of the consistent interpolation element, for a uniformly distributed transverse load, is equal to that of the Hermite cubic element of the classical beam theory.

The plane truss element and frame elements of the classical and Timoshenko beam theories have also been discussed. A plane truss element is a bar element that carries only axial loads and is oriented, in general, at an angle from the horizontal axis (the global x axis is taken to be horizontal). The bar element is first modified by adding columns and rows of zeros corresponding to the transverse displacements, so that it has two degrees of freedom (axial and transverse displacements) per node. The arbitrary plane truss element is then obtained by transforming the stiffness matrix and force vector from element coordinates to global coordinates, which are taken to be horizontal and vertical. Thus, a plane truss element has two degrees of freedom (horizontal and vertical displacements) per node, and carries only axial loads. The frame element is a superposition of the beam and bar elements, and has three degrees of freedom (axial displacement, transverse deflection, and rotation about an axis perpendicular to the plane of axial and transverse coordinates). The general plane frame element is oriented at an angle from the horizontal position, and its equations are obtained by transforming the equations of the frame element in local coordinates.

Finally, a procedure for including constraint conditions among the displacements and/or forces has been presented.

PROBLEMS

4.1–4.16. For the beam problems shown in Figs. P4.1–P4.16, use the Euler–Bernoulli beam element, and give:

(*a*) the assembled stiffness matrix and force vector;

(*b*) the specified global displacements and forces, and the equilibrium conditions;

(*c*) the condensed matrix equations for the primary unknowns (i.e., generalized displacements) and the secondary unknowns (i.e., generalized forces) separately.

Solve for the unknown displacements if there are less than four unknown displacements (use Cramer's rule), and evaluate the bending moment $M^c = EI\, d^2w/dx^2$ at point C using the finite element interpolation of w (when the condensed equations are not solved, express M^c in terms of the nodal values of

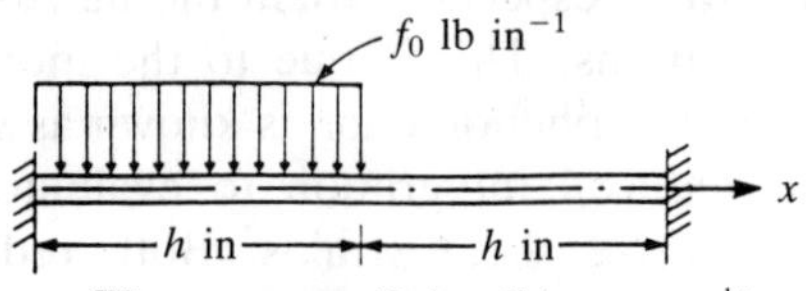

EI = constant; Point C is at $x = \frac{1}{2}h$

FIGURE P4.1

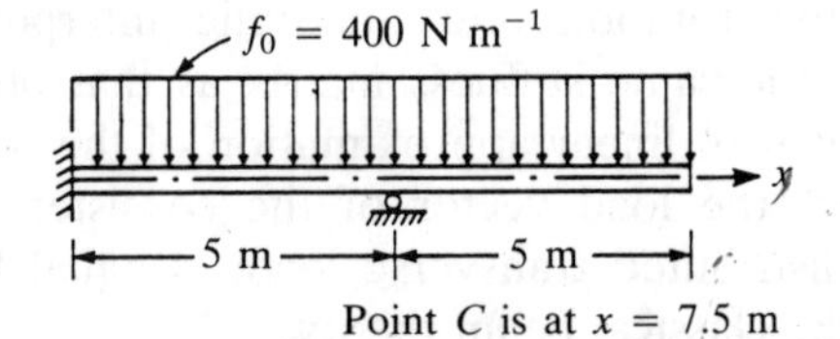

Point C is at x = 7.5 m

E = 200 GPa, G = 79 GPa, $A = 2.86 \times 10^{-3}$ m^2, $I = 20 \times 10^{-6}$ m^4

FIGURE P4.2

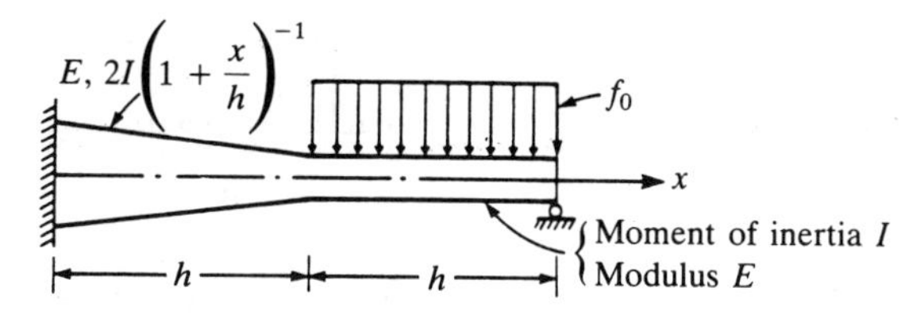

FIGURE P4.3

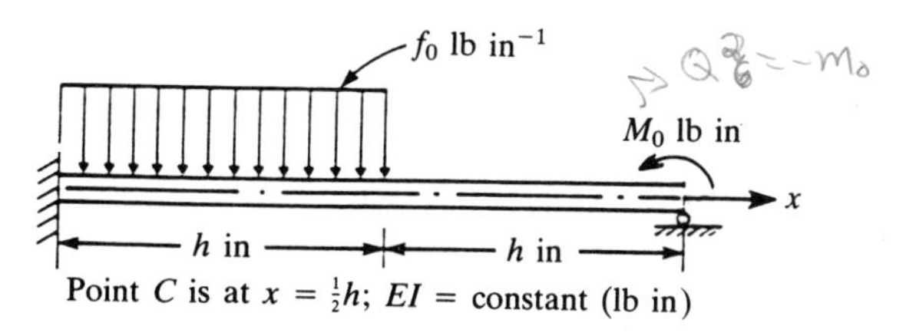

FIGURE P4.4

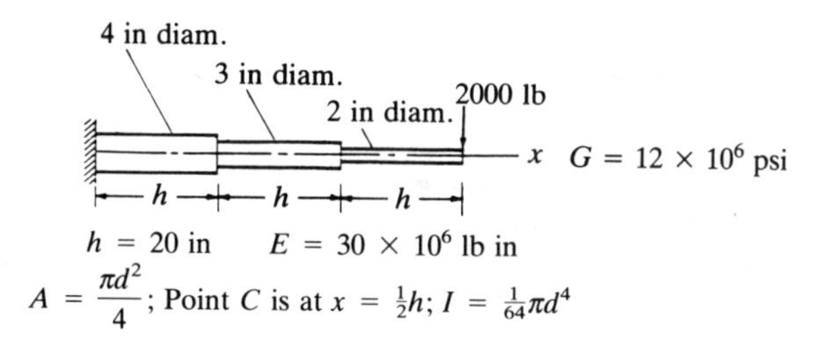

FIGURE P4.5

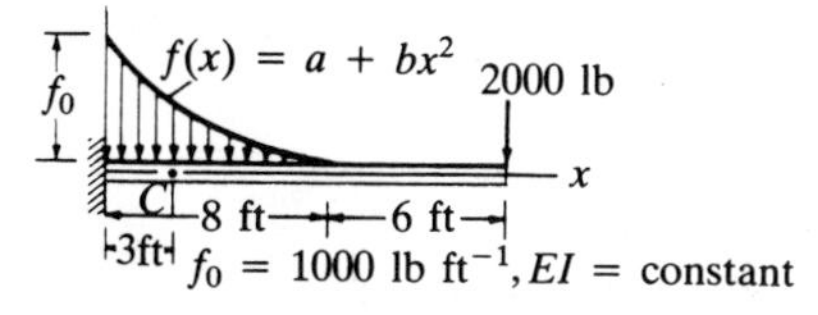

FIGURE P4.6

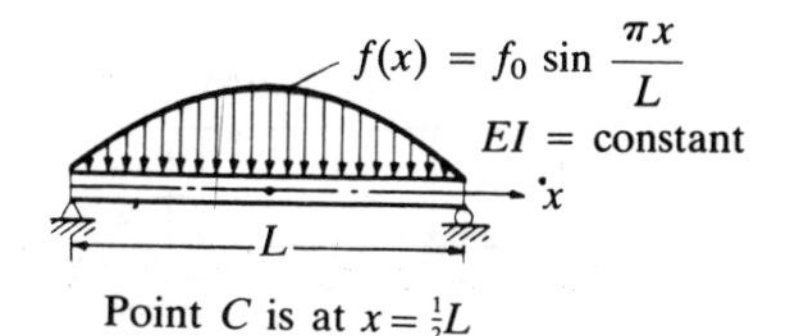

FIGURE P4.7

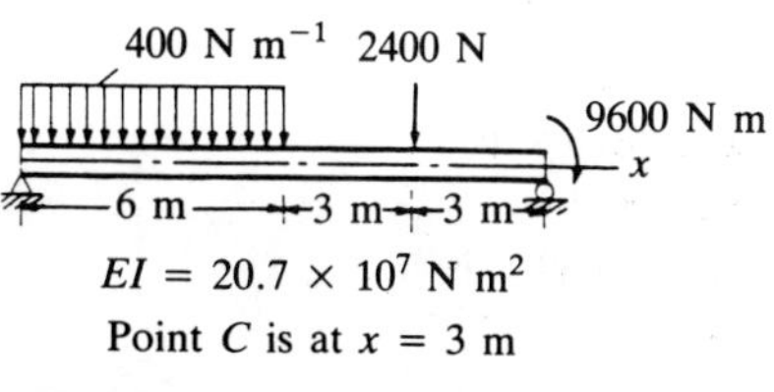

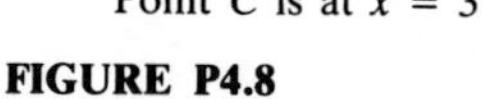
FIGURE P4.8

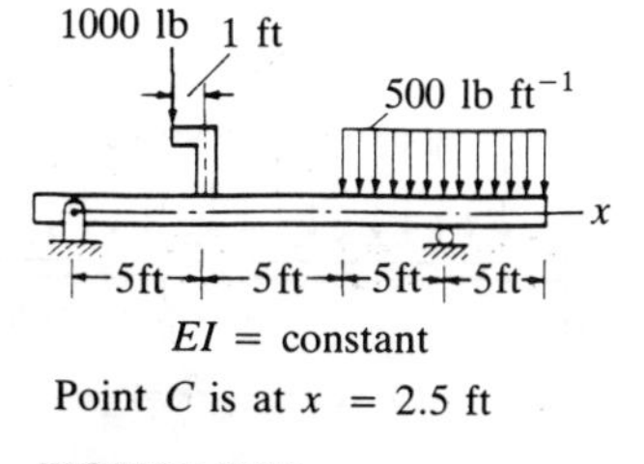

FIGURE P4.9

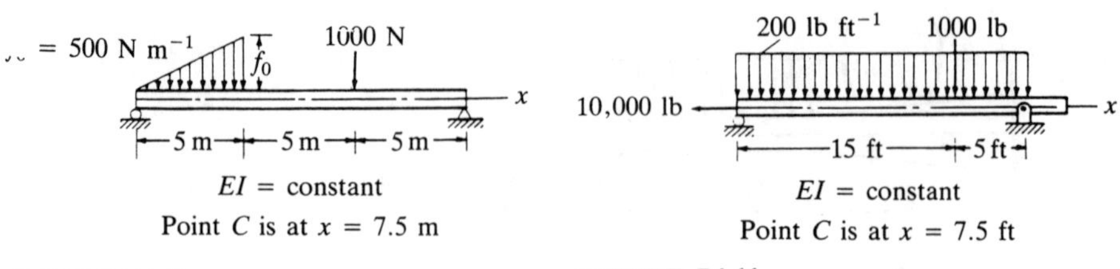

FIGURE P4.10

FIGURE P4.11

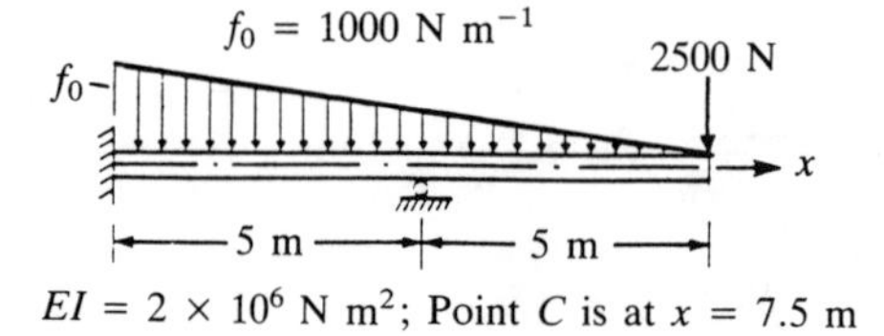

$EI = 2 \times 10^6$ N m^2; Point C is at $x = 7.5$ m

FIGURE P4.12

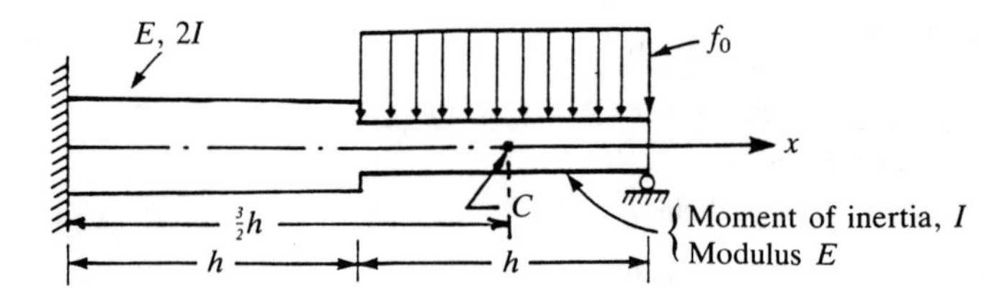

FIGURE P4.13

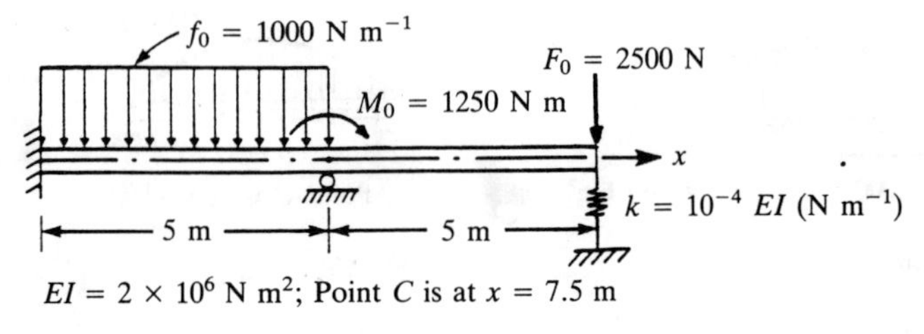

$EI = 2 \times 10^6$ N m^2; Point C is at $x = 7.5$ m

FIGURE P4.14

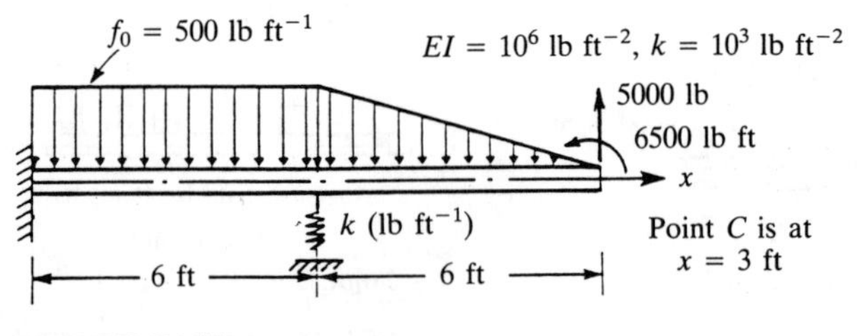

FIGURE P4.15

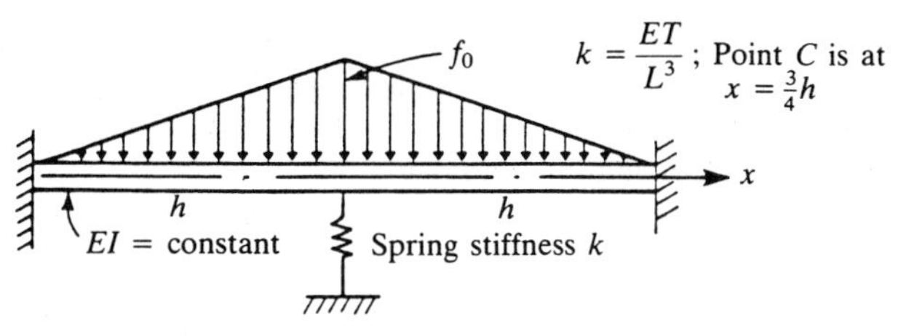

FIGURE P4.16

the appropriate element). Use the minimum number of elements required in each problem.

Answer: (Problem 4.1) $U_3 = -f_0h^4/48EI$, $U_4 = -(f_0h^3/96EI)$.

4.17–4.21. Repeat Problems 4.1, 4.2, 4.4–4.6 using the Timoshenko beam element (with reduced integration). Use a value of $\frac{5}{6}$ for the shear correction factor. Note that accurate results can be obtained only with a sufficiently large number of elements, when compared with the Euler–Bernoulli element.

4.22. Consider a simply supported beam on an elastic foundation (with foundation modulus k) and subjected to uniform transverse loading. Determine the transverse displacement at midspan using one Euler–Bernoulli beam element.

4.23. Consider the axisymmetric bending of a linear elastic circular plate of constant thickness. The governing differential equation according to the thin plate assumption is (see Section 7.3 for additional details)

$$\frac{1}{r}\frac{d^2}{dr^2}\left(rD_{11}\frac{d^2w}{dr^2} + D_{12}\frac{dw}{dr}\right) - \frac{1}{r}\frac{d}{dr}\left(D_{12}\frac{d^2w}{dr^2} + \frac{D_{22}}{r}\frac{dw}{dr}\right) = f$$

where D_{11}, D_{12}, and D_{22} are the plate material stiffnesses (constant), w is the transverse deflection, and r is the radial coordinate. Develop

(a) the weak form of the equation over a typical element $\Omega^e = (r_A, r_B)$; the quantities in parentheses should not be integrated by parts;

(b) the finite element model of the equation in the form

$$[K^e]\{u^e\} = \{f^e\} + \{Q^e\}$$

Make sure that $[K^e]$ is symmetric (i.e., the bilinear form in (a) should be symmetric). Comment on the interpolation functions that are admissible for the element.

4.24. The differential equations governing axisymmetric bending of circular plates according to thick plate theory are

$$-\frac{1}{r}\frac{d}{dr}\left[rA_{55}\left(\Psi + \frac{dw}{dr}\right)\right] = f$$

$$-\frac{1}{r}\frac{d}{dr}\left[r\left(D_{11}\frac{d\Psi}{dr} + \frac{D_{12}}{r}\Psi\right)\right] + \frac{1}{r}\left(D_{12}\frac{d\Psi}{dr} + \frac{D_{22}}{r}\Psi\right) + A_{55}\left(\Psi + \frac{dw}{dr}\right) = 0$$

where A_{55}, D_{11}, D_{12}, and D_{22} are plate material stiffnesses, Ψ is the rotation function, w is the transverse deflection, and f is the transverse load. Develop

(a) the weak form of the equations over an element;

(b) the finite element model of the equations.

4.25. Solve the problem of a thin (radius-to-thickness ratio $r_0/t = 100$) clamped isotropic ($\nu = 0.3$) circular plate with uniformly distributed load using the element developed in Problem 4.23. Exploit symmetry, and use two thin plate (Euler–Bernoulli) elements in the computational domain.

4.26. Solve the circular plate problem of Problem 4.25 using a two-element mesh of Timoshenko elements (use $k = \frac{5}{6}$).

4.27. Consider the fourth-order equation (4.1) and its weak form (4.4). Suppose that a two-node element is employed, with *three* primary variables at each node: (w, θ, κ), where $\theta = dw/dx$ and $\kappa = d^2w/dx^2$. Show that the associated interpolation (Hermite) functions are given by

$$\phi_1 = \frac{2h^5 - 20x^3h^2 + 30x^4h - 12x^5}{2h^5}, \qquad \phi_2 = \frac{2h^5x - 12x^3h^3 + 16x^4h^2 - 6x^5h}{2h^5}$$

$$\phi_3 = \frac{x^2h^5 - 3x^3h^4 + 3x^4h^3 - x^5h^2}{2h^5}, \qquad \phi_4 = \frac{-20x^3h^2 + 30x^4h - 12x^5}{2h^5}$$

$$\phi_5 = \frac{8x^3h^3 - 14x^4h^2 + 6x^5h}{2h^5}, \qquad \phi_6 = \frac{x^3h^4 - 2x^4h^3 + x^5h^2}{2h^5}$$

where x is the element coordinate with the origin at node 1. Also compute the element stiffness matrix and force vector.

4.28. Consider the weak form (4.4*a*) of the Euler–Bernoulli beam element. Use a three-node element with *two* degrees of freedom (w, θ), where $\theta \equiv -dw/dx$. Derive the Hermite interpolation functions for the element. Compute the element stiffness matrix and force vector.

4.29–4.36. For the truss and frame problems shown in Figs. P4.29–P4.36, give (*a*) the transformed element matrices; (*b*) the assembled element matrices; (*c*) the condensed matrix equations for the unknown generalized displacements and forces.

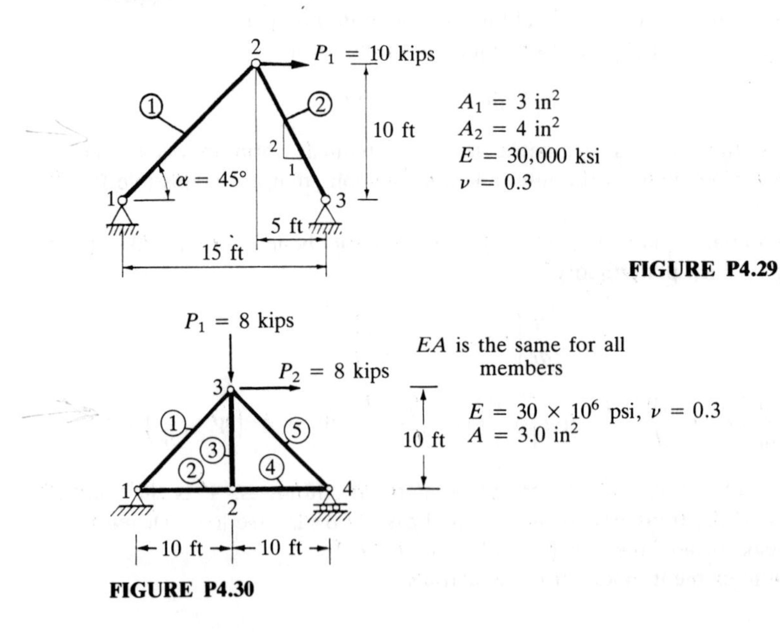

FIGURE P4.29

FIGURE P4.30

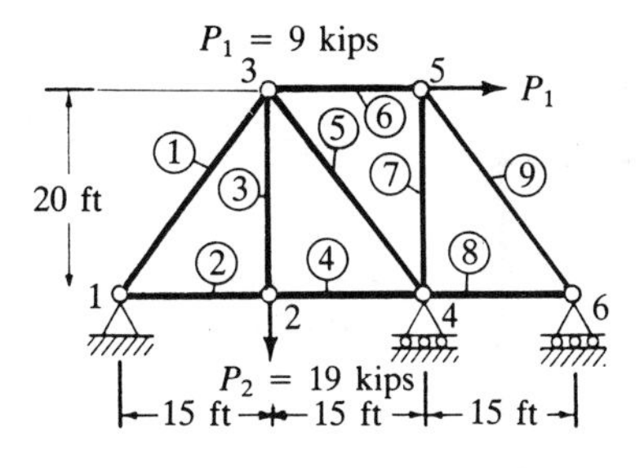

$E = 30 \times 10^6$ psi, $\nu = 0.3$
$A_1 = A_5 = A_9 = 5$ in^2
$A_2 = A_4 = A_6 = A_8 = 6$ in^2
$A_3 = A_7 = 2$ in^2

FIGURE P4.31

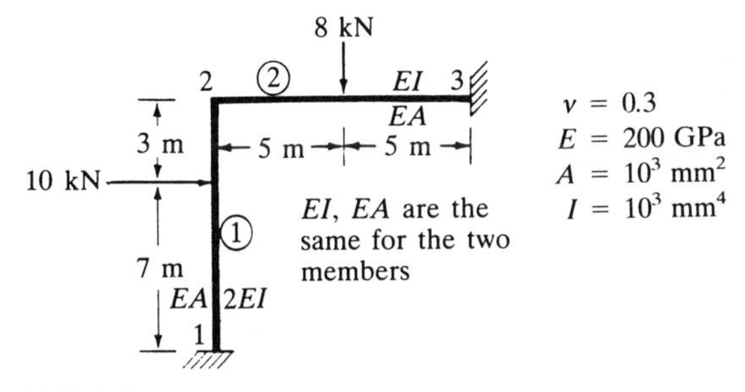

$\nu = 0.3$
$E = 200$ GPa
$A = 10^3$ mm^2
$I = 10^3$ mm^4

FIGURE P4.32

$\nu = 0.3$
$E = 30 \times 10^6$ psi
$A = 10$ in^2
$I = 10$ in^4

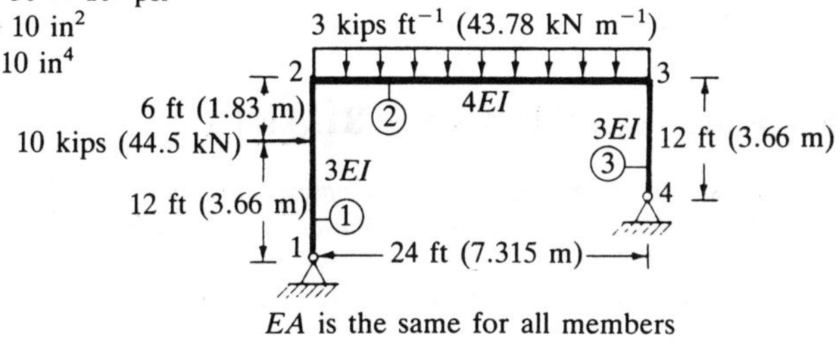

3 kips ft^{-1} (43.78 kN m^{-1})
2
3
6 ft (1.83 m)
4EI
10 kips (44.5 kN)
3EI
12 ft (3.66 m)
3EI
12 ft (3.66 m)
4
1
24 ft (7.315 m)

EA is the same for all members

FIGURE P4.33

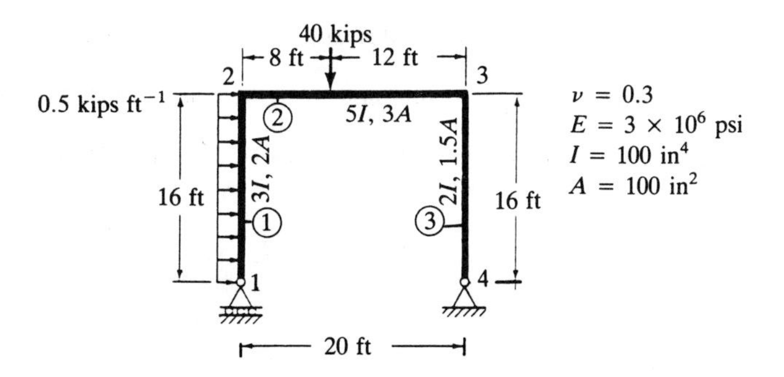

$\nu = 0.3$
$E = 3 \times 10^6$ psi
$I = 100$ in^4
$A = 100$ in^2

FIGURE P4.34

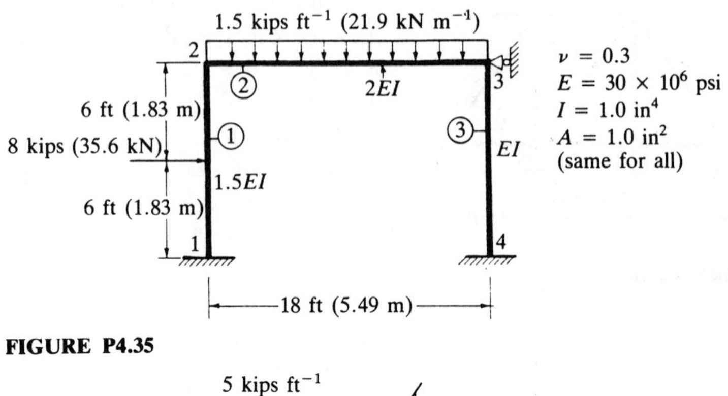

FIGURE P4.35

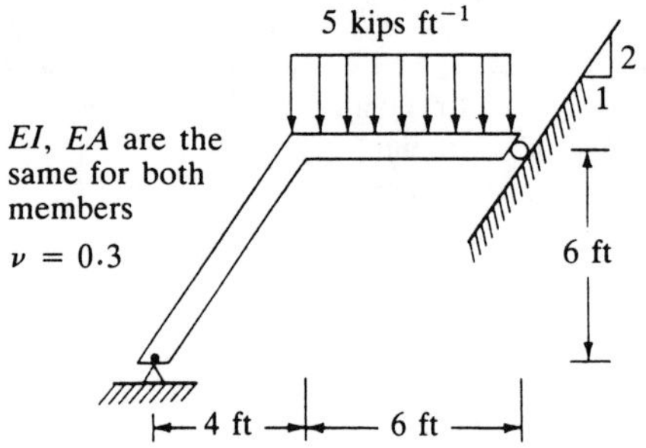

5 kips ft^{-1}

EI, *EA* are the same for both members

$\nu = 0.3$

2
1
6 ft
4 ft
6 ft

FIGURE P4.36

Answer: (Problem 4.29) $U_3 = 0.022973$ in, $U_4 = 0.002169$ in, $\bar{Q}_1^1 = -9428$ lb, $\bar{Q}_1^2 = 7454$ lb.

REFERENCES FOR ADDITIONAL READING

Budynas, R. G.: *Advanced Strength and Applied Stress Analysis,* McGraw-Hill, New York, 1977.

Dym, C. L., and I. H. Shames: *Solid Mechanics: A Variational Approach,* McGraw-Hill, New York, 1973.

Reddy, J. N.: *Energy and Variational Methods in Applied Mechanics,* John Wiley, New York, 1984.

Timoshenko, S. P., and J. N. Goodier: *Theory of Elasticity,* McGraw-Hill, New York, 1970.

Ugural, A. C., and S. K. Fenster: *Advanced Strength and Applied Elasticity,* Elsevier, New York, 1975.

Volterra, E., and J. H. Gaines: *Advanced Strength of Materials,* Prentice-Hall, Englewood Cliffs, NJ, 1971.

Wang, Chu-Kia, and C. G. Salmon: *Introductory Structural Analysis,* Prentice-Hall, Englewood Cliffs, NJ, 1984.

Willems, N., and W. M. Lucas, Jr.: *Structural Analysis for Engineers,* McGraw-Hill, New York, 1978.

CHAPTER 5

FINITE ELEMENT ERROR ANALYSIS

5.1 APPROXIMATION ERRORS

The errors introduced into the finite element solution of a given differential equation can be attributed to three basic sources:

1. *Domain approximation error,* which is due to the approximation of the domain.
2. *Quadrature and finite arithmetic errors,* which are due to the numerical evaluation of integrals and the numerical computation on a computer.
3. *Approximation error,* which is due to the approximation of the solution (see (iv) in the Note to Remark 10 in Section 3.2.7):

$$u \approx u_h \equiv \sum_{I=1}^{N} U_I \Phi_I \tag{5.1}$$

where U_I denotes the value of u at global node I, and Φ_I denotes the global interpolation function associated with global node I (see Fig. 3.3b).

In the one-dimensional problems discussed thus far, the domains considered have been straight lines. Therefore, no approximation of the domain has been necessary. In two-dimensional problems involving nonrectangular domains (as will be seen in Chapter 8), domain (or boundary) approximation errors are introduced into the finite element solutions. In general, these can be interpreted as errors in the specification of the data of the problem because we are now solving the given differential equation on a modified domain. As we refine the mesh, the domain is more accurately represented, and, therefore, the boundary approximation errors are expected to approach zero.

When finite element computations are performed on a computer, round-off errors in the computation of numbers and errors due to the numerical evaluation of integrals are introduced into the solution. In most linear problems with a reasonably small number of total degrees of freedom in the system, these errors are expected to be small (or zero when only a certain decimal point accuracy is desired).

The error introduced into the finite element solution U^e because of the approximation of the dependent variable u in an element Ω^e is inherent to any problem

$$u \approx u_h = \sum_{e=1}^{N} \sum_{i=1}^{n} u_i^e \psi_i^e = \sum_{I=1}^{M} U_I \Phi_I \tag{5.2}$$

where u_h is the finite element solution over the domain ($u_h = U^e$ in Ω^e), N is the number of elements in the mesh, M is the total number of global nodes, and n is the number of nodes in an element. We wish to know how the error $E = u - u_h$, measured in a meaningful way, behaves as the number of elements in the mesh is increased. It can be shown that the approximation error is zero for the single second-order and fourth-order equations with element-wise-constant coefficients [see (5.30)–(5.35)].

5.2 VARIOUS MEASURES OF ERRORS

There are several ways in which one can measure the "difference" (or distance) between any two functions u and u_h. The *pointwise error* is the difference of u and u_h at each point of the domain. One can also define the difference of u and u_h to be the maximum of all absolute values of the differences of u and u_h in the domain $\Omega = (a, b)$:

$$\|u - u_h\|_\infty \equiv \max_{a \leq x \leq b} |u(x) - u_h(x)| \tag{5.3}$$

This measure of difference is called the *supmetric*. Note that the supmetric is a real number, whereas the pointwise error is a function and does not qualify as a distance or *norm* in a strict mathematical sense. The norm of a function is a non-negative real number.

More generally used measures (or norms) of the difference of two functions are the *energy norm* and the L_2 *norm* (pronounced "L-two norm"). For any square-integrable functions u and u_h defined on the domain $\Omega =$

(a, b), the two norms are defined by

$$\text{energy norm} \qquad \|u - u_h\|_m = \left(\int_a^b \sum_{i=0}^{m} \left|\frac{d^i u}{dx^i} - \frac{d^i u_h}{dx^i}\right|^2 dx\right)^{1/2} \tag{5.4}$$

$$L_2 \text{ norm} \qquad \|u - u_h\|_0 = \left(\int_a^b |u - u_h|^2\, dx\right)^{1/2} \tag{5.5}$$

where $2m$ is the order of the differential equation being solved. The term "energy norm" is used to indicate that this norm contains the same-order derivatives as the quadratic functional (which, for most solid mechanics problems, denotes the energy) associated with the equation. Various measures of the distance between two functions are illustrated in Fig. 5.1. These definitions can easily be modified for two-dimensional domains.

5.3 CONVERGENCE OF SOLUTION

The finite element solution u_h in (5.1) is said to *converge in the energy norm* to the true solution u if

$$\|u - u_h\|_m \leq ch^p \quad \text{for } p > 0 \tag{5.6}$$

where c is a constant independent of u and u_h, and h is the characteristic length of an element. The constant p is called the *rate of convergence*. Note that the convergence depends on h as well as on p; p depends on the order of the derivative of u in the weak form and the degree of the polynomials used to approximate u [see (5.15) below]. Therefore, the error in the approximation can be reduced either by reducing the size of the elements or increasing the degree of approximation. Convergence of the finite element solutions with mesh refinements (i.e., more of the same kind of elements are used) is termed *h-convergence*. Convergence with increasing degree of polynomials is called *p-convergence*.

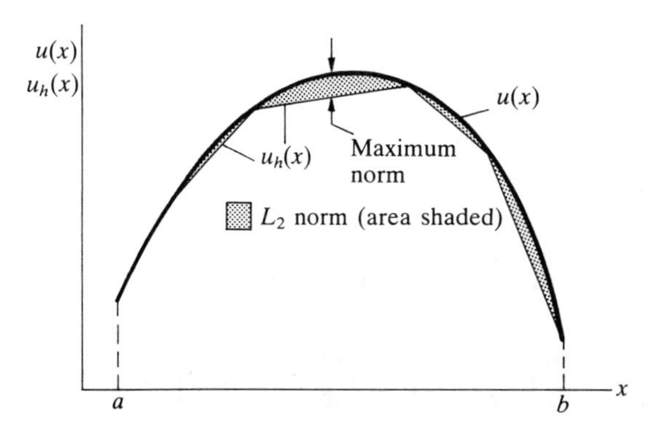

FIGURE 5.1
Different measures of error $E = u - u_h$ between the exact solution u and the finite element solution u_h. The maximum norm and the L_2 norm are illustrated.

5.4 ACCURACY OF THE SOLUTION

Returning to the question of estimating the approximation error, we consider a $2m$th-order differential equation in one dimension ($m = 1$, second-order equations; $m = 2$, fourth-order equations):

$$\sum_{i=1}^{m} (-1)^i \frac{d^i}{dx^i}\left(a_i \frac{d^i u}{dx^i}\right) = f \quad \text{for } 0 < x < L \tag{5.7}$$

where the coefficients $a_1(x)$ and $a_2(x)$ are assumed to be positive. Suppose that the essential boundary conditions of the problem are

$$u(0) = u(L) = 0 \quad (m = 1, 2) \tag{5.8}$$

$$\left(\frac{du}{dx}\right)\bigg|_{x=0} = \left(\frac{du}{dx}\right)\bigg|_{x=L} = 0 \quad (m = 2) \tag{5.9}$$

The variational formulation of (5.7) and (5.9) is given by

$$0 = \int_0^L \left(\sum_{i=1}^{m} a_i \frac{d^i v}{dx^i} \frac{d^i u}{dx^i} - vf\right) dx \tag{5.10}$$

The quadratic functional corresponding to the variational form is

$$I(u) = \int_0^L \frac{1}{2}\left[\sum_{i=1}^{m} a_i \left(\frac{d^i u}{dx^i}\right)^2\right] dx - \int_0^L uf\, dx \tag{5.11}$$

Now consider a finite element discretization of the domain by N elements of equal length h. If u_h denotes the finite element solution in (5.1), we have, from (5.11),

$$I(u_h) = \int_0^L \frac{1}{2}\left[\sum_{i=1}^{m} a_i \left(\frac{d^i u_h}{dx^i}\right)^2\right] dx - \int_0^L u_h f\, dx \tag{5.12}$$

In the following paragraphs, we show that the energy I associated with the finite element solution approaches the true energy from above, and we then give an error estimate. We confine our discussion, for the sake of simplicity, to the second-order equation ($m = 1$).

From (5.11) and (5.12), and

$$f = -\frac{d}{dx}\left(a_1 \frac{du}{dx}\right)$$

we have

$$I(u_h) - I(u) = \int_0^L \frac{1}{2}\left[a_1\left(\frac{du_h}{dx}\right)^2 - a_1\left(\frac{du}{dx}\right)^2 + 2f(u - u_h)\right] dx$$

$$= \int_0^L \left[\frac{a_1}{2}\left(\frac{du_h}{dx}\right)^2 - \frac{a_1}{2}\left(\frac{du}{dx}\right)^2 - \frac{d}{dx}\left(a_1 \frac{du}{dx}\right)(u - u_h)\right] dx$$

$$= \int_0^L \left\{ \frac{a_1}{2} \left[\left(\frac{du_h}{dx} \right)^2 - \left(\frac{du}{dx} \right)^2 \right] + a_1 \frac{du}{dx} \frac{d}{dx}(u - u_h) \right\} dx$$

$$= \int_0^L \frac{a_1}{2} \left[\left(\frac{du_h}{dx} \right)^2 + \left(\frac{du}{dx} \right)^2 - 2 \frac{du}{dx} \frac{du_h}{dx} \right] dx$$

$$= \int_0^L \frac{a_1}{2} \left(\frac{du_h}{dx} - \frac{du}{dx} \right)^2 dx \geq 0 \tag{5.13}$$

Thus,

$$\boxed{I(u_h) \geq I(u)} \tag{5.14}$$

The equality holds only for $u = u_h$. Equation (5.14) implies that the convergence of the energy of the finite element solution to the true energy is from above. Since the relation in (5.14) holds for any u_h, the inequality also indicates that the true solution u minimizes the energy. A similar relation can be established for the fourth-order equation ($m = 2$).

Now suppose that the finite element interpolation functions Φ_I ($I = 1, 2, \ldots, M$) are complete polynomials of degree k. Then the error in the energy norm can be shown to satisfy the inequality [see Reddy (1986, 1991), p. 401]

$$\boxed{\|e\|_m \equiv \|u - u_h\|_m \leq ch^p, \qquad p = k + 1 - m > 0} \tag{5.15}$$

where c is a constant. This estimate implies that the error goes to zero as the pth power of h as h is decreased (or the number of elements is increased). In other words, the logarithm of the error in the energy norm versus the logarithm of h is a straight line whose slope is $k + 1 - m$. The greater the degree of the interpolation functions, the more rapid the rate of convergence. Note also that the error in the energy goes to zero at the rate of $k + 1 - m$; the error in the L_2 norm will decrease even more rapidly, namely, at the rate of $k + 1$, i.e., derivatives converge more slowly than the solution itself.

Error estimates of the type in (5.15) are very useful because they give an idea of the accuracy of the approximate solution, whether or not we know the true solution. While the estimate gives an idea of how rapidly the finite element solution converges to the true solution, it does not tell us when to stop refining the mesh. This decision rests with the analysts, because only they know what a reasonable tolerance is for the problems they are solving.

As an example of estimating the error in the approximation, i.e. (5.15), consider the linear (two-node) element for a second-order equation ($m = 1$). We have for an element

$$u_h = u_1(1 - s) + u_2 s \tag{5.16}$$

where $s = \bar{x}/h$ and $\bar{x}$ is the local coordinate. Since u_2 can be viewed as a function of u_1 via (5.16), one can expand u_2 in a Taylor series around node 1 to obtain

$$u_2 = u_1 + u_1' + \tfrac{1}{2}u_1'' + \ldots \tag{5.17}$$

where $u' \equiv du/ds$. Substituting this into (5.16), we obtain

$$u_h = u_1 + u_1's + \tfrac{1}{2}u_1''s + \ldots \tag{5.18}$$

Expanding the true solution in a Taylor series about node 1, we obtain

$$u = u_1 + u_1's + \tfrac{1}{2}u_1''s^2 + \ldots \tag{5.19}$$

Therefore, we have, from (5.18) and (5.19),

$$|u_h - u| \leqslant \frac{1}{2}(s - s^2) \max_{0 \leqslant s \leqslant 1} \left|\frac{d^2u}{ds^2}\right| = \frac{1}{2}(s - s^2)h^2 \max_{0 \leqslant \bar{x} \leqslant h} \left|\frac{d^2u}{d\bar{x}^2}\right| \tag{5.20}$$

$$\left|\frac{d}{d\bar{x}}(u_h - u)\right| \leqslant \frac{1}{2}h \max_{0 \leqslant \bar{x} \leqslant h} \left|\frac{d^2u}{d\bar{x}^2}\right| \tag{5.21}$$

These lead to

$$\|u - u_h\|_0 \leqslant c_1h^2, \qquad \|u - u_h\|_1 \leqslant c_2h \tag{5.22}$$

where the constants c_1 and c_2 depend only on the length L of the domain.

The reader may carry a similar error analysis for the fourth-order equation.

Example 5.1. Here we consider a computational example to verify the error estimates in (5.22). Consider the differential equation

$$-\frac{d^2u}{dx^2} = 2 \quad \text{for } 0 < x < 1 \tag{5.23}$$

with

$$u(0) = u(1) = 0$$

The exact solution is

$$u(x) = x(1 - x) \tag{5.24}$$

while the finite element solutions are, for $N = 2$,

$$u_h = \begin{cases} h^2(x/h) & \text{for } 0 \leqslant x \leqslant h \\ h^2(2 - x/h) & \text{for } h \leqslant x \leqslant 2h \end{cases}$$

for $N = 3$,

$$u_h = \begin{cases} 2h^2(x/h) & \text{for } 0 \leqslant x \leqslant h \\ 2h^2(2 - x/h) + 2h^2(x/h - 1) & \text{for } h \leqslant x \leqslant 2h \\ 2h^2(3 - x/h) & \text{for } 2h \leqslant x \leqslant 3h \end{cases} \tag{5.25}$$

and, for $N = 4$,

$$u_h = \begin{cases} 3h^2(x/h) & \text{for } 0 \leqslant x \leqslant h \\ 3h^2(2 - x/h) + 4h^2(x/h - 1) & \text{for } h \leqslant x \leqslant 2h \\ 4h^2(3 - x/h) + 3h^2(x/h - 2) & \text{for } 2h \leqslant x \leqslant 3h \\ 3h^2(4 - x/h) & \text{for } 3h \leqslant x \leqslant 4h \end{cases}$$

TABLE 5.1
The L_2 error and error in the energy norm of the solution to (5.23) (Example 5.1)

h	$\log_{10} h$	$\|e\|_0$	$\log_{10} \|e\|_0$	$\|e\|_1$	$\log_{10} \|e\|_1$
$\frac{1}{2}$	−0.301	0.04564	−1.341	0.2887	−0.5396
$\frac{1}{3}$	−0.477	0.02028	−1.693	0.1925	−0.7157
$\frac{1}{4}$	−0.601	0.01141	−1.943	0.1443	−0.8406

For the two-element case ($h = 0.5$), the errors are given by

$$\begin{aligned} \|u - u_h\|_0^2 &= \int_0^h (x - x^2 - hx)^2\,dx + \int_h^{2h} (x - x^2 - 2h^2 + xh)^2\,dx \\ &= 0.002083 \\ \left\|\frac{du}{dx} - \frac{du_h}{dx}\right\|_0^2 &= \int_0^h (1 - 2x - h)^2\,dx + \int_h^{2h} (1 - 2x + h)^2\,dx \\ &= 0.08333 \end{aligned} \tag{5.26}$$

Similar calculations can be performed for $N = 3$ and $N = 4$. Table 5.1 gives the errors for $N = 2$, 3, and 4.

Plots of $\log \|e\|_0$ and $\log \|e\|_1$ versus $\log h$ show that (see Fig. 5.2)

$$\log \|e\|_0 = 2 \log h + \log c_1, \qquad \log \|e\|_1 = \log h + \log c_2 \tag{5.27}$$

In other words, the rate of convergence of the finite element solution is 2 in the L_2 norm and 1 in the energy norm, verifying the estimates in (5.22).

Much of the discussion presented in this section can be carried over to curved elements and two-dimensional elements. When the former, i.e.,

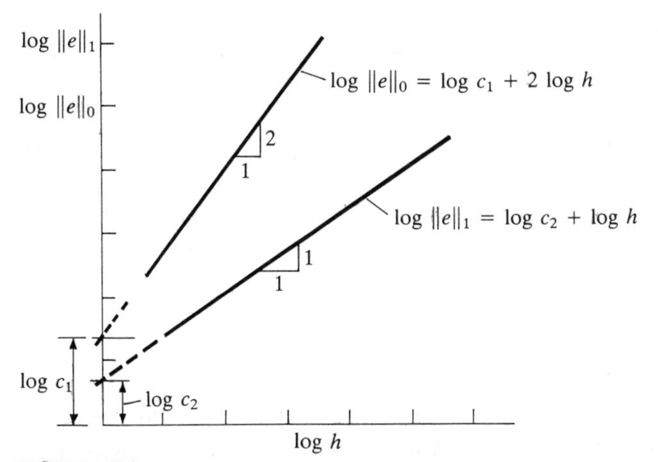

FIGURE 5.2
Plots of the L_2 and energy norms of errors versus the mesh size. The log–log plots give the rates of convergence in the respective norms. The rates of convergence are given by the slopes of the lines (the plots shown are for linear elements).

elements with nonstraight sides, are involved, the error estimate also depends on the Jacobian of the transformation. Because of the introductory nature of the present study, these topics are not discussed here. Interested readers can consult Cairlet (1978), Wait and Mitchell (1985), Oden and Reddy (1982), Strang and Fix (1973), and Reddy (1986, 1991).

As noted earlier, in the case of both second- and fourth-order equations in a single unknown and with constant coefficients, the error between the exact solution and the finite element solution at the nodes is zero. This is not accidental. We can prove that when the coefficients a and b are constant, the finite element solutions of the equations

$$-\frac{d}{dx}\left(a\frac{du}{dx}\right)=f(x) \tag{5.28}$$

$$\frac{d^2}{dx^2}\left(b\frac{d^2u}{dx^2}\right)=f(x) \tag{5.29}$$

coincide with the exact solutions at the nodes. The proof is presented below for the second-order equation.

Consider the equation

$$-a\frac{d^2u}{dx^2}=f \quad \text{for } 0<x<L \tag{5.30}$$

with

$$u(0)=0, \qquad u(L)=0 \tag{5.31}$$

The global finite element solution is given by ($U_1 = U_N = 0$)

$$u_h = \sum_{I=2}^{N-1} U_I\Phi_I \tag{5.32}$$

where Φ_I are the linear global interpolation functions shown in Fig. 3.3(b). From the definition of the variational problem, we have

$$\int_0^1 \left(\frac{d\Phi_I}{dx}\frac{du_h}{dx} - \Phi_I\hat{f}\right) dx = 0 \quad \text{for each } I=2,\ldots,N-1 \tag{5.33}$$

where $\hat{f}=f/a$. The exact solution also satisfies this equation. Hence, by subtracting the finite element equation (5.33) from the exact solution, we obtain

$$\int_0^L \left(\frac{du}{dx}-\frac{du_h}{dx}\right)\frac{d\Phi_I}{dx}\,dx = 0 \quad (I=2,\ldots,N-1)$$

Since we have $\Phi_I=0$ for $x \geqslant (I+1)h$ and $x \leqslant (I-1)h$, and $d\Phi_I/dx = 1/h$ for $(I-1)h \leqslant x \leqslant Ih$ and $d\Phi_I/dx = -1/h$ for $Ih \leqslant x \leqslant (I+1)h$, it follows that

$$\int_{(I-1)h}^{Ih} \left(\frac{du}{dx}-\frac{du_h}{dx}\right)\frac{1}{h}\,dx + \int_{Ih}^{(I+1)h} \left(\frac{du}{dx}-\frac{du_h}{dx}\right)\left(-\frac{1}{h}\right) dx = 0 \tag{5.34}$$

for $I = 2, 3, \ldots, N-1$. Denoting $\epsilon(x) = u(x) - u_h(x)$, we have

$$\frac{1}{h}(\epsilon_I - \epsilon_{I-1}) + \left(-\frac{1}{h}\right)(\epsilon_{I+1} - \epsilon_I) = 0$$

or

$$\frac{1}{h}(-\epsilon_{I-1} + 2\epsilon_I - \epsilon_{I+1}) = 0 \quad (I = 2, 3, \ldots, N-1) \tag{5.35}$$

where $\epsilon_I = \epsilon(Ih)$ (i.e., the value of ϵ at $x = Ih$). Since $\epsilon_0 = \epsilon_N = 0$ (because both u and u_h satisfy the essential boundary conditions), it follows from the above homogeneous equations that the solution is trivial: $\epsilon_1 = \epsilon_2 = \ldots = \epsilon_{N-1} = 0$. This implies that the finite element solution coincides with the exact solution at the nodes.

5.5 SUMMARY

Various types of errors in the finite element approximation of differential equations have been discussed and different measures of the error (or difference between two functions) have been defined. Error estimates for the second-order differential equations have been presented. It has been shown that the finite element solutions of differential equations with constant coefficients are exact at the nodes. The proof has been presented for a single second-order differential equation. This result does not hold for coupled second-order differential equations with constant coefficients.

PROBLEMS

5.1. Show that the error estimate for the fourth-order equation (4.1) is given by

$$\|w - w_h\|_2 \leq ch^2$$

where c is a constant, w_h is the finite element solution obtained by using the Hermite cubic interpolation, and w is the exact solution of the problem.

5.2. Consider a Lagrange quadratic element extending between $x = -h$ and $x = h$ and having the three nodes at $x = -h$, $x = \alpha h$, and $x = h$. The transformation between x and a normalized coordinate ξ is given by

$$x = h[\xi + \alpha(1 - \xi^2)]$$

If the dependent variable u is interpolated by a quadratic polynomial in ξ, show that the error $\epsilon = u - u_h$ is given by

$$\|\epsilon\|_1 \leq \frac{c\alpha}{1 - 2\alpha} h$$

Hint: First show that

$$|u - u_h| \leq c \max_{-1 \leq \xi \leq 1} \left|\frac{d^3u}{d\xi^3}\right|$$

and then

$$\frac{d^3u}{d\xi^3} = -6h^2\alpha(1 - 2\alpha\xi)\frac{d^2u}{dx^2} + h^3(1 - 2\alpha\xi)^3\frac{d^3u}{dx^3}, \qquad \frac{d}{dx} = \frac{1}{h(1 - 2\alpha\xi)}\frac{d}{d\xi}$$

5.3. The *error* in the finite element approximation is often measured in terms of the *energy* associated with the problem under consideration. For the second-order problem considered in (3.1), the energy of the problem is given by the functional

$$I = \sum_{e=1}^{N} I_e(u_e) \tag{i}$$

where N is the total number of elements in the finite element mesh and I_e is the functional given in (3.11). The error in the solution is defined to be the difference between the true or exact solution u_0 and the finite element solution u_h, and the error in the energy is defined by

$$E = u_0 - u_h, \qquad \|E\|_1 \equiv \sum_{e=1}^{N} I_e(E) \tag{ii, iii}$$

If the finite element approximation u_h is an interpolant of the true solution u_0, determine the error in the solution and the energy of (3.1) when (a) u_h is a linear interpolant and (b) u_h is a quadratic interpolant.

Hint: Use pertinent references at the end of the chapter to determine the interpolation error in terms of the mesh parameter (spacing) h, and use the result (iii) to determine the error in the energy in the form

$$\|E_1\| = c_0 h^p$$

where c_0 and p are constants; p determines the rate of convergence $(p > 0)$ or divergence $(p < 0)$.

REFERENCES FOR ADDITIONAL READING

Ciarlet, P. G.: *The Finite Element Method for Elliptic Problems,* North-Holland, Amsterdam, 1978.

Davies, A. L.: *The Finite Element Method, A First Approach,* Clarendon Press, Oxford, 1980.

Oden, J. T., and J. N. Reddy: *An Introduction to the Mathematical Theory of Finite Elements,* Wiley-Interscience, New York, 1982.

Reddy, J. N.: *Applied Functional Analysis and Variational Methods in Engineering,* McGraw-Hill, New York, 1986; Krieger, Melbourne, FL, 1991.

Rektorys, K.: *Variational Methods in Mathematics, Science and Engineering,* Reidel, Boston, 1977.

Strang, G., and G. J. Fix: *An Analysis of the Finite Element Method,* Prentice-Hall, Englewood Cliffs, NJ, 1973.

Wait, R., and A. R. Mitchell: *Finite Element Analysis and Applications,* John Wiley, New York, 1985.

CHAPTER 6

EIGENVALUE AND TIME-DEPENDENT PROBLEMS

6.1 EIGENVALUE PROBLEMS

6.1.1 Introduction

Determination of the values of the parameter λ such that the equation

$$A(u) = \lambda B(u) \tag{6.1}$$

where A and B denote linear differential operators, has nontrivial solutions u is called an *eigenvalue problem*. The values of λ are called *eigenvalues* and the associated functions u are called *eigenfunctions*. For example, the equation

$$-\frac{d^2u}{dx^2} = \lambda u, \quad \text{with } A = \frac{d^2}{dx^2}, \qquad B = 1$$

which arises in connection with the axial oscillations of a bar or the transverse oscillations of a cable, constitutes an eigenvalue problem. Here λ denotes the square of the frequency of vibration ω.

In general, the determination of the eigenvalues is of engineering as well as mathematical importance. In structural problems, the eigenvalues denote either natural frequencies or buckling loads. In fluid mechanics and heat transfer, eigenvalue problems arise in connection with the determination of the homogeneous parts of the solution. In these cases, eigenvalues often denote amplitudes of the Fourier components making up the solution. Eigenvalues are also useful in determining the stability characteristics of temporal schemes, as discussed in Section 6.2.

In this section, we develop finite element models of eigenvalue problems. In view of the close similarity between the equations of eigenvalue and boundary value problems, the steps involved in the construction of their finite element models are entirely analogous. Differential eigenvalue problems are reduced to algebraic eigenvalue problems by means of the finite element approximation. The methods of solution of algebraic eigenvalue problems are then used to solve for the eigenvalues and eigenvectors.

6.1.2 Formulation of Eigenvalue Problems

Consider the parabolic partial differential equation

$$\rho c A \frac{\partial u}{\partial t} - \frac{\partial}{\partial x}\left(kA \frac{\partial u}{\partial x}\right) = q(x, t) \tag{6.2}$$

which governs transient heat transfer in one-dimensional systems (e.g., a plane wall). Here u denotes the temperature, k the thermal conductivity, ρ the density, A the cross-sectional area, c the specific heat, and q the heat generation per unit length. The homogeneous solution (i.e., the solution when $q = 0$) of (6.2) is often sought in the form of a product of a function of x and a function of t (i.e., through the *separation of variables* technique):

$$u(x, t) = S(x)T(t) \tag{6.3}$$

Substitution of this assumed form of solution into the homogeneous form of (6.2) gives

$$\rho c A S \frac{dT}{dt} - \frac{d}{dx}\left(kA \frac{dS}{dx}\right)T = 0$$

Separating the variables (assuming that $\rho c A$ and kA are functions of x only),

$$\frac{1}{T}\frac{dT}{dt} = \frac{1}{\rho c A}\frac{1}{S}\frac{d}{dx}\left(kA \frac{dS}{dx}\right) \tag{6.4}$$

Note that the left-hand side of this equation is a function of t only while the right-hand side is a function of x only. For two functions of two independent variables to always be equal, both must be equal to the same constant, say $-\lambda$:

$$\frac{1}{T}\frac{dT}{dt} = \frac{1}{\rho c A}\frac{1}{S}\frac{d}{dx}\left(kA \frac{dS}{dx}\right) = -\lambda \tag{6.5a}$$

or

$$\frac{dT}{dt} = -\lambda T, \qquad -\frac{d}{dx}\left(kA \frac{dS}{dx}\right) - \lambda \rho c A S = 0 \tag{6.5b}$$

The negative sign in the constant λ is based on the physical requirement that the solution $S(x)$ be harmonic in x while $T(t)$ must decay exponentially with increasing t. The solution of the first equation is $T = T_0 e^{-\lambda t}$. When k, A, ρ, and c are constants, the solution of the second equation is

$$S(x) = B_1 \sin \sqrt{\bar{\lambda}} x + B_2 \cos \sqrt{\bar{\lambda}} x, \qquad \bar{\lambda} = \frac{\rho c}{k} \lambda$$

The constants λ, T_0, B_1, and B_2 are determined with the help of initial and boundary conditions.

In view of the above discussion, the solution of (6.2) is of the form

$$u(x, t) = U(x) e^{-\lambda t} \tag{6.6}$$

This form is consistent with the solutions we derived above, with $U(x) \equiv S(x) T_0$. Substituting (6.6) into the homogeneous form of (6.2), we obtain

$$-\frac{d}{dx}\left(kA \frac{dU}{dx}\right) e^{-\lambda t} - \lambda \rho c A U(x) e^{-\lambda t} = 0$$

or

$$\boxed{-\frac{d}{dx}\left(kA \frac{dU}{dx}\right) - \lambda \rho c A U = 0} \tag{6.7}$$

We wish to determine λ and nonzero $U(x)$ such that (6.7) holds and the boundary conditions of the problem are met. Equation (6.7) describes an eigenvalue problem, λ being the eigenvalue and $U(x)$ the eigenfunction.

The axial motion of a bar can be described by the hyperbolic equation

$$\rho A \frac{\partial^2 u}{\partial t^2} - \frac{\partial}{\partial x}\left(EA \frac{\partial u}{\partial x}\right) = q(x, t) \tag{6.8}$$

Here u denotes the axial displacement, E the modulus of elasticity, A the cross-sectional area, ρ the density, and q the axial force per unit length.

The natural axial oscillations of the bar are periodic, and they can be determined by assuming a solution of the form

$$u(x, t) = U(x) e^{-i\omega t}, \quad \text{with } i = \sqrt{-1} \tag{6.9}$$

where ω denotes the frequency of natural axial motion (or vibration), and $U(x)$ denotes the configuration of the bar, called the *mode shape*, during the vibration. For each value of ω, there is an associated mode shape. Substitution of (6.9) into the homogeneous form of (6.8) gives

$$\left[-\rho A \omega^2 U - \frac{d}{dx}\left(EA \frac{dU}{dx}\right)\right] e^{-i\omega t} = 0$$

or

$$-\frac{d}{dx}\left(EA\frac{dU}{dx}\right) - \lambda\rho AU = 0 \tag{6.10}$$

where $\lambda = \omega^2$. Equation (6.10) is an eigenvalue problem, which involves determining the square of natural frequencies λ and mode shapes U. This equation also arises in the solution of (6.8) by means of separation of variables, as discussed in connection with the parabolic equation. Note that (6.7) and (6.10) are of the same form. Only the coefficients are different.

Equations similar to (6.10) can be derived for the transverse vibrations of a beam using the Euler–Bernoulli or the Timoshenko beam theories. For the Euler–Bernoulli beam theory, we assume

$$w(x, t) = W(x)e^{-i\omega t}$$

where ω is the frequency of natural transverse motion and $W(x)$ is the mode shape of the transverse motion. Substitution of this form into the equation of motion of the Euler–Bernoulli beam theory

$$\rho A\frac{\partial^2 w}{\partial t^2} + \frac{\partial^2}{\partial x^2}\left(EI\frac{\partial^2 w}{\partial x^2}\right) = 0 \tag{6.11}$$

gives

$$\frac{d^2}{dx^2}\left(EI\frac{d^2W}{dx^2}\right) - \lambda\rho AW = 0 \tag{6.12}$$

where $\lambda = \omega^2$. For the Timoshenko beam theory, we assume

$$w(x, t) = W(x)e^{-i\omega t}, \qquad \Psi(x, t) = S(x)e^{-i\omega t}$$

and substitute into the equations of motion of the theory

$$\begin{aligned}\rho A\frac{\partial^2 w}{\partial t^2} - \frac{\partial}{\partial x}\left[GAK\left(\frac{\partial w}{\partial x} + \Psi\right)\right] &= 0\\ \rho I\frac{\partial^2 \Psi}{\partial t^2} - \frac{\partial}{\partial x}\left(EI\frac{\partial \Psi}{\partial x}\right) + GAK\left(\frac{\partial w}{\partial x} + \Psi\right) &= 0\end{aligned} \tag{6.13}$$

to obtain the eigenvalue problem

$$\begin{aligned}-\frac{d}{dx}\left[GAK\left(\frac{dW}{dx} + S\right)\right] - \lambda\rho AW &= 0\\ -\frac{d}{dx}\left(EI\frac{dS}{dx}\right) + GAK\left(\frac{dW}{dx} + S\right) - \lambda\rho IS &= 0\end{aligned} \tag{6.14}$$

The study of buckling of beam-columns also leads to an eigenvalue problem. For example, the equation governing the equilibrium of a beam subjected to an axial force P, according to the Euler–Bernoulli beam theory, is

$$\frac{d^2}{dx^2}\left(EI\frac{d^2w}{dx^2}\right) + P\frac{d^2w}{dx^2} = 0 \tag{6.15}$$

which is an eigenvalue equation with $\lambda = P$ as the eigenvalue, which represents the buckling load. Often one is interested only in the smallest value of λ, called the *critical buckling load.* For the Timoshenko beam theory, the buckling equations are

$$\begin{aligned} -\frac{d}{dx}\left[GAK\left(\frac{dw}{dx} + \Psi\right)\right] + P\frac{d^2w}{dx^2} = 0 \\ -\frac{d}{dx}\left(EI\frac{d\Psi}{dx}\right) + GAK\left(\frac{dw}{dx} + \Psi\right) = 0 \end{aligned} \tag{6.16}$$

This completes the formulation of eigenvalue problems associated with the model problems studied in this book. In the next section, we develop the finite element models of (6.7), (6.10), (6.12), and (6.14)–(6.16).

6.1.3 Finite Element Models

An examination of the eigenvalue equations derived in the previous section shows that they are a special case of the equations studied in Chapters 3 and 4. For example, consider

$$-\frac{d}{dx}\left(a\frac{du}{dx}\right) = f \tag{6.17}$$

The eigenvalue equation associated with this is

$$-\frac{d}{dx}\left(a\frac{dU}{dx}\right) = \lambda c_0 U \tag{6.18}$$

where a and c_0 are quantities that depend on the physical problem: for heat transfer,

$$a = kA, \qquad c_0 = \rho c A$$

where c is the specific heat, while, for a bar,

$$a = EA, \qquad c_0 = \rho A$$

Similarly, the eigenvalue equations associated with the transverse vibrations and buckling of beams are special cases of their static counterparts. Therefore, the finite element models of the eigenvalue equations can readily be developed. It is important to note that the spatial derivative operators of the static (i.e., non-time-dependent) and eigenvalue equations are the same. The difference between (6.17) and (6.18) is that, in place of the source term f, we have $\lambda c_0 U$ in the eigenvalue equations. This difference is responsible for another coefficient matrix, in addition to the usual coefficient matrix $[K^e]$, in the eigenvalue problems. The derivation of the finite element models of eigenvalue equations is presented next.

Over a typical element Ω^e, we seek a finite element approximation of u in the form

$$U = \sum_{j=1}^{n} u_j^e \psi_j^e(x) \tag{6.19}$$

The weak form of (6.18) is

$$0 = \int_{x_A}^{x_B} \left(\frac{dw}{dx} a \frac{dU}{dx} - \lambda c_0 w U \right) dx - Q_1 w(x_A) - Q_n w(x_B) \tag{6.20a}$$

where w is the weight function, and Q_1 and Q_n are the usual secondary variables ($Q_i^e = 0,\ i \neq 1, n$)

$$Q_1 = -\left(a \frac{dU}{dx} \right)\bigg|_{x_A}, \quad Q_n = \left(a \frac{dU}{dx} \right)\bigg|_{x_B} \tag{6.20b}$$

Substitution of the finite element approximation into the weak form gives the finite element model of the eigenvalue equation (6.18):

$$[K^e]\{u^e\} - \lambda[M^e]\{u^e\} = \{Q^e\} \tag{6.21a}$$

where

$$K_{ij}^e = \int_{x_A}^{x_B} a \frac{d\psi_i^e}{dx} \frac{d\psi_j^e}{dx}\, dx, \qquad M_{ij}^e = \int_{x_A}^{x_B} c_0 \psi_i^e \psi_j^e \, dx \tag{6.21b}$$

Equation (6.21*a*) represents the finite element model of the eigenvalue equations (6.18) and (6.10).

The finite element model of (6.12) is

$$[K^e]\{u^e\} - \lambda[M^e]\{u^e\} = \{Q^e\} \tag{6.22a}$$

where $\{u^e\}$ and $\{Q^e\}$ are the columns of nodal generalized displacement and force degrees of freedom of the Euler–Bernoulli beam element:

$$\{u^e\} = \begin{Bmatrix} W_1^e \\ \left(-\dfrac{dW}{dx}\right)_1^e \\ W_2^e \\ \left(-\dfrac{dW}{dx}\right)_2^e \end{Bmatrix}, \qquad \{Q^e\} = \begin{Bmatrix} \left[\dfrac{d}{dx}\left(EI\dfrac{d^2W}{dx^2}\right)\right]_1 \\ \left(EI\dfrac{d^2W}{dx^2}\right)_1 \\ \left[-\dfrac{d}{dx}\left(EI\dfrac{d^2W}{dx^2}\right)\right]_2 \\ \left(-EI\dfrac{d^2W}{dx^2}\right)_2 \end{Bmatrix} \tag{6.22b}$$

where the subscripts 1 and 2 refer to element nodes 1 and 2 (at $x = x_A$ and $x = x_B$). The matrices $[K^e]$ and $[M^e]$, known as the *stiffness* and *mass matrices*, are defined by

$$K_{ij}^e = \int_{x_A}^{x_B} EI \frac{d^2\phi_i^e}{dx^2}\frac{d^2\phi_j^e}{dx^2}\,dx, \qquad M_{ij}^e = \int_{x_A}^{x_B} \rho A \phi_i^e \phi_j^e\,dx \tag{6.22c}$$

where ϕ_i^e are the Hermite cubic interpolation functions (see Chapter 4 for details).

The finite element model of (6.14), with equal interpolation of w and S, is

$$\begin{bmatrix} [K^{11}] & [K^{12}] \\ [K^{12}]^T & [K^{22}] \end{bmatrix} \begin{Bmatrix} \{w^e\} \\ \{S^e\} \end{Bmatrix} - \lambda \begin{bmatrix} [M^{11}] & [0] \\ [0] & [M^{22}] \end{bmatrix} \begin{Bmatrix} \{w^e\} \\ \{S^e\} \end{Bmatrix} = \begin{Bmatrix} \{V^e\} \\ \{M^e\} \end{Bmatrix} \tag{6.23a}$$

where

$$K_{ij}^{11} = \int_{x_A}^{x_B} GAK \frac{d\psi_i^e}{dx}\frac{d\psi_j^e}{dx}\,dx, \qquad K_{ij}^{12} = \int_{x_A}^{x_B} GAK \frac{d\psi_i^e}{dx}\psi_j^e\,dx$$

$$K_{ij}^{22} = \int_{x_A}^{x_B} \left(GAK\,\psi_i^e\psi_j^e + EI \frac{d\psi_i^e}{dx}\frac{d\psi_j^e}{dx} \right) dx$$

$$M_{ij}^{11} = \int_{x_A}^{x_B} \rho A \psi_i^e \psi_j^e\,dx, \qquad M_{ij}^{22} = \int_{x_A}^{x_B} \rho I \psi_i^e \psi_j^e\,dx \tag{6.23b}$$

$$V_1^e = \left[-GAK\left(\frac{dW}{dx} + S\right) \right]_1, \qquad V_2^e = \left[GAK\left(\frac{dW}{dx} + S\right) \right]_2$$

$$M_1^e = \left(-EI \frac{dS}{dx} \right)_1, \qquad M_2^e = \left(EI \frac{dS}{dx} \right)_2$$

Note, that for sufficiently large ratios of length L to height H of the beam, the Timoshenko beam element gives the results of the Euler–Bernoulli beam element. For example, for $L/H \geqslant 100$, the effect of shear deformation is negligible, and both elements give approximately the same solution.

The finite element models of (6.15) and (6.16) are the same as those in (6.22a) and (6.23a), respectively, with $\lambda = P$, and $[M^e]$ and $[M^{11}]$ (and $[M^{22}] = [0]$) replaced by $[G^e]$ and $[G^{11}]$, respectively, where

$$G_{ij}^e = \int_{x_A}^{x_B} \frac{d\phi_i^e}{dx}\frac{d\phi_j^e}{dx}\,dx, \qquad G_{ij}^{11} = \int_{x_A}^{x_B} \frac{d\psi_i^e}{dx}\frac{d\psi_j^e}{dx}\,dx \tag{6.24}$$

The coefficient matrix $[G^e]$ is known as the *stability matrix*.

The numerical form of the stiffness matrices $[K^e]$ were given in the previous sections [see, e.g., (3.34a), (3.40a), (4.15), and (4.98)]. Expressions for the mass matrices M_{ij}^e and stability matrices G_{ij}^e for the Lagrange elements are also available from the previous derivations [see, e.g., (3.34a) and (3.35)]. The mass and stability matrices for the Hermite cubic interpolation,

$$M_{ij}^e = \int_{x_A}^{x_B} c_e \phi_i^e \phi_j^e\,dx, \qquad G_{ij}^e = \int_{x_A}^{x_B} a_e \frac{d\phi_i^e}{dx}\frac{d\phi_j^e}{dx}\,dx \tag{6.25}$$

can be evaluated numerically for element-wise-constant values of c_e and a_e:

$$[M^e] = \frac{ch}{420}\begin{bmatrix} 156 & -22h & 54 & 13h \\ -22h & 4h^2 & -13h & -3h^2 \\ 54 & -13h & 156 & 22h \\ 13h & -3h^2 & 22h & 4h^2 \end{bmatrix} \tag{6.26a}$$

$$[G^e] = \frac{a}{30h}\begin{bmatrix} 36 & -3h & -36 & -3h \\ -3h & 4h^2 & 3h & -h^2 \\ -36 & 3h & 36 & 3h \\ -3h & -h^2 & 3h & 4h^2 \end{bmatrix} \tag{6.26b}$$

The assembly procedure and imposition of boundary conditions in the eigenvalue analysis remains the same as for static problems. The boundary conditions are necessarily homogeneous.

The standard error estimates for the fundamental eigenvalues and eigenfunctions of the problems discussed here are [see Strang and Fix (1973)]

$$\boxed{\begin{gathered} \lambda^{(0)} \leq \lambda^h \leq \lambda^{(0)} + ch^{2(k+1-m)}[\lambda^{(0)}]^{(k+1)/m} \\ \|\mathbf{u}^0 - \mathbf{u}^h\|_m \leq ch^{k+1-m}[\lambda^{(0)}]^{(k+1)/2m} \end{gathered}} \tag{6.27}$$

where $(\lambda^{(0)}, \mathbf{u}^0)$ is the exact solution, $(\lambda^h, \mathbf{u}^h)$ is the finite element solution, m is the order of derivatives appearing in the weak form, k is the degree of polynomials used in $\mathbf{u}^h$, and h is the characteristic length of the element (see Chapter 5 for the notation and definitions of errors).

6.1.4 Applications

Here we consider a couple of examples of eigenvalue problems to illustrate the concepts described in the previous section. We consider one example of a heat-transfer-type problem and one of free vibration of beams.

Example 6.1. Consider a plane wall, initially at a uniform temperature T_0, which has both surfaces suddenly exposed to a fluid at temperature T_∞. The governing differential equation

$$\frac{\partial^2 T}{\partial x^2} = \frac{1}{\alpha}\frac{\partial T}{\partial t}, \qquad \alpha = \frac{k}{\rho c_p} \tag{6.28}$$

and the initial condition is

$$T(x, 0) = T_0$$

where α is the diffusion coefficient, k the thermal conductivity, ρ the density, and c_p the specific heat at constant pressure. Equation (6.28) is also known as the *diffusion equation.*

We consider two sets of boundary conditions.

Set 1. If the heat transfer coefficient at the surfaces of the wall is assumed to be infinite, the boundary conditions can be expressed as

$$T(0, t) = T_\infty, \qquad T(L, t) = T_\infty \quad \text{for } t > 0 \tag{6.29a}$$

Set 2. If we assume that the wall at $x = L$ is subjected to ambient temperature, we have

$$T(0, t) = T_\infty, \qquad \left[kA \frac{\partial T}{\partial x} + \beta A (T - T_\infty) \right] \bigg|_{x=L} = 0 \tag{6.29b}$$

Equation (6.28) can be normalized while making the boundary conditions homogeneous. Let

$$\bar{x} = \frac{x}{L}, \qquad \bar{t} = \frac{\alpha t}{L^2}, \qquad u = \frac{T - T_\infty}{T_0 - T_\infty}$$

The differential equation (6.28), boundary conditions (6.29), and the initial condition become

$$-\frac{\partial^2 u}{\partial x^2} + \frac{\partial u}{\partial t} = 0 \tag{6.30a}$$

$$u(0, t) = 0, \qquad u(1, t) = 0, \qquad u(x, 0) = 1 \tag{6.30b}$$

$$u(0, t) = 0, \qquad \left(\frac{du}{dx} + Hu \right) \bigg|_{x=1} = 0, \qquad H = \frac{\beta L}{k} \tag{6.30c}$$

where the bars over x and t are omitted in the interest of brevity.

Solution of (6.30a,b) by separation of variables leads to solution of the eigenvalue problem

$$-\frac{d^2 U}{dx^2} - \lambda U = 0, \qquad U(0) = 0, \qquad U(1) = 0 \tag{6.31}$$

The finite element model of this equation is given by (6.21), with $a = 1$ and $c_0 = 1$. For the choice of linear and quadratic interpolation functions, the element equations (6.21a) become [see (3.34) and (3.40)]

$$\left(\frac{1}{h_e} \begin{bmatrix} 1 & -1 \\ -1 & 1 \end{bmatrix} - \frac{\lambda h_e}{6} \begin{bmatrix} 2 & 1 \\ 1 & 2 \end{bmatrix} \right) \begin{Bmatrix} u_1^e \\ u_2^e \end{Bmatrix} = \begin{Bmatrix} Q_1^e \\ Q_2^e \end{Bmatrix} \quad \text{(linear element)}$$

$$\left(\frac{1}{3h_e} \begin{bmatrix} 7 & -8 & 1 \\ -8 & 16 & -8 \\ 1 & -8 & 7 \end{bmatrix} - \lambda \frac{h_e}{30} \begin{bmatrix} 4 & 2 & -1 \\ 2 & 16 & 2 \\ -1 & 2 & 4 \end{bmatrix} \right) \begin{Bmatrix} u_1^e \\ u_2^e \\ u_3^e \end{Bmatrix} = \begin{Bmatrix} Q_1^e \\ 0 \\ Q_3^e \end{Bmatrix} \quad \text{(quadratic element)}$$

For a mesh of two linear elements (the minimum needed), with $h_1 = h_2 = 0.5$, the assembled equations are

$$\left(2 \begin{bmatrix} 1 & -1 & 0 \\ -1 & 2 & -1 \\ 0 & -1 & 1 \end{bmatrix} - \lambda \frac{1}{12} \begin{bmatrix} 2 & 1 & 0 \\ 1 & 4 & 1 \\ 0 & 1 & 2 \end{bmatrix} \right) \begin{Bmatrix} U_1 \\ U_2 \\ U_3 \end{Bmatrix} = \begin{Bmatrix} Q_1^1 \\ 0 \\ Q_2^2 \end{Bmatrix}$$

The Set 1 boundary conditions $U(0) = 0$ and $U(1) = 0$ require $U_1 = U_3 = 0$. Hence, the eigenvalue problem reduces to the single equation

$$(4 - \tfrac{4}{12}\lambda) U_2 = 0, \qquad \text{or} \quad \lambda_1 = 12.0, \qquad U_2 \neq 0$$

For a mesh of one quadratic element, we have ($h = 1.0$)

$$\tfrac{16}{3} - \tfrac{1}{30}\lambda 16 = 0, \qquad \text{or} \quad \lambda_1 = 10.0$$

The corresponding eigenfunction amplitude is $U_2 = 1.0$ (or any nonzero constant), so that the eigenfunctions are as follows: for linear elements ($h = 0.5$),

$$U(x) = \begin{cases} U_2\psi_2^1 = \dfrac{x}{h} & \text{for } 0 \leqslant x \leqslant 0.5 \\[2ex] U_2\psi_1^2 = \dfrac{1-x}{h} & \text{for } 0.5 \leqslant x \leqslant 1.0 \end{cases}$$

while, for a quadratic element ($h = 1.0$),

$$U(x) = U_2\psi_2^1 = 4\frac{x}{h}\left(1 - \frac{x}{h}\right) \quad \text{for } 0 \leqslant x \leqslant 1.0$$

The exact eigenvalues are $\lambda_n = (n\pi)^2$ and $\lambda_1 = \pi^2 = 9.8696$. The exact eigenfunctions for Set 1 boundary conditions are $U_n(x) = \sin n\pi x$ and $U_1 = \sin \pi x$.

The eigenvalues and eigenfunctions can be used to construct the solution of the transient problem. For example, the solution of (6.28) with Set 1 boundary conditions is

$$u(x, t) = \sum_{n=1}^{\infty} T_n U_n(x) e^{-\lambda_n t} = \sum_{n=1}^{\infty} T_n \sin n\pi x\, e^{-(n\pi)^2 t}$$

where T_n are constants to be determined using the initial condition of the problem. The finite element solution of the same problem, when one quadratic element is used, is given by

$$u_h(x, t) = 4x(1 - x)e^{-10t}T_0$$

For a mesh of two linear elements, the Set 2 boundary conditions translate into $U_1 = 0$ and $Q_2^2 + U_3 = 0$. The condensed equations are

$$\left(2\begin{bmatrix} 2 & -1 \\ -1 & 1 \end{bmatrix} - \frac{\lambda}{12}\begin{bmatrix} 4 & 1 \\ 1 & 2 \end{bmatrix}\right)\begin{Bmatrix} U_2 \\ U_3 \end{Bmatrix} = \begin{Bmatrix} 0 \\ -U_3 \end{Bmatrix}$$

or

$$\left(\begin{bmatrix} 4 & -2 \\ -2 & 3 \end{bmatrix} - \frac{\lambda}{12}\begin{bmatrix} 4 & 1 \\ 1 & 2 \end{bmatrix}\right)\begin{Bmatrix} U_2 \\ U_3 \end{Bmatrix} = \begin{Bmatrix} 0 \\ 0 \end{Bmatrix}$$

The characteristic polynomial of the above eigenvalue problem is obtained by setting the determinant of the coefficient matrix equal to zero:

$$\begin{vmatrix} 4 - 4\bar{\lambda} & -(2 + \bar{\lambda}) \\ -(2 + \bar{\lambda}) & 3 - 2\bar{\lambda} \end{vmatrix} = 0, \qquad \bar{\lambda} = \tfrac{1}{12}\lambda$$

or

$$7\bar{\lambda}^2 - 24\bar{\lambda} + 8 = 0$$

The roots of this equation are

$$\bar{\lambda}_{1,2} = \frac{12 \pm \sqrt{88}}{7} \qquad (\bar{\lambda}_1 = 0.374167,\ \bar{\lambda}_2 = 3.0544)$$

and the eigenvalues are ($\lambda = 12\bar{\lambda}$)

$$\lambda_1 = 4.4900, \qquad \lambda_2 = 36.6529$$

The eigenvectors are computed from the equations

$$\begin{bmatrix} 4-\frac{4}{12}\lambda_i & -(2+\frac{1}{12}\lambda_i) \\ -(2+\frac{1}{12}\lambda_i) & 3-\frac{2}{12}\lambda_i \end{bmatrix} \begin{Bmatrix} U_2 \\ U_3 \end{Bmatrix}^{(i)} = \begin{Bmatrix} 0 \\ 0 \end{Bmatrix}$$

For example, for $\lambda_1 = 4.4900$, we have

$$2.5033U_2^{(1)} - 2.3742U_3^{(1)} = 0$$

or

$$\{U^{(1)}\} = (1.000, 1.0544) = (0.6881, 0.7256) \text{ (normalized)}$$

Hence, the eigenfunction corresponding to $\lambda_1 = 4.4900$ $(h = 0.5)$ is

$$U^{(1)}(x) = \begin{cases} 0.6881x/h & \text{for } 0 \le x \le 0.5 \\ 0.6881(1-x)/h + 0.7256(2x-1)/2h & \text{for } 0.5 \le x \le 1.0 \end{cases}$$

The exact eigenfunctions for Set 2 boundary conditions are

$$U_n(x) = \sin\sqrt{\lambda_n}x$$

and the eigenvalues λ_n are computed from the equation

$$1 + \sqrt{\lambda_n}\cot\sqrt{\lambda_n} = 0$$

Table 6.1 gives a comparison of the eigenvalues obtained using meshes of linear and quadratic elements with the exact values. Note that the number of eigenvalues we can obtain is equal to the number of unknown nodal values. As the mesh is refined, not only do we increase the number of eigenvalues but we also improve the accuracy of the preceding eigenvalues. Note also that the convergence of the numerical eigenvalues to

TABLE 6.1
Eigenvalues of the heat conduction equation (6.28) for two sets of boundary conditions
Set 1: $U(0) = 0, \quad U(1) = 0$
Set 2 (in parentheses): $U(0) = 0, \quad (dU/dx + U)|_{x=1} = 0$

Mesh	λ_1	λ_2	λ_3	λ_4	λ_5	λ_6	λ_7
2L	12.0000						
	(4.4900)	(36.6529)					
4L	10.3866	48.0000	126.756				
	(4.2054)	(27.3318)	(85.7864)	(177.604)			
8L	9.9971	41.5466	99.4855	192.000	328.291	507.025	686.512
	(4.1380)	(24.9088)	(69.1036)	(143.530)	(257.580)	(417.701)	(607.022)
1Q	10.000						
	(4.1545)	(38.5121)					
2Q	9.9439	40.000	128.723				
	(4.1196)	(24.8995)	(81.4446)	(207.653)			
4Q	9.8747	39.7754	91.7847	160.000	308.253	514.891	794.794
	(4.1161)	(24.2039)	(64.7705)	(129.261)	(240.539)	(405.253)	(658.137)
Exact	9.8696	39.4784	88.8264	157.9137	246.740	355.306	483.611
	(4.1159)	(24.1393)	(63.6591)	(122.889)	(201.851)	(300.550)	(418.987)

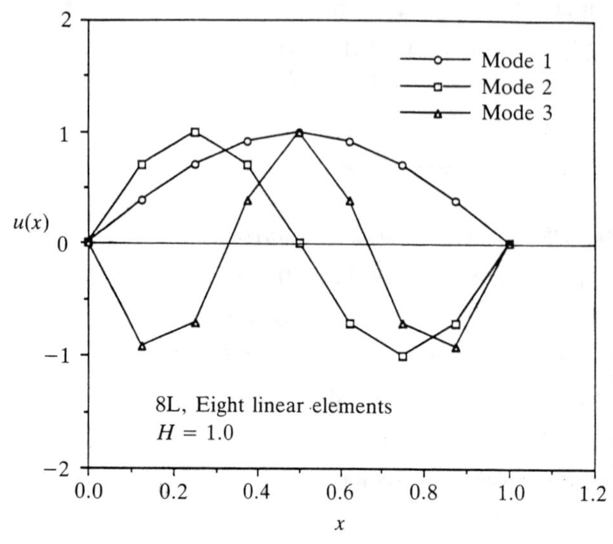

FIGURE 6.1
The first three mode shapes (or eigenvectors) as predicted by the mesh of eight linear elements for the heat transfer problem in Example 6.1, Set 1 boundary conditions.

the exact ones is from the above, i.e., the finite element solution provides an upper bound to the exact eigenvalues. For structural systems, this can be interpreted as follows. According to the principle of total minimum potential energy, any approximate displacement field would overestimate the total potential energy of the system. This is equivalent to approximating the stiffness of the system with a larger value than the actual one. A stiffer system will have larger eigenvalues (or frequencies). The first three mode shapes of the system are shown in Figs. 6.1 and 6.2.

We close this example by noting that the eigenvalue equation (6.31) can also be interpreted as that arising in connection with the axial vibrations of a constant cross-section member. In that case, U denotes the axial displacement and $\lambda = \omega^2 \rho / E$, ω being the frequency of vibration. The boundary conditions in (6.30c) can be interpreted as the left end of the bar being fixed and the right end being connected to a linear

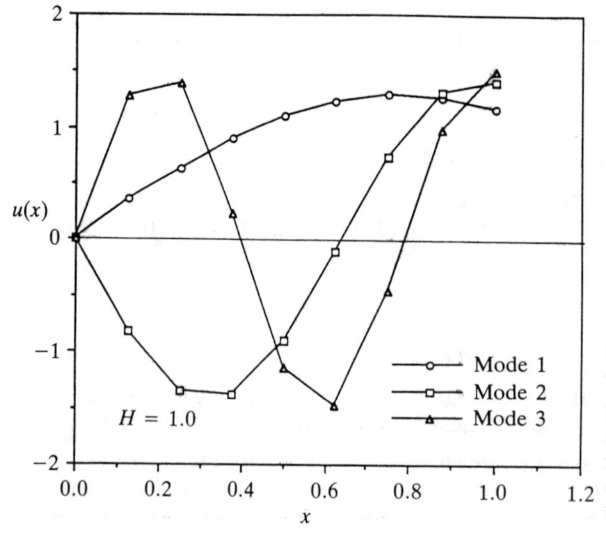

FIGURE 6.2
The first three mode shapes as predicted by a mesh of eight linear elements for the heat transfer problem with convection ($H = 1.0$). See Example 6.1, Set 2 boundary conditions.

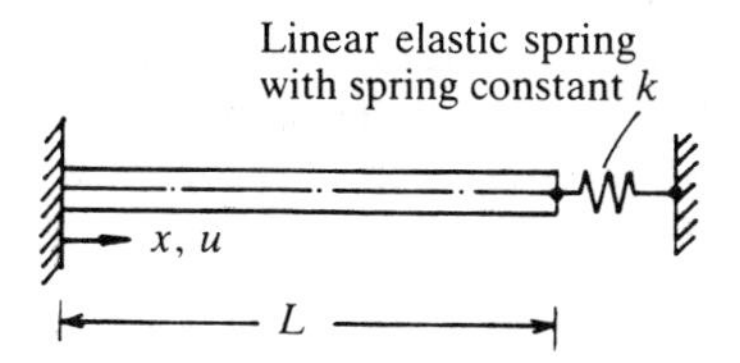

FIGURE 6.3
An elastic bar with an end spring. The axial vibrations in this case can be shown to be governed by the same eigenvalue problem as that used for the heat transfer problem with convective boundary condition at $x = L$ (Set 2 boundary conditions in Example 6.1).

elastic spring (see Fig. 6.3). The constant H is equal to k/EA, k being the spring constant. Thus, the results presented in Table 6.1 can be interpreted as the square of the natural frequencies of a uniform bar (multiplied by ρ/E).

In the next example we study transverse vibrations of cantilever beams using the Timoshenko beam theory. The effect of shear deformation is brought out by considering two different ratios of length to height of the beam. In problems with more than one independent variable and certain coarse meshes, the computed eigenvalues are not always the lowest eigenvalues of the problem. This is due to the restrictions placed by the particular mesh on the mode shapes it can model.

Example 6.2. Consider a uniform beam of rectangular cross-section, length L, width B, and height H. The beam is fixed at one end, say at $x = 0$, and free at the other, $x = L$. We assume that Poisson's ratio is $\nu = 0.25$. We wish to determine the first four natural frequencies associated with the transverse deflection w.

The finite element model of the Timoshenko beam theory is given by (6.23a). The number of eigenvalues we wish to determine dictates the minimum number of elements to be used. If we use one linear element, we shall have four degrees of freedom, with two degrees of freedom at the fixed end being specified as zeros. Thus, we can only obtain two eigenvalues. If two linear elements are used, there are six degrees of freedom, with two of them known to be zero, and we can obtain four eigenvalues. However, the four computed eigenvalues may not be the lowest four. Indeed, a mesh of two linear elements can only represent the first two mode shapes of the cantilever beam (see Fig. 6.4). In order to represent the first four mode shapes, we must use at least four linear elements.

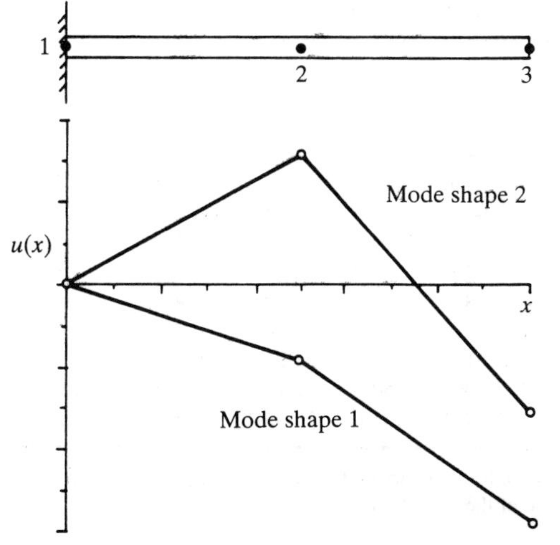

FIGURE 6.4
Two possible nonzero mode shapes that can be represented by a mesh of two linear (Lagrange) elements.

Because of the algebraic complexity of the element matrices for meshes of 4, 8, and 16 linear elements and 2, 4, and 8 quadratic elements that are used to study the convergence characteristics of the eigenvalues of the beam, the element matrices and assembled equations are not presented here. For all meshes used, the boundary conditions at the fixed end require $w(0)=0$ and $\Psi(0)=0$, or

$$U_1 = 0, \qquad U_2 = 0$$

The frequencies obtained by solving the resulting eigenvalue problems are shown in Table 6.2 for two different values of the length-to-height ratio L/H. The value $L/H = 100$ makes the effect of shear deformation negligible and yields essentially the Euler–Bernoulli beam theory results. Since the frequencies are normalized, $\bar{\omega} = \omega L^2(\rho A/EI)^{1/2}$, it is necessary only to select the values of ν and L/H. For computational purposes, we take $\nu = 0.25$, $\rho I = 1.0$, $EI = 1.0$, and $L = 1.0$, and compute

$$GAK = \frac{E}{2(1+\nu)} BH \frac{5}{6} = \frac{4EI}{H^2}, \qquad \rho A = \frac{12\rho I}{H^2}$$

where B and H are the width and height, respectively, of the beam. Thus for $L/H = 100$ (or $H = 10^{-2}$), we take $GAK = 4 \times 10^4$ and $\rho A = 12 \times 10^4$, and, for $L/H = 10$, we have $GAK = 4 \times 10^2$ and $\rho A = 12 \times 10^2$.

We note from Table 6.2 that the finite element results converge with h-refinement (i.e., when more of the same kind of elements are used) and also with p-refinement (i.e., when higher-order elements are used). The p-refinement shows more rapid convergence of the fundamental (i.e., first) frequency (see Fig. 6.5). The rates of convergence are consistent with the error estimates in (6.27). The mesh refinements with either more or higher-order elements refine the higher frequencies more than the fundamental frequency (see Table 6.2). Note also that the effect of shear deformation is to reduce the natural frequencies. This is because of the increased flexibility in the Timoshenko beam theory compared with the Euler–Bernoulli beam theory. The first four mode shapes of the cantilever beam, as obtained using the 16-element mesh of linear elements, are shown in Fig. 6.6.

TABLE 6.2
Natural frequencies of a cantilever beam according to the Timoshenko and Euler–Bernoulli beam theories $[\bar{\omega} = \omega L^2(\rho A/EI)^{1/2}]$

	$L/H = 100$				$L/H = 10$			
Mesh	$\bar{\omega}_1$	$\bar{\omega}_2$	$\bar{\omega}_3$	$\bar{\omega}_4$	$\bar{\omega}_1$	$\bar{\omega}_2$	$\bar{\omega}_3$	$\bar{\omega}_4$
4L	3.5406	25.6726	98.3953	417.1330	3.5137	24.1345	80.2244	189.9288
8L	3.5223	22.8851	68.8937	151.8431	3.4956	21.7004	60.6297	119.2798
16L	3.5174	22.2350	63.3413	127.5434	3.4908	21.1257	56.4714	104.6799
2Q	3.5214	23.3226	78.3115	328.3250	3.4947	22.0762	67.0884	181.0682
4Q	3.5161	22.1054	63.3271	133.9828	3.4895	21.0103	56.4572	108.6060
8Q	3.5158	22.0280	61.7325	121.4458	3.4892	20.4421	55.2405	100.7496
Exact TBT†	3.5158	22.0226	61.6179	120.6152	3.4892	20.9374	55.1530	100.2116
Exact EBT†	3.5158	22.0315	61.6774	120.8300	3.5092	21.7425	59.8013	114.2898
Exact EBT‡	3.5160	22.0345	61.6972	120.9019	3.5160	22.0345	61.6972	120.9019

† TBT, Timoshenko beam theory; EBT, Euler–Bernoulli beam theory (with rotary inertia).
‡ Rotory inertia neglected; the results are independent of the ratio L/H.

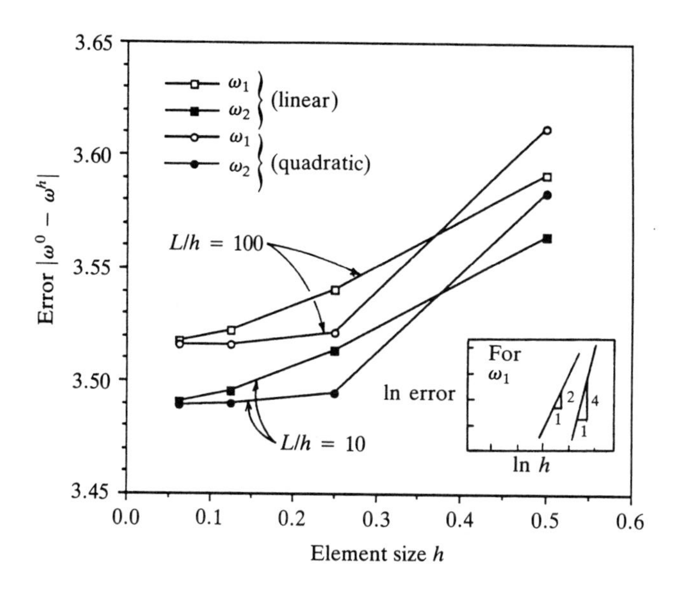

FIGURE 6.5
Plots of error in the frequencies computed using linear and quadratic finite elements (for length-to-height ratios $L/H = 10$ and 100).

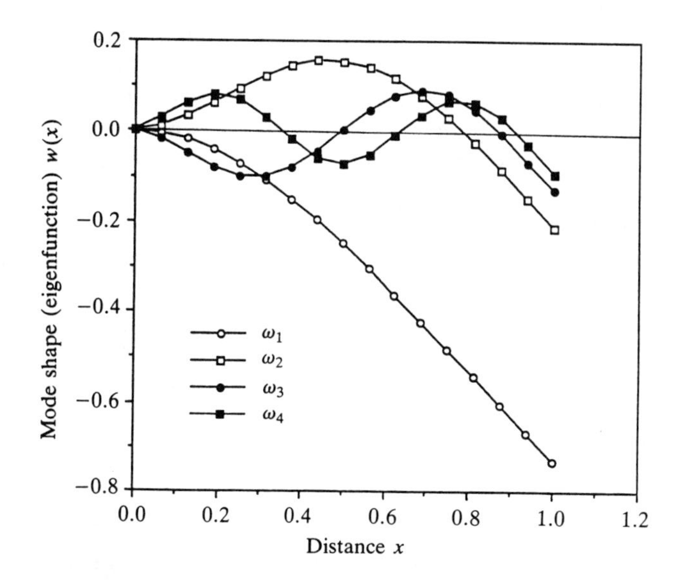

FIGURE 6.6
First four natural mode shapes of a cantilever beam, as predicted using a 16-element mesh of linear Timoshenko beam elements ($L/H = 10$).

6.2 TIME-DEPENDENT PROBLEMS

6.2.1 Introduction

In this section, we develop the finite element models of one-dimensional time-dependent problems and describe time approximation schemes to convert ordinary differential equations in time to algebraic equations. We consider finite element models of the time-dependent version of the differential equations studied in Chapters 3 and 4. These include the second-order (in space) parabolic (i.e., first time derivative) and hyperbolic (i.e., second time derivative) equations and fourth-order hyperbolic equations arising in connection with the bending of beams. Recall that second-order parabolic equations arise in heat transfer and fluid mechanics (see Sections 3.3.1 and 3.3.2), while second- and fourth-order hyperbolic equations arise in solid mechanics problems.

The finite element formulation of time-dependent problems involves two steps:

1. *Spatial approximation,* where the solution u of the equation under consideration is approximated by expressions of the form

$$u(x, t) \approx U^e(x, t) = \sum_{j=1}^{n} u_j^e(t)\psi_j^e(x) \tag{6.32}$$

   and the spatial finite element model of the equation is developed using the procedures of static or steady-state problems, while carrying all time-dependent terms in the formulation. This step results in a set of ordinary differential equations (i.e., a semidiscrete system of equations) in time for the nodal variables $u_j^e(t)$ of the element. Equation (6.32) represents the spatial approximation of u for any time t. When the solution is separable into functions of time only and space only, $u(x, t) = T(t)X(x)$, the approximation (6.32) is clearly justified. Even when the solution is not separable, (6.32) can represent a good approximation of the actual solution, provided a sufficiently small time step is used.
2. *Temporal approximation,* where the system of ordinary differential equations are further approximated in time, often using finite difference formulae for the time derivatives. This step allows conversion of the system of ordinary differential equations into a set of algebraic equations among u_j^e at time $t_{s+1} = (s+1)\,\Delta t$, where Δt is the time increment and s is an integer.

All time approximation schemes seek to find u_j at time t_{s+1} using the known values of u_j from previous times:

$$\text{compute} \quad \{u\}_{s+1} \quad \text{using} \quad \{u\}_s,\ \{u\}_{s-1},\ \ldots$$

Thus, at the end of the two-stage approximation, one has a continuous spatial solution at discrete intervals of time:

$$U^e(x, t_s) = \sum_{j=1}^{n} u_j^e(t_s)\psi_j^e(x) \quad (s = 0, 1, \ldots)$$

Here we study the details of these two steps by considering a model differential equation that contains both second- and fourth-order spatial derivatives and first- and second-order time derivatives:

$$\boxed{-\frac{\partial}{\partial x}\left(a\frac{\partial u}{\partial x}\right) + \frac{\partial^2}{\partial x^2}\left(b\frac{\partial^2 u}{\partial x^2}\right) + c_0 u + c_1\frac{\partial u}{\partial t} + c_2\frac{\partial^2 u}{\partial t^2} = f(x, t)} \qquad (6.33a)$$

subject to appropriate boundary and initial conditions. The boundary conditions are of the form

$$\text{specify} \quad u(x, t) \quad \text{or} \quad -a\frac{\partial u}{\partial x}(x, t) + \frac{\partial}{\partial x}\left(b\frac{\partial^2 u}{\partial x^2}\right)$$

and $\qquad (6.33b)$

$$\text{specify} \quad \frac{\partial u}{\partial x}(x, t) \quad \text{or} \quad b\frac{\partial^2 u}{\partial x^2}$$

at $x = 0, L$, and the initial conditions involve specify

$$c_2 u(x, 0) \quad \text{and} \quad c_2\dot{u}(x, 0) + c_1 u(x, 0) \qquad (6.33c)$$

where $\dot{u} \equiv \partial u/\partial t$. Equation (6.33$a$) contains, as special cases, time-dependent second- and fourth-order equations. Second-order equations arise, for example, in determining the transverse motion of a cable ($a = T$; $b = 0$, $c_1 = \rho$, $c_2 = 0$, $c_0 = 0$), the longitudinal motion of a rod ($a = EA$, $b = 0$, $c_1 = 0$ if damping is not considered, $c_2 = \rho A$, $c_0 = 0$), and the temperature transients in a fin ($a = kA$, $b = 0$, $c_1 = \rho A$, $c_2 = 0$). A fourth-order equation arises in determining the transverse motion of a beam ($a = 0$, $b = EI$, $c_0 = k$, $c_1 = 0$, $c_2 = \rho A$).

6.2.2 Semidiscrete Finite Element Models

The semidiscrete formulation involves approximation of the spatial variation of the dependent variable, which follows essentially the same steps as described in Section 3.2. The first step involves the construction of the weak form of the equation over a typical element.

Following the three-step procedure of constructing the weak form of a differential equation, we can develop the weak form of (6.33a) over an element. Integration by parts is used on the first term once and on the second term twice to distribute the spatial derivatives equally between the weight function w and the dependent variable u:

$$
\begin{aligned}
0 &= \int_{x_A}^{x_B} w\left[-\frac{\partial}{\partial x}\left(a\frac{\partial u}{\partial x}\right)+\frac{\partial^2}{\partial x^2}\left(b\frac{\partial^2 u}{\partial x^2}\right)+c_0 u+c_1\frac{\partial u}{\partial t}+c_2\frac{\partial^2 u}{\partial t^2}-f\right]dx \\
&= \int_{x_A}^{x_B}\left[\frac{\partial w}{\partial x}a\frac{\partial u}{\partial x}+\frac{\partial^2 w}{\partial x^2}b\frac{\partial^2 u}{\partial x^2}+c_0 wu+c_1 w\frac{\partial u}{\partial t}+c_2 w\frac{\partial^2 u}{\partial t^2}-wf\right]dx \\
&\quad +\left[w\left[\left(-a\frac{\partial u}{\partial x}\right)+\frac{\partial}{\partial x}\left(b\frac{\partial^2 u}{\partial x^2}\right)\right]+\frac{\partial w}{\partial x}\left(-b\frac{\partial^2 u}{\partial x^2}\right)\right]_{x_A}^{x_B} \\
&= \int_{x_A}^{x_B}\left(a\frac{\partial w}{\partial x}\frac{\partial u}{\partial x}+b\frac{\partial^2 w}{\partial x^2}\frac{\partial^2 u}{\partial x^2}+c_0 wu+c_1 w\frac{\partial u}{\partial t}+c_2 w\frac{\partial^2 u}{\partial t^2}-wf\right)dx \\
&\quad -\hat{Q}_1 w(x_A)-\hat{Q}_3 w(x_B)-\hat{Q}_2\left(-\frac{\partial w}{\partial x}\right)\bigg|_{x_A}-\hat{Q}_4\left(-\frac{\partial w}{\partial x}\right)\bigg|_{x_B}
\end{aligned} \tag{6.34a}
$$

where

$$
\hat{Q}_1=\left[-a\frac{\partial u}{\partial x}+\frac{\partial}{\partial x}\left(b\frac{\partial^2 u}{\partial x^2}\right)\right]\bigg|_{x_A},\qquad \hat{Q}_2=\left(b\frac{\partial^2 u}{\partial x^2}\right)\bigg|_{x_A}
$$

$$
\hat{Q}_3=-\left[-a\frac{\partial u}{\partial x}+\frac{\partial}{\partial x}\left(b\frac{\partial^2 u}{\partial x^2}\right)\right]\bigg|_{x_B},\qquad \hat{Q}_4=-\left(b\frac{\partial^2 u}{\partial x^2}\right)\bigg|_{x_B} \tag{6.34b}
$$

Next, we assume that u is interpolated by an expression of the form (6.32). Equation (6.32) implies that, at any arbitrarily fixed time $t>0$, the function u can be approximated by a linear combination of the ψ_j^e, with $u_j^e(t)$ being the value of u at time t at the jth node of the element Ω^e. In other words, the time and spatial variations of u are separable. This assumption is not valid, in general, because it may not be possible to write the solution $u(x, t)$ as the product of a function of time only and a function of space only. However, with sufficiently small time steps, it is possible to obtain accurate solutions to even those problems for which the solution is not separable in time and space. The finite element solution that we obtain at the end of the analysis is continuous in space but not in time. We only obtain the finite element solution in the form

$$
u(x, t_s)=\sum_{j=1}^{n} u_j^e(t_s)\psi_j^e(x)=\sum_{j=1}^{n}(u_j^s)^e\psi_j^e(x)\quad (s=1, 2, \ldots) \tag{6.35}
$$

where $(u_j^s)^e$ is the value of $u(x, t)$ at time $t=t_s$ and node j of the element Ω^e.

Substituting $w = \psi_i(x)$ and (6.32) into (6.34a), we obtain

$$0 = \int_{x_A}^{x_B} \left[a \frac{d\psi_i}{dx} \left(\sum_{j=1}^{n} u_j \frac{d\psi_j}{dx} \right) + b \frac{d^2\psi_i}{dx^2} \left(\sum_{j=1}^{n} u_j \frac{d^2\psi_j}{dx^2} \right) \right.$$
$$\left. + c_0\psi_i \left(\sum_{j=1}^{n} u_j \psi_j \right) + c_1 \psi_i \left(\sum_{j=1}^{n} \frac{du_j}{dt} \psi_j \right) + c_2 \psi_i \left(\sum_{j=1}^{n} \frac{d^2u_j}{dt^2} \psi_j \right) - \psi_i f \right] dx$$
$$- \hat{Q}_1 \psi_i(x_A) - \hat{Q}_3 \psi_i(x_B) - \hat{Q}_2 \left(-\frac{d\psi_i}{dx} \right) \bigg|_{x_A} - \hat{Q}_4 \left(-\frac{d\psi_i}{dx} \right) \bigg|_{x_B}$$
$$= \sum_{j=1}^{n} \left[(K_{ij}^1 + K_{ij}^2 + M_{ij}^0) u_j + M_{ij}^1 \frac{du_j}{dt} + M_{ij}^2 \frac{d^2u_j}{dt^2} \right] - F_i \tag{6.36}$$

In matrix form, we have

$$\boxed{[K]\{u\} + [M^1]\{\dot{u}\} + [M^2]\{\ddot{u}\} = \{F\}} \tag{6.37a}$$

where

$$[K] = [K^1] + [K^2] + [M^0] \tag{6.37b}$$

$$M_{ij}^0 = \int_{x_A}^{x_B} c_0 \psi_i \psi_j \, dx$$

$$M_{ij}^1 = \int_{x_A}^{x_B} c_1 \psi_i \psi_j \, dx, \qquad M_{ij}^2 = \int_{x_A}^{x_B} c_2 \psi_i \psi_j \, dx$$

$$K_{ij}^1 = \int_{x_A}^{x_B} a \frac{d\psi_i}{dx} \frac{d\psi_j}{dx} dx, \qquad K_{ij}^2 = \int_{x_A}^{x_B} b \frac{d^2\psi_i}{dx^2} \frac{d^2\psi_j}{dx^2} dx \tag{6.37c}$$

$$F_i = \int_{x_A}^{x_B} \psi_i f \, dx + \hat{Q}_i$$

This completes the semidiscrete finite element formulation of (6.33) over an element.

6.2.3 Time Approximations

As special cases, (6.37a) contains the parabolic equation (set $[M^2] = [0]$) and the hyperbolic equation (set $[M^1] = [0]$). The time approximation of (6.37a) for these two cases will be considered separately: in Case 1, $c_2 = 0$; in Case 2, $c_1 = 0$.

CASE 1: PARABOLIC EQUATIONS. Consider the parabolic equation [set $[M^2]$ equal to zero in (6.37a)]

$$[M^1]\{\dot{u}\} + [K]\{u\} = \{F\} \tag{6.38a}$$

subject to the initial condition

$$\{u\}_0 = \{u_0\} \tag{6.38b}$$

where $\{u\}_0$ denotes the value of the enclosed quantity u at time $t=0$, whereas $\{u_0\}$ denotes the column of values u_{j0}.

The most commonly used method for solving (6.38*a*) is the *α family of approximation,* in which a weighted average of the time derivative of a dependent variable is approximated at two consecutive time steps by linear interpolation of the values of the variable at the two steps:

$$(1-\alpha)\{\dot{u}\}_s + \alpha\{\dot{u}\}_{s+1} = \frac{\{u\}_{s+1} - \{u\}_s}{\Delta t_{s+1}} \quad \text{for } 0 \leq \alpha \leq 1 \tag{6.39a}$$

where $\{\ \}_s$ refers to the value of the enclosed quantity at time $t = t_s = \sum_{i=1}^{s} \Delta t_i$, and $\Delta t_s = t_s - t_{s-1}$ is the sth time step. If the time interval $[0, T_0]$ is divided into equal time steps then $t_s = s\,\Delta t$. Equation (6.39*a*) can be interpreted as

$$\boxed{\begin{aligned} \{u\}_{s+1} &= \{u\}_s + \Delta t\{\dot{u}\}_{s+\alpha} \\ \{\dot{u}\}_{s+\alpha} &= (1-\alpha)\{\dot{u}\}_s + \alpha\{\dot{u}\}_{s+1} \quad \text{for } 0 \leq \alpha \leq 1 \end{aligned}} \tag{6.39b}$$

For different values of α, we obtain the following well-known numerical integration schemes:

$$\boxed{\alpha = \begin{cases} 0, & \text{the forward difference (or Euler) scheme (conditionally stable); order of accuracy} = O(\Delta t) \\ \frac{1}{2}, & \text{the Crank–Nicolson scheme (stable); } O((\Delta t)^2) \\ \frac{2}{3}, & \text{the Galerkin method (stable); } O((\Delta t)^2) \\ 1, & \text{the backward difference scheme (stable); } O(\Delta t) \end{cases}} \tag{6.40}$$

Equation (6.39*a*) can be used to reduce the ordinary differential equations (6.38*a*) to algebraic equations among the u_j at time t_{s+1}. Since (6.38*a*) is valid for any $t > 0$, it is valid for $t = t_s$ and $t = t_{s+1}$ ($[M^1] = [M]$):

$$[M]\{\dot{u}\}_s + [K]_s\{u\}_s = \{F\}_s \tag{6.41a}$$

$$[M]\{\dot{u}\}_{s+1} + [K]_{s+1}\{u\}_{s+1} = \{F\}_{s+1} \tag{6.41b}$$

where it is assumed that the mass matrix is independent of time. After multiplying (6.39*a*) throughout by Δt_{s+1}, we premultiply both sides of the equation with $[M]$ and obtain

$$\Delta t_{s+1}\,\alpha[M]\{\dot{u}\}_{s+1} + \Delta t_{s+1}(1-\alpha)[M]\{\dot{u}\}_s = [M](\{u\}_{s+1} - \{u\}_s)$$

Substituting for $[M]\{\dot{u}\}_{s+1}$ and $[M]\{\dot{u}\}_s$ from (6.41*a*) and (6.41*b*), respectively, in the above equation, we obtain

$$\Delta t_{s+1}\,\alpha(\{F\}_{s+1} - [K]_{s+1}\{u\}_{s+1}) + \Delta t_{s+1}(1-\alpha)(\{F\}_s - [K]_s\{u\}_s) = [M](\{u\}_{s+1} - \{u\}_s)$$

Rearranging the terms into known and unknown ones, we obtain

$$[\hat{K}]_{s+1}\{u\}_{s+1} = [\bar{K}]_s\{u\}_s + \{\hat{F}\}_{s,s+1} \tag{6.42a}$$

where

$$\begin{gathered} [\hat{K}]_{s+1} = [M^1] + a_1[K]_{s+1}, \qquad [\bar{K}]_s = [M^1] - a_2[K]_s \\ \{\hat{F}\}_{s,s+1} = \Delta t_{s+1}\,[\alpha\{F\}_{s+1} + (1-\alpha)\{F\}_s] \\ a_1 = \alpha\,\Delta t_{s+1}, \qquad a_2 = (1-\alpha)\,\Delta t_{s+1} \end{gathered} \tag{6.42b}$$

Note that, in deriving (6.42), it has been assumed that $[M^1]$ is independent of time and that the time step is nonuniform.

Equations (6.42) are valid for a typical element. The assembly, imposition of boundary conditions, and solution of the assembled equations are the same as described before for static (or steady-state) problems. Calculation of $[\hat{K}]$ and $\{\hat{F}\}$ at time $t=0$ requires knowledge of the initial conditions $\{u\}_0$ and the time variation of $\{F\}$. Note that, for $\alpha=0$ (the forward difference scheme), we obtain $[\hat{K}]=[M^1]$. When the mass matrix $[M^1]$ is diagonal, (6.42*a*) become *explicit,* and one can solve for $\{u\}_{s+1}$ directly without inverting $[\hat{K}]$. The mass matrix obtained according to the weak form, called the *consistent mass matrix,* is not diagonal. There are several ways to diagonalize mass matrices; these will be described in Section 6.2.4.

Stability and accuracy. Since (6.39*a*) represents an approximation, which is used to derive (6.42*a*), error is introduced into the solution $\{u\}_{s+1}$ at each time step. In addition to the truncation error introduced in approximating the derivative, round-off errors can be introduced because of the finite arithmetic used in our computations. Since the solution at time t_{s+1} depends on the solution at time t_s, the error can grow with time. If it grows unboundedly with time, the solution scheme is said to be *unstable.* If it is bounded (i.e., it increases for one time step and decreases for another time step, but never exceeds a certain finite value), the solution scheme is said to be *stable.* The numerical scheme (6.42) is said to be *consistent* with the continuous problem (6.38) if the round-off and truncation errors go to zero as $\Delta t \rightarrow 0$. *Accuracy* of a numerical scheme is a measure of the closeness between the approximate solution and the exact solution, whereas *stability* of a solution is a measure of the boundedness of the approximate solution with time. As one might expect, the size of the time step can influence both accuracy and stability. When we construct an approximate solution, we like it to converge to the true solution when the number of elements or the degree of approximation is increased and the time step Δt is decreased. A time approximation scheme is said to be *convergent* if, for fixed t_s and Δt, the numerical value $\{u\}_s$ converges to its true value $\{u(t_s)\}$ as $\Delta t \rightarrow 0$. Accuracy is measured in terms of the rate at which the approximate solution converges. If a numerical scheme is stable and consistent, it is also convergent [see Isaacson and Keller (1966)].

A numerical scheme is said to be *conditionally stable* if it is stable only when certain restrictions on the time step are satisfied. For all numerical schemes in which $\alpha < \frac{1}{2}$, the α family of approximations is stable only if the time step satisfies the following (stability) condition:

$$\Delta t < \Delta t_{cr} \equiv \frac{2}{(1-2\alpha)\lambda} \tag{6.43}$$

where λ is the largest eigenvalue of the finite element equations (6.38). Note that the same mesh as that used for the transient analysis must be used to calculate the eigenvalues.

When $\alpha = 0$, the numerical scheme is called an *explicit* scheme. The name comes from the fact that when the mass matrix $[M]$ is diagonal, (6.42) can be solved without inverting $[\hat{K}]$ (because $[\hat{K}] = [M]$ is diagonal). When $\alpha \neq 0$ the scheme is said to be *implicit,* indicating that $[\hat{K}]$ has to be inverted (whether or not $[M]$ is diagonal). Such a classicfication of time integration schemes seems to have originated with finite difference methods. When these methods are used for spatial discretization, the mass matrix $[M]$ ends up as a diagonal matrix. However, in the finite element method, irrespective of the time integration scheme used, the consistent mass matrix $[M]$ is not diagonal and hence it is necessary to invert $[\hat{K}]$. When the consistent mass matrix is replaced by an equivalent diagonal mass matrix (see Section 6.2.4), an explicit scheme results in an explicit set of algebraic equations that can be solved without inverting the coefficient matrix $[\hat{K}]$.

CASE 2: HYPERBOLIC EQUATIONS. For this case $c_1 = 0$ (i.e., $[M^1] = [0]$), and (6.37*a*) takes the form

$$[M^2]\{\ddot{u}\} + [K]\{u\} = \{F\} \tag{6.44}$$

The hyperbolic nature is dictated by $c_2 \neq 0$, not by $c_1 = 0$; in fact, in structural systems, both c_1 and c_2 can be nonzero. For example, consider the case where $[M^1]$ is the damping matrix in a structural system; see Problem 6.24.

There are several numerical integration methods available to integrate second-order (i.e., hyperbolic) equations. Among these, the Newmark family of time integration schemes is widely used in structural dynamics. Other methods, such as the Wilson method and the Houbolt method, can be used to develop the algebraic equations from the second-order differential equations (6.44).

In the Newmark method, the function and its first time derivative are approximated according to

$$\begin{aligned} \{u\}_{s+1} &= \{u\}_s + \Delta t\, \{\dot{u}\}_s + \tfrac{1}{2}(\Delta t)^2\{\ddot{u}\}_{s+\gamma} \\ \{\dot{u}\}_{s+1} &= \{\dot{u}\}_s + \{\ddot{u}\}_{s+\alpha}\, \Delta t \end{aligned} \tag{6.45a}$$

where

$$\{\ddot{u}\}_{s+\theta} = (1-\theta)\{\ddot{u}\}_s + \theta\{\ddot{u}\}_{s+1} \tag{6.45b}$$

and α and γ $(=2\beta)$ are parameters that determine the stability and accuracy of the scheme. The following schemes are special cases of (6.45a,b):

$\alpha = \frac{1}{2}$,	$\gamma = 2\beta = \frac{1}{2}$,	the constant-average acceleration method (stable)
$\alpha = \frac{1}{2}$,	$\gamma = 2\beta = \frac{1}{3}$,	the linear acceleration method (conditionally stable)
$\alpha = \frac{1}{2}$,	$\gamma = 2\beta = 0$,	the central difference method (conditionally stable)
$\alpha = \frac{3}{2}$,	$\gamma = 2\beta = \frac{8}{5}$,	the Galerkin method (stable)
$\alpha = \frac{3}{2}$,	$\gamma = 2\beta = 2$,	the backward difference method (stable)

(6.45c)

For all schemes in which $\gamma < \alpha$ and $\alpha \geqslant \frac{1}{2}$, the stability requirement is

$$\Delta t \leqslant \Delta t_{cr} = [\tfrac{1}{2}\omega_{max}^2(\alpha - \gamma)]^{-1/2} \tag{6.46}$$

where ω_{max} is the maximum natural frequency of the system (6.44).

The use of (6.45) in (6.44) gives the following system of algebraic equations:

$$[\hat{K}]_{s+1}\{u\}_{s+1} = \{\hat{F}\}_{s,s+1} \tag{6.47a}$$

where

$$\begin{gathered} [\hat{K}]_{s+1} = [K]_{s+1} + a_3[M]_{s+1} \\ \{\hat{F}\}_{s,s+1} = \{F\}_{s+1} + [M]_{s+1}(a_3\{u\}_s + a_4\{\dot{u}\}_s + a_5\{\ddot{u}\}_s) \\ a_3 = \frac{2}{\gamma(\Delta t)^2}, \qquad a_4 = \frac{2}{\gamma\,\Delta t}, \qquad a_5 = \frac{1}{\gamma} - 1 \end{gathered} \tag{6.47b}$$

Again, (6.47) are valid over a typical element.

Note that the calculation of $[\hat{K}]$ and $\{\hat{F}\}$ requires knowledge of the initial conditions $\{u\}_0$, $\{\dot{u}\}_0$, and $\{\ddot{u}\}_0$. In practice, one does not know $\{\ddot{u}\}_0$. As an approximation, it can be calculated from (6.44) (we often assume that the applied force is zero at $t = 0$):

$$\{\ddot{u}\}_0 = [M^2]^{-1}(\{F\}_0 - [K]\{u\}_0) \tag{6.48}$$

At the end of each time step, the new velocity vector $\{\dot{u}\}_{s+1}$ and acceleration vector $\{\ddot{u}\}_{s+1}$ are computed using the equations

$$\begin{gathered} \{\ddot{u}\}_{s+1} = a_3(\{u\}_{s+1} - \{u\}_s) - a_4\{\dot{u}\} - a_5\{\ddot{u}\}_s \\ \{\dot{u}\}_{s+1} = \{\dot{u}\}_s + a_2\{\ddot{u}\}_s + a_1\{\ddot{u}\}_{s+1} \\ a_1 = \alpha\,\Delta t, \qquad a_2 = (1 - \alpha)\,\Delta t \end{gathered} \tag{6.49}$$

The remaining procedure stays the same as in static (i.e., non-time-dependent) problems.

6.2.4 Mass Lumping

Recall from the time approximation of parabolic equations that use of the forward difference scheme (i.e., $\alpha = 0$) results in the following time marching scheme [see (6.42)]:

$$[M^e]\{u\}_{s+1} = ([M^e] - \Delta t\,[K^e])\{u\}_s + \Delta t\,\{F^e\}_s \tag{6.50}$$

The mass matrix $[M^e]$ derived from the weighted-integral formulation of the governing equation is called the *consistent mass matrix,* and it is symmetric, positive-definite, and nondiagonal. Solution of the global equations associated with (6.50) requires inversion of the assembled mass matrix. If the mass matrix is diagonal then the assembled equations can be solved explicitly,

$$(U_I)_{s+1} = M_{II}^{-1}\left[M_{II}(U_I)_s - \Delta t \sum_{J=1}^{N} K_{IJ}(U_J)_s + \Delta t\,(F_I)_s\right] \tag{6.51}$$

thus saving computational time. The explicit nature of (6.51) motivated analysts to find rational ways of diagonalizing the mass matrix.

There are several ways of constructing diagonal mass matrices [see Hughes (1987)]. Diagonal mass matrices are known as *lumped mass matrices.* The error estimates in (6.27) are generally not valid for lumped mass matrices. The *row-sum* and *proportional lumping* techniques are discussed here.

ROW-SUM LUMPING. The sum of the elements of each row of the consistent mass matrix is used as the diagonal element:

$$\boxed{M_{ii}^e = \sum_{j=1}^{n} \int_{x_A}^{x_B} \rho\psi_i^e\psi_j^e\,dx = \int_{x_A}^{x_B} \rho\psi_i^e\,dx} \tag{6.52}$$

where the property $\sum_{j=1}^{n} \psi_j^e = 1$ of the interpolation functions is used. When ρ is constant, (6.52) gives

$$[M^e]_L = \frac{\rho h_e}{2}\begin{bmatrix} 1 & 0 \\ 0 & 1 \end{bmatrix} \quad \text{for the Lagrange linear element}$$

$$[M^e]_L = \frac{\rho h_e}{6}\begin{bmatrix} 1 & 0 & 0 \\ 0 & 4 & 0 \\ 0 & 0 & 1 \end{bmatrix} \quad \text{for the Lagrange quadratic element} \tag{6.53a}$$

Compare these lumped mass matrices with the consistent mass matrices

$$[M^e]_C = \frac{\rho h_e}{6}\begin{bmatrix} 2 & 1 \\ 1 & 2 \end{bmatrix} \quad \text{for the Lagrange linear element}$$

$$[M^e]_C = \frac{\rho h_e}{30}\begin{bmatrix} 4 & 2 & -1 \\ 2 & 16 & 2 \\ -1 & 2 & 4 \end{bmatrix} \quad \text{for the Lagrange quadratic element} \tag{6.53b}$$

Here subscripts L and C refer to lumped and consistent mass matrices, respectively.

PROPORTIONAL LUMPING. Here the diagonal elements of the lumped mass matrix are computed to be proportional to the diagonal elements of the consistent mass matrix while conserving the total mass of the element:

$$M_{ii}^e = \alpha \int_{x_A}^{x_B} \rho \psi_i^e \psi_i^e \, dx, \qquad \alpha = \left(\int_{x_A}^{x_B} \rho \, dx \right) \Big/ \left(\sum_{i=1}^{n} \int_{x_A}^{x_B} \rho \psi_i^e \psi_i^e \, dx \right) \tag{6.54}$$

For constant ρ, the proportional lumping gives the same lumped mass matrices as those obtained in the row-sum technique for the Lagrange linear and quadratic elements.

The use of a lumped mass matrix in transient analyses can save computational time in two ways. First, for forward difference schemes, lumped mass matrices result in explicit algebraic equations, not requiring matrix inversions. Second, the critical time step required for conditionally stable schemes is larger, and hence less computational time is required when lumped mass matrices are used. To see this, consider the stability criterion in (6.46) for the case $\alpha = \frac{1}{2}$, $\beta = 0$. For a one linear element model of a uniform bar of stiffness EA and mass ρA, fixed at the left end, the eigenvalue problem with a consistent mass matrix is

$$\left(\frac{EA}{h} \begin{bmatrix} 1 & -1 \\ -1 & 1 \end{bmatrix} - \omega^2 \frac{\rho A h}{6} \begin{bmatrix} 2 & 1 \\ 1 & 2 \end{bmatrix} \right) \begin{Bmatrix} U_1 \\ U_2 \end{Bmatrix} = \begin{Bmatrix} Q_1^1 \\ Q_2^1 \end{Bmatrix}$$

Since $U_1 = 0$ and $Q_2^1 = 0$, we have

$$\omega^2 = \frac{EA}{h} \Big/ \frac{\rho A h}{3} = \frac{3E}{\rho h^2}$$

Substituting this into the critical time step relation (6.46), we have,

$$(\Delta t_{\text{cr}})_C = 2/\omega_{\max} = h(4\rho/3E)^{1/2}$$

If we use the lumped matrix, ω is given by

$$\omega = (2E/\rho)^{1/2}/h$$

and the critical time step is

$$(\Delta t_{\text{cr}})_L = h(2\rho/E)^{1/2} > (\Delta t_{\text{cr}})_C \tag{6.55}$$

6.2.5 Applications

Here we consider two examples of applications of finite element models of one-dimensional problems. Problems are taken from heat transfer and solid mechanics. Other field problems can be related to heat-transfer-type problems.

Example 6.3. Consider the transient heat conduction problem with equation

$$\frac{\partial u}{\partial t} - \frac{\partial^2 u}{\partial x^2} = 0 \quad \text{for } 0 < x < 1 \tag{6.56a}$$

boundary conditions

$$u(0, t) = 0, \qquad \frac{\partial u}{\partial x}(1, t) = 0 \tag{6.56b}$$

and initial condition

$$u(x, 0) = 1.0 \tag{6.56c}$$

The problem at hand is a special case of (6.33a) with $a = 1$, $b = 0$, $c_0 = 0$, $c_1 = 1$, $c_2 = 0$, and $f = 0$. The finite element model of (6.56a) is given by (6.38a):

$$[M^e]\{\dot{u}\} + [K^e]\{u\} = \{Q^e\} \tag{6.57a}$$

where

$$M_{ij}^e = \int_{x_A}^{x_B} \psi_i^e \psi_j^e \, dx, \qquad K_{ij}^e = \int_{x_A}^{x_B} \frac{d\psi_i^e}{dx}\frac{d\psi_j^e}{dx}\, dx \tag{6.57b}$$

For the choice of linear interpolation functions, (6.57a) becomes

$$\frac{h}{6}\begin{bmatrix} 2 & 1 \\ 1 & 2 \end{bmatrix}\begin{Bmatrix} \dot{u}_1 \\ \dot{u}_2 \end{Bmatrix} + \frac{1}{h}\begin{bmatrix} 1 & -1 \\ -1 & 1 \end{bmatrix}\begin{Bmatrix} u_1 \\ u_2 \end{Bmatrix} = \begin{Bmatrix} Q_1 \\ Q_2 \end{Bmatrix} \tag{6.57c}$$

where h is the length of the element. Use of the α-family of approximation (6.39) results in the equation [see (6.42)]

$$\begin{aligned}([M^e] + \Delta t\, \alpha[K^e])\{u^e\}_{s+1} = ([M^e] - \Delta t\,(1-\alpha)[K^e])\{u^e\}_s \\ + \Delta t(\alpha\{Q^e\}_{s+1} + (1-\alpha)\{Q^e\}_s)\end{aligned} \tag{6.58a}$$

where Δt is the time step.

For a one-element model, we have

$$\begin{aligned}\begin{bmatrix} \frac{1}{3}h + \alpha\frac{\Delta t}{h} & \frac{1}{6}h - \alpha\frac{\Delta t}{h} \\ \frac{1}{6}h - \alpha\frac{\Delta t}{h} & \frac{1}{3}h + \alpha\frac{\Delta t}{h} \end{bmatrix}\begin{Bmatrix} U_1 \\ U_2 \end{Bmatrix}_{s+1} \\ = \begin{bmatrix} \frac{1}{3}h - (1-\alpha)\frac{\Delta t}{h} & \frac{1}{6}h + (1-\alpha)\frac{\Delta t}{h} \\ \frac{1}{6}h + (1-\alpha)\frac{\Delta t}{h} & \frac{1}{3}h - (1-\alpha)\frac{\Delta t}{h} \end{bmatrix}\begin{Bmatrix} U_1 \\ U_2 \end{Bmatrix}_s + \Delta t\begin{Bmatrix} \bar{Q}_1 \\ \bar{Q}_2 \end{Bmatrix}\end{aligned} \tag{6.58b}$$

where $\bar{Q}_i = \alpha(Q_i^1)_{s+1} + (1-\alpha)(Q_i^1)_s$. The boundary conditions of the problem require

$$(U_1)_s = 0, \qquad (Q_2^1)_s = 0 \quad \text{for all } s > 0 \text{ (i.e., } t > 0)$$

while the initial condition requires

$$(U_1)_0 = 1, \qquad (U_2)_0 = 1 \quad \text{for } t = 0$$

Since the initial condition should be consistent with the boundary conditions, we take $(U_1)_0 = 0.0$. Using the boundary conditions, we can write for the one-element model ($h = 1.0$)

$$\left(\tfrac{1}{3}h + \alpha\frac{\Delta t}{h}\right)(U_2)_{s+1} = \left[\tfrac{1}{3}h - (1-\alpha)\frac{\Delta t}{h}\right](U_2)_s \tag{6.59}$$

which can be solved repeatedly for U_2 at different times, $s = 0, 1, \ldots$.

Repeated use of (6.59) can cause the temporal approximation error to grow with time, depending on the value of α. As noted earlier, the forward difference scheme ($\alpha = 0$) is a conditionally stable scheme. The critical time step is given by

$$\Delta t_{cr} = 2/\lambda_{max}$$

where λ_{max} is the maximum eigenvalue associated with (6.57c):

$$-\lambda[M]\{U\} + [K]\{U\} = \{0\}$$

For the model at hand, this reduces to

$$-\tfrac{1}{3}\lambda h U_2 + h^{-1} U_2 = 0, \quad \text{or } \lambda = 3/h^2 = 3$$

Hence $\Delta t_{cr} = 0.6667$. Thus, in order for the forward difference solution of (6.59) to be stable, the time step should be smaller than $\Delta t_{cr} = 0.6667$; otherwise, the solution will be unstable, as shown in Fig. 6.7.

For unconditionally stable schemes ($\alpha \geq \frac{1}{2}$), there is no restriction on the time step. However, to obtain a sufficiently accurate solution, the time step must be taken as a fraction of Δt_{cr}. Of course, the accuracy of the solution also depends on the mesh size h. As this is decreased (i.e., the number of elements is increased), Δt_{cr} decreases.

Figure 6.8 shows plots of $u(1, t)$ versus time for $\alpha = 0.5$ and $\Delta t = 0.05$. Solutions predicted by meshes of one and two linear elements and the mesh of one quadratic element are compared with the exact solution. The convergence of the solution with increasing number of elements is clear. The finite element solutions obtained with different methods, time steps, and meshes are compared with the exact solution in Table 6.3.

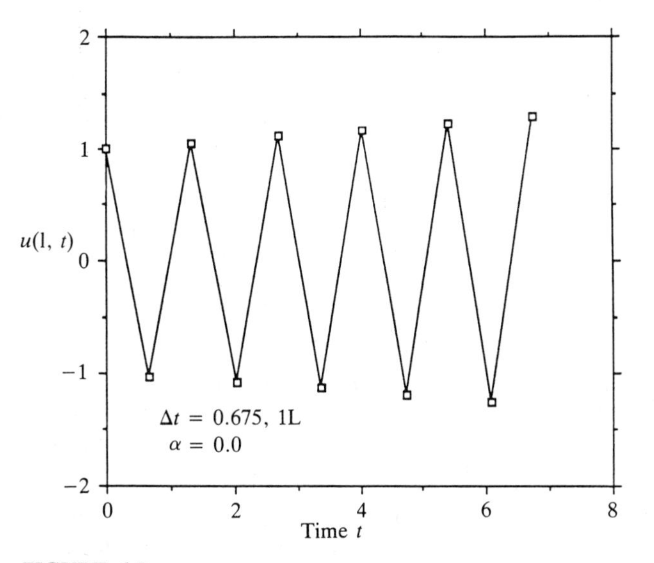

FIGURE 6.7
Transient solution of a parabolic equation according to the forward difference scheme ($\Delta t = 0.675$, one linear element). The solution is unstable because the time step is larger than the critical time step.

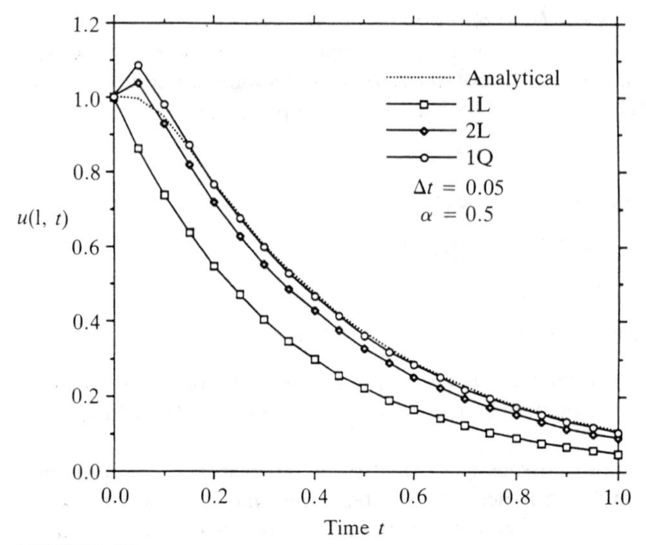

FIGURE 6.8
Transient solution of a parabolic equation according to linear and quadratic finite elements.

TABLE 6.3
A comparison of the finite element solutions obtained using various time approximation schemes and meshes with the analytical solution of a parabolic equation arising in conductive heat transfer

Time	1L		Linear elements ($\alpha = 0.5$)				Quadratic elements ($\alpha = 0.5$, $\Delta t = 0.05$)			Exact
t	$\alpha = 0$	$\alpha = 1$	1L	2L	4L	8L	1Q	2Q	4Q	solution
0.00	1.0000	1.0000	1.0000	1.0000	1.0000	1.0000	1.0000	1.0000	1.0000	1.0000
0.05	0.8500	0.8696	0.8605	1.0359	0.9951	0.9933	1.0870	0.9942	0.9928	0.9969
0.10	0.7225	0.7561	0.7404	0.9279	0.9588	0.9554	0.9819	0.9550	0.9549	0.9493
0.15	0.6141	0.6575	0.6371	0.8169	0.8639	0.8707	0.8693	0.8831	0.8725	0.8642
0.20	0.5220	0.5718	0.5482	0.7176	0.7557	0.7694	0.7679	0.7633	0.7731	0.7723
0.25	0.4437	0.4972	0.4717	0.6300	0.6759	0.6824	0.6780	0.6933	0.6855	0.6854
0.30	0.3771	0.4323	0.4059	0.5533	0.5906	0.6037	0.5987	0.6006	0.6070	0.6068
0.35	0.3206	0.3759	0.3492	0.4858	0.5250	0.5325	0.5286	0.5394	0.5358	0.5367
0.40	0.2725	0.3269	0.3005	0.4266	0.4608	0.4713	0.4668	0.4710	0.4741	0.4745
0.45	0.2316	0.2843	0.2586	0.3746	0.4083	0.4158	0.4121	0.4201	0.4188	0.4194
0.50	0.1969	0.2472	0.2225	0.3289	0.3592	0.3676	0.3639	0.3687	0.3701	0.3708
0.55	0.1673	0.2149	0.1914	0.2888	0.3176	0.3247	0.3213	0.3275	0.3273	0.3277
0.60	0.1422	0.1869	0.1647	0.2536	0.2798	0.2868	0.2837	0.2883	0.2890	0.2897
0.65	0.1209	0.1625	0.1418	0.2227	0.2472	0.2535	0.2505	0.2556	0.2556	0.2561
0.70	0.1028	0.1413	0.1220	0.1955	0.2180	0.2238	0.2212	0.2253	0.2258	0.2264
0.75	0.0874	0.1229	0.1050	0.1717	0.1924	0.1979	0.1953	0.1995	0.1996	0.2001
0.80	0.0743	0.1069	0.0903	0.1508	0.1697	0.1747	0.1725	0.1761	0.1764	0.1769
0.85	0.0631	0.0929	0.0777	0.1324	0.1498	0.1544	0.1523	0.1557	0.1559	0.1563
0.90	0.0536	0.0808	0.0669	0.1162	0.1322	0.1363	0.1345	0.1375	0.1378	0.1382
0.95	0.0456	0.0703	0.0575	0.1020	0.1166	0.1205	0.1187	0.1216	0.1218	0.1222
1.00	0.0388	0.0611	0.0495	0.0896	0.1029	0.1065	0.1048	0.1074	0.1076	0.1080

The next example concerns a hyperbolic equation.

Example 6.4. Consider the transverse motion of a beam clamped at both ends, according to the Euler–Bernoulli beam theory,

$$\frac{\partial^2 w}{\partial t^2}+\frac{\partial^4 w}{\partial x^4}=0 \quad \text{for } 0<x<1 \tag{6.60a}$$

$$w(0,t)=0, \quad \frac{\partial w}{\partial x}(0,t)=0, \quad w(1,t)=0, \quad \frac{\partial w}{\partial x}(1,t)=0 \tag{6.60b}$$

$$w(x,0)=\sin \pi x-\pi x(1-x), \quad \frac{\partial w}{\partial t}(x,0)=0 \tag{6.60c}$$

Note that the initial deflection of the beam is consistent with the boundary conditions. The initial slope is given by

$$\frac{\partial w}{\partial x}(x,0)=\pi\cos \pi x-\pi(1-2x) \tag{6.60d}$$

Because of symmetry about $x=0.5$, we consider only half of the beam for finite element modeling. In this case, the boundary condition at $x=0.5$ is $(\partial w/\partial x)(0.5,t)=0$. For a one-element model with the Euler–Bernoulli beam element, we have

$$\frac{h}{420}\begin{bmatrix}156 & -22h & 54 & 13h\\ -22h & 4h^2 & -13h & -3h^2\\ 54 & -13h & 156 & 22h\\ 13h & -3h^2 & 22h & 4h^2\end{bmatrix}\begin{Bmatrix}\ddot{u}_1\\ \ddot{u}_2\\ \ddot{u}_3\\ \ddot{u}_4\end{Bmatrix}$$

$$+\frac{2}{h^3}\begin{bmatrix}6 & -3h & -6 & -3h\\ -3h & 2h^2 & 3h & h^2\\ -6 & 3h & 6 & 3h\\ -3h & h^2 & 3h & 2h^2\end{bmatrix}\begin{Bmatrix}u_1\\ u_2\\ u_3\\ u_4\end{Bmatrix}=\begin{Bmatrix}Q_1\\ Q_2\\ Q_3\\ Q_4\end{Bmatrix}$$

The boundary conditions for the one-element model translate into

$$U_1=0, \quad U_2=0, \quad U_4=0, \quad Q_3=0 \quad \text{for all } t>0$$

while the initial conditions

$$\left.\begin{matrix}U_1=0, & U_2=0, & U_3=0.2146, & U_4=0\\ \dot{U}_1=0, & \dot{U}_2=0, & \dot{U}_3=0, & \dot{U}_4=0\end{matrix}\right\} \quad \text{for } t=0$$

The time marching scheme (6.47a) for this case takes the form

$$(K_{33}+a_3M_{33})(U_3)_{s+1}=(\hat{F}_3)_{s,s+1}\equiv M_{33}(a_3U_3+a_4\dot{U}_3+a_5\ddot{U}_3)_s$$

where a_3, a_4, and a_5 are defined in (6.47b). The second derivative $\ddot{U}_3$ for time $t=0$ (i.e., when $s=0$) is computed from the equation of motion:

$$(\ddot{U}_3)_0=-\frac{K_{33}(U_3)_0}{M_{33}}=-\left(\frac{12}{h^3}\times 0.2146\right)\frac{420}{156h}=-110.932$$

The stability criterion (6.46) requires, for $\gamma<\frac{1}{2}$, that the time step Δt be less than Δt_{cr}. For the present model, ω_{max} is computed from the eigenvalue problem

$$(K_{33}-\omega^2M_{33})U_3=0 \quad \text{or} \quad \omega^2=M_{33}/K_{33}=516.923$$

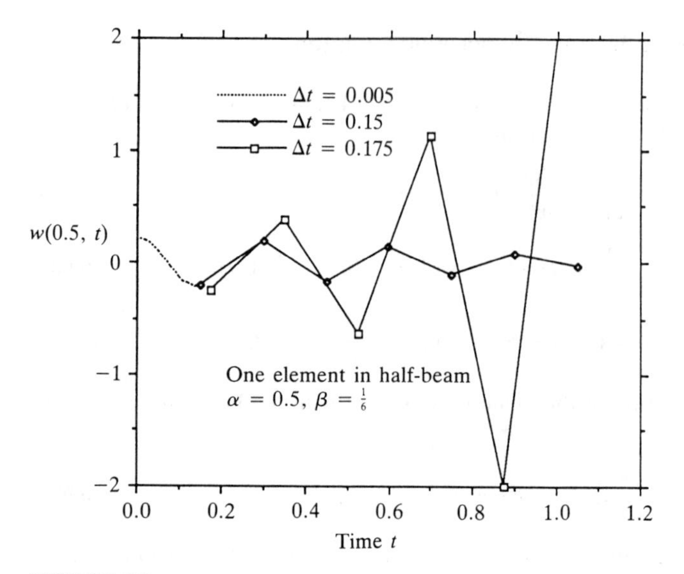

FIGURE 6.9
Transient response of a beam clamped at both ends, according to the linear acceleration method ($\alpha = 0.5$ and $\beta = \frac{1}{6}$). The critical time step for the method using one element is 0.1524. The remainder of the $\Delta t = 0.005$ plot is not shown.

Hence, the critical time step for $\alpha = 0.5$ and $\gamma = \frac{1}{3}$ (i.e., the linear acceleration scheme) is

$$\Delta t_{cr} = \sqrt{12}/\omega_{max} = 0.15236$$

Although there is no restriction on time integration schemes with $\alpha = 0.5$ and $\gamma > 0.5$, the critical time step provides an estimate of the time step to be used.

Figure 6.9 shows plots of $w(0.5, t)$ versus time for the scheme $\alpha = 0.5$, $\gamma = \frac{1}{3}$. Three different time steps, $\Delta t = 0.175$, 0.150, and 0.005, are used to illustrate the accuracy. For $\Delta t = 0.175 > \Delta t_{cr}$, the solution is unstable, whereas for $\Delta t = 0.15 < \Delta t_{cr}$, it is stable but inaccurate. The period of the solution is given by

$$T = 2\pi/\omega = 0.27635$$

For a time step $\Delta t = 0.005$, the solution is predicted very accurately.

For two- and four-element meshes, the critical steps are computed by computing the maximum eigenvalues of the corresponding discrete systems. The critical time steps for the two meshes are

$$(\Delta t_{cr})_2 = 0.00897, \qquad (\Delta t_{cr})_4 = 0.00135$$

where the subscripts refer to the number of elements in the mesh. Figure 6.10 shows a comparison of the transverse deflections obtained with the two Euler–Bernoulli elements ($\Delta t = 0.005$) and the Galerkin method (see Chapter 2) for a complete period $(0, 0.28)$.

The problem can also be analyzed using the Timoshenko beam element, which requires us to select the coefficients GAK, EI, ρA, and ρI consistent with those in the differential equation (6.11). Comparing (6.60*a*) with (6.11), we have $EI = 1.0$ and

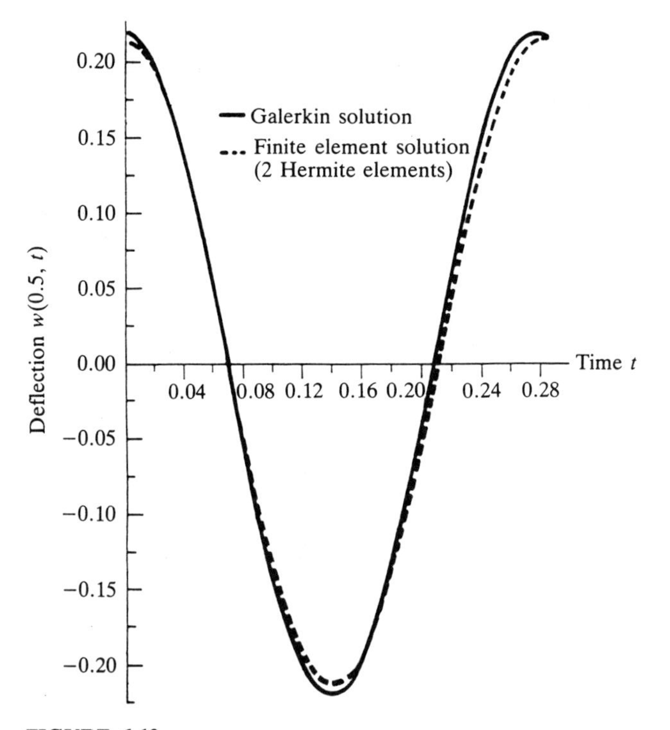

FIGURE 6.10
Transient response of a beam clamped at both ends and subjected to an initial transverse deflection ($\Delta t = 0.005$, $\alpha = 0.5$, and $\beta = 0.25$).

$\rho A = 1.0$. Therefore, GAK can be computed as

$$GAK = \frac{E}{2(1+\nu)} BHK = \frac{EI}{2(1+\nu)} \frac{12}{H^2} \frac{5}{6} = \frac{4}{H^2} EI \tag{6.61}$$

where B is the width and H the height of the beam, and $I = \frac{1}{12}BH^3$, $\nu = 0.25$, and $K = \frac{5}{6}$ are used in arriving at the last expression. Similarly,

$$\rho I = \rho \tfrac{1}{12} BH^3 = \tfrac{1}{12} \rho A H^2 \tag{6.62}$$

For $L/H = 100$ (since $L = 1.0$, $H = 0.01$), the shear deformation effect is small; while for $L/H = 10$ (or $H = 0.1$), it is significant enough to warrant its inclusion in the analysis.

Table 6.4 gives values of $w(0.5, t)$ as obtained using the Euler–Bernoulli and Timoshenko elements for various numbers of elements. The time step is taken to be $\Delta t = 0.005$, which is larger than that required for stability of the four-element mesh of the Euler–Bernoulli beam element when $\gamma = \frac{1}{3}$. Figure 6.11 shows plots of $w(0.5, t)$ obtained with two quadratic Timoshenko elements for $L/H = 100$ and 10. The Timoshenko elements have better stability characteristics (i.e., larger Δt_{cr}) than the Euler–Bernoulli beam element for small ratios L/H. This is because, as L/H is decreased, the ω_{max} predicted by the Timoshenko beam theory is smaller than that predicted by the Euler–Bernoulli beam theory. The critical time steps for the mesh of

TABLE 6.4

Effect of mesh and shear deformation on the transient response of a beam clamped at both ends ($\Delta t = 0.005$)

	Euler–Bernoulli beam element						Timoshenko element (2Q)		
	$\alpha = 0.25,\ \beta = 1/6$			$\alpha = 0.5,\ \beta = 0.25$			$\alpha = 0.5,\ \beta = 0.25$		Galerkin in space and exact in time
Time t	1	2	4†	1	2	4	$L/H = 100$	$L/H = 10$	
0.00	0.2146	0.2146	0.2146	0.2146	0.2146	0.2146	0.2146	0.2146	0.2146
0.01	0.2091	0.2098	0.2098	0.2091	0.2098	0.2098	0.2100	0.2116	0.2157
0.02	0.1928	0.1951	0.1950	0.1928	0.1951	0.1951	0.1953	0.1951	0.1988
0.03	0.1666	0.1696	0.1681	0.1667	0.1696	0.1698	0.1695	0.1690	0.1716
0.04	0.1319	0.1346	0.1181	0.1320	0.1348	0.1350	0.1342	0.1427	0.1356
0.05	0.0904	0.0930	−0.0795	0.0905	0.0931	0.0935	0.0929	0.1067	0.0925
0.06	0.0442	0.0481	−1.7165	−0.0443	0.0482	0.0483	0.0484	0.0657	0.0447
0.07	−0.0043	0.0014	−17.877	−0.0041	0.0014	0.0018	0.0016	0.0234	−0.0055
0.08	−0.0525	−0.0462	−179.92	−0.0523	−0.0459	−0.0455	−0.0469	−0.0267	−0.0553
0.09	−0.0980	−0.0926	−1796.9	−0.0978	−0.0923	−0.0916	−0.0937	−0.0706	−0.1023
0.10	−0.1385	−0.1345	·	−0.1383	−0.1342	−0.1336	−0.1349	−0.1100	−0.1441
0.11	−0.1719	−0.1685	·	−0.1717	−0.1685	−0.1682	−0.1680	−0.1461	−0.1783
0.12	−0.1964	−0.1933	·	−0.1963	−0.1933	−0.1932	−0.1931	−0.1717	−0.2034
0.13	−0.2108	−0.2088	·	−0.2108	−0.2088	−0.2087	−0.2100	−0.1969	−0.2179
0.14	−0.2144	−0.2153	·	−0.2144	−0.2150	−0.2148	−0.2169	−0.2110	−0.2211
0.15	−0.2070	−0.2113	Diverged	−0.2071	−0.2112	−0.2111	−0.2117	−0.2146	−0.2129

† $\Delta t_{cr} = 0.00135$

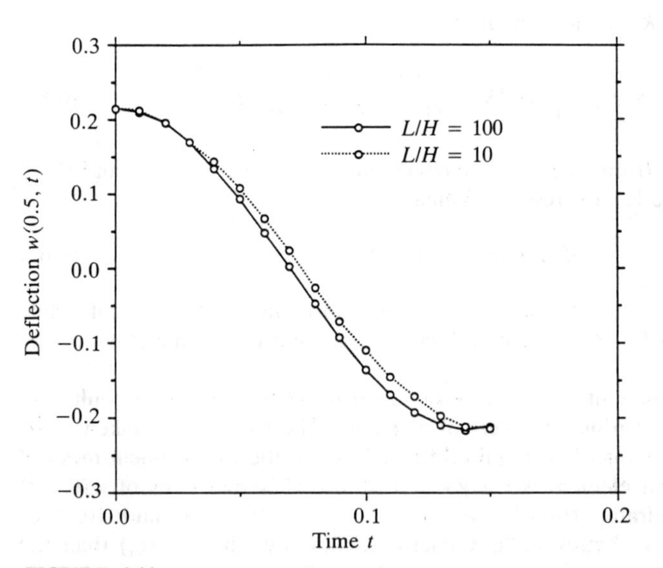

FIGURE 6.11

Transient response of a beam clamped at both ends, according to the Timoshenko beam theory ($\Delta t = 0.005$, two quadratic elements are used, $\alpha = 0.5$, and $\beta = 0.25$).

two quadratic elements are

$$(\Delta t_{cr})_{100} = 0.00665, \qquad (\Delta t_{cr})_{10} = 0.00644$$

where the subscripts refer to the ratio L/H. This completes the example.

6.3 SUMMARY

In this chapter eigenvalue problems and their formulation for one-dimensional second- and fourth-order equations (beams) have been discussed, finite element models of the equations have been developed, and applications to heat transfer and the natural vibration of beams have been presented. Except for the solution procedure, the finite element formulation of eigenvalue problems is entirely analogous to boundary value problems.

Finite element models of time-dependent problems described by parabolic and hyperbolic equations have also been presented. A two-step procedure to derive finite element models from differential equations has been described. In the first step, we seek spatial approximations of the dependent variables of the problem as linear combinations of nodal values that are functions of time and interpolation functions that are functions of space. This procedure is entirely analogous to the finite element formulation presented for boundary value problems in Chapters 3 and 4. The end result of this step is a set of ordinary differential equations (in time) among the nodal values. In the second step, the ordinary differential equations are further approximated to replace the time derivatives with the values of the functions at different times.

PROBLEMS

Section 6.1

6.1. Determine the first two longitudinal frequencies of a rod (E, A, L) fixed at one end and spring-supported at the other:

$$-EA\frac{\partial^2 u}{\partial x^2} + \rho A\frac{\partial^2 u}{\partial t^2} = 0 \quad \text{for } 0 < x < L$$

$$u(0) = 0, \qquad \left(EA\frac{du}{dx} + ku\right)\Big|_{x=L} = 0$$

Use (*a*) two linear finite elements and (*b*) one quadratic element.

Answer: (a) The characteristic equation is $7\lambda^2 - (10 + 4c)\lambda + (1 + 2c) = 0$, $c = kh/EA$, $\lambda = (\rho h^2/6E)\omega^2$.

6.2. Determine the smallest natural frequency of a beam with clamped ends, and of constant cross-sectional area A, moment of inertia I, and length L. Use the symmetry and two Euler–Bernoulli beam elements in the half-beam.

6.3. Re-solve the above problem with two linear Timoshenko beam elements in the half-beam.

6.4. Consider a beam (A, I, E, L) with its left end clamped and its right end supported by an elastic spring. Determine the fundamental natural frequency using (*a*) one Euler–Bernoulli beam and (*b*) one Timoshenko beam element.

6.5. Determine the critical buckling load of a cantilever beam (A, I, L, E) using (a) one Euler–Bernoulli beam element and (b) one Timoshenko beam element.

6.6. Determine the fundamental natural frequency of the truss shown in Fig. P4.29.

6.7. Determine the fundamental natural frequency of the frame shown in Fig. P4.32.

6.8. Determine the first two longitudinal natural frequencies of a rod (A, E, L, m), fixed at one end and with an attached mass m_2 at the other. Use two linear elements.

Hint: Note that the boundary conditions for the problem are $u(0) = 0$ and $(EA\, \partial u/\partial x + m_2\, \partial^2 u/\partial t^2)|_{x=L} = 0$.

6.9. The equation governing the torsional vibration of a circular rod is

$$-GJ\frac{\partial^2 \phi}{\partial x^2} + mJ\frac{\partial^2 \phi}{\partial t^2} = 0$$

where ϕ is the angular displacement, J the moment of inertia, G the shear modulus, and m the density. Determine the fundamental torsional frequency of a rod with disk (J_1) attached at each end. Use the symmetry and (a) two linear elements, (b) one quadratic element.

6.10. The equations governing the motion of a beam according to the Timoshenko beam theory can be written as

$$a^2\frac{\partial^4 w}{\partial x^4} + \frac{\partial^2 w}{\partial t^2} - b^2\left(1 + \frac{E}{kG}\right)\frac{\partial^4 w}{\partial x^2\,\partial t^2} + b^2\frac{m}{kG}\frac{\partial^4 w}{\partial t^4} = 0$$

where $a^2 = EI/mA$ and $b^2 = I/A$. Assuming that $b^2 m/kG \ll 1$ (i.e., neglecting the last term in the governing equation), formulate the eigenvalue problem for the determination of natural frequencies, and develop the finite element model of the eigenvalue problem.

6.11. Use the finite element model of Problem 6.10 to determine the fundamental frequency of a simply supported beam.

6.12. Find the critical buckling load P_{cr} by determining the eigenvalues of the equation

$$EI\frac{d^4 w}{dx^4} + P_{cr}\frac{d^2 w}{dx^2} = 0 \quad \text{for } 0 < x < L$$

$$w(0) = w(L) = 0, \qquad \left(EI\frac{d^2 w}{dx^2}\right)\bigg|_{x=0} = \left(EI\frac{d^2 w}{dx^2}\right)\bigg|_{x=L} = 0$$

Use one Euler–Bernoulli element in the half-beam.

Answer: $P_{cr} = 9.9439EI/L^2$.

Section 6.2

6.13. Consider the partial differential equation arising in connection with unsteady heat transfer in an insulated rod:

$$\frac{\partial u}{\partial t} - \frac{\partial}{\partial x}\left(a\frac{\partial u}{\partial x}\right) = f \quad \text{for } 0 < x < L$$

$$u(0, t) = 0, \qquad u(x, 0) = u_0, \qquad \left[a\frac{\partial u}{\partial x} + \beta(u - u_\infty) + \hat{q}\right]\bigg|_{x=L} = 0$$

Following the procedure outlined in Section 6.2, derive the semidiscrete variational form, the semidiscrete finite element model, and the fully discretized finite element equations for a typical element.

6.14. Using a two-element (linear) model and the semidiscrete finite element equations derived in Problem 6.13, determine the nodal temperatures as functions of time for the case in which $a = 1$, $f = 0$, $u_0 = 1$, and $\hat{q} = 0$. Use the Laplace transform technique [see Reddy (1986)] to solve the ordinary differential equations in time.

6.15. Consider a uniform bar of cross-sectional area A, modulus of elasticity E, mass density m, and length L. The axial displacement under the action of time-dependent axial forces is governed by the wave equation

$$\frac{\partial^2 u}{\partial t^2} = a^2 \frac{\partial^2 u}{\partial x^2}, \qquad a = \left(\frac{E}{m}\right)^{1/2}$$

Determine the transient response [i.e., find $u(x, t)$] of the bar when the end $x = 0$ is fixed and the end $x = L$ is subjected to a force P_0. Assume zero initial conditions. Use one linear element to approximate the spatial variation of the solution, and solve the resulting ordinary differential equation in time exactly to obtain

$$u_2(x, t) = \frac{P_0 L}{AE} \frac{x}{L}(1 - \cos \alpha t), \qquad \alpha = \sqrt{3}\frac{a}{L}$$

6.16. Re-solve Problem 6.15 with a mesh of two linear elements. Use the Laplace transform method to solve the two ordinary differential equations in time.

6.17. Solve Problem 6.15 when the right end is subjected to an axial force F_0 and supported by an axial spring of stiffness k.

Answer:

$$u_2(t) = c(1 - \cos \beta t), \qquad c = \frac{3F_0}{mAL\beta^2}, \qquad \beta = \sqrt{3}\frac{a}{L}\left(1 + \frac{kL}{EA}\right)^{1/2}$$

6.18. A bar of length L moving with velocity v_0 strikes a spring of stiffness k. Determine the motion $u(x, t)$ from the instant the end $x = 0$ strikes the spring. Use one linear element.

6.19. A uniform rod of length L and mass m is fixed at $x = 0$ and loaded with a mass M at $x = L$. Determine the motion $u(x, t)$ of the system when the mass M is subjected to a force P_0. Use one linear element.

Answer:

$$u_2(t) = c(1 - \cos \lambda t), \qquad c = \frac{P_0 L}{AE}, \qquad \lambda = \sqrt{3}\frac{a}{L}\left(\frac{3M}{AL} + m\right)^{-1}$$

6.20. The flow of liquid in a pipe, subjected to a surge-of-pressure wave (i.e., a water hammer), experiences a surge pressure p, which is governed by the equation

$$\frac{\partial^2 p}{\partial t^2} - c^2 \frac{\partial^2 p}{\partial x^2} = 0, \qquad c^2 = \frac{1}{m}\left(\frac{1}{k} + \frac{D}{bE}\right)^{-1}$$

where m is the mass density and K the bulk modulus of the fluid, D is the diameter and b the thickness of the pipe, and E is the modulus of elasticity of the pipe material. Determine the pressure $p(x, t)$ using one linear finite element, for the following boundary and initial conditions:

$$p(0, t) = 0, \qquad \frac{\partial p}{\partial x}(L, t) = 0, \qquad p(x, 0) = p_0, \qquad \dot{p}(x, 0) = 0$$

6.21. Consider the problem of determining the temperature distribution of a solid cylinder, initially at a uniform temperature T_0 and cooled in a medium of zero temperature (i.e., $T_\infty = 0$). The governing equation of the problem is

$$\rho c \frac{\partial T}{\partial t} - \frac{1}{r}\frac{\partial}{\partial r}\left(rk\frac{\partial T}{\partial r}\right) = 0$$

The boundary conditions are

$$\frac{\partial T}{\partial r}(0, t) = 0, \qquad \left(rk\frac{\partial T}{\partial r} + \beta T\right)\Big|_{r=R} = 0$$

The initial condition is $T(r, t) = T_0$. Determine the temperature distribution $T(r, t)$ using one linear element. Take $R = 2.5$ cm, $T_0 = 130\,°\text{C}$, $k = 215\ \text{W m}^{-1}\,°\text{C}^{-1}$, $\beta = 525\ \text{W m}^{-2}\,°\text{C}^{-1}$, $\rho = 2700\ \text{kg m}^{-2}$, and $c = 0.9\ \text{kJ kg}^{-1}\,°\text{C}^{-1}$. What is the heat loss at the surface?

6.22. Determine the nondimensional temperature $\theta(r, t)$ in the region bounded by two long cylindrical surfaces of radii R_1 and R_2. The dimensionless heat conduction equation is

$$-\frac{1}{r}\frac{\partial}{\partial r}\left(r\frac{\partial \theta}{\partial r}\right) + \frac{\partial \theta}{\partial t} = 0$$

with boundary and initial conditions

$$\frac{\partial \theta}{\partial r}(R_1, t) = 0, \qquad \theta(R_2, t) = 1, \qquad \theta(r, 0) = 0$$

6.23. Show that (6.44) and (6.45) can be expressed in the alternative form [see (6.47)]

$$[H]\{\ddot{u}\}_{s+1} = \{F\}_{s+1} - [K]\{b\}_s$$

where

$$[H] = \beta(\Delta t)^2[K] + [M], \qquad \{b\} = \{u\}_s + \Delta t\,\{\dot{u}\}_s + (\tfrac{1}{2} - \beta)(\Delta t)^2\{\ddot{u}\}_s$$

6.24. Using the Newmark integration scheme (6.45), express the equation

$$[M]\{\ddot{u}\} + [C]\{\dot{u}\} + [K]\{u\} = \{F\}$$

in the form

$$[\hat{K}]\{u\}_{s+1} = \{\hat{F}\}_{s+1}$$

where

$$[\hat{K}] = [K] + a_3[M] + a_5[C]$$

$$\{\hat{F}\}_{s+1} = \{F\}_{s+1} + [M](a_3\{u\}_s + a_4\{\dot{u}\}_s + a_5\{\ddot{u}\}_s) + [C](a_5\{u\}_s + a_6\{\dot{u}\}_s + a_7\{\ddot{u}\}_s)$$

$$a_6 = \frac{\alpha}{\beta} - 1, \qquad a_7 = \frac{\Delta t}{2}\left(\frac{\alpha}{\beta} - 2\right)$$

6.25. A uniform cantilever beam of length L, moment of inertia I, modulus of elasticity E, and mass m begins to vibrate with initial displacement

$$w(x, 0) = w_0 x^2/L^2$$

and zero initial velocity. Find its displacement at the free end at any subsequent time. Use one Euler–Bernoulli beam element to determine the solution. Solve the resulting differential equations in time using the Laplace transform method.

6.26. Re-solve Problem 6.25 using one Timoshenko element.

6.27. Use the Newmark integration scheme to reduce the ordinary differential equations of time in Problem 6.10 to algebraic equations.

REFERENCES FOR ADDITIONAL READING

Bathe, K.J.: *Finite Element Procedures in Engineering Analysis,* Prentice-Hall, Englewood Cliffs, NJ, 1982.

—— and E. L. Wilson: "Stability and Accuracy Analysis of Direct Integration Methods," *International Journal of Earthquake Engineering and Structural Dynamics,* vol. 1, pp. 283–291, 1973.

Belytschko, T.: "An Overview of Semidiscretization and Time Integration Procedures," in *Computational Methods for Transient Analysis,* eds. T. Belytschko and T. J. R. Hughes, pp. 1–65, North-Holland, Amsterdam, 1983.

Goudreau, G. L., and R. L. Taylor: "Evaluation of Numerical Integration Methods in Elastodynamics," *Journal of Computer Methods in Applied Mechanics and Engineering,* vol. 2, No. 1, pp. 69–97, 1973.

Hilber, H. M.: "Analysis and Design of Numerical Integration Methods in Structural Dynamics," EERC Report no. 77–29, Earthquake Engineering Research Center, University of California, Berkeley, California, November 1976.

Houbolt, J. C.: "A Recurrence Matrix Solution for the Dynamic Response of Elastic Aircraft," *Journal of Aeronautical Science,* vol. 17, pp. 540–550, 1950.

Hughes, T. J. R.: "Analysis of Transient Algorithms with Particular Reference to Stability Behavior," in *Computational Methods for Transient Analysis,* eds. T. Belytschko and T. J. R. Hughes, pp. 67–155, North-Holland, Amsterdam, 1983.

———: *The Finite Element Method,* Prentice-Hall, Englewood Cliffs, NJ, 1987.

——— and W. K. Liu: "Implicit–Explicit Finite Elements in Transient Analysis: Stability Theory," *Journal of Applied Mechanics,* vol. 45, pp. 371–374, 1978.

——— and W. K. Liu: "Implicit–Explicit Finite Elements in Transient Analysis: Implementation and Numerical Examples," *Journal of Applied Mechanics,* vol. 45, pp. 375–378, 1978.

———, Pister, K. S., and Taylor, R. L.: "Implicit–Explicit Finite Elements in Nonlinear Transient Analysis," *Computer Methods in Applied Mechanics and Engineering,* vol. 17/18, pp. 159–182, 1979.

Isaacson, E., and H. B. Keller: *Analysis of Numerical Methods,* John Wiley, New York, 1966.

Krieg, R. D., "Unconditional Stability in Numerical Time Integration Methods," *Transactions of the ASME, Journal of Applied Mechanics,* vol. 40, pp. 417–421, June 1973.

Lanczos, C.: *Applied Analysis,* Prentice-Hall, Englewood Cliffs, NJ, 1956.

Newmark, N. M.: "A Method of Computation for Structural Dynamics," *Journal of Engineering Mechanics Division, ASCE,* vol. 85, pp. 67–94, 1959.

Nickell, R. E.: "On the Stability of Approximation Operators in Problems of Structural Dynamics," *International Journal of Solids and Structures,* vol. 7, pp. 301–319, 1971.

———: "Direct Integration in Structural Dynamics," *Journal of Engineering Mechanics Division, ASCE,* vol. 99, pp. 303–317, 1973.

———: "Nonlinear Dynamics by Mode Superposition," *Journal of Computer Methods in Applied Mechanics and Engineering,* vol. 7, pp. 107–129, 1976.

Park, K. C.: "An Improved Stiffly Stable Method for Direct Integration of Nonlinear Structural Dynamics Equations," *Journal of Applied Mechanics,* vol. 42, pp. 464–470, 1975.

Wilson, E. L., I. Farhoomand, and K. J. Bathe: "Nonlinear Dynamic Analysis of Complex Structures," *International Journal of Earthquake Engineering and Structural Dynamics,* vol. 1, pp. 241–252, 1973.

CHAPTER 7

NUMERICAL INTEGRATION AND COMPUTER IMPLEMENTATION

7.1 ISOPARAMETRIC FORMULATIONS AND NUMERICAL INTEGRATION

7.1.1 Background

Exact evaluation of the integrals appearing in element coefficient matrices and source vectors is not always possible because of the algebraic complexity of the coefficients *a, b,* and *c* in differential equations. In such cases, it is natural to seek numerical evaluation of these integral expressions. Numerical evaluation of the coefficient matrices is also useful in problems with constraints, where reduced integration techniques are used (see, e.g., the reduced integration element of the Timoshenko beam theory, Section 4.4).

Numerical evaluation of integrals, called *numerical integration* or *numerical quadrature,* involves approximation of the integrand by a polynomial of sufficient degree, because the integral of a polynomial can be evaluated exactly. For example, consider the integral,

$$\mathcal{I} = \int_{x_A}^{x_B} F(x)\, dx \tag{7.1}$$

We approximate the function $F(x)$ by a polynomial:

$$F(x) \approx \sum_{I=1}^{N} F_I \psi_I(x) \tag{7.2}$$

where F_I denotes the value of $F(x)$ at the Ith point of the interval $[x_A, x_B]$ and $\psi_I(x)$ are polynomials of degree $N-1$. The representation can be viewed as the finite element interpolation of $F(x)$, where F_I is the value of the function at the Ith node. The interpolation can be of the Lagrange type or the Hermite type.

Substitution of (7.2) into (7.1) and evaluation of the integral gives an approximate value of $\mathcal{I}$. For example, suppose that we choose linear interpolation of $F(x)$. Then $N=2$, $\psi_1 = (x_B - x)/h$, $\psi_2 = (x - x_A)/h$, and

$$\mathcal{I} = \tfrac{1}{2}h(F_1 + F_2), \qquad F_1 = F(x_A), \qquad F_2 = F(x_B) \tag{7.3}$$

Thus, the value of the integral is given by the area of a trapezoid used to approximate the area under the function $F(x)$ (see Fig. 7.1). Equation (7.3) is known as the *trapezoidal rule* of numerical integration. If we use the Lagrange quadratic interpolation of $F(x)$, we obtain

$$\begin{aligned} \mathcal{I} &= \tfrac{1}{6}h(F_1 + 4F_2 + F_3), \qquad F_1 = F(x_A), \\ F_2 &= F(x_A + \tfrac{1}{2}h), \qquad F_3 = F(x_B) \end{aligned} \tag{7.4}$$

which is known as *Simpson's one-third rule.*

Equations (7.3) and (7.4) represent the form of numerical quadrature formulae. In general, a quadrature formula has the form

$$\mathcal{I} = \int_{x_A}^{x_B} F(x)\,dx \approx \sum_{I=1}^{r} F(x_I)W_I \tag{7.5}$$

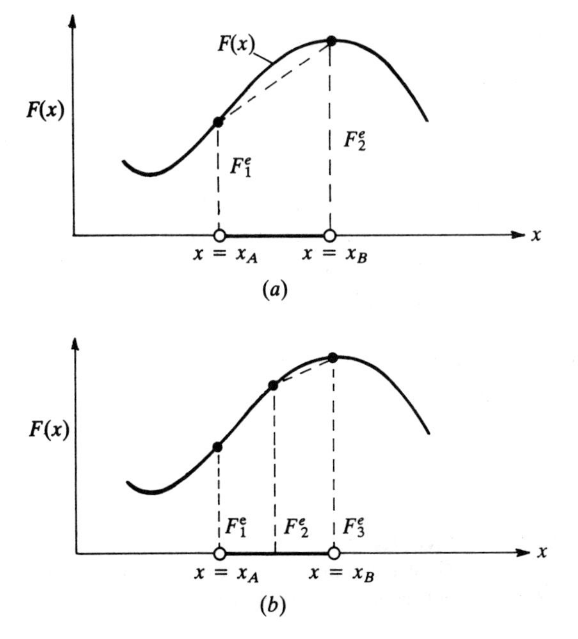

FIGURE 7.1
Approximate evaluation of an integral using the trapezoidal rule: (*a*) Two-point formula; (*b*) three-point formula.

where x_I are called the *quadrature points* and W_I are the *quadrature weights.* These formulae require functional evaluations, multiplications, and additions to obtain the numerical value of the integral. They yield exact values of the integral whenever $F(x)$ is a polynomial of order $r-1$.

In this section, we describe several numerical integration techniques and formulations in which the geometry as well as the dependent variables are approximated using different degrees of polynomials. We begin with the discussion of a local coordinate system.

7.1.2 Natural Coordinates

Of all the quadrature formulae, as will be discussed in the subsequent sections, the Gauss–Legendre one is the most commonly used. The details of the method itself will be discussed shortly. The formula requires the integral to be cast as one to be evaluated over the interval $[-1, 1]$. This requires the transformation of the problem coordinate x to a local coordinate ξ such that (see Fig. 7.2):

$$\text{when} \quad x = x_A, \quad \xi = -1; \qquad \text{when} \quad x = x_B, \quad \xi = 1$$

The transformation between x and ξ can be represented by the linear "stretch" transformation

$$x = a + b\xi$$

where a and b are constants to be determined such that the above conditions hold:

$$x_A = a + b(-1), \qquad x_B = a + b(1)$$

Solving for a and b, we obtain

$$b = \tfrac{1}{2}(x_B - x_A) = \tfrac{1}{2}h_e, \qquad a = \tfrac{1}{2}(x_B + x_A) = x_A + \tfrac{1}{2}h_e$$

Hence the transformation takes the form

$$x = x_A + \tfrac{1}{2}h_e(1 + \xi) \tag{7.6}$$

where x_A denotes the global coordinate of the left end node of the element Ω^e and h_e is the element length (see Fig. 7.2). The local coordinate ξ is called the *normal coordinate* or *natural coordinate,* and its values always lie between -1 and 1, with its origin at the center of the element.

The local coordinate ξ is useful in two ways: (i) it is convenient in constructing the interpolation functions; (ii) it is required in numerical integration using Gauss–Legendre quadrature.

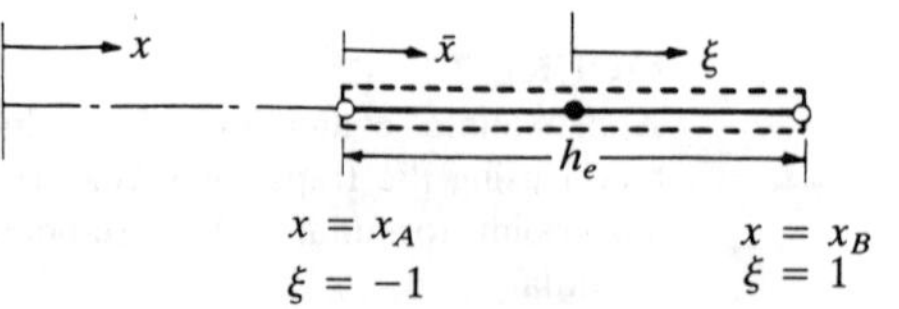

FIGURE 7,2
Global coordinate x, local coordinate $\bar{x}$, and normalized local coordinate ξ.

The derivation of the Lagrange family of interpolation functions in terms of the natural coordinate ξ is made easy by the property 1 in (3.28) of the interpolation functions:

$$\psi_i(\xi_j) = \begin{cases} 1 & \text{if } i = j \\ 0 & \text{if } i \neq j \end{cases} \tag{7.7}$$

where ξ_j is the ξ coordinate of the jth node in the element. For an element with n nodes, ψ_i $(i = 1, 2, \ldots, n)$ are functions of degree $n - 1$. To construct ψ_i satisfying (7.7), we proceed as follows. For each ψ_i, we form the product of $n - 1$ linear functions $\xi - \xi_j$ $(j = 1, 2, \ldots, i - 1, i + 1, \ldots, n, j \neq i)$:

$$\psi_i = c_i(\xi - \xi_1)(\xi - \xi_2) \cdots (\xi - \xi_{i-1})(\xi - \xi_{i+1}) \cdots (\xi - \xi_n)$$

Note that ψ_i is zero at all nodes except the ith. Next we determine the constant c_i such that $\psi_i = 1$ at $\xi = \xi_i$:

$$c_i = [(\xi_i - \xi_1)(\xi_i - \xi_2) \cdots (\xi_i - \xi_{i-1})(\xi_i - \xi_{i+1}) \cdots (\xi_i - \xi_n)]^{-1}$$

Thus, the interpolation function associated with node i is

$$\psi_i(\xi) = \frac{(\xi - \xi_1)(\xi - \xi_2) \cdots (\xi - \xi_{i-1})(\xi - \xi_{i+1}) \cdots (\xi - \xi_n)}{(\xi_i - \xi_1)(\xi_i - \xi_2) \cdots (\xi_i - \xi_{i-1})(\xi_i - \xi_{i+1}) \cdots (\xi_i - \xi_n)} \tag{7.8}$$

Interpolation functions that satisfy (7.7) are said to belong to the *Lagrange family of interpolation functions,* and the associated finite elements belong to the *Lagrange family of finite elements.* The interpolation functions ψ_i in (3.16*b*) and (3.18) provide an example of the Lagrange interpolation functions ($n = 2$). Figure 7.3 shows the linear, quadratic, and cubic Lagrange interpolation functions expressed in terms of the natural coordinate (for equally spaced nodes).

7.1.3 Approximation of Geometry

Recall from (3.31*c*) and (4.12*c*) that the element matrices involve the derivatives of the interpolation functions with respect to the global coordinate

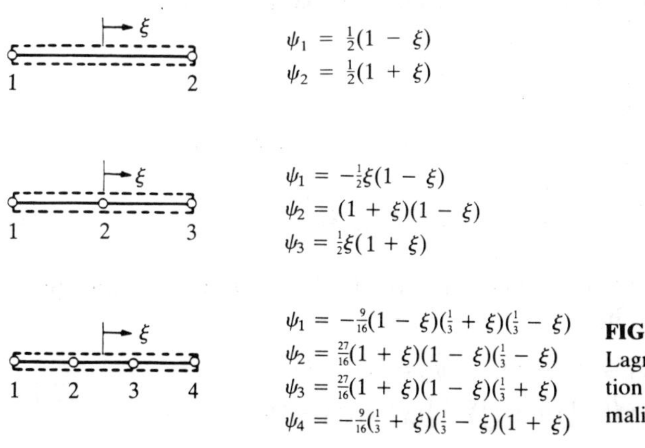

ξ
1 2
$\psi_1 = \frac{1}{2}(1 - \xi)$
$\psi_2 = \frac{1}{2}(1 + \xi)$
ξ
1 2 3
$\psi_1 = -\frac{1}{2}\xi(1 - \xi)$
$\psi_2 = (1 + \xi)(1 - \xi)$
$\psi_3 = \frac{1}{2}\xi(1 + \xi)$
ξ
1 2 3 4
$\psi_1 = -\frac{9}{16}(1 - \xi)(\frac{1}{3} + \xi)(\frac{1}{3} - \xi)$
$\psi_2 = \frac{27}{16}(1 + \xi)(1 - \xi)(\frac{1}{3} - \xi)$
$\psi_3 = \frac{27}{16}(1 + \xi)(1 - \xi)(\frac{1}{3} + \xi)$
$\psi_4 = -\frac{9}{16}(\frac{1}{3} + \xi)(\frac{1}{3} - \xi)(1 + \xi)$

FIGURE 7.3
Lagrange family of 1-D interpolation functions in terms of the normalized coordinate.

x. A transformation of the form

$$x = f(\xi) \tag{7.9}$$

is required in order to rewrite the integrals in terms of ξ $(-1 \leq \xi \leq 1)$. The function f is assumed to be a one-to-one transformation. An example of $f(\xi)$ is provided by (7.6):

$$f(\xi) = x_A + \tfrac{1}{2}h_e(1 + \xi),$$

In this case, $f(\xi)$ is a linear function of ξ. Hence, a straight line is transformed into a straight line. When f is a nonlinear function, a straight line is mapped into a curve of the same degree as the transformation.

It is natural to think of approximating the geometry in the same way as we approximated a dependent variable. In other words, the transformation $x = f(\xi)$ can be written as

$$x = \sum_{i=1}^{m} x_i^e \hat{\psi}_i^e(\xi) \tag{7.10}$$

where x_i^e is the global coordinate of the ith node of the element Ω^e and $\hat{\psi}_i^e$ are the Lagrange interpolation functions of degree $m-1$. Equation (7.10) represents the shape (or geometry) of an element, and the $\hat{\psi}_i^e$ are therefore called *shape functions*. Equation (7.10) maps a geometric shape from ξ space into x space, i.e., for any given ξ, (7.10) gives the corresponding x. When the element is a straight line, (7.10) is exactly the same as (7.6).

The transformation (7.10) is useful in evaluating integrals by the Gauss quadrature. It should be noted that the transformation is not used to change the actual geometry of the element; the transformation is used to write integral expressions involving x in terms of expressions involving ξ:

$$\int_{x_A}^{x_B} F(x)\, dx = \int_{-1}^{1} \hat{F}(\xi)\, d\xi \tag{7.11}$$

so that the Gauss quadrature can be used to evaluate the integral over $[-1, 1]$.

The differential element dx in the global coordinate system x is related to the differential element $d\xi$ in the natural coordinate system ξ by

$$dx = \frac{dx}{d\xi}\, d\xi = \mathcal{J}_e\, d\xi$$

where $\mathcal{J}_e$ is called the *Jacobian* of the transformation. We have

$$\mathcal{J}_e = \frac{dx}{d\xi} = \frac{d}{d\xi}\left(\sum_{i=1}^{m} x_i^e \hat{\psi}_i^e\right) = \sum_{i=1}^{m} x_i^e \frac{d\hat{\psi}_i^e}{d\xi} \tag{7.12}$$

For a linear transformation, i.e., when $\hat{\psi}_i^e$ are the linear Lagrange interpolation functions ($m = 2$), we have

$$\begin{gathered} \hat{\psi}_1 = \tfrac{1}{2}(1 - \xi), \qquad \hat{\psi}_2 = \tfrac{1}{2}(1 + \xi) \\ \mathcal{J}_e = x_1^e(-\tfrac{1}{2}) + x_2^e(\tfrac{1}{2}) = \tfrac{1}{2}(x_2^e - x_1^e) = \tfrac{1}{2}h_e \end{gathered} \tag{7.13}$$

It can be shown that $\mathcal{J}_e = \frac{1}{2}h_e$ whenever the element is a straight line, irrespective of the degree of interpolation used in the transformation (7.10).

7.1.4 Isoparametric Formulations

Recall that a dependent variable u is approximated in the element Ω^e by expressions of the form

$$u(x) = \sum_{j=1}^{n} u_j^e \psi_j^e(x) \tag{7.14}$$

In general, the degree of approximation used to describe the coordinate transformation (7.10) is not equal to the degree of approximation (7.14) used to represent a dependent variable, i.e., $\hat{\psi}_i^e \neq \psi_i^e$. In other words, two independent elements can be used in the finite element analysis: one for the approximation of the geometry x and the other for the interpolation of the dependent variable u. Depending on the relationship between the degree of approximation used for the coordinate transformation and that used for the dependent variable, the finite element formulations are classified into three categories:

1. Subparametric formulations: $m < n$
2. Isoparametric formulations: $m = n$ (7.15)
3. Superparametric formulations: $m > n$

In subparametric formulations, the geometry is represented by lower-order elements than those used to approximate the dependent variables. An example of this category is provided by the Euler–Bernoulli beam element, where the Hermite cubic element is used to approximate the transverse deflection, while the geometry is approximated, when straight beams are analyzed, with linear interpolation functions. In isoparametric formulations (the most common in practice), the same element is used to approximate the geometry as well as the dependent unknowns: $\psi_i(x) = \hat{\psi}_i(\xi)$. In the superparametric formulations, the geometry is represented with higher-order elements than those used to approximate the dependent variables. This formulation is seldom used in practice.

7.1.5 Numerical Integration

The evaluation of integrals of the form

$$\int_a^b F(x)\, dx \tag{7.16}$$

by exact means is either difficult or impossible owing to the complicated form of the integrand F. Numerical integration is also required when the integrand is to be evaluated inexactly (as in the Timoshenko beam element) and when the integrand depends on a quantity that is known only at discrete points (e.g., in nonlinear problems).

The basic idea behind all numerical integration techniques is to find a function $P(x)$ that is both a suitable approximation of $F(x)$ and simple to integrate. The interpolating polynomials of degree n, denoted by P_n, which interpolate the integrand at $n+1$ points of the interval $[a, b]$, often produce a suitable approximation and possess the desired property of simple integrability. An illustration of the approximation of the function $F(x)$ by the polynomial $P_4(x)$ that exactly matches the function $F(x)$ at the indicated base points is given in Fig. 7.4(a). The exact value of (7.16) is given by the area under the solid curve, while the approximate value

$$\int_a^b P_4(x)\, dx$$

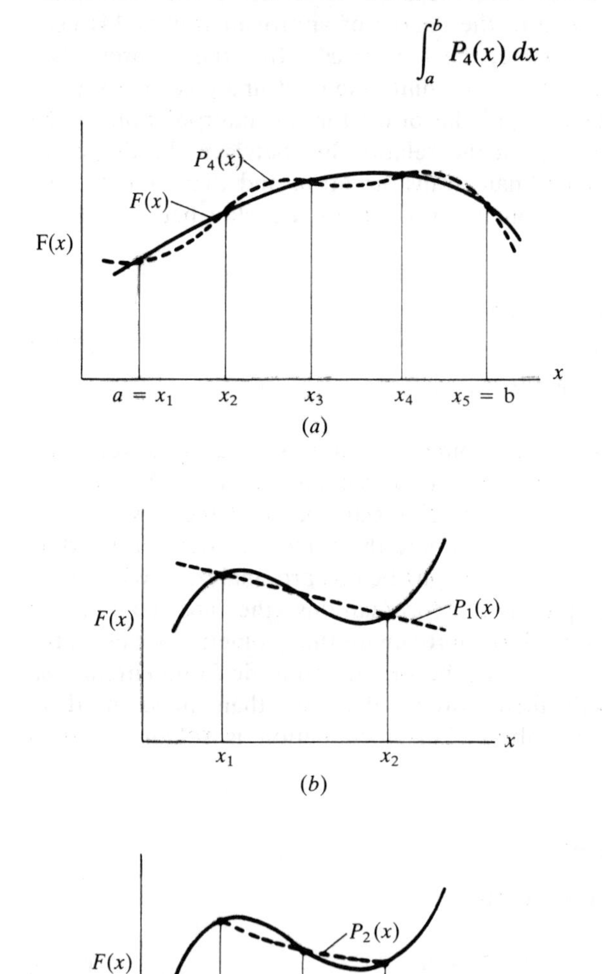

FIGURE 7.4
Numerical integration by the Newton–Cotes quadrature: (a) approximation of a function by $P_4(x)$; (b) the trapezoidal rule; (c) Simpson's rule.

is given by the area under the dashed curve. It should be noted that the difference (i.e., the error in the approximation) $E = F(x) - P_4(x)$ is not always of the same sign, and therefore the overall integration error may be small (because positive errors in one part cancel negative errors in other parts), even when P_4 is not a good approximation of F.

The commonly used integration methods can be classified into two basic groups:

1. The Newton–Cotes formulae that employ values of the function at equally spaced *base points.*
2. The Gauss quadrature formula that employs unequally spaced base points.

These are described here.

THE NEWTON–COTES QUADRATURE. For r equally spaced base points, the Newton–Cotes integration formula is given by

$$\boxed{\int_a^b F(x)\, dx = (b-a) \sum_{I=1}^{r} F(x_I) w_I} \tag{7.17}$$

where w_I are the weighting coefficients and x_I are the base points, which are equally spaced. For $r = 1$, (7.17) gives the rectangle formula. For $r = 2$, it gives the familiar *trapezoidal rule,* in which the required area under the solid curve in Fig. 7.4(b) is approximated by the area under the dotted straight line [i.e., $F(x)$ is approximated by $P_1(x)$]:

$$\int_{a=x_1}^{b=x_2} F(x)\, dx = \tfrac{1}{2}h[F(x_1) + F(x_2)], \qquad E = O(h^3) \tag{7.18}$$

where E denotes the error in the approximation and h is the spacing between two base points. The notation $O(h)$, read as "order of h", is used to indicate the order of the error in terms of the spacing h. For $r = 3$, (7.17) gives the familiar *Simpson's one-third rule* (see Fig. 7.4c):

$$\int_{a=x_1}^{b=x_3} F(x)\, dx = \tfrac{1}{3}h[F(x_1) + 4F(x_2) + F(x_3)], \qquad E = O(h^5) \tag{7.19}$$

The weighting coefficients for $r = 1, 2, \ldots, 7$ are given in Table 7.1. Note that $\sum_{I=1}^{r+1} w_I = 1$. The base point location for $r = 1$ is $x_1 = a + \frac{1}{2}(b-a) = \frac{1}{2}(a+b)$. For $r > 1$, the base point locations are

$$x_1 = a, \qquad x_2 = a + \Delta x, \qquad \ldots, \qquad x_r = a + (r-1)\,\Delta x = b$$

where $\Delta x = (b-a)/(r-1)$.

A comment is in order on the use of the Newton–Cotes integration formula (7.17). For odd r (i.e., when there is an even number of intervals or an odd number of base points), the formula is exact when $F(x)$ is a polynomial of degree r or less; for even r, the formula is exact for a polynomial of degree $r-1$ or less. Conversely, a pth-order polynomial is integrated exactly by choosing $r = p + 1$ base points.

TABLE 7.1
Weighting coefficients for the Newton–Cotes formula (7.17)

r	w_1	w_2	w_3	w_4	w_5	w_6	w_7
1	1						
2	$\frac{1}{2}$	$\frac{1}{2}$					
3	$\frac{1}{6}$	$\frac{4}{6}$	$\frac{1}{6}$				
4	$\frac{1}{8}$	$\frac{3}{8}$	$\frac{3}{8}$	$\frac{1}{8}$			
5	$\frac{7}{90}$	$\frac{32}{90}$	$\frac{12}{90}$	$\frac{32}{90}$	$\frac{7}{90}$		
6	$\frac{19}{288}$	$\frac{75}{288}$	$\frac{50}{288}$	$\frac{50}{288}$	$\frac{75}{288}$	$\frac{19}{288}$	
7	$\frac{41}{840}$	$\frac{216}{840}$	$\frac{27}{840}$	$\frac{272}{840}$	$\frac{27}{840}$	$\frac{216}{840}$	$\frac{41}{840}$

THE GAUSS–LEGENDRE QUADRATURE. In the Newton–Cotes quadrature, the base point locations have been specified. If the x_I are not specified then there will be $2r+2$ undetermined parameters, the weights w_I and base points x_I, which define a polynomial of degree $2r+1$. The Gauss–Legendre quadrature is based on this idea. The base points x_I and the weights w_I are chosen so that the sum of the $r+1$ appropriately weighted values of the function yields the integral exactly when $F(x)$ is a polynomial of degree $2r+1$ or less. The Gauss–Legendre quadrature formula is given by (see Fig. 7.5)

$$\int_a^b F(x)\,dx = \int_{-1}^{1} \hat{F}(\xi)\,d\xi \approx \sum_{I=1}^{r} \hat{F}(\xi_I)w_I \tag{7.20}$$

where w_I are the weight factors, ξ_I are the base points [roots of the Legendre

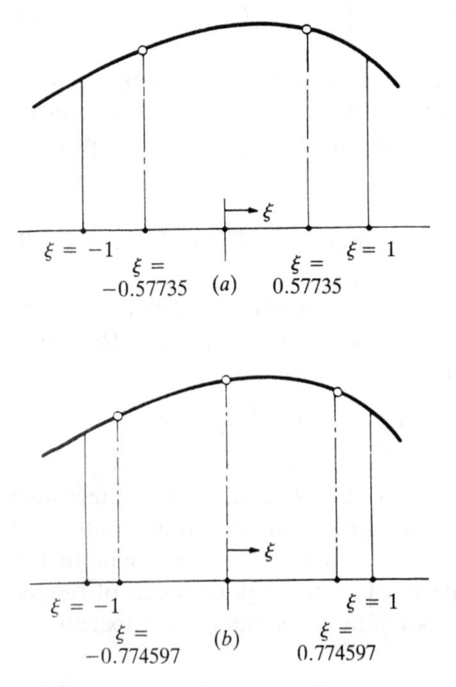

FIGURE 7.5
The two-point (a) and three-point (b) Gauss–Legendre quadratures.

TABLE 7.2
Weights and Gauss points for the Gauss–Legendre quadrature

$$\int_{-1}^{1} F(\xi)d\xi = \sum_{i=1}^{r} F(\xi_i)w_i$$

Points ξ_i	r	Weights w_i
0.0000000000	One-point formula	2.0000000000
±0.5773502692	Two-point formula	1.0000000000
0.0000000000	Three-point formula	0.8888888889
±0.7745966692		0.5555555555
±0.3399810435	Four-point formula	0.6521451548
±0.8611363116		0.3478548451
0.0000000000	Five-point formula	0.5688888889
±0.5384693101		0.4786286705
±0.9061798459		0.2369268850
±0.2386191861	Six-point formula	0.4679139346
±0.6612093865		0.3607615730
±0.9324695142		0.1713244924

polynomial $P_{r+1}(\xi)$], and $\hat{F}$ is the transformed integrand

$$\hat{F}(\xi) = F(x(\xi))\mathscr{J}(\xi) \tag{7.21}$$

The weight factors and Gauss points for the Gauss–Legendre quadrature (7.20) are given, for $r = 1, \ldots, 6$, in Table 7.2.

The Gauss–Legendre quadrature is more frequently used than the Newton–Cotes quadrature because it requires fewer base points (hence a saving in computation) to achieve the same accuracy. The error in the approximation is zero if the $(2r + 2)$th derivative of the integrand vanishes. In other words, a polynomial of degree p is integrated exactly by employing $r = \frac{1}{2}(p + 1)$ Gauss points. When $p + 1$ is odd, one should pick the nearest larger integer:

$$r = [\tfrac{1}{2}(p + 1)] \tag{7.22}$$

In finite element formulations, we encounter integrals whose integrands F are functions of x, $\psi_i(x)$ and derivatives of $\psi_i(x)$ with respect to x. For the Gauss–Legendre quadrature, we must transform $F(x)\,dx$ to $\hat{F}(\xi)\,d\xi$ to use (7.20). For example, consider the integral

$$K_{ij}^e = \int_{x_A}^{x_B} a(x) \frac{d\psi_i^e}{dx} \frac{d\psi_j^e}{dx}\, dx \tag{7.23}$$

Using the chain rule of differentiation, we have

$$\frac{d\psi_i^e(x)}{dx} = \frac{d\psi_i^e(\xi)}{d\xi} \frac{d\xi}{dx} = \mathscr{J}^{-1} \frac{d\psi_i^e(\xi)}{d\xi} \tag{7.24}$$

Therefore, the integral can be written, with the help of (7.10), as

$$K_{ij}^e = \int_{-1}^{1} a(x(\xi)) \frac{1}{\mathcal{J}} \frac{d\psi_i^e}{d\xi} \frac{1}{\mathcal{J}} \frac{d\psi_j^e}{d\xi} \mathcal{J} \, d\xi \tag{7.25}$$

$$\approx \sum_{I=1}^{r} \hat{F}_{ij}^e(\xi_I) w_I \tag{7.26}$$

where

$$\hat{F}_{ij}^e = a \frac{1}{\mathcal{J}} \frac{d\psi_i^e}{d\xi} \frac{d\psi_j^e}{d\xi}, \qquad \mathcal{J} = \sum_{i=1}^{m} x_i^e \frac{d\hat{\psi}_i^e}{d\xi} \tag{7.27}$$

For the isoparametric formulation, we take $\psi_i^e = \hat{\psi}_i^e$. The transformation from x to ξ is not required for the Newton–Cotes quadrature.

As noted earlier, the Jacobian matrix will be the same ($\mathcal{J}_e = \frac{1}{2}h_e$) when the element is straight, even if the coordinate transformation is quadratic or cubic. However, when the element is curved, the Jacobian is a function of ξ for transformations other than linear.

It is possible to determine the number of Gauss points required to evaluate the finite element matrices

$$K_{ij}^e = \int_{x_A}^{x_B} \frac{d\psi_i^e}{dx} \frac{d\psi_j^e}{dx} \, dx, \qquad M_{ij}^e = \int_{x_A}^{x_B} \psi_i^e \psi_j^e \, dx, \qquad F_i^e = \int_{x_A}^{x_B} \psi_i^e \, dx \tag{7.28}$$

exactly using linear, quadratic, and cubic interpolation functions. For linear interpolation functions, the integrand of K_{ij}^e is constant, requiring only one-point Gauss quadrature. The integrand of the mass matrix M_{ij}^e is quadratic ($p = 2$), requiring $[r = \frac{1}{2}(p +)1 = \frac{3}{2}]$ two-point quadrature. The coefficients f_i^e are evaluated exactly by one-point quadrature. Similarly, for quadratic and cubic elements, we can estimate the number of Gauss points needed to evaluate K_{ij}^e, M_{ij}^e, and f_i^e exactly. The results are summarized below. Note that, in estimating the quadrature points, it is assumed that the Jacobian is a constant, which holds true when the element is a straight line.

Element type	Number of Gauss quadrature points needed K_{ij}^e	M_{ij}^e	f_i^e
Linear	1	2	1
Quadratic	2	3	2
Cubic	3	4	3

If the matrices in (7.28) have variable coefficients or the elements are curved [and hence $\mathcal{J}_e = \mathcal{J}_e(\xi)$], the degree of the variation of the integrands changes and the number of Gauss points needed to evaluate the integral exactly changes. If the elements are straight, and the coefficients $a = a(x)$ and $c = c(x)$ together with $f = f(x)$ are no more than linear in x, then the number of Gauss points for evaluating the coefficients

$$K_{ij}^e = \int_{x_A}^{x_B} a \frac{d\psi_i^e}{dx} \frac{d\psi_j^e}{dx}\, dx, \qquad M_{ij}^e = \int_{x_A}^{x_B} c\psi_i^e \psi_j^e\, dx \tag{7.29}$$

remain the same as listed in the above table. However, the evaluation of f_i^e requires one point more than before. Conversely, the two-point quadrature for linear elements, three-point quadrature for quadratic elements, and four-point quadrature for cubic elements would exactly evaluate K_{ij}^e with a quadratic variation of $a(x)$, M_{ij}^e with linear variation of $c(x)$, and f_i^e with quadratic variation of $f(x)$.

The use of Gauss quadrature on (7.28) yields the following values (exact up to the fifth decimal place) when the element is straight and the isoparametric formulation is used:

Quadratic (three-point formula)

$$[K] = \frac{1}{h_e}\begin{bmatrix} 2.33333 & -2.66667 & 0.33333 \\ -2.66667 & 5.33333 & -2.66667 \\ 0.33333 & -2.66667 & 2.33333 \end{bmatrix}$$

$$[M] = \frac{h_e}{10}\begin{bmatrix} 1.33333 & 0.66667 & -0.33333 \\ 0.66667 & 5.33333 & 0.66667 \\ -0.33333 & 0.66667 & 1.333433 \end{bmatrix} \tag{7.30}$$

$$\{F\} = h_e \begin{Bmatrix} 0.166667 \\ 0.666667 \\ 0.166667 \end{Bmatrix}$$

Cubic (four-point formula)

$$[K] = \frac{1}{h_e}\begin{bmatrix} 3.700 & -4.725 & 1.350 & -0.325 \\ -4.725 & 10.800 & -7.425 & 1.350 \\ 1.350 & -7.425 & 10.800 & -4.725 \\ -0.325 & 1.350 & -4.725 & 3.700 \end{bmatrix}$$

$$[M] = \frac{h_e}{10}\begin{bmatrix} 0.761905 & 0.589286 & -0.214286 & 0.113095 \\ 0.589286 & 3.857143 & -0.482143 & -0.214286 \\ -0.214286 & -0.482143 & 3.857143 & 0.589286 \\ 0.113095 & -0.214286 & 0.589286 & 0.761905 \end{bmatrix} \tag{7.31}$$

$$\{F\} = h_e \begin{Bmatrix} 0.125 \\ 0.375 \\ 0.375 \\ 0.125 \end{Bmatrix}$$

In Section 7.2, we study the computer implementation of the steps involved in the finite element analysis of one-dimensional problems. As a part of the element calculations there, the computer implementation of the numerical integration ideas presented in this section will be studied. A model finite element program (FEM1DV2) for the solution of one-dimensional problems is also described, and its application is demonstrated via several examples. Appendix 1 contains a source listing of program FEM1DV2.

7.2 COMPUTER IMPLEMENTATION

7.2.1 Introductory Comments

Chapters 3–6 were devoted to the finite element formulations of two *classes* of boundary value, initial value, or eigenvalue problems in one dimension:

1. Second-order differential equations (e.g., heat transfer, fluid mechanics, 1-D elasticity, bars, and the Timoshenko beam theory);
2. Fourth-order differential equations governing the Euler–Bernoulli beam theory.

The frame element, obtained by superposing the bar element and the beam element, was discussed in Chapter 4.

By now, it should be clear to the reader that the steps involved in the finite element analysis of a general class of problems (e.g., single second-order, single fourth-order, and a pair of second-order equations) are systematic and can be implemented on a digital computer. Indeed, the success of the finite element method is largely due to the ease with which the analysis of a class of problems, without regard to a specific problem, can be implemented on a digital computer. For different geometries, boundary and initial conditions, and problem data, a specific problem from the general class can be solved by simply supplying the required input data to the program. For example, if we develop a general computer program to solve equations of the form

$$c_1 \frac{\partial u}{\partial t} + c_2 \frac{\partial^2 u}{\partial t^2} - \frac{\partial}{\partial x}\left(a \frac{\partial u}{\partial x}\right) + \frac{\partial^2}{\partial x^2}\left(b \frac{\partial^2 u}{\partial x^2}\right) + cu = f \tag{7.32}$$

then all physical problems described by (3.1) and (4.1) and their time-dependent versions can be solved for any compatible boundary and initial conditions.

The purpose of this section is to discuss the basic steps involved in the development of a computer program for second- and fourth-order one-dimensional differential equations studied in the preceding chapters. The ideas presented here are used in the development of the model program FEM1DV2 (a revised version of FEM1D from the first edition of this book), and they are meant to be illustrative of the steps used in a typical finite element program development. One can make use of the ideas presented here to develop a program of one's own.

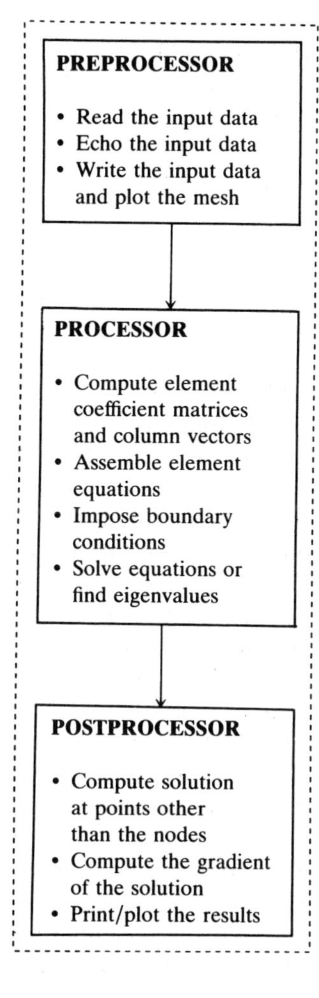

FIGURE 7.6
The three main functional units of a finite element program.

7.2.2 General Outline

A finite element program consists of three basic parts (see Fig. 7.6):

1. Preprocessor
2. Processor
3. Postprocessor

In the preprocessor part of the program, the input data of the problem are read in and/or generated. This includes the geometry (e.g., length of the domain and boundary conditions), the data of the problem (e.g., coefficients in the differential equation), finite element mesh information (e.g., element type, number of elements, element length, coordinates of the nodes, and connectivity matrix), and indicators for various options (e.g., print, no print, type of field problem analyzed, static analysis, eigenvalue analysis, transient analysis, and degree of interpolation).

In the processor part, all steps in the finite element method, except for postprocessing, discussed in the preceding chapters are performed. These include the following:

1. Generation of the element matrices using numerical integration.
2. Assembly of element equations.
3. Imposition of the boundary conditions.
4. Solution of the algebraic equations for the nodal values of the primary variables.

In the postprocessor part of the program, the solution is computed by interpolation at points other than nodes, secondary variables that are derivable from the solution are computed, and the output data are processed in a desired format for printout and/or plotting.

The preprocessor and postprocessors can be a few Fortran statements to read and print pertinent information, simple subroutines (e.g., subroutines to generate mesh and compute the gradient of the solution), or complex programs linked to other units via disk and tape files. The processor, where typically large amounts of computing time are spent, can consist of several subroutines, each having a special purpose (e.g., a subroutine for the calculation of element matrices, a subroutine for the imposition of boundary conditions, and a subroutine for the solution of the equations). The degree of sophistication and the complexity of a finite element program depend on the general class of problems being programmed, the generality of the data in the equation, and the intended user of the program. It is always desirable to describe, through comment statements, all variables used in the computer program. A flow chart of the computer program FEM1DV2 is presented in Fig. 7.7. In the following sections, a discussion of the basic components of a typical finite element program is presented, and then the basic ideas are illustrated via FORTRAN statements (see Appendix 1).

7.2.3 Preprocessor

The preprocessor unit consists of reading input data and generating finite element mesh, and printing the data and mesh information. The input data to a finite element program consist of element type IELEM (i.e., Lagrange element or Hermite element), number of elements used (NEM), specified boundary conditions on primary and secondary variables (number of boundary conditions, global node number and degree of freedom, and specified values of the degrees of freedom), the global coordinates of nodes X_I, and element properties (e.g., coefficients a_e, b_e, c_e, f_e, etc.). If a uniform mesh is used, the length of the domain should be read in, and global coordinates of the nodes can be generated in the program.

The preprocessor portion that deals with the generation of finite element mesh (when not supplied by the user) can be separated into a subroutine

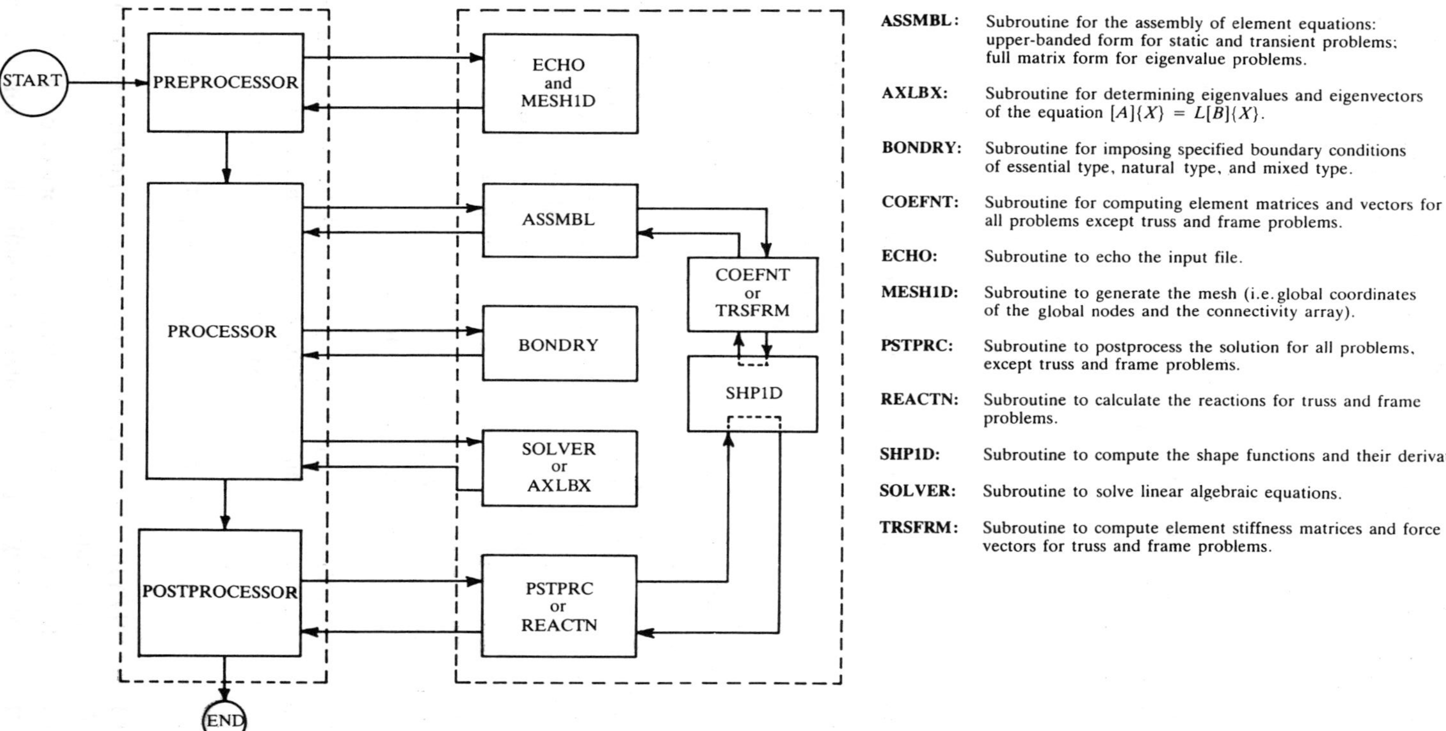

ASSMBL: Subroutine for the assembly of element equations: upper-banded form for static and transient problems; full matrix form for eigenvalue problems.

AXLBX: Subroutine for determining eigenvalues and eigenvectors of the equation $[A]\{X\} = L[B]\{X\}$.

BONDRY: Subroutine for imposing specified boundary conditions of essential type, natural type, and mixed type.

COEFNT: Subroutine for computing element matrices and vectors for all problems except truss and frame problems.

ECHO: Subroutine to echo the input file.

MESH1D: Subroutine to generate the mesh (i.e. global coordinates of the global nodes and the connectivity array).

PSTPRC: Subroutine to postprocess the solution for all problems, except truss and frame problems.

REACTN: Subroutine to calculate the reactions for truss and frame problems.

SHP1D: Subroutine to compute the shape functions and their derivatives.

SOLVER: Subroutine to solve linear algebraic equations.

TRSFRM: Subroutine to compute element stiffness matrices and force vectors for truss and frame problems.

FIGURE 7.7
Flow chart of the computer program FEM1DV2 along with the subroutines.

(MESH1D), depending on the convenience and complexity of the program. Mesh generation includes computation of the global coordinates X_I and the connectivity array $[B] = [\text{NOD}]$. Recall that the connectivity matrix describes the relationship between element nodes to global nodes:

NOD(I, J) = global node number corresponding to the Jth node of element I

This array is used in the assembly procedure as well as to transfer information from element to the global system and vice versa. For example, to extract the element nodal coordinates x_i^n of the element Ω^n from the global coordinates X_I, we can use the array NOD:

$$x_i^n = X_I, \qquad I = \text{NOD}(n, i)$$

The arrays {ELX} and {GLX} are used in FEM1DV2 for $\{x_i^n\}$ and $\{X_I\}$, respectively.

7.2.4 Calculation of Element Matrices (Processor)

The most significant part of a processor is where we generate element matrices. The element matrices are computed in various subroutines (COEFNT and TRSFRM), depending on the type of problem being solved. These subroutines typically involve numerical evaluations of the element matrices $[K^e]$ and $[M^e]$ (program variables ELK and ELM) and the element vector $\{f^e\}$ (program variable ELF) for various field problems. The Gauss quadrature described in Section 7.1.5 is used to evaluate element matrices and vectors, and the arrays are assembled as soon as they are computed. Thus, a loop on the number of elements in the mesh (NEM) is used to compute element matrices and assemble them (subroutine ASSMBL). It is here that the connectivity array NOD plays a crucial role. By putting element matrices into global locations one at a time, we avoid the computation of all element matrices at once.

Element matrices for different model equations (MODEL) and type of problem (NTYPE) are generated by assigning values as follows:

1. MODEL = 1, NTYPE = 0: all field problems described by the model equation (3.1), including radially symmetric heat-transfer-type problems.
2. MODEL = 1, NTYPE = 1: radially symmetric elastic disk problems (see Problem 3.33).
3. MODEL = 2, NTYPE = 0 (RIE) or 2 (CIE): the Timoshenko theory of beams.
4. MODEL = 2, NTYPE = 1 (RIE) or 3 (CIE): the Timoshenko theory for bending of circular plates.
5. MODEL = 3, NTYPE = 0: the Euler–Bernoulli theory of beams.
6. MODEL = 3, NTYPE = 1: the Euler–Bernoulli theory for bending of circular plates.

7. MODEL = 4, NTYPE = 0: the two-node truss element.

8. MODEL = 4, NTYPE = 1: the two-node Euler–Bernoulli frame element.

9. MODEL = 4, NTYPE = 2: the two-node, Timoshenko frame element.

The time-dependent option is exercised through variable ITEM:

ITEM = 0	static analysis
ITEM = 1	first-order time derivative (i.e., parabolic) problems
ITEM = 2	second-order time derivative (i.e., hyperbolic) problems

The element matrices are evaluated using the Gauss quadrature, except for MODEL = 4, where the explicit forms of element coefficients are programmed in the interest of computational efficiency.

The element shape functions SF and their derivatives GDSF are evaluated at the Gauss points in subroutine SHP1D. The Gaussian weights and points associated with two-, three-, four-, and five-point integration are stored in arrays GAUSWT and GAUSPT, respectively. The nth column of GAUSWT, for example, contains the weights corresponding to the n-point Gauss quadrature rule:

GAUSPT(i, n) = ith Gauss point corresponding to the n-point Gauss rule

The variable NGP is used to denote the number of Gauss points, which is selected to achieve good accuracy. As noted earlier, the linear, quadratic, and cubic interpolation functions require two, three, and four quadrature points, respectively, to evaluate the element coefficients exactly. Thus, if IELEM is the element type,

$$\text{IELEM} = \begin{cases} 1 & \text{for linear} \\ 2 & \text{for quadratic} \qquad \text{(Lagrange elements)} \\ 3 & \text{for cubic} \end{cases}$$

then NGP = IELEM + 1 would evaluate K_{ij}^e, M_{ij}^e, and f_i^e [see (7.29)] exactly when $c(x)$ is linear, and $a(x)$, $b(x)$, and $f(x)$ are quadratic functions. The Hermite cubic element is identified with IELEM = 0, for which case NGP is taken to be 4.

The coefficients $a(x)$ = AX, $b(x)$ = BX, and $c(x)$ = CX, together with $f(x)$ = FX in the differential equation (7.32) are assumed to vary with x as follows:

AX = AX0 + AX1 * X	$(a = a_0 + a_1 x)$
BX = BX0 + BX1 * X	$(b = b_0 + b_1 x)$
CX = CX0 + CX1 * X	$(c = c_0 + c_1 x)$
FX = FX0 + FX1 * X + FX2 * X * X	$(f = f_0 + f_1 x + f_2 x^2)$

For radially symmetric elasticity problems, (AX0, AX1) [or (BX0, BX1) for circular plates] are used to input Young's modulus E and Poisson's ratio ν.

The Gauss quadrature formula (7.20) can be implemented on the computer as follows. Consider K_{ij}^e of the form

$$K_{ij}^e = \int_{x_A}^{x_B} \left[a(x) \frac{d\psi_i^e}{dx} \frac{d\psi_j^e}{dx} + c(x)\psi_i^e \psi_j^e \right] dx \tag{7.33}$$

Let us use the following program variables for the quantities in (7.33):

$$\text{ELK(I, J)} = K_{ij}^e, \qquad \text{SF(I)} = \psi_i^e, \qquad \text{GDSF(I)} = \frac{d\psi_i^e}{dx}$$

$$\text{AX} = a(x), \qquad \text{CX} = c(x), \qquad \text{ELX(I)} = x_i^e$$

$$\text{NPE} = n, \quad \text{the number of nodes in the element}$$

After transforming x to ξ,

$$x = \sum_{i=1}^{n} x_i^e \psi_i^e \qquad [= x_A + \tfrac{1}{2}h_e(1 + \xi)] \tag{7.34}$$

the coefficients K_{ij}^e in (7.33) take the form [see (7.25)]

$$K_{ij}^e = \int_{-1}^{1} \left[a(\xi) \frac{1}{\mathscr{J}} \frac{d\psi_i^e}{d\xi} \frac{1}{\mathscr{J}} \frac{d\psi_j^e}{d\xi} + c(\xi)\psi_i^e \psi_j^e \right] \mathscr{J}\, d\xi \tag{7.35a}$$

$$= \sum_{I=1}^{\text{NGP}} F_{ij}^e(\xi_I) \mathscr{J} W_I \tag{7.35b}$$

where F_{ij}^e denotes the expression in the square brackets in (7.35a), $\mathscr{J}$ is the Jacobian, and (ξ_I, W_I) are the Gauss points and weights.

Examination of (7.35b) shows that there are three free indices: i, j, and I. We take the Gauss-point loop on I as the outermost one. Inside this loop, we evaluate F_{ij}^e at the Gauss point ξ_I for each i and j, multiply with the Jacobian $\mathscr{J} = \frac{1}{2}h_e$ and the weights W_I, and sum:

$$\text{ELK}(i, j) = \text{ELK}(i, j) + F_{ij}^e(\xi_I)\mathscr{J} W_I \tag{7.36}$$

To accomplish this, we must initialize all arrays that are being evaluated using the Gauss quadrature:

This initialization is made outside the do-loop on number of Gauss points.

The computation of coefficients F_{ij}^e in (7.35b) requires evaluation of a, c, ψ_i, and $d\psi_i/d\xi$ at the Gauss point ξ_I. Hence, inside the loop on I, we call subroutine SHP1D to evaluate ψ_i, $d\psi_i/dx = (d\psi_i/d\xi)/\mathscr{J}$. We now have all quantities needed to compute K_{ij}^e in (7.35b):

```
      DO 100 NI = 1,NGP
      XI = GAUSPT(NI,NGP)
C
C     Call subroutine SHP1D to evaluate the interpolation functions
C     (SF) and their global derivatives (GDSF) at the Gauss point XI
C
      CALL SHP1D(XI,NPE,SF,GDSF,GJ)
      CONST = GJ*GAUSWT(NI,NGP)
C
      X = 0.0
      DO 20 I=1,NPE
   20 X = X + SF(I)*ELX(I)
C
      AX=AX0 + AX1*X
      CX=CX0 + CX1*X
      DO 30 J = 1,NPE
      ELF(J)  = ELF(J)  + CONST*SF(J)*FX
      DO 30 I = 1,NPE
   30 ELK(I,J)=ELK(I,J)+CONST*(AX*GDSF(I)*GDSF(J)+CX*SF(I)*SF(J))
```

In the same way, all other coefficients (e.g., M_{ij}^e and f_i^e) can be evaluated. Recall that the element properties (i.e., K_{ij}^e, M_{ij}^e, and f_i^e) are calculated by calling a suitable subroutine (COEFNT or TRSFRM) for the field problem being analyzed within a loop on number of elements. As soon as the element properties are available for a particular element, they are put into their proper locations (i.e., assembled) with the help of array NOD. The assembly is explained in the next section.

7.2.5 Assembly of Element Equations (Processor)

The assembly of element equations should be carried out as soon as they are computed, rather than waiting till element coefficients of all elements are computed. The latter requires storage of the element coefficients of each element. In the former case, we can perform the assembly in the same loop in which a subroutine is called to calculate element matrices.

A feature of the finite element equations that enables us to save storage and computing time is the assembly of element matrices in upper-banded form. When element matrices are symmetric, as is the case in most problems of interest in this book, the resulting global (or assembled) matrix is also symmetric, with many zeros away from the main diagonal. Therefore, it is sufficient to store only the upper *half-band* of the assembled matrix. The half-bandwidth of a matrix is defined as follows. Let N_i be the number of matrix elements between the diagonal element and the last nonzero element in the ith row, after which all elements in that row are zero; the half-bandwidth is the maximum of $(N_i + 1) \times \text{NDF}$, where NDF is the number of degrees of freedom per node:

$$b_i = \max_{1 \leq i \leq n} [(N_i + 1) \times \text{NDF}]$$

where n is the number of rows in the matrix (or equations in the problem). General-purpose equation solvers are available for such banded systems of equations.

The half-bandwidth NHBW of the assembled (i.e., global) finite element matrix can be determined in the finite element program itself. The local nature of the finite element interpolation functions (i.e., ψ_i^e are defined to be nonzero only over the element Ω^e) is responsible for the banded character of the assembled matrix. Recall from earlier discussions that if two global nodes do not belong to the same element then the corresponding entries in the global matrix are zeros:

$$K_{IJ} = 0 \quad \text{if global nodes } I \text{ and } J \text{ do not correspond to local nodes of the same element}$$

This property enables us to determine the half-bandwidth NHBW of the assembled matrix:

$$\text{NHBW} = \max_{\substack{1 \leq N \leq \text{NEM} \\ 1 \leq I,\, J \leq \text{NPE}}} \{\text{abs}\,[\text{NOD}(N, I) - \text{NOD}(N, J)] + 1\} \times \text{NDF} \tag{7.37a}$$

where

$$\begin{aligned} \text{NEM} &= \text{number of elements in the mesh} \\ \text{NPE} &= \text{number of nodes per element} \\ \text{NDF} &= \text{number of degrees of freedom per element} \end{aligned} \tag{7.37b}$$

Clearly, for one-dimensional problems with elements connected in series, the maximum difference between nodes of an element is equal to NPE − 1. Hence,

$$\text{NHBW} = [(\text{NPE} - 1) + 1] \times \text{NDF} = \text{NPE} \times \text{NDF} \tag{7.38}$$

Of course, NHBW is always less than or equal to the total number of primary degrees of freedom, i.e., the number of equations, NEQ, in the finite element mesh of the problem.

The logic for assembling the element matrices K_{ij}^e into the upper-banded form of the global coefficients K_{IJ} is that the assembly can be skipped whenever $J < \text{I}$ and $J > \text{NHBW}$. The main diagonal, $I = J$, of the assembled square matrix (i.e., full storage form), becomes the first column of the assembled banded matrix (i.e., banded storage form), as shown in Fig. 7.8. The upper diagonals (parallel to the main diagonal) take the position of respective columns in the banded matrix. Thus, the banded matrix has dimension NEQ × NHBW, where NEQ denotes the total number of equations in the problem.

The element coefficients K_{ij}^n and f_i^n of a typical element Ω^n are to be assembled into the global coefficient matrix $[K]$ and source vector $\{F\}$, respectively. If the ith node of the element is equal to the Ith global node, and the jth node of the element is equal to the Jth global node, we have

$$K_{IJ} = K_{ij}^n, \qquad F_I = F_i^n \qquad (\text{for NDF} = 1) \tag{7.39a}$$

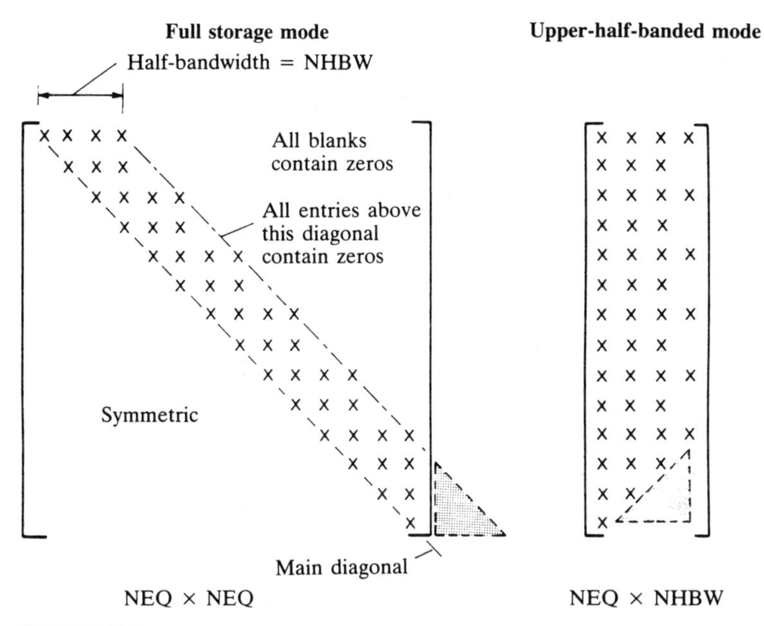

FIGURE 7.8
Finite element coefficient matrix storage in upper-half-banded form.

The values of I and J can be obtained with the help of array NOD:

$$I = \text{NOD}(n, i), \qquad J = \text{NOD}(n, j) \tag{7.39b}$$

Recall that it is possible that the same I and J may correspond to a pair of i and j of some other element Ω^m. In that case, K_{ij}^m will be added to existing coefficients K_{IJ} during the assembly. For NDF > 1, the logic still holds, with the change

$$K_{(\text{NR})(\text{NC})} = K^n_{(i+p-1)(j+q-1)} \quad (p, q = 1, 2, \ldots, \text{NDF}) \tag{7.40a}$$

where

$$\text{NR} = (I-1) \times \text{NDF} + p, \qquad \text{NC} = (J-1) \times \text{NDF} + q \tag{7.40b}$$

and I and J are related to i and j by (7.39b). A Fortran listing of the subroutine ASSMBL can be found in FEM1DV2 in Appendix 1.

7.2.6 Imposition of Boundary Conditions (Processor)

Imposition of boundary conditions on the primary and secondary global degrees of freedom can be carried out through a subroutine (BONDRY), which remains unchanged for 2-D or 3-D problems. There are three types of boundary conditions for any problem:

1. Essential boundary conditions, i.e., boundary conditions on primary variables (Dirichlet boundary conditions).

2. Natural boundary conditions, i.e., boundary conditions on secondary variables (Neumann boundary conditions).
3. Mixed boundary conditions (Newton boundary conditions).

The procedure for implementing the boundary conditions on the primary variables involves modifying the assembled coefficient matrix (GLK) and right-hand column vector (GLF) by three operations:

Step 1 moving the known products to the right-hand column of the matrix equation.

Step 2 replacing the columns and rows of GLK corresponding to the known primary variable by zeros, and setting the coefficient on the main diagonal to unity.

Step 3 replacing the corresponding component of the right-hand column by the specified value of the variable.

Consider the following N algebraic equations in full matrix form:

$$\begin{bmatrix} K_{11} & K_{12} & K_{13} & \cdots \\ K_{21} & K_{22} & K_{23} & \cdots \\ K_{31} & K_{32} & K_{33} & \cdots \\ \vdots & \vdots & \vdots & \end{bmatrix} \begin{Bmatrix} U_1 \\ U_2 \\ U_3 \\ \vdots \end{Bmatrix} = \begin{Bmatrix} F_1 \\ F_2 \\ F_3 \\ \vdots \end{Bmatrix}$$

where U_I and F_I are the global primary and secondary degrees of freedom, and K_{IJ} are the assembled coefficients. Suppose that $U_S = \hat{U}_S$ is specified. Recall that when the primary degree of freedom at a node is known, the corresponding secondary degree of freedom is unknown, and vice versa. Set $K_{SS} = 1$ and $F_S = \hat{U}_S$; further, set $K_{SI} = K_{IS} = 0$ for $I = 1, 2, \ldots, N$ and $I \neq S$. For $S = 2$, the modified equations are

$$\begin{bmatrix} K_{11} & 0 & K_{13} & K_{14} & \cdots & K_{1n} \\ 0 & 1 & 0 & 0 & \cdots & 0 \\ K_{31} & 0 & K_{33} & K_{34} & \cdots & K_{3n} \\ \vdots & \vdots & \vdots & \vdots & & \vdots \\ K_{n1} & 0 & K_{n3} & K_{n4} & \cdots & K_{nn} \end{bmatrix} \begin{Bmatrix} U_1 \\ U_2 \\ U_3 \\ \vdots \\ U_n \end{Bmatrix} = \begin{Bmatrix} \hat{F}_1 \\ \hat{U}_2 \\ \hat{F}_3 \\ \vdots \\ \hat{F}_n \end{Bmatrix}$$

where

$$\hat{F}_i = F_i - K_{i2}\hat{U}_2 \quad (i = 1, 3, 4, 5, \ldots, n; i \neq 2)$$

Thus, in general, if $U_S = \hat{U}_S$ is known, we have

$$K_{SS} = 1, \qquad F_S = \hat{U}_S; \qquad \hat{F}_i = F_i - K_{iS}\hat{U}_S; \qquad K_{Si} = K_{iS} = 0$$

where $i = 1, 2, \ldots, S-1, S+1, \ldots, n$ $(i \neq S)$. This procedure is repeated for every specified primary degree of freedom. It enables us to retain the original order of the matrix, and the specified boundary conditions on the primary degrees of freedom are printed as part of the solution. Of course, the logic should be implemented for a banded system of equations.

The specified secondary degrees of freedom (Q_i) are implemented directly by adding their specified values to the computed values. Suppose that the point source corresponding to the Rth secondary degree of freedom is specified to be $\hat{F}_R$. Then

$$F_R = f_R + \hat{F}_R$$

where f_R is the contribution due to the distributed source $f(x)$; f_R is computed as a part of the element computations and assembled.

Mixed-type boundary conditions are of the form

$$a\frac{du}{dx} + k(u - \bar{u}) = 0 \quad (\bar{u} \text{ and } k \text{ are known constants}) \tag{7.41}$$

which contains both the primary variable u and the secondary variable $a\,du/dx$. Thus $a\,du/dx$ at the node P is replaced by $-k_p(u_p - \bar{u}_p)$:

$$Q_p = -k_p(U_p - \bar{U}_p)$$

This amounts to modifying K_{pp} by adding k_p to its existing value,

$$K_{pp} \leftarrow K_{pp} + k_p$$

and adding $k_p\bar{U}_p$ to F_p,

$$F_p \leftarrow F_p + k_p\bar{U}_p$$

All three types of boundary conditions are implemented in subroutine BONDRY for boundary, initial, and eigenvalue problems. The following are used in subroutine BONDRY (see Appendix 1):

NSPV	number of specified primary variables	
NSSV	number of specified secondary variables	
NNBC	number of Newton boundary conditions	
VSPV	column of the specified values $\hat{U}_s$ of primary variables	
VSSV	column of the specified values $\hat{F}_R$ of secondary variables	(7.42)
VNBC	column of the specified values k_p	
UREF	column of the specified values $\bar{U}_p$	
ISPV	array of the global node and degree of freedom at the node that is specified [ISPV(I, 1) = global node of the Ith boundary condition, ISPV(I, 2) = degree of freedom specified at the global node, ISPV(I, 1)]	

Similar definitions are used for ISSV and INBC arrays.

7.2.7 Solution of Equations and Postprocessing

Subroutine SOLVE is used to solve a banded system of equations, and the solution is returned in array GLF. The program performs the Gaussian elimination and back-substitution to compute the solution. For a discussion of the Gaussian elimination used to solve a set of linear algebraic equations, the reader is referred to the book by Carnhan, Luther, and Wilkes (1969). On most computing systems, a variety of equation solvers are available, and one can use any of the programs that suits the needs. See Appendix 1 for a listing of subroutine SOLVER.

Postprocessing involves computation of the solution and its gradient at preselected points of the domain. Subroutine PSTPRC is used to evaluate the solution and its derivatives in any element:

$$u^e(x) = \sum_{j=1}^{n} u_j^e \psi_j^e(x), \qquad \left(\frac{du^e}{dx}\right)\bigg|_x = \sum_{j=1}^{n} u_j^e \left(\frac{d\psi_j^e}{dx}\right)\bigg|_x \tag{7.43a}$$

for the Lagrange elements and

$$w^e(x) = \sum_{j=1}^{n} u_j^e \phi_j^e(x), \qquad \frac{d^m w^e}{dx^m} = \sum_{j=1}^{n} u_j^e \left(\frac{d^m \phi_j^e}{dx^m}\right)\bigg|_x \quad (m = 1, 2, 3) \tag{7.43b}$$

for the Hermite cubic elements. The nodal values u_j^e of the element Ω^e are deduced from the global nodal values U_I as follows:

$$u_j^e = U_I, \qquad I = \text{NOD}(e, j), \qquad \text{when NDF} = 1$$

For NDF > 1, I is given by $I = [\text{NOD}(e, j) - 1] \times \text{NDF}$ and

$$u_{j+p}^e = U_{I+p} \quad (p = 1, 2, \ldots, \text{NDF})$$

The values computed using the derivatives of the solution are often inaccurate because the derivatives of the approximate solution become increasingly inaccurate with increasing order of differentiation. For example, the shear force computed in the Euler–Bernoulli beam theory,

$$V = \frac{d}{dx}\left(b\frac{d^2w}{dx^2}\right) = \sum_{j=1}^{n} u_j^e \frac{d}{dx}\left(b\frac{d^2\phi_j^e}{dx^2}\right) \tag{7.44}$$

will be in considerable error compared with the true value of V. The accuracy increases, rather slowly, with mesh refinement and higher-order elements. When accurate values of the secondary variables are desired, it is recommended that they be computed from the element equations:

$$Q_i^e = \sum_{j=1}^{n} K_{ij}^e u_j^e - f_i^e \quad (i = 1, 2, \ldots, n) \tag{7.45}$$

However, this requires recomputation or saving of element coefficients K_{ij}^e and f_i^e.

7.3 APPLICATIONS OF THE COMPUTER PROGRAM FEM1DV2

7.3.1 General Comments

The computer program FEM1DV2, which embodies the ideas presented in the previous section, is intended to illustrate the use of the finite element models developed in Chapters 3–6 to a variety of one-dimensional field problems, some of them not discussed in this book. FEM1DV2 is a modified version of

FEM1D from the first edition of this book. It was developed as a learning computational tool for students of the course. In the interest of simplicity and ease of understanding, only the model equations discussed in this book and their immediate extensions are included in the program (see Appendix 1).

Table 7.3 contains a summary of the definitions of coefficients of various model problems and their corresponding program variables. The table can be used as a ready reference to select proper values of AX0, AX1, and so on for different problems.

7.3.2 Illustrative Examples

Here we revisit some of the example problems considered earlier to illustrate the use of FEM1DV2 in their solution. Only certain key observations concerning the input data are made, but complete listings of input files for each problem are given. In the interest of brevity, the output files for most problems are not included. Table 7.4 contains a description of the input variables to program FEM1DV2.

Example 7.1 Steady heat transfer in a fin. The problem is governed by

$$-\frac{d^2\theta}{dx^2}+c\theta=0 \quad \text{for } 0<x<L$$

$$\theta(0)=\theta_0, \quad \left(\frac{d\theta}{dx}\right)\bigg|_{x=L}=0 \qquad \text{(Set 1)}$$

$$\theta(0)=\theta_0, \quad \left(\frac{d\theta}{dx}+\frac{\beta}{k}\theta\right)\bigg|_{x=L}=0 \quad \text{(Set 2)}$$

where θ is the nondimensional temperature, and L, c, θ_0, β, and k are

$$\begin{aligned} L&=0.25\text{ m}, \quad c=256\text{ m}^{-2}, \quad \theta_0=100°\text{C},\\ \beta&=64\text{ W m}^{-2}\,°\text{C}^{-1}, \quad k=50\text{ W m}^{-1}\,°\text{C}^{-1}\end{aligned} \tag{7.46}$$

Thus the problem is governed by the model equation (3.1).

Hence, MODEL = 1, NTYPE = 0, and ITEM = 0 (for a steady-state solution). Since $a=a_0=1.0$ and $c=c_0=P\beta/k=256$ are the same for all elements, we set ICONT = 1, AX0 = 1.0, and CX0 = 256.0. All other coefficients are zero [$b=0$ and f $(=q)=0$] for this problem. For a uniform mesh of four linear elements (NEM = 4, IELEM = 1), the increments DX(I) are [DX(1) is always the coordinate of node 1]:

$$\{DX\}=\{0.0, 0.0625, 0.0625, 0.0625, 0.0625\}$$

(because $\frac{1}{4}L=0.25/4=0.0625$).

Set 1 boundary conditions are $U_1=0$ and $Q_2^4=0$. Since the natural boundary condition ($Q_2^4=0$) is homogeneous, there is no need to add a zero to the corresponding entry of the source column ($F_5+Q_2^4\rightarrow F_5$). There are no mixed (i.e., convection) boundary conditions. Hence, NSPV = 1, NSSV = 0, and NNBC = 0. The specified boundary condition on the primary variable is at node 1 and degree of freedom 1: ISPV(1, 1) = 1 and ISPV(1, 2) = 1. The specified value is VSPV(1) = 100.0.

TABLE 7.3
Meaning of the program variables AX0, AX1, . . . , **for various field problems**
Subscripts 0, 1, and 2 on variables denote the constant, linear, and quadratic coefficients of the variables [e.g., f_0, f_1, and f_2 denote the coefficients in $f(x) = f_0 + f_1 x + f_2 x^2$]

Field problem	MODEL	NTYPE	ITEM†	AX0	AX1	BX0	BX1	CX0	CX1	FX0	FX1	FX2	CT0‡	CT1‡
1. Plane wall	1	0	1	k_0	k_1	0.0	0.0	0.0	0.0	f_0	f_1	f_2	ρ_0	ρ_1
2. Heat exchanger fin	1	0	1	$(kA)_0$	$(kA)_1$	0.0	0.0	c_0	c_1	f_0	f_1	f_2	ρ_0	ρ_1
3. Radially symmetric heat transfer	1	0	1	0.0	k_1	0.0	0.0	0.0	0.0	0.0	f_1	f_2	0.0	ρ_1
4. Viscous flow through channels	1	0	1	μ_0	μ_1	0.0	0.0	0.0	0.0	f_0	f_1	f_2	ρ_0	ρ_1
5. Viscous flow through pipes	1	0	1	0.0	μ_1	0.0	0.0	0.0	0.0	0.0	f_1	f_2	0.0	ρ_1
6. Unidirectional seepage	1	0	1	μ_0	μ_1	0.0	0.0	0.0	0.0	f_0	f_1	f_2	ρ_0	ρ_1
7. Radially symmetric seepage (ground water flow)	1	0	1	0.0	μ_1	0.0	0.0	0.0	0.0	0.0	f_1	f_2	0.0	ρ_1
8. Axial deformation of a bar	1	0	2	$(AE)_0$	$(AE)_1$	0.0	0.0	c_0	c_1	f_0	f_1	f_2	$(\rho A)_0$	$(\rho A)_1$
9. Radially symmetric deformation of a disk (plane stress)	1	1	2	E_1	E_2	ν_{12}	H	c_0	c_1	f_0	f_1	f_2	$(\rho A)_0$	$(\rho A)_1$

10. Radially symmetric deformation of a cylinder (plane strain)	1	2	2	E_1	E_2	ν_{12}	H	c_0	c_1	f_0	f_1	f_2	$(\rho A)_0$	$(\rho A)_1$
11. Euler–Bernoulli beam theory	3	0	2	0.0	0.0	$(EI)_0$	$(EI)_1$	c_0	c_1	f_0	f_1	f_2	ρA	ρI
12. Euler–Bernoulli theory for circular plates	3	1	2	E_1	E_2	ν_{12}	H	c_0	c_1	f_0	f_1	f_2	ρA	ρI
13. Timoshenko beam theory (RIE)	2	0	2	$(GAK)_0$	$(GAK)_1$	$(EI)_0$	$(EI)_1$	c_0	c_1	f_0	f_1	f_2	ρA	ρI
14. Timoshenko beam theory (CIE)	2	2	2	$(GAK)_0$	$(GAK)_1$	$(EI)_0$	$(EI)_1$	c_0	c_1	f_0	f_1	f_2	ρA	ρI
15. Timoshenko theory of circular plates (RIE)	2	1	2	E_1	E_2	ν_{12}	H	c_0	c_1	f_0	f_1	G_{13}	ρA	ρI
16. Timoshenko theory of circular plates (CIE)	2	3	2	E_1	E_2	ν_{12}	H	c_0	c_1	f_0	f_1	G_{13}	ρA	ρI
17. Truss	4	0	0	For field problems 17–19, these parameters are not read; instead, E = SE, A = SA, L = SL, and so on are read for each member of the structure: SE = modulus E, SA = cross-sectional area A, SI = moment of inertia I, SL = length L of the member, $CN = \cos\alpha$, $SN = \sin\alpha$, etc. (see Table 7.4)										
18. The Euler–Bernoulli frame element	4	1	0											
19. The Timoshenko frame element	4	2	0											

† For time-dependent problems only; when steady-state solutions are required, set ITEM = 0.
‡ For transient analysis only; no transient analysis option is included for truss and frame problems.

TABLE 7.4
Description of the input variables to the program FEM1DV2†

```
* Data Card 1:

        TITLE - Title of the problem being solved (80 characters)

* Data Card 2:

        MODEL - Model equation being solved (see below)
        NTYPE - Type of problem solved (see below)

                MODEL=1, NTYPE=0: A problem of MODEL EQUATION 1
                MODEL=1, NTYPE=1: A circular DISK (PLANE STRESS)
                MODEL=1, NTYPE>1: A circular DISK (PLANE STRAIN)
                MODEL=2, NTYPE=0: A Timoshenko BEAM (RIE#) problem
                MODEL=2, NTYPE=1: A Timoshenko PLATE (RIE) problem
                MODEL=2, NTYPE=2: A Timoshenko BEAM (CIE##) problem
                MODEL=2, NTYPE>2: A Timoshenko PLATE (CIE) problem
                MODEL=3, NTYPE=0: A Euler-Bernoulli BEAM problem
                MODEL=3, NTYPE>0: A Euler-Bernoulli Circular plate
                MODEL=4, NTYPE=0: A plane TRUSS problem
                MODEL=4, NTYPE=1: A Euler-Bernoulli FRAME problem
                MODEL=4, NTYPE=2: A Timoshenko (CIE) FRAME problem

                #  - Reduced Integration Element (RIE)
                ## - Consistent Interpolation Element (CIE)

        ITEM  - Indicator for transient analysis:

                ITEM=0, Steady-state solution
                ITEM=1, Transient analysis of PARABOLIC equations
                ITEM=2, Transient analysis of HYPERBOLIC equations
                ITEM=3, Eigenvalue analysis

* Data Card 3:

        IELEM - Type of finite element:

                IELEM=0, Hermite cubic finite element
                IELEM=1, Linear Lagrange finite element
                IELEM=2, Quadratic Lagrange finite element

        NEM   - Number of elements in the mesh

* Data Card 4:

        ICONT - Indicator for continuity of data for the problem:

                ICONT=1, Data (AX,BX,CX,FX and mesh) is continuous
                ICONT=0, Data is element dependent

        NPRNT - Indicator for printing of element/global matrices:

                NPRNT=0, Not print element or global matrices
                           but postprocess the solution and print
                NPRNT=1, Print Element 1 coefficient matrices only
                           but postprocess the solution and print
```

† SKIP means that you omit the input data. It *does not* mean that you leave a blank. In the "free format" used here, variables of each data card are read from the same line; if the values are not found on the same line, the computer will look for them on the next line(s). Each *data card* should start with a new line. Note that the variable names included in the data sets of Tables 7.5–7.14 are only for the convenience of the reader—they are *not* read by the computer.

TABLE 7.4 (Continued)

```
                  NPRNT=2, Print Element 1 and global matrices but
                           NOT postprocess the solution
                  NPRNT>2, Not print element or global matrices and
                           NOT postprocess the solution

   SKIP Cards 5 through 15 for TRUSS/FRAME problems (MODEL = 4), and
   read Cards 5 through 15 only if MODEL.NE.4. SKIP Cards 5 through 9
   if data is discontinuous (ICONT = 0) ______________________________

 * Data Card 5:

         DX(I) - Array of element lengths. DX(1) denotes the global
                 coordinate of Node 1 of the mesh; DX(I) (I=2,NEM1)
                 denotes the length of the (I-1)st element. Here
                 NEM1=NEM+1, and NEM=number of elements in the mesh.

   Cards 6 through 9 define the coefficients in the model equations.
   All coefficients are expressed in terms of GLOBAL coordinate x.
   See Table 7.3 for the meaning of the coefficients, especially for
   deformation of circular plates and Timoshenko elements.___________

 * Data Card 6:

         AX0   - Constant term of the coefficient AX
         AX1   - Linear   term of the coefficient AX

 * Data Card 7:

         BX0   - Constant term of the coefficient BX
         BX1   - Linear   term of the coefficient BX

 * Data Card 8:

         CX0   - Constant term of the coefficient CX
         CX1   - Linear   term of the coefficient CX

   SKIP Card 9 for eigenvalue problems (i.e. ITEM=3) ________________

 * Data Card 9:

         FX0   - Constant term of the coefficient FX
         FX1   - Linear   term of the coefficient FX
         FX2   - Quadratic term of the coefficient FX

   SKIP Cards 10 through 15 if data is continuous (ICONT.NE.0). Cards
   10 through 15 are read for each element (i.e., NEM times).     All
   coefficients are expressed in terms of the GLOBAL coord __________

 * Data Card 10:

         {GLX} - Global x-coordinates of the FIRST and LAST nodes
                 of the element

 * Data Card 11:

         {NOD} - Connectivity of the element:
                 NOD(N,I)=Global node number corresponding to the
                          I-th node of Element N (I=1,NPE)
                 NPE denotes the Number of nodes Per Element
 * Data Card 12:

         {DCAX}- Constant and linear terms of the coefficient AX

 * Data Card 13:

         {DCBX}- Constant and linear terms of the coefficient BX
```

TABLE 7.4 (Continued)

```
* Data Card 14:

        {DCCX}- Constant and linear terms of the coefficient CX

* Data Card 15:

        {DCFX}- Constant, linear and quadratic terms of FX

  READ Cards 16 through 21 for TRUSS/FRAME problems (MODEL = 4).
  SKIP Cards 16 through 21 if MODEL.NE.4 ____________________________

* Data Card 16:

        NNM    - Number of nodes in the finite element mesh

  SKIP Cards 17 through 19 for TRUSS problems (NTYPE = 0). Cards
  19 through 19 are read for each element, i.e., NEM times _________

* Data Card 17:

        PR     - Poisson's ratio of the material#
        SE     - Young's modulus of the material
        SL     - Length of the element
        SA     - Cross-ectional area of the element
        SI     - Moment of inertia of the element
        CS     - Cosine of the angle of orientation of the element
        SN     - Sine   of the angle of orientation of the element
                 Angle is measured counter-clock-wise from x axis

        # PR is not used for the Euler -Bernoulli element

* Data Card 18:

        HF     - Intensity of the horizontal distributed force
        VF     - Intensity of the transversely distributed force
        PF     - Point load on the element
        XB     - Distance from node 1, along the length of the
                 element, to the point of load application, PF
        CNT    - Cosine of the angle of orientation of the load PF
        SNT    - Sine   of the angle of orientation of the load PF
                 Angle is measured counter-clock-wise from the
                 element x axis
* Data Card 19:

        {NOD} - Connectivity of the element:
                NOD(N,I) = Global node number corresponding to the
                           I-th node of Element N (I=1,NPE)

  READ Cards 20  and 21 only for TRUSS problems (NTYPE = 0). Cards
  20 and 21  are read for each element;i.e. NEM times ___________

* Data Card 20:

        SE     - Young's modulus of the material
        SL     - Length of the element
        SA     - Cross-ectional area of the element
        CS     - Cosine of the angle of orientation of the element
        SN     - Sine   of the angle of orientation of the element
                 Angle is measured counter-clock-wise from x axis
        HF     - Intensity of the horizontal distributed force

* Data Card 21:

        {NOD} - Connectivity of the element:
                NOD(N,I) = Global node number corresponding to the
                           Ith node of Element N (I=1,NPE)
```

TABLE 7.4 (Continued)

```
* Data Card 22:

      NSPV  - Number of specified PRIMARY degrees of freedom

  SKIP Card 23 if no primary variables is specified (NSPV=0). Repeat
  Card 23 NSPV times ______________________________________________

* Data Card 23:    (I = 1 to NSPV)

      ISPV(I,1) - Node number at which the PV is specified
      ISPV(I,2) - Specified local primary DOF at the node
      VSPV(I)   - Specified value of the primary variable (PV)
                  (will not read for eigenvalue problems)

  SKIP Card 24 for eigenvalue problems (i.e. when ITEM=3) __________

* Data Card 24:

      NSSV  - Number of specified (nonzero) SECONDARY variables

  SKIP Card 25 if no secondary variable is specified (NSSV=0). Repeat
  Card 25 NSSV times ______________________________________________

* Data Card 25:    (I = 1 to NSSV)

      ISSV(I,1) - Node number at which the SV is specified
      ISSV(I,2) - Specified local secondary DOF at the node
      VSSV(I)   - Specified value of the secondary variable (PV)

* Data Card 26:

      NNBC  - Number of the Newton (mixed) boundary conditions

  SKIP Card 27 if no mixed boundary condition is specified (NNBC=0).
  The mixed boundary condition is assumed to be of the form: SV+VNBC
  *(PV-UREF) = 0. Repeat Card 28 NNBC times ________________________

* Data Card 27:    (I = 1 to NNBC)

      INBC(I,1) - Node number at which the mixed B.C. is specified
      INBC(I,2) - Local DOF of the PV and SV at the node
      VNBC(I)   - Value of the coefficient of the PV in the B.C.
      UREF(I)   - Reference value of the PV

  SKIP Card 28 if ITEM=0 (read  only for time-dependent and eigen-
  -value problems ___________________________________________________

* Data Card 28:
      CT0    - Constant part of CT = CT0 + CT1*X
      CT1    - Linear part of CT = CT0 + CT1*X

  SKIP remaining cards if steady-state or eigenvalue analysis is to
  be performed (ITEM=0 or ITEM=3) ___________________________________

* Data Card 29:
      DT     - Time increment (uniform)
      ALFA   - Parameter in the time approximation scheme
      GAMA   - Parameter in the time approximation scheme

* Data Card 30:
      INCOND- Indicator for initial conditions:
              INCOND=0, Homogeneous (zero) initial conditions
              INCOND>0, Nonhomogeneous initial conditions
      NTIME - Number of time steps for which solution is sought
      INTVL - Time step intervals at which solution is to printed
```

TABLE 7.4 (Continued)

```
    SKIP Cards 31 and 32 if initial conditions are zero (INCOND=0) ___

  * Data Card 31:

            {GU0} - Array of initial values of the primary variables

    SKIP Card 32 for parabolic equations (ITEM=1) _______________

  * Data Card 32:

            {GU1} - Array of initial values of the first time-deriva-
                    tives of the primary variables
```

For Set 2 boundary conditions, we have one essential boundary condition and one mixed boundary condition:

$$U_1 = 0, \qquad Q_2^4 + \frac{\beta}{k}(U_5 - U_\infty) = 0$$

Hence, NSPV = 1, NSSV = 0, and NNBC = 1, and the values in the mixed boundary conditions are input as

$$\text{VNBC}(1) = \beta/k = 1.28, \quad \text{UREF}(1) = U_\infty = 0.0, \quad \text{INBC}(1,1) = 5, \quad \text{INBC}(1,2) = 1.$$

The output file for Set 1 boundary conditions is presented in Table 7.5(a). The input files are echoed in the output. The input file for Set 2 boundary conditions is given in Table 7.5(b).

Example 7.2 (Example 3.2) Radially symmetric heat transfer in a solid cylinder. The governing equation of the problem is given by [see (3.78)]

$$-\frac{d}{dr}\left(2\pi kr\frac{dT}{dr}\right) = 2\pi q_0 \quad \text{for } 0 < r < R_o \tag{7.47}$$

$$\left(2\pi kr\frac{dT}{dr}\right)\bigg|_{r=0} = 0, \qquad T(R_o) = T_o \tag{7.48}$$

with $k = 20\ \text{W m}^{-1}\,°\text{C}^{-1}$, $q_0 = 2\times10^8\ \text{W m}^{-3}$, $T_o = 100°\text{C}$, and $R_o = 0.01$ m (see Table 3.3). We have MODEL = 1, NTYPE = 0, and ITEM = 0 (for a steady-state solution), and the data is continuous (ICONT = 1) in the domain for a mesh of two quadratic elements (IELEM = 2, NEM = 2). The data is

$$a = 2\pi kr \to a_0 = 0, \quad a_1 = 2\pi k; \quad b = 0 \to b_0 = 0.0, \quad b_1 = 0.0$$

$$c = 0 \to c_0 = 0.0, \quad c_1 = 0.0; \quad f = 2\pi q_0 r \to f_0 = 0.0, \quad f_1 = 2\pi q_0, \quad f_2 = 0.0$$

Thus, we have [for values $k = 20\ \text{W m}^{-1}\,°\text{C}^{-1}$, $q_0 = 2\times10^8\ \text{W m}^{-3}$

$$\text{AX0} = 0.0 \quad \text{AX1} = 125.6637, \quad \text{BX0} = 0.0, \quad \text{BX1} = 0.0$$

$$\text{CX0} = 0.0, \quad \text{CX1} = 0.0, \quad \text{FX0} = 0.0, \quad \text{FX1} = 12.5664\text{E}8, \quad \text{FX2} = 0.0$$

The array {DX} and boundary information are given by

$$\{\text{DX}\} = \{0.0, 0.0025, 0.0025, 0.0025, 0.0025\}$$

$$\text{NSPV} = 1, \quad \text{ISPV}(1,1) = 5, \quad \text{ISPV}(1,2) = 1, \quad \text{VSPV}(1) = 100$$

TABLE 7.5
Output from FEM1DV2 for the problem in Example 7.1

(*a*) Set 1 boundary conditions (edited output)

```
    *** ECHO OF THE INPUT DATA STARTS ***

Example 7.1: Steady heat transfer in a fin (Set 1 boundary conditions)
  1  0  0                                  MODEL, NTYPE, ITEM
  1  4                                     IELEM, NEM
  1  1                                     ICONT, NPRNT
    0.0  0.0625  0.0625  0.0625  0.0625    DX(I)
    1.0  0.0                               AX0, AX1
    0.0  0.0                               BX0, BX1
  256.0  0.0                               CX0, CX1
    0.0  0.0  0.0                          FX0,FX1,FX2
  1                                        NSPV
  1  1  100.0                              ISPV(1,J) [J=1,2], VSPV(1)
  0                                        NSSV
  0                                        NNBC
    **** ECHO OF THE INPUT DATA ENDS ****

  ________________________________________________________

   OUTPUT FROM PROGRAM   FEM1DV2  BY  J. N. REDDY
  ________________________________________________________

Example 7.1: Steady heat transfer in a fin (Set 1 boundary conditions)

   *** ANALYSIS OF MODEL 1, AND TYPE 0 PROBLEM ***

        Element type (0, Hermite,>0, Lagrange)..=   1
        No. of deg. of freedom per node, NDF....=   1
        No. of elements in the mesh, NEM........=   4
        No. of total DOF in the model, NEQ......=   5
        No. of specified primary DOF, NSPV......=   1
        No. of specified secondary DOF, NSSV....=   0
        No. of specified Newton B. C.: NNBC.....=   0

  Boundary information on primary variables:

      1    1    0.10000E+03

  Global coordinates of the nodes, {GLX}:

   0.00000E+00  0.62500E-01  0.12500E+00  0.18750E+00  0.25000E+00

  Coefficients of the differential equation:

        AX0 =  0.1000E+01     AX1 =  0.0000E+00
        BX0 =  0.0000E+00     BX1 =  0.0000E+00
        CX0 =  0.2560E+03     CX1 =  0.0000E+00
        FX0 =  0.0000E+00     FX1 =  0.0000E+00      FX2 =  0.0000E+00

 SOLUTION (values of PVs) at the NODES:

   0.10000E+03  0.35158E+02  0.12504E+02  0.48560E+01  0.30350E+01
```

TABLE 7.5 (Continued)

X	P. Variable	S. Variable
0.00000E+00	0.10000E+03	-0.10375E+04
0.78125E-02	0.91895E+02	-0.10375E+04
0.15625E-01	0.83789E+02	-0.10375E+04
0.23438E-01	0.75684E+02	-0.10375E+04
0.31250E-01	0.67579E+02	-0.10375E+04
0.39063E-01	0.59473E+02	-0.10375E+04
0.46875E-01	0.51368E+02	-0.10375E+04
0.54688E-01	0.43263E+02	-0.10375E+04
0.62500E-01	0.35158E+02	-0.10375E+04
0.62500E-01	0.35158E+02	-0.36245E+03
0.70313E-01	0.32326E+02	-0.36245E+03
0.78125E-01	0.29494E+02	-0.36245E+03
0.85938E-01	0.26663E+02	-0.36245E+03
0.93750E-01	0.23831E+02	-0.36245E+03
0.10156E+00	0.20999E+02	-0.36245E+03
0.10938E+00	0.18168E+02	-0.36245E+03
0.11719E+00	0.15336E+02	-0.36245E+03
0.12500E+00	0.12504E+02	-0.36245E+03
0.12500E+00	0.12504E+02	-0.12237E+03
0.13281E+00	0.11548E+02	-0.12237E+03
0.14063E+00	0.10592E+02	-0.12237E+03
0.14844E+00	0.96362E+01	-0.12237E+03
0.15625E+00	0.86801E+01	-0.12237E+03
0.16406E+00	0.77241E+01	-0.12237E+03
0.17188E+00	0.67681E+01	-0.12237E+03
0.17969E+00	0.58120E+01	-0.12237E+03
0.18750E+00	0.48560E+01	-0.12237E+03
0.18750E+00	0.48560E+01	-0.29136E+02
0.19531E+00	0.46284E+01	-0.29136E+02
0.20313E+00	0.44008E+01	-0.29136E+02
0.21094E+00	0.41731E+01	-0.29136E+02
0.21875E+00	0.39455E+01	-0.29136E+02
0.22656E+00	0.37179E+01	-0.29136E+02
0.23438E+00	0.34903E+01	-0.29136E+02
0.24219E+00	0.32626E+01	-0.29136E+02
0.25000E+00	0.30350E+01	-0.29136E+02

(*b*) Set 2 boundary conditions (input data only)

```
Example 7.1: Steady heat transfer in a fin (Set 2 boundary conditions)
  1  0  0                                          MODEL, NTYPE, ITEM
  1  4                                             IELEM, NEM
  1  1                                             ICONT, NPRNT
    0.0  0.0625  0.0625  0.0625  0.0625            DX(I)
    1.0  0.0                                       AX0, AX1
    0.0  0.0                                       BX0, BX1
  256.0  0.0                                       CX0, CX1
    0.0  0.0  0.0                                  FX0,FX1,FX2
  1                                                NSPV
  1  1  100.0                                      ISPV(1,J), VSPV(1)
  0                                                NSSV
  1                                                NNBC
  5  1  1.28  0.0                                  INBC(1,J), VNBC(1), UREF(1)
```

TABLE 7.6
Input file for the problem in Example 7.2

```
Example 7.2: Radially symmetric heat transfer in a solid cylinder
  1  0  0                                          MODEL, NTYPE, ITEM
  2  2                                             IELEM, NEM
  1  1                                             ICONT, NPRNT
    0.0  0.005   0.005                             DX(I)
    0.0  125.6637                                  AX0, AX1
    0.0  0.0                                       BX0, BX1
    0.0  0.0                                       CX0, CX1
    0.0  12.56637E8  0.0                           FX0,FX1,FX2
  1                                                NSPV
  5  1  100.0                                      ISPV(1,J), VSPV(1)
  0                                                NSSV
  0                                                NNBC
```

The input file of the problem is presented in Table 7.6 (cf. Table 3.3). Note that the finite element solution obtained with two quadratic elements is more accurate (essentially the same as the exact) than the solution obtained with four linear elements.

The fluid mechanics problem in Example 3.3 is very simple, and the reader should be able to generate the input file very easily to solve it with FEM1DV2.

Example 7.3 Deformation of a rotating disk. For this case, we have MODEL = 1, NTYPE = 1, and ITEM = 0. For a mesh of two quadratic elements (i.e., IEL = 2 and NEM = 2), we use ICONT = 1 and

$$\{\mathrm{DX}\} = \{0.0, 0.25R, 0.25R, 0.25R, 0.25R\}$$

where R is the radius of the disk. For a uniform and homogeneous disk of thickness H and made of isotropic material, we take the moduli $E_1 = E_2 = E$. Since we are seeking results in nondimensional form, we take

$$R = 1.0, \quad E = 1.0, \quad \nu = 0.3, \quad H = 1.0$$

Thus, we have

$$\mathrm{AX0}\ (= E_1) = 1.0, \quad \mathrm{AX1}\ (= E_2) = 1.0$$

$$\mathrm{BX0}\ (= \nu_{12}) = 0.3, \quad \mathrm{BX1}\ (= H) = 1.0$$

$$\mathrm{CX0} = 0.0, \quad \mathrm{CX1} = 0.0$$

The body force is $f_r = \rho\omega^2 r \equiv f_0 + f_1 r + f_2 0$. Hence, taking $\rho\omega^2 = 1.0$,

$$\mathrm{FX0} = 0.0, \quad \mathrm{FX1} = 1.0, \quad \mathrm{FX2} = 0.0$$

The boundary conditions are $u(0) = 0$ (by symmetry) and $r\sigma_r = 0$ at $r = R$ (stress-free condition). Since the secondary variable is homogeneous, there is no need to impose it—only the boundary condition on the primary variable is to be imposed. We have

$$\mathrm{NSPV} = 1, \quad \mathrm{NSSV} = 0, \quad \mathrm{NNBC} = 0$$

$$\mathrm{ISPV}(1,1) = 1, \quad \mathrm{ISPV}(1,2) = 1, \quad \mathrm{VSPV}(1) = 0.0$$

The input file for this problem is presented in Table 7.7.

TABLE 7.7
Input file for the problem in Example 7.3

```
Example 7.3: Deformation of a rotating disk
 1  1  0                                      MODEL, NTYPE, ITEM
 2  2                                         IELEM, NEM
 1  1                                         ICONT, NPRNT
   0.0  0.50    0.50                          DX(I)
   1.0  1.0                                   AX0, AX1
   0.3  1.0                                   BX0, BX1
   0.0  0.0                                   CX0, CX1
   0.0  1.0     0.0                           FX0, FX1, FX2
 1                                            NSPV
 1  1   0.0                                   ISPV(1,J), VSPV(1)
 0                                            NSSV
 0                                            NNBC
```

Example 7.4 (Example 4.2) Clamped and spring-supported beam in Fig. 4.7. We solve the problem using the Euler–Bernoulli beam element (MODEL = 3, NTYPE = 0, IELEM = 0) and the Timoshenko beam element (MODEL = 2, NTYPE = 0 or 2, IELEM = 1, 2, or 3).

Since the loading is discontinuous, we set ICONT = 0. A minimum of two elements are required to model the problem (i.e., NEM = 2).

If we take $EI = 1.0\text{E}6$ (i.e., 10^6 ft^2 lb) then

$$GAK = \frac{E}{2(1+\nu)} BH \frac{5}{6} = \frac{EI}{1+\nu} \frac{5}{H^2}$$

For $L/H = 10$, we have $H = 1.0$ because $L = 10$ ft. For the choice $\nu = 0.25$, we have

$$GAK = 4EI/H^2 = 4 \times 10^6 \text{ lb}$$

Thus, we use

$$\text{AX0} = 0.0, \quad \text{AX1} = 0.0, \quad \text{BX0}\ (= EI) = 1.0\text{E}6$$
$$\text{BX1} = 0.0, \quad \text{CX0} = 0.0, \quad \text{CX1} = 0.0$$

for the Euler–Bernoulli beam and

$$\text{AX0}\ (= GAK) = 4.0\text{E}6, \quad \text{AX1} = 0.0, \quad \text{BX0}\ (= EI) = 1.0\text{E}6$$
$$\text{BX1} = 0.0, \quad \text{CX0} = 0.0, \quad \text{CX1} = 0.0$$

for the Timoshenko beam.

The distributed transverse load is zero in element 1, and it is

$$f(x) = -\frac{100}{6}(x-4)$$

in element 2. Hence,

$$\text{FX0} = 0.0, \quad \text{FX1} = 0.0, \quad \text{FX2} = 0.0 \quad \text{in element 1}$$
$$\text{FX0} = 66.666, \quad \text{FX1} = -16.666, \quad \text{FX2} = 0.0 \quad \text{in element 2}$$

The global coordinates of nodes and the connectivity matrix entries for each element are obvious from the geometry. For the Euler–Bernoulli beam element the number of nodes is always equal to 2 (NPE = 2), whereas for the Timoshenko beam element, the number of nodes depend on the degree of interpolation (or element type) selected: NPE = IELEM + 1.

The boundary conditions for this problem are

$$w(0) = 0, \quad \left(\frac{dw}{dx}\right)\bigg|_{x=0} = 0, \quad (V + kw)\bigg|_{x=L} = 0$$

Therefore, we have (NSSV = 0)

$$\text{NSPV} = 2, \quad \text{ISPV}(1,1) = 1, \quad \text{ISPV}(1,2) = 1, \quad \text{VSPV}(1) = 0.0$$
$$\text{ISPV}(2,1) = 1, \quad \text{ISPV}(2,2) = 2, \quad \text{VSPV}(2) = 0.0$$
$$\text{NNBC} = 1, \quad \text{INBC}(1,1) = 3, \quad \text{INBC}(1,2) = 1$$
$$\text{VNBC}(1)\ (= k) = 1.0\text{E}6 \quad (\text{for } k/EI = 1) \text{ and } 0.0 \ (\text{for } k = 0.0)$$
$$\text{UREF}(1) = 0.0$$

Tables 7.8 and 7.9 give the input files for the Euler–Bernoulli and Timoshenko elements. Note that the Euler–Bernoulli element is a Hermite cubic element, whereas the Timoshenko element is only a Lagrange linear element.

Example 7.5 (Example 4.4) Analysis of a plane frame. Here we consider the two-member frame structure shown in Fig. 4.13(a). We shall analyze it using the Euler–Bernoulli frame element (MODEL = 4, NTYPE = 1) and the Timoshenko frame element (MODEL = 4, NTYPE = 2). The former gives an exact solution for all frame structures with constant cross-section members. The Timoshenko frame element, on the other hand, does not yield accurate results unless several elements per member of the structure are used.

The input files of the problem are presented in Tables 7.10 and 7.11. The eight-element mesh of Timoshenko elements gives results comparable to the two-element mesh with the Euler–Bernoulli frame elements (results are not included here).

The next example deals with the use of FEM1DV2 for time-dependent problems (i.e., ITEM = 1 or 2).

TABLE 7.8
Input file for the problem in Example 7.4

```
Example 7.4: Clamped and spring-supported beam (E-B element; k=1.0E4)
  3  0  0                                      MODEL, NTYPE, ITEM
  0  2                                         IELEM, NEM
  0  1                                         ICONT, NPRNT
    0.0     4.0                                GLX(I)           |
    1         2                                NOD(1,J)         |
    0.0     0.0                                AX0, AX1         |  Data for
    1.0E6   0.0                                BX0, BX1         |  Element 1
    0.0     0.0                                CX0, CX1         |
    0.0     0.0       0.0                      FX0,FX1,FX2      |
    4.0    10.0                                GLX(I)           |
    2         3                                NOD(2,J)         |
    0.0     0.0                                AX0, AX1         |  Data for
    1.0E6   0.0                                BX0, BX1         |  Element 2
    0.0     0.0                                CX0, CX1         |
    66.666667     -16.666667   0.0             FX0,FX1,FX2      |
  2                                            NSPV
  1     1    0.0                               ISPV(1,J), VSPV(1)
  1     2    0.0                               ISPV(2,J), VSPV(2)
  0                                            NSSV
  1                                            NNBC   (with transv. spring)
  3     1    1.0E04   0.0
```

TABLE 7.9
Input file for the problem in Example 7.4

```
Example 7.4:  Clamped and spring-supported beam (TIM element; k=1.0E4)
  2  0  0                                     MODEL, NTYPE, ITEM
  1  2                                        IELEM, NEM
  0  1                                        ICONT, NPRNT
    0.0     4.0                               GLX(I)          |
    1         2                               NOD(1,J)        |
    4.0E6   0.0                               AX0, AX1        |   Data for
    1.0E6   0.0              (L/H = 10)       BX0, BX1        |   Element 1
    0.0     0.0                               CX0, CX1        |
    0.0     0.0     0.0                       FX0,FX1,FX2     |
    4.0    10.0                               GLX(I)          |
    2         3                               NOD(2,J)        |
    4.0E6   0.0                               AX0, AX1        |   Data for
    1.0E6   0.0                               BX0, BX1        |   Element 2
    0.0     0.0                               CX0, CX1        |
    66.666667    -16.666667   0.0             FX0,FX1,FX2     |
  2                                           NSPV
  1    1    0.0                               ISPV(1,J), VSPV(1)
  1    2    0.0                               ISPV(2,J), VSPV(2)
  0                                           NSSV
  1                                           NNBC  (with tranv. spring)
  3    1    1.0E4   0.0                       INBC(1,J), VNBC(1), UREF(1)
```

Example 7.6 (Example 6.3) Transient heat conduction in a plane wall. Consider the transient heat conduction problem of Example 6.4 [see (6.56)]:

$$\frac{\partial u}{\partial t} - \frac{\partial^2 u}{\partial x^2} = 0 \quad \text{for } 0 < x < 1 \tag{7.49a}$$

$$u(0, t) = 0, \qquad \frac{\partial u}{\partial x}(1, t) = 0 \qquad \text{(Set 1)} \tag{7.49b}$$

TABLE 7.10
Input file for the problem in Example 7.5

```
Example 7.5:  Analysis of a plane frame  (E-B element)
  4  1  0                                        MODEL, NTYPE, ITEM
  0  2                                           IELEM, NEM
  0  1                                           ICONT, NPRNT
  3                                              NNM
  0.3    1.0E6   144.0  10.0  10.0  0.0   1.0    PR, SE, SL, SA, SI, CS, SN
  0.0   -0.0138888   0.0   0.0   0.0   0.0       HF, VF, PF, XB, CST, SNT
  1  2                                           NOD(1,J)
  0.3    1.0E6   180.0  10.0  10.0  0.8   0.6
  0.0    0.0      -4.0  90.0   0.6  0.8          Same as above for Element 2
    2      3
    6                                            NSPV
    1      1     0.0         <-----------
    1      2     0.0                     |
    1      3     0.0                     |
    3      1     0.0                     |       ISPV, VSPV
    3      2     0.0                     |
    3      3     0.0         <-----------
    1                                            NSSV
    2      2    -2.0                             ISSV, VSSV
    0                                            NNBC
```

TABLE 7.11
Input file for the problem in Example 7.5

```
Example 7.5:  Analysis of a plane frame  (TIM element)
 4  2  0                                    MODEL, NTYPE, ITEM
 0  2                                       IELEM, NEM
 0  1                                       ICONT, NPRNT
 3                                          NNM
0.3   1.0E6  144.0  10.0  10.0  0.0  1.0   PR, SE, SL, SA, SI, CS, SN
0.0  -0.0138888  0.0  0.0  0.0  0.0        HF, VF, PF, XB, CST, SNT
  1     2                                   NOD(1,J)
0.3   1.0E6  180.0  10.0  10.0  0.8  0.6
0.0   0.0     -4.0  90.0   0.6  0.8        Same as above for Element 2
  2     3
  6                                         NSPV
  1     1    0.0         <-----------
  1     2    0.0                      |
  1     3    0.0                      |
  3     1    0.0                      |     ISPV, VSPV
  3     2    0.0                      |
  3     3    0.0         <-----------
  1                                         NSSV
  2     2   -2.0                            ISSV, VSSV
  0                                         NNBC
```

$$u(0,t)=0, \qquad \left(\frac{\partial u}{\partial x}+u\right)\bigg|_{x=1}=0 \quad \text{(Set 2)} \tag{7.49c}$$

$$u(x,0)=1$$

In Example 6.3, only Set 1 boundary conditions were considered.

We have MODEL = 1, NTYPE = 0, ITEM = 1, ICONT = 1, NSPV = 1, NSSV = 0, and NNBC = 0 for Set 1 and NNBC = 1 for Set 2. The coefficients of the differential equations are ($a = 1.0$, $b = 0.0$, $c = 0.0$, $c_t = 1.0$ and $f = 0$)

$$\text{AX0}=1.0, \quad \text{AX1}=0.0, \quad \text{BX0}=0.0, \quad \text{BX1}=0.0$$
$$\text{CX0}=0.0, \quad \text{CX1}=0.0, \quad \text{FX0}=0.0, \quad \text{FX1}=0.0$$
$$\text{FX2}=0.0, \quad \text{CT0}=1.0, \quad \text{CT1}=0.0$$

The boundary and initial conditions (since INCOND = 1) are input as

$$\text{ISPV}(1,1)=1, \quad \text{ISPV}(1,2)=1, \quad \text{VSPV}(1)=0.0$$
$$\text{GU0(I)}=\{0.0, 1.0, 1.0, \ldots\}$$

From the discussions of Example 6.4, we use $\Delta t = 0.01$ (DT = 0.01) and print the solution for every time step (i.e., INTVL = 1).

The input file of the problem for Set 1 boundary conditions is presented in Table 7.12.

Example 7.7 (Example 6.2) Natural vibrations of a cantilever beam. For this problem, the boundary conditions are given by

$$w(0)=0, \quad \left(\frac{dw}{dx}\right)\bigg|_{x=0}=0 \quad \text{for the Euler–Bernoulli beam theory}$$

$$w(0)=0, \quad \Psi(0)=0 \quad \text{for the Timoshenko beam theory}$$

The input data for all variables is the same as in the static analysis. In addition, we must input c_{t0} and c_{t1}. For the Timoshenko beam theory [see (6.13)], c_{t0} denotes the inertia

TABLE 7.12
Input file for the problem in Example 7.6

```
Example 7.6: Transient heat conduction in a plane wall
  1  0  1                                                MODEL, NTYPE, ITEM
  1  2                                                   IELEM, NEM
  1  1                                                   ICONT, NPRNT
    0.0    0.5    0.5                                    DX(I)
    1.0    0.0                                           AX0, AX1
    0.0    0.0                                           BX0, BX1
    0.0    0.0                                           CX0, CX1
    0.0    0.0    0.0                                    FX0,FX1,FX2
    1                                                    NSPV
    1   1  0.0                                           ISPV(I,J), VSPV(I)
    0                                                    NSSV
    0                                                    NNBC
    1.0    0.0                                           CT0, CT1
    0.05   0.5    0.0                                    DT, ALFA, BETA
    1      20     2                                      INCOND, NTIME, INTVL
    0.0    1.0    1.0                                    GU0(I)
```

ρA, and c_{t1} denotes the rotatory inertia ρI. The eigenvalue solver used in FEM1DV2 requires the matrix $[B]$ in $[A]\{x\} = \lambda[B]\{x\}$ to be positive-definite. Hence, c_{t0} and c_{t1} should be nonzero, otherwise, the mass matrix coefficients associated with Ψ will be zero.

The input files for the natural vibrations of the cantilever beam by the two types of elements are given in Tables 7.13 and 7.14. The reader can investigate the convergence characteristics of the elements in improving the accuracy of the fundamental frequency with the use of FEM1DV2.

7.4 SUMMARY

In this chapter three main items have been discussed: numerical integration of finite element coefficient matrices and vectors, logical units of a typical finite element program and their contents, and applications of the finite element program FEM1DV2. The numerical evaluation of the coefficients is required because of (*a*) variable coefficients of the differential equations modeled and (*b*) special evaluation of the coefficients, as was required for the Timoshenko

TABLE 7.13
Input file for the problem in Example 7.7

```
Example 7.7: Natural vibrations of a cantilever beam (E-B; with RI)
  3  0  3                                                MODEL, NTYPE, ITEM
  0  4                                                   IELEM, NEM
  1  1                                                   ICONT, NPRNT
    0.0     0.25    0.25    0.25    0.25                 DX(I)
    0.0     0.0                                          AX0, AX1
    1.0     0.0                                          BX0, BX1
    0.0     0.0                                          CX0, CX1
    2                                                    NSPV
  1  1                                                   ISPV(1,J)
  1  2                                                   ISPV(2,J)
    0                                                    NNBC
    1200.0 1.0                                           CT0, CT1   (L/H=10)
```

TABLE 7.14
Input file for the problem in Example 7.7

```
Example 7.7: Natural vibrations of a cantilever beam (TIM element)
  2  0  3                                         MODEL, NTYPE, ITEM
  1  4                                            IELEM, NEM
  1  1                                            ICONT, NPRNT
     0.0     0.25    0.25    0.25    0.25         DX(I)
     4.0E2   0.0                                  AX0, AX1
     1.0     0.0                                  BX0, BX1
     0.0     0.0                                  CX0, CX1
     2                                            NSPV
  1  1                                            ISPV(1,J)
  1  2                                            ISPV(2,J)
     0                                            NNBC
    12.0E2   1.0                                  CT0, CT1   (L/H=10)
```

beam element with equal interpolation. The Newton–Cotes and Gauss–Legendre integration rules have been discussed. The integration rules require the transformation of the integral expressions from the global coordinate system to a local coordinate system. This transformation requires interpolation of the global coordinate x. Depending on the relative degrees of interpolation of the geometry and the dependent variables, the formulations are classified as subparametric, isoparametric, and superparametric.

The three logical units—preprocessor, processor, and postprocessor—have been discussed. The contents of processor, where most finite element calculations are carried out, have been considered in detail. Fortran statements for numerical evaluation of integral expressions, assembly of element coefficients, and imposition of boundry conditions have been discussed.

A description of the finite element computer program FEM1DV2 has been presented, and its application to problems of heat transfer and solid mechanics has been discussed.

PROBLEMS

Section 7.1

In Problems 7.1–7.5, compute the matrix coefficients using (*a*) the Newton–Cotes integration formula and (*b*) the Gauss–Legendre quadrature. Use the appropriate number of integration points, and verify the results with those obtained by the exact integration.

7.1.

$$K_{12} = \int_{x_A}^{x_B} (1+x)\frac{d\psi_1}{dx}\frac{d\psi_2}{dx}\,dx, \qquad G_{12} = \int_{x_A}^{x_B} (1+x)\psi_1\psi_2\,dx$$

where ψ_i are the linear (Lagrange) interpolation functions.
Answer: $K_{12} = -h^{-1}[1 + \frac{1}{2}(x_A + x_B)]$, $G_{12} = \frac{1}{6}h(1 + x_A + \frac{1}{2}h)$.

7.2. Repeat Problem 7.1 for the quadratic interpolation functions.

7.3.

$$K_{11} = \int_{x_A}^{x_B} \left(\frac{d^2\psi_1}{dx^2}\right)^2 dx, \quad G_{11} = \int_{x_A}^{x_B} (\psi_1)^2\, dx$$

where ψ_i are the Hermite cubic interpolation functions.

Answer: $r = 2$: $K_{11} = 12/h^3$ (exact), $G_{11} = 0.398148h$.

7.4. Repeat Problem 7.3 for the case in which the interpolation functions are the fifth-order Hermite polynomials of Problem 4.21.

Answer: $K_{11} = \frac{120}{7}h^3$, $G_{11} = \frac{181}{462}h$ (exact for five-point quadrature).

7.5.

$$K_{11} = \int_{x_A}^{x_B} \left(\frac{d\psi_1}{dx}\right)^2 dx$$

where $\psi_1 = \sin\left[\frac{1}{4}\pi(\xi + 3\xi^2)\right]$. Use three- and five-point Gauss quadrature to compute K_{11}.

Section 7.2

Computer exercises (use FEM1DV2)

7.6. Solve the problem in Example 3.1 (Set 2) using two, four, and six linear elements. Tabulate the results along with the exact solution. Use the following data: $L = 0.02$ m, $k = 20\ \text{W m}^{-1}\,°\text{C}^{-1}$, $g_0 = 10^5\ \text{W m}^{-2}$, $T_\infty = 50°\text{C}$, $\beta = 500\ \text{W m}^{-1}\,°\text{C}^{-1}$.

7.7. Solve the problem in Example 3.1 (Set 2) using one, two, and three quadratic elements. Compare the finite element results for the temperature and heat flux with the exact solutions at the nodes.

7.8. Solve the heat transfer problem in Example 3.2, using four linear elements and two quadratic elements, and tabulate them with the exact solution at nodes (see Table 3.3).

7.9. Solve the problem in Example 3.2 using four quadratic elements, and compare the solution with that obtained using eight linear elements and the exact solution.

7.10. Solve the one-dimensional flow problem of Example 3.3 (Set 1), for $dP/dx = -24$, using four linear elements and two quadratic elements. Compare the finite element results with the exact one.

7.11. Solve the Couette flow problem in Example 3.3 (Set 2) using (*a*) four linear elements and (*b*) two quadratic elements. Compare the finite element solution with the exact solution.

7.12. Solve the problem of heat flow in a rod (Problem 3.19) using (*a*) four linear elements and (*b*) two quadratic elements, and compare the results with the analytical solution.

7.13. Solve Problem 3.26 using (*a*) four linear elements and (*b*) two quadratic elements, and compare the finite element solution with the exact solution at nodes.

7.14. Solve the problem of axisymmetric deformation of a rotating circular disk using (*a*) four linear elements and (*b*) two quadratic elements (see Example 7.3). Assume that the body force is given by $f = \rho\omega^2 r$.

7.15. Solve Problem 3.28.

7.16. Solve Problem 3.30.

7.17. Solve the problem in Example 3.4 using (*a*) four linear and (*b*) two quadratic elements. Determine the stress at the fixed end.

7.18–7.29. Solve the beam problems 4.1–4.12 using the minimum number of Hermite cubic elements.

7.30. Analyze a clamped circular plate under a uniformly distributed transverse load using the Euler–Bernoulli plate element. Investigate the convergence using two, four, and six elements by comparing with the exact solution

$$w(r) = \frac{q_0 a^4}{64D}\left[1 - \left(\frac{r}{a}\right)^2\right]^2$$

where $D = EH^3/12(1-\nu^2)$, q_0 is the intensity of the distributed load, a is the radius of the plate, H is its thickness, and ν is Poisson's ratio ($\nu = 0.25$).

7.31. Repeat Problem 7.30 with (*a*) two, four, and six linear and (*b*) one, two, and three quadratic Timoshenko plate elements (RIE) for $a/H = 10$ and 100.

7.32. Repeat Problem 7.30 with the Timoshenko plate element (CIE) for $a/H = 10$ and 100.

7.33. Consider an annular plate of outer radius a and and inner radius b, and thickness H. If the plate is simply supported at the outer edge and subjected to a uniformly distributed moment M_0 at the inner edge (see Fig. P7.33), analyze the problem using the Euler–Bernoulli plate element. Investigate the accuracy using two and four elements and comparing with the analytical solution ($\nu = 0.25$)

$$w(r) = \frac{b^2 M_0 (a^2 - r^2)}{2(1+\nu)D(a^2 - b^2)} - \frac{a^2 b^2 M_0}{(1-\nu)D(a^2 - b^2)} \ln\left(\frac{r}{a}\right)$$

See Problem 7.30 for the definition of D.

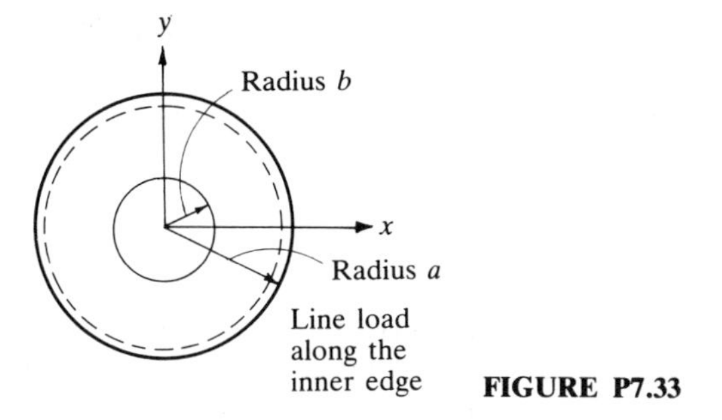

FIGURE P7.33

7.34. Repeat Problem 7.33 with (*a*) two and four linear and (*b*) one and two quadratic Timoshenko (RIE) elements for $a/H = 10$ and 100.

7.35. Consider the simply supported annular plate described in Problem 7.33. Suppose that the inner edge is loaded with a uniformly distributed shearing force Q_0. Use meshes of two and four Euler–Bernoulli plate elements to analyze the problem.

7.36. Analyze Problem 7.35 with (*a*) two and four linear and (*b*) one and two quadratic Timoshenko (RIE) plate elements for $a/H = 10$ and 100.

7.37. Repeat Problem 7.36 with the Timoshenko (CIE) plate element.

7.38. Consider a simply supported circular plate of radius a, loaded at the center with a load P. Analyze the problem with two, four, and six Euler–Bernoulli plate elements and compare with the analytical solution ($\nu = 0.25$)

$$w(r) = \frac{P}{8\pi D}\left[\frac{3+\nu}{2(1+\nu)}(a^2 - r^2) + r^2 \ln\left(\frac{r}{a}\right)\right]$$

Note that $Q_1^1 = P/2\pi$.

7.39. Analyze Problem 7.38 with four and eight linear Timoshenko (RIE) plate elements for $a/H = 10$ and 100.

7.40. Analyze the simply supported annular plate in Problem 7.33 when it is subjected to a uniformly distributed load of intensity q_0. Use the Euler–Bernoulli plate element.

7.41. Analyze the annular plate in Problem 7.40 using two and four linear Timoshenko (CIE) plate elements.

7.42–7.49. Solve the truss and frame problems in Figs. P4.29–P4.36.

7.50. Consider the axial motion of an elastic bar, governed by the second-order equation

$$EA\frac{\partial^2 u}{\partial x^2} = \rho A\frac{\partial^2 u}{\partial t^2} \quad \text{for } 0 < x < L$$

with the following data: length of bar $L = 500$ mm, cross-sectional area $A = 1\text{ mm}^2$, modulus of elasticity $E = 20{,}000\text{ N mm}^{-2}$, density $\rho = 0.008\text{ N s}^2\text{ mm}^{-4}$, boundary conditions

$$u(0, t) = 0, \qquad EA\frac{\partial u}{\partial x}(L, t) = 1$$

and zero initial conditions. Using 20 linear elements and $\Delta t = 0.002$ s, determine the axial displacement and plot the displacement as a function of position along the bar for $t = 0.8$ s.

7.51. Consider the following nondimensionalized differential equation governing the plane wall transient [see Myers (1971), p. 101]:

$$-\frac{\partial^2 T}{\partial x^2} + \frac{\partial T}{\partial t} = 0 \quad \text{for } 0 < x < 1$$

with boundary conditions $T(0, t) = 1$ and $T(1, t) = 0$, and initial condition $T(x, 0) = 0$. Solve the problem using (a) eight linear elements, and (b) four quadratic elements. Determine the critical time step, solve the problem with Crank-Nicholson's method ($\Delta t = 0.002$).

Note. Program FEM1DV2 cannnot be used without modification to solve Problems 7.52–7.54.

7.52. Consider a simply supported beam of length L subjected to a point load

$$P(t) = \begin{cases} P_0 \sin\dfrac{\pi t}{\tau} & \text{for } 0 \leqslant t \leqslant \tau \\ 0 & \text{for } t \geqslant \tau \end{cases}$$

at a distance c from the left end of the beam (assumed to be at rest at $t = 0$). The transverse deflection $w(x, t)$ is given by [see Harris and Crede (1961), pp. 8–53]

$$w(x,t) = \begin{cases} \dfrac{2P_0L^3}{\pi^4 EI} \displaystyle\sum_{i=1}^{\infty} \frac{1}{i^4} \sin\frac{i\pi c}{L} \sin\frac{i\pi x}{L} \left[\frac{1}{1 - T_i^2/4\tau^2}\left(\sin\frac{\pi t}{\tau} - \frac{T_i}{2\tau}\sin\omega_i t\right)\right] & \text{for } 0 \leq t \leq \tau \\ \dfrac{2P_0L^3}{\pi^4 EI} \displaystyle\sum_{i=1}^{\infty} \frac{1}{i^4} \sin\frac{i\pi c}{L} \sin\frac{i\pi x}{L} \left[\frac{(T_i/\tau)\cos(\pi\tau/T_i)}{T_i^2/4\tau^2 - 1} \sin\omega_i(t - \tfrac{1}{2}\tau)\right] & \text{for } t \geq \tau \end{cases}$$

where

$$T_i = \frac{2\pi}{\omega_i} = \frac{2L^2}{i^2\pi}\frac{A\rho}{EI} = \frac{T_1}{i^2}$$

Use the data $P_0 = 1000$ lb, $\tau = 20 \times 10^{-6}$ s, $L = 30$ in, $E = 30 \times 10^6$ lb in^{-2}, $\rho = 733 \times 10^{-6}$ lb in^{-3}, $\Delta t = 10^{-6}$ s, and assume that the beam is of square cross-section 0.5 in by 0.5 in. Using five Euler–Bernoulli elements in the half-beam, obtain the finite element solution and compare with the series solution at midspan for the case $c = \frac{1}{2}L$.

7.53. Repeat Problem 7.52 for $c = \frac{1}{4}L$ and eight elements in the full span.

7.54. Repeat Problem 7.52 for $P(t) = P_0$ at midspan and eight elements in the full span.

7.55. Consider a cantilevered beam with a point load at the free end. Using the load and data of Problem 7.52, find the finite element solution for the transverse deflection using Euler–Bernoulli beam elements.

7.56. Repeat Problem 7.52 for a clamped beam with the load at the midspan.

7.57–7.61. Solve Problems 7.52–7.56 using Timoshenko beam elements. Use $\nu = 0.3$.

REFERENCES FOR ADDITIONAL READING

Fluid mechanics

Bird, R. B., W. E. Stewart, and E. N. Lightfoot: *Transport Phenomena,* John Wiley, New York, 1960.

Duncan, W. J., A. S. Thom, and A. D. Young: *Mechanics of Fluids,* 2d ed., Elsevier, New York, 1970.

Eskinazi, S.: *Principles of Fluid Mechanics,* Allyn and Bacon, Boston, 1962.

Harr, M. E.: *Ground Water and Seepage,* McGraw-Hill, New York, 1962.

Nadai, A.: *Theory of Flow and Fracture of Solids,* vol. II, McGraw-Hill, New York, 1963.

Schlichting, H.: *Boundary-Layer Theory* (translated by J. Kestin), 7th ed., McGraw-Hill, New York, 1979.

Shames, I. H.: *Mechanics of Fluids,* McGraw-Hill, New York, 1962.

Vallentine, H. R.: *Applied Hydrodynamics,* Butterworths, London, 1959.

Verrujit, A.: *Theory of Groundwater Flow,* Gordon and Breach, New York, 1970.

Heat transfer

Carslaw, H. S., and J. C. Jaeger: *Conduction of Heat in Solids,* Clarendon Press, Oxford, 1959.

Holoman, J. P.: *Heat Transfer,* 6th ed., McGraw-Hill, New York, 1986.

Kreith, F.: *Principles of Heat Transfer,* 3rd ed., Harper & Row, New York, 1973.

Myers, G. G.: *Analytical Methods in Conduction Heat Transfer,* McGraw-Hill, New York, 1972.

Özisik, M. N.: *Heat Transfer: A Basic Approach,* McGraw-Hill, New York, 1985.

Plane elasticity

Budynas, R. G.: *Advanced Strength and Applied Stress Analysis,* McGraw-Hill, New York, 1977.
Harris, C. M., and C. E. Crede: *Shock and Vibration Handbook,* vol. 1, McGraw-Hill, New York, 1961.
Ugural, A. C., and S. K. Fenster: *Advanced Strength and Applied Elasticity,* Elsevier, New York, 1975.
Volterra, E., and J. H. Gaines: *Advanced Strength of Materials,* Prentice-Hall, Englewood Cliffs, NJ, 1971.

Time approximations

Bathe, K. J.: *Finite Element Procedures in Engineering Analysis,* Prentice-Hall, Englewood Cliffs, NJ, 1982.
Clough, R. W., and J. Penzien: *Dynamics of Structures,* McGraw-Hill, New York, 1975.
Hughes, T. J. R.: *The Finite Element Method,* Prentice-Hall, Englewood-Cliffs, NJ, 1987.

Numerical integration

Carnahan, B., H. A. Luther, and J. O. Wilker: *Applied Numerical Methods,* John Wiley, New York, 1969.
Loxan, A. N., N. Davids, and A. Levenson: "Table of the Zeros of the Legendre Polynomials of Order 1–16 and the Weight Coefficients for Gauss' Quadrature Formula," *Bulletin of the American Mathematical Society,* vol. 48, pp. 739–743, 1942.
Silvester, P.: "Newton–Cotes Quadrature Formulae for *N*-Dimensional Simplexes," *Proceedings of 2d Canadian Congress of Applied Mechanics, Waterloo, Ontario, 1969.* University of Waterloo, Waterloo, Ontario, 1969.
Stroud, A. H., and D. Secrest: *Gaussian Quadrature Formulas,* Prentice-Hall, Englewood Cliffs, NJ, 1966.
Zienkiewicz, O. C., and R. L. Taylor: *The Finite Element Method,* vols. 1 and 2, McGraw-Hill, 1989 and 1991.

PART 3

FINITE ELEMENT ANALYSIS OF TWO-DIMENSIONAL PROBLEMS

CHAPTER 8

SINGLE-VARIABLE PROBLEMS

8.1 INTRODUCTION

The finite element analysis of two-dimensional problems involves the same basic steps as those described for one-dimensional problems in Chapter 3. The analysis is somewhat complicated by the fact that two-dimensional problems are described by partial differential equations over geometrically complex regions. The boundary Γ of a two-dimensional domain Ω is, in general, a curve. Therefore, the finite elements are simple two-dimensional geometric shapes that can be used to approximate a given two-dimensional domain as well as the solution over it. Thus, in two-dimensional problems, we not only seek an approximate solution to a given problem on a domain but we also approximate the domain by a suitable finite element mesh. Consequently, in the finite element analysis of two-dimensional problems, we shall have approximation errors due to the approximation of the solution as well as discretization errors due to the approximation of the domain. The finite element mesh (discretization) consists of simple two-dimensional elements, such as triangles, rectangles, and/or quadrilaterals, that allow unique derivation of the interpolation functions. The elements are connected to each other at nodal points on the boundaries of the elements. The ability to represent domains with irregular geometries by a collection of finite elements makes the

TABLE 8.1
Some examples of the Poisson equation $-\nabla \cdot (k\nabla u) = f$
Natural boundary condition: $k\,\partial u/\partial n + \beta(u - u_\infty) = q$. Essential boundary condition: $u = \hat{u}$

Field of application	Primary variable u	Material constant k	Source variable f	Secondary variables $q, \frac{\partial u}{\partial x}, \frac{\partial u}{\partial y}$
1. Heat transfer	Temperature T	Conductivity k	Heat source Q	Heat flow q [comes from conduction $k\,\partial T/\partial n$ and convection $h(T - T_\infty)$]
2. Irrotational flow of an ideal fluid	Stream function ψ	Density ρ	Mass production σ (normally zero)	Velocities: $\frac{\partial \psi}{\partial x} = -v,\ \frac{\partial \psi}{\partial y} = u$
	Velocity potential ϕ	Density ρ	Mass production σ (normall zero)	$\frac{\partial \phi}{\partial x} = u,\ \frac{\partial \phi}{\partial y} = v$
3. Groundwater flow	Piezometric head ϕ	Permeability K	Recharge Q (or pumping, $-Q$)	Seepage $q = k\frac{\partial \phi}{\partial n}$ Velocities: $u = -k\frac{\partial \phi}{\partial x},\ v = -k\frac{\partial \phi}{\partial y}$
4. Torsion of members with constant cross-section	Stress function Ψ	$k = 1$	$f = 2$	$G\theta\frac{\partial \Psi}{\partial x} = -\sigma_{yz}$
		G = shear modulus	θ = angle of twist per unit length	$G\theta\frac{\partial \Psi}{\partial y} = \sigma_{xz}$
5. Electrostatics	Scalar potential ϕ	Dielectric constant ε	Charge density ρ	Displacement flux density D_n
6. Magnetostatics	Magnetic potential ϕ	Permeability μ	Charge density ρ	Magnetic flux density B_n
7. Transverse deflection of elastic membranes	Transverse deflection u	Tension T in membrane	Transversely distributed load	Normal force q

method a valuable practical tool for the solution of boundary, initial, and eigenvalue problems arising in various fields of engineering.

The objective of this chapter is to extend the basic steps discussed earlier for one-dimensional problems to two-dimensional boundary value problems involving a single dependent variable. Once again, we describe the basic steps of the finite element analysis with a model second-order partial differential equation, namely the Poisson equation, governing a single variable. This equation arises in a number of fields, including electrostatics, heat transfer, fluid mechanics, and solid mechanics (see Table 8.1).

8.2 BOUNDARY VALUE PROBLEMS

8.2.1 The Model Equation

Consider the problem of finding the solution u of the second-order partial differential equation

$$\boxed{-\frac{\partial}{\partial x}\left(a_{11}\frac{\partial u}{\partial x}+a_{12}\frac{\partial u}{\partial y}\right)-\frac{\partial}{\partial y}\left(a_{21}\frac{\partial u}{\partial x}+a_{22}\frac{\partial u}{\partial y}\right)+a_{00}u-f=0} \tag{8.1}$$

for given data a_{ij} $(i, j=1, 2)$, a_{00} and f, and specified boundary conditions. The form of the boundary conditions will be apparent from the weak formulation. As a special case, one can obtain the Poisson equation from (8.1) by setting $a_{11}=a_{22}=a$ and $a_{12}=a_{21}=a_{00}=0$:

$$-\nabla\cdot(a\,\nabla u)=f \quad \text{in } \Omega \tag{8.2}$$

where ∇ is the gradient operator. If $\hat{\mathbf{i}}$ and $\hat{\mathbf{j}}$ denote the unit vectors directed along the x and y axes, respectively, the gradient operator can be expressed as (see Section 2.2.2)

$$\nabla=\hat{\mathbf{i}}\frac{\partial}{\partial x}+\hat{\mathbf{j}}\frac{\partial}{\partial y}$$

and (8.2) in the cartesian coordinate system takes the form

$$-\frac{\partial}{\partial x}\left(a\frac{\partial u}{\partial x}\right)-\frac{\partial}{\partial y}\left(a\frac{\partial u}{\partial y}\right)=f \tag{8.3}$$

In the following, we shall develop the finite element model of (8.1). The major steps are as follows:

1. Discretization of the domain into a set of finite elements.
2. Weak (or weighted-integral) formulation of the governing differential equation.
3. Derivation of finite element interpolation functions.
4. Development of the finite element model using the weak form.
5. Assembly of finite elements to obtain the global system of algebraic equations.

6. Imposition of boundary conditions.
7. Solution of equations.
8. Post-computation of solution and quantities of interest.

Steps 6 and 7 remain unchanged from one-dimensional finite element analysis because at the end of Step 5 we have a set of algebraic equations whose form is independent of the dimension of the domain or nature of the problem. In the following sections, we discuss each step in detail.

8.2.2 Finite Element Discretization

In two dimensions, there is more than one simple geometric shape that can be used as a finite element (see Fig. 8.1). As we shall see shortly, the interpolation functions depend not only on the number of nodes in the element but also on the shape of the element. The shape of the element must be such that its geometry is uniquely defined by a set of points, which serve as the element nodes in the development of the interpolation functions. As will be discussed later in this section, a triangle is the simplest geometric shape, followed by a rectangle.

The representation of a given region by a set of elements (i.e., discretization or *mesh generation*) is an important step in finite element analysis. The choice of element type, number of elements, and density of elements depends on the geometry of the domain, the problem to be analyzed,

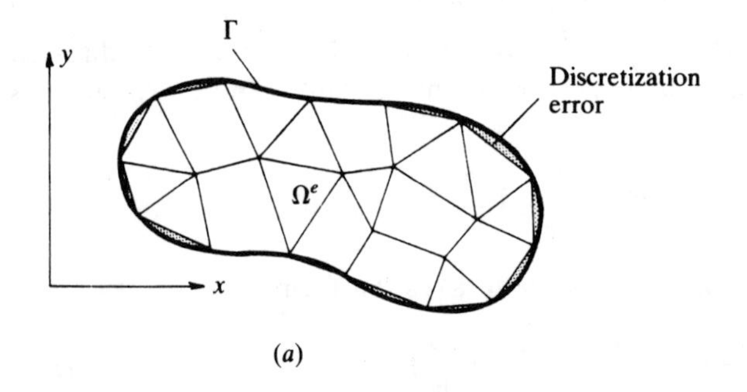

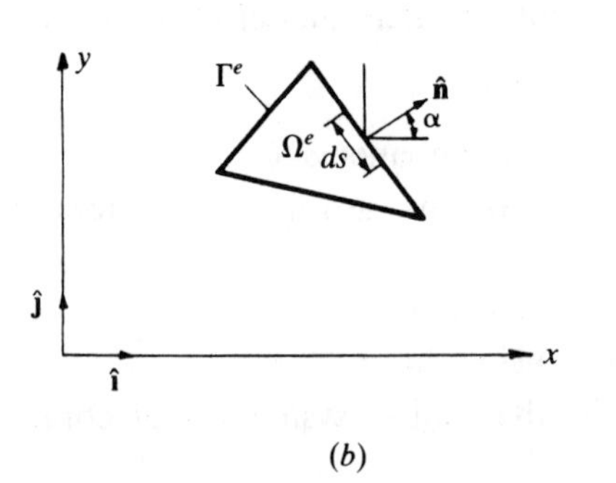

FIGURE 8.1
Finite element discretization of an irregular domain: (*a*) discretization of a domain by triangular and quadrilateral elements; (*b*) a typical triangular element (boundary Γ^e, the unit normal $\hat{\mathbf{n}}$ on the boundary of the element).

and the degree of accuracy desired. Of course, there are no specific formulae for obtaining this information. In general, the analyst is guided by his or her technical background, insight into the physics of the problem being modeled (e.g., a qualitative understanding of the solution), and experience with finite element modeling. The general rules of mesh generation for finite element formulations include the following:

1. Select elements that characterize the governing equations of the problem.
2. The number, shape, and type (i.e., linear or quadratic) of elements should be such that the geometry of the domain is represented as accurately as desired.
3. The density of elements should be such that regions of large gradients of the solution are adequately modeled (i.e., more or higher-order elements should be used in regions of large gradients).
4. Mesh refinements should vary gradually from high-density regions to low-density regions. If *transition elements* are used, they should be used away from critical regions (i.e., regions of large gradients). Transition elements are those that connect lower-order elements to higher-order elements (e.g., linear to quadratic).

Additional discussion of finite element meshes and types of elements (linear and higher-order Lagrange elements, transition elements, etc.) will be presented in Chapter 9.

8.2.3 Weak Form

In the development of the weak form, we need only consider an arbitrarily typical element. We assume that Ω^e is such an element, whether triangular or quadrilateral, of the finite element mesh, and we develop the finite element model of (8.1) over Ω^e. Various two-dimensional elements will be discussed in the sequel.

Following the three-step procedure presented in Chapters 2 and 3, we develop the weak form of (8.1) over the typical element Ω^e. The first step is to multiply (8.1) with a weight function w, which is assumed to be once-differentiable with respect to x and y, and then integrate the resulting equation over the element domain Ω^e:

$$0 = \int_{\Omega^e} w\left[-\frac{\partial}{\partial x}(F_1) - \frac{\partial}{\partial y}(F_2) + a_{00}u - f\right] dx\, dy \tag{8.4a}$$

where

$$F_1 = a_{11}\frac{\partial u}{\partial x} + a_{12}\frac{\partial u}{\partial y}, \qquad F_2 = a_{21}\frac{\partial u}{\partial x} + a_{22}\frac{\partial u}{\partial y} \tag{8.4b}$$

In the second step, we distribute the differentiation equally between u and w. To achieve this, we integrate the first two terms in (8.4a) by parts. First we note the identities

$$\frac{\partial}{\partial x}(wF_1) = \frac{\partial w}{\partial x} F_1 + w\frac{\partial F_1}{dx} \quad \text{or} \quad -w\frac{\partial F_1}{\partial x} = \frac{\partial w}{\partial x} F_1 - \frac{\partial}{\partial x}(wF_1) \tag{8.5a}$$

$$\frac{\partial}{\partial y}(wF_2) = \frac{\partial w}{\partial y} F_2 + w\frac{\partial F_2}{\partial y} \quad \text{or} \quad -w\frac{\partial F_2}{\partial y} = \frac{\partial w}{\partial y} F_2 - \frac{\partial}{\partial y}(wF_2) \tag{8.5b}$$

Next, we recall the component form of the gradient (or divergence) theorem in (2.17b):

$$\int_{\Omega^e} \frac{\partial}{\partial x}(wF_1)\,dx\,dy = \oint_{\Gamma^e} wF_1 n_x\,ds \tag{8.6a}$$

$$\int_{\Omega^e} \frac{\partial}{\partial y}(wF_2)\,dx\,dy = \oint_{\Gamma^e} wF_2 n_y\,ds \tag{8.6b}$$

where n_x and n_y are the components (i.e., direction cosines) of the unit normal vector

$$\hat{\mathbf{n}} = n_x\hat{\mathbf{i}} + n_y\hat{\mathbf{j}} = \cos\alpha\hat{\mathbf{i}} + \sin\alpha\hat{\mathbf{j}} \tag{8.7}$$

on the boundary Γ^e, and ds is the arclength of an infinitesimal line element along the boundary (see Fig. 8.1*b*). Using (8.5) and (8.6) in (8.4*a*), we obtain

$$0 = \int_{\Omega^e}\left[\frac{\partial w}{\partial x}\left(a_{11}\frac{\partial u}{\partial x} + a_{12}\frac{\partial u}{\partial y}\right) + \frac{\partial w}{\partial y}\left(a_{21}\frac{\partial u}{\partial x} + a_{22}\frac{\partial u}{\partial y}\right) + a_{00}wu - wf\right]dx\,dy$$
$$- \oint_{\Gamma^e} w\left[n_x\left(a_{11}\frac{\partial u}{\partial x} + a_{12}\frac{\partial u}{\partial y}\right) + n_y\left(a_{21}\frac{\partial u}{\partial x} + a_{22}\frac{\partial u}{\partial y}\right)\right]ds \tag{8.8}$$

From an inspection of the boundary term in this equation, we note that the specification of u constitutes the essential boundary condition, and hence u is the primary variable. The specification of the coefficient of the weight function in the boundary expression

$$q_n \equiv n_x\left(a_{11}\frac{\partial u}{\partial x} + a_{12}\frac{\partial u}{\partial y}\right) + n_y\left(a_{21}\frac{\partial u}{\partial x} + a_{22}\frac{\partial u}{\partial y}\right) \tag{8.9}$$

constitutes the natural boundary condition; thus, q_n is the secondary variable of the formulation. The function $q_n = q_n(s)$ denotes the projection of the vector $\mathbf{a} \cdot \nabla u$ along the unit normal $\hat{\mathbf{n}}$. By definition, q_n is positive outward from the surface as we move counter-clockwise along the boundary Γ^e. The secondary variable q_n is of physical interest in most problems. For example, in the case of heat transfer through an anisotropic medium, a_{ij} are the conductivities of the medium, and q_n is the heat flux normal to the boundary of the element (into the element).

The third and last step of the formulation is to use the definition (8.9) in (8.8) and write the weak form of (8.1) as

$$0 = \int_{\Omega^e} \left[\frac{\partial w}{\partial x} \left(a_{11} \frac{\partial u}{\partial x} + a_{12} \frac{\partial u}{\partial y} \right) + \frac{\partial w}{\partial y} \left(a_{21} \frac{\partial u}{\partial x} + a_{22} \frac{\partial u}{\partial y} \right) + a_{00} wu - wf \right] dx\, dy - \oint_{\Gamma^e} w q_n \, ds \tag{8.10a}$$

or

$$B(w, u) = l(w) \tag{8.10b}$$

where the bilinear form $B(\cdot, \cdot)$ and linear form $l(\cdot)$ are

$$B(w, u) = \int_{\Omega^e} \left[\frac{\partial w}{\partial x} \left(a_{11} \frac{\partial u}{\partial x} + a_{12} \frac{\partial u}{\partial y} \right) + \frac{\partial w}{\partial y} \left(a_{21} \frac{\partial u}{\partial x} + a_{22} \frac{\partial u}{\partial y} \right) + a_{00} wu \right] dx\, dy \tag{8.10c}$$

$$l(w) = \int_{\Omega^e} wf \, dx\, dy + \oint_{\Gamma^e} w q_n \, ds$$

The weak form (also called the *variational problem*) in (8.10) forms the basis of the finite element model of (8.1). The quadratic functional associated with the problem can be obtained from (2.43*b*) when $B(\cdot, \cdot)$ is symmetric [see Mikhlin (1964), Reddy and Rasmussen (1982), and Reddy (1986)]:

$$I(u) = \tfrac{1}{2} B(u, u) - l(u)$$

The bilinear form in (8.10*c*) is symmetric when $a_{12} = a_{21}$.

8.2.4 Finite Element Model

The weak form in (8.10) requires that the approximation chosen for u should be at least linear in both x and y so that there are no terms in (8.10) that are identically zero. Since the primary variable is simply the function itself, the Lagrange family of interpolation functions is admissible. Suppose that u is approximated over a typical finite element Ω^e by the expression

$$u(x, y) \approx U^e(x, y) = \sum_{j=1}^{n} u_j^e \psi_j^e(x, y) \tag{8.11}$$

where u_j^e is the value of U^e at the jth node (x_j, y_j) of the element, and ψ_j^e are the Lagrange interpolation functions, with the property

$$\psi_i^e(x_j, y_j) = \delta_{ij} \tag{8.12}$$

The specific form of ψ_i^e will be derived for linear triangular and rectangular elements in Section 8.2.5, and higher-order interpolation functions will be derived in Chapter 9.

Substituting the finite element approximation (8.11) for u into the weak form (8.10), we obtain

$$0=\int_{\Omega^e}\left[\frac{\partial w}{\partial x}\left(a_{11}\sum_{j=1}^{n}u_j\frac{\partial\psi_j}{\partial x}+a_{12}\sum_{j=1}^{n}u_j\frac{\partial\psi_j}{\partial y}\right)+\frac{\partial w}{\partial y}\left(a_{21}\sum_{j=1}^{n}u_j\frac{\partial\psi_j}{\partial x}+a_{22}\sum_{j=1}^{n}u_j\frac{\partial\psi_j}{\partial y}\right)+a_{00}w\sum_{j=1}^{n}u_j\psi_j-wf\right]dx\,dy-\oint_{\Gamma^e}wq_n\,ds \tag{8.13}$$

This equation must hold for any weight function w. Since we need n independent algebraic equations to solve for the n unknowns $u_1, u_2, \ldots, u_n$, we choose n independent functions for w: $w=\psi_1, \psi_2, \ldots, \psi_n$. This particular choice is a natural one when the weight function is viewed as a virtual variation of the dependent unknown (i.e., $w=\delta u=\sum_{i=1}^n \delta u_i\,\psi_i$). For each choice of w, we obtain an algebraic relation among $(u_1, u_2, \ldots, u_n)$. We label the algebraic equation resulting from substitution of ψ_1 for w into (8.13) as the first algebraic equation. The ith algebraic equation is obtained by substituting $w=\psi_i$ into (8.13):

$$0=\sum_{j=1}^{n}\left\{\int_{\Omega^e}\left[\frac{\partial\psi_i}{\partial x}\left(a_{11}\frac{\partial\psi_j}{\partial x}+a_{12}\frac{\partial\psi_j}{\partial y}\right)+\frac{\partial\psi_i}{\partial y}\left(a_{21}\frac{\partial\psi_j}{\partial x}+a_{22}\frac{\partial\psi_j}{\partial y}\right)+a_{00}\psi_i\psi_j\right]dx\,dy\right\}u_j-\int_{\Omega^e}f\psi_i\,dx\,dy-\oint_{\Gamma^e}\psi_i q_n\,ds \quad (i=1,2,\ldots,n)$$

or

$$\boxed{\sum_{j=1}^{n}K_{ij}^e u_j^e=f_i^e+Q_i^e} \tag{8.14a}$$

where

$$\boxed{\begin{aligned}K_{ij}^e&=\int_{\Omega^e}\left[\frac{\partial\psi_i}{\partial x}\left(a_{11}\frac{\partial\psi_j}{\partial x}+a_{12}\frac{\partial\psi_j}{\partial y}\right)+\frac{\partial\psi_i}{\partial y}\left(a_{21}\frac{\partial\psi_j}{\partial x}+a_{22}\frac{\partial\psi_j}{\partial y}\right)+a_{00}\psi_i\psi_j\right]dx\,dy\\ f_i^e&=\int_{\Omega^e}f\psi_i^e\,dx\,dy,\qquad Q_i^e=\oint_{\Gamma^e}q_n\psi_i^e\,ds\end{aligned}} \tag{8.14b}$$

In matrix notation, (8.14a) takes the form

$$[K^e]\{u^e\}=\{f^e\}+\{Q^e\} \tag{8.14c}$$

Note that $K_{ij}^e=K_{ji}^e$ (i.e., $[K]$ is symmetric) only when $a_{12}=a_{21}$. Equation (8.14c) represents the finite element model of (8.1). This completes the development of the finite element model. Before we discuss assembly of elements, it is informative to consider how the interpolations ψ_i^e are derived for certain basic elements and evaluate the element matrices in (8.14*b*).

8.2.5 Interpolation Functions

The finite element approximation $U^e(x, y)$ of $u(x, y)$ over an element Ω^e must satisfy the following conditions in order for the approximate solution to be convergent to the true one:

1. U^e must be differentiable, as required in the weak form of the problem (i.e., all terms in the weak form are represented as nonzero values).
2. The polynomials used to represent U^e must be complete (i.e., all terms, beginning with a constant term up to the highest order used in the polynomial, should be included in U^e).
3. All terms in the polynomial should be linearly independent.

The number of linearly independent terms in the representation of U^e dictates the shape and number of degrees of freedom of the element. Next, we discuss some of the basic polynomials and associated elements for problems with a single degree of freedom per node.

An examination of the variational form (8.10) and the finite element matrices in (8.14*b*) shows that ψ_i^e should be at least linear functions of x and y. For example, the polynomial

$$U^e(x, y) = c_1 + c_2 x + c_3 y \tag{8.15}$$

contains three linearly independent terms, and it is linear in both x and y. To write the constants c_i $(i = 1, 2, 3)$ in terms of the nodal values of U^e, we must identify three points or nodes in the element Ω^e. These three nodes must also uniquely define the geometry of the element. Obviously, the geometric shape defined by three points in a two-dimensional domain is a triangle. Thus the polynomial (8.15) is associated with a triangular element, and there are three nodes identified, namely, the vertices of the triangle.

On the other hand, the polynomial

$$U^e(x, y) = c_1 + c_2 x + c_3 y + c_4 xy \tag{8.16}$$

contains four linearly independent terms, and is linear in x and y, with a bilinear term in x and y. This polynomial requires an element with four nodes. There are two possible geometric shapes: a triangle with the fourth node at its center (or centroid), or a rectangle with the nodes at the vertices. A triangle with a fourth node at the center does not provide a single-valued variation of u at interelement boundaries, resulting in *incompatible* variations of u there, and is therefore not admissible. The linear rectangular element is a compatible element.

A polynomial with five constants is the incomplete quadratic polynomial

$$U^e(x, y) = c_1 + c_2x + c_3y + c_4xy + c_5(x^2 + y^2) \tag{8.17}$$

which can be used to construct an element with five nodes (e.g., a rectangle with a node at each corner and at its midpoint); however, the element does not give single-valued variation of u. Also, note that the x^2 and y^2 terms cannot be varied independently of each other.

The quadratic polynomial

$$U^e(x, y) = c_1 + c_2x + c_3y + c_4xy + c_5x^2 + c_6y^2 \tag{8.18}$$

with six constants can be used to construct an element with six nodes. For example, a triangle with a node at each vertex and a node at the midpoint of each side is admissible. It is known as the quadratic triangular element. Examples of three-, four-, five-, and six-node elements are shown in Fig. 8.2.

Here we derive the linear interpolation functions for the three-node triangular element and the four-node rectangular element. The procedure used is the same as that for one-dimensional elements. The interpolation functions for linear, quadratic and cubic triangular and rectangular elements can be found in many books, and the present study is intended to illustrate the procedure used to derive the interpolation functions. Additional discussion on the generation of interpolation functions is presented in Chapter 9.

LINEAR TRIANGULAR ELEMENT. Consider the linear approximation (8.15). The set $\{1, x, y\}$ is linearly independent and complete. We must rewrite the approximation (8.15) such that it satisfies the conditions

$$U^e(x_i^e, y_i^e) = u_i^e \quad (i = 1, 2, 3) \tag{8.19}$$

where (x_i^e, y_i^e) $(i = 1, 2, 3)$ are the global coordinates of the three vertices of the triangle Ω^e. In other words, we determine the three constants c_i in (8.15) in

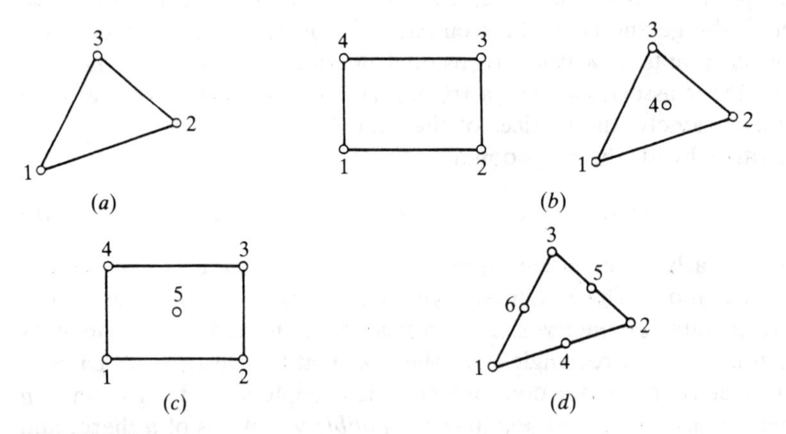

FIGURE 8.2
Finite elements in two dimensions: (*a*) a three-node element; (*b*) four-node elements; (*c*) a five-node element; (*d*) a six-node element.

terms of u_i^e from (8.19):

$$
\begin{aligned}
u_1 &\equiv u(x_1, y_1) = c_1 + c_2 x_1 + c_3 y_1 \\
u_2 &\equiv u(x_2, y_2) = c_1 + c_2 x_2 + c_3 y_2 \\
u_3 &\equiv u(x_3, y_3) = c_1 + c_2 x_3 + c_3 y_3
\end{aligned}
\tag{8.20}
$$

where the element label e on the us, xs, ys, and cs is omitted for simplicity. Throughout the following discussion, this format will be followed. In matrix form, we have

$$
\begin{Bmatrix} u_1 \\ u_2 \\ u_3 \end{Bmatrix} = \begin{bmatrix} 1 & x_1 & y_1 \\ 1 & x_2 & y_2 \\ 1 & x_3 & y_3 \end{bmatrix} \begin{Bmatrix} c_1 \\ c_2 \\ c_3 \end{Bmatrix}
\tag{8.21}
$$

Solution of (8.21) for c_i ($i = 1, 2, 3$) requires inversion of the coefficient matrix in (8.21). The inverse does not exist when any two rows or columns of (8.21) are the same. This happens only when all three nodes lie on the same line. Thus, in theory, as long as the three vertices of the triangle are distinct and do not lie on a line, the coefficient matrix is invertible. However, in actual computations, if any two of the three nodes are *very close* relative to the third, or the three nodes are almost on the same line, the coefficient matrix can be *nearly singular* and numerically noninvertible. Hence, one should avoid elements with narrow geometries (see Fig. 8.3) in finite element meshes.

Inverting the coefficient matrix in (8.21), we obtain

$$
[A]^{-1} = \frac{1}{2A_e} \begin{bmatrix} \alpha_1 & \alpha_2 & \alpha_3 \\ \beta_1 & \beta_2 & \beta_3 \\ \gamma_1 & \gamma_2 & \gamma_3 \end{bmatrix}, \qquad 2A_e = \alpha_1 + \alpha_2 + \alpha_3
\tag{8.22}
$$

and, solving for c_i in terms of u_i,

$$
\{c\} = [A]^{-1}\{u\},
$$

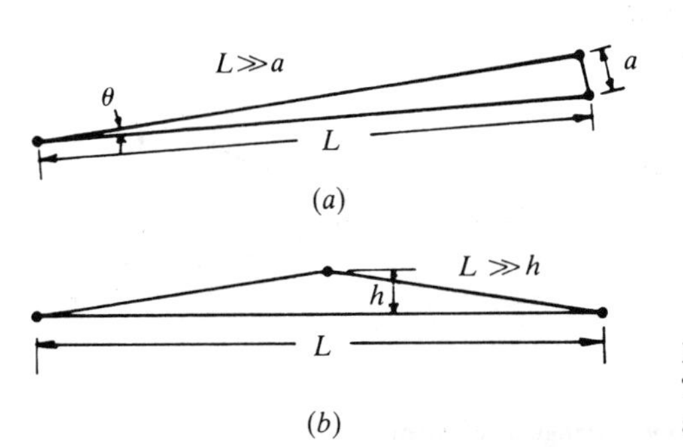

FIGURE 8.3
Triangular geometries that should be avoided in finite element meshes.

we obtain

$$c_1 = \frac{1}{2A_e}(\alpha_1 u_1 + \alpha_2 u_2 + \alpha_3 u_3)$$

$$c_2 = \frac{1}{2A_e}(\beta_1 u_1 + \beta_2 u_2 + \beta_3 u_3) \tag{8.23a}$$

$$c_3 = \frac{1}{2A_e}(\gamma_1 u_1 + \gamma_2 u_2 + \gamma_3 u_3),$$

where A_e is the area of the triangle (or $2A_e$ is the determinant of $[A]$), and α_i, β_i, and γ_i are the geometric constants

$$\left.\begin{aligned} \alpha_i &= x_j y_k - x_k y_j \\ \beta_i &= y_j - y_k \\ \gamma_i &= -(x_j - x_k) \end{aligned}\right\} \quad (i \neq j \neq k;\ \text{and } i,\ j \text{ and } k \text{ permute in a natural order}) \tag{8.23b}$$

Substituting for c_i from (8.23*a*) into (8.15), we obtain

$$\begin{aligned} U^e(x, y) &= \frac{1}{2A_e}[(\alpha_1 u_1 + \alpha_2 u_2 + \alpha_3 u_3) + (\beta_1 u_1 + \beta_2 u_2 + \beta_3 u_3)x \\ &\quad + (\gamma_1 u_1 + \gamma_2 u_2 + \gamma_3 u_3)y] \\ &= \sum_{i=1}^{3} u_i^e \psi_i^e(x, y) \end{aligned} \tag{8.24}$$

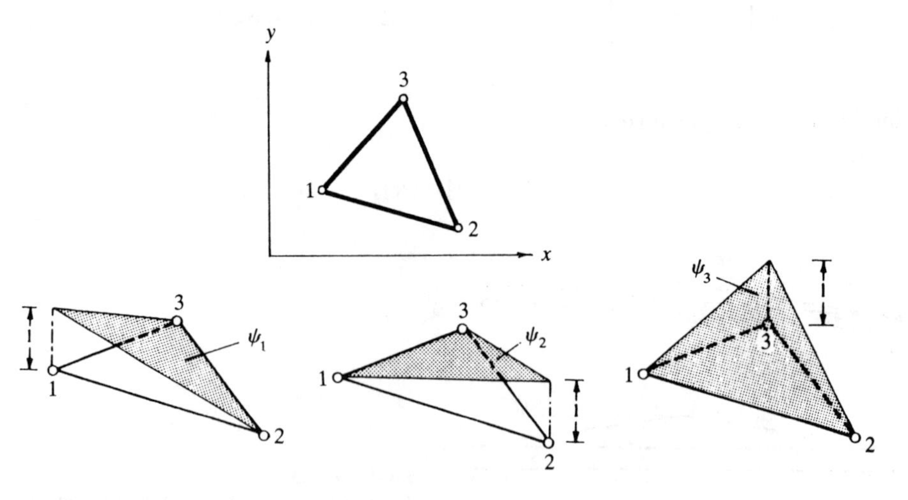

FIGURE 8.4
Linear interpolation functions for the three-node triangular element.

where ψ_i^e are the linear interpolation functions for the triangular element,

$$\boxed{\psi_i^e = \frac{1}{2A_e}(\alpha_i^e + \beta_i^e x + \gamma_i^e y) \quad (i = 1, 2, 3)} \tag{8.25}$$

and α_i, β_i, and γ_i are constants defined in (8.23*b*). The linear interpolation functions ψ_i^e are shown in Fig. 8.4. They have the properties

$$\begin{gathered}\psi_i^e(x_j^e, y_j^e) = \delta_{ij} \quad (i, j = 1, 2, 3) \\ \sum_{i=1}^{3} \psi_i^e = 1, \qquad \sum_{i=1}^{3} \frac{\partial \psi_i^e}{\partial x} = 0, \qquad \sum_{i=1}^{3} \frac{\partial \psi_i^e}{\partial y} = 0\end{gathered} \tag{8.26}$$

Note that (8.24) determines a plane surface passing through u_1, u_2, and u_3. Hence, use of the linear interpolation functions ψ_i^e of a triangle will result in the approximation of the curved surface $u(x, y)$ by a planar function $U^e = \sum_{i=1}^{3} u_i^e \psi_i^e$ as shown in Fig. 8.5. We consider an example of the computation of ψ_i^e.

Example 8.1. Consider the triangular element shown in Fig. 8.6. Let

$$U(x, y) = c_1 + c_2 x + c_3 y = \{1 \quad x \quad y\} \begin{Bmatrix} c_1 \\ c_2 \\ c_3 \end{Bmatrix}$$

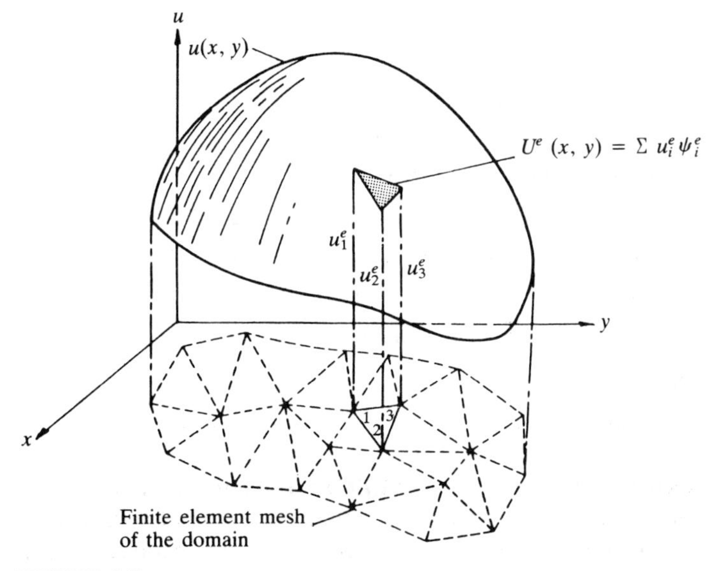

FIGURE 8.5
Representation of a continuous function $u(x, y)$ by linear interpolation functions of three-node triangular elements.

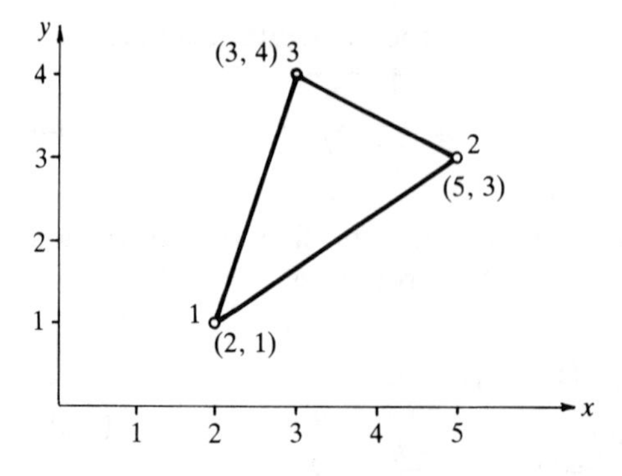

FIGURE 8.6
A triangular element with element nodes and coordinates (for Example 8.1).

Evaluating this polynomial at nodes 1, 2, and 3, we obtain the equations, from (8.21)

$$\begin{Bmatrix} u_1 \\ u_2 \\ u_3 \end{Bmatrix} = \begin{bmatrix} 1 & 2 & 1 \\ 1 & 5 & 3 \\ 1 & 3 & 4 \end{bmatrix} \begin{Bmatrix} c_1 \\ c_2 \\ c_3 \end{Bmatrix}, \quad \begin{Bmatrix} c_1 \\ c_2 \\ c_3 \end{Bmatrix} = [A]^{-1} \begin{Bmatrix} u_1 \\ u_2 \\ u_3 \end{Bmatrix}$$

where

$$[A]^{-1} = \frac{1}{7} \begin{bmatrix} 11 & -5 & 1 \\ -1 & 3 & -2 \\ -2 & -1 & 3 \end{bmatrix}$$

Substituting into U, we obtain ($[A]^{-1} = [A^*]$)

$$U^e(x, y) = \{1 \quad x \quad y\}[A^*] \begin{Bmatrix} u_1 \\ u_2 \\ u_3 \end{Bmatrix} = \{\psi_1 \quad \psi_2 \quad \psi_3\} \begin{Bmatrix} u_1 \\ u_2 \\ u_3 \end{Bmatrix} = \sum_{i=1}^{3} \psi_i^e u_i^e$$

where

$$\psi_i^e = A_{1i}^* + A_{2i}^* x + A_{3i}^* y$$

and A_{ij}^* are the elements of the inverse matrix $[A]^{-1}$.

Alternatively, from the definitions (8.23*b*), we have

$$\alpha_1 = 5 \times 4 - 3 \times 3 = 11, \quad \alpha_2 = 3 \times 1 - 2 \times 4 = -5, \quad \alpha_3 = 2 \times 3 - 5 \times 1 = 1$$
$$\beta_1 = 3 - 4 = -1, \quad \beta_2 = 4 - 1 = 3, \quad \beta_3 = 1 - 3 = -2$$
$$\gamma_1 = -(5 - 3) = -2, \quad \gamma_2 = -(3 - 2) = -1, \quad \gamma_3 = -(2 - 5) = 3$$
$$2A_e = \alpha_1 + \alpha_2 + \alpha_3 = 7$$

The interpolation functions are

$$\psi_1^e = \tfrac{1}{7}(11 - x - 2y), \quad \psi_2^e = \tfrac{1}{7}(-5 + 3x - y), \quad \psi_3^e = \tfrac{1}{7}(1 - 2x + 3y)$$

LINEAR (FOUR-NODE) RECTANGULAR ELEMENT. Here we consider an approximation of the form (8.16) and use a rectangular element with sides a and b (see Fig. 8.7*a*). For the sake of convenience, we choose a local coordinate system $(\bar{x}, \bar{y})$ to derive the interpolation functions. We assume that

$$U(\bar{x}, \bar{y}) = c_1 + c_2\bar{x} + c_3\bar{y} + c_4\bar{x}\bar{y} \tag{8.27}$$

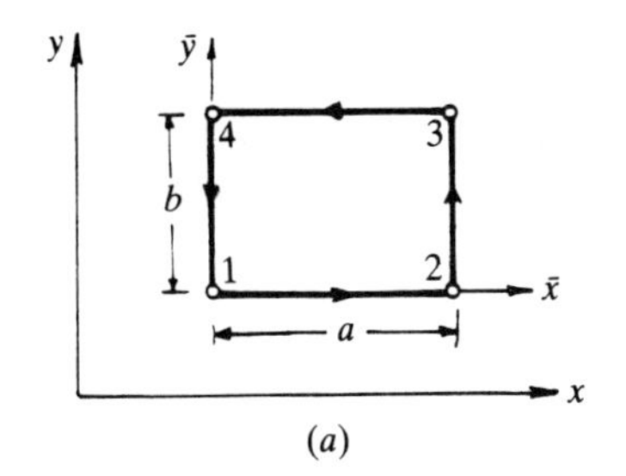

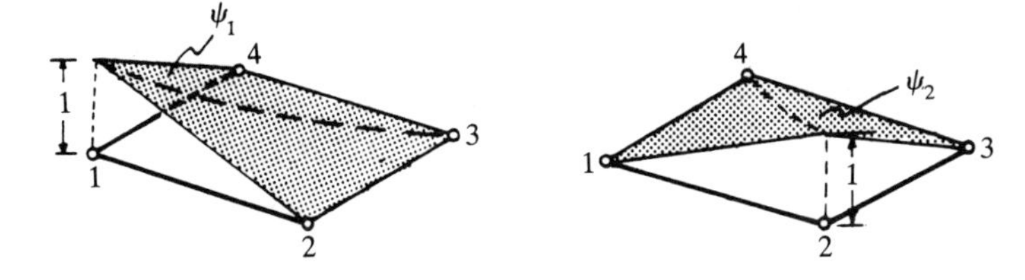

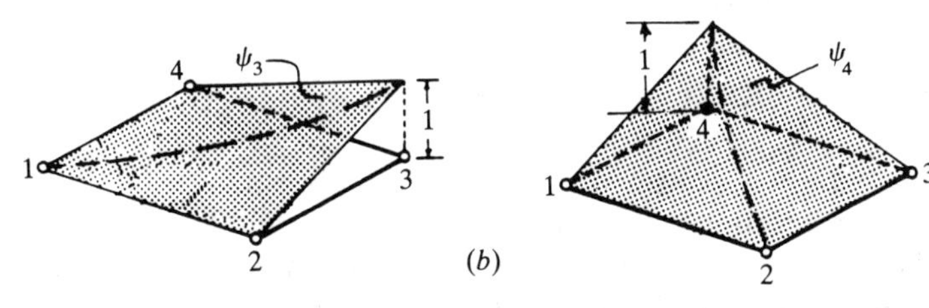

FIGURE 8.7
Linear rectangular element and its interpolation functions: (*a*) geometry of the element; (*b*) interpolation functions.

and require

$$
\begin{aligned}
u_1 &= U(0, 0) = c_1 \\
u_2 &= U(a, 0) = c_1 + c_2 a \\
u_3 &= U(a, b) = c_1 + c_2 a + c_3 b + c_4 ab \\
u_4 &= U(0, b) = c_1 + c_3 b
\end{aligned}
\tag{8.28}
$$

Solving for c_i ($i = 1, \ldots, 4$), we obtain

$$
\begin{aligned}
c_1 &= u_1, \qquad c_2 = \frac{u_2 - u_1}{a} \\
c_3 &= \frac{u_4 - u_1}{b}, \qquad c_4 = \frac{u_3 - u_4 + u_1 - u_2}{ab},
\end{aligned}
\tag{8.29}
$$

Substituting these into (8.27), we obtain

$$U(\bar{x}, \bar{y}) = u_1\left(1 - \frac{\bar{x}}{a} - \frac{\bar{y}}{b} + \frac{\bar{x}}{a}\frac{\bar{y}}{b}\right) + u_2\left(\frac{\bar{x}}{a} - \frac{\bar{x}}{a}\frac{\bar{y}}{b}\right) + u_3\frac{\bar{x}}{a}\frac{\bar{y}}{b} + u_4\left(\frac{\bar{y}}{b} - \frac{\bar{x}}{a}\frac{\bar{y}}{b}\right)$$
$$= u_1\psi_1 + u_2\psi_2 + u_3\psi_3 + u_4\psi_4 = \sum_{i=1}^{4} u_i^e\psi_i^e \tag{8.30}$$

where

$$\boxed{\begin{aligned} \psi_1^e &= \left(1 - \frac{\bar{x}}{a}\right)\left(1 - \frac{\bar{y}}{b}\right), \quad \psi_2^e = \frac{\bar{x}}{a}\left(1 - \frac{\bar{y}}{b}\right) \\ \psi_3^e &= \frac{\bar{x}}{a}\frac{\bar{y}}{b}, \quad \psi_4^e = \left(1 - \frac{\bar{x}}{a}\right)\frac{\bar{y}}{b} \end{aligned}} \tag{8.31a}$$

or, in concise form,

$$\psi_i^e(\bar{x}, \bar{y}) = (-1)^{i+1}\left(1 - \frac{\bar{x} + \bar{x}_i}{a}\right)\left(1 - \frac{\bar{y} + \bar{y}_i}{b}\right) \tag{8.31b}$$

where $(\bar{x}_i, \bar{y}_i)$ are the $(\bar{x}, \bar{y})$ coordinates of node i. The interpolation functions are shown in Fig. 8.7(b). We again have

$$\psi_i^e(\bar{x}_j, \bar{y}_j) = \delta_{ij} \quad (i, j = 1, \ldots, 4), \qquad \sum_{i=1}^{4} \psi_i^e = 1 \tag{8.32a,b}$$

where $(\bar{x}_j, \bar{y}_j)$ are the coordinates of node j in the $(\bar{x}, \bar{y})$ system.

Note that the linear interpolation functions for the four-node rectangular element can also be obtained by taking the tensor product of the 1-D linear interpolation functions (3.18) associated with sides 1–2 and 2–3:

$$\begin{Bmatrix} 1 - \dfrac{\bar{x}}{a} \\ \dfrac{\bar{x}}{a} \end{Bmatrix} \begin{Bmatrix} 1 - \dfrac{\bar{y}}{b} & \dfrac{\bar{y}}{b} \end{Bmatrix} = \begin{bmatrix} \psi_1 & \psi_4 \\ \psi_2 & \psi_3 \end{bmatrix} \tag{8.33}$$

The procedure given above for the construction of the Lagrange interpolation functions involves the inversion of an $n \times n$ matrix, where n is the number of nodes in the element. When n is large, the inversion becomes very tedious. The alternative procedure discussed in Chapter 3 for one-dimensional elements proves to be algebraically simple. Here we illustrate the alternative procedure for the four-node rectangular element. Equation (8.32a) requires that

$$\psi_1^e(\bar{x}_i, \bar{y}_i) = 0 \quad (i = 2, 3, 4), \qquad \psi_1^e(x_1, y_1) = 1$$

That is, ψ_1^e is identically zero on lines $\bar{x} = a$ and $\bar{y} = b$. Hence, $\psi_1^e(\bar{x}, \bar{y})$ must be of the form

$$\psi_1^e(\bar{x}, \bar{y}) = c_1(a - \bar{x})(b - \bar{y})$$

Using the condition $\psi_1^e(x_1, y_1) = \psi_1^e(0, 0) = 1$, we obtain $c_1 = 1/ab$. Hence,

$$\psi_1^e(\bar{x}, \bar{y}) = \frac{1}{ab}(a - \bar{x})(b - \bar{y}) = \left(1 - \frac{\bar{x}}{a}\right)\left(1 - \frac{\bar{y}}{b}\right)$$

Likewise, one can obtain the remaining three interpolation functions.

8.2.6 Evaluation of Element Matrices, and Vectors

The exact evaluation of the element matrices $[K^e]$ and $\{f^e\}$ in (8.14*b*) is, in general, not easy. Therefore, they are evaluated using numerical integration techniques described in Section 7.8.5. However, when a_{ij}, a_{00}, and f are element-wise constant, it is possible to evaluate the integrals exactly over the linear triangular and rectangular elements discussed in the previous section. The boundary integral in $\{Q^e\}$ of (8.14*b*) can be evaluated whenever q_n is known. For an interior element (i.e., one that does not have any of its sides on the boundary of the problem), the contribution from the boundary integral cancels with similar contributions from adjoining elements of the mesh (analogous to the Q_i^e in one-dimensional problems). A more detailed discussion is given below.

For the sake of brevity, we rewrite $[K^e]$ in (8.14) as the sum of four basic matrices $[S^{\alpha\beta}]$ $(\alpha, \beta = 0, 1, 2)$:

$$[K^e] = a_{00}[S^{00}] + a_{11}[S^{11}] + a_{12}[S^{12}] + a_{21}[S^{12}]^T + a_{22}[S^{22}] \tag{8.34}$$

where $[\ ]^T$ denotes the transpose of the enclosed matrix, and

$$S_{ij}^{\alpha\beta} = \int_{\Omega^e} \psi_{i,\alpha}\psi_{j,\beta}\, dx\, dy \tag{8.35}$$

with $\psi_{i,\alpha} \equiv \partial\psi_i/\partial x_\alpha$, $x_1 = x$, and $x_2 = y$; $\psi_{i,0} = \psi_i$. All matrices in (8.34) and interpolation functions in (8.35) are understood to be defined over an element; i.e., all expressions and quantities should have the element label *e*, but these are omitted in the interest of brevity. We now proceed to compute the matrices in (8.34) and (8.14*b*) using the linear interpolation functions derived in the previous section.

ELEMENT MATRICES FOR A LINEAR TRIANGULAR ELEMENT. For a triangle, the following exact integral formulae are available for evaluating the integrals. Let

$$I_{mn} \equiv \int_\Delta x^m y^n\, dx\, dy \tag{8.36}$$

Then we have (the element label on A is omitted)

$$I_{00} = A \quad \text{(area of the triangle)}$$

$$I_{10} = A\hat{x}, \qquad \hat{x} = \tfrac{1}{3}\sum_{i=1}^{3} x_i, \qquad I_{01} = A\hat{y}, \qquad \hat{y} = \tfrac{1}{3}\sum_{i=1}^{3} y_i$$

$$I_{11} = \frac{A}{12}\left(\sum_{i=1}^{3} x_i y_i + 9\hat{x}\hat{y}\right), \qquad I_{20} = \frac{A}{12}\left(\sum_{i=1}^{3} x_i^2 + 9\hat{x}^2\right), \qquad I_{02} = \frac{A}{12}\left(\sum_{i=1}^{3} y_i^2 + 9\hat{y}^2\right) \tag{8.37}$$

Using the linear interpolation functions (8.25) in (8.35), and noting that

$$\frac{\partial \psi_i}{\partial x} = \frac{\beta_i}{2A}, \qquad \frac{\partial \psi_i}{\partial y} = \frac{\gamma_i}{2A} \tag{8.38}$$

we obtain

$$S_{ij}^{11} = \frac{1}{4A}\beta_i\beta_j, \qquad S_{ij}^{12} = \frac{1}{4A}\beta_i\gamma_j, \qquad S_{ij}^{22} = \frac{1}{4A}\gamma_i\gamma_j$$

$$S_{ij}^{00} = \frac{1}{4A}\Big\{[\alpha_i\alpha_j + (\alpha_i\beta_j + \alpha_j\beta_i)\hat{x} + (\alpha_i\gamma_j + \alpha_j\gamma_i)\hat{y}] \\ + \frac{1}{A}[I_{20}\beta_i\beta_j + I_{11}(\gamma_i\beta_j + \gamma_j\beta_i) + I_{02}\gamma_i\gamma_j]\Big\} \tag{8.39}$$

In view of the identity $\alpha_i + \beta_i\hat{x} + \gamma_i\hat{y} = \tfrac{2}{3}A$ [which follows from (8.23b) and (8.37)], for an element-wise-constant value of $f = f_e$, we have

$$f_i^e = \tfrac{1}{2}f_e(\alpha_i + \beta_i\hat{x} + \gamma_i\hat{y}) = \tfrac{1}{3}f_e A_e \tag{8.40}$$

This should be obvious, because for a constant source f_e, the total magnitude of the source on the element is equal to $f_e A_e$, which is distributed equally among the nodes. Once the coordinates of the element nodes are known, one can compute α_i, β_i, and γ_i from (8.23b) and substitute into (8.39) to obtain the element matrices, which in turn can be used in (8.34) to obtain the element matrix $[K^e]$. For example, when a_{12}, a_{21}, and a_{00} are zero, and a_{11} and a_{22} are element–wise-constant, we have

$$\boxed{K_{ij}^e = \frac{1}{4A_e}(a_{11}^e\beta_i^e\beta_j^e + a_{22}^e\gamma_i^e\gamma_j^e)} \tag{8.41}$$

ELEMENT MATRICES FOR A LINEAR RECTANGULAR ELEMENT. When a_{ij} $(i, j = 0, 1, 2)$ and f are constants, we can use the interpolation functions of (8.31a) expressed in the local coordinates $\bar{x}$ and $\bar{y}$, which are related to the global coordinates by

$$x = \bar{x} + x_1^e, \qquad y = \bar{y} + y_1^e, \qquad dx = d\bar{x}, \qquad dy = d\bar{y} \tag{8.42}$$

where (x_1^e, y_1^e) are the global coordinates of node 1 of the element Ω^e with respect to the global (x, y) coordinate system. For example, we have

$$S_{ij}^{00} = \int_{x_1^e}^{x_1^e+a} \int_{y_1^e}^{y_1^e+b} \psi_i \psi_j \, dx\, dy = \int_0^a \int_0^b \psi_i \psi_j \, d\bar{x}\, d\bar{y}$$

where a and b are the lengths along the $\bar{x}$ and $\bar{y}$ axes of the element. Consider the coefficient

$$\begin{aligned} S_{11}^{00} &= \int_0^a \int_0^b \psi_1 \psi_1 \, dx\, dy = \int_0^a \int_0^b \left(1 - \frac{\bar{x}}{a}\right)\left(1 - \frac{\bar{y}}{b}\right)\left(1 - \frac{\bar{x}}{a}\right)\left(1 - \frac{\bar{y}}{b}\right) d\bar{x}\, d\bar{y} \\ &= \int_0^a \left(1 - \frac{\bar{x}}{a}\right)^2 d\bar{x} \int_0^b \left(1 - \frac{\bar{y}}{b}\right)^2 d\bar{y} = \frac{a}{3}\frac{b}{3} = \frac{ab}{9} \end{aligned}$$

Similarly, we can evaluate all the matrices $[S^{\alpha\beta}]$ with the aid of the integral identities

$$\begin{aligned} &\int_0^a \left(1 - \frac{\bar{x}}{a}\right)^2 d\bar{x} = \tfrac{1}{3}a, \qquad \int_0^a \frac{\bar{x}}{a}\left(1 - \frac{\bar{x}}{a}\right) d\bar{x} = \tfrac{1}{6}a \\ &\int_0^a \left(1 - \frac{\bar{x}}{a}\right) d\bar{x} = \tfrac{1}{2}a, \qquad \int_0^a \frac{\bar{x}}{a}\, d\bar{x} = \tfrac{1}{2}a \end{aligned} \tag{8.43}$$

We have

$$[S^{11}] = \frac{b}{6a}\begin{bmatrix} 2 & -2 & -1 & 1 \\ -2 & 2 & 1 & -1 \\ -1 & 1 & 2 & -2 \\ 1 & -1 & -2 & 2 \end{bmatrix}, \qquad [S^{12}] = \frac{1}{4}\begin{bmatrix} 1 & 1 & -1 & -1 \\ -1 & -1 & 1 & 1 \\ -1 & -1 & 1 & 1 \\ 1 & 1 & -1 & -1 \end{bmatrix}$$

$$[S^{22}] = \frac{a}{6b}\begin{bmatrix} 2 & 1 & -1 & -2 \\ 1 & 2 & -2 & -1 \\ -1 & -2 & 2 & 1 \\ -2 & -1 & 1 & 2 \end{bmatrix}, \qquad [S^{00}] = \frac{ab}{36}\begin{bmatrix} 4 & 2 & 1 & 2 \\ 2 & 4 & 2 & 1 \\ 1 & 2 & 4 & 2 \\ 2 & 1 & 2 & 4 \end{bmatrix} \tag{8.44}$$

$$\{f\} = \tfrac{1}{4}fab\{1 \quad 1 \quad 1 \quad 1\}^T$$

EVALUATION OF THE BOUNDARY INTEGRALS. Here we consider the evaluation of boundary integrals of the type

$$Q_i^e = \oint_{\Gamma^e} q_n^e \psi_i^e(s)\, ds \tag{8.45}$$

where q_n^e is a known function of the distance s along the boundary Γ^e. It is not necessary to compute such integrals when a portion of Γ^e does not coincide with the boundary Γ of the total domain Ω. On portions of Γ^e that are in the interior of the domain Ω, q_n^e on side (i, j) of the element Ω^e cancels with q_n^f on side (p, q) of the element Ω^f when sides (i, j) of Ω^e and (p, q) of Ω^f are the same (i.e., at the interface of Ω^e and Ω^f). This can be viewed as the equilibrium of the internal "flux" (see Figs. 8.8b,c). When Γ^e falls on the

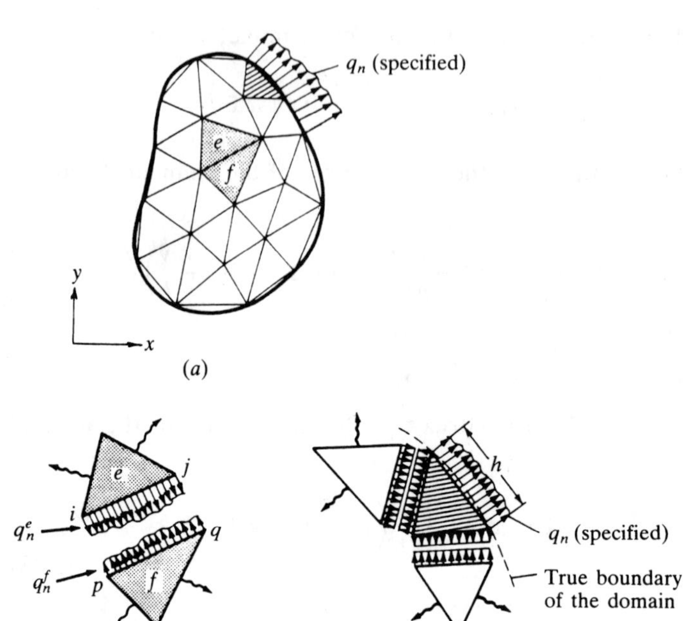

FIGURE 8.8
Computation of boundary forces and equilibrium of secondary variables at interelement boundaries: (*a*) finite element discretization; (*b*) equilibrium of forces at interfaces; (*c*) computation of forces on the true boundary.

boundary of the domain Ω, q_n^e is either known, in general, as a function of s, or is to be determined in the post-computation. In the latter case, the primary variable will be specified on the portion of the boundary where q_n is not specified.

The boundary Γ^e of linear two-dimensional elements is a set of linear one-dimensional elements. Therefore, the evaluation of the boundary integrals in two-dimensional problems amounts to evaluating line integrals. It should not be surprising to the reader that when two-dimensional interpolation functions are evaluated on the boundary, we obtain the corresponding one-dimensional interpolation functions. For example, consider the linear triangular element shown in Fig. 8.9. The linear interpolation functions for this element are given by (8.25). Now let us choose a coordinate system (s, t) with its origin at node 1 and the coordinate s parallel to the side connecting nodes 1 and 2. The two coordinate systems (x, y) and (s, t) are related by

$$x = a_1 + b_1 s + c_1 t$$

$$y = a_2 + b_2 s + c_2 t$$

The constants a_1, b_1, c_1, a_2, b_2, and c_2 can be determined with the following conditions:

$$\text{when } s=0,\ t=0, \quad x=x_1,\ y=y_1$$
$$\text{when } s=a,\ t=0, \quad x=x_2,\ y=y_2$$
$$\text{when } s=c,\ t=b, \quad x=x_3,\ y=y_3$$

We obtain

$$\begin{aligned} x(s,t) &= x_1+(x_2-x_1)\frac{s}{a}+\left[\left(\frac{c}{a}-1\right)x_1-\frac{c}{a}x_2+x_3\right]\frac{t}{b} \\ y(s,t) &= y_1+(y_2-y_1)\frac{s}{a}+\left[\left(\frac{c}{a}-1\right)y_1-\frac{c}{a}y_2+y_3\right]\frac{t}{b} \end{aligned} \tag{8.46}$$

These equations allow us to express $\psi_i(x, y)$ as $\psi_i(s, t)$, which can be evaluated on the side connecting nodes 1 and 2 by setting $t=0$ in $\psi_i(s, t)$:

$$\psi_i(s) \equiv \psi_i(s,0) = \psi_i(x(s,0), y(s,0))$$
$$x(s) = x_1+(x_2-x_1)\frac{s}{a}, \qquad y(s) = y_1+(y_2-y_1)\frac{s}{a}$$

For instance,

$$\begin{aligned} \psi_1(s) &= \frac{1}{2A}\left\{\alpha_1+\beta_1\left[\left(1-\frac{s}{a}\right)x_1+\frac{s}{a}x_2\right]+\gamma_1\left[\left(1-\frac{s}{a}\right)y_1+\frac{s}{a}y_2\right]\right\} \\ &= \frac{1}{2A}(\alpha_1+\alpha_2+\alpha_3)\left(1-\frac{s}{a}\right) = 1-\frac{s}{a} \end{aligned}$$

where the definitions of α_1, β_1, and γ_1 have been used to rewrite the entire expression. Similarly,

$$\psi_2(s) = \frac{s}{a}, \qquad \psi_3(s) = 0$$

where $a = h_{12}$ is the length of side 1–2. We note that $\psi_1(s)$ and $\psi_2(s)$ are precisely the linear one-dimensional interpolation functions associated with the line element connecting nodes 1 and 2. Similarly, when $\psi_i(x, y)$ are evaluated on side 3–1 of the element, we obtain

$$\psi_1(s) = \frac{s}{h_{13}}, \qquad \psi_2 = 0, \qquad \psi_3(s) = 1-\frac{s}{h_{13}}$$

where the s coordinate is taken along the side 3–1, with origin at node 3, and h_{13} is the length of side 1–3. Thus evaluation of Q_i^e involves the use of appropriate 1-D interpolation functions and the known variation of q_n on the boundary:

$$\begin{aligned} Q_i^e &= \int_{1-2} \psi_i(s)q_n(s)\,ds + \int_{2-3} \psi_i(s)q_n(s)\,ds + \int_{3-1} \psi_i(s)q_n(s)\,ds \\ &\equiv Q_{i1}^e + Q_{i2}^e + Q_{i3}^e \end{aligned} \tag{8.47}$$

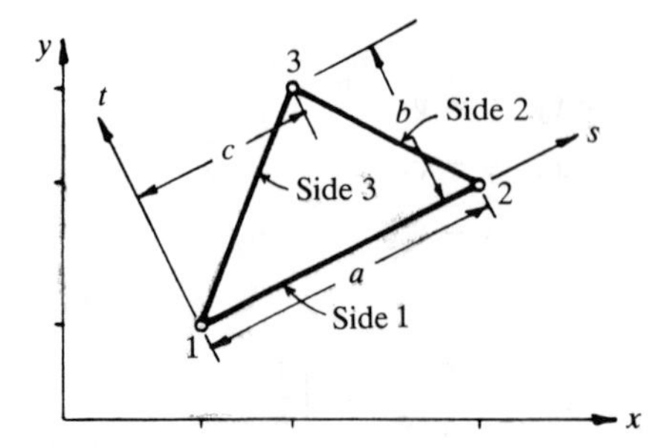

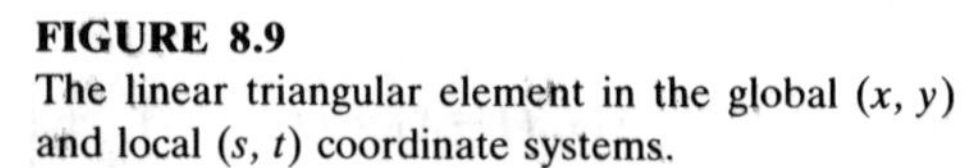
FIGURE 8.9
The linear triangular element in the global (x, y) and local (s, t) coordinate systems.

where $\int_{i-j}$ denotes the integral over the line connecting node i to node j, the s coordinate is taken from node i to node j, with origin at node i, and Q_{iJ}^e is defined as the contribution to Q_i^e from q_n^e on side J (see Fig. 8.9) of the element Ω^e:

$$Q_{iJ}^e = \int_{\text{side } J} \psi_i q_n \, ds \tag{8.48}$$

For example,

$$Q_1^e = \oint_{\Gamma^e} q_n \psi_1(s)\, ds = \int_{1-2} (q_n)_{1-2} \psi_1 \, ds + 0 + \int_{3-1} (q_n)_{3-1} \psi_1 \, ds$$

The contribution from side 2–3 is zero, because ψ_1 is zero on side 2–3 of a triangular element. For a rectangular element, Q_1^e has contributions from sides 1–2 and 4–1, because ψ_1 is zero on sides 2–3 and 3–4.

Example 8.2. Consider the evaluation of the boundary integral Q_i^e in (8.45) for the four cases of $q(s)$ and finite elements shown in Fig. 8.10. For each case, we must use the $q(s)$ and the interpolation functions associated with the type of boundary element (i.e., linear or quadratic).

Case 1. $q(s) = q_0 = \text{constant}$; linear element:

$$Q_i^e = \oint_{\Gamma^e} q_0 \psi_i \, ds = q_0 \int_0^{h_e} \psi_i \, ds + 0 + 0 \quad (i = 1, 2, 3)$$

where

$$\psi_1 = 1 - \frac{s}{h_e}, \qquad \psi_2 = \frac{s}{h_e}, \qquad \psi_3 = 0$$

We have

$$Q_1^e = \tfrac{1}{2} q_0 h_e \; (= Q_{11}^e), \qquad Q_2^e = \tfrac{1}{2} q_0 h_e \; (= Q_{21}^e), \qquad Q_3^e = 0$$

Case 2. $q(s) = q_0 s / h_e$ (linear variation); linear element:

$$Q_i^e = \oint_{\Gamma^e} q_0 \frac{s}{h_e} \psi_i \, ds = \frac{q_0}{h_e} \int_0^{h_e} s \psi_i \, ds \quad (i = 1, 2, 3)$$

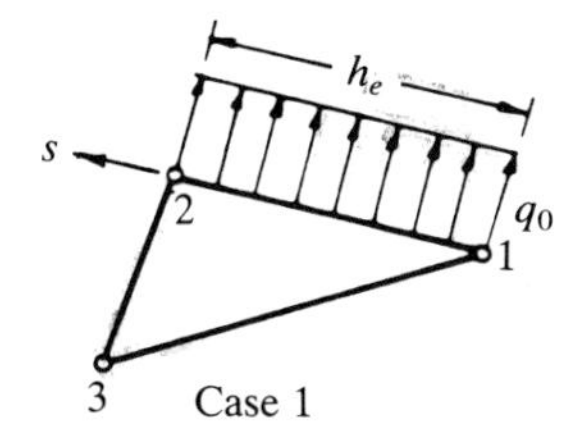

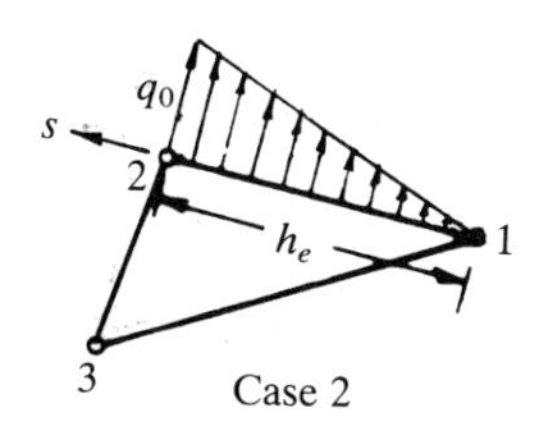

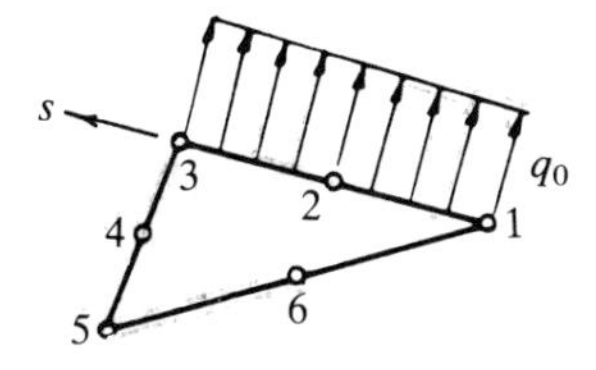

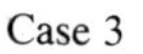

Case 4

FIGURE 8.10
Evaluation of boundary integrals in the finite element analysis (Example 8.2).

where

$$\psi_1 = 1 - \frac{s}{h_e}, \qquad \psi_2 = \frac{s}{h_e}, \qquad \psi_3 = 0$$

We have

$$Q_1^e = \tfrac{1}{6} q_0 h_e \ (= Q_{11}^e), \qquad Q_2^e = \tfrac{1}{3} q_0 h_e \ (= Q_{21}^e), \qquad Q_3^e = 0$$

Case 3. $q(s) = q_0 =$ constant; quadratic element:

$$Q_i^e = \oint_{\Gamma^e} q_0 \psi_i \, ds \quad (i = 1, 2, \ldots, 6)$$

$$\psi_1 = \left(1 - \frac{s}{h_e}\right)\left(1 - \frac{2s}{h_e}\right), \qquad \psi_2 = \frac{4s}{h_e}\left(1 - \frac{s}{h_e}\right), \qquad \psi_3 = -\frac{s}{h_e}\left(1 - \frac{2s}{h_e}\right)$$

and ψ_4, ψ_5, and ψ_6 are zero on side 1–2–3. We have

$$Q_1^e = \tfrac{1}{6} q_0 h_e \ (= Q_{11}^e), \qquad Q_2^e = \tfrac{4}{6} q_0 h_e \ (= Q_{21}^e), \qquad Q_3^e = \tfrac{1}{6} q_0 h_e \ (= Q_{31}^e)$$

Case 4. $q(s)$ as shown in Fig. 8.10; linear element:

$$Q_i^e = \oint_{\Gamma^e} q(s) \psi_i \, ds = \int_{1-2} q_0 \frac{s}{h_{12}} \psi_i \, ds + \int_{2-3} q_1 \psi_i \, ds + 0$$

$$= Q_{i1}^e + Q_{i2}^e \qquad (Q_{i3}^e = 0)$$

We obtain,

$$Q_1^e = \int_{1-2} q_0 \frac{s}{h_{12}}\left(1 - \frac{s}{h_{12}}\right) ds + 0 + 0 = \tfrac{1}{6} q_0 h_{12} \quad (= Q_{11}^e)$$

$$Q_2^e = \int_{1-2} q_0 \frac{s}{h_{12}} \frac{s}{h_{12}}\, ds + \int_{2-3} q_1\left(1 - \frac{s}{h_{23}}\right) ds + 0$$

$$= \tfrac{1}{3} q_0 h_{12} + \tfrac{1}{2} q_1 h_{23} \quad (= Q_{21}^e + Q_{22}^e)$$

$$Q_3^e = 0 + \int_{2-3} q_1 \frac{s}{h_{23}}\, ds + 0 = \tfrac{1}{2} q_1 h_{23} \quad (= Q_{32}^e)$$

8.2.7 Assembly of Element Equations

The assembly of finite element equations is based on the same two principles that were used in one-dimensional problems:

1. Continuity of primary variables
2. "Equilibrium" (or "balance") of secondary variables

We illustrate the procedure by considering a finite element mesh consisting of a triangular element and a quadrilateral element (see Fig. 8.11a). Let K_{ij}^1 $(i, j = 1, 2, 3)$ denote the coefficient matrix corresponding to the triangular element, and let K_{ij}^2 $(i, j = 1, \ldots, 4)$ denote the coefficient matrix corresponding to the quadrilateral element. From the finite element mesh shown in Fig. 8.11(a), we note the following correspondence (i.e., connectivity relations)

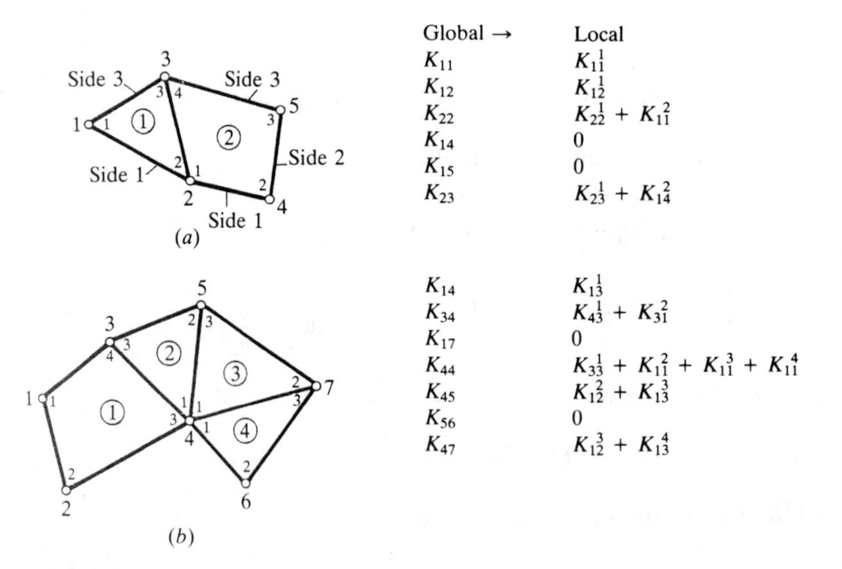

Global →
Local
K_{11} K_{11}^1
K_{12} K_{12}^1
K_{22} $K_{22}^1 + K_{11}^2$
K_{14} 0
K_{15} 0
K_{23} $K_{23}^1 + K_{14}^2$
Side 3
Side 3
Side 2
Side 1
Side 1
(a)
K_{14} K_{13}^1
K_{34} $K_{43}^1 + K_{31}^2$
K_{17} 0
K_{44} $K_{33}^1 + K_{11}^2 + K_{11}^3 + K_{11}^4$
K_{45} $K_{12}^2 + K_{13}^3$
K_{56} 0
K_{47} $K_{12}^3 + K_{13}^4$
(b)

FIGURE 8.11
Assembly of finite element coefficient matrices using the correspondence between global and element nodes (one unknown per node): (a) assembly of two elements; (b) assembly of several elements.

between the global and element nodes:

$$[B] = \begin{bmatrix} 1 & 2 & 3 & \times \\ 2 & 4 & 5 & 3 \end{bmatrix} \tag{8.49}$$

where $\times$ indicates that there is no entry. The correspondence between the local and global nodal values is (see Fig. 8.11*a*)

$$u_1^1 = U_1, \quad u_2^1 = u_1^2 = U_2, \quad u_3^1 = u_4^2 = U_3, \quad u_2^2 = U_4, \quad u_3^2 = U_5 \tag{8.50}$$

which amounts to imposing the continuity of the primary variables at the nodes common to elements 1 and 2.

Note that the continuity of the primary variables at the interelement nodes guarantees the continuity of the primary variable along the entire interelement boundary. For the case in Fig. 8.11(*a*), the requirement

$$u_2^1 = u_1^2 \text{ and } u_3^1 = u_4^2 \quad \text{guarantees} \quad U^1(s) = U^2(s)$$

on the side connecting global nodes 2 and 3. This can be shown as follows. The solution $U^1(s)$ along the line connecting global nodes 2 and 3 is linear, and is given by

$$U^1(s) = u_2^1\left(1 - \frac{s}{h}\right) + u_3^1 \frac{s}{h}$$

where s is the local coordinate along side 2–3 with its origin at global node 2 and h is the length of side 2–3 (or side 2). Similarly, the finite element solution along the same line but from element 2 is

$$U^2(s) = u_1^2\left(1 - \frac{s}{h}\right) + u_4^2 \frac{s}{h}$$

Since $u_1^2 = u_2^1$ and $u_4^2 = u_3^1$, it follows that $U^1(s) = U^2(s)$ for every value of s along the interface of the two elements.

Next we use the balance of secondary variables. At the interface between the two elements, the fluxes from them should be equal in magnitude and opposite in sign. For the two elements in Fig. 8.11(*a*), the interface is along the side connecting global nodes 2 and 3. Hence, the internal flux q_n^1 on side 2–3 of element 1 should balance the flux q_n^2 on side 4–1 of element 2 (recall the sign convention on q_n^e):

$$(q_n^1)_{2\text{–}3} = (q_n^2)_{4\text{–}1} \quad \text{or} \quad (q_n^1)_{2\text{–}3} = (-q_n^2)_{1\text{–}4} \tag{8.51}$$

In the finite element method, we impose the above relation in a weighted-integral sense:

$$\int_{h_{23}^1} q_n^1 \psi_2^1 \, ds = -\int_{h_{14}^2} q_n^2 \psi_1^2 \, ds, \quad \int_{h_{23}^1} q_n^1 \psi_3^1 \, ds = -\int_{h_{14}^2} q_n^2 \psi_4^2 \, ds \tag{8.52a}$$

where h^e_{pq} denotes the length of the side connecting node p to node q of the element Ω^e. The above equations can be written in the form

$$\int_{h^1_{23}} q^1_n \psi^1_2 \, ds + \int_{h^2_{14}} q^2_n \psi^2_1 \, ds = 0, \qquad \int_{h^1_{23}} q^1_n \psi^1_3 \, ds + \int_{h^2_{14}} q^2_n \psi^2_4 \, ds = 0 \tag{8.52b}$$

or

$$Q^1_{22} + Q^2_{14} = 0, \qquad Q^1_{32} + Q^2_{44} = 0 \tag{8.52c}$$

where Q^e_{iJ} denotes the part of Q^e_i that comes from side J of element e [see (8.48)]:

$$Q^e_{iJ} = \int_{\text{side } J} q^e_n \psi^e_i \, ds$$

The sides of triangular and rectangular elements are numbered as shown in Fig. 8.11. These balance relations must be imposed in assembling the element equations. We note that Q^e_{iJ} is only a portion of Q^e_i [see (8.47)].

The element equations of the two elements are written first. For the model problem at hand, there is only one primary degree of freedom (NDF = 1) per node. For the triangular element, the element equations are of the form

$$\begin{aligned}
K^1_{11}u^1_1 + K^1_{12}u^1_2 + K^1_{13}u^1_3 &= f^1_1 + Q^1_1 \\
K^1_{21}u^1_1 + K^1_{22}u^1_2 + K^1_{23}u^1_3 &= f^1_2 + Q^1_2 \\
K^1_{31}u^1_1 + K^1_{32}u^1_2 + K^1_{33}u^1_3 &= f^1_3 + Q^1_3
\end{aligned} \tag{8.53a}$$

For the rectangular element, the element equations are given by

$$\begin{aligned}
K^2_{11}u^2_1 + K^2_{12}u^2_2 + K^2_{13}u^2_3 + K^2_{14}u^2_4 &= f^2_1 + Q^2_1 \\
K^2_{21}u^2_1 + K^2_{22}u^2_2 + K^2_{23}u^2_3 + K^2_{24}u^2_4 &= f^2_2 + Q^2_2 \\
K^2_{31}u^2_1 + K^2_{32}u^2_2 + K^2_{33}u^2_3 + K^2_{34}u^2_4 &= f^2_3 + Q^2_3 \\
K^2_{41}u^2_1 + K^2_{42}u^2_2 + K^2_{43}u^2_3 + K^2_{44}u^2_4 &= f^2_4 + Q^2_4
\end{aligned} \tag{8.53b}$$

In order to impose the balance of secondary variables in (8.52), we must add the second equation of element 1 to the first equation of element 2, and also add the third equation of element 1 to the fourth equation of element 2:

$$\begin{aligned}
&(K^1_{21}u^1_1 + K^1_{22}u^1_2 + K^1_{23}u^1_3) + (K^2_{11}u^2_1 + K^2_{12}u^2_2 + K^2_{13}u^2_3 + K^2_{14}u^2_4) \\
&\qquad = (f^1_2 + Q^1_2) + (f^2_1 + Q^2_1) \\
&(K^1_{31}u^1_1 + K^1_{32}u^1_2 + K^1_{33}u^1_3) + (K^2_{41}u^2_1 + K^2_{42}u^2_2 + K^2_{43}u^2_3 + K^2_{44}u^2_4) \\
&\qquad = (f^1_3 + Q^1_3) + (f^2_4 + Q^2_4)
\end{aligned}$$

Using the global-variable notation in (8.50), we can rewrite the above equations as follows [this amounts to imposing continuity of the primary variables in (8.50)]:

$$\begin{aligned}
K^1_{21}U_1 + (K^1_{22} + K^2_{11})U_2 + (K^1_{23} + K^2_{14})U_3 + K^2_{12}U_4 + K^2_{13}U_5 &= f^1_2 + f^2_1 + (Q^1_2 + Q^2_1) \\
K^1_{31}U_1 + (K^1_{32} + K^2_{41})U_2 + (K^1_{33} + K^2_{44})U_3 + K^2_{42}U_4 + K^2_{43}U_5 &= f^1_3 + f^2_4 + (Q^1_3 + Q^2_4)
\end{aligned}$$

Now we can impose the conditions in (8.52) by setting appropriate portions of the expressions in parentheses on the right-hand sides of the above equations equal to zero:

$$\begin{aligned} Q_2^1 + Q_1^2 &= (Q_{21}^1 + Q_{22}^1 + Q_{23}^1) + (Q_{11}^2 + Q_{12}^2 + Q_{13}^2 + Q_{14}^2) \\ &= Q_{21}^1 + Q_{23}^1 + \underline{(Q_{22}^1 + Q_{14}^2)} + Q_{11}^2 + Q_{12}^2 + Q_{13}^2 \\ Q_3^1 + Q_4^2 &= (Q_{31}^1 + Q_{32}^1 + Q_{33}^1) + (Q_{41}^2 + Q_{42}^2 + Q_{43}^2 + Q_{44}^2) \\ &= Q_{31}^1 + Q_{33}^1 + \underline{(Q_{32}^1 + Q_{44}^2)} + Q_{41}^2 + Q_{42}^2 + Q_{43}^2 \end{aligned}$$

The underlined terms are zero, by the balance requirement (8.52*c*). The remaining terms of each equation will be either known because q_n is known on the boundary or will remain unknown because the primary variable is specified on the boundary.

In general, when several elements are connected, the assembly of the elements is carried out by putting element coefficients K_{ij}^e, f_i^e, and Q_i^e into proper locations of the global coefficient matrix and right-hand column vectors. This is done by means of the connectivity relations, i.e., the correspondence of the local node number to the global node number. For example, if global node number 3 corresponds to node 3 of element 1 and node 4 of element 2 then we have

$$F_3 = F_3^1 + F_4^2 \equiv f_3^1 + f_4^2 + Q_3^1 + Q_4^2, \qquad K_{33} = K_{33}^1 + K_{44}^2$$

If global node numbers 2 and 3 correspond, respectively, to nodes 2 and 3 of element 1 and nodes 1 and 4 of element 2 then the global coefficients K_{22}, K_{23} and K_{33} are given by

$$K_{22} = K_{22}^1 + K_{11}^2, \qquad K_{23} = K_{23}^1 + K_{14}^2, \qquad K_{33} = K_{33}^1 + K_{44}^2$$

Similarly, the source components of global nodes 2 and 3 are added:

$$F_2 = F_2^1 + F_1^2, \qquad F_3 = F_3^1 + F_4^2$$

For the two-element mesh shown in Fig. 8.11(*a*), the assembled equations are

$$\begin{bmatrix} K_{11}^1 & K_{12}^1 & K_{13}^1 & 0 & 0 \\ K_{21}^1 & K_{22}^1 + K_{11}^2 & K_{23}^1 + K_{14}^2 & K_{12}^2 & K_{13}^2 \\ K_{31}^1 & K_{32}^1 + K_{41}^2 & K_{33}^1 + K_{44}^2 & K_{42}^2 & K_{43}^2 \\ 0 & K_{21}^2 & K_{24}^2 & K_{22}^2 & K_{23}^2 \\ 0 & K_{31}^2 & K_{34}^2 & K_{32}^2 & K_{33}^2 \end{bmatrix} \begin{Bmatrix} U_1 \\ U_2 \\ U_3 \\ U_4 \\ U_5 \end{Bmatrix} = \begin{Bmatrix} F_1^1 \\ F_2^1 + F_1^2 \\ F_3^1 + F_4^2 \\ F_2^2 \\ F_3^2 \end{Bmatrix} \tag{8.54}$$

The assembly procedure described above can be used to assemble elements of any shape and type. The procedure can be implemented on a computer, as described for one-dimensional problems, with the help of the array [*B*] (program variable is NOD). For hand calculations, the procedure described above must be used. For example, consider the finite element mesh shown in Fig. 8.11(*b*). Location (4, 4) of the global coefficient matrix contains $K_{33}^1 + K_{11}^2 + K_{11}^3 + K_{11}^4$. Location 4 in the assembled column vector contains $F_3^1 + F_1^2 + F_1^3 + F_1^4$. Locations (1,5), (1,6), (1,7), (2,5), (2,6), (2,7), (3,6), (3,7), and (5,6) of the global matrix contain zeros because $K_{IJ} = 0$ when global nodes I and J do not correspond to nodes of the same element in the mesh.

This completes the first five steps in the finite element modeling of the model equation (8.1). The next two steps of the analysis, namely, the imposition of boundary conditions and solution of equations, remain the same as for one-dimensional problems. Postprocessing of the solution for two-dimensional problems is discussed next.

8.2.8 Postprocessing

The finite element solution at any point (x, y) in an element Ω^e is given by

$$U^e(x, y) = \sum_{j=1}^{n} u_j^e \psi_j^e(x, y) \tag{8.55}$$

and its derivatives are computed as

$$\frac{\partial U^e}{\partial x} = \sum_{j=1}^{n} u_j^e \frac{\partial \psi_j^e}{\partial x}, \qquad \frac{\partial U^e}{\partial y} = \sum_{j=1}^{n} u_j^e \frac{\partial \psi_j^e}{\partial y} \tag{8.56}$$

Equations (8.55) and (8.56) can be used to compute the solution and its derivatives at any point (x, y) in the element. It is useful to generate, by interpolation from (8.55), information needed to plot contours of U^e and its gradient.

The derivatives of U^e will not be continuous at interelement boundaries, because continuity of the derivatives is not imposed during the assembly procedure. The weak form of the equations suggests that the primary variable is u, which is to be carried as the nodal variable. If additional variables, such as higher-order derivatives of the dependent unknown, are carried as nodal variables in the interest of making them continuous across interelement boundaries, the degree of interpolation (or order of the element) increases. In addition, the continuity of higher-order derivatives that are not identified as the primary variables may violate the physical features of the problem. For example, making $\partial u/\partial x$ continuous will violate the requirement that q_x $(= a_{11}\, \partial u/\partial x)$ be continuous at the interface of two dissimilar materials, because a_{11} is different for the two materials at the interface.

For linear triangular elements, the derivatives are constants within each element:

$$\psi_j^e = \frac{1}{2A_e}(\alpha_j + \beta_j x + \gamma_j y)$$

$$\frac{\partial \psi_j^e}{\partial x} = \frac{1}{2A}\beta_j, \qquad \frac{\partial \psi_j^e}{\partial y} = \frac{1}{2A}\gamma_j \tag{8.57}$$

$$\frac{\partial U^e}{\partial x} = \sum_{j=1}^{n} \frac{u_j^e \beta_j}{2A_e}, \qquad \frac{\partial U^e}{\partial y} = \sum_{j=1}^{n} \frac{u_j^e \gamma_j}{2A_e}$$

For linear rectangular elements, $\partial U^e/\partial x$ is linear in $\bar{y}$, and $\partial U^e/\partial y$ is linear in $\bar{x}$ [see (8.31*b*)]:

$$\frac{\partial \psi_j^e}{\partial \bar{x}} = -\frac{1}{a}\left(1 - \frac{\bar{y} + \bar{y}_j}{b}\right), \qquad \frac{\partial \psi_j^e}{\partial \bar{y}} = -\frac{1}{b}\left(1 - \frac{\bar{x} + \bar{x}_j}{a}\right)$$
$$\frac{\partial U^e}{\partial \bar{x}} = -\frac{1}{a}\sum_{j=1}^{n} u_j^e\left(1 - \frac{\bar{y} + \bar{y}_j}{b}\right), \qquad \frac{\partial U^e}{\partial \bar{y}} = -\frac{1}{b}\sum_{j=1}^{n} u_j^e\left(1 - \frac{\bar{x} + \bar{x}_j}{a}\right) \tag{8.58}$$

where $\bar{x}$ and $\bar{y}$ are the local coordinates (see Fig. 8.7*a*). Although $\partial U^e/\partial \bar{x}$ and $\partial U^e/\partial \bar{y}$ are linear functions of y and x, respectively, in each element, they are discontinuous at interelement boundaries. Consequently, quantities computed using derivatives of the finite element solution U^e are discontinuous there. For example, if one computes $q_x^e = a_{11}^e\, \partial U^e/\partial x$ at a node shared by three different elements, three different values of q_x^e are expected. The differences between the three values will diminish as the mesh is refined. Some commercial finite element softwares give a single value of q_x at the node by averaging the values obtained from various elements connected at the node.

8.2.9 Axisymmetric Problems

In studying problems involving cylindrical geometries, it is convenient to use the cylindrical coordinate system (r, θ, z) to formulate the problem. If the geometry, material properties, boundary conditions, and loading (or source) of the problem are independent of the angular coordinate θ, the problem solution will also be independent of θ. Consequently, a three-dimensional problem is reduced to a two-dimensional one in (r, z) coordinates. Here we consider a model axisymmetric problem, develop its weak form, and formulate the finite element model.

MODEL EQUATION. Consider the partial differential equation

$$-\frac{1}{r}\frac{\partial}{\partial r}\left(r\hat{a}_{11}\frac{\partial u}{\partial r}\right) - \frac{\partial}{\partial z}\left(\hat{a}_{22}\frac{\partial u}{\partial z}\right) + \hat{a}_{00}u = \hat{f}(r, z) \tag{8.59}$$

where $\hat{a}_{00}$, $\hat{a}_{11}$, $\hat{a}_{22}$, and $\hat{f}$ are given functions of r and z. This equation arises in the study of heat transfer in cylindrical geometries, as well as in other fields of engineering and applied science. Our objective is to develop the finite element model of the equation based on the weak form of (8.59).

WEAK FORM. Following the three-step procedure, we write the weak form of (8.59):

$$\text{(i)} \quad 0 = \int_{\Omega^e} w\left[-\frac{1}{r}\frac{\partial}{\partial r}\left(r\hat{a}_{11}\frac{\partial u}{\partial r}\right) - \frac{\partial}{\partial z}\left(\hat{a}_{22}\frac{\partial u}{\partial z}\right) + \hat{a}_{00}u - \hat{f}\right] r\,dr\,dz$$
$$\text{(ii)} \quad 0 = \int_{\Omega^e}\left(\frac{\partial w}{\partial r}r\hat{a}_{11}\frac{\partial u}{\partial r} + \frac{\partial w}{\partial z}r\hat{a}_{22}\frac{\partial u}{\partial z} + wr\hat{a}_{00}u - wr\hat{f}\right) dr\,dz - \oint_{\Gamma^e} w\left(r\hat{a}_{11}\frac{\partial u}{\partial r}n_r + r\hat{a}_{22}\frac{\partial u}{\partial z}n_z\right) ds$$
$$\text{(iii)} \quad 0 = \int_{\Omega^e}\left(\hat{a}_{11}\frac{\partial w}{\partial r}\frac{\partial u}{\partial r} + \hat{a}_{22}\frac{\partial w}{\partial z}\frac{\partial u}{\partial z} + \hat{a}_{00}wu - w\hat{f}\right) r\,dr\,dz - \oint_{\Gamma^e} wq_n\,ds \tag{8.60}$$

where w is the weight function and q_n is the normal flux,

$$q_n = r\left(\hat{a}_{11}\frac{\partial u}{\partial r}n_r + \hat{a}_{22}\frac{\partial u}{\partial z}n_z\right) \tag{8.61}$$

Note that the weak form (8.60) does not differ significantly from that developed for the model equation (8.1) when $a_{12} = a_{21} = 0$. The only difference is the presence of r in the integrand. Consequently, (8.60) can be obtained as a special case of (8.10) for $a_{00} = \hat{a}_{00}x$, $a_{11} = \hat{a}_{11}x$, $a_{22} = \hat{a}_{22}x$, and $f = \hat{f}x$; the coordinates r and z are treated like x and y, respectively.

FINITE ELEMENT MODEL. Let us assume that $u(r, z)$ is approximated by the finite element interpolation U^e over the element Ω^e:

$$u \approx U^e(r, z) = \sum_{j=1}^{n} u_j^e \psi_j^e(r, z) \tag{8.62}$$

The interpolation functions $\psi_j^e(r, z)$ are the same as those developed in (8.25) and (8.31*a*) for linear triangular and rectangular elements, with $x = r$ and $y = z$. Substitution of (8.62) for u and ψ_i^e for w into the weak form gives the ith equation of the finite element model:

$$\begin{aligned} 0 = \sum_{j=1}^{n}\left[\int_{\Omega^e}\left(\hat{a}_{11}\frac{\partial \psi_i^e}{\partial r}\frac{\partial \psi_j^e}{\partial r} + \hat{a}_{22}\frac{\partial \psi_i^e}{\partial z}\frac{\partial \psi_j^e}{\partial z} + \hat{a}_{00}\psi_i^e\psi_j^e\right) r\,dr\,dz\right]u_j^e \\ - \int_{\Omega^e}\psi_i^e\hat{f}r\,dr\,dz - \oint_{\Gamma^e}\psi_i^e q_n\,ds \end{aligned} \tag{8.63}$$

or

$$0 = \sum_{j=1}^{n} K_{ij}^e u_j^e - f_i^e - Q_i^e \tag{8.64a}$$

where

$$\begin{aligned} K_{ij}^e &= \int_{\Omega^e}\left(\hat{a}_{11}\frac{\partial \psi_i^e}{\partial r}\frac{\partial \psi_j^e}{\partial r} + \hat{a}_{22}\frac{\partial \psi_i^e}{\partial z}\frac{\partial \psi_j^e}{\partial z} + \hat{a}_{00}\psi_i^e\psi_j^e\right) r\,dr\,dz \\ f_i^e &= \int_{\Omega^e}\psi_i^e\hat{f}r\,dr\,dz, \qquad Q_i^e = \oint_{\Gamma^e}\psi_i^e q_n\,ds \end{aligned} \tag{8.64b}$$

Exact evaluation of the integrals in K_{ij}^e and f_i^e for polynomial forms of $\hat{a}_{ij}$ and $\hat{f}$ is possible. However, we evaluate them numerically using the numerical integration methods discussed in Chapter 7 (see Section 7.1.5), and reviewed in Chapter 9. This completes the development of the finite element model of an axisymmetric problem.

8.2.10 An Example

The model equation in (8.1) arises in many fields of engineering and applied sciences, and some examples are given in Table 8.1. The application of the finite element model developed in Sections 8.2.2–8.2.8 to a problem is discussed here. This example will be particularly useful for readers who are interested in the mathematical aspects rather than the physical background of the problem.

Example 8.3. Consider the Poisson equation

$$-\nabla^2 u = f_0 \quad \text{or} \quad -\left(\frac{\partial^2 u}{\partial x^2} + \frac{\partial^2 u}{\partial y^2}\right) = f_0 \quad \text{in } \Omega \tag{8.65}$$

in a square region Ω (see Fig. 8.12*a*). The boundary condition of the problem is

$$u = 0 \quad \text{on } \Gamma \tag{8.66}$$

We wish to solve the problem using the finite element method.

A problem possesses symmetry of the solution about a line only when there is symmetry of (*a*) the geometry, (*b*) the material properties, (*c*) the source variation, and

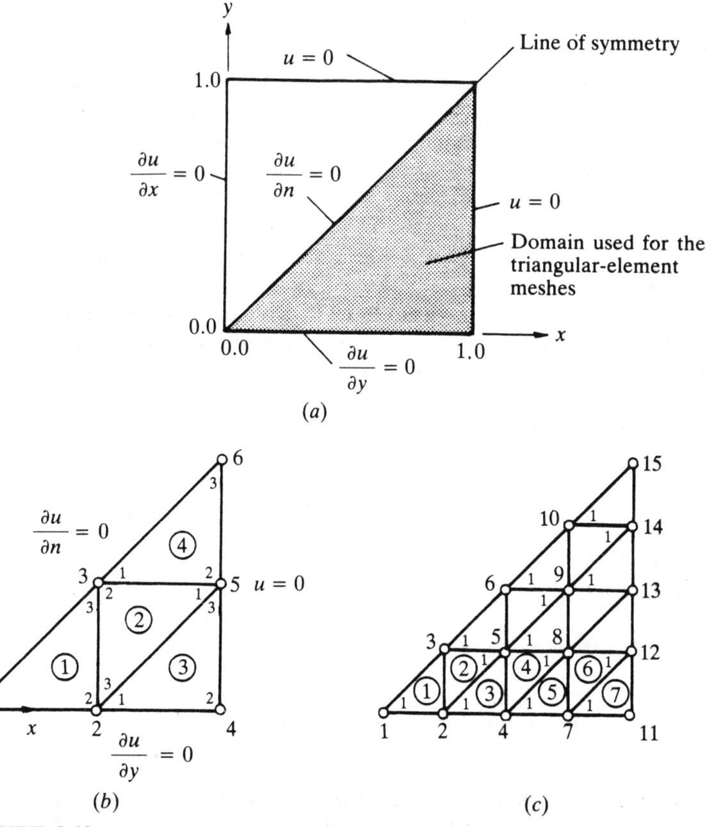

FIGURE 8.12
Finite element analysis of the Poisson equation in a square region: (*a*) geometry and computational domain, and boundary conditions of the problem; (*b*) a coarse finite element mesh of linear triangular elements; (*c*) a refined mesh of linear triangular elements (Example 8.3).

(*d*) the boundary conditions about the line. Whenever a portion of the domain is modeled to exploit symmetries available in the problem, a portion of the boundary of the computational domain is a line of symmetry. On lines of symmetry, the normal derivative of the solution (i.e., the derivative with respect to the coordinate normal to the line of symmetry) is zero:

$$q_n \equiv \frac{\partial u}{\partial n} = \frac{\partial u}{\partial x} n_x + \frac{\partial u}{\partial y} n_y = 0 \tag{8.67}$$

The problem at hand has symmetry about the $x = 0$ and $y = 0$ axes; it is also symmetric about the diagonal line $x = y$ (see Fig. 8.12*a*). We can exploit such symmetries in modeling the problem. Thus, we can use a quadrant for meshes of rectangular elements and an octant for meshes of triangular elements of the domain to analyze the problem. Of course, it is possible to mix triangular and rectangular elements to represent the domain as well as the solution.

Solution by linear triangular elements. Owing to the symmetry along the diagonal $x = y$, we model the triangular domain shown in Fig. 8.12(*a*). As a first choice, we use a uniform mesh of four linear triangular elements to represent the domain (see Fig. 8.12*b*), and then a refined mesh (see Fig. 8.12*c*) to compare the solutions. In the present case, there is no discretization error involved in the problem because the geometry is exactly represented.

Elements 1, 3, and 4 are identical in orientation as well as geometry. Element 2 is geometrically identical with element 1, except that it is oriented differently. If we number the local nodes of element 2 to match those of element 1 then all four elements have the same element matrices, and it is necessary to compute them only for element 1. When the element matrices are calculated on a computer, such considerations are not taken into account. In solving the problem by hand, we use the correspondence between a master element (element 1) and the other elements in the mesh to avoid unnecessary calculations.

We consider element 1 as the typical element, with its local coordinate system $(\bar{x}, \bar{y})$. Suppose that the element dimensions, i.e., length and height, are a and b, respectively. The coordinates of the element nodes are

$$(\bar{x}_1, \bar{y}_1) = (0, 0), \qquad (\bar{x}_2, \bar{y}_2) = (a, 0), \qquad (\bar{x}_3, \bar{y}_3) = (a, b)$$

Hence, the parameters α_i, β_i, and γ_i are given by

$$\begin{aligned}
&\alpha_1 = \bar{x}_2\bar{y}_3 - \bar{x}_3\bar{y}_2 = ab, \qquad \alpha_2 = \bar{x}_3\bar{y}_1 - \bar{x}_1\bar{y}_3 = 0, \qquad \alpha_3 = \bar{x}_1\bar{y}_2 - \bar{x}_2\bar{y}_1 = 0 \\
&\beta_1 = \bar{y}_2 - \bar{y}_3 = -b, \qquad \beta_2 = \bar{y}_3 - \bar{y}_1 = b, \qquad \beta_3 = \bar{y}_1 - \bar{y}_2 = 0 \\
&\gamma_1 = -(\bar{x}_2 - \bar{x}_3) = 0, \qquad \gamma_2 = -(\bar{x}_3 - \bar{x}_1) = -a, \qquad \gamma_3 = -(\bar{x}_1 - \bar{x}_2) = a
\end{aligned} \tag{8.68}$$

The element coefficients K_{ij}^e and f_i^e are given by

$$[K^1] = \frac{1}{2ab}\begin{bmatrix} b^2 & -b^2 & 0 \\ -b^2 & a^2 + b^2 & -a^2 \\ 0 & -a^2 & a^2 \end{bmatrix}, \qquad \{f^1\} = \frac{f_0 ab}{6}\begin{Bmatrix} 1 \\ 1 \\ 1 \end{Bmatrix} \tag{8.69a}$$

The element matrix in (8.69*a*) is valid for the Laplace operator $-\nabla^2$ on any right-angled triangle with sides a and b in which the right-angle is at node 2, and the diagonal line of the triangle connects node 3 to node 1. Note that the off-diagonal coefficient associated with the nodes on the diagonal line is zero for a right-angled triangle. These observations can be used to write the element matrix associated with the

Laplace operator on *any* right-angled triangle, i.e., for any element-node numbering system. For example, if the right-angled corner is numbered as node 1, and the diagonal-line nodes are numbered as 2 and 3 (following the counter-clockwise numbering scheme), we have (note that a denotes the length of side connecting nodes 1 and 2)

$$[K^e] = \frac{1}{2ab}\begin{bmatrix} a^2+b^2 & -b^2 & -a^2 \\ -b^2 & b^2 & 0 \\ -a^2 & 0 & a^2 \end{bmatrix} \tag{8.69b}$$

For the mesh shown in Fig. 8.12(b), we have

$$[K^1]=[K^2]=[K^3]=[K^4], \qquad \{f^1\}=\{f^2\}=\{f^3\}=\{f^4\}$$

For $a=b$, the coefficient matrix in (8.69a) takes the form

$$[K^e] = \frac{1}{2}\begin{bmatrix} 1 & -1 & 0 \\ -1 & 2 & -1 \\ 0 & -1 & 1 \end{bmatrix} \tag{8.70}$$

The assembled coefficient matrix for the finite element mesh is 6×6, because there are six global nodes, with one unknown per node. The assembled matrix can be obtained directly by using the correspondence between the global nodes and the local nodes, expressed through the connectivity matrix

$$[B] = \begin{bmatrix} 1 & 2 & 3 \\ 5 & 3 & 2 \\ 2 & 4 & 5 \\ 3 & 5 & 6 \end{bmatrix} \tag{8.71}$$

The assembled system of equations is

$$\frac{1}{2}\left[\begin{array}{ccc:ccc} 1 & -1 & 0 & 0 & 0 & 0 \\ -1 & 4 & -2 & -1 & 0 & 0 \\ 0 & -2 & 4 & 0 & -2 & 0 \\ \hdashline 0 & -1 & 0 & 2 & -1 & 0 \\ 0 & 0 & -2 & -1 & 4 & -1 \\ 0 & 0 & 0 & 0 & -1 & 1 \end{array}\right]\begin{Bmatrix} U_1 \\ U_2 \\ U_3 \\ U_4 \\ U_5 \\ U_6 \end{Bmatrix} = \frac{f_0}{24}\begin{Bmatrix} 1 \\ 3 \\ 3 \\ 1 \\ 3 \\ 1 \end{Bmatrix} + \begin{Bmatrix} Q_1^1 \\ Q_2^1+Q_3^2+Q_1^3 \\ Q_3^1+Q_2^2+Q_1^4 \\ \hdashline Q_2^3 \\ Q_1^2+Q_3^3+Q_2^4 \\ Q_3^4 \end{Bmatrix} \tag{8.72}$$

The sums of the secondary variables at global nodes 2, 3, and 5 are

$$\begin{aligned} Q_2^1+Q_3^2+Q_1^3 &\equiv \hat{Q}_2 \\ Q_3^1+Q_2^2+Q_1^4 &\equiv \hat{Q}_3 \\ Q_1^2+Q_3^3+Q_2^4 &\equiv \hat{Q}_5 \end{aligned} \tag{8.73}$$

At nodes 1, 4, and 6, we have $Q_1^1\equiv\hat{Q}_1$, $Q_2^3\equiv\hat{Q}_4$, and $Q_3^4\equiv\hat{Q}_6$.

The specified boundary conditions on the primary degrees of freedom of the problem are

$$U_4=U_5=U_6=0 \tag{8.74}$$

The specified secondary degrees of freedom are (all due to symmetry)

$$\hat{Q}_1=0, \qquad \hat{Q}_2=0, \qquad \hat{Q}_3=0 \tag{8.75}$$

Since U_4, U_5, and U_6 are known, the secondary variables at these nodes, i.e., $\hat{Q}_4$, $\hat{Q}_5$, and $\hat{Q}_6$, are unknown, and can be obtained in the post-computation.

Since the only unknown primary variables are (U_1, U_2, and U_3), and (U_4, U_5, and U_6) are specified to be zero, the condensed equations for the primary unknowns can be obtained by deleting rows and columns 4, 5, and 6 from the system (8.72). In retrospect, it would have been sufficient to assemble the element coefficients associated with the global nodes 1, 2, and 3, i.e., writing out equations 1, 2 and 3:

$$\begin{bmatrix} K_{11}^1 & K_{12}^1 & K_{13}^1 \\ K_{21}^1 & K_{22}^1 + K_{33}^2 + K_{11}^3 & K_{23}^1 + K_{32}^2 \\ K_{31}^1 & K_{32}^1 + K_{23}^2 & K_{33}^1 + K_{22}^2 + K_{11}^4 \end{bmatrix} \begin{Bmatrix} U_1 \\ U_2 \\ U_3 \end{Bmatrix} = \begin{Bmatrix} f_1^1 \\ f_2^1 + f_3^2 + f_1^3 \\ f_3^1 + f_2^2 + f_1^4 \end{Bmatrix} + \begin{Bmatrix} 0 \\ 0 \\ 0 \end{Bmatrix} \tag{8.76}$$

The unknown secondary variables $\hat{Q}_4$, $\hat{Q}_5$, and $\hat{Q}_6$ can be computed either from the equations (i.e., from equilibrium)

$$\begin{Bmatrix} \hat{Q}_4 \\ \hat{Q}_5 \\ \hat{Q}_6 \end{Bmatrix} = -\begin{Bmatrix} f_2^3 \\ f_1^2 + f_3^3 + f_2^4 \\ f_3^4 \end{Bmatrix} + \begin{bmatrix} 0 & K_{21}^3 & 0 \\ 0 & K_{13}^2 + K_{31}^3 & K_{12}^2 + K_{21}^4 \\ 0 & 0 & K_{31}^4 \end{bmatrix} \begin{Bmatrix} U_1 \\ U_2 \\ U_3 \end{Bmatrix} \tag{8.77}$$

or from their definitions (8.73) and (8.47). For example, we have

$$\hat{Q}_4 = Q_2^3 = \int_{1-2} q_n^3 \psi_2^3 \, dx + \int_{2-3} q_n^3 \psi_2^3 \, dy + \int_{3-1} q_n^3 \psi_2^3 \, ds \tag{8.78a}$$

where

$$(q_n^3)_{1-2} = \left(\frac{\partial u}{\partial x} n_x + \frac{\partial u}{\partial y} n_y \right)_{1-2} = 0 \quad \left(n_x = 0, \quad \frac{\partial u}{\partial y} = 0 \right)$$

$$(q_n^3)_{2-3} = \left(\frac{\partial u}{\partial x} n_x + \frac{\partial u}{\partial y} n_y \right)_{2-3} = \frac{\partial u}{\partial x} \quad (n_x = 1, \ n_y = 0)$$

$$(\psi_2^3)_{2-3} = 1 - \frac{y}{h_{23}}, \qquad (\psi_2^3)_{1-3} = 0$$

Thus,

$$\hat{Q}_4 = Q_{22}^3 = \int_0^{h_{23}} \frac{\partial u}{\partial x} \left(1 - \frac{y}{h_{23}} \right) dy$$

where $\partial u / \partial x$ from the finite element interpolation is

$$\frac{\partial u}{\partial x} = \sum_{j=1}^{3} u_j^3 \frac{\beta_j^3}{2A_3}$$

We obtain ($h_{23} = a$, $\beta_1^3 = -a$, $2A_3 = a^2$, $U_4 = U_5 = 0$)

$$\hat{Q}_4 = \frac{h_{23}}{4A_3} \sum_{j=1}^{3} u_j^3 \beta_j^3 = -0.5 U_2 \tag{8.78b}$$

Using the numerical values of the coefficients K_{ij}^e and f_i^e (with $f_0 = 1$), we write the condensed equations for U_1, U_2, and U_3 as

$$\begin{bmatrix} 0.5 & -0.5 & 0 \\ -0.5 & 2.0 & -1.0 \\ 0 & -1.0 & 2.0 \end{bmatrix} \begin{Bmatrix} U_1 \\ U_2 \\ U_3 \end{Bmatrix} = \frac{1}{24} \begin{Bmatrix} 1 \\ 3 \\ 3 \end{Bmatrix} \tag{8.79}$$

Solving (8.79) for U_i $(i = 1, 2, 3)$, we obtain

$$\begin{Bmatrix} U_1 \\ U_2 \\ U_3 \end{Bmatrix} = \frac{1}{24}\begin{bmatrix} 3 & 1 & 0.5 \\ 1 & 1 & 0.5 \\ 0.5 & 0.5 & 0.75 \end{bmatrix}\begin{Bmatrix} 1 \\ 3 \\ 3 \end{Bmatrix} = \frac{1}{24}\begin{Bmatrix} 7.5 \\ 5.5 \\ 4.25 \end{Bmatrix} = \begin{Bmatrix} 0.31250 \\ 0.22917 \\ 0.17708 \end{Bmatrix} \tag{8.80}$$

and, from (8.77), we have

$$\begin{Bmatrix} Q^3_{22} \\ Q^3_{32} + Q^4_{22} \\ Q^4_{32} \end{Bmatrix} = -\frac{1}{24}\begin{Bmatrix} 1 \\ 3 \\ 1 \end{Bmatrix} + \begin{bmatrix} 0 & -0.5 & 0 \\ 0 & 0 & -1 \\ 0 & 0 & 0 \end{bmatrix}\begin{Bmatrix} U_1 \\ U_2 \\ U_3 \end{Bmatrix} = \begin{Bmatrix} -0.197917 \\ -0.302083 \\ -0.041667 \end{Bmatrix} \tag{8.81}$$

By interpolation, Q^3_{22}, for example, is equal to $-0.5U_2$, and it differs from Q^3_{22} computed from equilibrium by the amount f^3_2 $(=\frac{1}{24})$.

Solution by linear rectangular elements. Note that we cannot exploit the symmetry along the diagonal $x = y$ to our advantage when we use a mesh of rectangular elements. Therefore, we use a 2×2 uniform mesh of four linear rectangular elements (see Fig. 8.13) to discretize a quadrant of the domain. Once again, no discretization error is introduced in the present case.

Since all elements in the mesh are identical, we shall compute the element matrices for only one element, say element 1. We have

$$\psi_1 = (1 - 2\bar{x})(1 - 2\bar{y}), \quad \psi_2 = 2\bar{x}(1 - 2\bar{y}), \quad \psi_3 = 4\bar{x}\bar{y}, \quad \psi_4 = (1 - 2\bar{x})2\bar{y}$$

$$K^e_{ij} = \int_0^{0.5}\int_0^{0.5}\left(\frac{\partial \psi_i}{\partial \bar{x}}\frac{\partial \psi_j}{\partial \bar{x}} + \frac{\partial \psi_i}{\partial \bar{y}}\frac{\partial \psi_j}{\partial \bar{y}}\right) d\bar{x}\, d\bar{y} \tag{8.82}$$

$$f^e_i = \int_0^{0.5}\int_0^{0.5} f_0 \psi_i \, d\bar{x}\, d\bar{y}$$

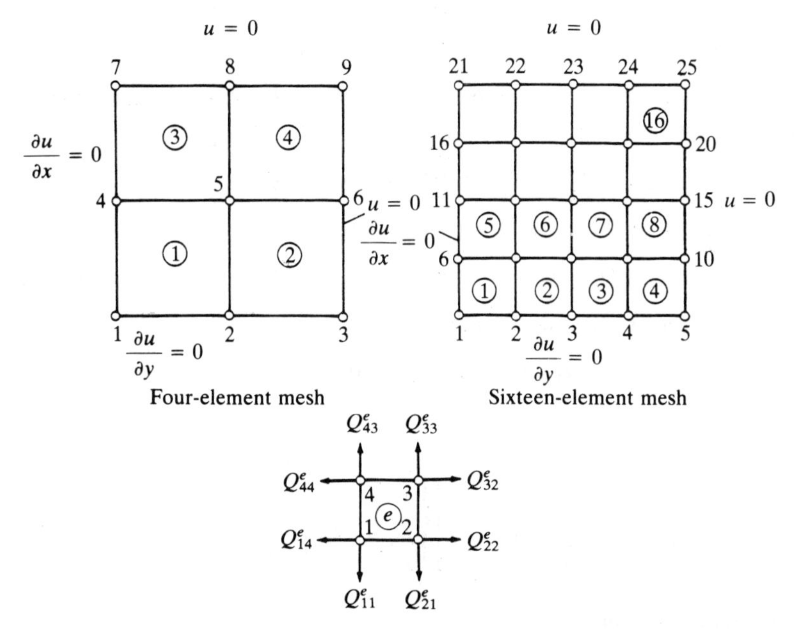

FIGURE 8.13
Finite element discretization of the domain of Example 8.3 by linear rectangular elements.

Evaluating these integrals, we obtain {see (8.44): $[K^e]=[S^{11}]+[S^{22}]$}

$$[K^e]=\frac{1}{6}\begin{bmatrix} 4 & -1 & -2 & -1 \\ -1 & 4 & -1 & -2 \\ -2 & -1 & 4 & -1 \\ -1 & -2 & -1 & 4 \end{bmatrix}, \qquad \{F^e\}=\frac{1}{16}\begin{Bmatrix} 1 \\ 1 \\ 1 \\ 1 \end{Bmatrix}+\begin{Bmatrix} Q_1^e \\ Q_2^e \\ Q_3^e \\ Q_4^e \end{Bmatrix} \tag{8.83a}$$

where

$$Q_i^e=\int_{\bar{x}_1}^{\bar{x}_2}[q_n^{(e)}\psi_i(\bar{x},\bar{y})]_{\bar{y}=0}\,d\bar{x}+\int_{\bar{y}_2}^{\bar{y}_3}[q_n^{(e)}\psi_i(\bar{x},\bar{y})]_{\bar{x}=a}\,d\bar{y}$$
$$+\int_{\bar{x}_3}^{\bar{x}_4}[q_n^{(e)}\psi_i(\bar{x},\bar{y})]_{\bar{y}=b}\,d\bar{x}+\int_{\bar{y}_4}^{\bar{y}_1}[q_n^{(e)}\psi_i(\bar{x},\bar{y})]_{\bar{x}=0}\,d\bar{y} \tag{8.83b}$$

and $(\bar{x}_i, \bar{y}_i)$ denote the local coordinates of the element nodes (and $a=\bar{x}_2-\bar{x}_1=\bar{x}_3-\bar{x}_4$ and $b=\bar{y}_4-\bar{y}_1=\bar{y}_3-\bar{y}_2$).

The coefficient matrix of the condensed equations for the primary unknowns can be directly assembled. There are four unknowns (at nodes 1, 2, 4, and 5). The condensed equations are

$$\begin{bmatrix} K_{11} & K_{12} & K_{14} & K_{15} \\ K_{21} & K_{22} & K_{24} & K_{25} \\ K_{41} & K_{42} & K_{44} & K_{45} \\ K_{51} & K_{52} & K_{54} & K_{55} \end{bmatrix}\begin{Bmatrix} U_1 \\ U_2 \\ U_4 \\ U_5 \end{Bmatrix}=\begin{Bmatrix} F_1 \\ F_2 \\ F_4 \\ F_5 \end{Bmatrix} \tag{8.84a}$$

where K_{IJ} and F_I are the global coefficients

$$K_{11}=K_{11}^1,\quad K_{12}=K_{12}^1,\quad K_{14}=K_{14}^1,\quad K_{15}=K_{13}^1$$
$$K_{22}=K_{22}^1+K_{11}^2,\quad K_{24}=K_{24}^1,\quad K_{25}=K_{23}^1+K_{14}^2$$
$$K_{44}=K_{44}^1+K_{11}^3,\quad K_{45}=K_{43}^1+K_{12}^3,\quad K_{55}=K_{33}^1+K_{44}^2+K_{11}^4+K_{22}^3 \tag{8.84b}$$
$$F_1=f_1^1+Q_1^1,\quad F_2=f_2^1+f_1^2+Q_2^1+Q_1^2,\quad F_4=f_4^1+f_1^3+Q_4^1+Q_1^3$$
$$F_5=f_3^1+f_4^2+f_1^4+f_2^3+Q_3^1+Q_4^2+Q_1^4+Q_2^3$$

The boundary conditions on the secondary variables are

$$Q_1^1=0,\quad Q_2^1+Q_1^2=0,\quad Q_4^1+Q_1^3=0 \tag{8.85a}$$

and the balance of secondary variables at global node 5 requires

$$Q_3^1+Q_4^2+Q_2^3+Q_1^4=0 \tag{8.85b}$$

Thus, we have

$$\frac{1}{6}\begin{bmatrix} 4 & -1 & -1 & -2 \\ -1 & 8 & -2 & -2 \\ -1 & -2 & 8 & -2 \\ -2 & -2 & -2 & 16 \end{bmatrix}\begin{Bmatrix} U_1 \\ U_2 \\ U_4 \\ U_5 \end{Bmatrix}=\frac{1}{16}\begin{Bmatrix} 1 \\ 2 \\ 2 \\ 4 \end{Bmatrix} \tag{8.86}$$

The solution of these equations is

$$U_1=0.31071,\quad U_2=0.24107,\quad U_4=0.24107,\quad U_5=0.19286 \tag{8.87}$$

The secondary variables $\hat{Q}_3$, $\hat{Q}_6$, and $\hat{Q}_9$ at nodes 3, 6, and 9, respectively, can be computed from the equations ($\hat{Q}_3 = Q_2^2$, $\hat{Q}_6 = Q_3^2 + Q_2^4$, $\hat{Q}_9 = Q_3^4$)

$$\begin{Bmatrix} \hat{Q}_3 \\ \hat{Q}_6 \\ \hat{Q}_9 \end{Bmatrix} = -\begin{Bmatrix} f_2^2 \\ f_3^2 + f_2^4 \\ f_3^4 \end{Bmatrix} + \begin{bmatrix} K_{31} & K_{32} & K_{34} & K_{35} \\ K_{61} & K_{62} & K_{64} & K_{65} \\ K_{91} & K_{92} & K_{94} & K_{95} \end{bmatrix} \begin{Bmatrix} U_1 \\ U_2 \\ U_4 \\ U_5 \end{Bmatrix} \tag{8.88a}$$

where

$$\begin{aligned} K_{31} &= 0, \quad K_{32} = K_{21}^2, \quad K_{34} = 0, \quad K_{35} = K_{34}^2 \\ K_{61} &= 0, \quad K_{62} = K_{31}^2, \quad K_{64} = 0, \quad K_{65} = K_{34}^2 + K_{21}^4 \\ K_{91} &= 0, \quad K_{92} = 0, \quad K_{94} = 0, \quad K_{95} = K_{31}^4 \end{aligned} \tag{8.88b}$$

We have

$$\begin{Bmatrix} \hat{Q}_3 \\ \hat{Q}_6 \\ \hat{Q}_9 \end{Bmatrix} = -\frac{1}{16}\begin{Bmatrix} 1 \\ 2 \\ 1 \end{Bmatrix} + \frac{1}{6}\begin{bmatrix} 0 & -1 & 0 & -2 \\ 0 & -2 & 0 & -2 \\ 0 & 0 & 0 & -2 \end{bmatrix} \begin{Bmatrix} U_1 \\ U_2 \\ U_4 \\ U_5 \end{Bmatrix} = -\begin{Bmatrix} 0.16697 \\ 0.26964 \\ 0.12679 \end{Bmatrix} \tag{8.89}$$

The finite element solutions obtained using two different meshes of triangular elements and two different meshes of rectangular elements are compared in Table 8.2 with the 50-term series solution (at $x = 0$ for varying y) in (2.102) (set $k = 1$, $q_0 = f_0 = 1$) and the one-parameter Ritz solution in (2.101); see also Fig. 8.14. The finite element solution obtained using 16 triangular elements (in an octant) is the most accurate one when compared with the series solution. The accuracy of the triangular element mesh is due to the large number of elements it has compared with the number of elements in the rectangular element mesh for the same size of domain.

The solution u and its gradient can be computed at any interior point of the domain. For a point (x, y) in the element Ω^e, we have

$$U(x, y) = \sum_{j=1}^{n} u_j^e \psi_j^e(x, y) \tag{8.90a}$$

$$q_y(x, y) = \frac{\partial U}{\partial y} = \sum_{i=1}^{n} u_i^e \frac{\partial \psi_i^e}{\partial y}, \qquad q_x(x, y) = \frac{\partial U}{\partial x} = \sum_{i=1}^{n} u_i^e \frac{\partial \psi_i^e}{\partial x} \tag{8.90b}$$

TABLE 8.2
Comparison of the finite element solutions $u(0, y)$ with the series solution and the Ritz solution of (8.65) (Example 8.3)

	Triangular elements		**Rectangular elements**			**Series**
					Ritz	**solution**
y	**4 elements**	**16 elements**	**4 elements**	**16 elements**	**(2.101)**	**(2.102)**
0.0	0.3125	0.3013	0.3107	0.2984	0.3125	0.2947
0.25	0.2709†	0.2805	0.2759†	0.2824	0.2930	0.2789
0.50	0.2292	0.2292	0.2411	0.2322	0.2344	0.2293
0.75	0.1146†	0.1393	0.1205†	0.1414	0.1367	0.1397
1.0	0.0000	0.0000	0.0000	0.0000	0.0000	0.0000

† Interpolated values.

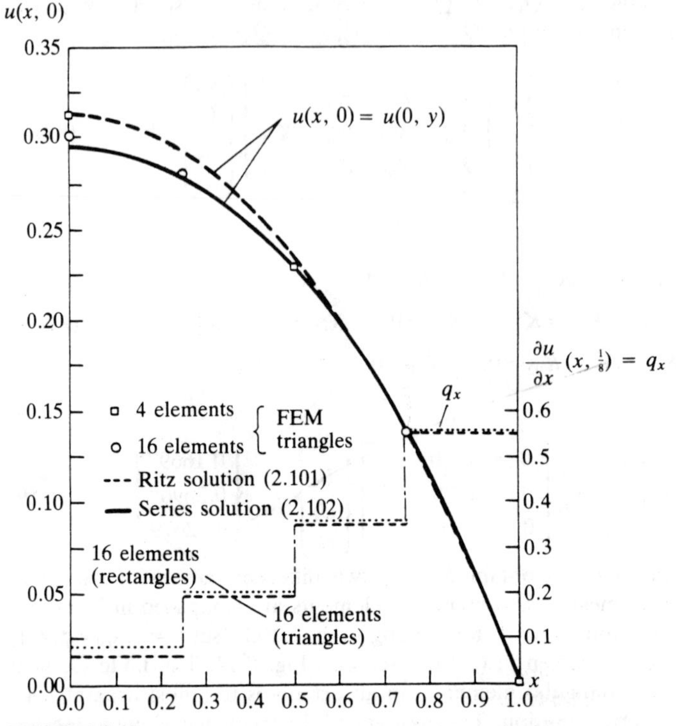

FIGURE 8.14
Comparison of the finite element solution with the two-parameter Ritz solution and the series solution (8.65) and (8.66) (Example 8.3).

Note that for a linear triangular element, q_x and q_y are constants over an entire element, whereas q_x is linear in y and q_y is linear in x for a linear rectangular element. For example, consider element 1. For triangular elements (4 elements),

$$
\begin{aligned}
q_x^1 &= \frac{1}{2A}\sum_{i=1}^{3} u_i^1 \beta_i^1 = 2(U_2 - U_1) = -0.166\,67 \\
q_y^1 &= \frac{1}{2A}\sum_{i=1}^{3} u_i^1 \gamma_i^1 = 2(U_3 - U_2) = -0.104\,17
\end{aligned}
\tag{8.91a}
$$

while for rectangular elements (4 elements),

$$
\begin{aligned}
q_x^1 &= \sum_{i=1}^{4} u_i^1 \frac{\partial \psi_i^1}{\partial x} = -2U_1(1-2y) + 2U_2(1-2y) + 4yU_5 - 4yU_4 \\
q_x^1&(0.25, 0.25) = -0.11785 \\
q_y^1 &= \sum_{i=1}^{4} u_i^1 \frac{\partial \psi_i^1}{\partial y} = -2U_1(1-2x) + 2U_2(1-2x) + 4xU_5 - 4xU_4 \\
q_y^1&(0.25, 0.25) = -0.11785
\end{aligned}
\tag{8.91b}
$$

Plots of q_x, obtained using the 16-element meshes of linear triangular and rectangular elements, as a function of x (for $y = 0.125$) are also shown in Fig. 8.14.

The computation of *isolines*, i.e., lines of constant u, for linear finite elements is straightforward. Suppose that we wish to find the $u = u_0$ (constant) isoline. On a side of a linear triangle or rectangular element, the solution u varies according to the equation

$$U^e(s) = u_1^e + \frac{u_3^e - u_1^e}{h} s$$

where s is the local coordinate with its origin at node 1 of the side, (u_1^e, u_3^e) are the nodal values (see Fig. 8.15), and h is the length of the side. Then, if $u \equiv u_0$ lies on the line (i.e., $u_1^e < u_0 < u_3^e$ or $u_3^e < u_0 < u_1^e$), the point s_0 at which $U^e(s_0) = u_0$ is given by

$$s_0 = \frac{(u_0 - u_1^e)h}{u_3^e - u_1^e} \tag{8.92}$$

Similar equations apply for other sides of the element. Since the solution varies linearly between any two points of linear elements, the isoline is determined by joining two points on any two sides of the element for which (8.92) gives a positive value (and $s_0 < h$).

For quadratic elements, isolines are determined by finding three points s_i in the element at which $U^e(s_i) = u_0$ $(i = 1, 2, 3)$:

$$\frac{s_0}{h} = \frac{-b \pm (b^2 - 4ac)^{1/2}}{2a} > 0 \tag{8.93a}$$

where

$$c = u_1^e - u_0, \qquad b = -3u_1^e + 4u_2^e - u_3^e, \qquad a = 2(u_1^e - 2u_2^e + u_3^e) \tag{8.93b}$$

Equation (8.93*a*) is to be applied on any three lines in the element until three different values $h > s_0 > 0$ are found.

The computational problem considered here [i.e., (8.65)] has several physical interpretations (see Table 8.1). The problem can be viewed as one of finding the temperature u in a unit square with uniform internal heat generation, where the sides $x = 0$ and $y = 0$ are insulated and the other two sides are at zero temperature (see Section 8.4.1). Another interpretation of the equation is that it defines the torsion of a cylindrical bar of 2 in square cross-section (see Section 8.4.3). In this case, u denotes the stress function Ψ,

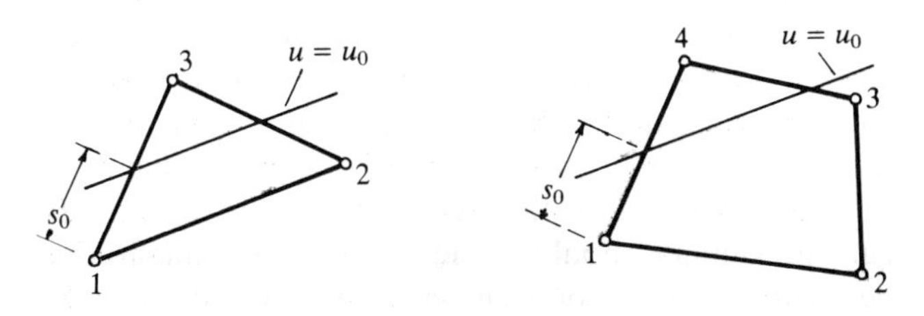

FIGURE 8.15
Isolines for triangular and quadrilateral elements (linear elements).

and the components of the gradient of the solution are the stresses (which are of primary interest):

$$\sigma_{xz} = G\theta \frac{\partial \Psi}{\partial y}, \qquad \sigma_{yz} = -G\theta \frac{\partial \Psi}{\partial x}$$

where G is the shear modulus and θ is the angle of twist per unit length of the bar.

A third interpretation of (8.65) is provided by groundwater (seepage) and potential flow problems. In this case, u is the piezometric head ϕ, stream function ψ or velocity potential ϕ (see Section 8.4.2). The x and y components of the velocity for the groundwater flow are defined as

$$u_1 = -a_{11} \frac{\partial \phi}{\partial x}, \qquad u_2 = -a_{22} \frac{\partial \phi}{\partial y}$$

where a_{11} and a_{22} are the permeabilities of the soil along the x and y directions, respectively.

Examples of each of these field problems will be considered in Section 8.4.

8.3 SOME COMMENTS ON MESH GENERATION AND IMPOSITION OF BOUNDARY CONDITIONS

8.3.1 Discretization of a Domain

The representation of a given domain by a collection of finite elements requires engineering judgement on the part of the finite element practitioner. The number, type (e.g., linear or quadratic), shape (e.g., triangular or rectangular), and density (i.e., mesh refinement) of elements used in a given problem depend on a number of considerations. The first is to discretize the domain as closely as possible with elements that are admissible. As we shall see later, one can use one set of elements for the approximation of a domain and another set for the solution. In discretizing a domain, consideration must be given to an accurate representation of the domain, point sources, distributed sources with discontinuities (i.e., sudden change in the intensity of the source), and material and geometric discontinuities, including a re-entrant corner. The discretization should include, for example, nodes at point sources (so that the point source is accurately lumped at the node), re-entrant corners, and element interfaces where abrupt changes in geometry and material properties occur. A second consideration, which requires some engineering judgement, is to discretize the body or portions of the body into sufficiently small elements so that steep gradients of the solution can be accurately calculated. The engineering judgement should come from both a qualitative understanding of the behavior of the solution and an estimate of the computational costs involved in the mesh refinement (i.e., reducing the size of the elements). For example, consider inviscid flow around a cylinder in a

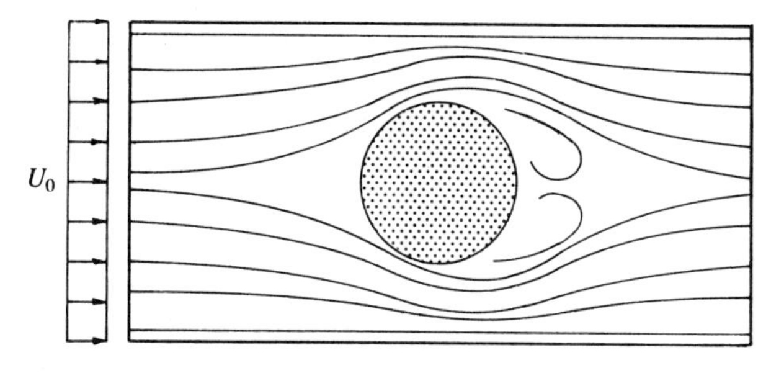

FIGURE 8.16
Flow of an inviscid fluid around a cylinder (streamlines).

channel. The flow entering the channel at the left goes around the cylinder and exits the channel at the right (see Fig. 8.16). Since the section at the cylinder is smaller than the inlet section, it is expected that the flow accelerates in the vicinity of the cylinder. On the other hand, the velocity field far from the cylinder (e.g., at the inlet) is essentially uniform. Such knowledge of the qualitative behavior of the flow allows us to employ a coarse mesh (i.e., elements that are relatively large in size) at sites sufficiently far from the cylinder, and a fine one at closer distances to the cylinder (see Fig. 8.17). Another purpose of using a refined mesh near the cylinder is to accurately represent the curved boundary of the domain there. In general, a refined mesh is required in places where acute changes in geometry, boundary conditions, loading, material properties or solution occur.

A mesh refinement should meet three conditions: (1) all previous meshes should be contained in the refined mesh; (2) every point in the body can be included within an arbitrarily small element at any stage of the mesh refinement; and (3) the same order of approximation for the solution may be retained through all stages of the refinement process. The last requirement eliminates comparison of two different approximations in two different meshes. When a mesh is refined, care should be taken to avoid elements with very large aspect ratios (i.e., the ratio of the smallest to the largest side of the element)

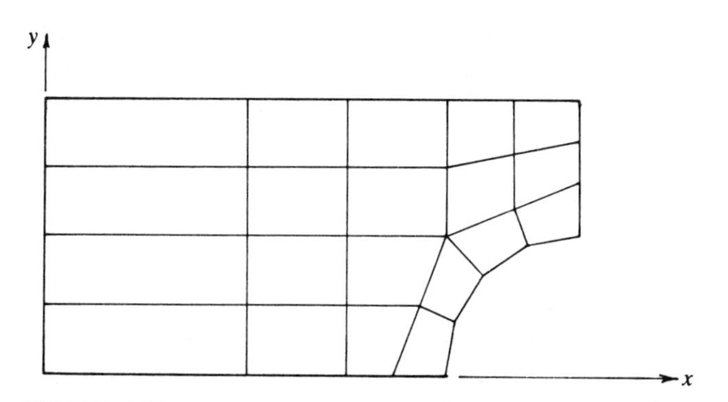

FIGURE 8.17
Finite element mesh considerations for inviscid flow around a cylinder. A typical mesh for a quadrant of the domain.

or small angles. Recall from the element matrices (8.41) and (8.68) that the coefficient matrices depend on the ratios of a to b and b to a. If the value of a/b or b/a is very large, the resulting coefficient matrices are ill-conditioned (i.e., numerically not invertible). Although the safe lower and upper limits on b/a are believed to be 0.1 and 10, respectively, the actual values are much more extreme (say, 1000), and they depend on the nature of the physical phenomenon being modeled. For example, in the inviscid flow problem discussed above, large aspect ratios are allowed at the entrance of the channel.

The words "coarse" and "fine" are relative. In any given problem, one begins with a finite element mesh that is believed to be adequate (based on experience and engineering judgement) to solve the problem at hand. Then, as a second choice, one selects a mesh that consists of a larger number of elements (and includes the first one as a subset) to solve the problem once again. If there is a significant difference between the two solutions, one sees the benefit of mesh refinement, and further refinement may be warranted. If the difference is negligibly small, further refinements are not necessary. Such numerical experiments with mesh refinements are not always feasible in practice, mostly because of the computational costs involved. In cases where computational cost is the prime concern, one must depend on one's judgement concerning what is a reasonably good mesh, which is often dictated by the geometry and qualitative understanding of the variations of the solution and its gradient. Since most practical problems are approximated in their engineering formulations, one should not be overly concerned with the numerical accuracy of the solution. A feel for the relative proportions and directions of various errors introduced into the analysis helps the finite element practitioner to make a decision on when to stop refining a mesh. In summary, scientific (or engineering) knowledge and experience with a given class of problems is an essential part of any approximate analysis.

8.3.2 Generation of Finite Element Data

An important part of finite element modeling is mesh generation, which involves numbering the nodes and elements, and the generation of nodal coordinates and the connectivity matrix. While the task of generating such data is quite simple, the type of the data has an effect on the computational efficiency as well as on accuracy. More specifically, the numbering of the nodes directly affects the bandwidth of the final assembled equations, which in turn increases the storage requirement and computational cost if equation solvers with the Gauss elimination procedure are used. The elements can be numbered arbitrarily, because this has no effect on the half-bandwidth. In a general-purpose program with a preprocessor, options to minimize the bandwidth are included. The saving of computational cost due to a smaller bandwidth in the solution of equations can be substantial, especially in problems where a large number of nodes and degrees of freedom per node are involved. While element numbering does not affect the half-bandwidth, it may affect the computer time required to assemble the global coefficient matrix—usually, a very small percentage of the time required to solve the equations.

The accuracy of the finite element solution can also depend on the choice of the finite element mesh. For instance, if the selected mesh violates the symmetry of the problem, the resulting solution will be less accurate than one obtained using a mesh that agrees with the physical symmetry of the problem. Geometrically, a triangular element has fewer (or no) lines of symmetry compared with a rectangular element, and therefore one should use meshes of triangular elements with care (e.g., one should select a mesh that does not violate the mathematical symmetry present in the problem).

The effect of the finite element meshes shown in Fig. 8.18 on the solution of the Poisson equation in Example 8.3 is investigated. The finite element solutions obtained by the three meshes are compared with the series solution in Table 8.3. Clearly, the solution obtained using mesh 3 is less accurate. This is to be expected, because mesh 3 is symmetric about the diagonal line connecting node 3 to node 7, whereas the mathematical symmetry is about the diagonal line connecting node 1 to node 9 (see Fig. 8.18). Mesh 1 is the most

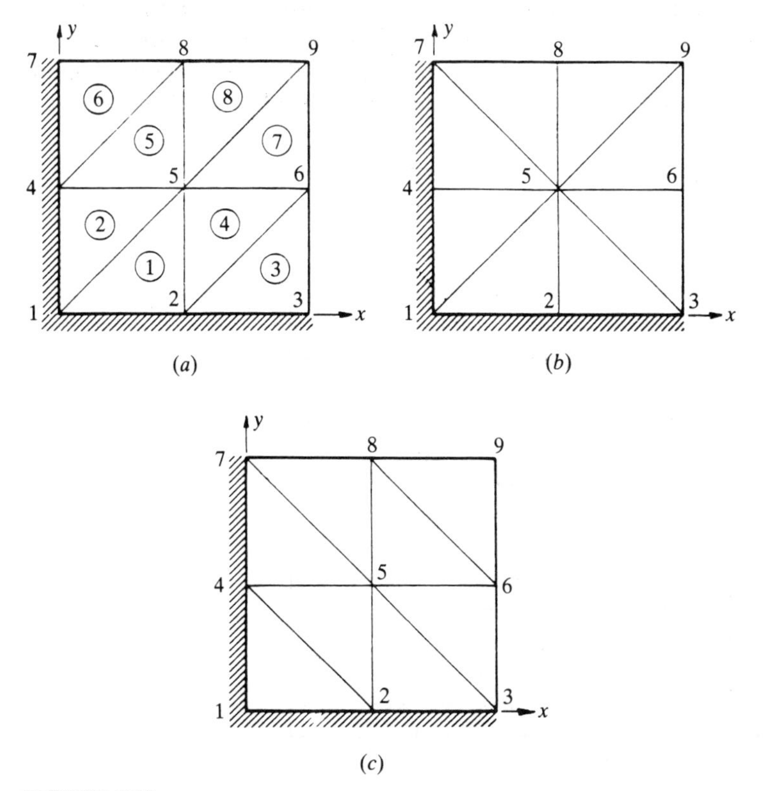

FIGURE 8.18
Various types of triangular-element meshes for the domain of Example 8.3: (*a*) mesh 1; (*b*) mesh 2; (*c*) mesh 3.

TABLE 8.3
Comparison of the finite element solutions obtained using various linear triangular-element meshes† with the series solution of the problem in Example 8.3

Node	Finite element solution Mesh 1	Mesh 2	Mesh 3	Series solution
1	0.31250	0.29167	0.25000	0.29469
2	0.22917	0.20833	0.20833	0.22934
4	0.22917	0.20833	0.20833	0.22934
5	0.17708	0.18750	0.16667	0.18114

† See Fig. 8.18 for the finite element meshes.

desirable of the three, because it does not violate the mathematical symmetry of the problem.

Next, the effect of mesh refinement with rectangular elements is investigated. Four different meshes of rectangular elements are shown in Fig. 8.19. Each mesh contains the previous mesh as a subset. The mesh shown in Fig. 8.19(*c*) is nonuniform; it is obtained by subdividing the first two rows and columns of elements of the mesh shown in Fig. 8.19(*b*). The finite element solutions obtained by these meshes are compared in Table 8.4. The numerical

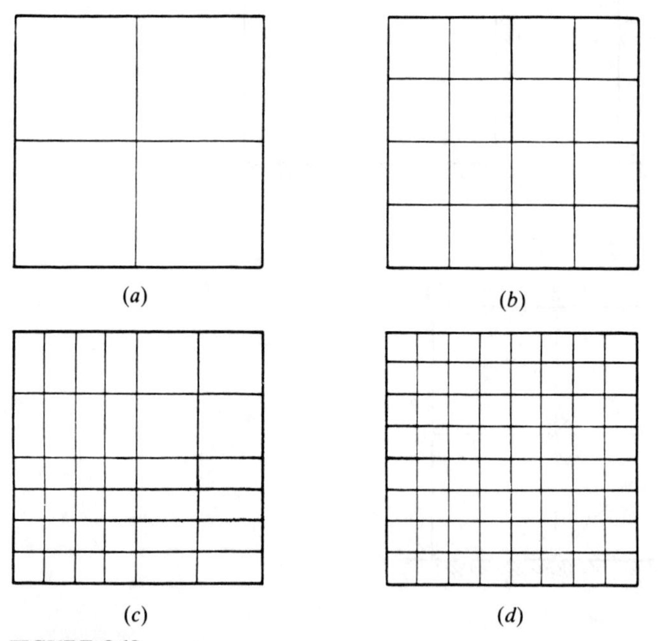

FIGURE 8.19
Mesh refinement; the meshes in (*a*), (*b*), and (*d*) are uniform; the mesh in (*c*) is nonuniform: (*a*) 2×2 mesh; (*b*) 4×4 mesh; (*c*) 6×6 mesh; (*d*) 8×8 mesh.

TABLE 8.4
Convergence of the finite element solution (with mesh refinement†) of the problem in Example 8.3

Location		Finite element solution				
x	y	2×2	4×4	6×6	8×8	Series solution
0.0	0.0	0.31071	0.29839	0.29641	0.29560	0.29469
0.125	0.0	—	—	0.29248	0.29167	0.29077
0.250	0.0	—	0.28239	0.28055	0.27975	0.27888
0.375	0.0	—	—	0.26022	0.24943	0.25863
0.50	0.0	0.24107	0.23220	0.23081	0.23005	0.22934
0.625	0.0	—	—	—	0.19067	0.19009
0.750	0.0	—	0.14137	0.14064	0.14014	0.13973
0.875	0.0	—	—	—	0.07709	0.07687
0.125	0.125	—	—	0.28862	0.28781	0.28692
0.250	0.250	—	0.26752	0.26580	0.26498	0.26415
0.375	0.375	—	—	0.22960	0.22873	0.22799
0.50	0.50	0.19286	0.18381	0.18282	0.18179	0.18114
0.625	0.625	—	—	—	0.12813	0.12757
0.750	0.750	—	0.07506	0.07481	0.07332	0.07282
0.875	0.875	—	—	—	0.02561	0.02510

† See Fig. 8.19 for the finite-element meshes.

convergence of the finite element solution of the refined meshes to the series solution is apparent from the results presented.

8.3.3 Imposition of Boundary Conditions

In most problems of interest, one encounters situations where the portion of the boundary on which natural boundary conditions are specified has points in common with the portion of the boundary on which the essential boundary conditions are specified. In other words, at a few nodal points of the mesh, both the primary and secondary degrees of freedom may be specified. Such points are called *singular points*. Obviously, one cannot impose boundary conditions on both the primary and secondary variables at the same point. As a general rule, one should impose the essential boundary condition (i.e., the boundary condition on the primary variables) at the singular points and disregard the natural boundary condition (i.e., the boundary condition on the secondary variables), because the essential boundary conditions are often maintained more strictly than the natural ones. Of course, if the true situation in a problem is that the natural boundary conditions are imposed and the essential boundary conditions are a result of this then consideration must be given to the former.

Another type of singularity that one encounters in the solution of boundary value problems is the specification of two different values of a primary variable at the same boundary point. An example of such a case is provided by the problem in Fig. 8.20, where u is specified to be zero on the

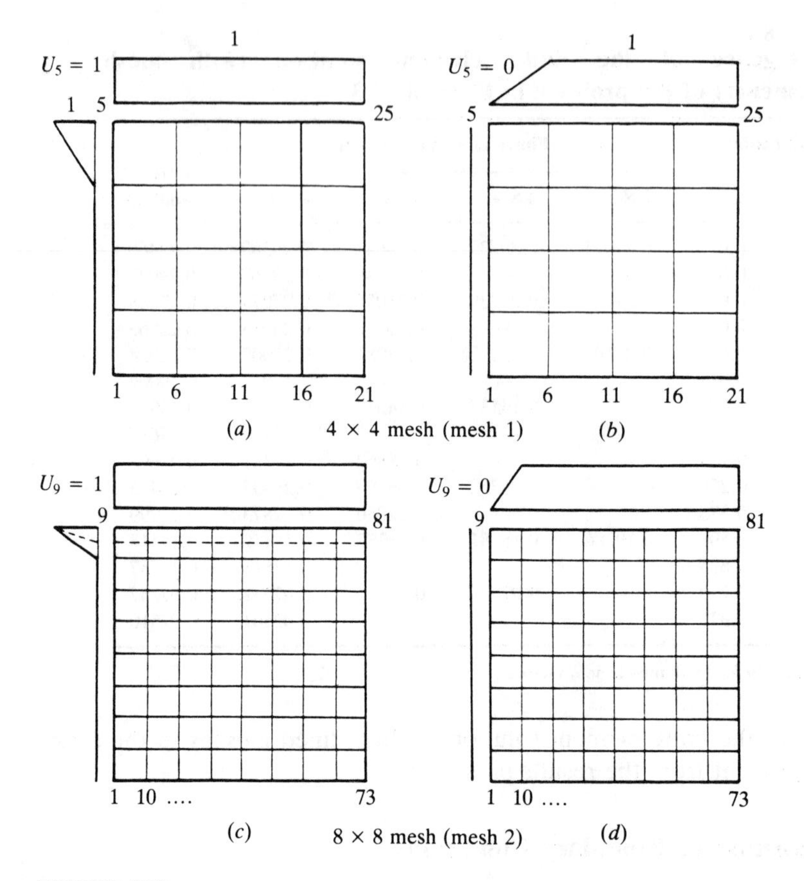

FIGURE 8.20
Effect of specifying (either of the) two values of a primary variable at a boundary node [node 5 in (*a*) and (*b*) and node 9 in (*c*) and (*d*)].

boundary defined by the line $x = 0$ (for any y), and to be unity on the boundary defined by the line $y = 1$ (for any x). Consequently, at $x = 0$ and $y = 1$, u has two different values. In the finite element analysis, one must make a choice between the two values, or a weighted average of the two values can be used. When a choice is made between two values, often the larger one is imposed. In any case, the true boundary condition is replaced by an approximate condition. The closeness of the approximate boundary condition to the true one depends on the size of the element containing the point (see Fig. 8.20). A mesh refinement in the vicinity of the singular point often yields an acceptable solution.

8.4 APPLICATIONS

8.4.1 Heat Transfer

In Chapter 3, Section 3.3.1, heat transfer (by conduction and convection) in one-dimensional (axial and radially symmetric) systems was considered. Here we consider heat transfer in two-dimensional plane and axisymmetric systems. The derivation of two-dimensional heat transfer equations in plane and axisymmetric geometries follows the same procedure as in one dimension, but considers heat transfer in the two directions. Details of such derivations can be found in textbooks on heat transfer (see the references at the end of the chapter). Here we record the governing equations for various cases, construct their finite element models, and present typical applications.

For heat conduction in plane or axisymmetric geometries, the finite element models developed in Sections 8.2 and 8.3 are immediately applicable with the following interpretation of the variables:

$$\begin{aligned} u = T &\equiv \text{temperature (in °C)} \\ q_n &\equiv \text{negative of heat flux (in W m}^{-2}\,°\text{C}^{-1}) \\ a_{11}, a_{22} &\equiv \text{conductivities (in W m}^{-1}\,°\text{C}^{-1}) \text{ of an orthotropic medium} \\ &\quad\ \text{whose principal material axes coincide with the } (x, y) \text{ axes} \\ f &\equiv \text{internal heat generation (in W m}^{-3}\,°\text{C}^{-1}) \\ a_{00} &= 0 \end{aligned} \tag{8.94}$$

For convective heat transfer, i.e., when heat is transferred from one medium to the surrounding medium (often a fluid) by convection, the finite element model developed earlier requires some modification. This is because, in two-dimensional problems, the convective boundary is a curve as opposed to a point in one-dimensional problems. Therefore, the contributions of the convection (or Newton-type) boundary condition to the coefficient matrix and source vector are to be computed by evaluating boundary integrals involving the interpolation functions of elements with convective boundaries. The model to be presented allows the computation of the additional contributions to the coefficient matrix and source vector whenever the element has the convection boundary condition.

PLANE SYSTEMS. The governing equation for steady-state heat transfer in plane systems is a special case of (8.1), and is given by

$$-\frac{\partial}{\partial x}\left(k_x \frac{\partial T}{\partial x}\right) - \frac{\partial}{\partial y}\left(k_y \frac{\partial T}{\partial y}\right) = f(x, y) \tag{8.95}$$

where T is the temperature (in °C), k_x and k_y are the thermal conductivities (in W m^{-1} °C^{-1}) along the x and y directions, respectively, and f is the internal heat generation per unit volume (in W m^{-3}). For a convective boundary, the natural boundary condition is a balance of energy transfer across the boundary due to conduction and/or convection (i.e., Newton's law of cooling):

$$k_x \frac{\partial T}{\partial x} n_x + k_y \frac{\partial T}{\partial y} n_y + \beta(T - T_\infty) = \hat{q}_n \tag{8.96}$$

where β is the convective conductance (or the convective heat transfer coefficient) (in W m^{-2} °C^{-1}), T_∞ is the (ambient) temperature of the surrounding fluid medium, and $\hat{q}_n$ is the specified heat flow. The first two terms account for heat transfer by conduction, the third for heat transfer by convection, while the term on the right-hand side accounts for the specified heat flux, if any. It is the presence of the term $\beta(T - T_\infty)$ that requires some modification of (8.10).

The weak form of (8.95) can be obtained from (8.8). The boundary integral should be modified to account for the convective heat transfer term in (8.96). Instead of replacing the coefficient of w in the boundary integral by q_n, we use (8.96) ($\hat{q}_n$ is replaced by q_n, which is obtained on the element boundary):

$$
\begin{aligned}
0 &= \int_{\Omega^e} \left(k_x \frac{\partial w}{\partial x}\frac{\partial T}{\partial x} + k_y \frac{\partial w}{\partial y}\frac{\partial T}{\partial y} - wf \right) dx\,dy - \oint_{\Gamma^e} w\left(k_x \frac{\partial T}{\partial x} n_x + k_y \frac{\partial T}{\partial y} n_y \right) ds \\
&= \int_{\Omega^e} \left(k_x \frac{\partial w}{\partial x}\frac{\partial T}{\partial x} + k_y \frac{\partial w}{\partial y}\frac{\partial T}{\partial y} - wf \right) dx\,dy - \oint_{\Gamma^e} w[q_n - \beta(T - T_\infty)]\, ds \\
&= B(w, T) - l(w)
\end{aligned}
\tag{8.97a}
$$

where w is the weight function, and $B(\cdot, \cdot)$ and $l(\cdot)$ are the bilinear and linear forms

$$
\begin{aligned}
B(w, T) &= \int_{\Omega^e} \left(k_x \frac{\partial w}{\partial x}\frac{\partial T}{\partial x} + k_y \frac{\partial w}{\partial y}\frac{\partial T}{\partial y} \right) dx\,dy + \oint_{\Gamma^e} \beta w T\, ds \\
l(w) &= \int_{\Omega^e} wf\, dx\,dy + \oint_{\Gamma^e} \beta w T_\infty\, ds + \oint_{\Gamma^e} w q_n\, ds
\end{aligned}
\tag{8.97b}
$$

The finite element model of (8.58) is obtained by substituting the finite element approximation of the form

$$
T = \sum_{j=1}^{n} T_j^e \psi_j^e(x, y) \tag{8.98}
$$

for T and ψ_i^e for w into (8.97):

$$
\boxed{\sum_{j=1}^{n} (K_{ij}^e + H_{ij}^e) T_j^e = F_i^e + P_i^e} \tag{8.99a}
$$

where

$$
\boxed{
\begin{aligned}
K_{ij}^e &= \int_{\Omega^e} \left(k_x \frac{\partial \psi_i}{\partial x}\frac{\partial \psi_j}{\partial x} + k_y \frac{\partial \psi_i}{\partial y}\frac{\partial \psi_j}{\partial y} \right) dx\,dy \\
F_i^e &= \int_{\Omega^e} f\psi_i\, dx\,dy + \oint_{\Gamma^e} q_n^e \psi_i\, ds \equiv f_i^e + Q_i^e \\
H_{ij}^e &= \beta^e \oint_{\Gamma^e} \psi_i \psi_j\, ds, \qquad P_i^e = \beta^e \oint_{\Gamma^e} \psi_i T_\infty\, ds
\end{aligned}
}
\tag{8.99b}
$$

Note that by setting the heat transfer coefficient β equal to zero, we obtain the heat conduction model with no account taken of convection.

The additional coefficients H_{ij}^e and P_i^e due to the convective boundary conditions can be computed by evaluating boundary integrals. These coefficients must be computed only for those elements and boundaries that are subject to the convective boundary condition. The computation of the coefficients for the linear triangular and rectangular elements is presented in the following paragraphs. The coefficients H_{ij}^e and P_i^e for a linear triangular element are defined by

$$\begin{aligned} H_{ij}^e &= \beta_{12}^e \int_0^{h_{12}^e} \psi_i^e \psi_j^e \, ds + \beta_{23}^e \int_0^{h_{23}^e} \psi_i^e \psi_j^e \, ds + \beta_{31}^e \int_0^{h_{31}^e} \psi_i^e \psi_j^e \, ds \\ P_i^e &= \beta_{12}^e T_\infty^{12} \int_0^{h_{12}^e} \psi_i^e \, ds + \beta_{23}^e T_\infty^{23} \int_0^{h_{23}^e} \psi_i^e \, ds + \beta_{31}^e T_\infty^{31} \int_0^{h_{31}^e} \psi_i^e \, ds \end{aligned} \tag{8.100}$$

where β_{ij}^e is the film coefficient (assumed to be constant) for the side connecting nodes i and j of the element Ω^e, T_∞^{ij} is the ambient temperature on that side, and h_{ij}^e is the length of the side. For a rectangular element, the expressions in (8.100) must be modified to account for four line integrals on four sides of the element.

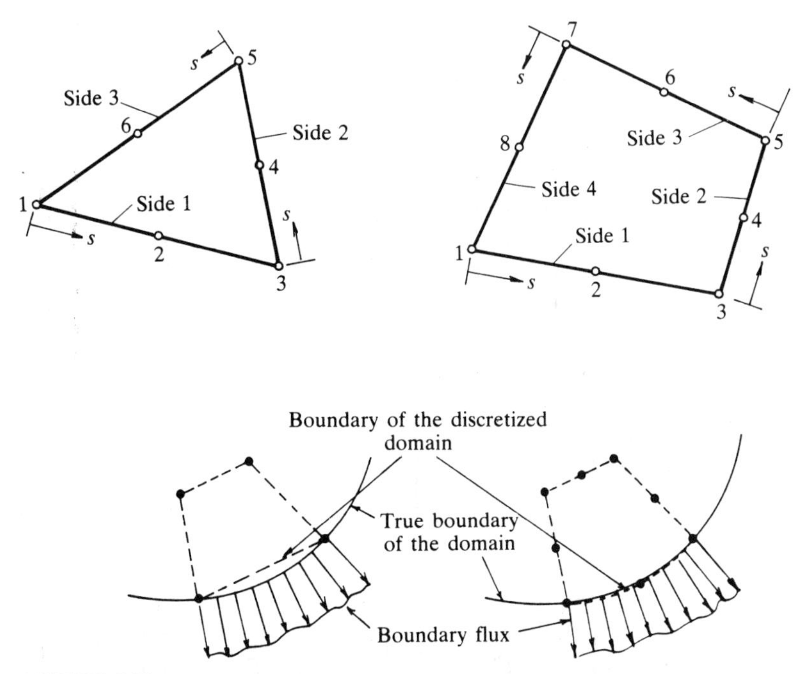

FIGURE 8.21
Triangular and quadrilateral elements, with node numbers and local coordinates for the evaluation of the boundary integrals. Also shown are the boundary approximation and flux representation using linear and quadratic elements.

The boundary integrals are line integrals involving the interpolation functions. The local coordinate s is taken along the side, with its origin at the first node of the side (see Fig. 8.21). As noted earlier, the interpolation functions on any given side are the one-dimensional interpolation functions. Therefore, the evaluation of integrals is made easy. Indeed, the integrals

$$\int_0^{h_{ij}^e} \psi_i^e \psi_j^e \, ds, \qquad \int_0^{h_{ij}^e} \psi_i^e \, ds$$

have been evaluated in Chapter 3 in connection with mass matrix coefficients and source vector coefficients for linear and quadratic elements. We summarize the results here.

For a linear triangular element, the matrices $[H^e]$ and $\{P^e\}$ are given by

$$[H^e] = \frac{\beta_{12}^e h_{12}^e}{6}\begin{bmatrix} 2 & 1 & 0 \\ 1 & 2 & 0 \\ 0 & 0 & 0 \end{bmatrix} + \frac{\beta_{23}^e h_{23}^e}{6}\begin{bmatrix} 0 & 0 & 0 \\ 0 & 2 & 1 \\ 0 & 1 & 2 \end{bmatrix} + \frac{\beta_{31}^e h_{31}^e}{6}\begin{bmatrix} 2 & 0 & 1 \\ 0 & 0 & 0 \\ 1 & 0 & 2 \end{bmatrix} \tag{8.101a}$$

$$\{P^e\} = \frac{\beta_{12}^e T_\infty^{12} h_{12}^e}{2}\begin{Bmatrix} 1 \\ 1 \\ 0 \end{Bmatrix} + \frac{\beta_{23}^e T_\infty^{23} h_{23}^e}{2}\begin{Bmatrix} 0 \\ 1 \\ 1 \end{Bmatrix} + \frac{\beta_{31}^e T_\infty^{31} h_{31}^e}{2}\begin{Bmatrix} 1 \\ 0 \\ 1 \end{Bmatrix} \tag{8.101b}$$

For a quadratic triangular element, we have

$$[H^e] = \frac{\beta_{13}^e h_{13}^e}{30}\begin{bmatrix} 4 & 2 & -1 & 0 & 0 & 0 \\ 2 & 16 & 2 & 0 & 0 & 0 \\ -1 & 2 & 4 & 0 & 0 & 0 \\ 0 & 0 & 0 & 0 & 0 & 0 \\ 0 & 0 & 0 & 0 & 0 & 0 \\ 0 & 0 & 0 & 0 & 0 & 0 \end{bmatrix} + \frac{\beta_{35}^e h_{35}^e}{30}\begin{bmatrix} 0 & 0 & 0 & 0 & 0 & 0 \\ 0 & 0 & 0 & 0 & 0 & 0 \\ 0 & 0 & 4 & 2 & -1 & 0 \\ 0 & 0 & 2 & 16 & 2 & 0 \\ 0 & 0 & -1 & 2 & 4 & 0 \\ 0 & 0 & 0 & 0 & 0 & 0 \end{bmatrix}$$

$$+ \frac{\beta_{51} h_{51}^e}{30}\begin{bmatrix} 4 & 0 & 0 & 0 & -1 & 2 \\ 0 & 0 & 0 & 0 & 0 & 0 \\ 0 & 0 & 0 & 0 & 0 & 0 \\ 0 & 0 & 0 & 0 & 0 & 0 \\ -1 & 0 & 0 & 0 & 4 & 2 \\ 2 & 0 & 0 & 0 & 2 & 16 \end{bmatrix} \tag{8.102a}$$

$$\{P^e\} = \frac{\beta_{13}^2 T_\infty^{13} h_{13}^e}{6}\begin{Bmatrix} 1 \\ 4 \\ 1 \\ 0 \\ 0 \\ 0 \end{Bmatrix} + \frac{\beta_{35}^e T_\infty^{35} h_{35}^e}{6}\begin{Bmatrix} 0 \\ 0 \\ 1 \\ 4 \\ 1 \\ 0 \end{Bmatrix} + \frac{\beta_{51}^e T_\infty^{51} h_{51}^e}{6}\begin{Bmatrix} 1 \\ 0 \\ 0 \\ 0 \\ 1 \\ 4 \end{Bmatrix} \tag{8.102b}$$

For a linear rectangular element, the matrix $[H^e]$ is of the form

$$[H^e] = \frac{\beta_{12}^e h_{12}^e}{6}\begin{bmatrix} 2 & 1 & 0 & 0 \\ 1 & 2 & 0 & 0 \\ 0 & 0 & 0 & 0 \\ 0 & 0 & 0 & 0 \end{bmatrix} + \frac{\beta_{23}^e h_{23}^e}{6}\begin{bmatrix} 0 & 0 & 0 & 0 \\ 0 & 2 & 1 & 0 \\ 0 & 1 & 2 & 0 \\ 0 & 0 & 0 & 0 \end{bmatrix}$$
$$+ \frac{\beta_{34}^e h_{34}^e}{6}\begin{bmatrix} 0 & 0 & 0 & 0 \\ 0 & 0 & 0 & 0 \\ 0 & 0 & 2 & 1 \\ 0 & 0 & 1 & 2 \end{bmatrix} + \frac{\beta_{41}^e h_{41}^e}{6}\begin{bmatrix} 2 & 0 & 0 & 1 \\ 0 & 0 & 0 & 0 \\ 0 & 0 & 0 & 0 \\ 1 & 0 & 0 & 2 \end{bmatrix} \tag{8.103a}$$

and $\{P^e\}$ is given by

$$\{P^e\} = \frac{\beta_{12}^e T_\infty^{12} h_{12}^e}{2}\begin{Bmatrix} 1 \\ 1 \\ 0 \\ 0 \end{Bmatrix} + \frac{\beta_{23}^e T_\infty^{23} h_{23}^e}{2}\begin{Bmatrix} 0 \\ 1 \\ 1 \\ 0 \end{Bmatrix} + \frac{\beta_{34}^e T_\infty^{34} h_{34}^e}{2}\begin{Bmatrix} 0 \\ 0 \\ 1 \\ 1 \end{Bmatrix} + \frac{\beta_{41}^e T_\infty^{41} h_{41}^e}{2}\begin{Bmatrix} 1 \\ 0 \\ 0 \\ 1 \end{Bmatrix} \tag{8.103b}$$

Similar expressions hold for a quadratic rectangular element.

AXISYMMETRIC SYSTEMS. For symmetric heat transfer about the z axis (i.e., independent of the circumferential coordinate), the governing equation is given by

$$-\left[\frac{1}{r}\frac{\partial}{\partial r}\left(rk_r\frac{\partial T}{\partial r}\right) + \frac{\partial}{\partial z}\left(k_z\frac{\partial T}{\partial z}\right)\right] = f(r, z) \tag{8.104}$$

where r is the radial coordinate and z is the axial coordinate. The temperature gradient vector is defined by

$$\mathbf{q} = r\left(k_r\frac{\partial T}{\partial r}\hat{\mathbf{i}} + k_z\frac{\partial T}{\partial z}\hat{\mathbf{j}}\right)$$

and the normal derivative of T (i.e., the negative of the heat flux) across the surface is

$$q_n = r\left(k_r\frac{\partial T}{\partial r}n_r + k_z\frac{\partial T}{\partial z}n_z\right) \tag{8.105}$$

where n_r and n_z are the direction cosines of the unit normal $\hat{\mathbf{n}}$,

$$\hat{\mathbf{n}} = n_r\hat{\mathbf{i}} + n_z\hat{\mathbf{j}}$$

The weak form of (8.104) is given by

$$0 = 2\pi\int_{\Omega^e} w\left\{-\left[\frac{1}{r}\frac{\partial}{\partial r}\left(k_r r\frac{\partial T}{\partial r}\right) + \frac{\partial}{\partial z}\left(k_z\frac{\partial T}{\partial z}\right)\right] - f\right\} r\,dr\,dz$$
$$= 2\pi\int_{\Omega^e}\left(k_r\frac{\partial w}{\partial r}\frac{\partial T}{\partial r} + k_z\frac{\partial w}{\partial z}\frac{\partial T}{\partial z} - wf\right) r\,dr\,dz - 2\pi\oint_{\Gamma^e} wq_n\,ds \tag{8.106}$$

where the factors of 2π come from the integration with respect to the circumferential coordinate over $(0, 2\pi)$, and q_n is given by (8.105). The convective boundary condition is of the form

$$q_n + r\beta(T - T_\infty) = \hat{q}_n$$

Substituting for $q_n = -r\beta(T - T_\infty) + \hat{q}_n$ in (8.106), we obtain

$$0 = 2\pi \int_{\Omega^e} \left(k_r \frac{\partial w}{\partial r} \frac{\partial T}{\partial r} + k_z \frac{\partial w}{\partial z} \frac{\partial T}{\partial z} - wf \right) r\, dr\, dz - 2\pi \oint_{\Gamma^e} w[-r\beta(T - T_\infty) + \hat{q}_n]\, ds \tag{8.107}$$

The finite element model of this equation is

$$[K^e + H^e]\{T^e\} = \{f^e\} + \{P^e\} + \{Q^e\} \tag{8.108a}$$

where

$$K_{ij}^e = 2\pi \int_{\Omega^e} \left(k_r \frac{\partial \psi_i^e}{\partial r} \frac{\partial \psi_j^e}{\partial r} + k_z \frac{\partial \psi_i^e}{\partial z} \frac{\partial \psi_j^e}{\partial z} \right) r\, dr\, dz$$

$$H_{ij}^e = 2\pi \oint_{\Gamma^e} \beta^e \psi_i^e \psi_j^e r\, ds, \qquad f_i^e = 2\pi \int_{\Omega^e} \psi_i^e f r\, dr\, dz \tag{8.108b}$$

$$Q_i^e = 2\pi \oint_{\Gamma^e} \hat{q}_n \psi_i^e\, ds, \qquad P_i^e = 2\pi \oint_{\Gamma^e} \beta^e T_\infty^e \psi_i^e r\, ds$$

Evaluation of the line integrals in $[H^e]$ and $\{P^e\}$ once again follows along the lines of Example 8.3.

The finite element models in (8.99) and (8.108) are valid for conductive and convective heat transfer boundary conditions. Radiative heat transfer boundary conditions are nonlinear, and therefore are not considered here. For problems with no convective boundary conditions, the convective contributions $[H^e]$ and $\{P^e\}$ to the element coefficients are omitted. Indeed, these contributions have to be included only for those elements whose sides fall on the problem boundary with convective heat transfer specified. For example, if side 2–3 of the element Ω^e is on the boundary with convective boundary conditions then the only contribution to $[H^e]$ and $\{P^e\}$ comes from the second integrals in (8.100).

Next we consider a couple of examples of heat transfer.

Example 8.4. Consider steady-state heat conduction in an isotropic rectangular region of dimensions $3a$ by $2a$ (see Fig. 8.22a). The origin of the x and y coordinates is taken at the lower left corner such that x is parallel to the side $3a$ and y is parallel to the side $2a$. The boundaries $x = 0$ and $y = 0$ are insulated, the boundary $x = 3a$ is maintained at zero temperature, and the boundary $y = 2a$ is maintained at a temperature $T = T_0 \cos(\pi x/6a)$. We wish to determine the temperature distribution using the finite element method in the region and the heat required at the boundary $x = 3a$ to maintain it at zero temperature.

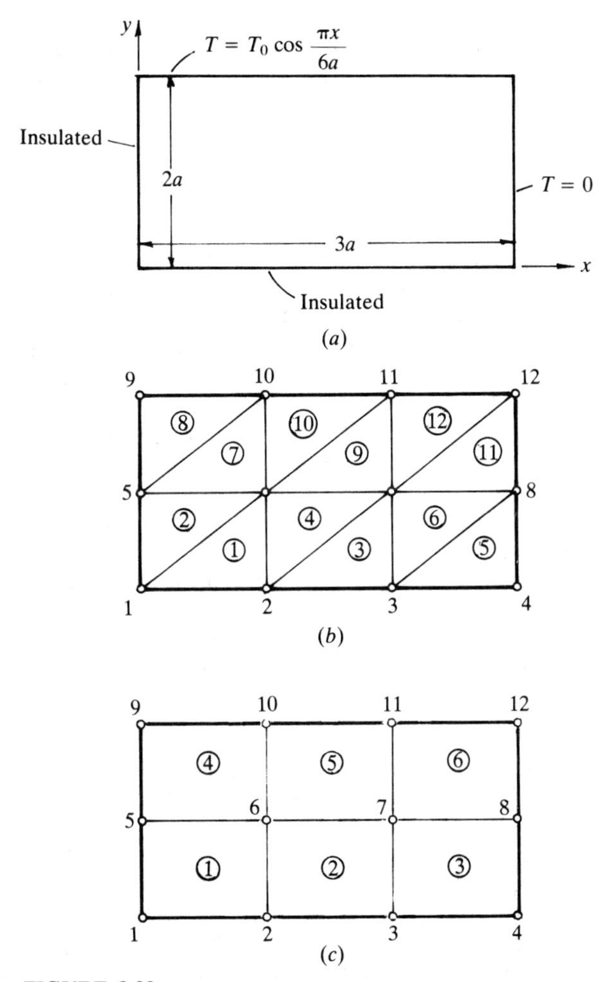

FIGURE 8.22
Finite element analysis of a heat conduction problem over a rectangular domain: (*a*) domain; (*b*) mesh of linear triangular elements; (*c*) mesh of linear rectangular elements.

To analyze the problem, we first note that it is governed by (8.95) with zero internal heat generation, $f = 0$, and no-convection boundary conditions:

$$-k\nabla^2 T = 0 \tag{8.109}$$

Hence, the finite element model of the problem is given by (8.99) with $[H^e]$ and $\{P^e\}$ omitted:

$$[K^e]\{u^e\} = \{Q^e\} \qquad (\{f^e\} = \{0\}) \tag{8.110a}$$

where u_i^e is the temperature at node i of the element Ω^e, and

$$K_{ij}^e = \int_{\Omega^e} k\left(\frac{\partial \psi_i}{\partial x}\frac{\partial \psi_j}{\partial x} + \frac{\partial \psi_i}{\partial y}\frac{\partial \psi_j}{\partial y}\right) dx\, dy, \qquad Q_i^e = \oint_{\Gamma^e} q_n \psi_i \, ds \tag{8.110b}$$

Suppose that we use a 3×2 mesh (i.e., 3 subdivisions along the x axis and 2 along the y axis) of linear triangular elements and then a 3×2 mesh of linear rectangular elements. Both meshes have the same number of global nodes (12) but differing numbers of elements.

Triangular element mesh (12 elements). The global node numbers, element numbers, and element node numbers used are shown in Fig. 8.22(b). Of course, the global node numbering and element numbering are arbitrary (they do not have to follow any particular pattern), although the global node numbering dictates the size of the half-bandwidth of the assembled equations, which in turn affects the computational time of Gauss elimination methods used in the solution of algebraic equations on a computer. The element node numbering scheme should be that used in the development of element interpolation functions. In the present study, a counter-clockwise numbering system was adopted [see (8.19) and Fig. 8.4]. According to the element node numbering scheme used in Fig. 8.22(a), all elements in the mesh fall into one of two geometric shapes: one with its base at the bottom of the element and another with its base at the top of the element. By renumbering the element nodes, as shown in Fig. 8.12(b), all elements can be made to have a common geometric shape, and thus the element coefficients need to be computed only for a single element. Such considerations are important only when hand calculations are carried out.

For a typical element of the mesh of triangles in Fig. 8.22(b), the element coefficient matrix is [see (8.69a) and (8.70)]

$$[K^e] = \frac{k}{2}\begin{bmatrix} 1 & -1 & 0 \\ -1 & 2 & -1 \\ 0 & -1 & 1 \end{bmatrix}$$

where k is the conductivity of the medium. Note that the element matrix is independent of the size of the element, as long as the latter is a right-angled triangle with base equal to height.

The assembly of the elements (on a computer) follows the logic discussed earlier. The boundary conditions require that

$$\begin{aligned} &U_4 = U_8 = U_{12} = 0, \qquad U_9 = T_0, \qquad U_{10} = \tfrac{1}{2}\sqrt{3}\,T_0, \qquad U_{11} = \tfrac{1}{2}T_0 \\ &F_1 = F_2 = F_3 = F_5 = 0 \quad \text{(zero heat flow due to insulated boundary)} \end{aligned} \tag{8.111}$$

We first write the six finite element equations for the six unknown primary variables. These equations come from nodes 1, 2, 3, 5, 6, and 7:

$$\frac{k}{2}\begin{bmatrix} 2 & -1 & 0 & -1 & 0 & 0 \\ -1 & 4 & -1 & 0 & -2 & 0 \\ 0 & -1 & 4 & 0 & 0 & -2 \\ -1 & 0 & 0 & 4 & -2 & 0 \\ 0 & -2 & 0 & -2 & 8 & -2 \\ 0 & 0 & -2 & 0 & -2 & 8 \end{bmatrix} \begin{Bmatrix} U_1 \\ U_2 \\ U_3 \\ U_5 \\ U_6 \\ U_7 \end{Bmatrix} = \frac{k}{2}\begin{Bmatrix} 0 \\ 0 \\ 0 \\ T_0 \\ \sqrt{3}\,T_0 \\ T_0 \end{Bmatrix} \tag{8.112}$$

The solution of these equations is (in °C)

$$\begin{aligned} &U_1 = 0.6362T_0, \qquad U_2 = 0.5510T_0, \qquad U_3 = 0.3181T_0 \\ &U_5 = 0.7214T_0, \qquad U_6 = 0.6248T_0, \qquad U_7 = 0.3607T_0 \end{aligned} \tag{8.113}$$

The exact solution of (8.109) for the boundary conditions shown in Fig. 8.22(a) is

$$T(x, y) = T_0 \frac{\cosh(\pi y/6a)\cos(\pi x/6a)}{\cosh \frac{1}{3}\pi} \tag{8.114a}$$

Evaluating the exact solution at the nodes, we have (in °C)

$$
\begin{aligned}
T_1 &= 0.6249T_0, \quad & T_2 &= 0.5412T_0, \quad & T_3 &= 0.3124T_0 \\
T_5 &= 0.7125T_0, \quad & T_6 &= 0.6171T_0, \quad & T_7 &= 0.3563T_0
\end{aligned}
\tag{8.114b}
$$

The heat at node 4, for example, can be compared from the fourth finite element equation:

$$
F_4 = Q_2^5 = K_{41}U_1 + K_{42}U_2 + K_{43}U_3 + K_{44}U_4 + K_{45}U_5 + K_{46}U_6 + K_{47}U_7 + K_{48}U_8 + \ldots \tag{8.115a}
$$

Noting that $K_{41} = K_{42} = K_{45} = \ldots = K_{4(12)} = 0$ and $U_4 = U_8 = 0$, we obtain

$$
Q_2^5 = -\tfrac{1}{2}k\, U_3 = -0.1591kT_0 \text{ (in W)} \tag{8.115b}
$$

Rectangular element mesh (6 elements). For a 3×2 mesh of linear rectangular elements (see Fig. 8.22c), the element coefficient matrix is given by (8.34) and (8.44) with $a_{00} = 0$, $a_{11} = a_{22} = k$, $a_{12} = 0$, and $a = b = 1$:

$$
[K^e] = \frac{k}{6}\begin{bmatrix} 4 & -1 & -2 & -1 \\ -1 & 4 & -1 & -2 \\ -2 & -1 & 4 & -1 \\ -1 & -2 & -1 & 4 \end{bmatrix}, \qquad \{f^e\} = \{0\} \tag{8.116}
$$

The present mesh of rectangular elements is node-wise-equivalent to the triangular element mesh considered in Fig. 8.22(b). Hence the boundary conditions in (8.111) are valid for the present case. The six finite element equations for the unknowns U_1, U_2, U_3, U_5, U_6, and U_7 have the same form as before.

The equations for the unknown temperatures (i.e., the condensed equations for the unknown primary variables) are given by

$$
\frac{k}{6}\begin{bmatrix} 4 & -1 & 0 & -1 & -2 & 0 \\ -1 & 8 & -1 & -2 & -2 & -2 \\ 0 & -1 & 8 & 0 & -2 & -2 \\ -1 & -2 & 0 & 8 & -2 & 0 \\ -2 & -2 & -2 & -2 & 16 & -2 \\ 0 & -2 & -2 & 0 & -2 & 16 \end{bmatrix}\begin{Bmatrix} U_1 \\ U_2 \\ U_3 \\ U_5 \\ U_6 \\ U_7 \end{Bmatrix} = \frac{k}{6}\begin{Bmatrix} 0 \\ 0 \\ 0 \\ T_0 + \sqrt{3}\,T_0 \\ 2T_0 + \sqrt{3}\,T_0 + T_0 \\ \sqrt{3}\,T_0 + T_0 \end{Bmatrix} \tag{8.117}
$$

Their solution is

$$
\begin{aligned}
U_1 &= 0.6128T_0, \quad & U_2 &= 0.5307T_0, \quad & U_3 &= 0.3064T_0 \\
U_5 &= 0.7030T_0, \quad & U_6 &= 0.6088T_0, \quad & U_7 &= 0.3515T_0
\end{aligned}
\tag{8.118}
$$

The value of the heat at node 4 is given by

$$
Q_2^3 = K_{43}U_3 + K_{47}U_7 = -\frac{k}{6}U_3 - \frac{2k}{6}U_7 = -0.1682kT_0 \text{ (in W)}
$$

We note that the results obtained using the 3×2 mesh of rectangular elements is not as accurate as that obtained with 3×2 mesh of triangular elements. This is due to the fact that there are only half as many elements in the former case as in the latter.

Table 8.5 gives a comparison of the finite element solutions with the analytical solution (8.114a) for two different meshes of linear triangular and rectangular elements, and Fig. 8.23 shows plots of $T(x, 0)$ and $q_x(x, 0)/T_0$, computed using various meshes of

TABLE 8.5
Comparison of the nodal temperatures $T(x, y)/T_0$, obtained using various finite element meshes,† with the analytical solution of (8.109) (Example 8.4)

		Triangles		Rectangles		Analytical solution
x	y	3×2	6×4	3×2	6×4	(8.114a)
0.0	0.0	0.6362	0.6278	0.6128	0.6219	0.6249
0.5	0.0	—	0.6064	—	0.6007	0.6036
1.0	0.0	0.5510	0.5437	0.5307	0.5386	0.5412
1.5	0.0	—	0.4439	—	0.4398	0.4419
2.0	0.0	0.3181	0.3139	0.3064	0.3110	0.3124
2.5	0.0	—	0.1625	—	0.1610	0.1617
0.0	1.0	0.7214	0.7148	0.7030	0.7102	0.7125
0.5	1.0	—	0.6904	—	0.6860	0.6882
1.0	1.0	0.6248	0.6190	0.6088	0.6150	0.6171
1.5	1.0	—	0.5054	—	0.5022	0.5038
2.0	1.0	0.3607	0.3574	0.3515	0.3551	0.3563
2.5	1.0	—	0.1850	—	0.1838	0.1844

† See Fig. 8.22 for the geometry and meshes.

triangular and rectangular elements. Note that Q_i^e are heats (in W) whereas q_x is the flux (in W m^{-1}) in the x direction ($q_x = -k\,\partial T/\partial x$).

Next we consider an example involving convective heat transfer.

Example 8.5. Consider heat transfer in a rectangular region of dimensions a by b, subject to the boundary conditions shown in Fig. 8.24. We wish to write the finite element algebraic equations for the unknown nodal temperatures and heats. For illustrative purposes, a 4 × 2 mesh of rectangular elements is chosen. We assume that the medium is orthotropic, with conductivities k_x and k_y in the x and y directions, respectively. It is assumed that there is no internal heat generation.

The heat transfer in the region is governed by the equation

$$-\frac{\partial}{\partial x}\left(k_x\frac{\partial T}{\partial x}\right)-\frac{\partial}{\partial y}\left(k_y\frac{\partial T}{\partial y}\right)=0$$

The finite element model of the equation is given by

$$[K^e + H^e]\{u^e\} = \{Q^e\} + \{P^e\} \qquad (\{f^e\} = \{0\}) \tag{8.119}$$

where u_i^e denotes the temperature at node i of the element Ω^e.

The element matrices are

$$[K^e]=\frac{k_x r}{6}\begin{bmatrix}2&-2&-1&1\\-2&2&1&-1\\-1&1&2&-2\\1&-1&-2&2\end{bmatrix}+\frac{k_y}{6r}\begin{bmatrix}2&1&-1&-2\\1&2&-2&-1\\-1&-2&2&1\\-2&-1&1&2\end{bmatrix}$$

$$[H^e]=\frac{\beta_{23}^e h_{23}^e}{6}\begin{bmatrix}0&0&0&0\\0&2&1&0\\0&1&2&0\\0&0&0&0\end{bmatrix} \qquad (e = 4, 8)$$

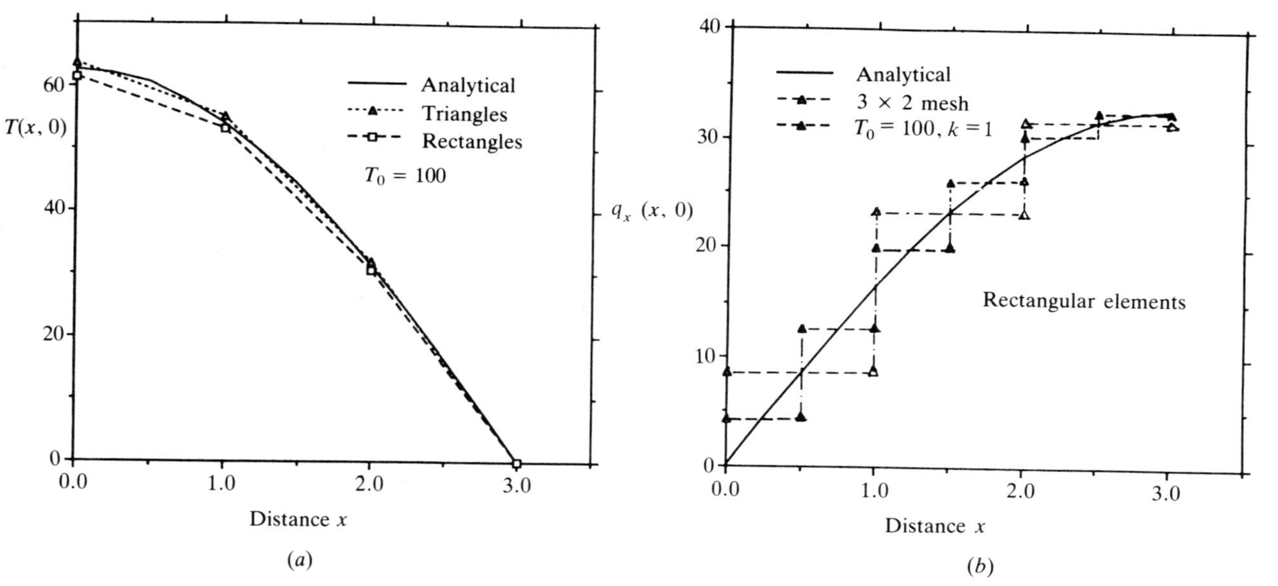

FIGURE 8.23
Comparison of the finite element solution with the analytical solution of the heat conduction problem of Example 8.4: (*a*) $T(x, 0)$ versus x; (*b*) $\partial T/\partial x$ versus x.

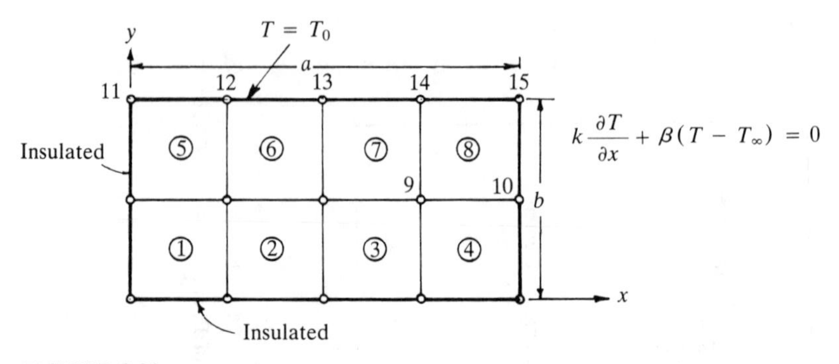

FIGURE 8.24
Domain and boundary conditions for convective heat transfer in a rectangular domain. A mesh of linear rectangular elements is also shown (Example 8.5).

$$\{P^e\} = \frac{\beta_{23}^e T_\infty^{23} h_{23}^e}{2}\begin{Bmatrix}0\\1\\1\\0\end{Bmatrix} \quad (e = 4, 8) \tag{8.120}$$

where

$$r = \tfrac{1}{2}b/\tfrac{1}{4}a = 2b/a$$

There are 10 nodal temperatures that are to be determined, and heats at all nodes except nodes 7, 8, 9 and 10 are to be computed. To illustrate the procedure, we write algebraic equations for only representative temperatures and heats.

Node 10 (for temperatures)

$$K_{31}^4 U_4 + (K_{32}^4 + H_{32}^4)U_5 + (K_{43}^4 + K_{21}^8)U_9$$
$$+ (K_{33}^4 + H_{33}^4 + K_{22}^8 + H_{22}^8)U_{10} + K_{24}^8 U_{14} + (K_{23}^8 + H_{23}^8)U_{15}$$
$$= (Q_3^4 + P_3^4) + (Q_2^8 + P_2^8) = P_3^4 + P_2^8 \quad \text{(known)}$$

Node 14 (for heat Q_{14})

$$Q_{14} \equiv Q_3^7 + Q_4^8 = K_{31}^7 U_8 + (K_{32}^7 + K_{41}^8)U_9 + K_{42}^8 U_{10} + K_{34}^7 U_{13} + (K_{33}^7 + K_{44}^8)U_{14} + K_{43}^8 U_{15}$$

From the boundary conditions, we know the temperatures at nodes 11–15 (i.e., $U_{11}, U_{12}, \ldots, U_{15}$ are known). Substituting the values of K_{ij}^e, H_{ij}^e, and P_i^e, we obtain explicit form of the algebraic equations. For example, the algebraic equation corresponding to node 10 is

$$-\frac{1}{6}\left(k_x r + \frac{k_y}{r}\right)U_4 + \left[\frac{1}{6}\left(k_x r - \frac{2k_y}{r}\right) + \frac{1}{12}\beta b\right]U_5 + \frac{1}{6}\left[\left(-2k_x r + \frac{k_y}{r}\right) + \left(-2k_x r + \frac{k_y}{r}\right)\right]U_9$$

$$+\frac{2}{3}\left[\left(k_x r + \frac{k_y}{r}\right) + \frac{\beta b}{2}\right]U_{10} + \frac{1}{6}\left(-k_x r - \frac{k_y}{r}\right)U_{14} + \frac{1}{6}\left[k_x r - \frac{2k_y}{r} + \frac{1}{2}\beta b\right]U_{15} = \tfrac{1}{2}\beta b T_\infty$$

8.4.2 Fluid Mechanics

Recall from Section 3.3.2 that there are three basic differential equations of fluid motion. They are as follows [see (3.100), (3.103), and (3.108)]:

Conservation of mass

$$\frac{\partial \rho}{\partial t} + \boldsymbol{\nabla} \cdot \rho(\mathbf{v}) = 0 \tag{8.121}$$

Conservation of linear momentum

$$\rho \frac{D\mathbf{v}}{Dt} = \mathbf{f} - \boldsymbol{\nabla} P + \boldsymbol{\nabla} \cdot \boldsymbol{\tau} \tag{8.122}$$

Conservation of energy

$$\rho \frac{De}{Dt} + P(\boldsymbol{\nabla} \cdot \mathbf{v}) = \boldsymbol{\nabla} \cdot (k\, \boldsymbol{\nabla} T) + \Phi \tag{8.123}$$

Here ρ is the density, $\mathbf{v}$ is the velocity vector, $\mathbf{f}$ is the body force vector, P is the pressure, $\boldsymbol{\tau}$ is the viscous stress tensor, e is the internal energy, k is the thermal conductivity, T is the temperature and Φ is the dissipation,

$$\Phi = \boldsymbol{\nabla} \mathbf{v} : \boldsymbol{\tau}$$

The operator D/Dt denotes the material time derivative,

$$\frac{D}{Dt} = \frac{\partial}{\partial t} + \mathbf{v} \cdot \boldsymbol{\nabla}$$

Equations (8.121)–(8.123) are supplemented with constitutive equations.

A fluid is said to be *incompressible* if the volume change is zero,

$$\boldsymbol{\nabla} \cdot \mathbf{v} = 0 \tag{8.124}$$

and it is termed *inviscid* if the viscosity is zero, $\mu = 0$. A flow with negligible angular velocity is called *irrotational,*

$$\boldsymbol{\nabla} \times \mathbf{v} = \mathbf{0} \tag{8.125}$$

The irrotational flow of an ideal fluid (i.e., $\rho = \text{constant}$ and $\mu = 0$) is called a *potential flow*.

For an ideal fluid ($\boldsymbol{\tau} = \mathbf{0}$), the continuity and momentum equations can be written as

$$\boldsymbol{\nabla} \cdot \mathbf{v} = 0 \tag{8.126a}$$

$$\tfrac{1}{2}\rho \boldsymbol{\nabla}(\mathbf{v} \cdot \mathbf{v}) - \rho[\mathbf{v} \times (\boldsymbol{\nabla} \times \mathbf{v})] = -\boldsymbol{\nabla} \hat{P} \tag{8.126b}$$

where $\boldsymbol{\nabla} \hat{P} = \boldsymbol{\nabla} P - \mathbf{f}$. For irrotational flow, the velocity field $\mathbf{v}$ satisfies (8.125).

For two-dimensional irrotational flows, these equations have the forms

$$\frac{\partial u}{\partial x}+\frac{\partial v}{\partial y}=0 \tag{8.126c}$$

$$\tfrac{1}{2}\rho(u^2+v^2)+\hat{P}=\text{constant} \tag{8.126d}$$

$$\frac{\partial u}{\partial y}-\frac{\partial v}{\partial x}=0 \tag{8.126e}$$

Equations (8.132), (8.133) and (8.126d) are used to determine u, v, and $\hat{P}$.

The problem of determining u, v, and $\hat{P}$ is simplified by introducing a function $\psi(x, y)$ such that the continuity equation is identically satisfied:

$$\boxed{u=\frac{\partial \psi}{\partial y}, \qquad v=-\frac{\partial \psi}{\partial x}} \tag{8.127}$$

Then the irrotational flow condition in terms of ψ takes the form

$$\boxed{\frac{\partial^2 \psi}{\partial y^2}+\frac{\partial^2 \psi}{\partial x^2}\equiv \nabla^2\psi=0} \tag{8.128}$$

This equation is used to determine ψ, and then the velocities u and v are determined from (8.127) and $\hat{P}$ from (8.126d).

The function ψ has the physical significance that lines of constant ψ are lines across which there is no flow, i.e., they are streamlines of the flow. Hence, $\psi(x, y)$ is called the *stream function.*

In cylindrical coordinates, the continuity equation takes the form

$$\frac{\partial u}{\partial r}+\frac{1}{r}\frac{\partial v}{\partial \theta}=0 \tag{8.129}$$

where u and v are the radial and circumferential velocity components. The stream function $\psi(r, \theta)$ is defined by

$$u=\frac{1}{r}\frac{\partial \psi}{\partial \theta}, \qquad v=-\frac{\partial \psi}{\partial r} \tag{8.130}$$

and (8.128) takes the form

$$\nabla^2\psi\equiv\frac{\partial^2 \psi}{\partial r^2}+\frac{1}{r}\frac{\partial \psi}{\partial r}+\frac{1}{r^2}\frac{\partial^2 \psi}{\partial \theta^2}=0 \tag{8.131}$$

There exists an alternative formulation of the potential flow equations (8.126). We introduce the function $\phi(x, y)$, called the *velocity potential,* such that the condition of irrotational flow is identically satisfied:

$$u = -\frac{\partial \phi}{\partial x}, \qquad v = -\frac{\partial \phi}{\partial y} \tag{8.132}$$

Then the continuity equation (8.126c) in terms of the velocity potential takes the form

$$-\nabla^2 \phi = 0 \tag{8.133}$$

Comparing (8.127) with (8.132), we note that

$$-\frac{\partial \phi}{\partial x} = \frac{\partial \psi}{\partial y}, \qquad -\frac{\partial \phi}{\partial y} = -\frac{\partial \psi}{\partial x} \tag{8.134}$$

The velocity potential has the physical significance that lines of constant ϕ are lines along which there is no change in velocity. The equipotential lines and streamlines intersect at right-angles.

Although both ψ and ϕ are governed by the Laplace equation, the boundary conditions on them are different in a flow problem, as should be evident from the definitions (8.127) and (8.132). In this section, we consider applications of the finite element method to potential flows, i.e., the solution of (8.128) and (8.133).

We consider two examples of fluid flow. The first deals with a groundwater flow problem and the second with the flow around a cylindrical body. In discussing these problems, emphasis is placed on certain modeling aspects, data generation, and postprocessing of solutions. Evaluation of element matrices and assembly is amply illustrated in previous examples, and will not be discussed, since it takes substantial space to write the assembled equations even for the crude meshes used in these examples.

Example 8.6 Groundwater flow or seepage. The governing differential equation for a homogeneous aquifer (i.e., one where material properties do not vary with position) of unit depth, with flow in the (x, y) plane, is given by

$$-\frac{\partial}{\partial x}\left(a_{11}\frac{\partial \phi}{\partial x}\right) - \frac{\partial}{\partial y}\left(a_{22}\frac{\partial \phi}{\partial y}\right) = f \tag{8.135}$$

where a_{11} and a_{22} are the coefficients of permeability (in m day^{-1}) along the x and y directions, respectively, ϕ is the piezometric head (in m), measured from a reference level (usually the bottom of the aquifer), and f is the rate of pumping (in m^3 day^{-1} m^{-3}). We know from the previous discussions that the natural and essential boundary conditions associated with (8.135) are

Natural

$$a_{11}\frac{\partial \phi}{\partial x} n_x + a_{22}\frac{\partial \phi}{\partial y} n_y = \phi_n \quad \text{on } \Gamma_1 \tag{8.136a}$$

Essential

$$\phi = \phi_0 \quad \text{on } \Gamma_2 \tag{8.136b}$$

where Γ_1 and Γ_2 are the portions of the boundary Γ of Ω such that $\Gamma_1 + \Gamma_2 = \Gamma$.

Here we consider the following specific problem: find the lines of constant potential ϕ (equipotential lines) in a $3000\text{ m} \times 1500\text{ m}$ rectangular aquifer Ω (see Fig. 8.25) bounded on the long sides by an impermeable material (i.e., $\partial\phi/\partial n = 0$) and on the short sides by a constant head of 200 m ($\phi_0 = 200$ m). In the way of sources, suppose that a river is passing through the aquifer, infiltrating it at a rate of $0.24\text{ m}^3\text{ day}^{-1}\text{ m}^{-2}$, and that two pumps are located at (1000, 670) and (1900, 900), pumping at rates $Q_1 = 1200\text{ m}^3\text{ day}^{-1}\text{ m}^{-1}$ and $Q_2 = 2400\text{ m}^3\text{ day}^{-1}\text{ m}^{-1}$, respectively. Pumping is considered as a negative point source.

A mesh of 64 triangular elements and 45 nodes is used to model the domain (see Fig. 8.26*a*). The river forms the interelement boundary between elements (33, 35, 37, 39) and (26, 28, 30, 32). In the mesh selected, neither pump is located at a node. This is done intentionally for the purpose of illustrating the calculation of the generalized forces due to a point source within an element. If the pumps are located at a node then the rate of pumping Q_0 is input as the specified secondary variable of the node. When a source (or sink) is located at a point other than a node, we must calculate its contribution to the nodes. Similarly, the source components due to the distributed line source (i.e., the river) should be computed.

First, consider the line source. We can view the river as a line source of constant intensity, $0.24\text{ m}^3\text{ day}^{-1}\text{ m}^{-2}$. Since the length of the river is equally divided by nodes 21–25 (into four parts), we can compute the contribution of the infiltration of the river at each of the nodes 21–25 by evaluating the integrals (see Fig. 8.26*b*):

node 25: $\displaystyle\int_0^h (0.24)\psi_1^1\, ds$

node 24: $\displaystyle\int_0^h (0.24)\psi_2^1\, ds + \int_0^h (0.24)\psi_1^2\, ds$

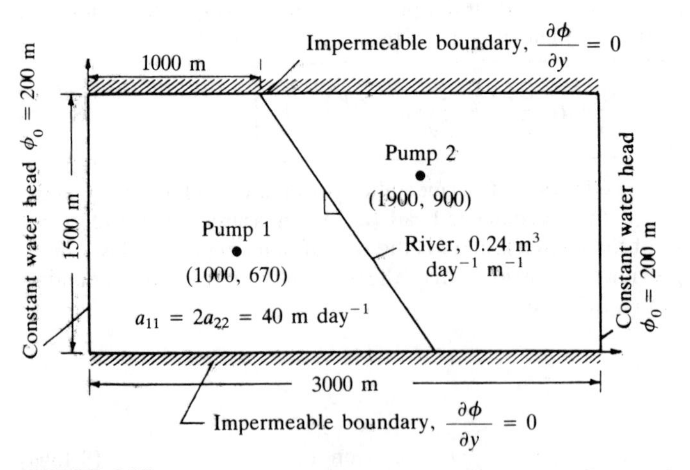

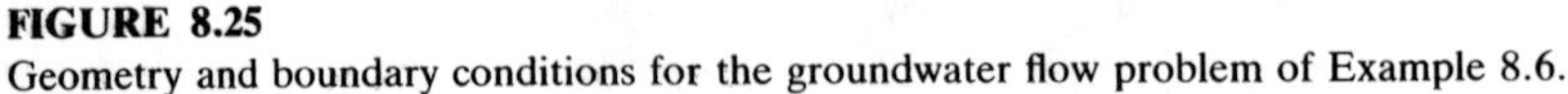
FIGURE 8.25
Geometry and boundary conditions for the groundwater flow problem of Example 8.6.

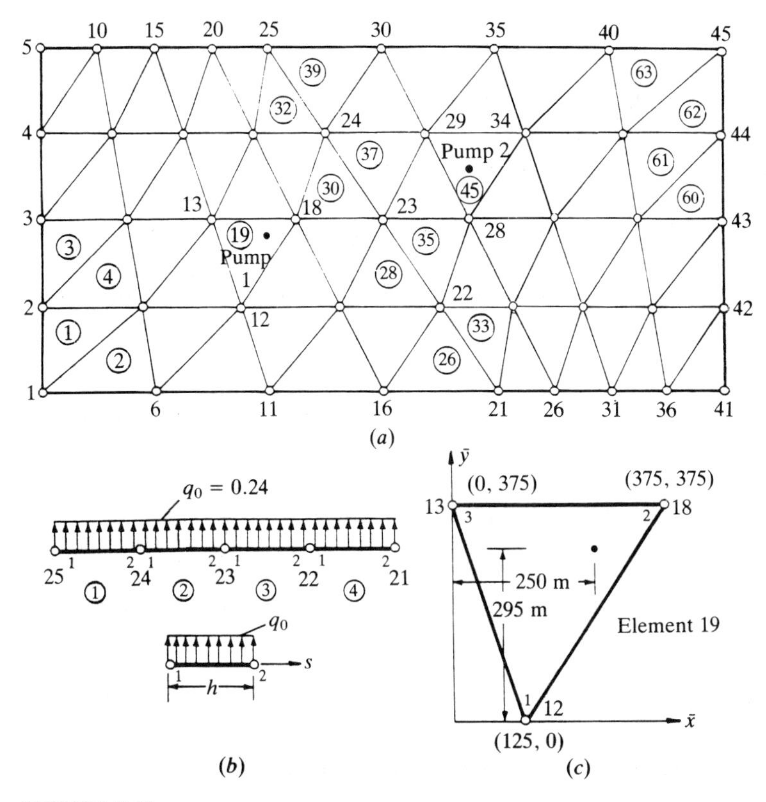

FIGURE 8.26.
Finite element mesh, and computation of force components for the groundwater (seepage) flow: (*a*) finite element mesh of triangular elements (45 nodes and 64 elements); (*b*) computation of global forces due to the infiltration of the river; (*c*) computation of global forces for pump 1, located inside element 19 (Example 8.6).

node 23: $\displaystyle\int_0^h (0.24)\psi_2^2\,ds + \int_0^h (0.24)\psi_1^3\,ds$

node 22: $\displaystyle\int_0^h (0.24)\psi_2^3\,ds + \int_0^h (0.24)\psi_1^4\,ds$

node 21: $\displaystyle\int_0^h (0.24)\psi_2^4\,ds$

For constant intensity q_0 and the linear interpolation functions $\psi_1^e(s) = 1 - s/h$ and $\psi_2^e(s) = s/h$, the contribution of these integrals is well known:

$$\int_0^h q_0\psi_i\,ds = \tfrac{1}{2}q_0 h, \qquad h = \tfrac{1}{4}[(1000)^2 + (1500)^2]^{1/2}, \qquad q_0 = 0.24$$

Hence, we have

$$F_{21} = \tfrac{1}{2}q_0 h, \qquad F_{22} = F_{23} = F_{24} = q_0 h, \qquad F_{25} = \tfrac{1}{2}q_0 h.$$

Next, we consider the contribution of the point sources. Since these are located inside an element, we distribute them to the nodes of the element by interpolation (see

Fig. 8.26c):

$$\boxed{f_i^e = \int_{\Omega^e} Q_0 \delta(x - x_0, y - y_0) \psi_i^e(x, y)\, dx\, dy = Q_0 \psi_i^e(x_0, y_0)}$$

where the two-dimensional Dirac delta function is defined by

$$\int_{-\infty}^{\infty} \int_{-\infty}^{\infty} F(x, y) \delta(x - x_0, y - y_0)\, dx\, dy = F(x_0, y_0)$$

For example, the source at pump 1 (located at $x_0 = 1000$ m, $y_0 = 670$ m) can be expressed as

$$Q_1(x, y) = -1200\delta(x - 1000, y - 670)$$

The interpolation functions ψ_i^e for element 19 are (in terms of the local coordinates $\bar{x}$ and $\bar{y}$; see Fig. 8.26c)

$$\psi_i(\bar{x}, \bar{y}) = \frac{1}{2A}(\alpha_i + \beta_i \bar{x} + \gamma_i \bar{y})$$

$$A = \tfrac{1}{2}(375)^2, \quad \alpha_1 = (375)^2, \quad \alpha_2 = -375(125), \quad \alpha_3 = 375(125)$$

$$\beta_1 = 0, \quad \beta_2 = 375, \quad \beta_3 = -375, \quad \gamma_1 = -375, \quad \gamma_2 = 125, \quad \gamma_3 = 250$$

We have $\bar{x} = x - 750$ and $\bar{y} = y - 375$, and, therefore,

$$\psi_1(250, 295) = 0.2133, \quad \psi_2 = (250, 295) = 0.5956, \quad \psi_3 = (250, 295) = 0.1911$$

Similar computations can be performed for pump 2.

In summary, the primary variables and nonzero secondary variables are

$$U_1 = U_2 = U_3 = U_4 = U_5 = U_{41} = U_{42} = U_{43} = U_{44} = U_{45} = 200.0$$

$$F_{21} = 54.08, \quad F_{22} = F_{23} = F_{24} = 108.17, \quad F_{25} = 54.08$$

$$F_{12} = -255.6, \quad F_{13} = -229.2, \quad F_{18} = -715.2, \quad F_{28} = -1440.0$$

$$F_{29} = -410.4, \quad F_{34} = -549.6$$

The secondary variables at nodes 6–11, 14–17, 19, 20, 26, 27, 30–33, and 35–40 are zero. This completes the data generation for the problem.

The assembled equations are solved, after imposing the specified boundary conditions, for the values of ϕ at the nodes. The equipotential lines can be determined using (8.92) or a postprocessor. The lines of constant ϕ are shown in Fig. 8.27(*a*) and the velocity vectors in Fig. 8.27(*b*). The greatest drawdown of water occurs at node 28, which has the largest portion of discharge from pump 2.

The solution of the same problem by an alternative mesh that puts pumps 1 and 2 at nodal points is left as an exercise.

Next, we consider an example of irrotational flow of an ideal fluid (i.e., a nonviscous fluid). Examples of physical problems that can be approximated by such flows are provided by flow around bodies such as weirs, airfoils, buildings, and so on, and by flow of water through the earth and dams. The Laplace equations (8.128) and (8.133) governing these flows are special cases of (8.1), and therefore one can use the finite element equations developed earlier to model these problems.

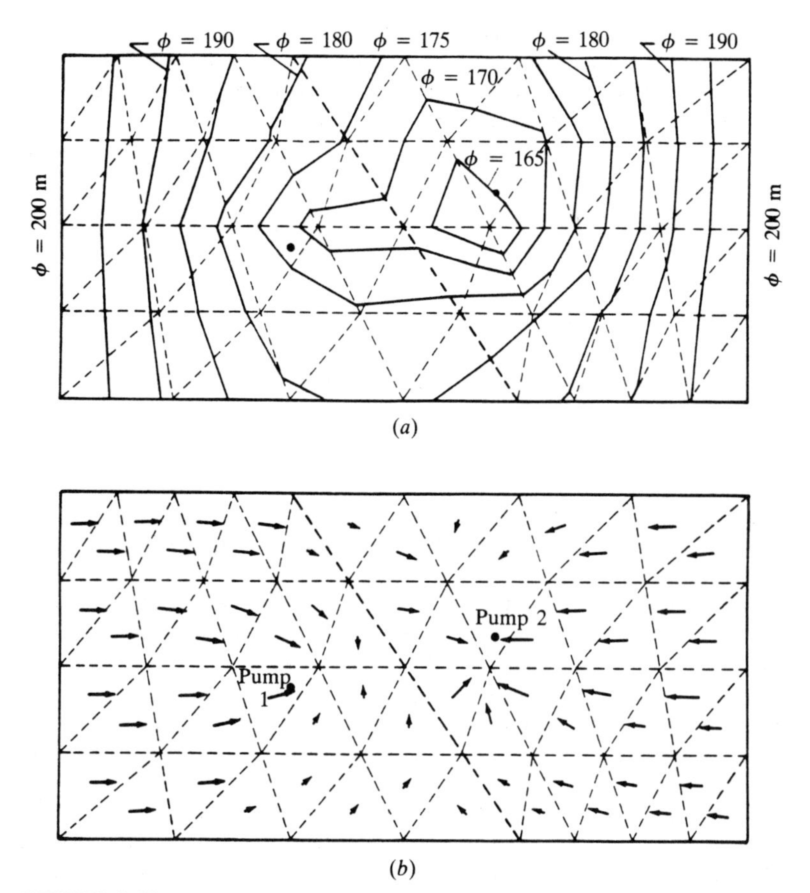

FIGURE 8.27
Plots of constant piezometric head and velocity vector for the groundwater flow: (*a*) lines of constant ϕ; (*b*) plot of velocity vectors (Example 8.6).

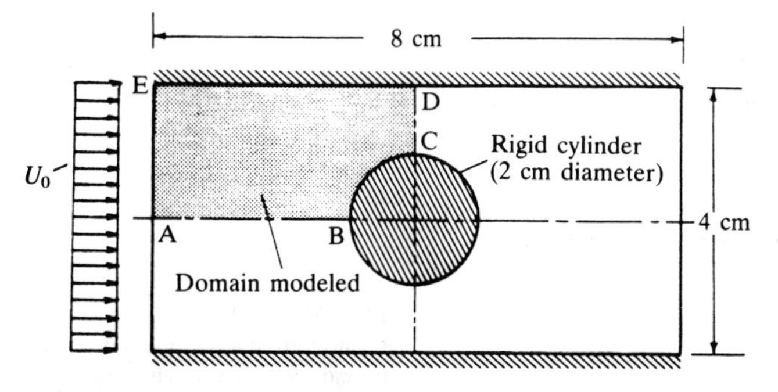

FIGURE 8.28
Domain and boundary conditions for the stream function and velocity potential formulations of irrotational flow about a cylinder (Example 8.7).

Example 8.7 Confined flow around a circular cylinder. The irrotational flow of an ideal fluid about a circular cylinder, placed with its axis perpendicular to the plane of the flow between two *long* horizontal walls (see Fig. 8.28) is to be analyzed using the finite element method. The equation governing the flow is

$$-\nabla^2 u = 0 \quad \text{in } \Omega$$

where u is either (1) the stream function ψ or (2) the velocity potential ϕ. If it is the former, the velocity components $\mathbf{u} = (u_1, u_2)$ of the flow field are given by

$$u_1 = \frac{\partial \psi}{\partial y}, \qquad u_2 = -\frac{\partial \psi}{\partial x}$$

If u is the velocity potential ϕ, the velocity components can be computed from

$$u_1 = -\frac{\partial \phi}{\partial x}, \qquad u_2 = -\frac{\partial \phi}{\partial y}$$

In either case, the velocity field is not affected by a constant term in the solution u. To determine the constant state of the solution, which does not affect the velocity field, we arbitrarily set the functions ψ and ϕ equal to zero (or a constant) on appropriate boundary lines. We analyze the problem using both formulations. For both, symmetry exists about the horizontal and vertical centerlines; therefore, only a quadrant of the flow region is used as the computational domain.

Stream function formulation. The boundary conditions on the stream function ψ can be determined as follows. Streamlines have the property that flow perpendicular to them is zero. Therefore, the fixed walls correspond to streamlines. Note that, for inviscid flows, fluid does not stick to rigid walls. Because of the biaxial symmetry about the horizontal and vertical centerlines, only a quadrant (say, ABCDE in Fig. 8.29) of the domain need to be used in the analysis. The fact that the velocity component perpendicular to the horizontal line of symmetry is zero allows us to use that line as a streamline. Since the velocity field depends on the relative difference of two streamlines, we take the value of the stream function that coincides with the horizontal axis of symmetry (i.e., on ABC) to be zero, and then determine the value of ψ on the upper wall from the condition

$$\frac{\partial \psi}{\partial y} = U_0$$

where U_0 is the inlet horizontal velocity of the field. We determine the value of the stream function on the boundary $x = 0$ by integrating the above equation with respect

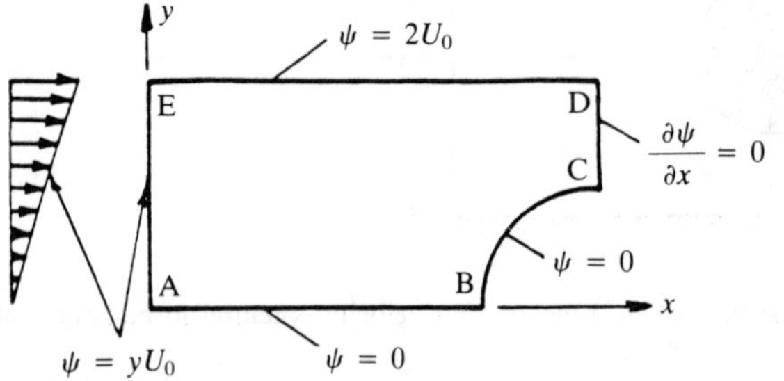

FIGURE 8.29
Computational domain and boundary conditions for the stream function formulation of inviscid flow around a cylinder (see Fig. 8.28).

to y:

$$\int_0^y \frac{d\psi}{dy}\,dy = \int_0^y U_0\,dy + \psi_A = U_0 y + 0 \tag{8.137}$$

because $\psi_A = 0$ from the previous discussion. This gives the boundary condition on AE. Since the line ED is a streamline and its value at the point E is $2U_0$, it follows that $\psi = 2U_0$ on the line ED. Lastly, we assume that the vertical velocity is zero on the line CD (i.e., $U_2 = 0$); hence $\partial\psi/\partial x = 0$ on CD. The boundary conditions are shown on the computational domain in Fig. 8.29.

In selecting a mesh, we should note that the velocity field is uniform (i.e., streamlines are horizontal) at the inlet, and that it has a parabolic profile at the exit (along CD). Therefore, the mesh at the inlet should be uniform, and the mesh close to the cylinder should be relatively more refined to be able to model the curved boundary and capture the rapid change in ψ. Two coarse finite element meshes are used to discuss the boundary conditions, and results for refined meshes will be discussed subsequently. Mesh T1 consists of 32 triangular elements and mesh Q1 consist of 16 quadrilateral elements. Both contain 25 nodes (see Fig. 8.30). The mesh with the solid lines in Fig. 8.30 corresponds to mesh Q1, and that with both the solid and dashed lines corresponds to mesh T1. It should be noted that the discretization error is not zero for this case.

The specified primary degrees of freedom (i.e., nodal values of ψ) for meshes T1 and Q1 are

$$\begin{gathered} U_1 = U_2 = \ldots = U_5 = U_{10} = U_{15} = U_{20} = U_{25} = 0.0 \\ U_6 = 1.0, \qquad U_{11} = U_{16} = U_{21} = 2.0 \end{gathered} \tag{8.138}$$

There are no nonzero specified secondary variables; the secondary variables are specified to be zero at the nodes on the line CD:

$$F_{22} = F_{23} = F_{24} = 0$$

Although the secondary variable is specified to be zero at nodes 21 and 25, where the primary variable is also specified, we choose to impose the boundary conditions on the primary variable instead of the secondary variables.

Table 8.6 gives the values of the stream function and its derivative $\partial\psi/\partial y$ $(= u_1)$ at selected points/elements of the meshes. The finite element program FEM2DV2 (see Chapter 13 for details) is used in the analysis. The stream function values obtained with meshes T1 and Q1 are very close to each other. Recall that the derivative $\partial\psi/\partial y$ is constant in a linear triangular element, whereas it varies linearly with x in a linear

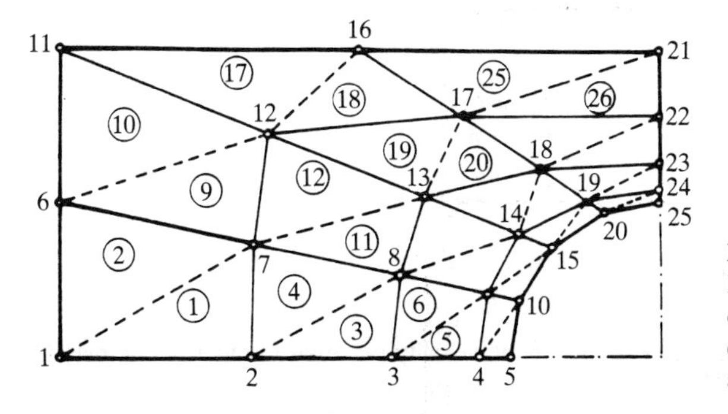

FIGURE 8.30
Meshes T1 and Q1 (remove the dashed lines for the mesh of quadrilaterals) used for inviscid flow around a cylinder.

TABLE 8.6
Finite element results from the stream function formulation of inviscid flow around a cylinder (Example 8.7)

		Stream function		Velocity $u_1 = \partial\psi/\partial y$		Velocity $u_1 = -\partial\phi/\partial x$	
x	y	Mesh T1	Mesh Q1	Mesh T1	Mesh Q1	Mesh T1	Mesh Q1
1.3183	0.7354	0.7092	0.7095	0.9643(1)†	0.9852(1)	0.9922(1)	0.9989(1)
2.2705	0.5444	0.4372	0.4379	0.8032(3)	0.9002(2)	0.9371(3)	0.9408(2)
2.8564	0.4268	0.1667	0.1650	0.3906(5)	0.6432(3)	0.7047(5)	0.7018(3)
1.4112	1.4459	1.4241	1.4270	0.0000(7)	0.2679(4)	0.2999(7)	0.3197(4)
2.4305	1.0457	0.8730	0.8823	0.4469(15)	0.8746(8)	0.6469(15)	0.8364(8)
3.0577	0.7995	0.3357	0.3384	1.636(24)	1.586(12)	1.873(24)	1.453(12)
2.6931	1.5388	1.3758	1.4010	2.544(32)	2.4551(16)	2.163(32)	2.075(16)
3.1937	1.2057	0.7706	0.7980				
3.5018	1.0007	0.2520	0.2658				
4.0000	1.5714	1.2395	1.2065				
4.0000	1.2619	0.6191	0.5796				
4.0000	1.0714	0.1817	0.1588				

† Denotes element number; the derivatives of ψ and ϕ are evaluated at the center of this element.

rectangular element. Therefore, the results for meshes T1 and Q1 will not be the same. The velocities included in Table 8.6 correspond to elements closest to the symmetry line (i.e., the $y = 0$ line) and the surface of the cylinder.

The tangential velocity u_t on the cylinder surface can be computed from the relation

$$u_t(\theta) = u_1 \sin\theta + u_2 \cos\theta = \frac{\partial\psi}{\partial y}\sin\theta - \frac{\partial\psi}{\partial x}\cos\theta \tag{8.139}$$

where θ is the angular distance along the cylinder surface.

Velocity potential formulation. The boundary conditions on the velocity potential ϕ can be derived as follows (see Fig. 8.31). The fact that $u_2 = -\partial\phi/\partial y = 0$ (no penetration) on the upper wall as well as on the horizontal line of symmetry gives the

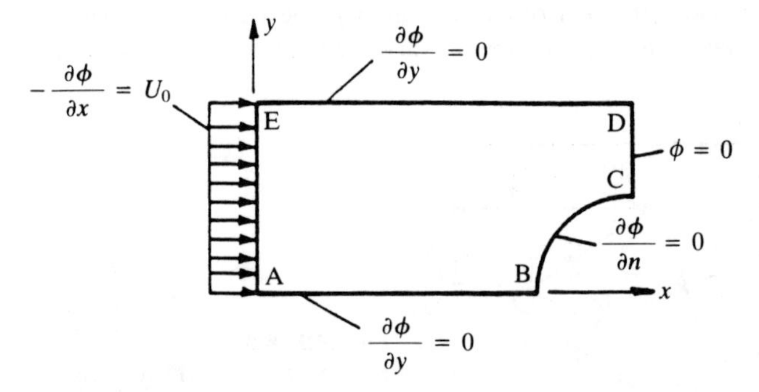

FIGURE 8.31
Computational domain and boundary conditions for the velocity potential formulation of inviscid flow around a cylinder (see Fig. 8.28).

boundary conditions there. Along AE, the velocity $u_1 = -\partial\phi/\partial x$ is specified to be U_0. On the surface of the cylinder, the normal velocity, $u_n = -\partial\phi/\partial n$ is zero. Thus all boundary conditions, so far, are of the flux type. On the boundary CD, we must know either ϕ or $\partial\phi/\partial n = \partial\phi/\partial x$. It is clear that $-\partial\phi/\partial x = u_1$ is not known on CD. Therefore, we assume that ϕ is known, and we set it equal to ϕ_0 = constant. The constant ϕ_0 is arbitrary and does not contribute to the velocity field (because $-\partial\phi/\partial x = u_1$ and $-\partial\phi/\partial y = u_2$ are independent of ϕ_0). It should be noted that specification of ϕ at least one point of the mesh is necessary to determine the constant part in the solution for ϕ (i.e., eliminate the rigid body motion). We take $\phi = \phi_0 = 0$ on CD.

The mathematical boundary conditions of the problem must be translated into finite element data. The boundary conditions on the primary variables are from the boundary CD. We have

$$U_{21} = U_{22} = U_{23} = U_{24} = U_{25} = 0.0$$

The only nonzero boundary conditions on the secondary variables come from the boundary AE. There we must evaluate the boundary integral

$$\int_{\Gamma^e} \frac{\partial\phi}{\partial n}\psi_i\, ds = U_0 \int_{\text{AE}} \psi_i(y)\, dy$$

We obtain

$$Q_1 = U_0 \int_0^{h_1} \left(1 - \frac{\bar{y}}{h_1}\right) d\bar{y} = \tfrac{1}{2}U_0 h_1 = 0.5U_0$$

$$Q_2 = U_0 \int_0^{h_1} \frac{\bar{y}}{h}\, d\bar{y} + U_0 \int_0^{h_2} \left(1 - \frac{\bar{y}}{h_2}\right) d\bar{y} = \tfrac{1}{2}U_0(h_1 + h_2) = U_0$$

$$Q_3 = U_0 \int_0^{h_2} \frac{\bar{y}}{h_2}\, d\bar{y} = \tfrac{1}{2}h_2 U_0 = 0.5U_0$$

The finite element solutions for $u_1 = -\partial\phi/\partial x$ obtained with meshes T1 and Q1 are listed in Table 8.6. Note that there is a difference between the velocities obtained with the two formulations (for either mesh). This is primarily due to the nature of the boundary value problems in the two formulations. In the stream function formulation, there are more boundary conditions on the primary variable than in the velocity potential formulation.

Contour plots of streamlines, velocity potential, and horizontal velocity $u_1 = \partial\psi/\partial y$ obtained with mesh Q1 (and a plotter routine) are shown in Figs 8.32(a–c). A plot of the variation of the tangential velocity with the angular distance along the cylinder surface is shown in Fig. 8.33, along with the analytical potential solution

$$u_t = U_0(1 + R^2/r^2)\sin\theta \tag{8.140a}$$

valid on the cylinder surface. The finite element solution of a refined mesh, mesh Q2, is also included in the figure. The angle θ, radial distance r, and tangential velocity u_t can be computed from the relations

$$\theta = \tan^{-1}\left(\frac{y}{4-x}\right), \qquad r = [(4-x)^2 + y^2]^{1/2}, \qquad u_t = u_1 \sin\theta + u_2 \cos\theta \tag{8.140b}$$

The finite element solution is in general agreement with the potential solution of the problem. However, the former is not expected to agree closely, because u_t is evaluated at a radial distance $r > R$, whereas the potential solution is evaluated at $r = R$ only.

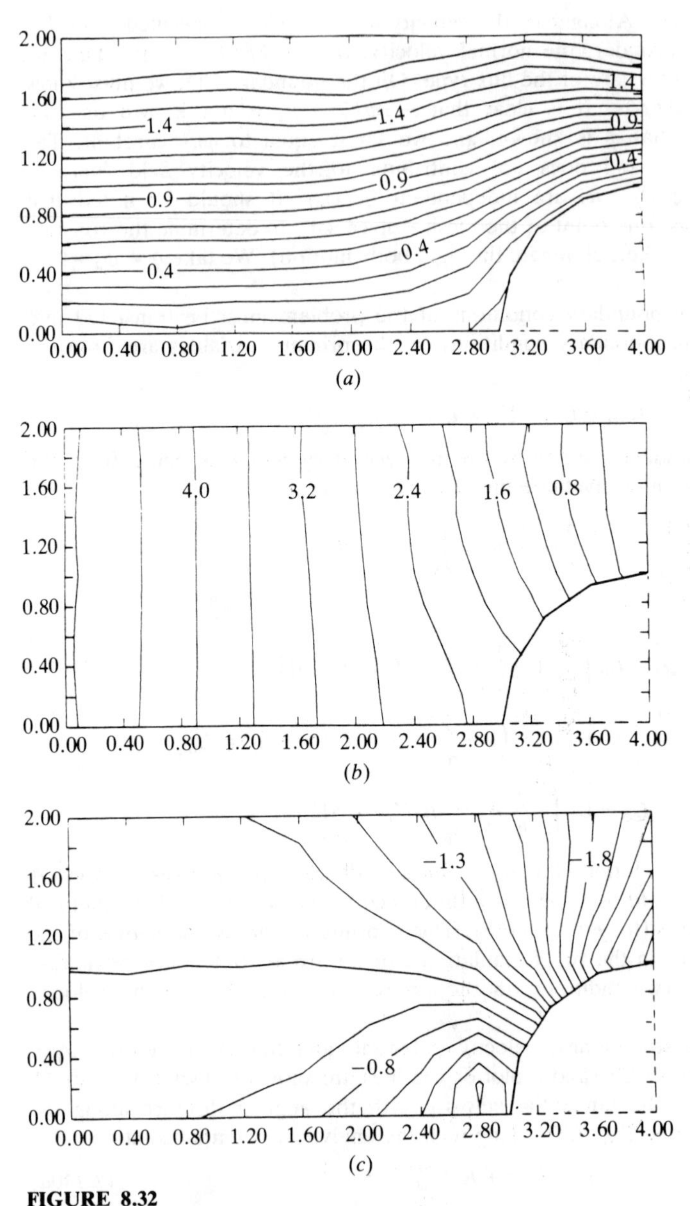

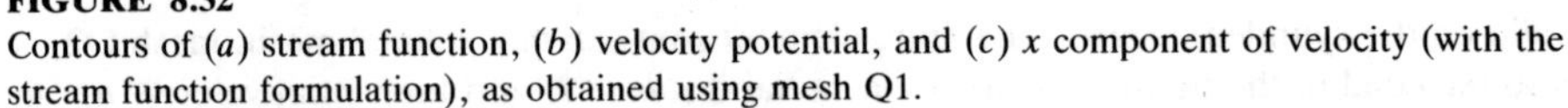
FIGURE 8.32
Contours of (*a*) stream function, (*b*) velocity potential, and (*c*) x component of velocity (with the stream function formulation), as obtained using mesh Q1.

This completes the section on fluid mechanics problems that are cast in terms of a single dependent unknown, such as the stream function or velocity potential. We return to fluid mechanics later in this book to consider two-dimensional flows of viscous incompressible fluids. The governing equations of such problems consist of several dependent variables and as many differential equations (see Chapter 11).

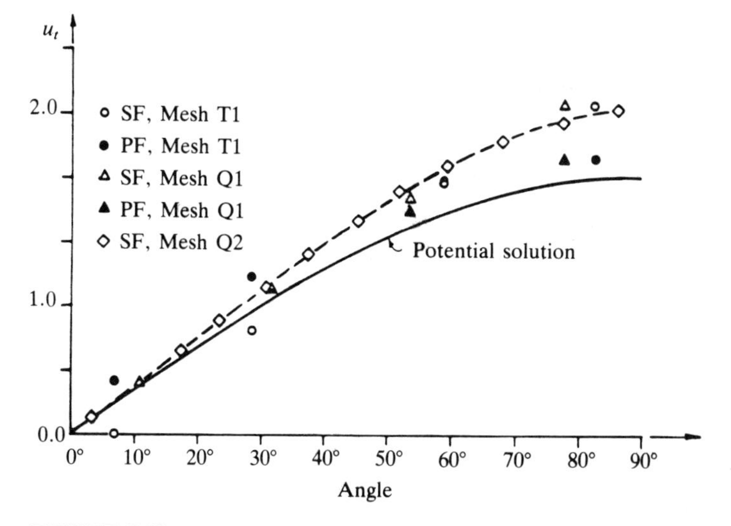

FIGURE 8.33
Variation of the tangential velocity along the cylinder surface: comparison of the finite element results with the potential theory solution (mesh Q2 contains 96 elements and 117 nodes).

8.4.3 Solid Mechanics

In this section, we consider two-dimensional boundary value problems of solid mechanics that are cast in terms of a single dependent unknown. These problems include torsion of cylindrical members and transverse deflection of membranes. The discussion is restricted to small deformations.

TORSION OF CYLINDRICAL BARS. Consider a cylindrical bar (i.e., a long member of uniform cross-section), fixed at one end and twisted by a couple (i.e., a moment or torque) of magnitude M directed along the axis (z) of the bar (see Fig. 8.34a). Suppose that the bar is not subjected to any body forces and is free from external forces on the lateral surface. We wish to determine the amount of twist and the associated stress field in the bar. To this end, we first describe the deformation of the bar analytically, and then analyze the equation using the finite element method.

In general, a member of noncircular cross-section subjected to a torsional moment experiences warping at any section. We assume that all cross-sections warp in the same way (which holds true for small twisting moments and deformation). This allows us to take the displacements (u, v, w) along the coordinates (x, y, z) to be of the form (see Fig. 8.34b)

$$u = -\theta zy, \qquad v = \theta zx, \qquad w = \theta\phi(x, y) \tag{8.141}$$

where $\phi(x, y)$ is a function to be determined and θ is the angle of twist per unit length of the bar.

The displacement field in (8.141) can be used to compute the strains, and stresses are computed using an assumed constitutive law. The stresses thus

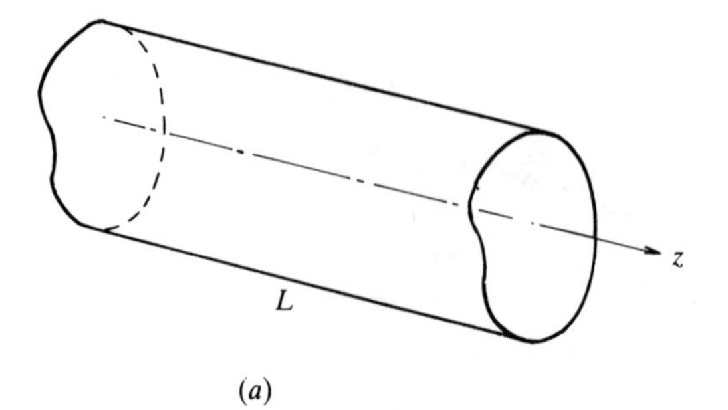

(a)

y
Moment
M
$\alpha = \theta z$
x
Ω

(b)

FIGURE 8.34
Torsion of cylindrical members: (a) a cylindrical member; (b) domain of analysis.

determined must satisfy the three-dimensional equations of stress equilibrium

$$
\begin{aligned}
&\frac{\partial \sigma_x}{\partial x} + \frac{\partial \sigma_{xy}}{\partial y} + \frac{\partial \sigma_{xz}}{\partial z} = 0 \\
&\frac{\partial \sigma_{xy}}{\partial x} + \frac{\partial \sigma_y}{\partial y} + \frac{\partial \sigma_{yz}}{\partial z} = 0 \qquad (8.142) \\
&\frac{\partial \sigma_{xz}}{\partial x} + \frac{\partial \sigma_{yz}}{\partial y} + \frac{\partial \sigma_z}{\partial z} = 0.
\end{aligned}
$$

and the stress boundary conditions on the lateral surface and at the ends of the cylindrical bar. Calculation of strains and then stresses using the generalized Hooke's law gives the expressions

$$
\boxed{\sigma_{xz} = G\theta\left(\frac{\partial \phi}{\partial x} - y\right), \qquad \sigma_{yz} = G\theta\left(\frac{\partial \phi}{\partial y} + x\right)} \qquad (8.143)
$$

and all other stresses are identically zero. Here G denotes the shear modulus of the material of the bar. Substitution of these stresses into (8.142) gives [the first two equations in (8.142) are identically satisfied, and the third leads to this result]

$$\boxed{\frac{\partial}{\partial x}\left(G\theta\frac{\partial\phi}{\partial x}\right)+\frac{\partial}{\partial y}\left(G\theta\frac{\partial\phi}{\partial y}\right)=0} \tag{8.144}$$

throughout the cross-section Ω of the cylinder. The boundary conditions on the lateral surfaces Γ require that

$$\sigma_{xz}n_x+\sigma_{yz}n_y=0$$

$$\left(\frac{\partial\phi}{\partial x}-y\right)n_x+\left(\frac{\partial\phi}{\partial y}+x\right)n_y=0$$

or

$$\frac{\partial\phi}{\partial n}=yn_x-xn_y \quad \text{on } \Gamma \tag{8.145}$$

Here (n_x, n_y) denote the direction cosines of the unit normal at a point on Γ.

In summary, the torsion of a cylindrical bar is governed by (8.144) and (8.145). The function $\phi(x, y)$ is called the *torsion function* or *warping function.* Since the boundary condition in (8.145) is of the flux type, this function can be determined to within an additive constant. The stresses in (8.143), however, are independent of this constant. The additive constant has the meaning of rigid-body movement of the cylinder as a whole in the z direction. For additional discussion of this topic, the reader is referred to Sokolnikoff (1956).

The Laplace equation (8.144) and the Neumann boundary condition (8.145) governing ϕ are not convenient in the analysis because of the nature and form of the boundary condition, especially for members of irregular cross-section. The theory of analytic functions can be used to rewrite these equations in terms of the function $\Psi(x, y)$, called the *stress function,* which is related to $\phi(x, y)$ by the equations

$$\frac{\partial\Psi}{\partial x}=-\frac{\partial\phi}{\partial y}-x, \qquad \frac{\partial\Psi}{\partial y}=\frac{\partial\phi}{\partial x}-y \tag{8.146}$$

Eliminating ϕ from (8.144) and (8.145) gives, respectively, the results

$$-\left(\frac{\partial^2\Psi}{\partial x^2}+\frac{\partial^2\Psi}{\partial y^2}\right)=2 \tag{8.147}$$

$$\frac{\partial\Psi}{\partial y}n_x-\frac{\partial\Psi}{\partial x}n_y=0 \tag{8.148}$$

The left-hand side of (8.148) denotes the tangential derivative $d\Psi/ds$, and $d\Psi/ds=0$ implies that

$$\Psi=\text{constant} \quad \text{on } \Gamma$$

Since the constant part of Ψ does not contribute to the stress field,

$$\boxed{\sigma_{xz} = G\theta \frac{\partial \Psi}{\partial y}, \qquad \sigma_{yz} = -G\theta \frac{\partial \Psi}{\partial x}} \tag{8.149}$$

we take $\Psi = 0$ on the boundary.

Now the torsion problem can be stated as one of determining the stress function Ψ such that

$$\boxed{\begin{aligned} -\nabla^2 \Psi &= 2 \quad \text{in } \Omega \\ \Psi &= 0 \quad \text{on } \Gamma \end{aligned}} \tag{8.150}$$

Once Ψ has been determined, the stresses can be computed from (8.149) for a given angle of twist per unit length, θ, and shear modulus G.

The finite element model of (8.150) follows immediately from that of (8.1):

$$[K^e]\{u^e\} = \{f^e\} + \{Q^e\} \tag{8.151a}$$

where u_i^e is the value of Ψ at the ith node of Ω^e and

$$\begin{aligned} K_{ij}^e &= \int_{\Omega^e} \left(\frac{\partial \psi_i}{\partial x} \frac{\partial \psi_j}{\partial x} + \frac{\partial \psi_i}{\partial y} \frac{\partial \psi_j}{\partial y} \right) dx\, dy \\ f_i^e &= \int_{\Omega^e} 2\psi_i \, dx\, dy, \qquad Q_i^e = \oint_{\Gamma^e} \frac{\partial \Psi}{\partial n} \psi_i \, ds \end{aligned} \tag{8.151b}$$

Next we consider an example of a torsion problem.

Example 8.8 Torsion of a square cross-section bar. It should be recalled that (8.1) can also be interpreted as the equation associated with the torsion of a square cross-section cylindrical bar of cross-sectional dimensions a by b. Note that the actual problem is antisymmetric as far as the loading and stress distribution are concerned; however, the stress function, being a scalar function governed by the Poisson equation, is symmetric about the x and y axes as well as the diagonal lines. When using rectangular elements, one quadrant of the bar cross-section can be used in the finite element analysis. The biaxial symmetry about the x and y axes requires imposition of the following boundary conditions on Ψ:

$$\frac{\partial \Psi}{\partial x} = 0 \quad \text{on the line } x = 0$$

$$\frac{\partial \Psi}{\partial y} = 0 \quad \text{on the line } y = 0$$

In addition, on the actual boundary we have the boundary condition

$$\Psi = 0 \quad \text{on the lines } x = a,\ y = b$$

The analytical solution of the problem is given by

$$\Psi(x, y) = \tfrac{1}{4}a^2 - x^2 - \frac{8a^2}{\pi^3} \sum_{n=0}^{\infty} \frac{(-1)^n}{(2n+1)^3} \frac{\cosh k_n y \cos k_n x}{\cosh \frac{1}{2}k_n b} \tag{8.152a}$$

$$\sigma_{xz} = -\frac{8aG\theta}{\pi^2} \sum_{n=0}^{\infty} \frac{(-1)^n}{(2n+1)^2} \frac{\sinh k_n y \cos k_n x}{\cosh \frac{1}{2}k_n b} \tag{8.152b}$$

$$\sigma_{yz} = G\theta\left[2x - \frac{8a}{\pi^2} \sum_{n=0}^{\infty} \frac{(-1)^n}{(2n+1)^2} \frac{\cosh k_n y \sin k_n x}{\cosh \frac{1}{2}k_n b}\right] \tag{8.152c}$$

where $k_n = \frac{1}{2}(2n+1)\pi$. The problem is analyzed here for shear stresses σ_{xz} and σ_{yz} [see (8.149)].

Here we investigate the convergence of the finite element solutions using gradually refined meshes of linear and quadratic rectangular elements. The results of this study are summarized in Tables 8.7 and 8.8. The convergence of the finite element solutions for the stress function and stresses to the analytical solutions (8.152) can be seen from these results. The contour lines of the surface $\Psi(x, y) = \text{constant}$, $\sigma_{xz} = \text{constant}$, and $\sigma_{yz} = \text{constant}$ are shown in Fig. 8.35.

TRANSVERSE DEFLECTION OF MEMBRANES. Suppose that a membrane, with fixed edges, occupies a region Ω in the (x, y) plane. Initially, the membrane is stretched so that the tension a in the membrane is uniform and that a is so large that it is not appreciably altered when the membrane is deflected by a distributed normal force $f(x, y)$. The equation governing the transverse deflection u of the membrane is

$$-a\left(\frac{\partial^2 u}{\partial x^2} + \frac{\partial^2 u}{\partial y^2}\right) = f(x, y) \quad \text{in } \Omega \tag{8.153a}$$

TABLE 8.7
Convergence of the finite element solutions for Ψ using linear and quadratic rectangular elements (four-node and nine-node elements) in Example 8.8

		Linear elements			Quadratic elements†		
x	y	2×2	4×4	8×8	1×1	2×2	4×4
0.0000	0.0000	0.15536	0.14920	0.14780	0.14744	0.14730	0.14734
0.0625	0.0000	—	—	0.14583	—	—	0.14538
0.1250	0.0000	—	0.14120	0.13987	—	0.13941	0.13944
0.1875	0.0000	—	—	0.12972	—	—	0.12931
0.2500	0.0000	0.12054	0.11610	0.11502	0.11378	0.11463	0.11467
0.3125	0.0000	—	—	0.09534	—	—	0.09505
0.3750	0.0000	—	0.07069	0.07007	—	0.06983	0.06986
0.4375	0.0000	—	—	0.03854	—	—	0.03844
0.1250	0.2500	—	0.11031	0.10925	—	0.10887	0.10890
0.2500	0.2500	0.09643	0.09191	0.09090	0.09095	0.09056	0.09057
0.3750	0.2500	—	0.05729	0.05660	—	0.05626	0.05636

† The 4 × 4 mesh of nine-node quadratic elements gives a solution that coincides with the analytical solution to five significant decimal places.

TABLE 8.8
Comparison of finite element solutions for the shear stress $\sigma_{yz}(x, y)$ $[= -\sigma_{xz}(y, x)]$, computed using various meshes, with the analytical solution (Example 8.8)

		Mesh			
x	y	2×2	4×4	8×8	Analytical solution
0.03125	↑	—	—	0.0312	0.0312
0.09375		—	—	0.0946	0.0946
0.15625		—	—	0.1612	0.1611
0.21875		—	—	0.2332	0.2331
0.28125	0.03125	—	—	0.3127	0.3124
0.34375		—	—	0.4015	0.4011
0.40625		—	—	0.5013	0.5008
0.46875	↓	—	—	0.6135	0.6128
0.06250	↑	—	0.06175	—	0.0618
0.1875		—	0.1942	—	0.1939
0.3125	0.0625	—	0.3529	—	0.3516
0.4375	↓	—	0.5528	—	0.5504
0.1250	0.1250	0.1179	—	—	0.1193
0.3750	0.1250	0.4339	—	—	0.4272

with

$$u = 0 \quad \text{on } \Gamma \tag{8.153b}$$

Note that this equation and boundary conditions are of the same form as those for the torsion of cylindrical bars [see (8.150)]. The finite element model of the equation is obvious. In view of the close analogy between this problem and the torsion of cylindrical bars, we shall not consider any numerical examples here.

8.5 EIGENVALUE AND TIME-DEPENDENT PROBLEMS

8.5.1 Introduction

This section deals with the finite element analysis of two-dimensional eigenvalue and time-dependent problems involving a single variable. We use the results of Section 6.2 to develop finite element algebraic equations from the semidiscrete finite element models of time-dependent problems. Since the weak form and temporal approximations have already been discussed in detail in Section 6.2, attention is focussed here first on the development of the semidiscrete finite element models and then on the associated eigenvalue and fully discretized models. The examples presented are very simple, because they are designed to illustrate the procedure for eigenvalue and time-dependent problems; solution of two-dimensional problems with complicated geometries requires the use of numerical integration. Chapter 9 is devoted to the discussion of various two-dimensional elements and their interpolation functions and numerical integration methods.

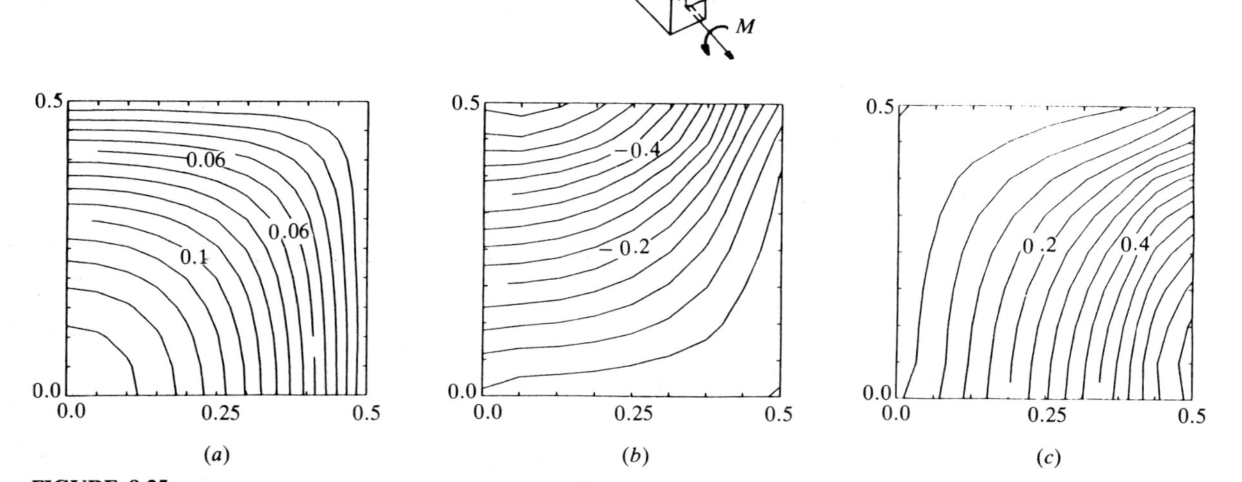

FIGURE 8.35
Contour plots of the stress function and shear stresses obtained using the 8×8 mesh of linear rectangular elements in a quadrant of square cross-section in Example 8.8: (*a*) stress function; (*b*) stress σ_{xz}; (*c*) stress σ_{yz}.

The development of finite element models of eigenvalue and time-dependent problems involves, as described in Section 6.2, two main stages. The first, called *semidiscretization,* is to develop the weak form of the equations over an element and to seek spatial approximations of the dependent variables of the problem. The end result of this step is a set of ordinary differential equations in time among the nodal values of the dependent variables. For transient problems, the second stage consists in time approximation of the ordinary differential equations (i.e., numerical integration of the equations) by finite difference schemes. This step leads to a set of algebraic equations involving the nodal values at time t_{s+1} $[=(s+1)\,\Delta t$, where s is an integer and Δt is the time increment] in terms of known values from the previous time step(s). For eigenvalue problems, the second stage consists in seeking a solution of the form $u_j(t)=U_j e^{-\lambda t}$ for nodal values and determining the eigenvalues λ and eigenfunctions $U_j\psi_j(x, y)$ (no sum on j). The two-stage procedure was clearly illustrated for one-dimensional problems in Sections 6.1 and 6.2. It will be applied here to two-dimensional problems involving a single equation in a single variable. Since the emphasis in this section is on the time approximations, the development of the weak form and spatial finite element model will not be covered explicitly here, and the reader is referred to Sections 8.2 and 8.3 for details.

8.5.2 Parabolic Equations

Consider the partial differential equation governing transient heat transfer and similar problems in a two-dimensional region Ω with total boundary Γ,

$$c\frac{\partial u}{\partial t}-\frac{\partial}{\partial x}\left(a_{11}\frac{\partial u}{\partial x}\right)-\frac{\partial}{\partial y}\left(a_{22}\frac{\partial u}{\partial y}\right)+a_0u=f(x, y, t) \tag{8.154}$$

with the boundary conditions

$$u=\hat{u} \text{ or } q_n=\hat{q}_n \quad \text{on } \Gamma \quad (t\geqslant 0) \tag{8.155a}$$

where

$$q_n=a_{11}\frac{\partial u}{\partial x}n_x+a_{22}\frac{\partial u}{\partial y}n_y \tag{8.155b}$$

The initial conditions (i.e., at $t=0$) are of the form

$$u(x, y, 0)=u_0(x, y) \quad \text{in } \Omega \tag{8.156}$$

Here t denotes time, and c, a_{11}, a_{22}, a_0, $\hat{u}$, u_0, f, and $\hat{q}_n$ are given functions of position and/or time. Equation (8.154) is a modification of (8.1) in that it contains a time-derivative term, which accounts for time variations of the physical process represented by (8.1).

The weak form of (8.154) and (8.155) over an element Ω^e is obtained by the standard procedure: multiply (8.154) with the weight function $v(x, y)$ and integrate over the element, integrate by parts (spatially) those terms that involve higher-order derivatives using the gradient or divergence theorem, and replace the coefficient of the weight function in the boundary integral with the secondary variable [i.e., use (8.155*b*)]. We obtain

$$0=\int_{\Omega^e}\left[v\left(c\frac{\partial u}{\partial t}+a_0u-f\right)+a_{11}\frac{\partial v}{\partial x}\frac{\partial u}{\partial x}+a_{22}\frac{\partial v}{\partial y}\frac{\partial u}{\partial y}\right]dx\,dy-\oint_{\Gamma^e}q_nv\,ds \tag{8.157}$$

Note that the procedure for obtaining the weak form for time-dependent problems is not very different from that used for steady-state problems in Section 8.2.3. The difference is that all terms of the equations may be functions of time. Also, there is no integration by parts with respect to time, and the weight function v is not a function of time.

The *semidiscrete* finite element model is obtained from (8.157) by substituting a finite element approximation for the dependent variable u. In selecting the approximation for u, once again we assume that the time dependence can be separated from the spatial variation:

$$u(x,y,t)\approx\sum_{j=1}^{n}u_j^e(t)\psi_j^e(x,y) \tag{8.158}$$

where u_j denotes the value of $u(x, y, t)$ at the spatial location (x_j, y_j) at time t. The ith differential equation (in time) of the finite element model is obtained by substituting $v=\psi_i^e(x,y)$ and replacing u by (8.158) in (8.157):

$$0=\sum_{j=1}^{n}\left(M_{ij}^e\frac{du_j^e}{dt}+K_{ij}^eu_j^e\right)-f_i^e-Q_i^e \tag{8.159a}$$

or, in matrix form,

$$[M^e]\{\dot{u}^e\}+[K^e]\{u^e\}=\{f^e\}+\{Q^e\} \tag{8.159b}$$

where a superposed dot on u denotes the time derivative ($\dot{u}=\partial u/\partial t$), and

$$M_{ij}^e=\int_{\Omega^e}c\psi_i\psi_j\,dx\,dy,\qquad K_{ij}^e=\int_{\Omega^e}\left(a_{11}\frac{\partial\psi_i}{\partial x}\frac{\partial\psi_j}{\partial x}+a_{22}\frac{\partial\psi_i}{\partial y}\frac{\partial\psi_j}{\partial y}+a_0\psi_i\psi_j\right)dx\,dy \tag{8.159c}$$

$$f_i^e=\int_{\Omega^e}f(x,y,t)\psi_i\,dx\,dy$$

This completes the semidiscretization step.

EIGENVALUE ANALYSIS. The problem of finding $u_j^e(t)=U_j^ee^{-\lambda t}$ such that (8.159) holds for homogeneous boundary and initial conditions and $f=0$ is called an *eigenvalue problem*. Substituting for $u_j^e(t)$ in (8.159*b*), we obtain

$$\boxed{(-\lambda[M^e]+[K^e])\{u^e\}=\{Q^e\}} \tag{8.160}$$

Upon assembly of the element equations (8.160), the right-hand-side column vector of the condensed equations is zero (because of the homogeneous boundary conditions), giving rise to the global eigenvalue problem

$$([K] - \lambda[M])\{U\} = \{0\} \tag{8.161}$$

The order of the matrix equations is $N \times N$, where N is the number of nodes at which the solution is not known. A nontrivial solution to (8.161) exists only if the determinant of the coefficient matrix is zero:

$$|[K] - \lambda[M]| = 0$$

which when expanded, results in an Nth-degree polynomial in λ. The N roots λ_j $(j = 1, 2, \ldots, N)$ of this polynomial give the first N eigenvalues of the discretized system (the continuous system, in general, has an infinite number of eigenvalues). There exist standard eigenvalue routines for solving (8.161), which give the N eigenvalues and eigenvectors.

TRANSIENT ANALYSIS. Note that the form of (8.159) is the same as the parabolic equation discussed in Section 6.2. The spatial dimension does not appear in (8.159*a,b*) because it is taken into account in the spatial approximation. Thus, irrespective of the spatial dimension, the finite element model of all problems with a first time derivative is the same. Therefore, the time approximation schemes discussed in Section 6.2.3 for parabolic equations can be readily applied.

Using the α-family of approximation (6.39),

$$\{u\}_{s+1} = \{u\}_s + \Delta t\,[(1-\alpha)\{\dot{u}\}_s + \alpha\{\dot{u}\}_{s+1}] \quad (0 \leq \alpha \leq 1) \tag{8.162}$$

we can transform the ordinary differential equations (8.159) into a set of algebraic equations at time t_{s+1}:

$$\boxed{[\hat{K}]_{s+1}\{u\}_{s+1} = \{\hat{F}\}_{s,s+1}} \tag{8.163a}$$

where

$$\boxed{\begin{aligned} [\hat{K}]_{s+1} &= [M] + a_1[K]_{s+1} \\ \{\hat{F}\} &= \Delta t\left(\alpha\{F\}_{s+1} + (1-\alpha)\{F\}_s\right) + ([M] - a_2[K]_s)\{u\}_s \\ a_1 = \alpha\,\Delta t,& \qquad a_2 = (1-\alpha)\,\Delta t \end{aligned}} \tag{8.163b}$$

Equation (8.163*a*), after assembly and imposition of boundary conditions, is solved at each time step for the nodal values u_j at time $t_s = (s+1)\,\Delta t$. At time $t = 0$ (i.e., $s = 0$), the right-hand side is computed using the initial values $\{u\}_0$; the vector $\{F\}$, which is the sum of the source vector $\{f\}$ and the internal flux vector $\{Q\}$, is always known, for both times t_s and t_{s+1}, at all nodes at which the solution is unknown [because $f(x, t)$ is a *known* function of time, and the sum of Q_j^e at these nodes is zero].

It should be recalled from Section 6.2.3 that, for different values of α, we obtain different well-known approximation schemes, with given order of accuracy:

$$\alpha = \begin{cases} 0, & \text{the forward difference scheme (conditionally stable); } O(\Delta t) \\ \frac{1}{2}, & \text{the Crank–Nicolson scheme (unconditionally stable); } O(\Delta t)^2 \\ \frac{2}{3}, & \text{the Galerkin scheme (unconditionally stable); } O(\Delta t)^2 \\ 1, & \text{the backward difference scheme (unconditionally stable); } O(\Delta t) \end{cases}$$

For the forward difference scheme, the stability requirement is

$$\boxed{\Delta t < \Delta t_{cr} = \frac{2}{(1-2\alpha)\lambda_{max}}, \qquad \alpha < \tfrac{1}{2}} \tag{8.164}$$

where λ_{max} is the largest eigenvalue of the finite element equations (8.161).

We next consider examples of eigenvalue and time-dependent problems.

Example 8.9 Eigenvalue analysis. Consider the differential equation

$$\frac{\partial u}{\partial t} - \left(\frac{\partial^2 u}{\partial x^2} + \frac{\partial^2 u}{\partial y^2}\right) = f \tag{8.165a}$$

in a square region, subject to the boundary conditions

$$\frac{\partial u}{\partial x}(0, y, t) = 0, \qquad \frac{\partial u}{\partial y}(x, 0, t) = 0, \qquad u(x, 1, t) = 0, \qquad u(1, y, t) = 0 \tag{8.165b}$$

and initial condition

$$u(x, y, 0) = 0 \tag{8.165c}$$

As a first choice, we take a 1×1 mesh of triangular (2 elements) and rectangular elements. Alternatively, for the choice of triangles, we can use the diagonal symmetry and model the domain with one triangular element. The triangular element matrices are

$$[K^e] = \frac{1}{2}\begin{bmatrix} 1 & -1 & 0 \\ -1 & 2 & -1 \\ 0 & -1 & 1 \end{bmatrix}, \qquad [M^e] = \frac{1}{24}\begin{bmatrix} 2 & 1 & 1 \\ 1 & 2 & 1 \\ 1 & 1 & 2 \end{bmatrix}$$

The eigenvalue problem becomes

$$\left(-\frac{\lambda}{24}\begin{bmatrix} 2 & 1 & 1 \\ 1 & 2 & 1 \\ 1 & 1 & 2 \end{bmatrix} + \frac{1}{2}\begin{bmatrix} 1 & -1 & 0 \\ -1 & 2 & -1 \\ 0 & -1 & 1 \end{bmatrix}\right)\begin{Bmatrix} U_1 \\ U_2 \\ U_3 \end{Bmatrix} = \begin{Bmatrix} 0 \\ 0 \\ 0 \end{Bmatrix}$$

The boundary conditions require $U_2 = U_3 = 0$. Hence, we have

$$(-\tfrac{1}{12}\lambda + \tfrac{1}{2})U_1 = 0, \qquad \text{or} \quad \lambda = 6$$

The eigenfunction becomes

$$U(x, y) = \psi_1(x, y) = 1 - x$$

which is defined over the octant of the domain. For a quadrant of the domain, by symmetry, it becomes $U(x, y) = (1 - x)(1 - y)$.

For one rectangular element, we have

$$[K^e] = \frac{1}{6}\begin{bmatrix} 4 & -1 & -2 & -1 \\ -1 & 4 & -1 & -2 \\ -2 & -1 & 4 & -1 \\ -1 & -2 & -1 & 4 \end{bmatrix}, \quad [M^e] = \frac{1}{36}\begin{bmatrix} 4 & 2 & 1 & 2 \\ 2 & 4 & 2 & 1 \\ 1 & 2 & 4 & 2 \\ 2 & 1 & 2 & 4 \end{bmatrix}$$

$$\left(-\frac{\lambda}{36}\begin{bmatrix} 4 & 2 & 1 & 2 \\ 2 & 4 & 2 & 1 \\ 1 & 2 & 4 & 2 \\ 2 & 1 & 2 & 4 \end{bmatrix} + \frac{1}{6}\begin{bmatrix} 4 & -1 & -2 & -1 \\ -1 & 4 & -1 & -2 \\ -2 & -1 & 4 & -1 \\ -1 & -2 & -1 & 4 \end{bmatrix} \right) \begin{Bmatrix} U_1 \\ U_2 \\ U_3 \\ U_4 \end{Bmatrix} = \begin{Bmatrix} 0 \\ 0 \\ 0 \\ 0 \end{Bmatrix}$$

Using the boundary conditions $U_2 = U_3 = U_4 = 0$, we obtain

$$(-\tfrac{1}{36}\lambda \times 4 + \tfrac{4}{6})U_1 = 0, \qquad \text{or} \quad \lambda = 6$$

The eigenfunction over the quadrant of the domain is given by

$$U(x, y) = \psi_1(x, y) = (1 - x)(1 - y)$$

For this problem, the one-element mesh of triangles in an octant of the domain gives the same solution as the one-element mesh of rectangular elements in a quadrant of the domain.

Table 8.9 gives eigenvalues obtained with various meshes of triangular and rectangular elements, along with the analytical solution of the problem. It is clear that the convergence of the minimum eigenvalue obtained using the finite element method to the analytical value is rapid compared with the convergence of the higher eigenvalues; i.e., the errors in the higher eigenvalues are always larger than that in the minimum eigenvalue. Also, the minimum eigenvalue converges faster with mesh refinements. The mesh used must be such that the required eigenmodes are represented accurately.

Example 8.10. We wish to solve the transient heat conduction equation

$$\frac{\partial T}{\partial t} - \left(\frac{\partial^2 T}{\partial x^2} + \frac{\partial^2 T}{\partial y^2}\right) = 1 \tag{8.166a}$$

TABLE 8.9
Comparison of finite element solutions for eigenvalues, obtained using various meshes, with the analytical solution (Example 8.9)

	Triangles				Rectangles				Analytical†
λ	1×1	2×2	4×4	8×8	1×1	2×2	4×4	8×8	λ_{mn}
$\lambda_1\ (\lambda_{11})$	6.000	5.415	5.068	4.969	6.000	5.193	4.999	4.951	4.935
$\lambda_2\ (\lambda_{13})$	—	32.000	27.250	25.340	—	34.290	27.370	25.330	24.674
$\lambda_3\ (\lambda_{31})$	—	38.200	28.920	25.730	—	34.290	27.370	25.330	24.674
$\lambda_4\ (\lambda_{33})$	—	76.390	58.220	48.080	—	63.380	49.740	45.710	44.413
$\lambda_4\ (\lambda_{15})$	—	—	85.350	69.780	—	—	84.570	69.260	64.152
$\lambda_5\ (\lambda_{51})$	—	—	86.790	69.830	—	—	84.570	69.260	64.152

† The analytical solution is $\lambda_{mn} = \frac{1}{4}\pi^2(m^2 + n^2)$ $(m, n = 1, 3, 5, \ldots)$.

subject to the boundary conditions, for $t \geqslant 0$,

$$\begin{aligned} \frac{\partial T}{\partial x}(0, y, t) = 0, \quad & \frac{\partial T}{\partial y}(x, 0, t) = 0 \\ T(1, y, t) = 0, \quad & T(x, 1, t) = 0 \end{aligned} \tag{8.166b}$$

and the initial condition

$$T(x, y, 0) = 0 \quad \text{for all } (x, y) \text{ in } \Omega \tag{8.166c}$$

We choose a 4×4 mesh of linear triangular elements (see Fig. 8.36) to model the domain, and investigate the stability and accuracy of the Crank–Nicolson method (i.e., $\alpha = 0.5$) and the forward difference scheme ($\alpha = 0.0$) for the temporal approximation. Since the Crank–Nicolson method is unconditionally stable, one can choose any value of Δt. However, for large values of Δt, the solution may not be accurate. The forward difference scheme is conditionally stable; it is stable if $\Delta t < \Delta t_{cr}$, where

$$\Delta t_{cr} = \frac{2}{\lambda_{max}} = \frac{2}{386.4} = 0.00518$$

where the maximum eigenvalue of (8.166a) for the 4×4 mesh of triangles is 386.4.

The element equations are given by (8.163), with $[M^e]$, $[K^e]$, and $\{f^e\}$ defined by (8.159c), wherein $c = 1$, $a_{11} = 1$, $a_{22} = 1$, $a_0 = 0$, and $f = 1$. The boundary conditions of the problem for the 4×4 mesh are given by

$$U_5 = U_{10} = U_{15} = U_{20} = U_{21} = U_{22} = U_{23} = U_{24} = U_{25} = 0.0$$

Beginning with the initial conditions $U_i = 0$ ($i = 1, 2, \ldots, 25$), we solve the assembled set of equations associated with (8.163).

The forward difference scheme would be unstable for $\Delta t > 0.00518$. To illustrate this point, the equations are solved using $\alpha = 0$, $\Delta t = 0.01$ and $\alpha = 0.5$, $\Delta t = 0.01$. The Crank–Nicolson method gives a stable and accurate solution, while the forward difference scheme yields an unstable solution (i.e., the solution error grows unboundedly with time), as can be seen from Fig. 8.37. For $\Delta t = 0.005$, the forward difference scheme yields a stable solution.

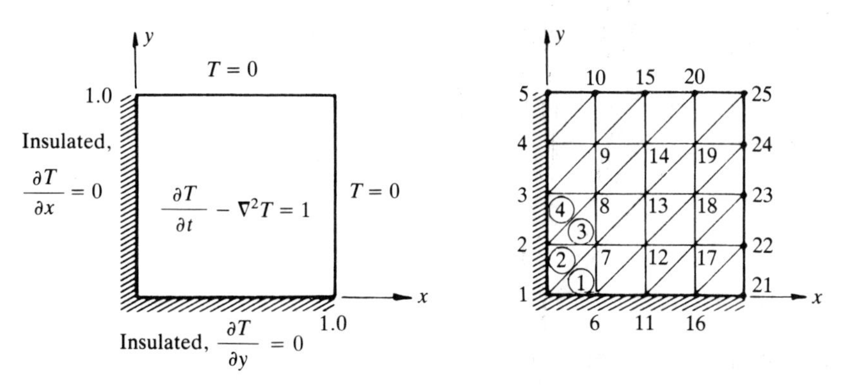

FIGURE 8.36
Domain, boundary conditions, and finite element mesh for the transient heat conduction problem of Example 8.10.

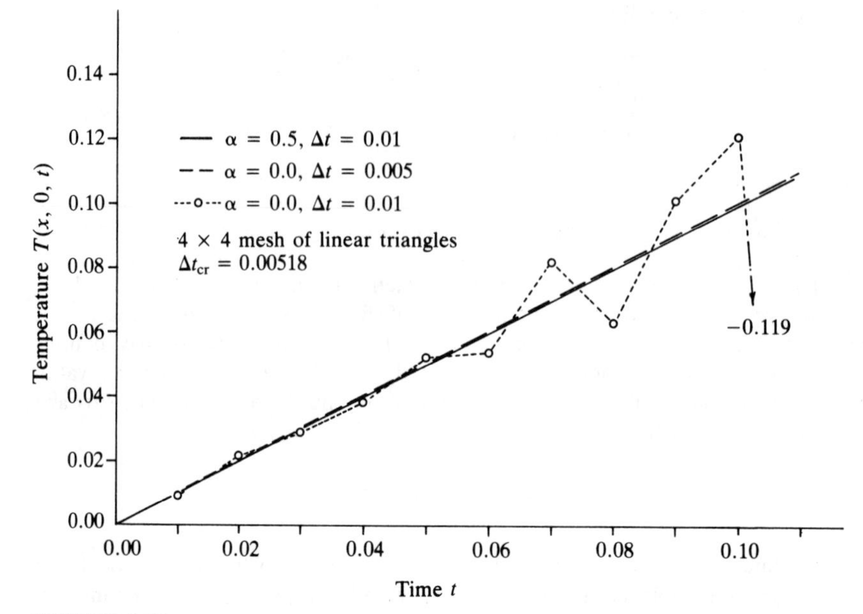

FIGURE 8.37
Stability of the transient solutions of the heat conduction problem in Example 8.10 analyzed using a 4×4 mesh of linear triangular elements and the Crank–Nicolson ($\alpha = 0.5$) and forward difference ($\alpha = 0.0$) time integration schemes.

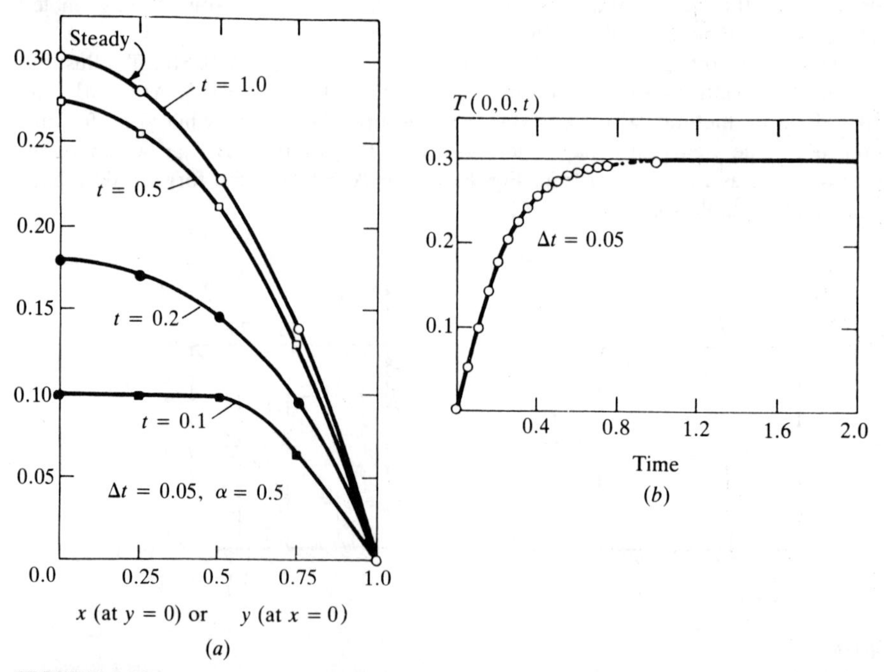

FIGURE 8.38
Variation of the temperature as a function of position x and time t for the transient heat conduction problem of Example 8.10.

TABLE 8.10
Comparison of finite difference and finite element solutions with the exact solution of the heat conduction problem in Example 8.10

Node	Exact solution	Finite difference solution	Error	Finite element solution (steady)	Error	Finite element solution (unsteady) at $t = 1.0$
1	0.2947	0.2911	0.0036	0.3013	−0.0066	0.2993
2	0.2789	0.2755	0.0034	0.2805	−0.0016	0.2786
3	0.2293	0.2266	0.0027	0.2292	0.0001	0.2278
4	0.1397	0.1381	0.0016	0.1392	0.0005	0.1385
5	0.0000	0.0000	0.0000	0.0000	0.0000	0.0000
7	0.2642	0.2609	0.0033	0.2645	−0.0003	0.2628
8	0.2178	0.2151	0.0027	0.2172	0.0006	0.2159
9	0.1333	0.1317	0.0016	0.1327	0.0006	0.1320
10	0.0000	0.0000	0.0000	0.0000	0.0000	0.0000
13	0.1811	0.1787	0.0024	0.1801	0.0010	0.1791
14	0.1127	0.1110	0.0017	0.1117	0.0010	0.1111
15	0.0000	0.0000	0.0000	0.0000	0.0000	0.0000
19	0.0728	0.0711	0.0017	0.0715	0.0013	0.0712
20	0.0000	0.0000	0.0000	0.0000	0.0000	0.0000
25	0.0000	0.0000	0.0000	0.0000	0.0000	0.0000

The Crank–Nicolson method gives a stable and accurate solution even for $\Delta t = 0.05$. The temperature $T(x, 0, t)$ is plotted versus x for various values of the time in Fig. 8.38(*a*). The steady state is reached at time $t = 1.0$. The temperature $T(0, 0, t)$ predicted by the Crank–Nicolson method is plotted versus time in Fig. 8.38(*b*), which indicates the evolution of the temperature from zero to the steady state. Table 8.10 gives a comparison of the transient solution at $t = 1.0$ with the steady-state finite element, the finite difference, and the analytical solutions. Table 8.11 gives the finite element solutions for temperature predicted by 4×4 meshes of triangles and rectangles and various values of Δt and $\alpha = 0.5$.

8.5.3 Hyperbolic Equations

The transverse motion of a membrane, for example, is governed by a partial differential equation of the form

$$c\frac{\partial^2 u}{\partial t^2} - \frac{\partial}{\partial x}\left(a_{11}\frac{\partial u}{\partial x}\right) - \frac{\partial}{\partial y}\left(a_{22}\frac{\partial u}{\partial y}\right) + a_0 u = f(x, y, t) \qquad (8.167a)$$

where $u(x, y, t)$ denotes the transverse deflection, c is the material density of the membrane, a_{11} and a_{22} are the tensions in the x and y directions of the membrane, a_0 is the modulus of the elastic foundation on which the membrane is stretched (often $a_0 = 0$, i.e., there is no foundation), and $f(x, y, t)$ is the transversely distributed force. Equation (8.167*a*) is known as the *wave equation,* and is classified mathematically as a hyperbolic equation. The function u must be determined such that it satisfies (8.167*a*) in a region Ω,

TABLE 8.11
Comparison of the transient solutions of (8.166) obtained using a mesh of triangular and rectangular elements ($\alpha = 0.5$) (Example 8.10)

Time		Temperature along the line $y = 0$: $T(x, 0, t) \times 10$			
t	Element†	$x = 0.0$	$x = 0.25$	$x = 0.5$	$x = 0.75$
0.1	T1	0.9758	0.9610	0.9063	0.7104
	R1	0.9684	0.9556	0.8956	0.6887
	T2	0.9928	0.9798	0.9168	0.6415
	R2	0.9841	0.9718	0.9020	0.6323
0.2	T1	1.8003	1.7238	1.4891	0.9321
	R1	1.7723	1.7216	1.4829	0.9367
	T2	1.7979	1.7060	1.4644	0.9462
	R2	1.7681	1.6990	1.4626	0.9469
0.3	T1	2.3130	2.1671	1.7961	1.1466
	R1	2.2747	2.1650	1.8084	1.1499
	T2	2.2829	2.1448	1.7943	1.1249
	R2	2.2479	2.1432	1.8018	1.1319
1.0	T1	2.9960	2.7871	2.2804	1.3843
	R1	2.9648	2.8053	2.3090	1.4059
	T2	2.9925	2.7862	2.2776	1.3849
	R2	2.9621	2.8037	2.3065	1.4053

† T1, triangular element mesh with $\Delta t = 0.1$; T2, triangular element mesh with $\Delta t = 0.05$; R1, rectangular element mesh with $\Delta t = 0.1$; R2, rectangular element mesh with $\Delta t = 0.05$.

together with the following boundary and initial conditions:

$$u = \hat{u} \quad \text{or} \quad q_n = \hat{q}_n \quad \text{on } \Gamma \quad (t \geq 0) \tag{8.167b}$$

$$u(x, y, 0) = u_0(x, y), \qquad \frac{\partial u}{\partial t}(x, y, 0) = v_0(x, y) \tag{8.167c}$$

where $\hat{u}$ and $\hat{q}_n$ are specified boundary values of u and q_n [see (8.155b)], and u_0 and v_0 are specified initial values of u and its time derivative, respectively.

The weak form of (8.167a,b) over a typical element Ω^e is similar to that of (8.154) [see (8.157)], except that here we have the second time derivative of u:

$$0 = \int_{\Omega^e} \left[v\left(c\frac{\partial^2 u}{\partial t^2} + a_0 u - f \right) + a_{11}\frac{\partial v}{\partial x}\frac{\partial u}{\partial x} + a_{22}\frac{\partial v}{\partial y}\frac{\partial u}{\partial y} \right] dx\, dy - \oint_{\Gamma^e} q_n v\, ds \tag{8.168}$$

where $v = v(x, y)$ is the weight function.

The semidiscrete finite element model of (8.167a) is obtained by substituting the finite element approximation (8.158) for u and $v = \psi_i$ into

(8.168):

$$0=\sum_{j=1}^{n}\left(M_{ij}^{e}\frac{d^{2}u_{j}^{e}}{dt^{2}}+K_{ij}^{e}u_{j}^{e}\right)-f_{i}^{e}-Q_{i}^{e} \tag{8.169a}$$

or, in matrix form,

$$[M^e]\{\ddot{u}^e\}+[K^e]\{u^e\}=\{f^e\}+\{Q^e\} \tag{8.169b}$$

The coefficients M_{ij}^e, K_{ij}^e, and f_i^e are the same as those in (8.159c).

EIGENVALUE ANALYSIS. The problem of finding $u_j(t)=U_je^{-i\omega t}$ $(i=\sqrt{-1})$ such that (8.169) holds for homogeneous boundary and initial conditions and $f=0$ is called an eigenvalue problem of (8.167). We obtain

$$\boxed{(-\omega^2[M^e]+[K^e])\{u^e\}=\{Q^e\}} \tag{8.170}$$

The eigenvalues ω^2 and eigenfunctions $\sum_{j=1}^{n} U_j\psi_j(x, y)$ are determined from the assembled equations associated with (8.170), after imposing the homogeneous boundary conditions. For a membrane problem, ω denotes the frequency of natural vibration. The number of eigenvalues of the discrete system (8.170) of the problem is equal to the number of unknown nodal values of U in the mesh.

Example 8.11. Consider the free vibrations of a rectangular membrane of homogeneous material, of dimensions a by b (in ft), material density ρ (in slugs ft^{-2}), and fixed on all its edges, i.e., $u=0$ on Γ. Although the problem has symmetry about the horizontal centerline and vertical centerlines of the domain (see Fig. 8.39), use of any symmetry in the finite element analysis will eliminate the unsymmetric modes of vibration of the membrane. For example, if we consider a quadrant of the domain in the finite element analysis, the frequencies ω_{mn} $(m, n\neq 1, 3, 5, \ldots)$ and associated eigenfunctions will be missed in the results [i.e., we can only obtain ω_{mn} $(m, n = 1, 3, 5, \ldots)$]. By considering the full domain, the first N frequencies allowed by the mesh can be computed, where N is the number of unknown nodal values in the mesh.

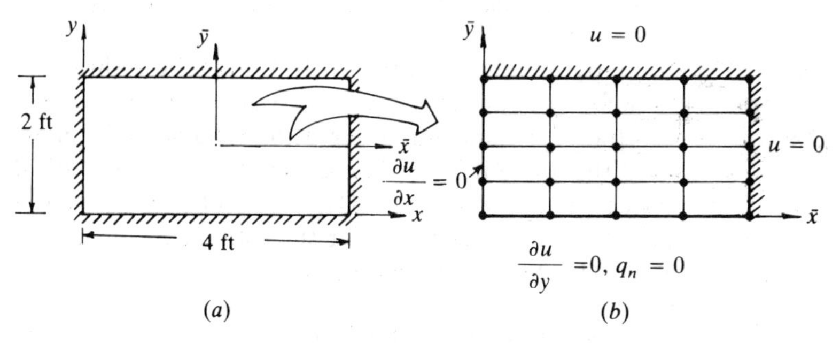

FIGURE 8.39
Geometry, computational domain, finite element mesh, and boundary conditions for the transient analysis of a rectangular membrane with initial deflection: (a) actual geometry; (b) computational domain, finite element mesh of rectangular elements, and boundary conditions (4×4 mesh of linear elements or 2×2 mesh of nine-node quadratic elements).

TABLE 8.12
Comparison of natural frequencies computed using various meshes of linear triangular and rectangular elements with the analytical solution of a rectangular membrane fixed on all its sides ($a_{11} = a_{22} = 12.5$, $\rho = T = 2.5$)

	Triangular (linear)			Rectangular (linear)			
ω_{mn}	2×2	4×4	8×8	2×2	4×4	8×8	Analytical
ω_{11}	5.0000	4.2266	4.0025	4.3303	4.0285	3.9522	3.9270
ω_{21}	—	5.9083	5.2068	—	5.2899	5.0478	4.9673
ω_{31}	—	8.2392	6.8788	—	7.2522	6.6020	6.3321
ω_{12}	—	8.3578	7.5271	—	7.9527	7.4200	7.2410
ω_{22}	—	10.0618	8.4565	—	8.6603	8.0571	7.8540
ω_{41}	—	12.1021	8.8856	—	9.9805	8.5145	7.8540
ω_{32}	—	13.2011	9.9280	—	12.7157	9.1117	8.7810
ω_{51}	—	14.6942	11.1193	—	13.1700	10.5797	9.4574
ω_{42}	—	15.8117	11.4425	—	14.0734	10.7280	9.9346

If only the first eigenvalue ω_{11} is of interest or only symmetric frequencies are required, one can use a quadrant of the domain in the analysis. Indeed the results of Example 8.9 are applicable here, with $\lambda_{mn} = \omega_{mn}^2$. The results presented in Table 8.9 can be interpreted as the squares of the symmetric natural frequencies of a square $a = b = 2$) membrane with $\rho = 1$ and $a_{11} = a_{22} \equiv T = 1$. The exact natural frequencies of a rectangular membrane of dimensions a by b, with tensions $a_{11} = a_{22} = T$ and density ρ are

$$\omega_{mn} = \pi\left(\frac{T}{\rho}\right)^{1/2}\left(\frac{m^2}{a^2} + \frac{n^2}{b^2}\right)^{1/2} \quad (m, n = 1, 2, \ldots)$$

To obtain all frequencies, the full domain must be modeled.

Table 8.12 contains the first nine frequencies of a rectangular membrane of 4 ft by 2 ft, tension $T = 12.5$ lb ft^{-1}, and density $\rho = 2.5$ slugs ft^{-2}, computed using various meshes of linear triangular and rectangular elements. The convergence of the finite element results to the analytical solution is clear. The mesh of linear rectangular element yields more accurate results than the mesh of linear triangular elements.

TRANSIENT ANALYSIS. The hyperbolic equation (8.169*b*) can be reduced to a system of algebraic equations by approximating the second time derivative. As discussed in Section 6.2, the Newmark time integration scheme is the most commonly used method, and therefore it is used here. Since the mathematical form of (8.169b) is exactly the same as that in (6.44), the results in (6.47) and (6.49) are immediately applicable to the former. For the sake of convenience, the results are repeated here. The Newmark scheme is

$$\boxed{\begin{aligned} \{u\}_{s+1} &= \{u\}_s + \Delta t\{\dot{u}\}_s + \tfrac{1}{2}(\Delta t)^2\{\ddot{u}\}_{s+\gamma} \\ \{\dot{u}\}_{s+1} &= \{\dot{u}\}_s + \Delta t\{\ddot{u}\}_{s+\alpha} \\ \{\ddot{u}\}_{s+\theta} &= (1-\theta)\{\ddot{u}\}_s + \theta\{\ddot{u}\}_{s+1} \end{aligned}} \tag{8.171a}$$

where, for example,

$\alpha = \frac{1}{2}$,	$\gamma = \frac{1}{2}$,	the constant-average acceleration method (stable)
$\alpha = \frac{1}{2}$,	$\gamma = \frac{1}{3}$,	the linear acceleration method (conditionally stable)
$\alpha = \frac{1}{2}$,	$\gamma = 0$,	the central difference method (conditionally stable)

(8.171*b*)

The stability criterion is

$$\Delta t < \Delta t_{cr} = [\tfrac{1}{2}\omega_{max}^2(\alpha - \gamma)]^{-1/2}, \qquad \alpha \geqslant \tfrac{1}{2}, \qquad \gamma < \alpha \tag{8.172}$$

where ω_{max}^2 is the maximum eigenvalue of the corresponding discrete eigenvalue problem (8.170) (i.e., the same mesh and element type used in the transient analysis must be used in the eigenvalue analysis). Note that a more refined mesh will yield a lower maximum eigenvalue and a higher Δt_{cr}.

Time marching scheme

$$[\hat{K}^e]_{s+1}\{u^e\}_{s+1} = \{\hat{F}^e\}_{s,s+1} \tag{8.173a}$$

where (the superscript e is omitted for brevity in the following)

$$\begin{gathered}
[\hat{K}]_{s+1} = [K]_{s+1} + a_3[M]_{s+1} \\
\{\hat{F}\}_{s,s+1} = \{F\}_{s+1} + [M]_{s+1}(a_3\{u\}_s + a_4\{\dot{u}\}_s + a_5\{\ddot{u}\}_s) \\
a_3 = \frac{2}{\gamma(\Delta t)^2}, \qquad a_4 = \Delta t\, a_3, \qquad a_5 = \frac{1}{\gamma} - 1
\end{gathered} \tag{8.173b}$$

Once $\{u\}_{s+1}$ has been calculated from (8.173*a*), the velocities and accelerations at time $t_{s+1} = \Delta t\,(s+1)$ are calculated from (6.49):

$$\begin{gathered}
\{\ddot{u}\}_{s+1} = a_3(\{u\}_{s+1} - \{u\}_s) - a_4\{\dot{u}\}_s - a_5\{\ddot{u}\}_s \\
\{\dot{u}\}_{s+1} = \{\dot{u}\}_s + a_2\{\ddot{u}\}_s + a_1\{\ddot{u}\}_{s+1} \\
a_1 = \alpha\,\Delta t, \qquad a_2 = (1-\alpha)\,\Delta t
\end{gathered} \tag{8.174}$$

Note that (8.173*a*) is valid for an element. Therefore, the operations indicated in (8.173*b*) are carried out for an element, and $[\hat{K}^e] = \{\hat{F}^e\}$ are assembled as in a static analysis. For the first time step, the initial conditions on u and $\partial u/\partial t$ are used to compute $\{u\}_0$ and $\{\dot{u}\}_0$ for each element of the entire mesh. The acceleration vector $\{\ddot{u}\}_0$ is computed from (8.169*b*) at $t = 0$:

$$\{\ddot{u}\}_0 = [M]^{-1}(\{F\}_0 - [K]\{U\}_0) \tag{8.175}$$

It is often assumed that $\{F\}_0 = \{0\}$. If the initial conditions are zero, $\{u\}_0 = \{0\}$, and the applied force is assumed to be zero at $t = 0$, we then take $\{\ddot{u}\}_0 = \{0\}$.

Example 8.12. Consider a homogeneous rectangular membrane of sides $a = 4$ ft and $b = 2$ ft, fixed on all its four edges. Assume that the tension in the membrane is 12.5 lb ft^{-1} (i.e., $a_{11} = a_{22} = 12.5$) and the density is $\rho = c = 2.5$ slugs ft^{-2}. The initial deflection of the membrane is assumed to be

$$u_0(x, y) = 0.1(4x - x^2)(2y - y^2) \tag{8.176}$$

and the initial velocity is $v_0 = 0$. We wish to determine the deflection $u(x, y, t)$ of the membrane as a function of time using the finite element method. The analytical solution of this problem is [see Kreyszig (1988), p. 684],

$$u(x, y, t) = \frac{409.6}{\pi^6} \sum_{m,n=1,3,\ldots} \frac{1}{m^3 n^3} \cos \omega_{mn} t \sin \tfrac{1}{4} m\pi x \sin \tfrac{1}{2} n\pi y \tag{8.177a}$$

$$\omega_{mn} = \tfrac{1}{4}\pi[5(m^2 + 4n^2)]^{1/2} \tag{8.177b}$$

where the origin of the (x, y) coordinate system is located at the lower corner of the domain (see Fig. 8.39*a*).

In the finite element analysis, we can utilize the biaxial symmetry of the problem and model one quadrant of the domain (see Fig. 8.39*b*). We set up a new coordinate system $(\bar{x}, \bar{y})$ for the computational domain. The inital displacement in the new coordinates is given by (8.176), with x and y replaced in terms of $\bar{x}$ and $\bar{y}$:

$$x = \bar{x} + 2, \qquad y = \bar{y} + 1$$

The initial values of $\ddot{u}$ are calculated using (8.175), with $\{F\}_0 = \{0\}$ and $\{u\}_0$ as given in (8.176) by $u_0(x, y)$. At $\bar{x} = 2$ and $\bar{y} = 1$, all nodal values for the function u and its time derivatives are zero.

As for the critical time step, we calculate λ_{max} from the solution of (8.170) using the same mesh as that used for the transient analysis, and then use (8.172) to compute Δt_{cr}. Of course, for $\alpha = \frac{1}{2}$ and $\gamma = \frac{1}{2}$, there is no restriction on the time step for a stable solution. For a 4×4 mesh of linear rectangular elements, $\lambda_{max} = (14.0734)^2$ (see Table 8.12), and $\Delta t_{cr} = 0.246$ for the linear acceleration scheme ($\alpha = 0.5$, $\gamma = \frac{1}{3}$).

Figure 8.40 shows plots of the center deflection $u(0, 0, t)$ versus time t, and Fig. 8.41 shows the stability of the solutions computed using the constant-average acceleration ($\alpha = 0.5$, $\gamma = 0.5$) and linear acceleration ($\alpha = 0.5$, $\gamma = \frac{1}{3}$) schemes for $\Delta t = 0.25 > \Delta t_{cr}$. Figure 8.40 also shows a comparison of the finite element solutions with the analytical solution (8.177). The finite element solutions are in good agreement with the analytical one.

8.6 SUMMARY

A step-by-step procedure for finite element formulation of second-order differential equations in two dimensions with a single dependent variable has been presented. The Poisson equation in two dimensions has been used to illustrate the steps involved. These include weak formulation of the equation, development of the finite element model, derivation of the interpolation functions for linear triangular and rectangular elements, evaluation of element matrices and vectors, assembly of element equations, solution of equations, and post-computation of the gradient of the solution. A number of illustrative problems of heat transfer (conduction and convection), fluid mechanics and solid mechanics have been discussed. Finally, the eigenvalue and time-

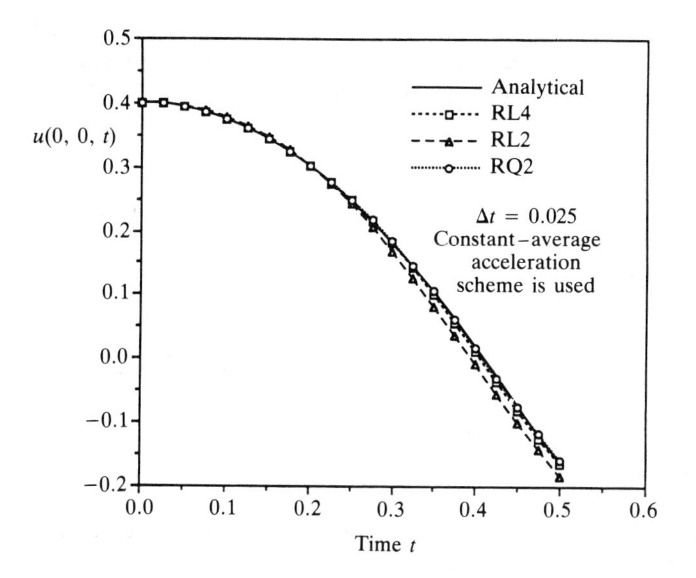

FIGURE 8.40
Comparison of the center deflection obtained using various meshes with the analytical solution of a rectangular membrane with initial deflection: RL4, 4×4 mesh of linear rectangular elements; RL2, 2×2 mesh of linear rectangular elements; RQ2, 2×2 mesh of nine-node quadratic elements.

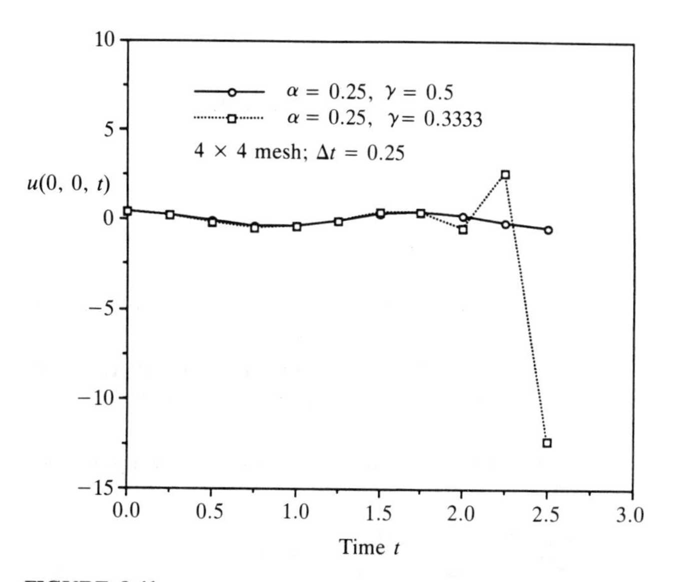

FIGURE 8.41
Stability characteristics of the constant-average acceleration and linear acceleration schemes (a 4×4 mesh of linear rectangular elements is used in a quadrant of the domain).

dependent problems associated with the model equation have also been discussed. This chapter constitutes the heart of the finite element analysis of two-dimensional problems to be discussed in Chapters 10–12.

PROBLEMS

Note that most of the problems given here require hand calculations only. When four or more simultaneous algebraic equations are to be solved, they should be left in matrix form. New problems can be created just by changing data and meshes.

Section 8.2

8.1. The electrostatic potential $\phi(x, y)$ of electrical charges in a region Ω is governed by the Poisson equation

$$-\nabla^2\phi = f(x, y)$$

where f is the distributed electric charge.

(a) Develop the finite element model of the equation.

(b) Identify the element coefficient matrices for linear triangular and rectangular elements from those available in this book.

(c) Write the specified primary and secondary variables at all boundary nodes and their specified values for the problem shown in Fig. P8.1.

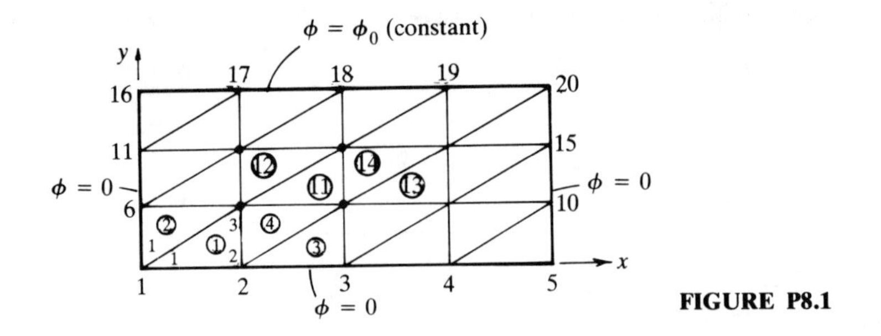

FIGURE P8.1

8.2. Consider the partial differential equation

$$-\nabla^2 u + cu = 0 \quad \text{in } \Omega^e, \qquad \text{with} \quad \frac{\partial u}{\partial n} + ku = 0 \quad \text{on } \Gamma^e$$

Develop the weak form and finite element model of the equation over an element Ω^e.

8.3. Assuming that c and k are constant in Problem 8.2, write the element coefficient matrix and source vector for (a) a linear rectangular element and (b) a linear triangular element.

8.4. Develop the finite element model of the following differential equation governing an axisymmetric problem,

$$-\frac{1}{r}\left[\frac{\partial}{\partial r}\left(r\frac{\partial u}{\partial r}\right) + \frac{\partial}{\partial z}\left(r\frac{\partial u}{\partial z}\right)\right] = f(r, z)$$

and compute the coefficients K_{ij}^e and f_i^e for a linear triangular element in terms of I_{mn} of (8.36).

8.5. Calculate the linear interpolation functions for the linear triangular and rectangular elements shown in Fig. P8.5.

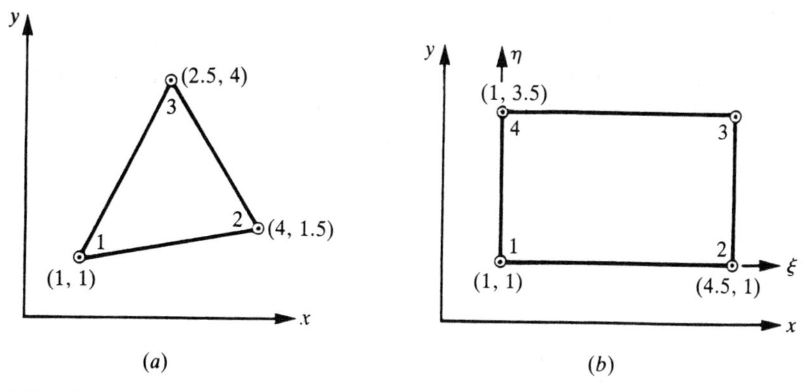

FIGURE P8.5

Answer: $\psi_1 = (12.25 - 2.5x - 1.5y)/8.25$.

8.6. For the linear triangular element shown in Fig. P8.6, show that the element coefficient matrix associated with the Laplace operator $-\nabla^2$,

$$K_{ij}^e = \int_{\Omega^e} \left(\frac{\partial \psi_i^e}{\partial x} \frac{\partial \psi_j^e}{\partial x} + \frac{\partial \psi_i^e}{\partial y} \frac{\partial \psi_j^e}{\partial y} \right) dx\, dy$$

is

$$[K^e] = \frac{1}{2} \begin{bmatrix} \alpha + \beta & -\alpha & -\beta \\ -\alpha & \alpha & 0 \\ -\beta & 0 & \beta \end{bmatrix}, \qquad \alpha = \frac{b}{a}, \qquad \beta = \frac{a}{b}$$

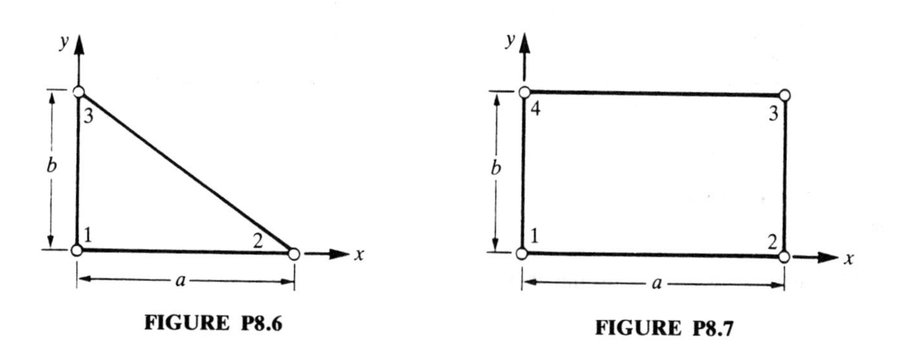

FIGURE P8.6 **FIGURE P8.7**

8.7. For the linear rectangular element shown in Fig. P8.7, show that the element coefficient matrix associated with the Laplace operator is

$$[K^e] = \frac{1}{6} \begin{bmatrix} 2(\alpha+\beta) & -2\alpha+\beta & -(\alpha+\beta) & \alpha-2\beta \\ -2\alpha+\beta & 2(\alpha+\beta) & \alpha-2\beta & -(\alpha+\beta) \\ -(\alpha+\beta) & \alpha-2\beta & 2(\alpha+\beta) & -2\alpha+\beta \\ \alpha-2\beta & -(\alpha+\beta) & -2\alpha+\beta & 2(\alpha+\beta) \end{bmatrix}, \qquad \alpha = \frac{b}{a}, \qquad \beta = \frac{a}{b}$$

8.8. Find the coefficient matrix associated with the Laplace operator when the rectangular element in Fig. P8.7 is divided into two triangles by joining node 1 to node 3.

8.9. The nodal values of an element in the finite element analysis of a field problem $-\nabla^2 u = f_0$ are $U_{10} = 389.79$, $U_{11} = 337.19$, and $U_{12} = 395.08$ (see Fig. P8.9). Find the gradient of the solution in the element. Also determine where the 392 isoline intersects the boundary of the element in Fig. P8.9.

Answer: $\boldsymbol{\nabla} T = 10.76\hat{\mathbf{e}}_1 - 105.02\hat{\mathbf{e}}_2$.

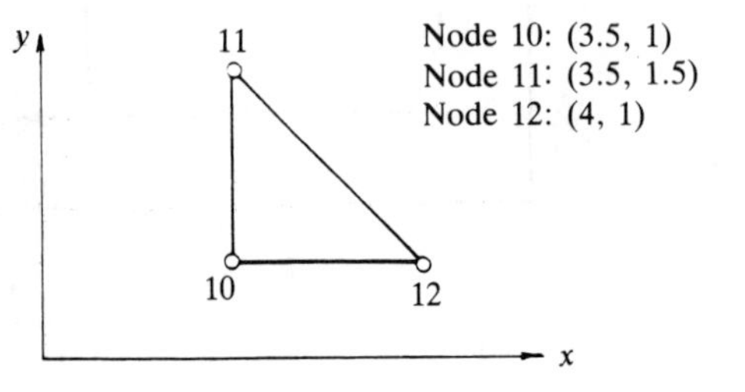

FIGURE P8.9

8.10. If the nodal values of the elements shown in Fig. P8.10 are $u_1 = 0.2645$, $u_2 = 0.2172$, $u_3 = 0.1800$ for the triangular element and $u_1 = 0.2173$, $u_3 = 0.1870$, $u_2 = u_4 = 0.2232$ for the rectangular element, compute u, $\partial u/\partial x$, and $\partial u/\partial y$ at the point $(x, y) = (0.375, 0.375)$.

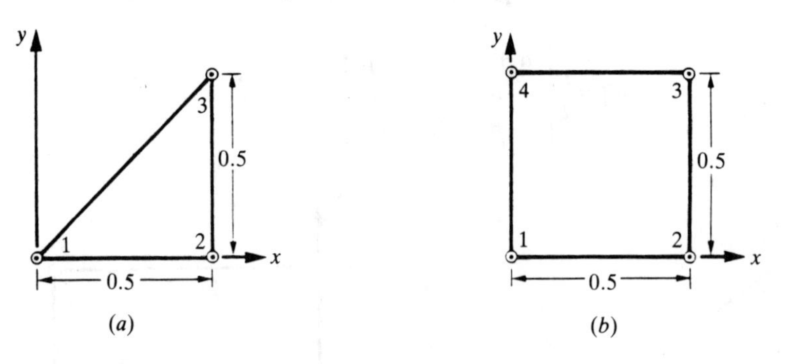

FIGURE P8.10

8.11. Compute the element matrices

$$S_{ij}^{01} = \int_0^a \int_0^b \psi_i \frac{d\psi_j}{dx}\, dx\, dy, \qquad S_{ij}^{02} = \int_0^a \int_0^b \psi_i \frac{d\psi_j}{dy}\, dx\, dy$$

where ψ_i are the linear interpolation functions of a rectangular element with sides a and b.

8.12. Give the assembled coefficient matrix for the finite element meshes shown in Fig. P8.12. Assume one degree of freedom per node and let $[K^e]$ denote the element coefficient matrix for the eth element. Your answer should be in terms of element matrices K_{ij}^e.

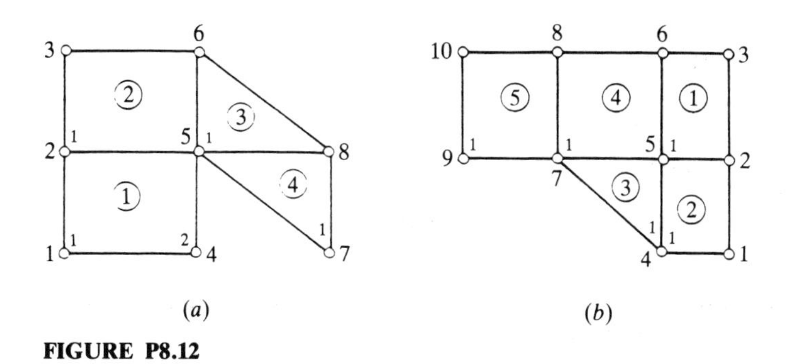

FIGURE P8.12

8.13. Compute the global source vector corresponding to the nonzero specified boundary flux for the finite element meshes of linear elements shown in Fig. P8.13.

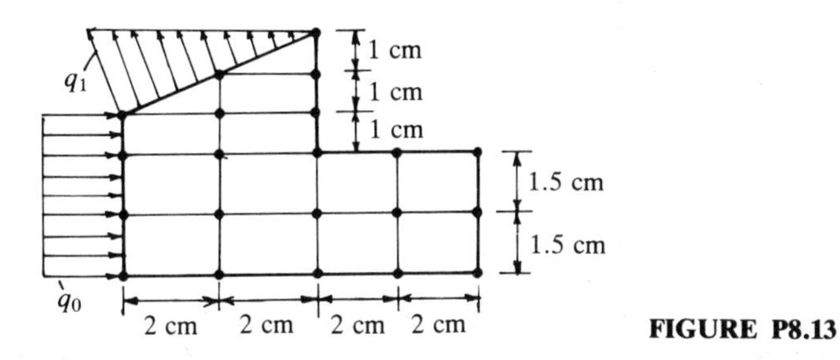

FIGURE P8.13

8.14. Repeat Problem 8.13 for the finite element mesh of quadratic elements shown in Fig. P8.14.

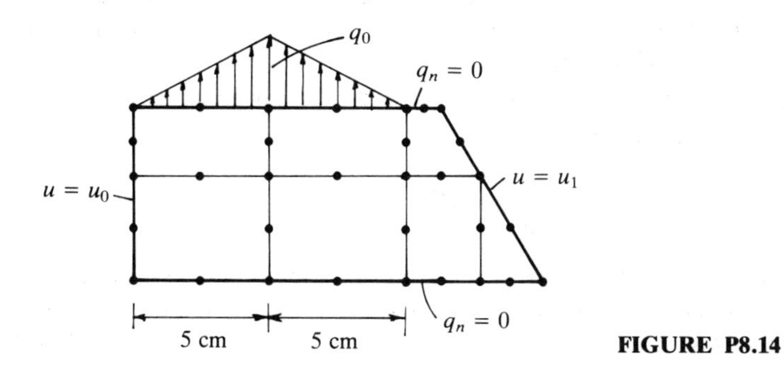

FIGURE P8.14

8.15. A point source of magnitude Q_0 is located at the point $(x, y) = (1.25, 2.5)$ inside the triangular element shown in Fig. P8.5(*a*). Determine the contribution of the point source to the element source vector.

8.16. Repeat Problem 8.15 for the rectangular element in Fig. P8.5(*b*).

8.17. A line source of intensity q_0 is located across the triangular element shown in Fig. 8.17. Compute the element source vector.

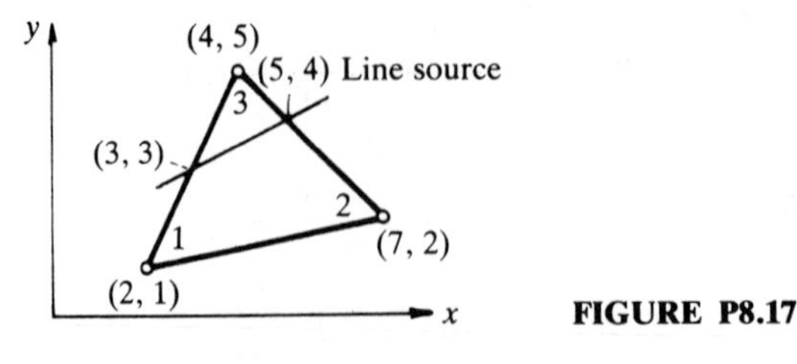

FIGURE P8.17

8.18. Repeat Problem 8.17 when the line source has varying intensity $q(s) = q_0 s/L$, where s is the coordinate along the line source.

8.19. When a_{11} and a_{22} in (8.1) are bilinear functions of the global coordinates x and y, $a_{ij} = a_{ij}^0 + a_{ij}^1 x + a_{ij}^2 y$, and $a_{12} = a_{21} = a_{00} = 0$, determine the explicit form of the coefficient matrix $[K^e]$ in (8.14*b*) for the linear triangular element. Express the result in terms of the integrals defined in (8.36).

8.20. Repeat Problem 8.19 for the linear rectangular element with local coordinate system $(\bar{x}, \bar{y})$. Use (8.42) and (8.43) to evaluate the integrals. You may also use the coefficient matrices in (8.44).

8.21. Determine the source vector

$$f_i^e = \int_{\Omega^e} f\psi_i \, dx \, dy$$

for linear rectangular elements when f is a function of the form $f = a_0 + a_1 x + a_2 y$, where x and y are the global coordinates.

8.22. Evaluate the coefficient matrix K_{ij}^e and source vector f_i^e in (8.64*b*) for a linear triangular element when $\hat{a}_{11}$, $\hat{a}_{22}$, and $\hat{f}$ are constant, and $\hat{a}_{00} = 0$. Note that the results of Problem 8.19 are useful here.

8.23. Repeat Problem 8.22 for a linear rectangular element.

8.24. For the mesh of linear triangular elements shown in Fig. P8.1, give the condensed set of equations for the unknown nodal values when $f = f_0 =$ constant. Use the symmetry of the problem. Write algebraic equations for the secondary variables Q_1^1 and Q_2^1 (from equilibrium).

8.25. Repeat Problem 8.24 with the equivalent mesh of linear rectangular elements.

8.26. Solve the Laplace equation

$$-\left(\frac{\partial^2 u}{\partial x^2} + \frac{\partial^2 u}{\partial y^2}\right) = 0 \quad \text{in } \Omega$$

on a rectangle, when $u(0, y) = u(a, y) = u(x, 0) = 0$ and $u(x, b) = u_0(x)$. Use the symmetry and (*a*) a mesh of 2×2 triangular elements and (*b*) a mesh of 2×2 rectangular elements (see Fig. P8.26). Compare the finite element solution with the exact solution

$$u(x, y) = \sum_{n=1}^{\infty} A_n \sin \frac{n\pi x}{a} \sinh \frac{n\pi y}{b}$$

where

$$A_n = \frac{2}{a \sinh (n\pi b/a)} \int_0^a u_0(x) \sin \frac{n\pi x}{a} \, dx$$

Take $a = b = 1$, and $u_0(x) = \sin \pi x$ in the computations. For this case, the exact solution becomes

$$u(x, y) = \frac{\sin \pi x \sinh \pi y}{\sinh \pi}$$

Answer: for a 2×2 mesh of triangles, $U_4 = 0.23025$ and $U_5 = 0.16281$; for a 2×2 mesh of rectangles, $U_4 = 0.15202$ and $U_5 = 0.10750$.

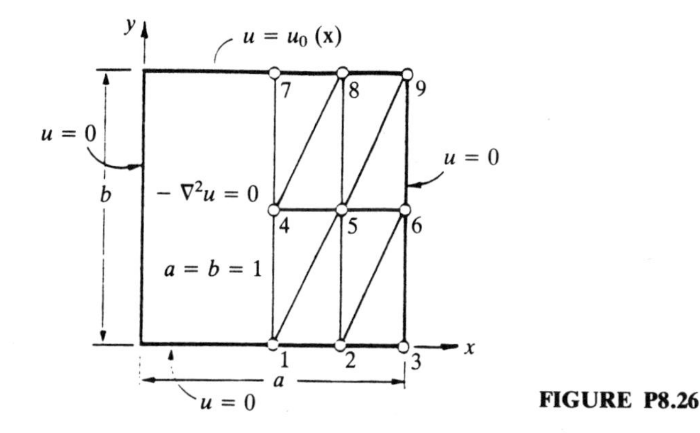

FIGURE P8.26

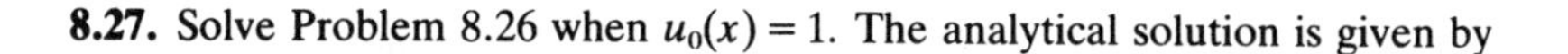

8.27. Solve Problem 8.26 when $u_0(x) = 1$. The analytical solution is given by

$$u(x, y) = \frac{4}{\pi} \sum_{n=0}^{\infty} \frac{\sin (2n+1)\pi x \sin (2n+1)\pi y}{(2n+1) \sinh (2n+1)\pi}$$

Answer: (*a*) $U_4 = 0.26471$ and $U_5 = 0.20588$.

8.28. Solve Problem 8.26 when $u_0(x) = 4(x - x^2)$.

Answer: $U_4 = 0.23529$ and $U_5 = 0.16912$.

8.29. Solve the Laplace equation for the unit square domain and boundary conditions given in Fig. P8.29. Use one rectangular element.

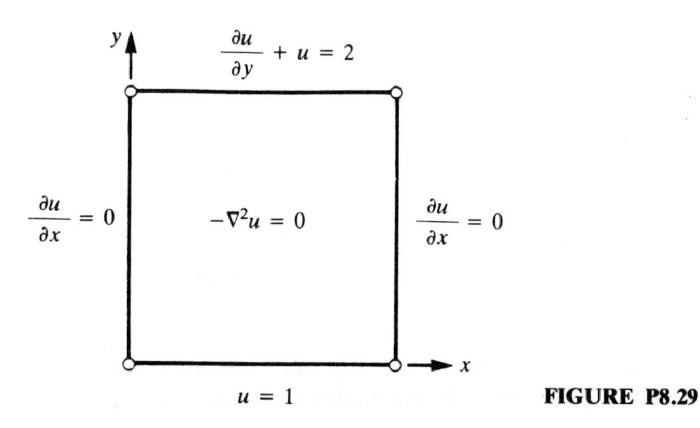

FIGURE P8.29

8.30. Use two triangular elements to solve the problem in Fig. P8.29. Use the mesh obtained by joining points (1, 0) and (0, 1).

8.31. Solve the Poisson equation $-\nabla^2 u = 2$ in the square whose vertices are at (0, 0), (1, 0), (1, 1), and (0, 1). The boundary conditions are $u(0, y) = y^2$, $u(x, 0) = x^2$, $u(1, y) = 1 - y$ and $u(x, 1) = 1 - x$. Use four linear rectangular elements (a 2×2 mesh).

8.32. Solve Problem 8.31 using four triangular elements in the upper half of the triangle because of the symmetry about the $x = y$ lines; join node 2 to 5 in Fig. P8.32.

Answer: $U_3 = 0.5$.

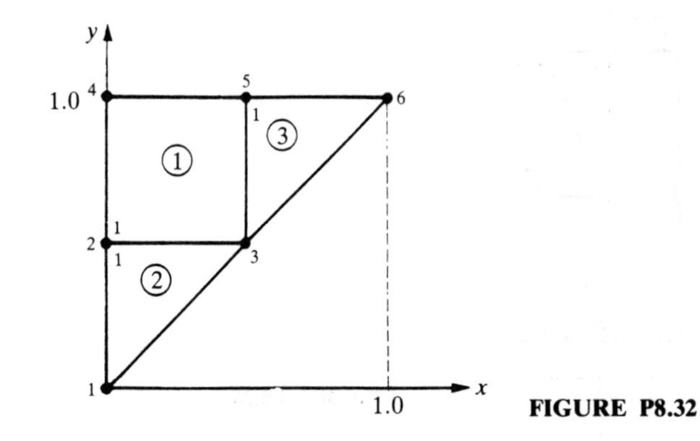

FIGURE P8.32

8.33. Solve Problem 8.31 using the mesh of a rectangle and two triangles, as shown in Fig. P8.32.

Answer: $U_3 = 0.675$.

8.34. Solve the Poisson equation $-\nabla^2 u = 2$ in Ω, with boundary conditions $u = 0$ on Γ_1 and $\partial u / \partial n = 0$ on Γ_2, where Ω is the first quadrant bounded by the parabola $y = 1 - x^2$ and the coordinate axes (see Fig. P8.34), and Γ_1 and Γ_2 are the boundaries shown in Fig. P8.34.

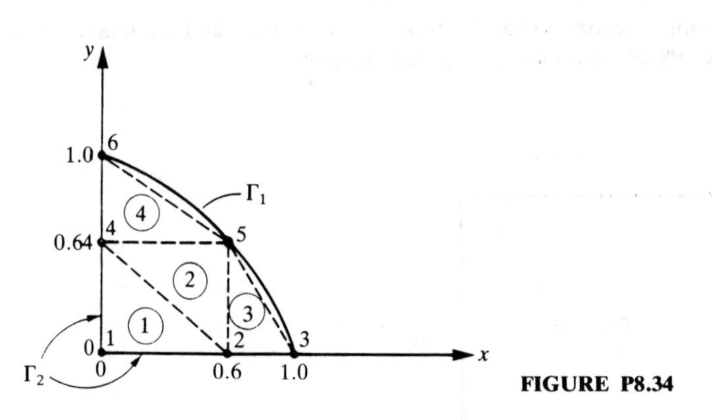

FIGURE P8.34

8.35. Show that a variable approximated by the quadratic interpolation functions of a rectangular element is continuous along interelement boundaries.

8.36. Solve the axisymmetric field problem shown in Fig. P8.36 for the mesh shown there. Note that the problem has symmetry about any line $z = \text{constant}$. Hence, the problem is essentially one-dimensional. Compare the finite element solution with the exact one.

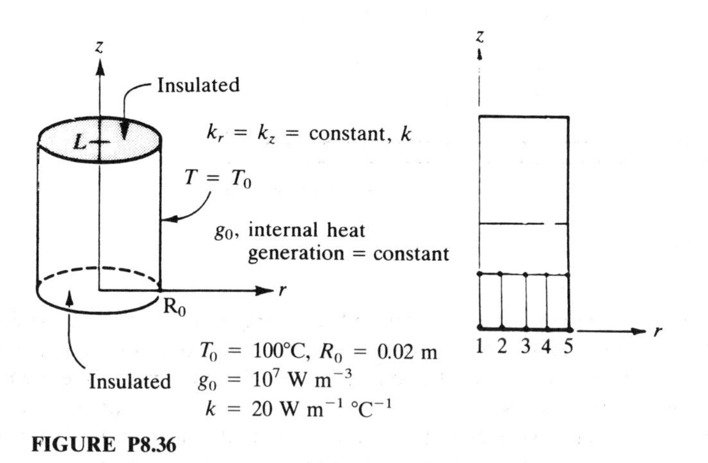

FIGURE P8.36

8.37. Formulate the axisymmetric field problem shown in Fig. P8.37 for the mesh shown. Compute the secondary variable at $r = \frac{1}{2}R_0$ using (*a*) equilibrium and (*b*) the definition. Use the element at the left of the node.

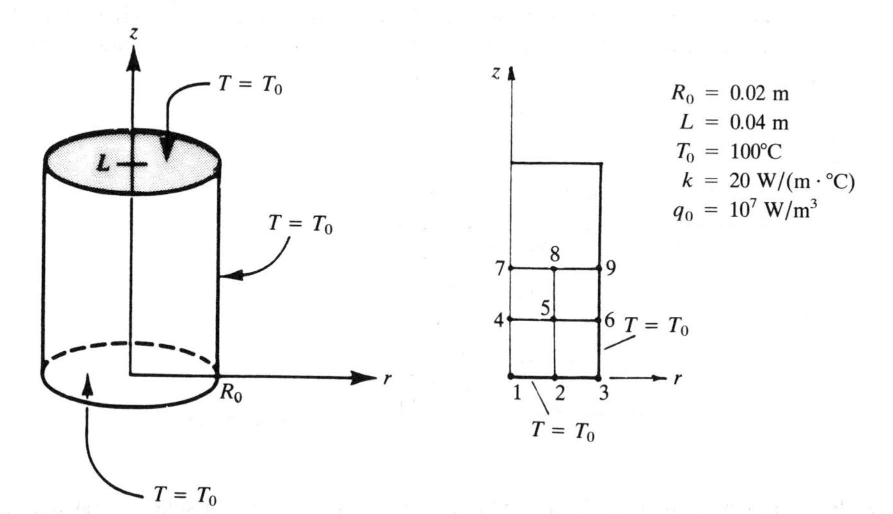

FIGURE P8.37

Section 8.4

8.38. A series of heating cables have been placed in a conducting medium, as shown in Fig. P8.38. The medium has conductivities of $k_x = 10$ W cm^{-1} °C^{-1} and $k_y =$

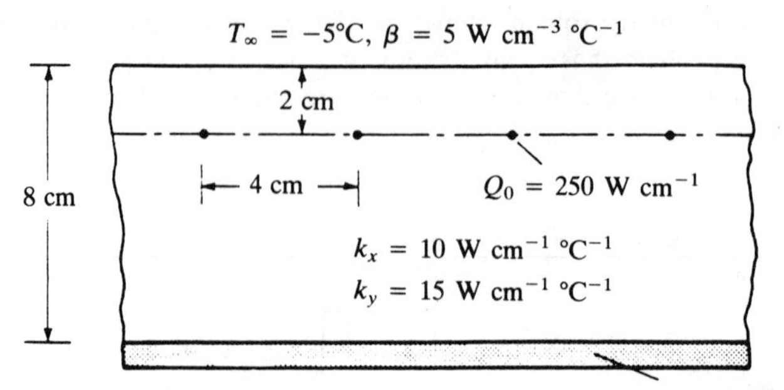

FIGURE P8.38

15 W cm^{-1} °C^{-1}, the upper surface is exposed to a temperature of −5 °C, and the lower surface is bounded by an insulating medium. Assume that each cable is a point source of 250 W cm^{-1}. Take the convection coefficient between the medium and the upper surface to be $\beta = 5$ W cm^{-2} °C^{-1}. Use a 2×8 mesh of linear rectangular (or triangular) elements in the computational domain (use the symmetry available in the problem), and formulate the problem (i.e., give element matrices for a typical element, give boundary conditions on primary and secondary variables, and compute convective boundary contributions).

8.39. Formulate the finite element analysis information to determine the temperature distribution in the molded asbestos insulation shown in Fig. P8.39. Use the symmetry to identify a computational domain and identify the specified boundary conditions at the nodes of the mesh. What is the size of the assembled coefficient matrix?

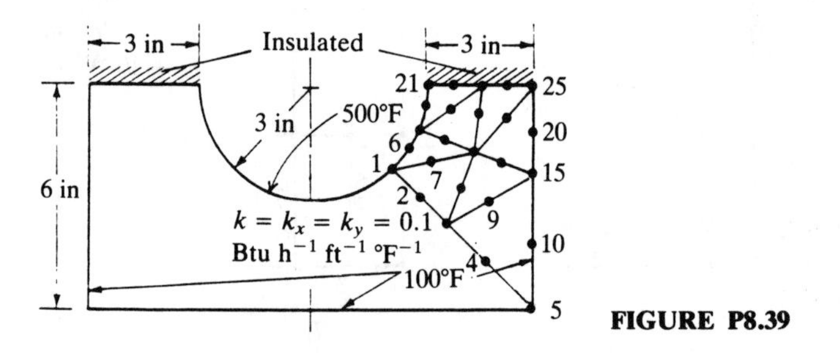

FIGURE P8.39

8.40. Consider steady-state heat conduction in a square region of side $2a$. Assume that the medium has conductivity k (in W m^{-1} °C^{-1}) and uniform heat (energy) generation f_0 (in W m^{-3}). For the boundary conditions and mesh shown in Fig. P8.40, write the finite element algebraic equations for nodes 1, 3, and 7.

8.41. Repeat Problem 8.40 for nodes 4, 5, and 9.

8.42. For the convective heat transfer problem shown in Fig. P8.42, write the four finite element equations for the unknown temperatures. Assume that the thermal conductivity of the material is $k = 5$ W m^{-1} °C^{-1}, the convective heat transfer

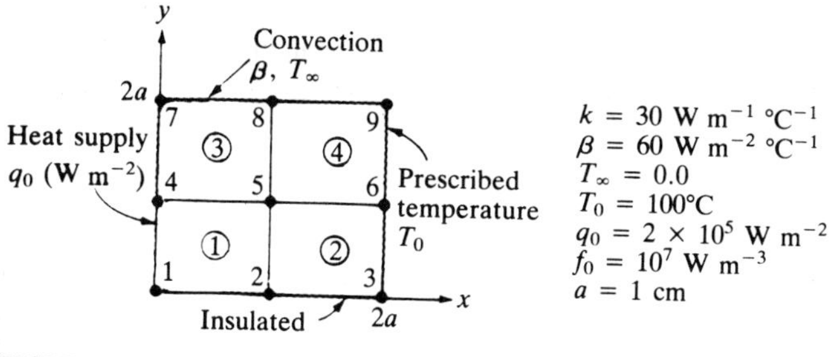

FIGURE P8.40

coefficient on the left surface is $\beta = 30\ \mathrm{W\,m^{-2}\,°C^{-1}}$, and the internal heat generation is zero. Compute the heats at nodes 2, 4, and 9 using (*a*) element equations (i.e., equilibrium) and (*b*) the definition (use the temperature field of elements 1 and 2).

8.43. Write the finite element equations for the unknown temperatures of the problem shown in Fig. P8.43.

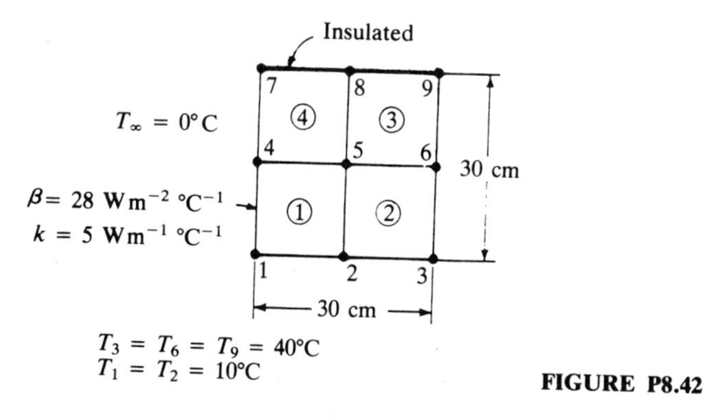

FIGURE P8.42

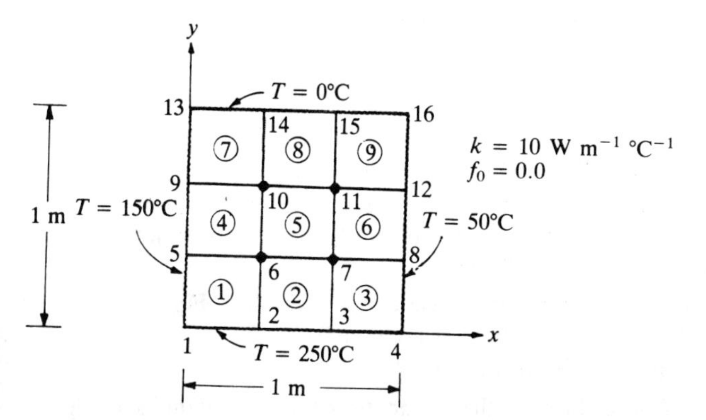

FIGURE P8.43

8.44. Write the finite element equations for the heats at nodes 1 and 13 of Problem 8.43. The answer should be in terms of the nodal temperatures $T_1, T_2, \ldots, T_{16}$.

8.45. Write the finite element equations associated with nodes 13, 16, and 19 for the problem shown in Fig. P8.45.

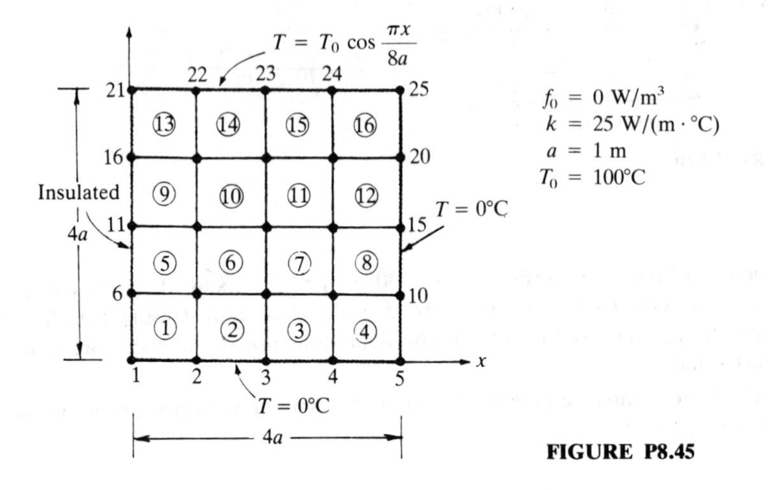

FIGURE P8.45

8.46. The fin shown in Fig. P8.46 has its base maintained at 300°C and is exposed to ambient temperature on its remaining boundary. Write the finite element equations at nodes 7 and 10.

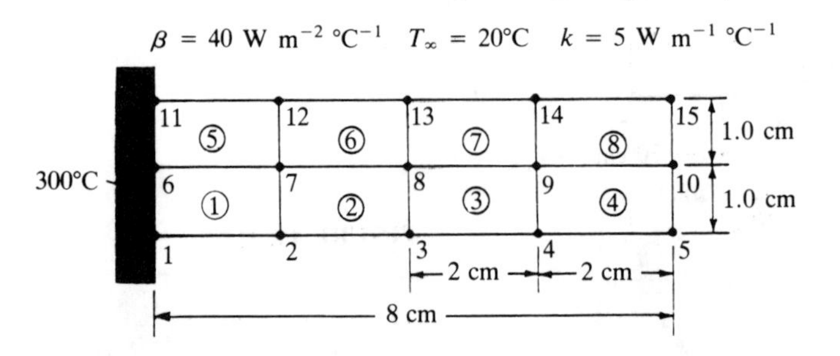

FIGURE P8.46

8.47. Compute the heat loss at nodes 10 and 13 of Problem 8.46.

8.48. Consider the problem of groundwater flow beneath a coffer dam. Formulate it using the velocity potential for finite element analysis. The geometry and boundary conditions are shown in Fig. P8.48.

8.49. Formulate the groundwater flow problem for the domain shown in Fig. P8.49 for finite element analysis. The pump is located at $(826\frac{2}{3}, 400)$.

8.50. Repeat Problem 8.49 for the domain shown in Fig. P8.50.

8.51. Consider steady confined flow through the foundation soil of a dam (see Fig. P8.51). Assuming that the soil is isotropic ($k_x = k_y$), formulate the problem for

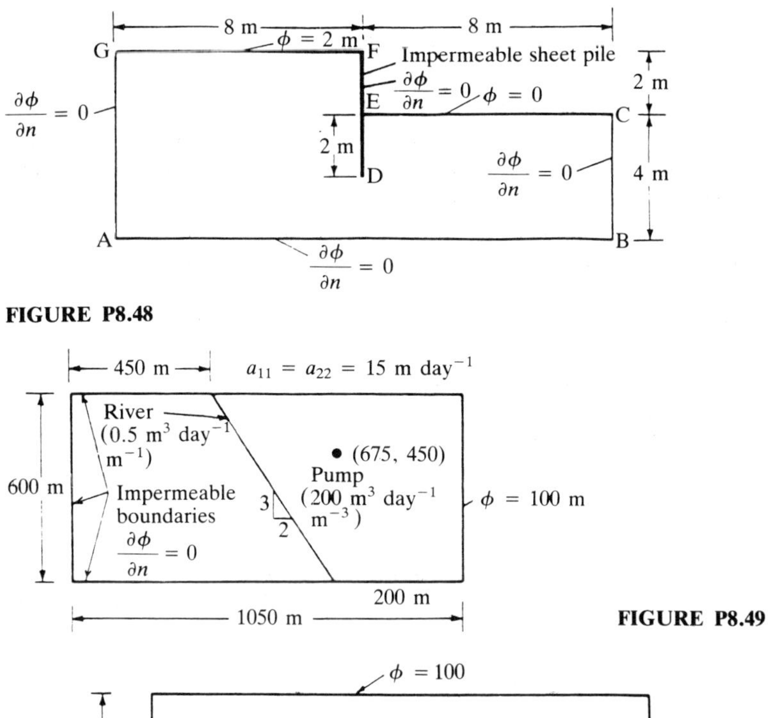

FIGURE P8.48

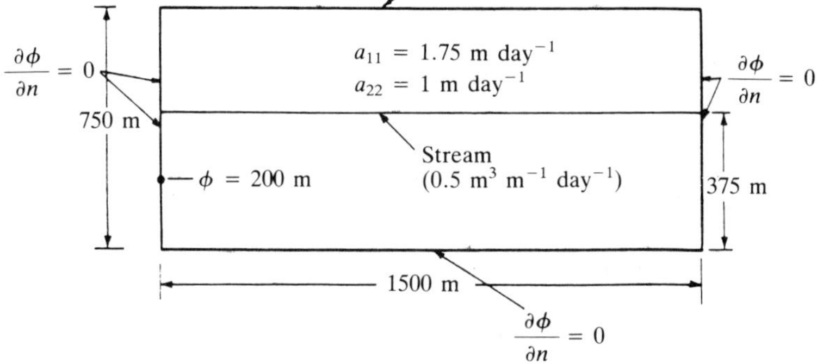

FIGURE P8.49

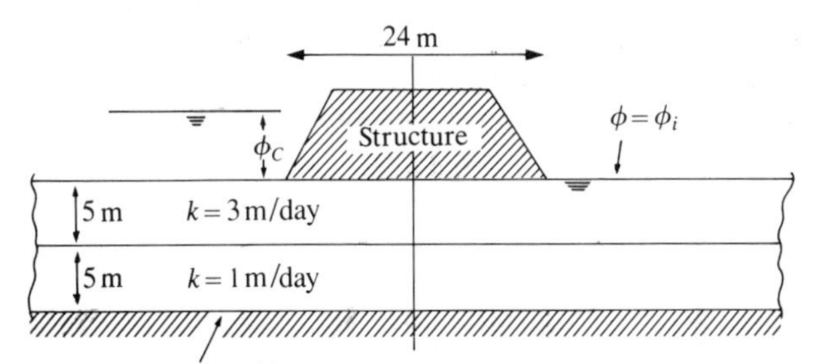

FIGURE P8.50

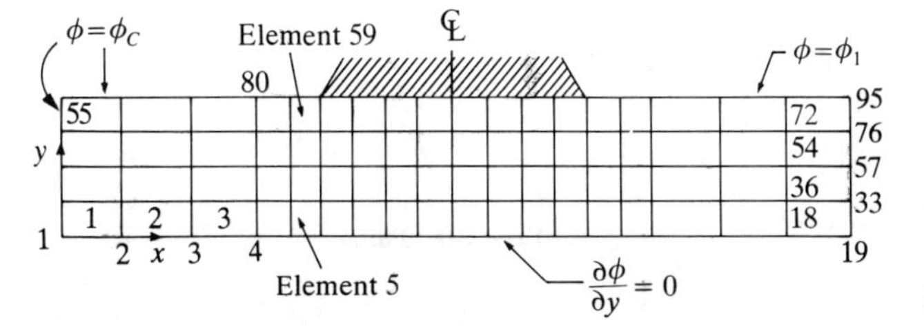

FIGURE P8.51

finite element analysis (identify the specified primary and secondary variables and their contribution to the nodes). In particular, write the finite element equations at nodes 8 and 11. Write the finite element equations for the horizontal velocity component in the fifth and tenth elements.

8.52. Formulate the problem of flow about an elliptical cylinder using (*a*) the stream function and (*b*) the velocity potential. The geometry and boundary conditions are shown in Fig. P8.52.

8.53. Repeat Problem 8.52 for the domain shown in Fig. P8.53.

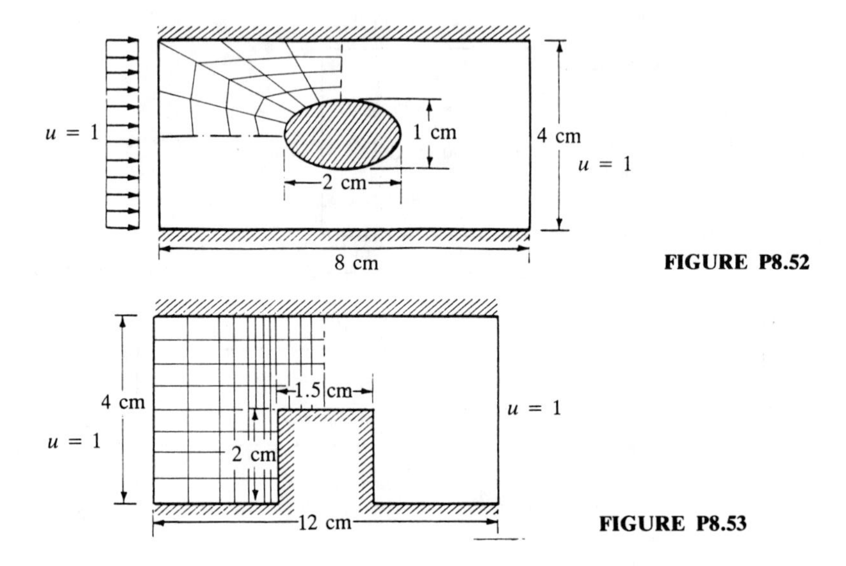

FIGURE P8.52

FIGURE P8.53

8.54. The Prandtl theory of torsion of a cylindrical member leads to

$$-\nabla^2 u = 2G\theta \quad \text{in } \Omega; \qquad u = 0 \quad \text{on } \Gamma$$

where Ω is the cross-section of the cylindrical member being twisted, Γ is the boundary of Ω, G is the shear modulus of the material of the member, θ is the angle of twist, and u is the stress function. Solve the equation for the case in which Ω is a circular section (see Fig. P8.54) using the mesh of linear triangular elements. Compare the finite element solution with the exact one (valid for

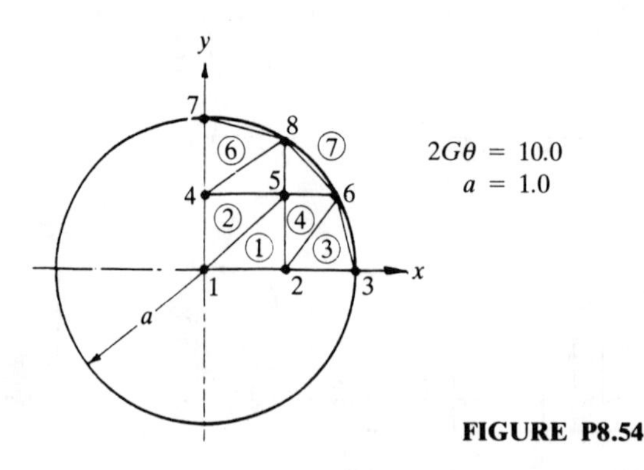

FIGURE P8.54

elliptical sections with axes a and b)

$$u = \frac{G\theta a^2 b^2}{a^2 + b^2}\left(1 - \frac{x^2}{a^2} - \frac{y^2}{b^2}\right)$$

Use $a = 1$, $b = 1$, and $2G\theta = 10$.

8.55. Repeat Problem 8.54 for a member with elliptical section. Use $a = 1$ and $b = 1.5$.

8.56. Repeat Problem 8.54 for the case in which Ω is an equilateral triangle (see Fig. P8.56). The exact solution is

$$u = -G\theta[\tfrac{1}{2}(x^2 + y^2) - \tfrac{1}{2}a(x^3 - 3xy^2) - \tfrac{2}{27}a^2]$$

Take $a = 1$ and $2G\theta = 10$.

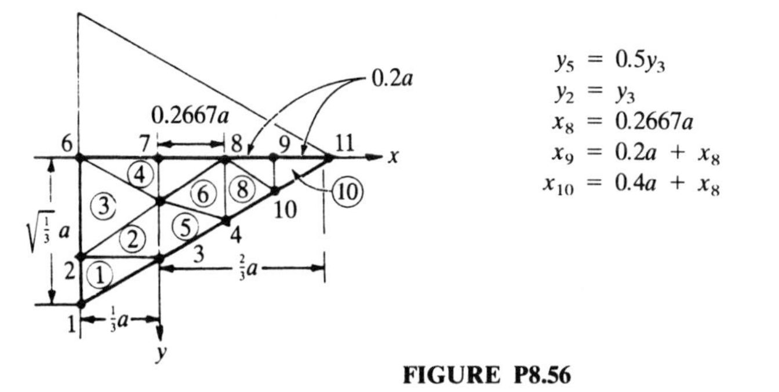

FIGURE P8.56

8.57. Consider the torsion of a hollow member with square cross-section. The stress function Ψ is required to satisfy the Poisson equation (8.150) and the following boundary conditions:

$$\Psi = 0 \quad \text{on the outer boundary;} \qquad \Psi = 2r^2 \quad \text{on the inner boundary}$$

where $r = b/c = 6a/2a = 3$ is the ratio of outside to inside dimensions. Formulate the problem for finite element analysis using the mesh shown in Fig. P8.57.

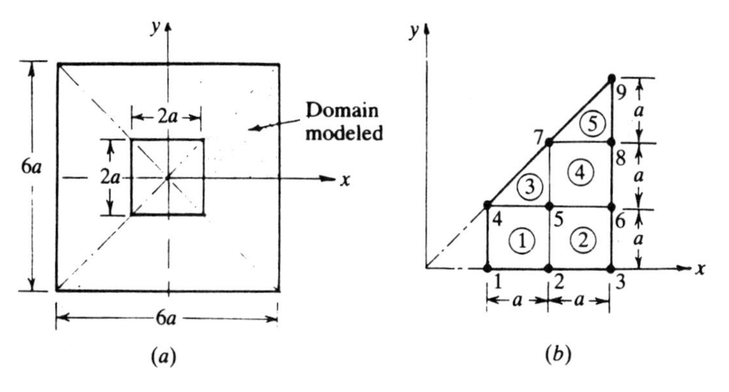

FIGURE P8.57

8.58. Repeat Problem 8.57 with the mesh of linear triangles (join nodes 1 and 5, 2 and 6, and 5 and 8 in Fig. P8.57b).

8.59. The membrane shown in Fig. P8.59 is subjected to a uniformly distributed load of intensity f_0 (in N m^{-2}). Formulate the problem for finite element analysis.

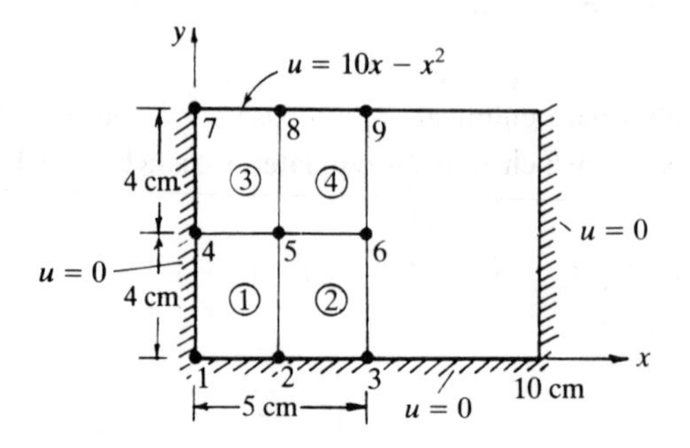

FIGURE P8.59

8.60. Repeat Problem 8.59 for the membrane shown in Fig. P8.60.

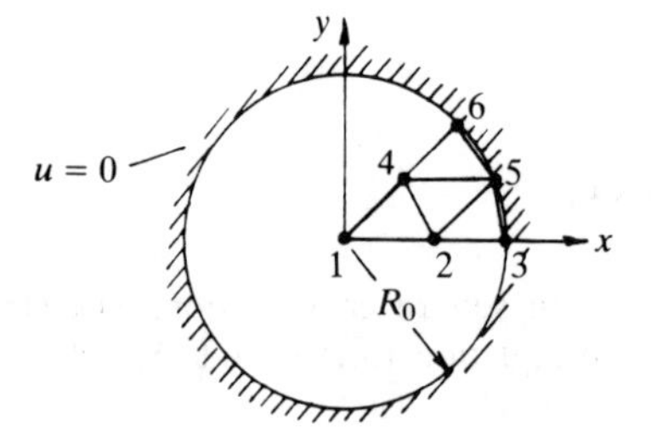

FIGURE P8.60

Section 8.5

8.61. Determine the time step for the transient analysis (with $\alpha \leq \frac{1}{2}$) of the problem

$$\frac{\partial u}{\partial t} - \nabla^2 u = 1 \quad \text{in } \Omega; \qquad u = 0 \quad \text{in } \Omega \text{ at } t = 0$$

by determining the maximum eigenvalue of the problem

$$-\nabla^2 u = \lambda u \quad \text{in } \Omega; \; u = 0 \quad \text{on } \Gamma$$

The geometry of the domain and mesh are shown in Fig. 8.12*b*.
Answer: $\lambda = 24$ for a 1×1 mesh in a quarter of the domain.

8.62. Set up the condensed equations for the transient problem in Problem 8.61 for the α family of approximation. Use the finite element mesh shown in Fig. P8.26.

8.63. Set up the condensed equations for the time-dependent analysis of the membrane in Problem 8.60.

8.64. (*Central difference method*) Consider the following matrix differential equation in time:

$$[M]\{\ddot{U}\} + [C]\{\dot{U}\} + [K]\{U\} = \{F\}$$

where the superposed dots indicate differentiation with respect to time. Assume

$$\{\ddot{U}\}_n = \frac{1}{(\Delta t)^2}(\{U\}_{n-1} - 2\{U\}_n + \{U\}_{n+1}), \quad \{\dot{U}\}_n = \frac{1}{2(\Delta t)}(\{U\}_{n+1} - \{U\}_{n-1})$$

and derive the algebraic equations for the solution of $\{U\}_{n+1}$ in the form

$$[A]\{U\}_{n+1} = \{F\}_n - [B]\{U\}_n - [D]\{U\}_{n-1}$$

Define $[A]$, $[B]$, and $[D]$ in terms of $[M]$, $[C]$, and $[K]$.

8.65. Repeat Problem 8.64 using the Newmark time approximation scheme.

8.66. Consider the first-order differential equation in time

$$a\frac{du}{dt} + bu = f$$

Using linear approximation, $u(t) = u_1\psi_1(t) + u_2\psi_2(t)$, $\psi_1 = 1 - t/\Delta t$, and $\psi_2 = t/\Delta t$, derive the associated algebraic equation and compare it with that obtained using the α family of approximation.

8.67. (*Space–time element*) Consider the differential equation

$$c\frac{\partial u}{\partial t} - \frac{\partial}{\partial x}\left(a\frac{\partial u}{\partial x}\right) = f \quad \text{for } 0 < x < L,\ 0 \leqslant t \leqslant T$$

with

$$u(0, t) = u(L, t) = 0 \quad \text{for } 0 \leqslant t \leqslant T; \qquad u(x, 0) = u_0(x) \quad \text{for } 0 < x < L$$

where $c = c(x)$, $a = a(x)$, $f = f(x, t)$, and u_0 are given functions. Consider the rectangular domain defined by

$$\Omega = \{(x, t): 0 < x < L,\ 0 \leqslant t \leqslant T\}$$

A finite element discretization of Ω by rectangles requires a space–time rectangular element (with y replaced by t). Give a finite element formulation of the equation over a space–time element, and discuss the mathematical and practical limitations of such a formulation. Compute the element matrices for a linear element.

8.68. (*Space-time finite elements*) Consider the time-dependent problem

$$\frac{\partial^2 u}{\partial x^2} = c\frac{\partial u}{\partial t} \quad \text{for } 0 < x < 1,\ t > 0$$

$$u(0, t) = 0, \qquad \frac{\partial u}{\partial x}(1, t) = 1, \qquad u(x, 0) = x$$

Use linear rectangular elements in the (x, t) plane to model the problem. Note that the finite element model is given by $[K^e]\{u^e\} = \{Q^e\}$, where

$$K_{ij}^e = \int_0^{\Delta t}\int_{x_A}^{x_B}\left(\frac{\partial\psi_i^e}{\partial x}\frac{\partial\psi_j^e}{\partial x} + c\psi_i\frac{\partial\psi_j}{\partial t}\right)dx\,dt$$

$$Q_1^e = \left(-\int_0^{\Delta t}\frac{\partial u}{\partial x}\,dt\right)\bigg|_{x=x_A}, \qquad Q_2^e = \left(\int_0^{\Delta t}\frac{\partial u}{\partial x}\,dt\right)\bigg|_{x=x_B}$$

8.69. For the heat transfer problem in Problem 8.45, set up the equations (for nodes 13, 16, and 19) for the transient case (see Fig. P8.45).

8.70. The collocation time approximation methods [see Hughes (1987), p. 530] are defined by the following relations:

$$\{\ddot{u}\}_{s+\alpha} = (1-\alpha)\{\ddot{u}\}_s + \alpha\{\ddot{u}\}_{s+1}$$

$$\{\dot{u}\}_{s+\alpha} = \{\dot{u}\}_s + \alpha\,\Delta t\,[(1-\gamma)\{\ddot{u}\}_s + \gamma\{\ddot{u}\}_{s+\alpha}]$$

$$\{u\}_{s+\alpha} = \{u\}_s + \alpha\,\Delta t\,\{\dot{u}\}_s + \tfrac{1}{2}\alpha(\Delta t)^2\,[(1-2\beta)\{\ddot{u}\}_s + 2\beta\{\ddot{u}\}_{s+\alpha}]$$

The collocation scheme contains two of the well-known schemes: $\alpha = 1$ gives the Newmark scheme; $\beta = \frac{1}{6}$, $\gamma = \frac{1}{2}$ gives the Wilson scheme. The collocation scheme is unconditionally stable and second-order-accurate for the following values of the parameters:

$$\alpha \geqslant 1, \qquad \gamma = \tfrac{1}{2}, \qquad \frac{\alpha}{2(1+\alpha)} \geqslant \beta \geqslant \frac{2\alpha^2 - 1}{4(2\alpha^3 - 1)}$$

Formulate the algebraic equations associated with the matrix differential equation

$$[M]\{\ddot{u}\} + [C]\{\dot{u}\} + [K]\{u\} = \{F\}$$

using the collocation scheme.

REFERENCES FOR ADDITIONAL READING

Bird, R. B., W. E. Stewart, and E. N. Lightfoot: *Transport Phenomena,* John Wiley, New York, 1960.

Budynas, R. G.: *Advanced Strength and Applied Stress Analysis,* McGraw-Hill, New York, 1977.

Carslaw, H. S., and J. C. Jaeger: *Conduction of Heat in Solids,* 2d ed., Oxford University Press, Fairlawn, NJ, 1959.

Clough, R. W., and J. Penzien, *Dynamics of Structures,* McGraw-Hill, New York, 1975.

Eskinazi, S.: *Principles of Fluid Mechanics,* Allyn and Bacon, Boston, 1962.

Holman, J. P.: *Heat Transfer,* 6th ed., McGraw-Hill, New York, 1986.

Hughes, J. T. R.: *The Finite Element Method (Linear, Static and Dynamic Finite Element Analysis),* Prentice-Hall, Englewood Cliffs, NJ, 1987.

Kohler, W., and J. Pittr: "Calculation of Transient Temperature Fields with Finite Elements in Space and Time Dimensions," *International Journal for Numerical Methods in Engineering,* vol. 8, pp. 625–631, 1974.

Kreyszig. E.: *Advanced Engineering Mathematics,* 6th ed., John Wiley, New York, 1988.

Myers, G. E.: *Analytical Methods in Conduction Heat Transfer,* McGraw-Hill, New York, 1971.

Mikhlin, S. G.: *Variational Methods in Mathematical Physics* (translated from the Russian by T. Boddington), Pergamon Press, Oxford, 1964.

———: *The Numerical Performance of Variational Methods* (translated from the Russian by R. S. Anderssen), Wolters–Noordhoff, The Netherlands, 1971.

Oden, J. T., and J. N. Reddy: *Variational Methods in Theoretical Mechanics,* 2d ed., Springer-Verlag, Berlin, 1983.

Özisik, M. N.: *Heat Transfer: A Basic Approach,* McGraw-Hill, New York, 1985.

Reddy, J. N.: *Energy and Variational Methods in Applied Mechanics,* John Wiley, New York, 1984.

———: *Applied Functional Analysis and Variational Methods in Engineering,* McGraw-Hill, New York, 1986; Krieger, Melbourne, FL, 1991.

——— and M. L. Rasmussen: *Advanced Engineering Analysis,* John Wiley, New York, 1982; Krieger, Melbourne, FL, 1990.

Rektorys, K.: *Variational Methods in Mathematics, Science and Engineering,* D. Reidel, Boston, MA, 1980.

———: *The Method of Discretization in Time,* D. Reidel, Boston, MA, 1982.

Schlichting, H.: *Boundary-Layer Theory* (translated by J. Kestin), 7th ed., McGraw-Hill, New York, 1979.

Sobey, R. J.: "Hermitian Space–Time Finite Elements for Estuarine Mass Transport." *International Journal for Numerical Methods in Fluids,* vol. 2, pp. 277–297, 1982.

Sokolnikoff, I. S.: *Mathematical Theory of Elasticity,* McGraw-Hill, New York, 1956.

Timoshenko, S. P., and J. N. Goodier: *Theory of Elasticity,* 3d ed., McGraw-Hill, New York, 1970.

Ugural, A. C., and S. K. Fenster: *Advanced Strength and Applied Elasticity,* Elsevier, New York, 1975.

Volterra, E., and J. H. Gaines: *Advanced Strength of Materials,* Prentice-Hall, Englewood Cliffs, NJ, 1971.

CHAPTER

9

INTERPOLATION FUNCTIONS, NUMERICAL INTEGRATION, AND MODELING CONSIDERATIONS

9.1 LIBRARY OF ELEMENTS AND INTERPOLATION FUNCTIONS

9.1.1 Introduction

In the previous chapter, we studied the finite element analysis of a model second-order equation and its analogues in the fields of heat transfer, fluid mechanics, and solid mechanics. During this study, we developed the interpolation functions for the basic elements, namely, the linear triangular and rectangular elements. These elements, which were developed in connection with the finite element analysis of a second-order partial differential equation in a single variable, are useful in all finite element models that admit Lagrange interpolation of the primary variables of the weak formulation.

The objective of this chapter is to develop a library of two-dimensional triangular and rectangular elements of the Lagrange family, i.e., elements over which only the function—not its derivatives—is interpolated. Once we have elements of different shapes and order at our disposal, we can choose appropriate elements and associated interpolation functions for a given problem. The regularly shaped elements, called *master elements,* for which interpolation functions are developed here, can be used for numerical evaluation of integrals defined on irregular elements. Of course, this requires a transformation of the geometry from the actual element shape to an associated master element. Section 9.2 deals with the transformation and numerical integration. With these preliminary comments, we now proceed to the discussion of interpolation functions for triangular and rectangular master elements.

9.1.2 Triangular Elements

The linear (three-node) triangular element was developed in Section 8.2.5. Higher-order triangular elements (i.e., those with interpolation functions of higher degree) can be systematically developed with the help of the so-called *Pascal's triangle,* which contains the terms of polynomials of various degrees in the two coordinates x and y, as shown in Fig. 9.1. Here x and y denote some local coordinates—they do not, in general, represent the global coordinates of the problem. One can view the position of the terms as the nodes of the triangle, with the constant term and the first and last terms of a given row being the vertices of the triangle. Of course, the shape of the triangle is arbitrary—it is not necessarily an equilateral triangle, as might appear from the position of the terms in Pascal's triangle. For example, a triangular element of order 2 (i.e., one for which the degree of the polynomial is 2) contains six nodes, as can be seen from the top three rows of Pascal's triangle. The position of the six nodes in the triangle is at the three vertices and at the midpoints of the three sides. The polynomial involves six constants, which can be expressed in terms of the nodal values of the variable being interpolated:

$$u = \sum_{i=1}^{6} u_i \psi_i(x, y) \tag{9.1}$$

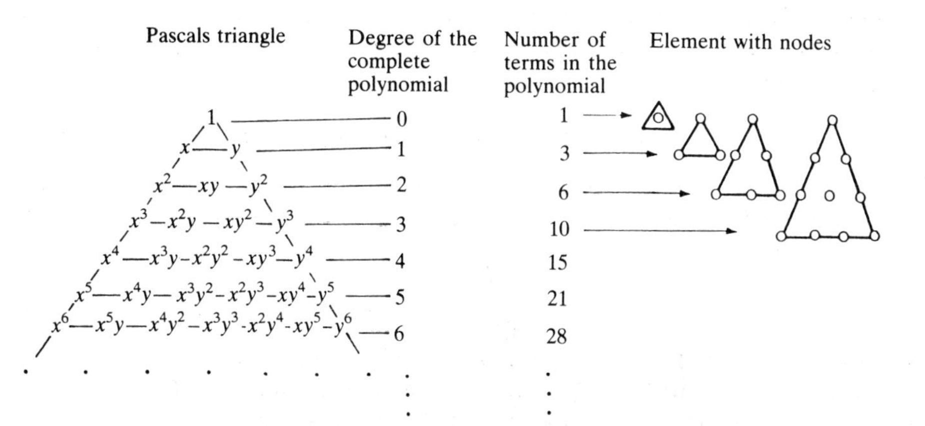

FIGURE 9.1
Pascal's triangle for the generation of the Lagrange family of triangular elements.

where ψ_i are the quadratic interpolation functions obtained following the same procedure as used for the linear element in Section 8.2.5. In general, a pth-order triangular element has n nodes, with

$$n = \tfrac{1}{2}(p+1)(p+2) \tag{9.2}$$

and a complete polynomial of the pth degree is given by

$$u(x, y) = \sum_{i=1}^{n} a_i x^r y^s = \sum_{j=1}^{n} u_j \psi_j, \qquad r + s \leq p \tag{9.3}$$

The location of the entries in Pascal's triangle gives a symmetric location of nodal points in elements that will produce exactly the right number of nodes to define a Lagrange interpolation of any degree. It should be noted that the Lagrange family of triangular elements (of order greater than zero) should be used for second-order problems that require only the dependent variables (not their derivatives) of the problem to be continuous at interelement boundaries. It can easily be seen that the pth-degree polynomial associated with the pth-order Lagrange element, when evaluated on the boundary of the latter, yields a pth-degree polynomial in the boundary coordinate. For example, the quadratic polynomial associated with the quadratic (six-node) triangular element shown in Fig. 9.2 is

$$u^e(x, y) = a_1 + a_2 x + a_3 y + a_4 xy + a_5 x^2 + a_6 y^2 \tag{9.4}$$

The derivatives of u^e are

$$\frac{\partial u^e}{\partial x} = a_2 + a_4 y + 2a_5 x, \qquad \frac{\partial u^e}{\partial y} = a_3 + a_4 x + 2a_6 y \tag{9.5}$$

The element shown in Fig. 9.2(a) is an arbitrary quadratic triangular element. By rotating and translating the (x, y) coordinate system, we obtain the (s, t) coordinate system. Since the transformation from the (x, y) system to the (s, t) system involves only rotation (which is linear) and translation, a kth-degree polynomial in the (x, y) coordinate system is still a kth-degree polynomial in the (s, t) system:

$$u^e(s, t) = \hat{a}_1 + \hat{a}_2 s + \hat{a}_3 t + \hat{a}_4 st + \hat{a}_5 s^2 + \hat{a}_6 t^2 \tag{9.6}$$

where $\hat{a}_i$ $(i = 1, 2, \ldots, 6)$ are constants depending on a_i and the angle of rotation α. Now, by setting $t = 0$, we obtain the restriction of u to side 1–2–3 of the element Ω^e:

$$u^e(s, 0) = \hat{a}_1 + \hat{a}_2 s + \hat{a}_5 s^2 \tag{9.7}$$

which is a quadratic polynomial in s. If a neighboring element Ω^f has its side 5–4–3 in common with side 1–2–3 of the element Ω^e then the function u on side 5–4–3 of the element Ω^f is also a quadratic polynomial:

$$u^f(s, 0) = \hat{b}_1 + \hat{b}_2 s + \hat{b}_5 s^2 \tag{9.8}$$

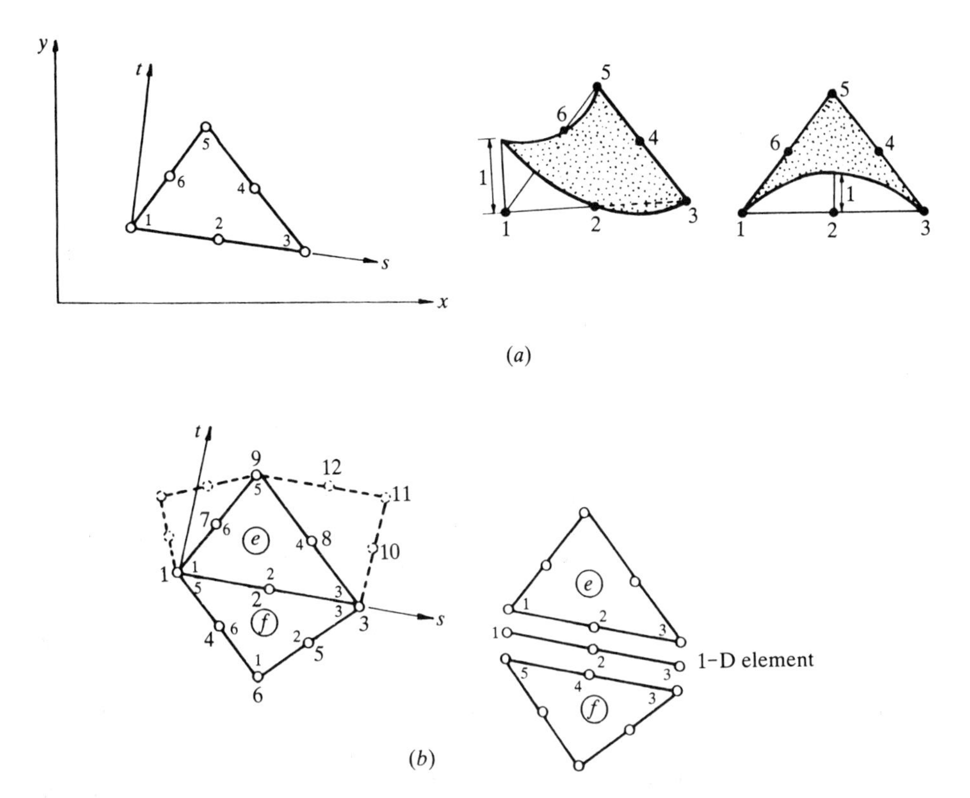

FIGURE 9.2
Variation of a function along the interelement boundaries of Lagrange (triangular) elements: (*a*) a typical higher-order element; (*b*) interelement continuity of a quadratic function.

Since the polynomials are uniquely defined by the same nodal values $U_1 = u_1^e = u_5^f$, $U_2 = u_2^e = u_4^f$, and $U_3 = u_3^e = u_3^f$, we have $u^e(s, 0) = u^f(s, 0)$, and hence the function u is uniquely defined on the interelement boundary of the elements Ω^e and Ω^f.

The above ideas can easily be extended to three dimensions, in which case Pascal's triangle takes the form of a Christmas tree and the elements are of a pyramidal shape, called tetrahedral elements. We shall not elaborate on this any further, because the scope of the present study is limited to two-dimensional elements only. An introduction to 3-D elements is presented in Chapter 14.

Recall from (8.19)–(8.25) that the procedure for deriving the interpolation functions involves the inversion of an $n \times n$ matrix, where n is the number of terms in the polynomial used to represent a function. When $n > 3$, this procedure is algebraically very tedious, and therefore one should devise an alternative way of developing the interpolation functions, as was discussed for one-dimensional elements in Chapter 3.

The alternative derivation of the interpolation functions for the higher-order Lagrange family of triangular elements is simplified by use of the *area*

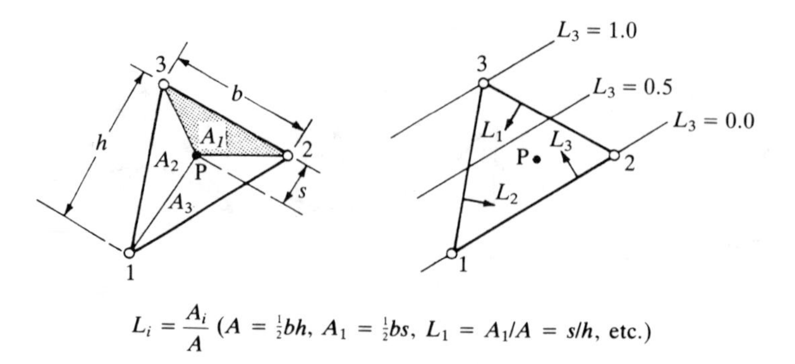

FIGURE 9.3
Definition of the natural coordinates of a triangular element.

coordinates L_i. For triangular elements, it is possible to construct three nondimensionalized coordinates L_i $(i = 1, 2, 3)$, which relate respectively to the sides directly opposite nodes 1, 2, and 3, such that (see Fig. 9.3)

$$\boxed{L_i = \frac{A_i}{A}, \qquad A = \sum_{i=1}^{3} A_i} \tag{9.9}$$

where A_i is the area of the triangle formed by nodes j and k and an arbitrary point P in the element, and A is the total area of the element. For example A_1 is the area of the shaded triangle, which is formed by nodes 2 and 3 and point P. The point P is at a perpendicular distance s from the side connecting nodes 2 and 3. We have $A_1 = \frac{1}{2}bs$ and $A = \frac{1}{2}bh$. Hence,

$$L_1 = A_1/A = s/h$$

Clearly, L_1 is zero on side 2–3 (and hence zero at nodes 2 and 3), and has a value of unity at node 1. Thus, L_1 is the interpolation function associated with node 1. Similarly, L_2 and L_3 are the interpolation functions associated with nodes 2 and 3, respectively. In summary, we have

$$\psi_i = L_i \tag{9.10}$$

for a linear triangular element. We shall use L_i to construct interpolation functions for higher-order triangular elements.

Consider a higher-order element with k nodes (equally spaced) per side (see Fig. 9.4a). Then the total number of nodes in the element is

$$n = \sum_{i=0}^{k-1} (k - i) = k + (k - 1) + \ldots + 1 = \tfrac{1}{2}k(k + 1) \tag{9.11}$$

and the degree of the interpolation functions is $k - 1$. For example, for the quadratic element, we have $k - 1 = 2$ and $n = 6$. Let the corner (i.e., vortex) nodes be denoted by I, J, and K, and let h_I be the perpendicular distance of node I from the side connecting J and K. Then the distance s_p to the pth row

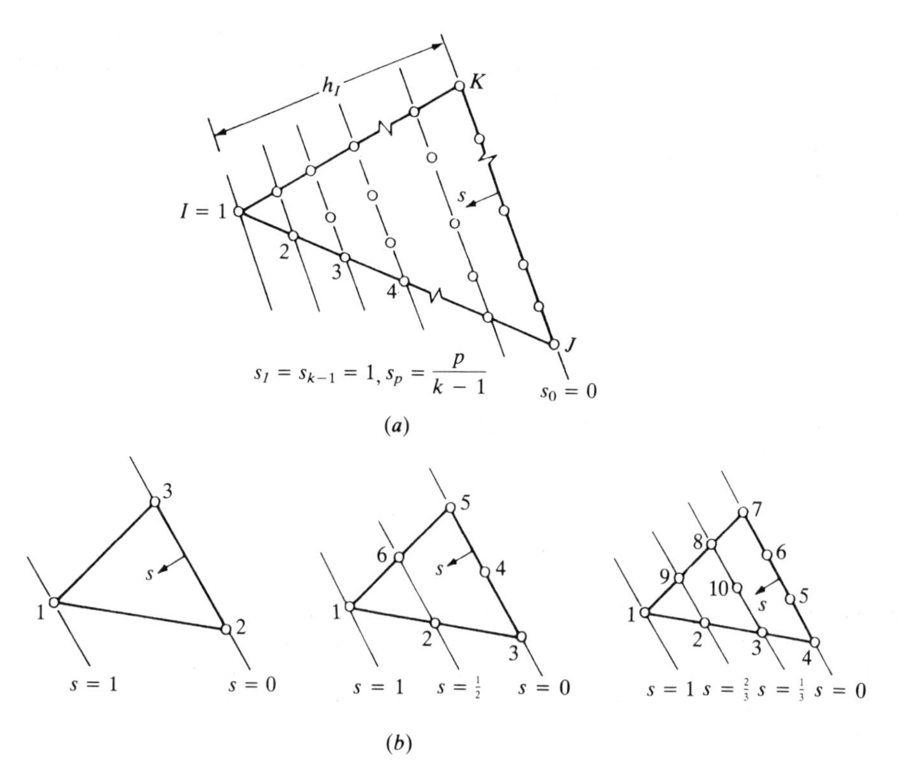

FIGURE 9.4
Construction of the element interpolation functions of the Lagrange triangular elements: (*a*) an arbitrary $(k-1)$th-order element; (*b*) linear, quadratic, and cubic elements.

parallel to side J–K (under the assumption that the nodes are equally spaced along the sides and the rows) is given in nondimensional form by

$$s_p = \frac{p}{k-1}, \qquad s_0 = 0, \qquad s_I = s_{k-1} = 1 \tag{9.12}$$

The interpolation function ψ_I should be zero at the nodes on the lines $L_I = 0$, $1/(k-1), \ldots, p/(k-1)$ $(p = 0, 1, \ldots, k-2)$, and ψ_I should be unity at $L_I = S_I$. Thus we have the necessary information for constructing the interpolation function ψ_I:

$$\boxed{\psi_I = \frac{(L_I - s_0)(L_I - s_1)(L_I - s_2)\cdots(L_I - s_{k-2})}{(s_I - s_0)(s_I - s_1)(s_I - s_2)\cdots(s_I - s_{k-2})} = \prod_{p=0}^{k-2} \frac{L_I - s_p}{s_I - s_p}} \tag{9.13}$$

Similar expressions can be derived for nodes located other than at the vertices. In general, ψ_i for node i is given by

$$\psi_i = \prod_{j=1}^{k-1} \frac{f_j}{f_j^i} \tag{9.14}$$

where f_j are functions of L_4, L_2, and L_3, and f_j^i is the value of f_j at node i. The functions f_j are derived from the equations of $k-1$ lines that pass through all the nodes except node i. The procedure is illustrated below via an example.

Example 9.1. First consider the triangular element that has two nodes per side (i.e., $k=2$). This is the linear triangular element with total number of nodes equal to 3 ($n=3$). For node 1 (see Fig. 9.4b), we have $k-2=0$ and

$$s_0 = 0, \quad s_1 = 1, \quad \psi_1 = \frac{L_1 - s_0}{s_1 - s_0} = L_1 \tag{9.15a}$$

Similarly, for ψ_2 and ψ_3, we obtain

$$\psi_2 = L_2, \quad \psi_3 = L_3 \tag{9.15b}$$

Next, consider the triangular element with three nodes per side ($k=3$). The total number of nodes is 6. For node 1, we have

$$s_0 = 0, \quad s_1 = \tfrac{1}{2}, \quad s_2 = 1$$

$$\psi_1 = \frac{L_1 - s_0}{s_2 - s_0}\frac{L_1 - s_1}{s_2 - s_1} = L_1(2L_1 - 1) \tag{9.16a}$$

The function ψ_2 (see Fig. 9.4b) should vanish at nodes 1, 3, 4, 5, and 6, and should be unity at node 2. Equivalently, ψ_2 should vanish along the lines connecting nodes 1 and 5, and 3 and 5. These two lines are given in terms of L_1, L_2, and L_3 (note that the subscripts on L refer to the nodes in the three-node triangular element) by $L_2 = 0$ and $L_1 = 0$. Hence,

$$\psi_2 = \frac{L_2 - s_0}{s_1 - s_0}\frac{L_1 - s_0}{s_1 - s_0} = \frac{L_2 - 0}{\frac{1}{2} - 0}\frac{L_1 - 0}{\frac{1}{2} - 0} = 4L_1L_2 \tag{9.16b}$$

Similarly,

$$\psi_3 = L_2(2L_2 - 1), \quad \psi_4 = 4L_2L_3, \quad \psi_5 = L_3(2L_3 - 1), \quad \psi_6 = 4L_1L_3 \tag{9.16c}$$

As a last example, consider the cubic element (i.e., $k-1=3$). For ψ_1, we note that it must vanish along the lines $L_1 = 0$, $L_1 = \frac{1}{3}$, and $L_1 = \frac{2}{3}$. Therefore,

$$\psi_1 = \frac{L_1 - 0}{1 - 0}\frac{L_1 - \frac{1}{3}}{1 - \frac{1}{3}}\frac{L_1 - \frac{2}{3}}{1 - \frac{2}{3}} = \tfrac{1}{2}L_1(3L_1 - 1)(3L_1 - 2)$$

Similarly,

$$\psi_2 = \frac{L_2 - 0}{\frac{1}{3} - 0}\frac{L_1 - 0}{\frac{2}{3} - 0}\frac{L_1 - \frac{1}{3}}{\frac{2}{3} - \frac{1}{3}} = \tfrac{9}{2}L_2L_1(3L_1 - 1)$$

and so on. We have

$$\begin{aligned}
\psi_3 &= \tfrac{9}{2}L_1L_2(3L_2 - 1), & \psi_4 &= \tfrac{1}{2}L_2(3L_2 - 1)(3L_2 - 2)\\
\psi_5 &= \tfrac{9}{2}L_2L_3(3L_2 - 1), & \psi_6 &= \tfrac{9}{2}L_2L_3(3L_3 - 1)\\
\psi_7 &= \tfrac{1}{2}L_3(3L_3 - 1)(3L_3 - 2), & \psi_8 &= \tfrac{9}{2}L_3L_1(3L_3 - 1)\\
\psi_9 &= \tfrac{9}{2}L_1L_3(3L_1 - 1), & \psi_{10} &= 27L_1L_2L_3
\end{aligned} \tag{9.17}$$

In closing this section, it should be pointed out that the area coordinates L_i facilitate not only the construction of the interpolation functions for the higher-order elements but also the integration of functions of L_i over line paths and areas. The following exact integration formulae prove to be useful:

$$\int_a^b L_1^m L_2^n \, ds = \frac{m!\,n!}{(m+n+1)!}(b-a) \tag{9.18a}$$

$$\iint_{\text{area}} L_1^m L_2^n L_3^p \, dA = \frac{m!\,n!\,p!}{(m+n+p+2)!} 2A \tag{9.18b}$$

where m, n, and p are arbitrary (positive) integers, A is the area of the domain of integration, and $m!$ denotes the factorial of m. Of course, one should transform the integrals from the x and y coordinates to L_i coordinates using

$$x = \sum_{i=1}^{n} x_i L_i, \qquad y = \sum_{i=1}^{n} y_i L_i \tag{9.19}$$

where (x_i, y_i) are the global coordinates of the ith node of the element.

9.1.3 Rectangular Elements

Analogous to the Lagrange family of triangular elements, the Lagrange family of rectangular elements can be developed from Pascal's triangle in Fig. 9.1 or the rectangular array shown in Fig. 9.5. Since a linear rectangular element has four corners (and hence four nodes), the polynomial should have the first four terms 1, x, y, and xy (which form a parallelogram in Pascal's triangle and a rectangle in the array given in Fig. 9.5). The coordinates (x, y) are usually taken to be the element (i.e., local) coordinates. In general, a pth-order Lagrange rectangular element has n nodes, with

$$n = (p+1)^2 \quad (p = 0, 1, \ldots)$$

and the associated polynomial contains the terms from the pth parallelogram or the pth rectangle in Fig. 9.5. When $p = 0$, it is understood (as in triangular elements) that the node is at the center of the element (i.e., the variable is a constant on the entire element). The Lagrange quadratic rectangular element has nine nodes, and the associated polynomial is given by

$$\begin{aligned} u(x, y) &= a_1 + a_2 x + a_3 y + a_4 xy + a_5 x^2 + a_6 y^2 \\ &\quad + a_7 x^2 y + a_8 xy^2 + a_9 x^2 y^2 \end{aligned} \tag{9.20a}$$

$$\begin{aligned} \frac{\partial u}{\partial x} &= a_2 + a_4 y + 2a_5 x + 2a_7 xy + a_8 y^2 + 2a_9 xy^2 \\ \frac{\partial u}{\partial y} &= a_3 + a_4 x + 2a_6 y + a_7 x^2 + 2a_8 xy + 2a_9 x^2 y \end{aligned} \tag{9.20b}$$

The polynomial contains the complete polynomial of the second degree plus the third-degree terms x^2y and xy^2 and also an x^2y^2 term. Four of the nine nodes are placed at the four corners, four at the midpoints of the sides, and one at the center of the element. The polynomial is uniquely determined by

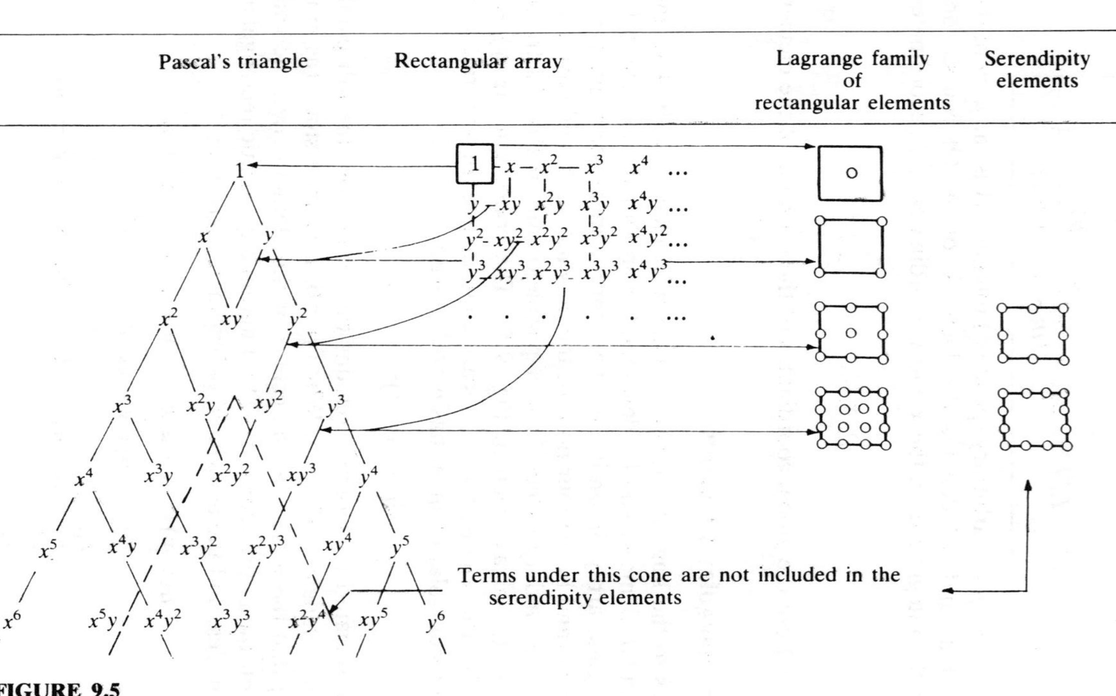

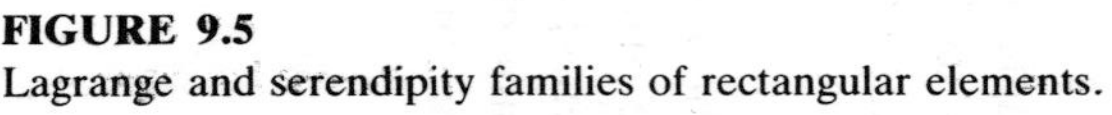

FIGURE 9.5
Lagrange and serendipity families of rectangular elements.

specifying its values at each of the nine nodes. Moreover, along the sides of the element, the polynomial is quadratic (with three terms—as can be seen by setting $y = 0$), and is determined by its values at the three nodes on that side. If two rectangular elements share a side, and the polynomial is required to have the same values from both elements at the three nodes of the elements, then u is uniquely defined along the entire side (shared by the two elements). Note that the normal derivative of u approximated by the quadratic Lagrange polynomials is quadratic in the tangential direction and linear in the normal direction (i.e., $\partial u/\partial x$ is quadratic in y and linear in x, and $\partial u/\partial y$ is quadratic in x and linear in y). Plots of ψ_1, ψ_2, and ψ_5 (the node numbers correspond to those in Fig. 9.6) of the nine-node rectangular element are shown in Fig. 9.7.

The pth-order Lagrange rectangular element has the pth-degree polynomial

$$u(x, y) = \sum_{i=1}^{n} a_i x^j y^k \quad (j + k \leq p + 1; i, j \leq p)$$
$$= \sum_{i=1}^{n} u_i \psi_i \tag{9.21}$$

and ψ_i are called the pth-order Lagrange interpolation functions.

The Lagrange interpolation functions associated with rectangular elements can be obtained from corresponding one-dimensional Lagrange inter-

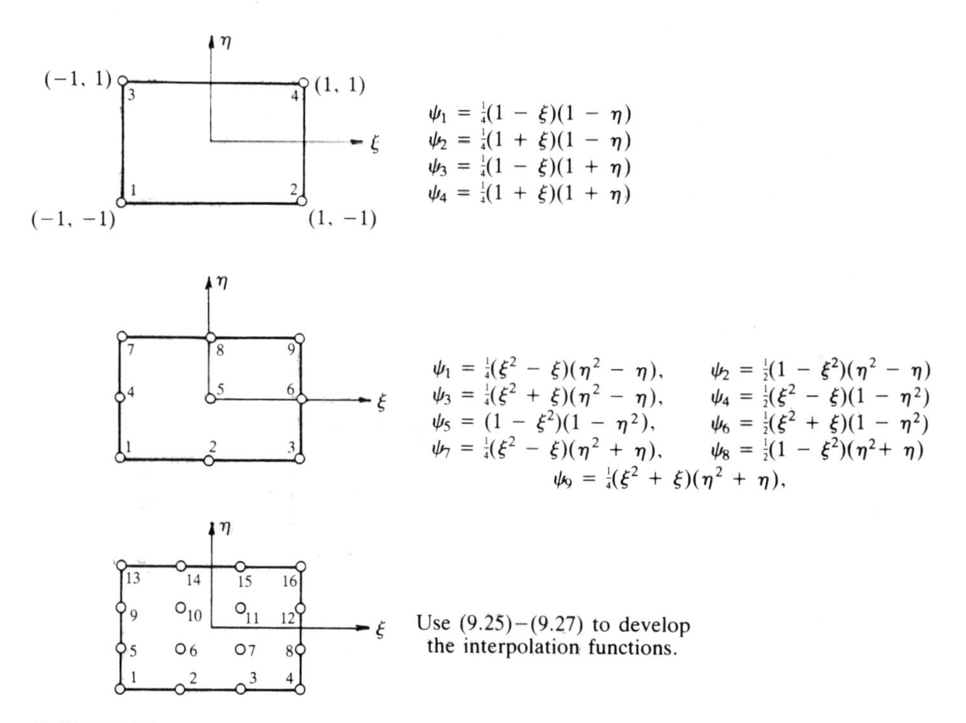

FIGURE 9.6
Node numbers and interpolation functions for the rectangular elements of the Lagrange family.

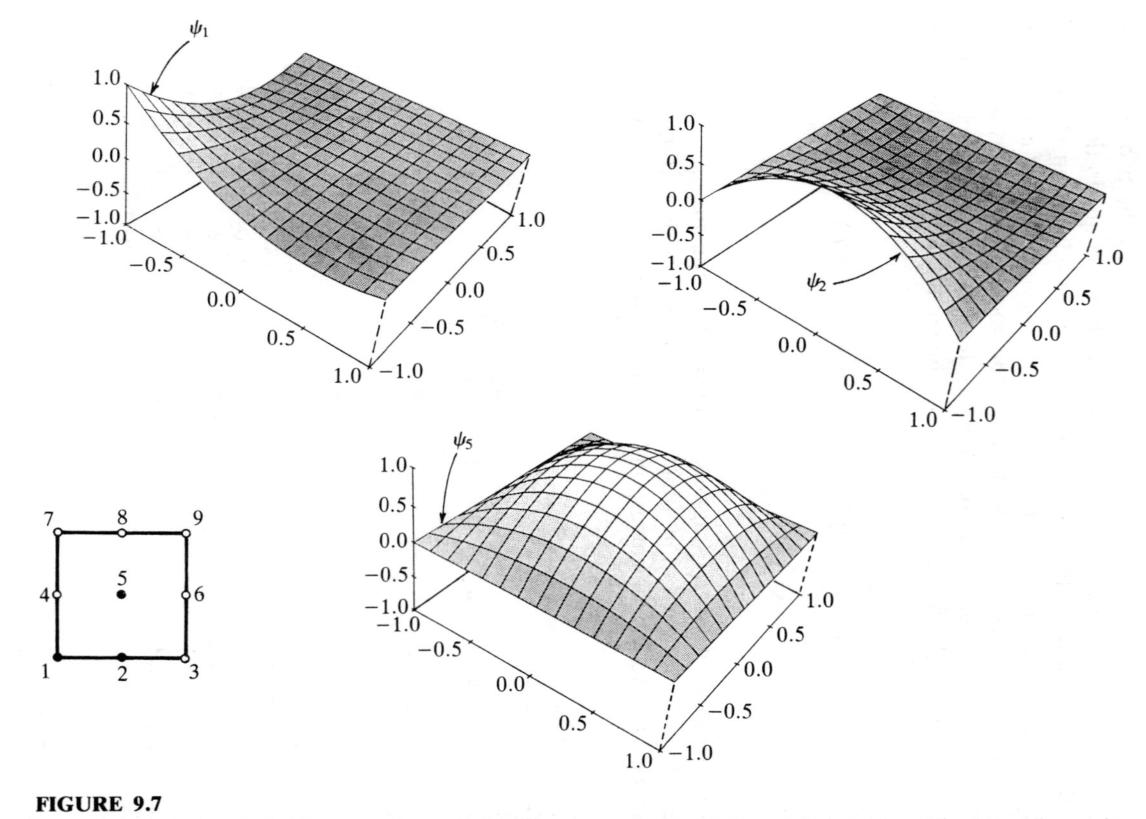

FIGURE 9.7
Geometric variation of the Lagrange interpolation functions at nodes 1, 2, and 5 (see Fig. 9.6) of the nine-node quadratic element.

polation functions by taking the tensor product of the x-direction (one-dimensional) interpolation functions with the y-direction (one-dimensional) interpolation functions. Let the x and y coordinates be taken along element sides, with the origin of the coordinate system at the lower left corner of the rectangle. Then for an element with dimensions a and b along the x and y directions, respectively, the interpolation functions are given as follows:

Linear ($p = 1$)

$$\begin{bmatrix} \psi_1 & \psi_3 \\ \psi_2 & \psi_4 \end{bmatrix} = \begin{Bmatrix} 1 - \dfrac{x}{a} \\ \dfrac{x}{a} \end{Bmatrix} \left\{ 1 - \frac{y}{b} \quad \frac{y}{b} \right\}$$
$$= \begin{bmatrix} \left(1 - \dfrac{x}{a}\right)\left(1 - \dfrac{y}{b}\right) & \left(1 - \dfrac{x}{a}\right)\dfrac{y}{b} \\ \dfrac{x}{a}\left(1 - \dfrac{y}{b}\right) & \dfrac{x}{a}\dfrac{y}{b} \end{bmatrix} \tag{9.22}$$

Quadratic ($p = 2$)

$$\begin{bmatrix} \psi_1 & \psi_4 & \psi_7 \\ \psi_2 & \psi_5 & \psi_8 \\ \psi_3 & \psi_6 & \psi_9 \end{bmatrix} = \begin{Bmatrix} \dfrac{(x - \frac{1}{2}a)(x - a)}{(-\frac{1}{2}a)(-a)} \\ \dfrac{x(x - a)}{\frac{1}{2}a(\frac{1}{2}a - a)} \\ \dfrac{x(x - \frac{1}{2}a)}{a(\frac{1}{2}a)} \end{Bmatrix} \begin{Bmatrix} \dfrac{(y - \frac{1}{2}b)(y - b)}{\frac{1}{2}b^2} \\ \dfrac{y(y - b)}{-\frac{1}{4}b^2} \\ \dfrac{y(y - b/2)}{\frac{1}{2}b^2} \end{Bmatrix}^T$$
$$\equiv \{f\}\{g\}^T = \begin{bmatrix} f_1 g_1 & f_1 g_2 & f_1 g_3 \\ f_2 g_1 & f_2 g_2 & f_2 g_3 \\ f_3 g_1 & f_3 g_2 & f_3 g_3 \end{bmatrix} \tag{9.23}$$

where $f_i(x)$ and $g_i(y)$ are the one-dimensional interpolation functions along the x and y directions, respectively. We obtain

$$\begin{aligned}
\psi_1 &= \left(1 - \frac{2x}{a}\right)\left(1 - \frac{x}{a}\right)\left(1 - \frac{2y}{b}\right)\left(1 - \frac{y}{b}\right) \\
\psi_2 &= \frac{4x}{a}\left(1 - \frac{x}{a}\right)\left(1 - \frac{2y}{b}\right)\left(1 - \frac{y}{b}\right), \quad \psi_3 = \frac{x}{a}\left(\frac{2x}{a} - 1\right)\left(1 - \frac{2y}{b}\right)\left(1 - \frac{y}{b}\right) \\
\psi_4 &= \left(1 - \frac{2x}{a}\right)\left(1 - \frac{x}{a}\right)\frac{4y}{b}\left(1 - \frac{y}{b}\right), \quad \psi_5 = \frac{4x}{a}\left(1 - \frac{x}{a}\right)\frac{4y}{b}\left(1 - \frac{y}{b}\right) \\
\psi_6 &= \frac{x}{a}\left(\frac{2x}{a} - 1\right)\frac{4y}{b}\left(1 - \frac{y}{b}\right), \quad \psi_7 = \left(1 - \frac{2x}{a}\right)\left(1 - \frac{x}{a}\right)\frac{y}{b}\left(\frac{2y}{b} - 1\right) \\
\psi_8 &= \frac{4x}{a}\left(1 - \frac{x}{a}\right)\frac{y}{b}\left(\frac{2y}{b} - 1\right), \quad \psi_9 = \frac{x}{a}\left(\frac{2x}{a} - 1\right)\frac{y}{b}\left(\frac{2y}{b} - 1\right)
\end{aligned} \tag{9.24}$$

At pth order

$$\begin{bmatrix} \psi_1 & \psi_{p+2} & \cdots & \psi_k \\ \psi_2 & & & \\ \vdots & \ddots & & \vdots \\ \psi_p & & \ddots & \\ \psi_{p+1} & \psi_{2p+2} & \cdots & \psi_n \end{bmatrix} = \begin{Bmatrix} f_1 \\ f_2 \\ \vdots \\ f_{p+1} \end{Bmatrix} \begin{Bmatrix} g_1 \\ g_2 \\ \vdots \\ g_{p+1} \end{Bmatrix}^T \tag{9.25}$$

$$k = (p+1)p + 1, \qquad n = (p+1)^2$$

where $f_i(x)$ and $g_i(y)$ are the pth-order interpolants in x and y, respectively. For example, the polynomial

$$f_i(\xi) = \frac{(\xi - \xi_1)(\xi - \xi_2)\cdots(\xi - \xi_{i-1})(\xi - \xi_{i+1})\cdots(\xi - \xi_{p+1})}{(\xi_i - \xi_1)(\xi_i - \xi_2)\cdots(\xi_i - \xi_{i-1})(\xi_i - \xi_{i+1})\cdots(\xi_i - \xi_{p+1})} \tag{9.26}$$

(where ξ_i is the ξ coordinate of node i) is the pth-degree interpolation polynomial in ξ that vanishes at the points $\xi_1, \xi_2, \ldots, \xi_{i-1}, \xi_{i+1}, \ldots, \xi_{p+1}$. We recall that (x, y) are the element coordinates.

It is convenient (for numerical integration purposes) to express the interpolation functions in (9.25) in terms of the natural coordinates ξ and η:

$$\xi = \frac{2(x - x_1) - a}{a}, \qquad \eta = \frac{2(y - y_1) - b}{b} \tag{9.27}$$

where x_1 and y_1 are the global coordinates of node 1 in the local x and y coordinates. For a coordinate system with origin fixed at node 1 and coordinates parallel to the sides of the element, we have $x_1 = y_1 = 0$. In this case, the quadratic interpolation functions in (9.24) can be written in terms of ξ and η as

$$\begin{aligned} &\psi_1 = \tfrac{1}{4}(1-\xi)(1-\eta)\xi\eta, && \psi_5 = (1-\xi^2)(1-\eta^2) \\ &\psi_2 = -\tfrac{1}{2}(1-\xi^2)(1-\eta)\eta, && \psi_6 = \tfrac{1}{2}(1+\xi)(1-\eta^2)\xi \\ &\psi_3 = -\tfrac{1}{4}(1+\xi)(1-\eta)\xi\eta, && \psi_7 = -\tfrac{1}{4}(1-\xi)(1+\eta)\xi\eta \\ &\psi_4 = -\tfrac{1}{2}(1-\xi)(1-\eta^2)\xi, && \psi_8 = \tfrac{1}{2}(1-\xi^2)(1+\eta)\eta \\ &\psi_9 = \tfrac{1}{4}(1+\xi)(1+\eta)\xi\eta \end{aligned} \tag{9.28}$$

The reader should be cautioned that the subscripts of ψ_i refer to the node numbering used in Fig. 9.6. For any renumbering of the nodes, the subscripts of the interpolation functions should be changed accordingly.

9.1.4 The Serendipity Elements

Since the internal nodes of the higher-order elements of the Lagrange family do not contribute to the interelement connectivity, they can be condensed out at the element level so that the size of the element matrices is reduced. The serendipity elements are those rectangular elements that have no interior nodes. In other words, all the node points are on the boundary of the element. The interpolation functions for serendipity elements cannot be obtained using tensor products of one-dimensional interpolation functions. Instead, an alternative procedure that employs the interpolation properties in (8.26) is used. Here we show how to construct the interpolation functions for the eight-node (quadratic) element using the natural coordinates (ξ, η).

The interpolation function for node 1 should be zero at nodes $2, 3, \ldots, 8$, and unity at node 1. Equivalently, ψ_1 should vanish on the sides defined by the equations $1-\xi=0$, $1-\eta=0$, and $1+\xi+\eta=0$ (see Fig. 9.8). Therefore, ψ_1 is of the form

$$\psi_1(\xi, \eta) = c(1-\xi)(1-\eta)(1+\xi+\eta) \tag{9.29a}$$

where c is a constant that should be determined so as to yield $\psi_1(-1, -1) = 1$. We obtain $c = -\frac{1}{4}$, and therefore

$$\psi_1(\xi, \eta) = -\tfrac{1}{4}(1-\xi)(1-\eta)(1+\xi+\eta) \tag{9.29b}$$

Similarly, we obtain

$$\begin{aligned} &\psi_2 = \tfrac{1}{2}(1-\xi^2)(1-\eta), && \psi_3 = \tfrac{1}{4}(1+\xi)(1-\eta)(-1+\xi-\eta) \\ &\psi_4 = \tfrac{1}{2}(1-\xi)(1-\eta^2), && \psi_5 = \tfrac{1}{2}(1+\xi)(1-\eta^2) \\ &\psi_6 = \tfrac{1}{4}(1-\xi)(1+\eta)(-1-\xi+\eta), && \psi_7 = \tfrac{1}{2}(1-\xi^2)(1+\eta) \\ &\psi_8 = \tfrac{1}{4}(1+\xi)(1+\eta)(-1+\xi+\eta) \end{aligned} \tag{9.30}$$

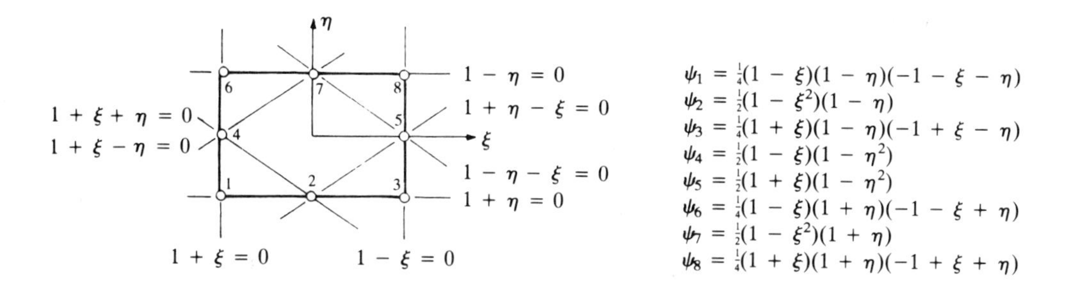

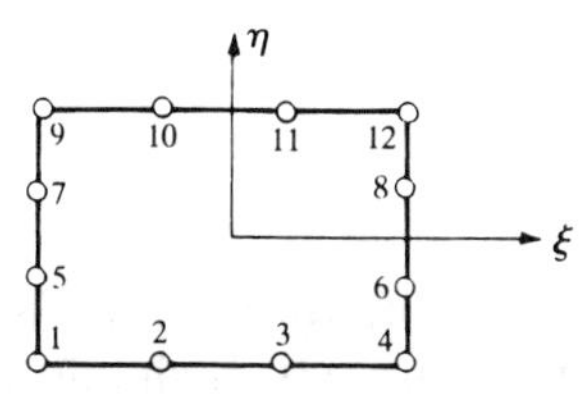

See (9.32) for the interpolation functions

FIGURE 9.8
Node numbers and interpolation functions for the serendipity family of elements.

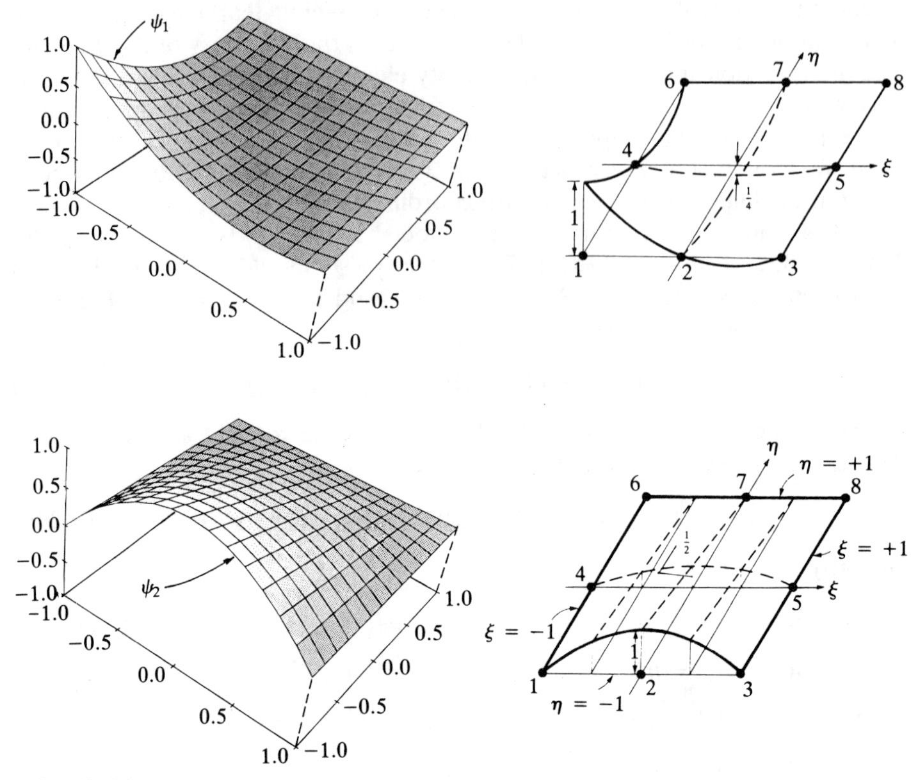

FIGURE 9.9
Geometric variation of the interpolation functions associated with nodes 1 and 2 of the eight-node serendipity element (see Fig. 9.8).

Note that all the ψ_i for the eight-node element have the form

$$\psi_i = c_1 + c_2\xi + c_3\eta + c_4\xi\eta + c_5\xi^2 + c_6\eta^2 + c_7\xi^2\eta + c_8\xi\eta^2 \tag{9.31a}$$

The derivatives of ψ_i with respect to ξ and η are of the form

$$\begin{aligned}\frac{\partial\psi_i}{\partial\xi} &= c_2 + c_4\eta + 2c_5\xi + 2c_7\xi\eta + c_8\eta^2 \\ \frac{\partial\psi_i}{\partial\eta} &= c_3 + c_4\xi + 2c_6\eta + c_7\xi^2 + 2c_8\xi\eta\end{aligned} \tag{9.31b}$$

Plots of ψ_1 and ψ_2 (the node numbers correspond to those in Fig. 9.8) for the eight-node serendipity element are shown in Fig. 9.9. It should be noted that ψ_2 of the nine-node element is zero at the element center, whereas ψ_2 of the eight-node element is nonzero there.

The interpolation functions for the cubic serendipity element, which has 12 nodes, are

$$\psi_1 = \tfrac{1}{32}(1-\xi)(1-\eta)[-10+9(\xi^2+\eta^2)], \qquad \psi_2 = \tfrac{9}{32}(1-\eta)(1-\xi^2)(1-3\xi)$$
$$\psi_3 = \tfrac{9}{32}(1-\eta)(1-\xi^2)(1+3\xi), \qquad \psi_4 = \tfrac{1}{32}(1+\xi)(1-\eta)[-10+9(\xi^2+\eta^2)]$$
$$\psi_5 = \tfrac{9}{32}(1-\xi)(1-\eta^2)(1-3\eta), \qquad \psi_6 = \tfrac{9}{32}(1+\xi)(1-\eta^2)(1-3\eta)$$
$$\psi_7 = \tfrac{9}{32}(1-\xi)(1-\eta^2)(1+3\eta), \qquad \psi_8 = \tfrac{9}{32}(1+\xi)(1-\eta^2)(1+3\eta)$$
$$\psi_9 = \tfrac{1}{32}(1-\xi)(1+\eta)[-10+9(\xi^2+\eta^2)], \qquad \psi_{10} = \tfrac{9}{32}(1+\eta)(1-\xi^2)(1-3\xi)$$
$$\psi_{11} = \tfrac{9}{32}(1+\eta)(1-\xi^2)(1+3\xi), \qquad \psi_{12} = \tfrac{1}{32}(1+\xi)(1+\eta)[-10+9(\xi^2+\eta^2)] \tag{9.32}$$

The interpolation functions ψ_i for the 12-node element are of the form

$$\psi_i = \text{terms of the form in (9.31}a) + c_9\xi^3 + c_{10}\eta^3 + c_{11}\xi^3\eta + c_{12}\xi\eta^3 \tag{9.33}$$

TABLE 9.1
Interpolation functions for the linear and quadratic Langrange rectangular elements, quadratic serendipity element and Hermite cubic rectangular element†

Element type	Interpolation functions	Remarks
Lagrange elements:		
Linear	$\frac{1}{4}(1+\xi\xi_i)(1+\eta\eta_i)$	Node i $(i = 1, \ldots, 4)$
Quadratic	$\frac{1}{4}\xi\xi_i(1+\xi\xi_i)\eta\eta_i(1+\eta\eta_i)$	Corner node
	$\frac{1}{2}\eta\eta_i(1+\eta\eta_i)(1-\xi^2)$	Side node, $\xi_i = 0$
	$\frac{1}{2}\xi\xi_i(1+\xi\xi_i)(1-\eta^2)$	Side node, $\eta_i = 0$
	$(1-\xi^2)(1-\eta^2)$	Interior node
Seredipity element:		
Quadratic	$\frac{1}{4}(1+\xi\xi_i)(1+\eta\eta_i)(\xi\xi_i+\eta\eta_i-1)$	Corner node
	$\frac{1}{2}(1-\xi^2)(1+\eta\eta_i)$	Side node, $\xi_i = 0$
	$\frac{1}{2}(1+\xi\xi_i)(1-\eta^2)$	Side node, $\eta_i = 0$
Hermite cubic element:		
Interpolation functions for		
variable u	$\frac{1}{16}(\xi+\xi_i)^2(\xi\xi_i-2)(\eta+\eta_i)^2(\eta\eta_i-2)$	For node i
derivative $\partial u/\partial\xi$	$-\frac{1}{16}\xi_i(\xi+\xi_i)^2(\xi\xi_i-1)(\eta+\eta_i)^2(\eta\eta_i-2)$	$(i = 1, \ldots, 4)$
derivative $\partial u/\partial\eta$	$-\frac{1}{16}(\xi+\xi_i)^2(\xi\xi_i-2)\eta_i(\eta+\eta_i)^2(\eta\eta_i-1)$	
derivative $\partial^2 u/\partial\xi\,\partial\eta$	$\frac{1}{16}\xi_i(\xi+\xi_i)^2(\xi\xi_i-1)\eta_i(\eta+\eta_i)^2(\eta\eta_i-1)$	
Interpolation functions for		
variable u	$\frac{1}{2}(\xi_0+1)(\eta_0+1)(2+\xi_0+\eta_0-\xi^2-\eta^2)$	For node i
derivative $\partial u/\partial\xi$	$\frac{1}{2}\xi_i(\xi_0+1)^2(\xi_0-1)(\eta_0+1)$	$(i = 1, \ldots, 4)$
derivative $\partial u/\partial\eta$	$\frac{1}{2}\eta_i(\xi_0+1)(\eta_0+1)^2(\eta_0-1)]$	
	$\xi = (x-x_c)/a, \quad \eta = (y-y_c)/b$	(2a and 2b are the
	$\xi_0 = \xi\xi_i, \quad \eta_0 = \eta\eta_i$	sides of the rectangular element)

† See Fig. 9.10 for the coordinate system; (ξ_i, η_i) denote the natural coordinates of the ith node of the element; (x_c, y_c) are the global coordinates of the center of the element.

In the above discussion, we developed only the Lagrange interpolation functions for triangular and rectangular elements. The Hermite family of interpolation functions (which interpolate the function and its derivatives) were not discussed. We recall that such functions are required in the finite element formulation of fourth-order (or higher-order) differential equations (e.g., the Euler–Bernoulli beam theory of Chapter 4 and the classical or Kirchhoff plate theory of Chapter 12). For the sake of completeness, while not presenting the details of the derivation, the Hermite cubic interpolation functions for two rectangular elements are summarized in Table 9.1. The first is based on the interpolation of $(u, \partial u/\partial x, \partial u/\partial y, \partial^2 u/\partial x\, \partial y)$ at each node, and the second is based on the interpolation of $(u, \partial u/\partial x, \partial u/\partial y)$ at each node. The node numbering system in Table 9.1 refers to that used in Fig. 9.10. The notation used in Table 9.1 and Fig. 9.10 is also followed in the computer program FEM2DV2, which will be discussed in Chapter 13.

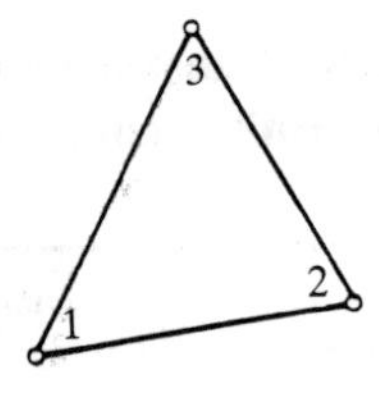

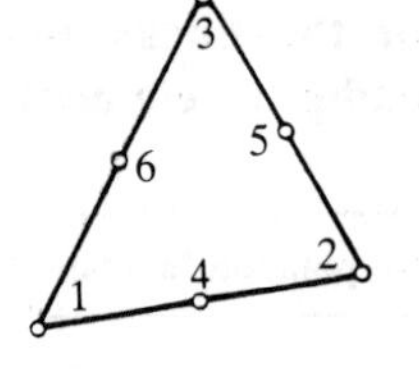

(a)

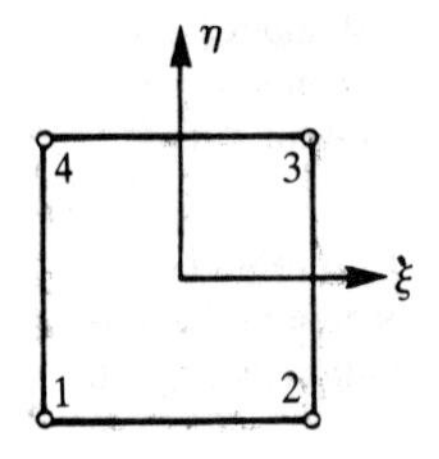

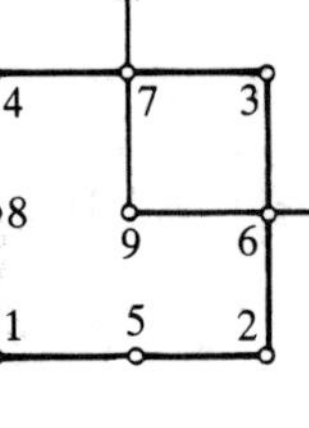

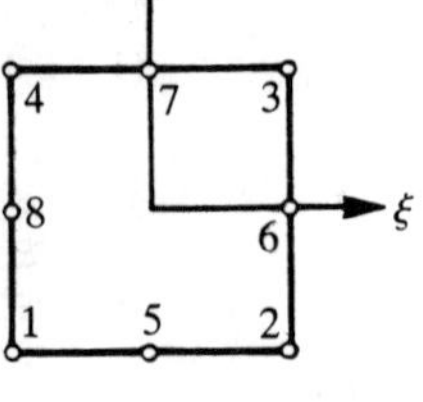

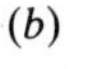

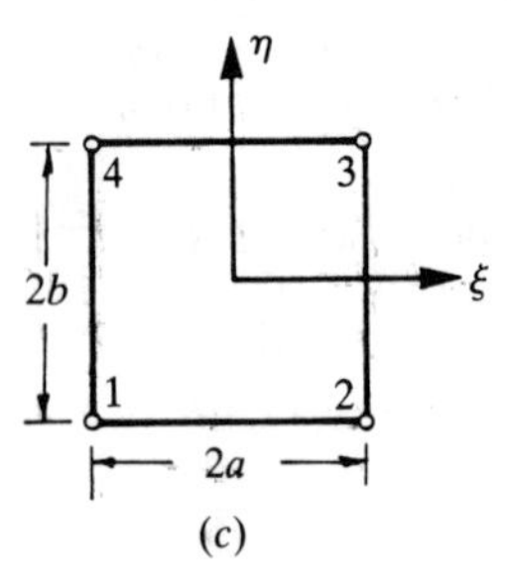

(c)

FIGURE 9.10
Triangular and rectangular elements: (a) linear and quadratic triangular elements; (b) linear and quadratic Lagrange elements; (c) Hermite cubic element.

9.2 NUMERICAL INTEGRATION

9.2.1 Preliminary Comments

An accurate representation of domains with curved boundaries can be accomplished by the use of refined meshes and/or curvilinear elements. For example, a nonrectangular region cannot be represented using rectangular elements; however, it can be represented by quadrilateral elements. Since the interpolation functions are easily derivable for a rectangular element, and it is easier to evaluate integrals over rectangular geometries, we transform the finite element integral statements defined over quadrilaterals to a rectangle. The transformation results in complicated expressions in terms of the coordinates used for the rectangular element. Therefore, numerical integration is used to evaluate such expressions. Numerical integration schemes, such as the Gauss–Legendre scheme, require that the integral be evaluated on a specfic domain or with respect to a specific coordinate system. Gauss quadrature, for example, requires that the integral be expressed over a square region $\hat{\Omega}$ of dimension 2 by 2 and that the coordinate system (ξ, η) be such that $-1 \leq (\xi, \eta) \leq 1$. The transformation of the geometry and the variable coefficients of the differential equation from the problem coordinates (x, y) to the coordinates (ξ, η) results in algebraically complex expressions, and this precludes analytical (i.e., exact) evaluation of the integrals. Thus, the transformation of a given integral expression, defined over the element Ω^e, to one on the domain $\hat{\Omega}$ must be such as to facilitate numerical integration. Each element of the finite element mesh is transformed to $\hat{\Omega}$, only for the purpose of numerically evaluating the integrals. The element $\hat{\Omega}$ is called a *master element.* For example, every quadrilateral element can be transformed to a square element with side 2 that facilitates the use of Gauss–Legendre quadrature to evaluate integrals defined over the quadrilateral element.

The transformation between Ω^e and $\hat{\Omega}$ [or, equivalently, between (x, y) and (ξ, η)] is accomplished by a coordinate transformation of the form

$$x = \sum_{j=1}^{m} x_j^e \hat{\psi}_j^e(\xi, \eta), \qquad y = \sum_{j=1}^{m} y_j^e \hat{\psi}_j^e(\xi, \eta) \tag{9.34}$$

where $\hat{\psi}_j^e$ denote the finite element interpolation functions of the master element $\hat{\Omega}$. Although the Lagrange interpolation of the geometry is implied by (9.34), one can also use Hermite interpolation. Consider, as an example, the master element shown in Fig. 9.11. The coordinates in the master element are chosen to be the natural coordinates (ξ, η) such that $-1 \leq (\xi, \eta) \leq 1$. This choice is dictated by the limits of integration in the Gauss quadrature rule that is used to evaluate the integrals. For this case, the $\hat{\psi}_j^e$ denote the interpolation functions of the four-node rectangular element shown in Fig. 9.11 (i.e., $m = 4$). The transformation (9.34) maps a point (ξ, η) in the master element $\hat{\Omega}$ onto a point (x, y) in the element Ω^e, and vice versa if the Jacobian of the transformation is positive definite. The transformation maps the line $\xi = 1$ in $\hat{\Omega}$ to the line defined parametrically by $x = x(1, \eta)$ and $y = y(1, \eta)$ in the (x, y)

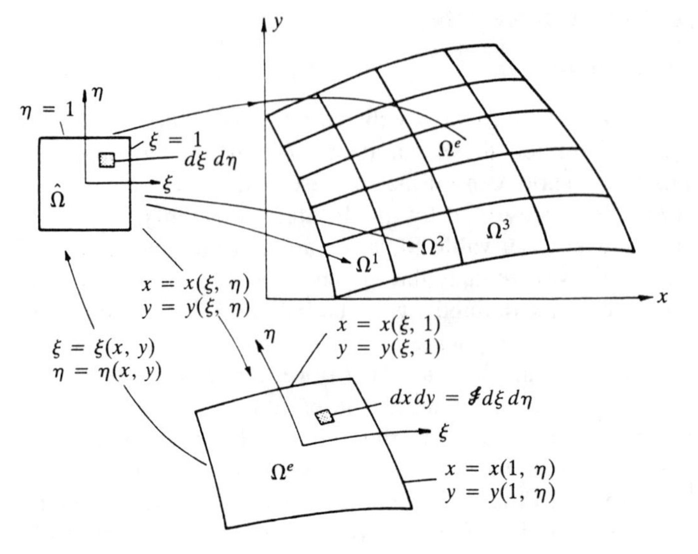

FIGURE 9.11
Generation of a finite element mesh from a master element.

plane. For instance, consider the line $\xi = 1$ in the master element $\hat{\Omega}$. We have

$$x(1, \eta) = \sum_{i=1}^{4} x_i \hat{\psi}_i(1, \eta) = x_1 0 + \tfrac{1}{2}x_2(1-\eta) + \tfrac{1}{2}x_3(1+\eta) + x_4 0$$
$$= \tfrac{1}{2}(x_2 + x_3) + \tfrac{1}{2}(x_3 - x_2)\eta \tag{9.35}$$
$$y(1, \eta) = \sum_{i=1}^{4} y_i \hat{\psi}_i(1, \eta) = \tfrac{1}{2}(y_2 + y_3) + \tfrac{1}{2}(y_3 - y_2)\,\eta$$

Clearly, x and y are linear functions on η. Therefore, they define a straight line. Similarly, the lines $\xi = -1$ and $\eta = \pm 1$ are mapped into straight lines in the element Ω^e. In other words, the master element $\hat{\Omega}$ is transformed, under the linear transformation, into a quadrilateral element (i.e., a four-sided element whose sides are not parallel) in the (x, y) plane. Conversely, every quadrilateral element of a mesh can be transformed to the same four-noded square (master) element $\hat{\Omega}$ in the (ξ, η) plane (see Fig. 9.6).

In general, the dependent variable(s) of the problem are approximated by expressions of the form

$$u(x, y) = \sum_{j=1}^{n} u_j^e \psi_j^e(x, y) \tag{9.36}$$

The interpolation functions ψ_j^e used for the approximation of the dependent variable are, in general, different from the $\hat{\psi}_j^e$ used in the approximation of the geometry. Depending on the relative degree of approximations used for the geometry [see (9.34)] and the dependent variable(s) [see (9.36)], the finite element formulations are classified into three categories.

1. *Superparametric* ($m > n$): the approximation used for the geometry is of higher order than that used for the dependent variable.
2. *Isoparametric* ($m = n$): equal degree of approximation is used for both geometry and dependent variables.
3. *Subparametric* ($m < n$): higher-order approximation of the dependent variable is used. (9.37)

For example, in the finite element analysis of the Euler–Bernoulli beams, we used linear Lagrange interpolation of the geometry,

$$x = \sum_{i=1}^{2} x_j \hat{\psi}_j(\xi) = x_A + \tfrac{1}{2}h_e(1 + \xi) \tag{9.38}$$

whereas Hermite cubic interpolation was used to approximate the transverse deflection. Such a formulation falls into the subparametric category. Since the axial displacement is approximated by the linear Lagrange interpolation functions, it can be said that the isoparametric formulation is used for the axial displacement. Superparametric formulations are rarely used. Also, the approximation of geometry by the Hermite family of interpolation functions is not common.

9.2.2 Coordinate Transformations

It should be noted that the transformation of a quadrilateral element of a finite element mesh to the master element $\hat{\Omega}$ is solely for the purpose of numerically evaluating the integrals. *No transformation of the physical domain or elements is involved in the finite element analysis.* The resulting algebraic equations of the finite element formulation are always among the nodal values of the physical domain. Different elements of the finite element mesh can be generated from the same master element by assigning the global coordinates of the elements (see Fig. 9.11). Master elements of different order define different transformations and hence different collections of finite element meshes. For example, a cubic-order master rectangular element can be used to generate a mesh of cubic curvilinear quadrilateral elements. Thus, with the help of an appropriate master element, any arbitrary element of a mesh can be generated. However, the transformations of a master element should be such that there are no spurious gaps between elements and no element overlaps. The elements in Figs. 9.6 and 9.8 can be used as master elements.

When a typical element of the finite element mesh is transformed to its master element for the purpose of numerically evaluating integrals, the integrand must also be expressed in terms of the coordinates (ξ, η) of the master element. For example, consider the element coefficients

$$K_{ij}^e = \int_{\Omega^e} \left[a(x, y) \frac{\partial \psi_i^e}{\partial x} \frac{\partial \psi_j^e}{\partial x} + b(x, y) \frac{\partial \psi_i^e}{\partial y} \frac{\partial \psi_j^e}{\partial y} + c(x, y) \psi_i^e \psi_j^e \right] dx\, dy \tag{9.39}$$

The integrand (i.e., the expression in the square brackets under the integral) is a function of the global coordinates x and y. We must rewrite it in terms of ξ and η using the transformation (9.34). Note that the integrand contains not only functions but also derivatives with respect to the global coordinates (x, y). Therefore, we must relate $\partial\psi_i^e/\partial x$ and $\partial\psi_i^e/\partial y$ to $\partial\psi_i^e/\partial\xi$ and $\partial\psi_i^e/\partial\eta$ using the transformation (9.34).

The functions $\psi_i^e(x, y)$ can be expressed in terms of the local coordinates ξ and η by means of (9.34). Hence, by the chain rule of partial differentiation, we have

$$\begin{aligned}\frac{\partial\psi_i^e}{\partial\xi} &= \frac{\partial\psi_i^e}{\partial x}\frac{\partial x}{\partial\xi} + \frac{\partial\psi_i^e}{\partial y}\frac{\partial y}{\partial\xi} \\ \frac{\partial\psi_i^e}{\partial\eta} &= \frac{\partial\psi_i^e}{\partial x}\frac{\partial x}{\partial\eta} + \frac{\partial\psi_i^e}{\partial y}\frac{\partial y}{\partial\eta}\end{aligned} \tag{9.40a}$$

or, in matrix notation,

$$\begin{Bmatrix}\dfrac{\partial\psi_i^e}{\partial\xi} \\ \dfrac{\partial\psi_i^e}{\partial\eta}\end{Bmatrix} = \begin{bmatrix}\dfrac{\partial x}{\partial\xi} & \dfrac{\partial y}{\partial\xi} \\ \dfrac{\partial x}{\partial\eta} & \dfrac{\partial y}{\partial\eta}\end{bmatrix}^e \begin{Bmatrix}\dfrac{\partial\psi_i^e}{\partial x} \\ \dfrac{\partial\psi_i^e}{\partial y}\end{Bmatrix} \tag{9.40b}$$

which gives the relation between the derivatives of ψ_i^e with respect to the global and local coordinates.

The matrix in (9.40*b*) is called the *Jacobian matrix* of the transformation (9.34):

$$[\mathscr{J}] = \begin{bmatrix}\dfrac{\partial x}{\partial\xi} & \dfrac{\partial y}{\partial\xi} \\ \dfrac{\partial x}{\partial\eta} & \dfrac{\partial y}{\partial\eta}\end{bmatrix}^e \tag{9.41}$$

Note from the expression given for K_{ij}^e in (9.39) that we must relate $\partial\psi_i^e/\partial x$ and $\partial\psi_i^e/\partial y$ to $\partial\psi_i^e/\partial\xi$ and $\partial\psi_i^e/\partial\eta$, whereas (9.40) provides the inverse relations. Therefore, (9.40*b*) must be inverted by inverting the Jacobian matrix:

$$\begin{Bmatrix}\dfrac{\partial\psi_i^e}{\partial x} \\ \dfrac{\partial\psi_i^e}{\partial y}\end{Bmatrix} = [\mathscr{J}]^{-1}\begin{Bmatrix}\dfrac{\partial\psi_i^e}{\partial\xi} \\ \dfrac{\partial\psi_i^e}{\partial\eta}\end{Bmatrix} \tag{9.42}$$

This requires that the Jacobian matrix $[\mathscr{J}]$ be nonsingular.

Although it is possible to write the relationship (9.42) directly by means of the chain rule,

$$\frac{\partial \psi_i^e}{\partial x} = \frac{\partial \psi_i^e}{\partial \xi}\frac{\partial \xi}{\partial x} + \frac{\partial \psi_i^e}{\partial \eta}\frac{\partial \eta}{\partial x}$$
$$\frac{\partial \psi_i^e}{\partial y} = \frac{\partial \psi_i^e}{\partial \xi}\frac{\partial \xi}{\partial y} + \frac{\partial \psi_i^e}{\partial \eta}\frac{\partial \eta}{\partial y} \tag{9.43}$$

it is not possible to evaluate $\partial\xi/\partial x$, $\partial\xi/\partial y$, $\partial\eta/\partial x$, and $\partial\eta/\partial y$ directly from the transformation equation (9.34). The transformation equation (9.34) allows direct evaluation of $\partial x/\partial\xi$, $\partial x/\partial\eta$, $\partial y/\partial\xi$ and $\partial y/\partial\eta$, and therefore $[\mathscr{J}]$.

Using the transformation (9.34), we can write

$$\frac{\partial x}{\partial \xi} = \sum_{j=1}^{m} x_j \frac{\partial \hat{\psi}_j^e}{\partial \xi}, \qquad \frac{\partial y}{\partial \xi} = \sum_{j=1}^{m} y_j \frac{\partial \hat{\psi}_j^e}{\partial \xi}$$
$$\frac{\partial x}{\partial \eta} = \sum_{j=1}^{m} x_j \frac{\partial \hat{\psi}_j^e}{\partial \eta}, \qquad \frac{\partial y}{\partial \eta} = \sum_{j=1}^{m} y_j \frac{\partial \hat{\psi}_j^e}{\partial \eta} \tag{9.44a}$$

and

$$[\mathscr{J}] = \begin{bmatrix} \dfrac{\partial x}{\partial \xi} & \dfrac{\partial y}{\partial \xi} \\ \dfrac{\partial x}{\partial \eta} & \dfrac{\partial y}{\partial \eta} \end{bmatrix} = \begin{bmatrix} \sum_{i=1}^{m} x_i \dfrac{\partial \hat{\psi}_i}{\partial \xi} & \sum_{i=1}^{m} y_i \dfrac{\partial \hat{\psi}_i}{\partial \xi} \\ \sum_{i=1}^{m} x_i \dfrac{\partial \hat{\psi}_i}{\partial \eta} & \sum_{i=1}^{m} y_i \dfrac{\partial \hat{\psi}_i}{\partial \eta} \end{bmatrix}$$
$$= \begin{bmatrix} \dfrac{\partial \hat{\psi}_1}{\partial \xi} & \dfrac{\partial \hat{\psi}_2}{\partial \xi} & \cdots & \dfrac{\partial \hat{\psi}_m}{\partial \xi} \\ \dfrac{\partial \hat{\psi}_1}{\partial \eta} & \dfrac{\partial \hat{\psi}_2}{\partial \eta} & \cdots & \dfrac{\partial \hat{\psi}_m}{\partial \eta} \end{bmatrix} \begin{bmatrix} x_1 & y_1 \\ x_2 & y_2 \\ \vdots & \vdots \\ x_m & y_m \end{bmatrix} \tag{9.44b}$$

Thus, given the glocal coordinates (x_j, y_j) of element nodes and the interpolation functions $\hat{\psi}_j^e$ used for geometry, the Jacobian matrix can be evaluated using (9.44*b*). Note that the $\hat{\psi}_j^e$ are different, in general, from the ψ_j^e used in the approximation of the dependent variables.

In order to compute the global derivatives of ψ_i^e (i.e., derivatives of ψ_i^e with respect to x and y) from (9.42), it is necessary to invert the Jacobian matrix. A necessary and sufficient condition for $[\mathscr{J}]^{-1}$ to exist is that the determinant $\mathscr{J}$, called the *Jacobian*, be non-negative at every point (ξ, η) in $\hat{\Omega}$:

$$\mathscr{J} \equiv \det[\mathscr{J}] = \frac{\partial x}{\partial \xi}\frac{\partial y}{\partial \eta} - \frac{\partial x}{\partial \eta}\frac{\partial y}{\partial \xi} > 0 \tag{9.45}$$

From this it is clear that the functions $\xi = \xi(x, y)$ and $\eta = \eta(x, y)$ must be continuous, differentiable, and invertible. Moreover, the transformation should be algebraically simple so that the Jacobian matrix can be easily evaluated. Transformations of the form (9.34) satisfy these requirements and the requirement that no spurious gaps between elements or overlapping of elements occur. We consider an example to illustrate the invertibility requirements.

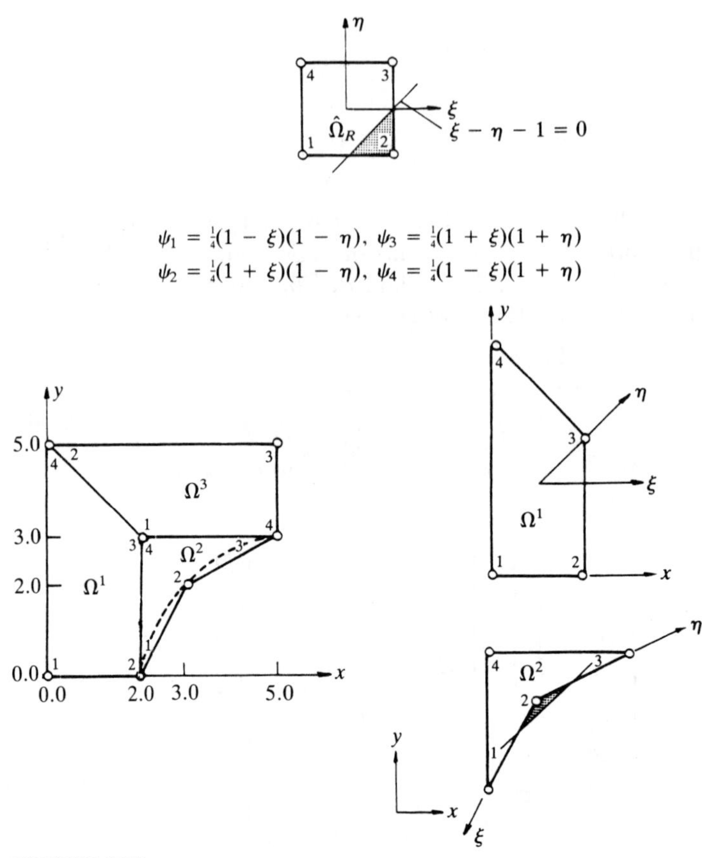

FIGURE 9.12
Examples of transformations of the master rectangular element $\hat{\Omega}_R$.

Example 9.2. Consider the three-element mesh of quadrilaterals shown in Fig. 9.12. The master element is the four-node square. Elements 1 and 2 have counter-clockwise element node numbering consistent with the node numbering in the master element, and element 3 has node numbering opposite to that of the master element. Elements 1 and 3 are *convex* domains in the sense that the line segment connecting any two arbitrary points of a convex domain lies entirely in the element. Clearly, element 2 is not convex, because, for example, the line segment joining nodes 1 and 3 is not entirely inside the element. In the following paragraphs, we investigate the effect of node numbering and element convexity on the transformations from the master element to each of the three elements.

First, we compute the elements of the Jacobian matrix (the interpolation functions are given in Fig. 9.12):

$$\frac{\partial x}{\partial \xi} = \sum_{i=1}^{4} x_i \frac{\partial \hat{\psi}_i}{\partial \xi} = \tfrac{1}{4}[-x_1(1-\eta) + x_2(1-\eta) + x_3(1+\eta) - x_4(1+\eta)]$$

$$\frac{\partial x}{\partial \eta} = \sum_{i=1}^{4} x_i \frac{\partial \hat{\psi}_i}{\partial \eta} = \tfrac{1}{4}[-x_1(1-\xi) - x_2(1+\xi) + x_3(1+\xi) + x_4(1-\xi)] \tag{9.46}$$

$$\frac{\partial y}{\partial \xi} = \sum_{i=1}^{4} y_i \frac{\partial \hat{\psi}_i}{\partial \xi} = \tfrac{1}{4}[-y_1(1-\eta) + y_2(1-\eta) + y_3(1+\eta) - y_4(1+\eta)]$$

$$\frac{\partial y}{\partial \eta} = \sum_{i=1}^{4} y_i \frac{\partial \hat{\psi}_i}{\partial \eta} = \tfrac{1}{4}[-y_1(1-\xi) - y_2(1+\xi) + y_3(1+\xi) + y_4(1-\xi)]$$

Next, we evaluate the Jacobian for each of the elements.

Element 1. We have $x_1 = x_4 = 0$, $x_2 = x_3 = 2$, $y_1 = y_2 = 0$, $y_3 = 3$, and $y_4 = 5$. The transformation and Jacobian are given by

$$x = 2\hat{\psi}_2 + 2\hat{\psi}_3 = 1 + \xi, \qquad y = 3\hat{\psi}_3 + 5\hat{\psi}_4 = (1+\eta)(2 - \tfrac{1}{2}\xi) \tag{9.47a}$$

$$\mathcal{J} = \det[\mathcal{J}] = \begin{vmatrix} 1 & -\frac{1}{2}(1+\eta) \\ 0 & 2 - \frac{1}{2}\xi \end{vmatrix} = \tfrac{1}{2}(4 - \xi) > 0 \tag{9.47b}$$

Clearly, the Jacobian is linear in ξ, and, for all values of ξ in $-1 \leq \xi \leq 1$, it is positive. Therefore, the transformation (9.47a) is invertible.

Element 2. Here we have $x_1 = x_4 = 2$, $x_2 = 3$, $x_3 = 5$, $y_1 = 0$, $y_2 = 2$, and $y_3 = y_4 = 3$. The transformation and Jacobian are given by

$$x = 3 + \xi + \tfrac{1}{2}\eta + \tfrac{1}{2}\xi\eta, \qquad y = 2 + \tfrac{1}{2}\xi + \eta - \tfrac{1}{2}\xi\eta \tag{9.48a}$$

$$\mathcal{J} = \begin{vmatrix} 1 + \frac{1}{2}\eta & \frac{1}{2}(1-\eta) \\ \frac{1}{2}(1+\xi) & 1 - \frac{1}{2}\xi \end{vmatrix} = \tfrac{3}{4}(1 + \eta - \xi) \tag{9.48b}$$

The Jacobian is *not* nonzero everywhere in the master element. It is zero along the line $\xi = 1 + \eta$ shown by the shaded area in the master element in Fig. 9.12. Moreover, this area is mapped into the shaded area outside element 2. Thus, elements with any interior angle greater than π should not be used in any finite element mesh.

Element 3. We have $x_1 = 2$, $x_2 = 0$, $x_3 = x_4 = 5$, $y_1 = y_4 = 3$, and $y_2 = y_3 = 5$. The transformation and Jacobian become (note that the nodes are numbered clockwise)

$$x = 3 - \tfrac{1}{2}\xi + 2\eta + \tfrac{1}{2}\xi\eta, \qquad y = 4 + \xi \tag{9.49a}$$

$$\mathcal{J} = \begin{vmatrix} -\frac{1}{2}(1-\eta) & 1 \\ 2 + \frac{1}{2}\xi & 0 \end{vmatrix} = -(2 + \tfrac{1}{2}\xi) < 0 \tag{9.49b}$$

The negative Jacobian indicates that a right-hand coordinate system is mapped into a left-hand coordinate system. Such coordinate transformations should be avoided.

The above example illustrates, for the four-node master element, that nonconvex elements are not admissible in finite element meshes. In general, any interior angle θ (see Fig. 9.13) should not be too small or too large, because the Jacobian $\mathcal{J} = (|d\mathbf{r}_1|\,|d\mathbf{r}_2| \sin\theta)/d\xi\,d\eta$ will be very small. Similar restrictions hold for higher-order master elements. Additional restrictions also exist for higher-order elements. For example, for higher-order triangular and rectangular elements, the placing of the side and interior nodes is restricted. For the eight-node rectangular element, it can be shown that the side nodes should be placed at a distance greater than a quarter of the length of the side from either corner node (see Fig. 9.13).

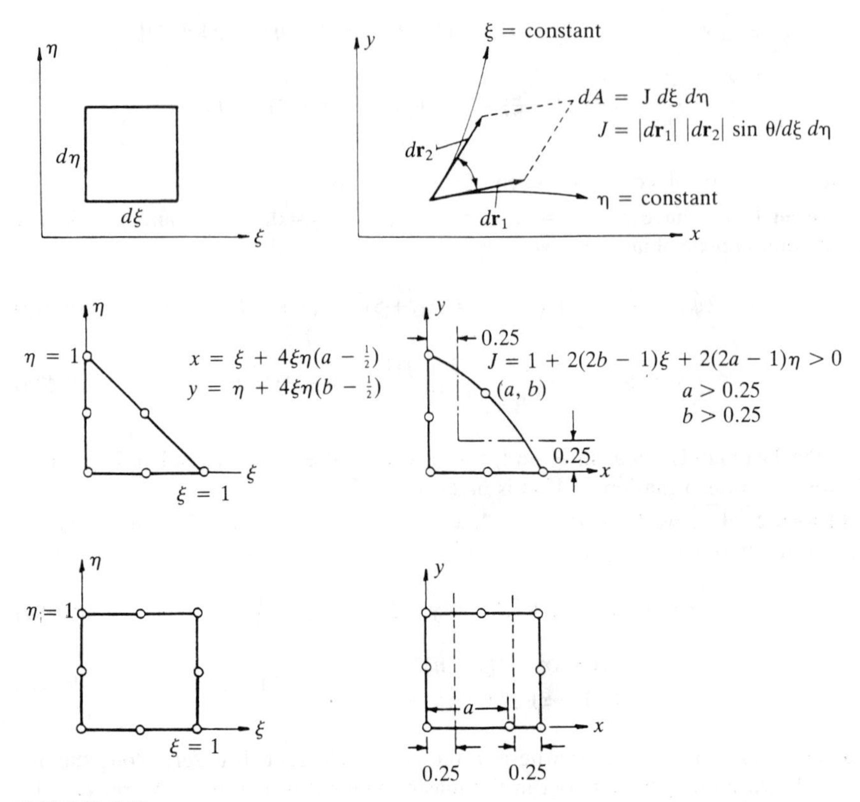

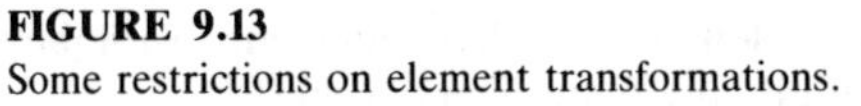

FIGURE 9.13
Some restrictions on element transformations.

Returning to numerical evaluation of integrals, we have, from (9.42),

$$\begin{Bmatrix} \dfrac{\partial \psi_i^e}{\partial x} \\ \dfrac{\partial \psi_i^e}{\partial y} \end{Bmatrix} = [\mathcal{J}]^{-1} \begin{Bmatrix} \dfrac{\partial \psi_i^e}{\partial \xi} \\ \dfrac{\partial \psi_i^e}{\partial \eta} \end{Bmatrix} \equiv [\mathcal{J}^*] \begin{Bmatrix} \dfrac{\partial \psi_i^e}{\partial \xi} \\ \dfrac{\partial \psi_i^e}{\partial \eta} \end{Bmatrix} \tag{9.50}$$

where $\mathcal{J}_{ij}^*$ is the element in position (i, j) of the inverse of the Jacobian matrix,

$$[\mathcal{J}]^{-1} \equiv [\mathcal{J}^*] = \begin{bmatrix} \mathcal{J}_{11}^* & \mathcal{J}_{12}^* \\ \mathcal{J}_{21}^* & \mathcal{J}_{22}^* \end{bmatrix} \tag{9.51}$$

The elemental area $dA = dx\,dy$ in the element Ω^e is transformed to

$$dA \equiv dx\,dy = \mathcal{J}\,d\xi\,d\eta \tag{9.52}$$

in the master element $\hat{\Omega}$.

Equations (9.42), (9.44), (9.51), and (9.52) provide the necessary relations to transform integral expressions on any element Ω^e to an associated master element $\hat{\Omega}$. For instance, consider the integral expression in (9.39) where $a = a(x, y)$, $b = b(x, y)$, and $c = c(x, y)$ are functions of x and y. Suppose that the mesh of finite elements is generated by a master element $\hat{\Omega}$. Under the transformation (9.34), we can write

$$
\begin{aligned}
K_{ij}^e &= \int_{\Omega^e} \left(a \frac{\partial \psi_i}{\partial x} \frac{\partial \psi_j}{\partial x} + b \frac{\partial \psi_i}{\partial y} \frac{\partial \psi_j}{\partial y} + c\psi_i \psi_j \right) dx\, dy \\
&= \int_{\hat{\Omega}} \left[\hat{a} \left(\mathcal{J}_{11}^* \frac{\partial \psi_i}{\partial \xi} + \mathcal{J}_{12}^* \frac{\partial \psi_i}{\partial \eta} \right) \left(\mathcal{J}_{11}^* \frac{\partial \psi_j}{\partial \xi} + \mathcal{J}_{12}^* \frac{\partial \psi_j}{\partial \eta} \right) \right. \\
&\quad \left. + \hat{b} \left(\mathcal{J}_{21}^* \frac{\partial \psi_i}{\partial \xi} + \mathcal{J}_{22}^* \frac{\partial \psi_i}{\partial \eta} \right) \left(\mathcal{J}_{21}^* \frac{\partial \psi_j}{\partial \xi} + \mathcal{J}_{22}^* \frac{\partial \psi_j}{\partial \eta} \right) + \hat{c}\psi_i \psi_j \right] \mathcal{J}\, d\xi\, d\eta \\
&\equiv \int_{\hat{\Omega}} F_{ij}(\xi, \eta)\, d\xi\, d\eta
\end{aligned}
\tag{9.53}
$$

where $\mathcal{J}_{ij}^*$ are the elements of the inverse of the Jacobian matrix in (9.51), and $\hat{a} = a(\xi, \eta)$, and so on. Equations (9.42), (9.44), and (9.51)–(9.53) are valid for master elements of both rectangular and triangular geometry. The master triangular and rectangular elements for linear and quadratic triangular and quadrilateral elements are shown in Fig. 9.14.

9.2.3 Integration over a Master Rectangular Element

Quadrature formulae for integrals defined over a rectangular master element $\hat{\Omega}_R$ (such as that shown in Fig. 9.14) can be derived from the one-dimensional quadrature formulae presented in Section 7.1.5. We have

$$
\begin{aligned}
\int_{\hat{\Omega}_R} F(\xi, \eta)\, d\xi\, d\eta &= \int_{-1}^{1} \left[\int_{-1}^{1} F(\xi, \eta)\, d\eta \right] d\xi \approx \int_{-1}^{1} \left[\sum_{J=1}^{N} F(\xi, \eta_J) W_J \right] d\xi \\
&\approx \sum_{I=1}^{M} \sum_{J=1}^{N} F(\xi_I, \eta_J) W_I W_J
\end{aligned}
\tag{9.54}
$$

where M and N denote the number of quadrature points in the ξ and η directions, (ξ_I, η_J) denote the Gauss points, and W_I and W_J denote the corresponding Gauss weights (see Table 7.2). The selection of the number of Gauss points is based on the same formula as that given in Section 7.1.5: a polynomial of degree p is integrated exactly employing $N = \text{int}\,[\frac{1}{2}(p+1)]$; that is, the smallest integer greater than $\frac{1}{2}(p+1)$. In most cases, the interpolation functions are of the same degree in both ξ and η, and therefore one has $M = N$. When the integrand is of different degree in ξ and η, the number of Gauss points is selected on the basis of the largest-degree polynomial. The minimum allowable quadrature rule is one that computes the mass of the element exactly when the density is constant.

Tables 9.2 and 9.3 give information on the selection of the integration order and the location of the Gauss points for linear, quadratic, and cubic elements. The maximum degree of the polynomial refers to the degree of the highest polynomial in ξ or η that is present in the integrands $F(\xi, \eta)$ of the

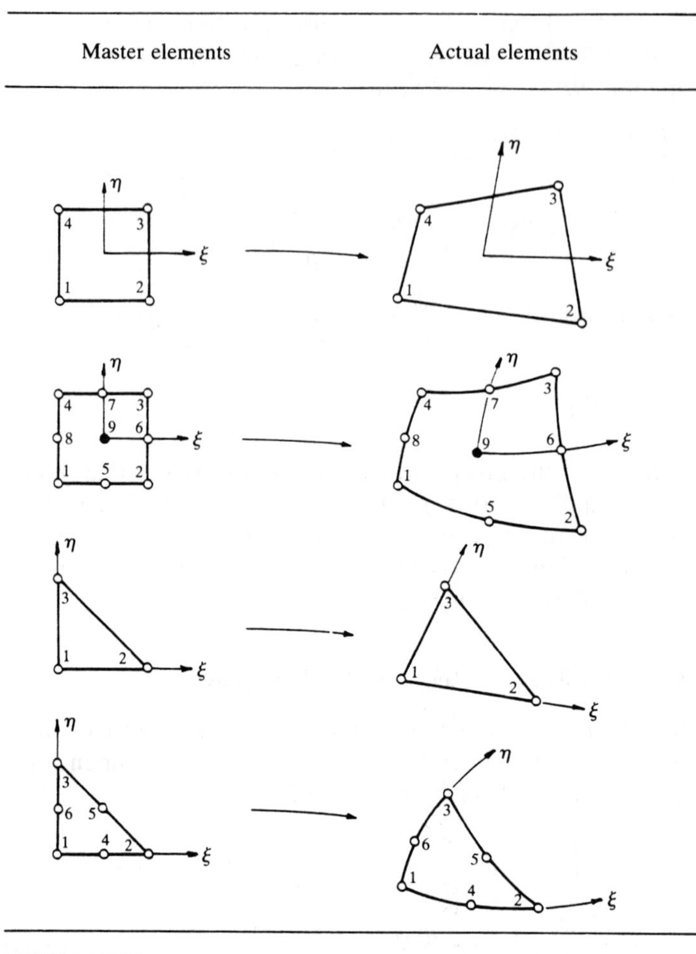

FIGURE 9.14
Linear and quadratic master elements and their transformations.

element matrices of the type in (9.53). Note that the polynomial degree of coefficients as well as $\mathscr{J}_{ij}^*$ and $\mathscr{J}$ should be accounted for in determining the total polynomial degree of the integrand. Of course, the coefficients a, b, and c, and $\mathscr{J}_{ij}^*$ in general, may not be polynomials. In those cases, their functional variations must be approximated by a suitable polynomial (for example, via a binomial series) in order to determine the polynomial degree of the integrand.

The $N \times N$ Gauss point locations are given by the *tensor product* of one-dimensional Gauss points ξ_I:

$$\begin{Bmatrix} \xi_1 \\ \xi_2 \\ \vdots \\ \xi_N \end{Bmatrix} \{\xi_1, \xi_2, \ldots, \xi_N\} = \begin{bmatrix} (\xi_1, \xi_1) & (\xi_1, \xi_2) & \ldots & (\xi_1, \xi_N) \\ (\xi_2, \xi_1) & & \ddots & \vdots \\ \vdots & & & \\ (\xi_N, \xi_1) & & \ldots & (\xi_N, \xi_N) \end{bmatrix} \tag{9.55}$$

The values of ξ_I $(I = 1, 2, \ldots, N)$ are presented in Table 7.2.

TABLE 9.2
Selection of the integration order and location of the Gauss points for linear, quadratic, and cubic quadrilateral elements (nodes not shown)

Element type	Maximum polynomial degree	Order of integration ($r \times r$)	Order of the residual	Location of integration points* in master element
Linear ($r = 2$)	2	2×2	$O(h^4)$	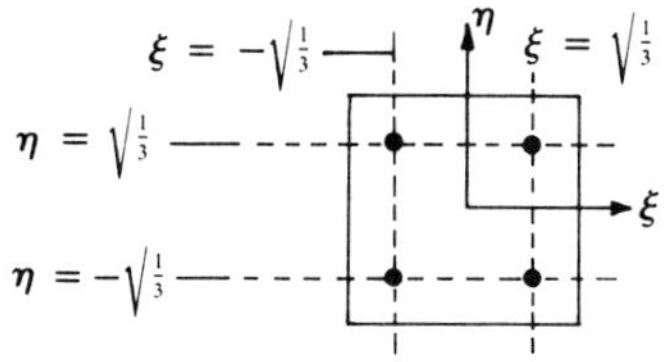 $\xi = -\sqrt{\frac{1}{3}}$, $\xi = \sqrt{\frac{1}{3}}$; $\eta = \sqrt{\frac{1}{3}}$, $\eta = -\sqrt{\frac{1}{3}}$
Quadratic ($r = 3$)	4	3×3	$O(h^6)$	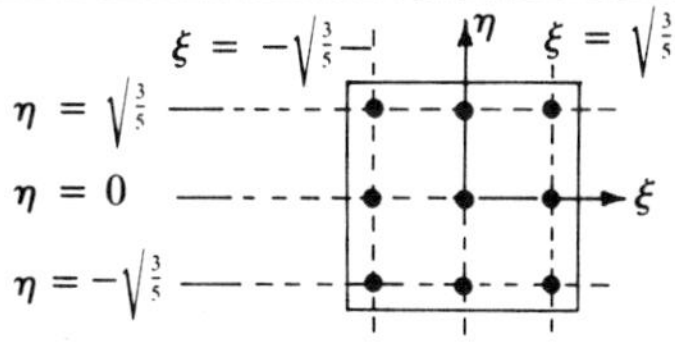 $\xi = -\sqrt{\frac{3}{5}}$, $\xi = \sqrt{\frac{3}{5}}$; $\eta = \sqrt{\frac{3}{5}}$, $\eta = 0$, $\eta = -\sqrt{\frac{3}{5}}$
Cubic ($r = 4$)	6	4×4	$O(h^8)$	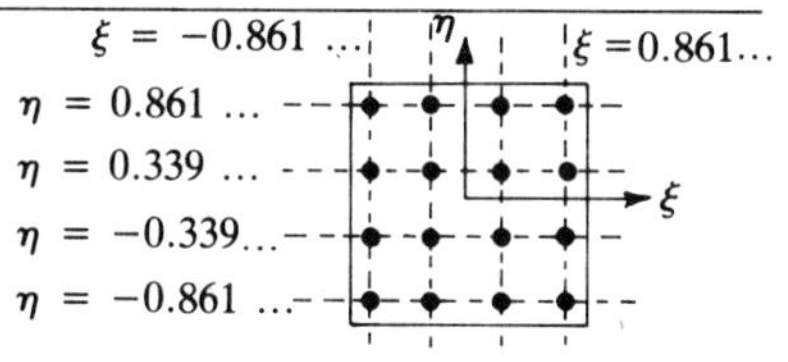 $\xi = -0.861\ldots$, $\xi = 0.861\ldots$; $\eta = 0.861\ldots$, $\eta = 0.339\ldots$, $\eta = -0.339\ldots$, $\eta = -0.861\ldots$

*See Table 7.2 for the integration points and weights for each coordinate direction.

The following examples illustrate the evaluation of the Jacobian and element matrices on rectangular elements.

Example 9.3. Consider the quadrilateral element Ω^1 shown in Fig. 9.12. We wish to evaluate $\partial\psi_i/\partial x$ and $\partial\psi_i/\partial y$ at $(\xi, \eta) = (0, 0)$ and $(\frac{1}{2}, \frac{1}{2})$ using the isoparametric formulation (i.e., $\hat{\psi}_i = \psi_i$). From (9.44b), we have

$$[\mathscr{J}] = \frac{1}{4}\begin{bmatrix} -(1-\eta) & 1-\eta & 1+\eta & -(1+\eta) \\ -(1-\xi) & -(1+\xi) & 1+\xi & 1-\xi \end{bmatrix}\begin{bmatrix} 0.0 & 0.0 \\ 2.0 & 0.0 \\ 2.0 & 3.0 \\ 0.0 & 5.0 \end{bmatrix}$$

$$= \begin{bmatrix} 1 & -\frac{1}{2}(1+\eta) \\ 0 & \frac{1}{2}(4-\xi) \end{bmatrix}$$

The inverse of the Jacobian matrix is given by

$$[\mathscr{J}]^{-1} = \begin{bmatrix} 1 & \dfrac{1+\eta}{4-\xi} \\ 0 & \dfrac{2}{4-\xi} \end{bmatrix}$$

$$\mathscr{J}^*_{11} = 1, \quad \mathscr{J}^*_{21} = 0, \quad \mathscr{J}^*_{12} = \frac{1+\eta}{4-\xi}, \quad \mathscr{J}^*_{22} = \frac{2}{4-\xi}$$

From (9.42), we have

$$\frac{\partial \psi_i}{\partial x} = \frac{\partial \psi_i}{\partial \xi} + \frac{1+\eta}{4-\xi}\frac{\partial \psi_i}{\partial \eta}, \qquad \frac{\partial \psi_i}{\partial y} = \frac{2}{4-\xi}\frac{\partial \psi_i}{\partial \eta}$$

where

$$\psi_i = \tfrac{1}{4}(1+\xi\xi_i)(1+\eta\eta_i), \qquad \frac{\partial \psi_i}{\partial \xi} = \tfrac{1}{4}\xi_i(1+\eta\eta_i), \qquad \frac{\partial \psi_i}{\partial \eta} = \tfrac{1}{4}\eta_i(1+\xi\xi_i) \tag{9.56}$$

(ξ_i, η_i) being the coordinates of the ith node in the master element (see Fig. 9.12):

Node	ξ_i	η_i
1	−1	−1
2	1	−1
3	1	1
4	−1	1

Then we have

$$\frac{\partial \psi_i}{\partial x} = \tfrac{1}{4}\xi_i(1+\eta\eta_i) + \frac{1}{4}\frac{1+\eta}{4-\xi}\eta_i(1+\xi\xi_i)$$

$$\frac{\partial \psi_i}{\partial y} = \frac{1}{4}\frac{2}{4-\xi}\eta_i(1+\xi\xi_i)$$

$(\xi, \eta) \rightarrow$	$(0, 0)$	$(\frac{1}{2}, \frac{1}{2})$
$\dfrac{\partial \psi_i}{\partial x}$	$\frac{1}{4}\xi_i + \frac{1}{16}\eta_i$	$\frac{1}{8}\xi_i(2+\eta_i) + \frac{3}{56}\eta_i(2+\xi_i)$
$\dfrac{\partial \psi_i}{\partial y}$	$\frac{1}{8}\eta_i$	$\frac{1}{14}\eta_i(2+\xi_i)$

Example 9.4. Consider the quadrilateral element in Fig. 9.15. We wish to compute the following element matrices using Gauss quadrature and the isoparametric formulation:

$$S_{ij} = \int_\Omega \psi_i \psi_j \, dx\, dy, \qquad S^{11}_{ij} = \int_\Omega \frac{\partial \psi_i}{\partial x}\frac{\partial \psi_j}{\partial x}\, dx\, dy$$

$$S^{22}_{ij} = \int_\Omega \frac{\partial \psi_i}{\partial y}\frac{\partial \psi_j}{\partial y}\, dx\, dy, \qquad S^{12}_{ij} = \int_\Omega \frac{\partial \psi_i}{\partial x}\frac{\partial \psi_j}{\partial y}\, dx\, dy \tag{9.57}$$

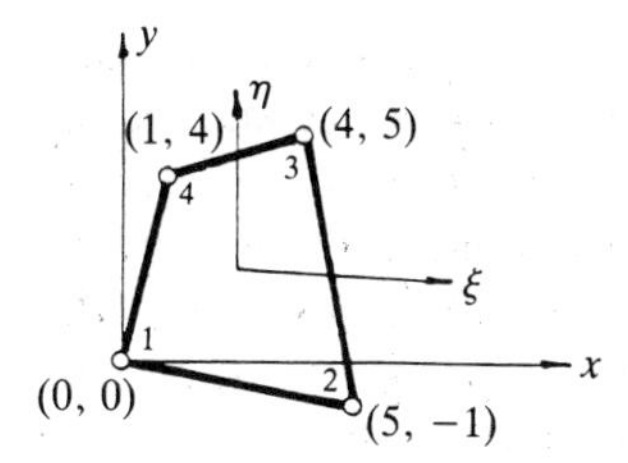

FIGURE 9.15
Geometry of the bilinear element used in Example 9.4

We have

$$[\mathscr{J}] = \frac{1}{4}\begin{bmatrix} 8-2\eta & 2\eta \\ -2\xi & 10+2\xi \end{bmatrix}$$

$$\mathscr{J} = (4-\eta)(5+\xi) + \xi\eta = 20 + 4\xi - 5\eta$$

$$S_{ij} = \int_\Omega \psi_i\psi_j \, dx\, dy = \int_{-1}^{1}\int_{-1}^{1} \psi_i\psi_j\mathscr{J}\, d\xi\, d\eta$$

$$S_{ij}^{11} = \int_\Omega \frac{\partial\psi_i}{\partial x}\frac{\partial\psi_j}{\partial x}\, dx\, dy \tag{9.58}$$

$$= \int_{-1}^{1}\int_{-1}^{1}\left(\mathscr{J}_{11}^{*}\frac{\partial\psi_i}{\partial\xi} + \mathscr{J}_{12}^{*}\frac{\partial\psi_i}{\partial\eta}\right)\left(\mathscr{J}_{11}^{*}\frac{\partial\psi_j}{\partial\xi} + \mathscr{J}_{12}^{*}\frac{\partial\psi_j}{\partial\eta}\right)\mathscr{J}\, d\xi\, d\eta$$

and so on, where $\partial\psi_i/\partial\xi$ and $\partial\psi_i/\partial\eta$ are given by (9.56). Note that the integrand of S_{ij} is a polynomial of order $p = 3$ in each coordinate. Hence, $N = M = 2$ will evaluate S_{ij} exactly. Evaluating the integrals in (9.57) using the 2×2 quadrative rule, we obtain

$$[S^{11}] = \begin{bmatrix} 0.40995 & -0.36892 & -0.20479 & 0.16376 \\ -0.36892 & 0.34516 & 0.25014 & -0.22639 \\ -0.20479 & 0.25014 & 0.43155 & -0.47690 \\ 0.16376 & -0.22639 & -0.47690 & 0.53953 \end{bmatrix}$$

$$[S^{22}] = \begin{bmatrix} 0.26237 & 0.16389 & -0.13107 & -0.29520 \\ 0.16389 & 0.22090 & -0.23991 & -0.14489 \\ -0.13107 & -0.23991 & 0.27619 & 0.09478 \\ -0.29520 & -0.14489 & 0.09478 & 0.34530 \end{bmatrix}$$

$$[S^{12}] = \begin{bmatrix} 0.24731 & 0.25156 & -0.25297 & -0.24589 \\ -0.24844 & -0.25090 & 0.25172 & 0.24762 \\ -0.25297 & -0.24828 & 0.24671 & 0.25454 \\ 0.25411 & 0.24762 & -0.24546 & -0.25627 \end{bmatrix}$$

$$[S] = \begin{bmatrix} 2.27780 & 1.25000 & 0.55556 & 1.00000 \\ 1.25000 & 2.72220 & 1.22220 & 0.55556 \\ 0.55556 & 1.22220 & 2.16670 & 0.97222 \\ 1.00000 & 0.55556 & 0.97222 & 1.72222 \end{bmatrix}$$

9.2.4 Integration over a Master Triangular Element

In the preceding section, we discussed numerical integration on quadrilateral elements, which can be used to represent very general geometries as well as field variables in a variety of problems. Here we discuss numerical integration on triangular elements. Since quadrilateral elements can be geometrically distorted, it is possible to distort a quadrilateral element to obtain a required triangular element by moving the position of the corner node to one of the neighboring nodes. In actual computation, this is achieved by assigning the same global node number to two corner nodes of the quadrilateral element. Thus, master triangular elements can be obtained in a natural way from associated master rectangular elements. Here we discuss the transformations from an arbitrary triangular element to a master triangular element.

We choose the unit right isosceles triangle (see Fig. 9.16a) as the master element. An arbitrary triangular element Ω^e can be generated from the master triangular element $\hat{\Omega}_T$ by a transformation of the form (9.34). The coordinate lines $\xi = 0$ and $\eta = 0$ in $\hat{\Omega}_T$ correspond to the skew curvilinear coordinate lines 1–3 and 1–2 in Ω^e. For the three-node triangular element, the transformation (9.34) is taken to be

$$x = \sum_{i=1}^{3} x_i \hat{\psi}_i(\xi, \eta), \qquad y = \sum_{i=1}^{3} y_i \hat{\psi}_i(\xi, \eta) \tag{9.59}$$

where $\hat{\psi}_i(\xi, \eta)$ are the interpolation functions of the master three-node triangular element (see Fig. 9.16b),

$$\hat{\psi}_1 = 1 - \xi - \eta, \qquad \hat{\psi}_2 = \xi, \qquad \hat{\psi}_3 = \eta \tag{9.60}$$

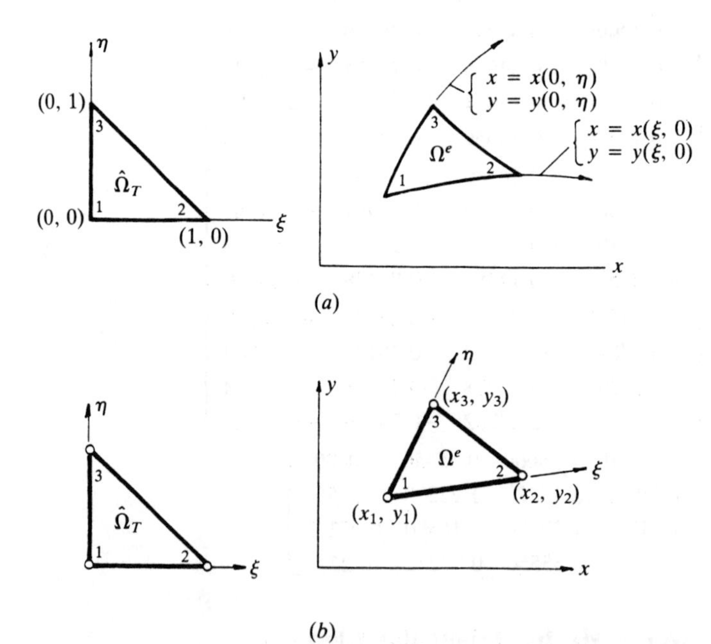

FIGURE 9.16
Triangular master element and its transformations: (a) general transformation; (b) linear transformation of a master element to a triangular element.

The inverse transformation from the element Ω^e to $\hat{\Omega}_T$ is given by inverting (9.59):

$$\xi = \frac{1}{2A}[(x - x_1)(y_3 - y_1) - (y - y_1)(x_3 - x_1)]$$
$$\eta = \frac{1}{2A}[(x - x_1)(y_1 - y_2) + (y - y_1)(x_2 - x_1)] \tag{9.61}$$

where A is the area of Ω^e.

With the help of (9.61), one can show that the interpolation functions in (8.25) are equivalent to the $\hat{\psi}_i$ in (9.60). Moreover, the area coordinates L_i in (9.9) are also equivalent to $\hat{\psi}_i$. The interpolation functions for the linear and higher-order triangular elements can be obtained from the area coordinates, as described in Section 9.1.

The Jacobian matrix for the linear triangular element is

$$[\mathscr{J}] = \begin{bmatrix} x_2 - x_1 & y_2 - y_1 \\ x_3 - x_1 & y_3 - y_1 \end{bmatrix} = \begin{bmatrix} \gamma_3 & -\beta_3 \\ -\gamma_2 & \beta_2 \end{bmatrix} \tag{9.62}$$

where β_i and γ_i are the constants defined in (8.23b). The inverse of the Jacobian matrix is given by

$$[\mathscr{J}]^{-1} = \frac{1}{\mathscr{J}}\begin{bmatrix} \beta_2 & \beta_3 \\ \gamma_2 & \gamma_3 \end{bmatrix}, \qquad \mathscr{J} = \beta_2\gamma_3 - \gamma_2\beta_3 = 2A \tag{9.63}$$

The relations (9.42) for the isoparametric formulation with linear triangular elements have the explicit form

$$\frac{\partial \psi_1}{\partial x} = -\frac{\beta_2 + \beta_3}{2A} = \frac{\beta_1}{2A}, \qquad \frac{\partial \psi_1}{\partial y} = -\frac{\gamma_2 + \gamma_3}{2A} = \frac{\gamma_1}{2A}$$
$$\frac{\partial \psi_2}{\partial x} = \frac{\beta_2}{2A}, \quad \frac{\partial \psi_2}{\partial y} = \frac{\gamma_2}{2A}, \quad \frac{\partial \psi_3}{\partial x} = \frac{\beta_3}{2A}, \quad \frac{\partial \psi_3}{\partial y} = \frac{\gamma_3}{2A} \tag{9.64}$$

In a general case, the derivatives of ψ_i with respect to the global coordinates can be computed from (9.43), which take the form

$$\frac{\partial \psi_i}{\partial x} = \frac{\partial \psi_i}{\partial L_1}\frac{\partial L_1}{\partial x} + \frac{\partial \psi_i}{\partial L_2}\frac{\partial L_2}{\partial x}$$
$$\frac{\partial \psi_i}{\partial y} = \frac{\partial \psi_i}{\partial L_1}\frac{\partial L_1}{\partial y} + \frac{\partial \psi_i}{\partial L_2}\frac{\partial L_2}{\partial y} \tag{9.65a}$$

or

$$\begin{Bmatrix} \dfrac{\partial \psi_i}{\partial x} \\ \dfrac{\partial \psi_i}{\partial y} \end{Bmatrix} = [\mathscr{J}]^{-1}\begin{Bmatrix} \dfrac{\partial \psi_i}{\partial L_1} \\ \dfrac{\partial \psi_i}{\partial L_2} \end{Bmatrix}, \qquad [\mathscr{J}] = \begin{bmatrix} \dfrac{\partial x}{\partial L_1} & \dfrac{\partial y}{\partial L_1} \\ \dfrac{\partial x}{\partial L_2} & \dfrac{\partial y}{\partial L_2} \end{bmatrix} \tag{9.65b}$$

Note that only L_1 and L_2 are treated as linearly independent coordinates, because $L_3 = 1 - L_1 - L_2$.

After transformation, integrals on $\hat{\Omega}_T$ have the form

$$\int_{\hat{\Omega}_T} G(\xi, \eta)\, d\xi\, d\eta = \int_{\hat{\Omega}_T} \hat{G}(L_1, L_2, L_3)\, dL_1\, dL_2 \tag{9.66}$$

which can be approximated by the quadrature formula

$$\int_{\hat{\Omega}_T} \hat{G}(L_1, L_2, L_3)\, dL_1\, dL_2 \approx \frac{1}{2} \sum_{I=1}^{N} W_I \hat{G}(\mathbf{S}_I) \tag{9.67}$$

TABLE 9.3
Quadrature points and weights for triangular elements

Number of integration points	Degree of polynomial and order of the residual	Location of integration points				
		L_1	L_2	L_3	W	Geometric locations
1	1 $O(h^2)$	$\frac{1}{3}$	$\frac{1}{3}$	$\frac{1}{3}$	1	a 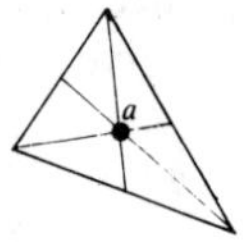
3	2 $O(h^3)$	$\frac{1}{2}$	0	$\frac{1}{2}$	$\frac{1}{3}$	a 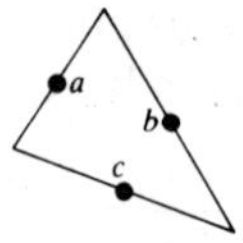
		$\frac{1}{2}$	$\frac{1}{2}$	0	$\frac{1}{3}$	b
		0	$\frac{1}{2}$	$\frac{1}{2}$	$\frac{1}{3}$	c
4	3 $O(h^4)$	$\frac{1}{3}$	$\frac{1}{3}$	$\frac{1}{3}$	$-\frac{27}{48}$	a 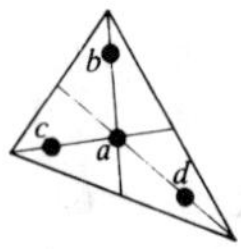
		0.6	0.2	0.2	$\frac{25}{48}$	b
		0.2	0.6	0.2	$\frac{25}{48}$	c
		0.2	0.2	0.6	$\frac{25}{48}$	d
7	5 $O(h^6)$	$\frac{1}{3}$	$\frac{1}{3}$	$\frac{1}{3}$	0.225	a 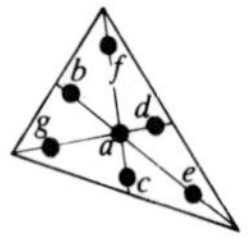
		α_1	β_1	β_1		b
		β_1	α_1	β_1	w_2	c
		β_1	β_1	α_1		d
		α_2	β_2	β_2		e
		β_2	α_2	β_2	w_3	f
		β_2	β_2	α_2		g

$\alpha_1 = 0.797\ 426\ 985\ 353$
$\beta_1 = 0.101\ 286\ 507\ 323$ $\quad W_2 = 0.125\ 939\ 180\ 544$
$\alpha_2 = 0.059\ 715\ 871\ 789$
$\beta_2 = 0.470\ 142\ 064\ 105$ $\quad W_3 = 0.132\ 394\ 152\ 788$

where W_I and $\mathbf{S}_I$ denote the weights and integration points of the quadrature rule. Table 9.3 gives the location of integration points and weights for one-, three-, and four-point quadrature rules over triangular elements. For the evaluation of integrals whose integrands are polynomials of degree higher than five (in any of the area coordinates), the reader should consult books on numerical integration.

Example 9.5. Consider the quadratic triangular element shown in Fig. 9.17. We wish to calculate $\partial\psi_1/\partial x$, $\partial\psi_1/\partial y$, $\partial\psi_4/\partial x$, and $\partial\psi_4/\partial y$ at the point $(x, y) = (2, 4)$ and evaluate the integral of the product $(\partial\psi_1/\partial x)(\partial\psi_4/\partial x)$.

Since the element has straight edges, its geometry is defined by the interpolation functions of the corner nodes (i.e., a subparametric formulation can be used). Note that if the element is curvilinear, we cannot use only three corner nodes to describe the geometry exactly (hence an isoparametric formulation must be used). For the element at hand, we have

$$x = \sum_{i=1}^{3} x_i L_i = 7L_2 + 2L_3 = 2 - 2L_1 + 5L_2$$

$$y = \sum_{i=1}^{3} y_i L_i = 2L_2 + 6L_3 = 6 - 6L_1 - 4L_2 \tag{9.68a}$$

$$[\mathcal{J}] = \begin{bmatrix} -2 & -6 \\ 5 & -4 \end{bmatrix}, \qquad [\mathcal{J}]^{-1} = \frac{1}{38}\begin{bmatrix} -4 & 6 \\ -5 & -2 \end{bmatrix}$$

The ψ_i for the quadratic triangular element are given by (9.16), with ψ_2 in (9.16c) being equal to ψ_4 here (compare the node numbering in Figs. 9.17 and 9.4b). The global derivatives of ψ_i are given by (9.65):

$$\begin{Bmatrix} \dfrac{\partial\psi_1}{\partial x} \\ \dfrac{\partial\psi_1}{\partial y} \end{Bmatrix} = \frac{1}{38}\begin{bmatrix} -4 & 6 \\ -5 & -2 \end{bmatrix} \begin{Bmatrix} \dfrac{\partial\psi_1}{\partial L_1} \\ \dfrac{\partial\psi_1}{\partial L_2} \end{Bmatrix} = \begin{Bmatrix} \dfrac{-4(4L_1 - 1)}{38} \\ \dfrac{-5(4L_1 - 1)}{38} \end{Bmatrix} \tag{9.68b}$$

$$\begin{Bmatrix} \dfrac{\partial\psi_4}{\partial x} \\ \dfrac{\partial\psi_4}{\partial y} \end{Bmatrix} = \frac{1}{38}\begin{bmatrix} -4 & 6 \\ -5 & -2 \end{bmatrix} \begin{Bmatrix} \dfrac{\partial\psi_4}{\partial L_1} \\ \dfrac{\partial\psi_4}{\partial L_2} \end{Bmatrix} = \frac{1}{38}\begin{Bmatrix} -16L_2 + 24L_1 \\ -20L_2 - 8L_1 \end{Bmatrix}$$

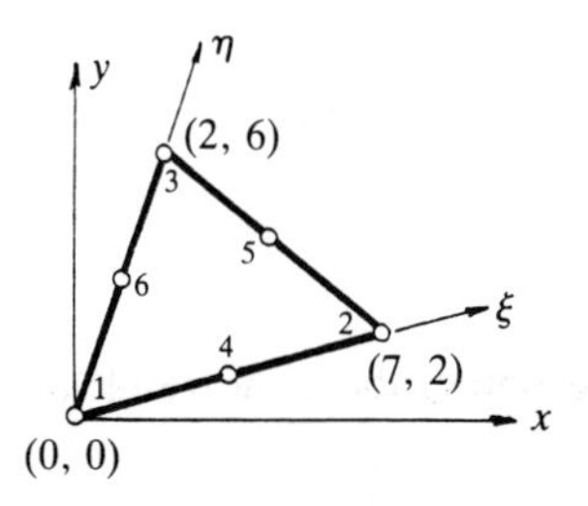

FIGURE 9.17
A quadratic triangular element in the global (x, y) and local coordinate systems (Example 9.5).

where $\psi_1 = L_1(2L_1 - 1)$ and $\psi_4 = 4L_1L_2$ [see (9.16) and Fig. 9.4*b*], so that

$$\frac{\partial \psi_1}{\partial L_1} = 4L_1 - 1, \quad \frac{\partial \psi_1}{\partial L_2} = 0, \quad \frac{\partial \psi_4}{\partial L_1} = 4L_2, \quad \frac{\partial \psi_4}{\partial L_2} = 4L_1$$

For the point (2, 4), the area coordinates L_i can be calculated from (9.68*a*):

$$2 = 7L_2 + 2L_3, \qquad 4 = 2L_2 + 6L_3$$

Once L_2 and L_3 have been computed from the above relations, L_1 is found from the relation $L_1 = 1 - L_2 - L_3$. We obtain

$$L_1 = \tfrac{5}{19}, \qquad L_2 = \tfrac{2}{19}, \qquad L_3 = \tfrac{12}{19}$$

Evaluating $\partial\psi_4/\partial x$ and $\partial\psi_4/\partial y$ at the point (2, 4), we obtain

$$\begin{aligned} \frac{\partial \psi_1}{\partial x} &= -\frac{2}{19}\left(\frac{20}{19} - 1\right) = -\frac{2}{361} \\ \frac{\partial \psi_1}{\partial y} &= -\frac{5}{38}\left(4 \times \frac{5}{19} - 1\right) = -\frac{5}{722} \\ \frac{\partial \psi_4}{\partial x} &= \frac{-16}{(19)^2} + \frac{60}{(19)^2} = \frac{44}{361} \\ \frac{\partial \psi_4}{\partial y} &= \frac{-20}{(19)^2} + \frac{-20}{(19)^2} = \frac{-40}{361} \end{aligned} \tag{9.69}$$

The integral of the product $(\partial\psi_1/\partial x)(\partial\psi_4/\partial x)$ over the quadratic element is

$$\int_{\hat{\Omega}_T} \frac{\partial \psi_1}{\partial x}\frac{\partial \psi_4}{\partial x}\,dx\,dy = \frac{-4\mathcal{J}}{361}\int_0^1 \int_0^{1-L_2} (4L_1 - 1)(6L_1 - 4L_2)\,dL_1\,dL_2$$

Since the integrand is quadratic in L_1 and bilinear in L_1 and L_2, we use the three-point quadrature (see Table 9.3) to evaluate the integral exactly:

$$\begin{aligned} &\frac{-4\mathcal{J}}{361}\int_0^1 \int_0^{1-L_2} (4L_1 - 1)(6L_1 - 4L_2)\,dL_1\,dL_2 \\ &= \frac{1}{2}\frac{(-4) \times 38}{361}\tfrac{1}{3}[(\tfrac{4}{2} - 1)(\tfrac{6}{2} - 0) + (\tfrac{4}{2} - 1)(\tfrac{6}{2} - \tfrac{4}{2}) + (0 - 1)(0 - \tfrac{4}{2})] \\ &= -\frac{8}{19} \end{aligned} \tag{9.70}$$

The result can be verified using the exact integration formula in (9.81):

$$\begin{aligned} \int_{\hat{\Omega}_T} \frac{\partial \psi_1}{\partial x}\frac{\partial \psi_4}{\partial x}\,dx\,dy &= \frac{4}{361}\left[6 \times \frac{1}{3!} - 4 \times \frac{1}{3!} - 24 \times \frac{2!}{4!} + 16 \times \frac{1}{4!}\right]2A \\ &= -\frac{8}{19} \end{aligned}$$

The area A of the triangle is equal to 19, and therfore we obtain the same result as above.

9.3 MODELING CONSIDERATIONS

9.3.1 Preliminary Comments

Numerical simulation of a physical process or a system requires (i) a mathematical model that describes the process and (ii) a numerical method to analyze the mathematical model. In the development of a mathematical model, we often make a set of assumptions about the process (e.g., constitutive behavior, loads, and boundary conditions) to derive the mathematical relationships governing the system. The mathematical model is used to gain an understanding of how the corresponding process works. If the relationships are simple, it is possible to obtain *exact* information on quantities of interest. This is known as the *analytical* solution. However, most practical problems are too complicated to allow analytical solutions of the models. Hence, these mathematical models must be studied by numerical methods, such as the finite element method. We use a computer to evaluate a mathematical model *numerically* to *estimate* the quantities that characterize the system.

Finite element analysis is a numerical simulation of a physical process. Therefore, finite element modeling involves assumptions concerning the representation of the system and/or its behavior. Valid assumptions can be made only if one has a qualitative understanding of how the process or system works. A good knowledge of the basic principles governing the process and the finite element theory enable the development of a good numerical model of the actual process.

Here we discuss several aspects of the development of finite element models. Guidelines concerning element geometries, mesh refinements, and load representations are given.

9.3.2 Element Geometries

Recall from Section 9.2 that the numerical evaluation of integrals over actual elements involves a coordinate transformation from the actual element to a master element. The transformation is acceptable if and only if every point in the actual element is mapped uniquely to a point in the master element, and vice versa. Such mappings are termed one-to-one. This requirement can be expressed as [see (9.45)]

$$\mathcal{J}^e \equiv \det [\mathcal{J}^e] > 0 \quad \text{everywhere in the element } \Omega^e \tag{9.71}$$

where $[\mathcal{J}^e]$ is the Jacobian matrix in (9.44*b*). Geometrically, the Jacobian $\mathcal{J}^e$ represents the ratio of an area element in the real element to the corresponding area element in the master element:

$$dA \equiv dx\, dy = \mathcal{J}^e\, d\xi\, d\eta$$

If $\mathcal{J}^e$ is zero then a nonzero area element in the real element is mapped into zero area in the master element, which is unacceptable. Also, if $\mathcal{J}^e < 0$, a right-handed coordinate system is mapped into a left-handed coordinate system.

In general, the Jacobian is a function of ξ and η, implying that the real element is nonuniformly mapped into the master element, i.e., the element is distorted. Excessive distortion of elements is not good, because a nonzero area element can be mapped into a zero or nearly zero area.

To ensure $\mathcal{J}^e > 0$ and keep within the extreme limits of acceptable distortion, certain geometric shapes of elements must be avoided. For example, the interior angle at each vertex of a triangular element should not be equal to either 0° or 180°. Indeed, in practice the angle should reasonably be larger than 0° and less than 180° to avoid numerical ill-conditioning of element matrices. Although the acceptable range depends on the problem, the

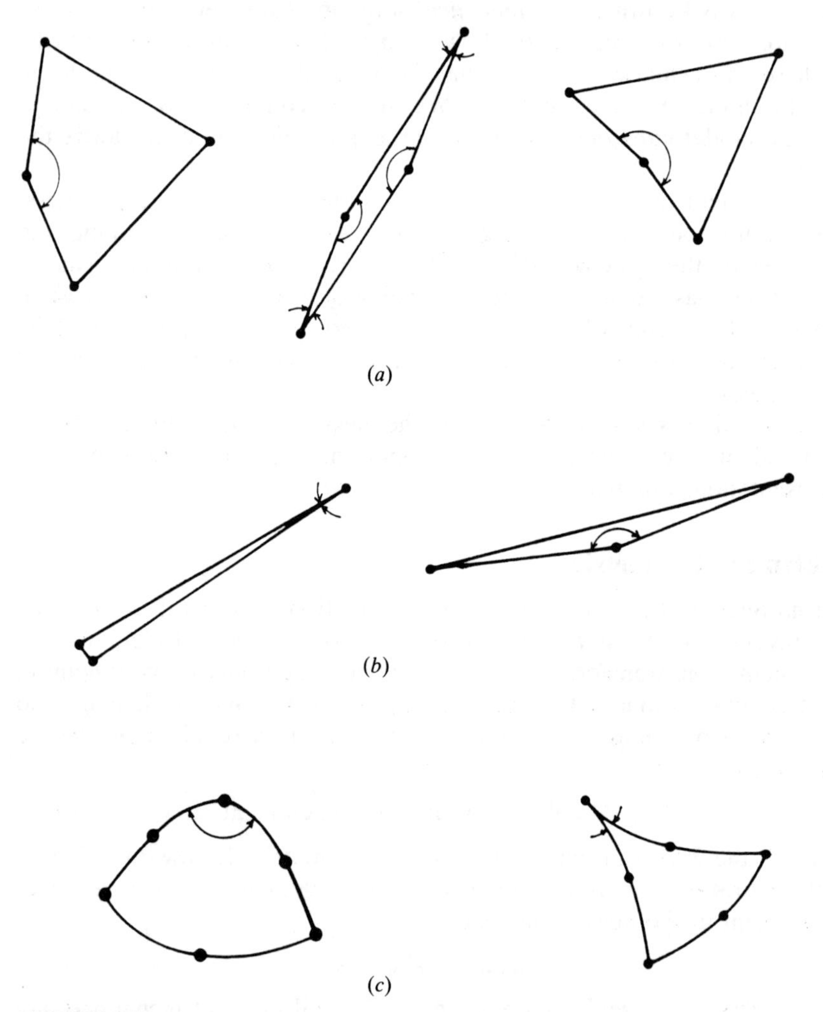

FIGURE 9.18
Finite elements with unacceptable vertex angles: (*a*) linear quadrilateral elements; (*b*) linear triangular elements; (*c*) quadratic quadrilateral and triangular elements. The angles marked are either too small compared with 0° or too large compared with 180°.

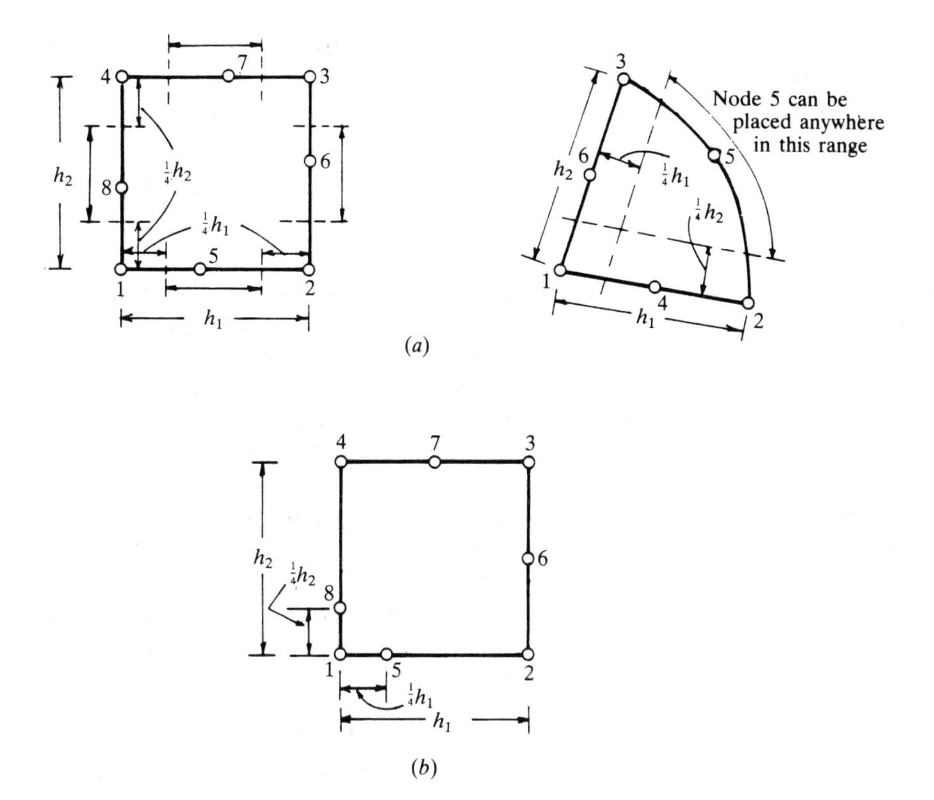

FIGURE 9.19
Range of acceptable locations of the midside nodes for quadratic elements: (*a*) eight-node quadratic element and six-node quadratic triangular element; (*b*) the quarter-point quadrilateral element.

range 15°–165° can be used as a guide. Figure 9.18 shows elements with unacceptable vertex angles for straight- and curved-sided elements.

For higher-order Lagrange elements (also called C^0 elements), the location of the interior nodes contributes to the element distortion, and therefore they are constrained to lie within a certain distance from the vertex nodes. For example, in the case of a quadratic element, the midside node should be at a distance not less than one-quarter of the length of the side from the vertex nodes (see Fig. 9.19). When the midside node is located exactly at a distance of one-quarter of the side length from a vertex, the element exhibits special properties (see Problem 9.20). Such elements, called *quarter-point elements,* are used in fracture mechanics problems to represent an inverse square-root singularity in the gradient of the solution at the nearest vertex node.

9.3.3 Mesh Generation

Generation of a finite element mesh for a given problem should follow the guidelines listed below:

1. The mesh should represent the geometry of the computational domain and load accurately.
2. It should be such that large gradients in the solution are adequately represented.
3. It should not contain elements with unacceptable geometries, especially in regions of large gradients.

Within the above guidelines, the mesh used can be *coarse* (i.e., have few elements) or *refined* (i.e., have many elements), and may consist of one or more orders and types of elements (e.g., linear and quadratic, triangular and quadrilateral). A judicious choice of element order and type could save computational cost while giving accurate results. It should be noted that the choice of elements and mesh is problem-dependent. What works well for one problem may not for another. An analyst with physical insight into the process being simulated can make a better choice of elements and mesh for the problem at hand. One should start with a coarse mesh that meets the three requirements listed above, exploit symmetries available in the problem, and evaluate the results thus obtained in the light of physical understanding and approximate analytical and/or experimental information. These results can be used to guide subsequent mesh refinements and analyses.

Generation of meshes of a single element type is easy because elements of the same degree are compatible with each other (see Fig. 9.20). Mesh refinements involve several options. The mesh can be refined by subdividing existing elements into two or more elements of the same type (see Fig. 9.21*a*). This is called *h-version mesh refinement.* Alternatively, existing elements can be replaced by elements of higher order (see Fig. 9.21*b*). This is called *p-version mesh refinement.* There is also *h, p-version mesh refinement,* in which elements are subdivided into two or more elements in some places and replaced with higher-order elements in other places. Generally, local mesh refinements should be such that elements of very small size are not placed adjacent to those of very large size (see Fig. 9.22).

Combining elements of different *kinds* naturally arises in solid and structural mechanics problems. For example, plate bending elements (2-D) can be connected to a beam element (1-D). If the plate element is based on classical plate theory (see Chapter 12), the beam element should be based on the Euler–Bernoulli beam theory so that they have the same degrees of freedom at the connecting node. When a plane elasticity element (see Chapter 10) is connected to a beam element, which is not compatible with the former in terms of the degrees of freedom at the nodes, one must construct a special element that makes the transition from the 2-D plane elasticity element to the 1-D beam element (see Problem 10.8). Such an element is called a *transition element*.

Combining elements of different order, say a linear to a quadratic element, may be necessary to accomplish local mesh refinements. There are two ways to do this. One way is to use a transition element that has a different number of nodes on different sides (see Fig. 9.23*a*). The other way is to

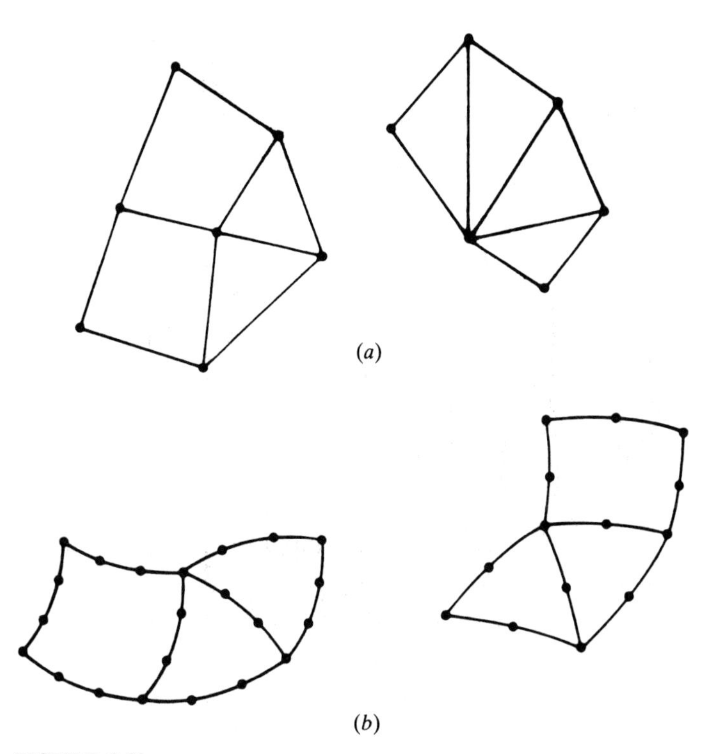

FIGURE 9.20
Connecting elements of the same order. The C^0 elements of the same order ensure the C^0 continuity along the element interfaces: (*a*) linear elements; (*b*) quadratic elements.

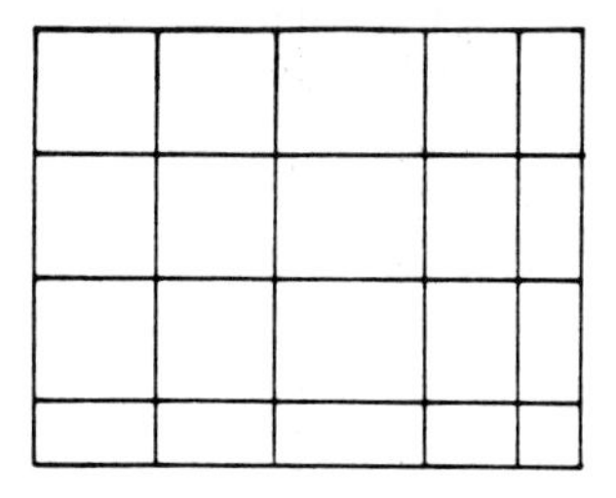

(*a*)

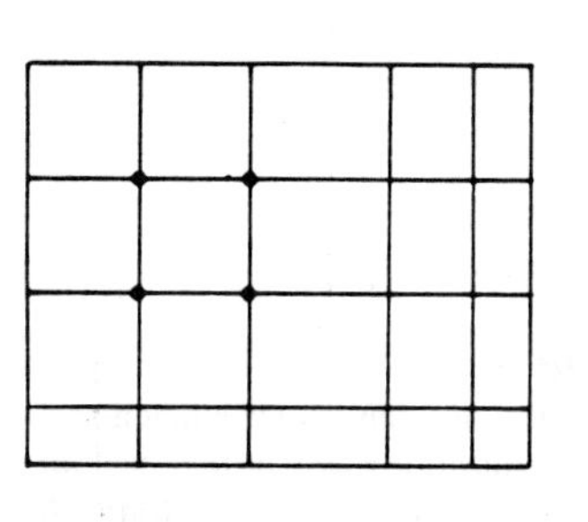

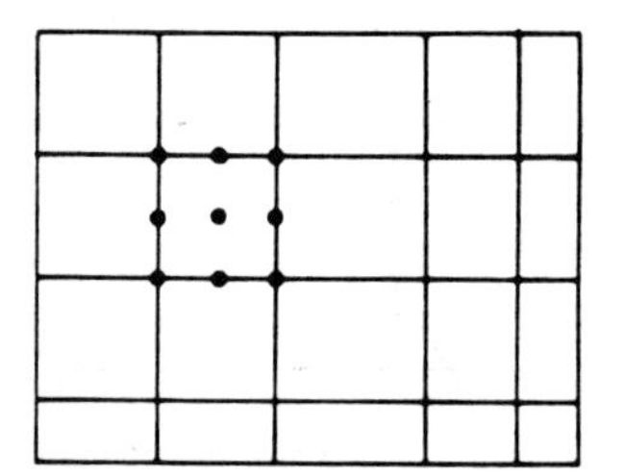

(*b*)

FIGURE 9.21
The h-version (*a*) and p-version (*b*) mesh refinements.

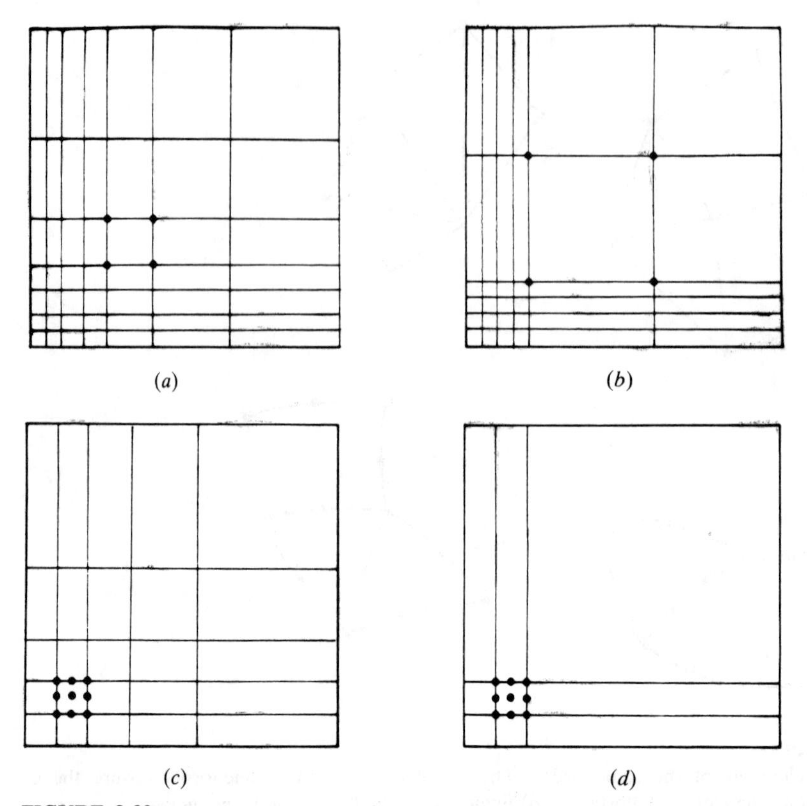

FIGURE 9.22
Finite element mesh refinements. Meshes shown in (*a*) and (*c*) are acceptable, and those shown in (*b*) and (*d*) are unacceptable. The typical element of the mesh is shown in each case.

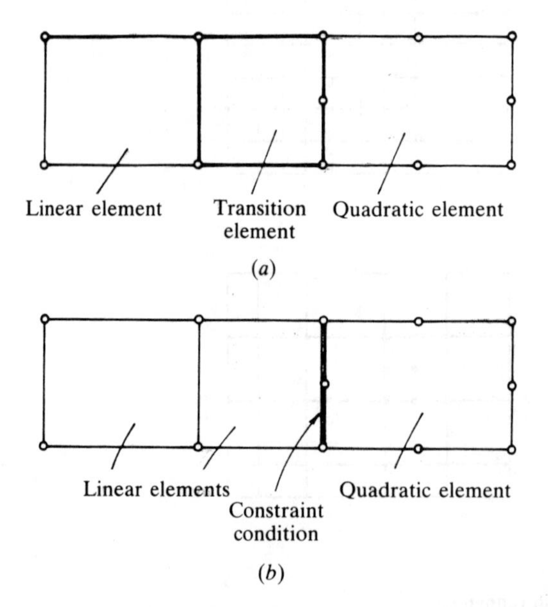

FIGURE 9.23
Combining different order elements: (*a*) use of a transition element that has three sides linear and one side quadratic; (*b*) use of a linear constraint equation to connect a linear side to a quadratic side.

impose a condition that constrains the midside node to have the same value as that experienced at the node by the lower-order element (see Fig. 9.23*b*). However, such combinations do not enforce interelement continuity of the solution. Figure 9.24 shows element connections that do not satisfy C^0 continuity along the connecting sides. Use of transition elements and constraint conditions in local mesh refinements is a common practice. Figure 9.25 shows a few examples of such refinements.

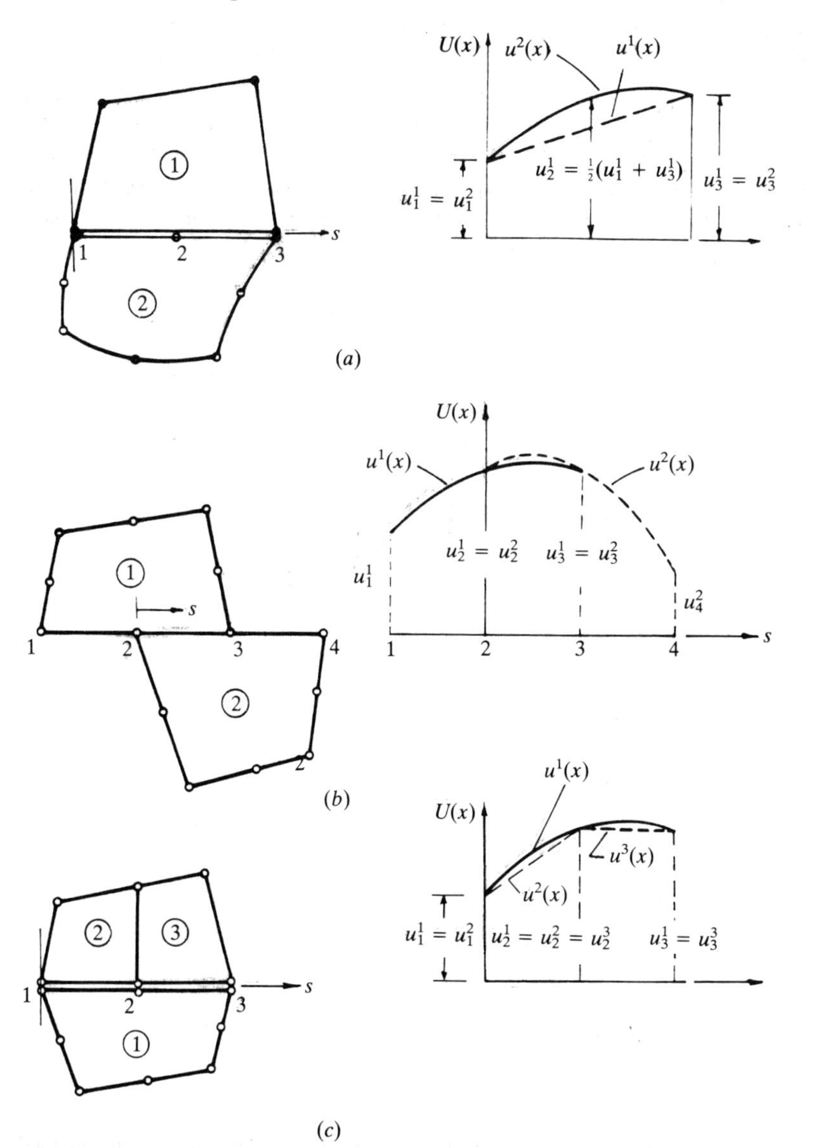

FIGURE 9.24
Various types of incompatible connections of finite elements. In all cases the interelement continuity of the function is violated along the connecting side.

9.3.4 Load Representation

Computation of the nodal contributions of a distributed boundary source was discussed in Chapter 8 [see (8.45)]. The accuracy of the result depends on the element and mesh used to represent the domain. For example, in heat transfer problems, the boundary source is the heat flux across (i.e., normal to) the boundary. Use of linear elements, for example, to represent the boundary will change the actual distribution (see Fig. 9.26). Of course, h-version or p-version mesh refinements will improve the representation of the boundary flux.

Another situation where boundary forces are subject to different approximations is the force due to contact between two bodies. For example, a solid plate in contact with a circular disk generates a reactive force that can be represented either as a point load or a locally distributed force. Representation of the contact force between deformable bodies as a point load is an approximation of the true distribution. A sine distribution might be a more realistic representation of the actual force (see Fig. 9.27).

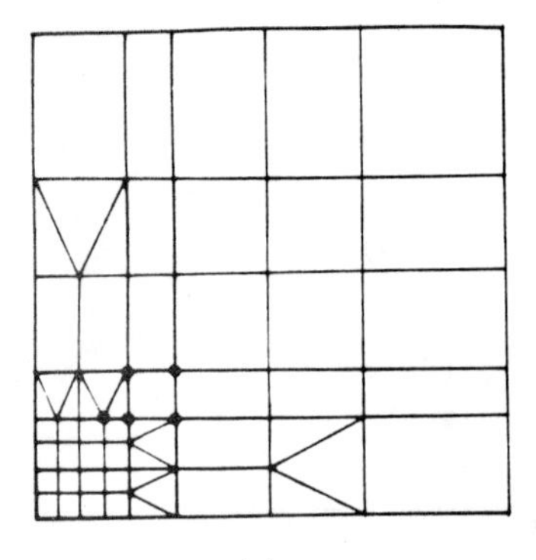

(*a*)

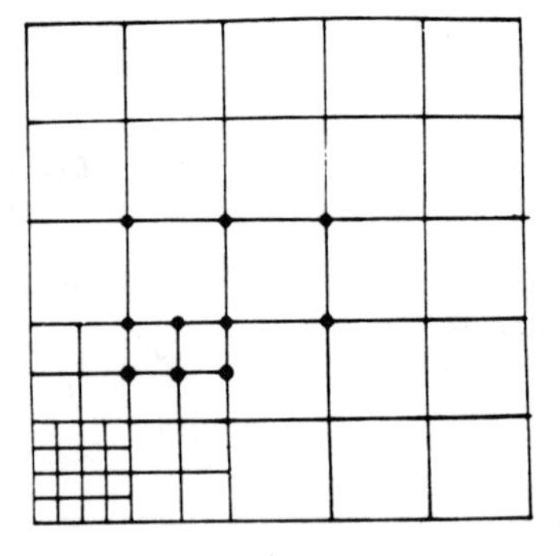

(*b*)

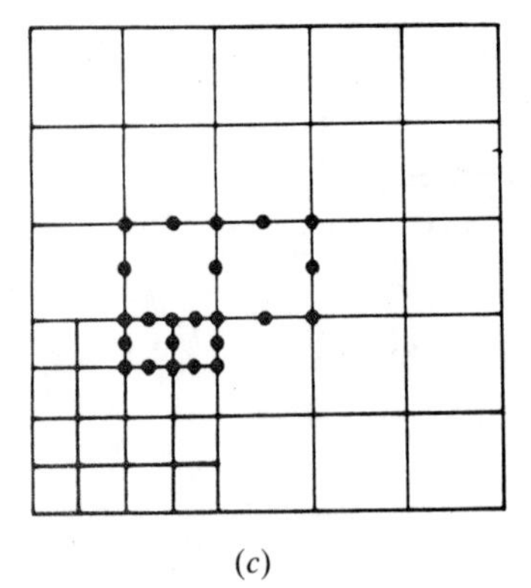

(*c*)

FIGURE 9.25
Some examples of local mesh refinements: (*a*) with compatible (C^0-continuous) elements; (*b*) with transition elements (or constraint conditions are imposed) between linear elements; (*c*) with transition elements between quadratic elements. In (*b*) and (*c*), the transition elements can be between linear and quadratic, and quadratic and cubic elements, respectively.

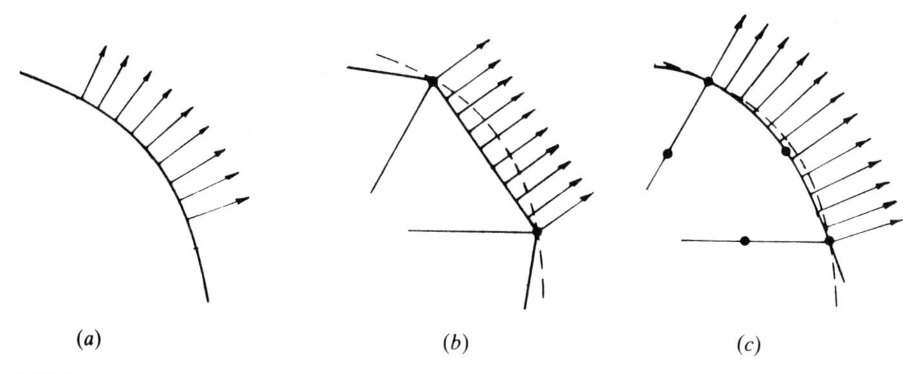

FIGURE 9.26
Approximation of the boundary fluxes in the finite element method: (*a*) actual geometry of the domain and distribution of flux; (*b*) approximation of the domain by linear finite elements and associated representation of the boundary flux; (*c*) approximation of the domain by quadratic finite elements and associated representation of the boundary flux.

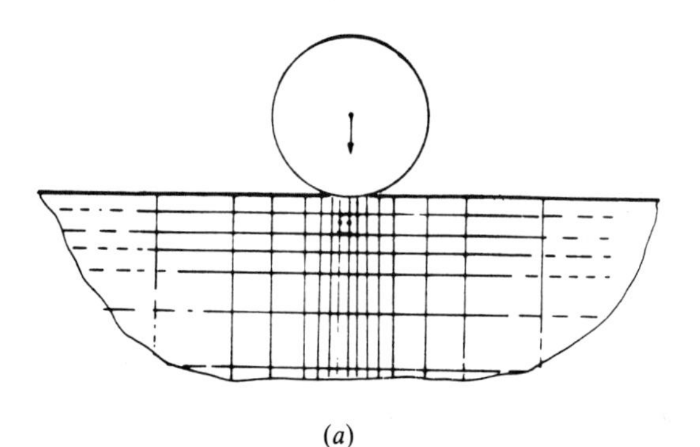

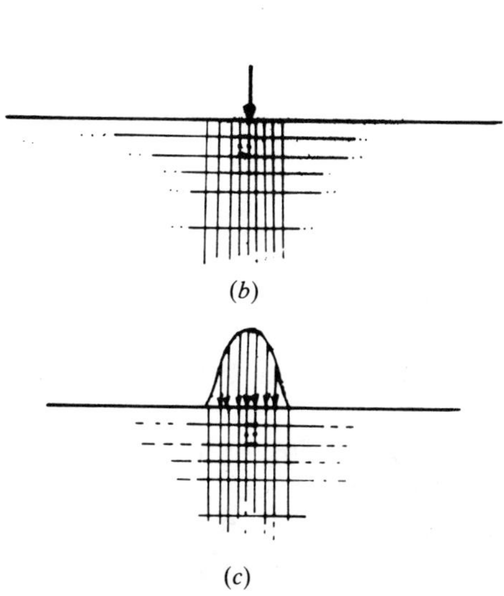

FIGURE 9.27
Representation of contact pressure developed between two bodies: (*a*) geometry of the bodies in contact; (*b*) representation of the contact pressure as a point load; (*c*) representation of the contact pressure as a distributed surface load. In the latter case, often the surface area of the distributed force is unknown.

9.4 SUMMARY

In this chapter three major topics have been discussed: (i) Lagrange interpolation functions for triangular and rectangular elements; (ii) numerical integration to evaluate integral expressions over triangular and rectangular elements; (iii) some modeling guidelines. Interpolation functions for linear, quadratic, and cubic triangular elements have been developed using the area coordinates. Linear, quadratic, and cubic interpolation functions for the Lagrange and serendipity families of rectangular elements have also been developed. A systematic description of the numerical evaluation of integral expressions involving interpolation functions and their derivatives with respect to global coordinates has been presented. This development is suitable for computer implementation, as will be seen in Chapter 13. Modeling is an art that can be improved by experience and understanding of the physical interactions involved in the process. It is necessary to critically evaluate the computed results before using them. The guidelines given are to encourage good modeling practice, and they should be followed to determine the actual "working" model.

PROBLEMS

9.1. Show that the bilinear interpolation functions for the four-node triangular element in Fig. P9.1 are of the form

$$\psi_i = a_i + b_i\xi + c_i\eta + d_i\xi\eta \quad (i = 1, \dots, 4)$$

where

$$a_1 = 1, \qquad a_2 = a_3 = a_4 = 0, \qquad -b_1 = b_2 = 1/a, \qquad b_3 = b_4 = 0$$

$$c_1 = \frac{6ab - a^2 - 2b^2}{ac(a-2b)}, \qquad c_2 = \frac{2b(a+b)}{ac(a-2b)}, \qquad c_3 = \frac{a+b}{c(a-2b)}, \qquad c_4 = \frac{-9b}{c(a-2b)}$$

$$d_1 = d_2 = d_3 = -\tfrac{1}{3}d_4 = -\frac{3}{c(a-2b)}$$

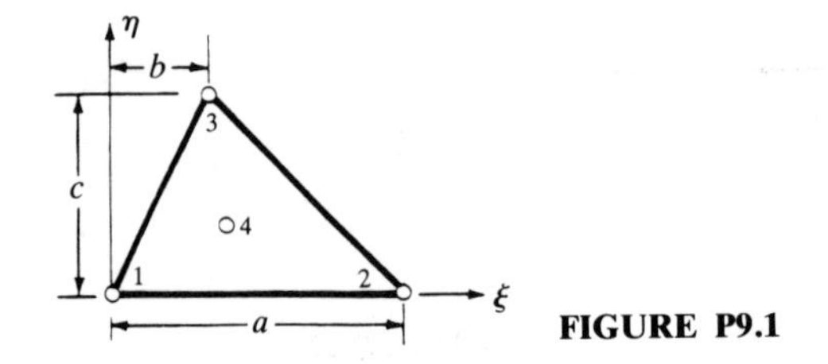

FIGURE P9.1

9.2. Show that the interpolation functions involving the term $\xi^2 + \eta^2$ for the five-node rectangular element shown in Fig. P9.2 are

$$\begin{aligned}
\psi_1 &= 0.25(-\xi - \eta + \xi\eta) + 0.125(\xi^2 + \eta^2) \\
\psi_2 &= 0.25(\xi - \eta - \xi\eta) + 0.125(\xi^2 + \eta^2) \\
\psi_3 &= 0.25(\xi + \eta + \xi\eta) + 0.125(\xi^2 + \eta^2) \\
\psi_4 &= 0.25(-\xi + \eta - \xi\eta) + 0.125(\xi^2 + \eta^2) \\
\psi_5 &= 1 - 0.5(\xi^2 + \eta^2)
\end{aligned}$$

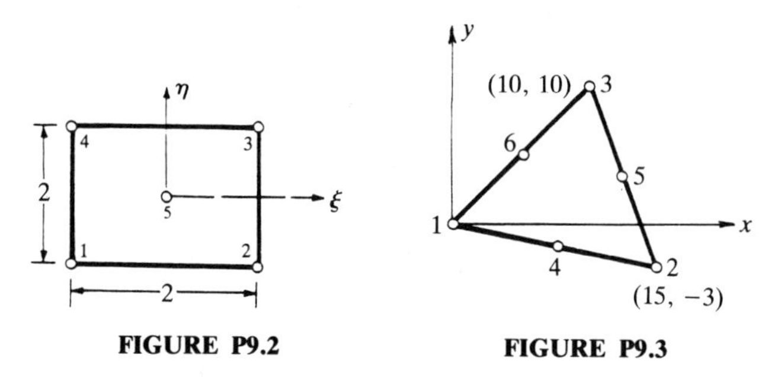

FIGURE P9.2

FIGURE P9.3

9.3. Calculate the interpolation functions for the quadratic triangular element shown in Fig. P9.3.

9.4. Determine the interpolation function ψ_{14} for the quartic triangular element shown in Fig. P9.4.

Answer: $32L_1L_2L_3(4L_2 - 1)$.

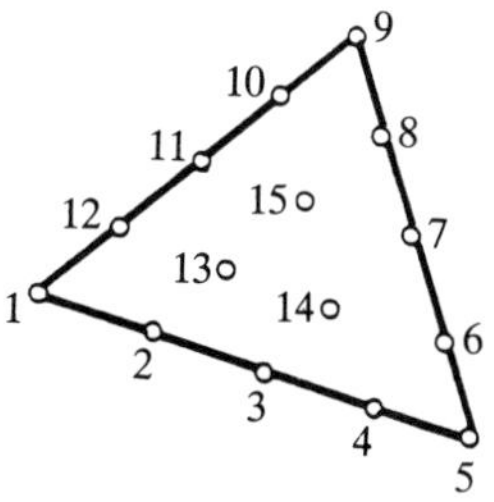

FIGURE P9.4

9.5. Derive the interpolation function of a corner node in a cubic serendipity element.

9.6. Consider the five-node element shown in Fig. P9.6. Using the basic linear and quadratic interpolations along the coordinate directions ξ and η, derive the interpolation functions for the element. Note that the element can be used as a transition element connecting four-node elements to eight- or nine-node elements.

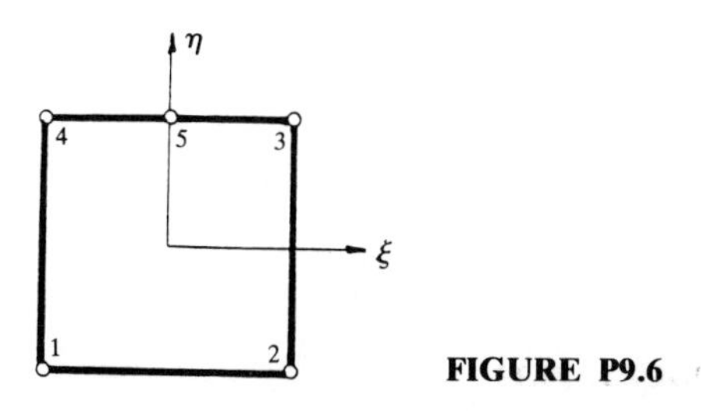

FIGURE P9.6

9.7. (*Nodeless variables*) Consider the four-node rectangular element with interpolation of the form

$$u = \sum_{i=1}^{4} u_i \psi_i + \sum_{i=1}^{4} c_i \phi_i$$

where u_i are the nodal values and c_i are arbitrary constants. Determine the form of ψ_i and ϕ_i for the element.

9.8–9.10. Determine the Jacobian matrix and the transformation equations for the elements shown in Figs. P9.8–P9.10.

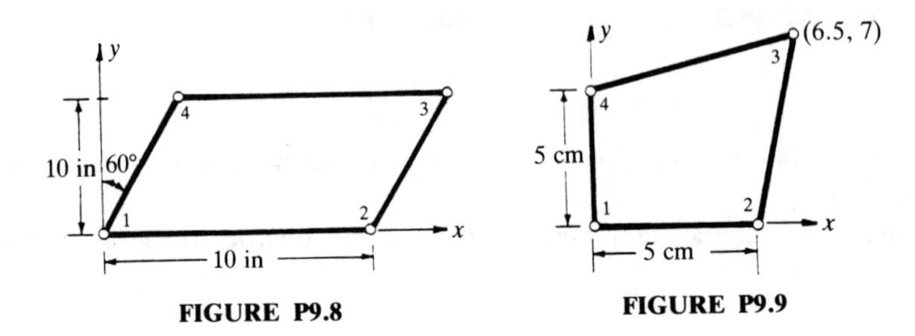

FIGURE P9.8

FIGURE P9.9

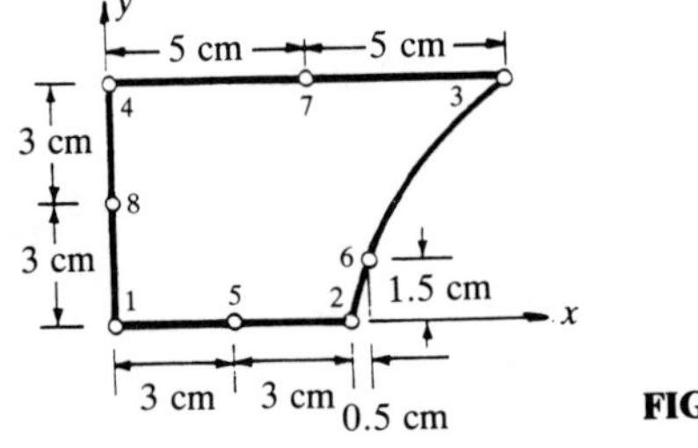

FIGURE P9.10

9.11. Using Gauss quadrature, determine the contribution of a constant distributed source to nodal points of the four-node finite element shown in Fig. P9.9.

Answer: $f_0(7.7083, 8.5417, 9.1667, 8.3333)$.

9.12. Show that the side nodes in the eight-node rectangular element of Fig. 9.13 should be located such that $0.25 < a < 0.75$.

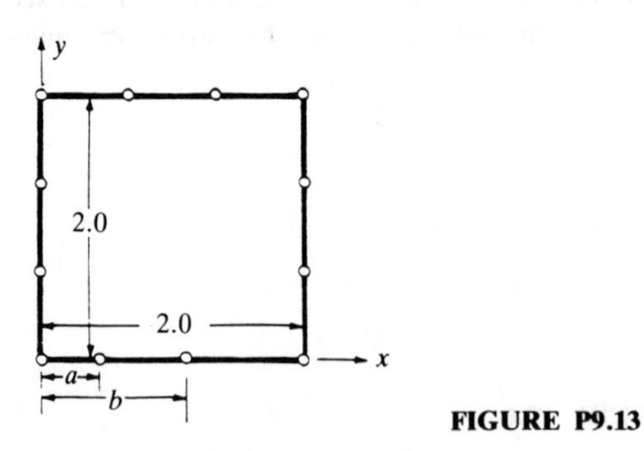

FIGURE P9.13

9.13. For a 12-node serendipity (cubic) element, as illustrated in Fig. P9.13, show that the determinant $\mathscr{J}$ is

$$\mathscr{J} = \tfrac{9}{16}(-3 - 2\xi + 9\xi 62)a + \tfrac{9}{2}(3 - 2\xi - 9\xi^2)b + \tfrac{1}{8}(-1 + 18\xi + 27\xi^2)$$

What can you conclude from the requirement $\mathscr{J} > 0$?

9.14. Determine the conditions on the location of node 3 of the quadrilateral element shown in Fig. P9.14. Show that the transformation equations are

$$x = \xi + \xi\eta(a - 2), \qquad y = \eta + \xi\eta(b - 2)$$

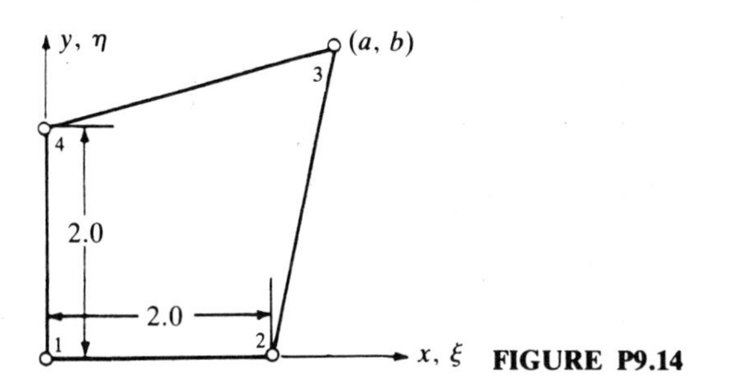

FIGURE P9.14

9.15. Determine the global derivatives of the interpolation functions for node 3 of the element shown in Fig. P9.9.

9.16. Let the transformation between the global coordinates (x, y) and local normalized coordinates (ξ, η) in a Lagrange element Ω^e be

$$x = \sum_{i=1}^{m} x_i \hat{\psi}_i(\xi, \eta), \qquad y = \sum_{i=1}^{m} y_i \hat{\psi}_i(\xi, \eta)$$

where (x_i^e, y_i^e) denote the global coordinates of the element nodes. The differential lengths in the two coordinate systems are related by

$$dx_e = \frac{\partial x_e}{\partial \xi} d\xi + \frac{\partial x_e}{\partial \eta} \partial\eta, \qquad dy_e = \frac{\partial y_e}{\partial \xi} \xi + \frac{\partial y_e}{\partial \eta} d\eta$$

or

$$\begin{Bmatrix} dx_e \\ dy_e \end{Bmatrix} = \begin{bmatrix} \dfrac{\partial x_e}{\partial \xi} & \dfrac{\partial x_e}{\partial \eta} \\ \dfrac{\partial y_e}{\partial \xi} & \dfrac{\partial y_e}{\partial \eta} \end{bmatrix} \begin{Bmatrix} d\xi \\ d\eta \end{Bmatrix} = [\mathscr{T}] \begin{Bmatrix} d\xi \\ d\eta \end{Bmatrix}$$

In the finite element literature the transpose of $[\mathscr{T}]$ is called the Jacobian matrix, $[\mathscr{J}]$. Show that the derivatives of the interpolation function $\psi_i^e(\xi, \eta)$ with respect to the global coordinates (x, y) are related to their derivatives with respect to the local coordinates (ξ, η) by

$$\begin{Bmatrix} \dfrac{\partial \psi_i^e}{\partial x} \\ \dfrac{\partial \psi_i^e}{\partial y} \end{Bmatrix} = [\mathscr{J}]^{-1} \begin{Bmatrix} \dfrac{\partial \psi_i^e}{\partial \xi} \\ \dfrac{\partial \psi_i^e}{\partial \eta} \end{Bmatrix}$$

and

$$\begin{Bmatrix} \dfrac{\partial^2 \psi_i^e}{\partial x^2} \\ \dfrac{\partial^2 \psi_i^e}{\partial y^2} \\ \dfrac{\partial^2 \psi_i^e}{\partial x\, \partial y} \end{Bmatrix} = \begin{bmatrix} \left(\dfrac{\partial x_e}{\partial \xi}\right)^2 & \left(\dfrac{\partial y_e}{\partial \xi}\right)^2 & 2\dfrac{\partial x_e}{\partial \xi}\dfrac{\partial y_e}{\partial \xi} \\ \left(\dfrac{\partial x_e}{\partial \eta}\right)^2 & \left(\dfrac{\partial y_e}{\partial \eta}\right)^2 & 2\dfrac{\partial x_e}{\partial \eta}\dfrac{\partial y_e}{\partial \eta} \\ \dfrac{\partial x_e}{\partial \xi}\dfrac{\partial x_e}{\partial \eta} & \dfrac{\partial y_e}{\partial \xi}\dfrac{\partial y_e}{\partial \eta} & \dfrac{\partial x_e}{\partial \eta}\dfrac{\partial y_e}{\partial \xi} + \dfrac{\partial x_e}{\partial \xi}\dfrac{\partial y_e}{\partial \eta} \end{bmatrix}^{-1}$$

$$\times \left(\begin{Bmatrix} \dfrac{\partial^2 \psi_i^e}{\partial \xi^2} \\ \dfrac{\partial^2 \psi_i^e}{\partial \eta^2} \\ \dfrac{\partial^2 \psi_i^e}{\partial \xi\, \partial \eta} \end{Bmatrix} - \begin{bmatrix} \dfrac{\partial^2 x_e}{\partial \xi^2} & \dfrac{\partial^2 y_e}{\partial \xi^2} \\ \dfrac{\partial^2 x_e}{\partial \eta^2} & \dfrac{\partial^2 y_e}{\partial \eta^2} \\ \dfrac{\partial^2 x_e}{\partial \xi\, \partial \eta} & \dfrac{\partial^2 y_e}{\partial \xi\, \partial \eta} \end{bmatrix} \begin{Bmatrix} \dfrac{\partial \psi_i^e}{\partial x} \\ \dfrac{\partial \psi_i^e}{\partial y} \end{Bmatrix} \right)$$

9.17. (*Continuation of Problem 9.16*) Show that the Jacobian can be computed from the equation

$$[\mathcal{J}] = \begin{bmatrix} \dfrac{\partial \psi_1^e}{\partial \xi} & \dfrac{\partial \psi_2^e}{\partial \xi} & \cdots & \dfrac{\partial \psi_n^e}{\partial \xi} \\ \dfrac{\partial \psi_1^e}{\partial \eta} & \dfrac{\partial \psi_2^e}{\partial \eta} & \cdots & \dfrac{\partial \psi_n^e}{\partial \eta} \end{bmatrix} \begin{bmatrix} x_1^e & y_1^e \\ x_2^e & y_2^e \\ \vdots & \vdots \\ x_n^e & y_n^e \end{bmatrix}$$

9.18. Consider the quadrilateral element shown in Fig. P9.18. Using the linear interpolation functions of a rectangular element, transform the element to the local coordinate system and sketch the transformed element.

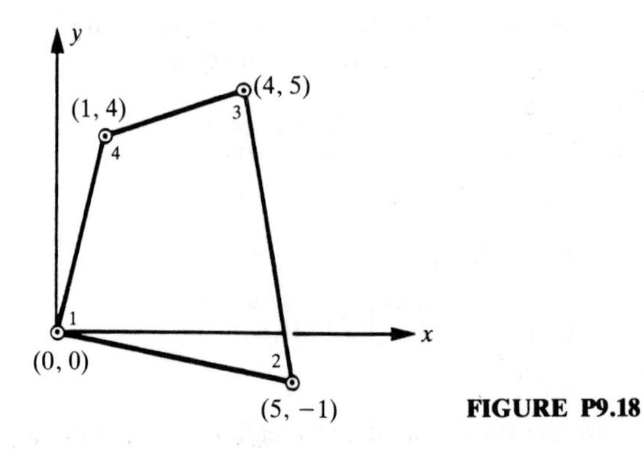

FIGURE P9.18

9.19. For the quadrilateral element Ω^e in Problem 9.18, express the following integral in the local coordinates:

$$K_{ij}^e = \int_{\Omega^e} \left[a\left(\frac{\partial \psi_i^e}{\partial x}\frac{\partial \psi_j^e}{\partial x} + \frac{\partial \psi_i^e}{\partial y}\frac{\partial \psi_j^e}{\partial y} \right) + b\psi_i^e \psi_j^e \right] dx\, dy$$

9.20. Find the Jacobian matrix for the nine-node rectangular element shown in Fig. P9.20. What is the determinant of the Jacobian matrix?

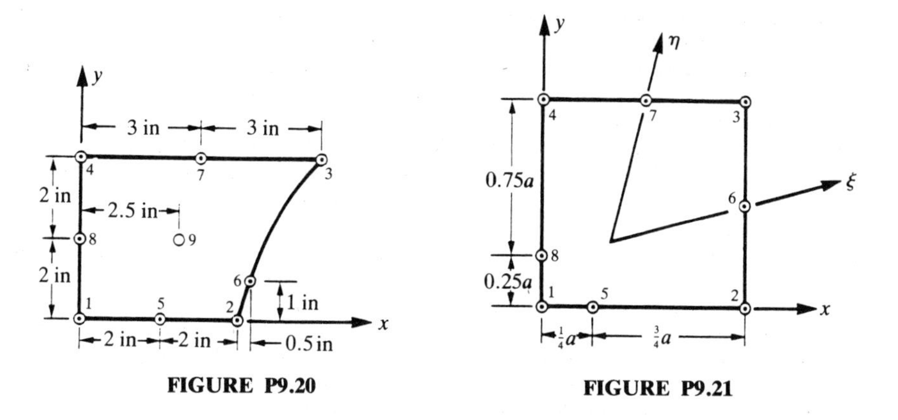

FIGURE P9.20 **FIGURE P9.21**

9.21. For the eight-node element shown in Fig. P9.21, show that the x coordinate along side 1–2 is related to the ξ coordinate by

$$x = -\tfrac{1}{2}\xi(1-\xi)x_1^e + \tfrac{1}{2}\xi(1+\xi)x_2^e + (1-\xi^2)x_5^e$$

and that the relations

$$\xi = 2\left(\frac{x}{a}\right)^{1/2} - 1, \qquad \frac{\partial x}{\partial \xi} = (xa)^{1/2}$$

hold. Also, show that

$$u_e(x, 0) = -\left[2\left(\frac{x}{a}\right)^{1/2} - 1\right]\left[1 - \left(\frac{x}{a}\right)^{1/2}\right]u_1^e + \left[-1 + 2\left(\frac{x}{a}\right)^{1/2}\right]\left(\frac{x}{a}\right)^{1/2} u_2^e + 4\left[\left(\frac{x}{a}\right)^{1/2} - \frac{x}{a}\right]u_5^e$$

$$\frac{\partial u_e(x, 0)}{\partial x} = \frac{1}{(xa)^{1/2}}\left\{\frac{1}{2}\left[3 - 4\left(\frac{x}{a}\right)^{1/2}\right]u_1^e + \frac{1}{2}\left[-1 + 4\left(\frac{x}{a}\right)^{1/2}\right]u_2^e + 2\left[1 - 2\left(\frac{x}{a}\right)^{1/2}\right]u_5^e\right\}$$

Thus, $\partial u_e/\partial x$ grows at a rate of $(xa)^{-1/2}$ as x approaches zero along side 1–2. In other words, we have a $x^{-1/2}$ singularity at node 1. Such elements are used in fracture mechanics problems.

9.22. Using the tensor product of the one-dimensional Hermite cubic interpolation functions, obtain the Hermite cubic interpolation functions (16 of them) for the four-node rectangular element.

REFERENCES FOR ADDITIONAL READING

Burnett, D. S.: *Finite Element Analysis from Concepts to Applications,* Addison-Wesley, Reading, MA, 1987.

Carnahan, B., H. A. Luther, and J. O. Wilkes: *Applied Numerical Methods,* John Wiley, New York, 1969.

Cowper, G. R.: "Gaussian Quadrature Formulas for Triangles," *International Journal for Numerical Methods in Engineering,* vol. 7, pp. 405–408, 1973.

Fellippa, C. A.: "Refined Finite Element Analysis of Linear and Non-linear Two-Dimensional Structures," Structures Materials Research Report no. 66-22, University of California, Berkeley, October 1966.

Froberg, C. E.: *Introduction to Numerical Analysis,* Addison-Wesley, Reading, MA, 1969.

Hammer, P. C., O. P. Marlowe, and A. H. Stroud: "Numerical Integration over Simplexes and Cones," *Mathematics Tables and Aids to Computation, National Research Council (Washington),* vol. 10, pp. 130–137 , 1956.

Irons, B. M.: "Quadrature Rules for Brick-Based Finite Elements," *International Journal for Numerical Methods in Engineering,* vol. 3, pp. 293–294, 1971.

Loxan, A. N., N. Davids, and A. Levenson: "Table of the Zeros of the Legendre Polynomials of Order 1–16 and the Weight Coefficients for Gauss' Mechanical Quadrature Formula," *Bulletin of the American Mathematical Society,* vol. 48, pp. 739–743, 1942.

Reddy, C. T., and D. J. Shippy: "Alternative Integration Formulae for Triangular Finite Elements," *International Journal for Numerical Methods in Engineering,* vol. 17, pp. 133–139, 1981.

Segerlind, L. J.: *Applied Finite Element Analysis,* John Wiley, New York, 1977.

Silvester, P.: "Newton–Cotes Quadrature Formulae for *N*-Dimensional Simplexes," *Proceedings of 2d Canadian Congress of Applied Mechanics, Waterloo, Ontario, 1969*, University of Waterloo, Waterloo, Ontario, 1969.

Stroud, A. H., and D. Secrest: *Gaussian Quadrature Formulas,* Prentice-Hall, Englewood Cliffs, NJ, 1966.

Zienkiewicz, O. C., and R. L. Taylor: *The Finite Element Method,* vol. 1: *Linear Problems,* McGraw-Hill, New York, 1989.

CHAPTER
10

PLANE ELASTICITY

10.1 INTRODUCTION

In Section 8.4.3, we considered the finite element analysis of second-order problems of solid mechanics that are described by only one dependent unknown. These include the torsion of a cylindrical bar and the deflection of a membrane. The governing equation in each case is the Laplace or Poisson equation for the dependent variable (e.g., stress function or deflection). Here we consider plane elasticity problems described by a pair of *coupled* partial differential equations expressed in terms of two dependent variables representing the two components of the displacement vector. The word "coupled" is used to imply that the same dependent variables appear in more than one equation of the set, and therefore no equation can be solved independent of the other(s) in the set.

The primary objective of this chapter is twofold: first, to describe the weak form and associated finite element model of the plane elasticity equations; and second, to describe how the linear Lagrange elements developed in Section 8.2 can be used in the solution of plane elasticity problems in two dimensions. The treatment of both these subjects proceeds along the same lines as in the one-dimensional problems.

10.2 GOVERNING EQUATIONS

10.2.1 Assumptions of Plane Elasticity

Consider a linear elastic solid Ω of uniform thickness h bounded by two parallel planes (say, by planes $z = -\frac{1}{2}h$ and $z = \frac{1}{2}h$) and any closed boundary Γ. If the thickness h is very large compared with the size of Ω then the problem is considered to be a *plane strain problem,* and if the thickness is small compared with the size of Ω then the problem is considered to be a *plane stress problem.* Both of these problems are simplifications of three-dimensional elasticity problems under the following assumptions on loading: the body forces, if any exist, cannot vary in the thickness direction and cannot have components in the z direction; the applied boundary forces must be uniformly distributed across the thickness (i.e., constant in the z direction); and no loads can be applied on the parallel planes bounding the top and bottom surfaces.

The assumption that the forces are zero on the parallel planes implies that for plane stress problems the stresses in the z direction are negligibly small (see Fig. 10.1):

$$\sigma_{xz} = 0, \qquad \sigma_{yz} = 0, \qquad \sigma_z = 0 \tag{10.1}$$

For plane strain problems, where the body is very thick in the z direction, the assumption is that the strains in the z direction are zero (see Fig. 10.2):

$$\epsilon_{xz} = 0, \qquad \epsilon_{yz} = 0, \qquad \epsilon_z = 0 \tag{10.2}$$

An example illustrating the difference between plane stress and plane strain problems is provided by the bending of a beam of rectangular cross-section. If the beam is narrow then the problem is a plane stress problem, and if it is very wide then the problem is a plane strain problem. The reader should consult the references on advanced strength of materials and elasticity listed at the end of the chapter.

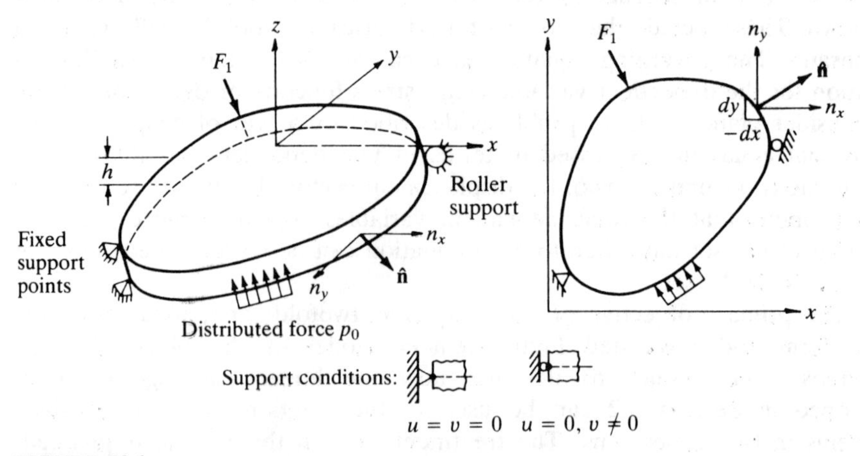

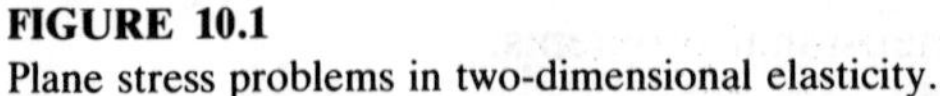

FIGURE 10.1
Plane stress problems in two-dimensional elasticity.

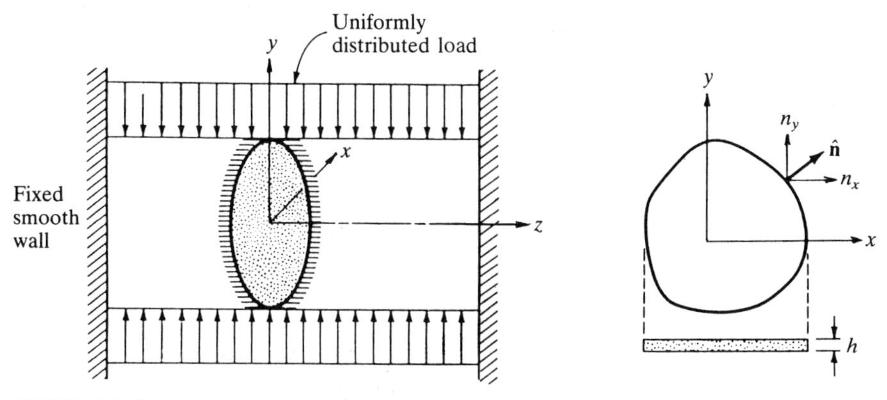

FIGURE 10.2
Plane strain problems in two-dimensional elasticity.

10.2.2 Basic Equations

The governing equations for the two types of plane elasticity problems discussed above are summarized below. The equations of the two classes of problems differ from each other only in the constitutive relations.

EQUATIONS OF MOTION

$$\begin{aligned}\frac{\partial \sigma_x}{\partial x} + \frac{\partial \sigma_{xy}}{\partial y} + f_x &= \rho \frac{\partial^2 u}{\partial t^2}\\ \frac{\partial \sigma_{xy}}{\partial x} + \frac{\partial \sigma_y}{\partial y} + f_y &= \rho \frac{\partial^2 v}{\partial t^2}\end{aligned} \tag{10.3}$$

where f_x and f_y denote the body forces per unit volume along the x and y directions, respectively, and ρ is the density of the material.

STRAIN–DISPLACEMENT RELATIONS

$$\epsilon_x = \frac{\partial u}{\partial x}, \qquad \epsilon_y = \frac{\partial v}{\partial y}, \qquad 2\epsilon_{xy} = \frac{\partial u}{\partial y} + \frac{\partial v}{\partial x} \tag{10.4}$$

STRESS–STRAIN (OR CONSTITUTIVE) RELATIONS

$$\begin{Bmatrix} \sigma_x \\ \sigma_y \\ \sigma_{xy} \end{Bmatrix} = \begin{bmatrix} c_{11} & c_{12} & 0 \\ c_{12} & c_{22} & 0 \\ 0 & 0 & c_{66} \end{bmatrix} \begin{Bmatrix} \epsilon_x \\ \epsilon_y \\ 2\epsilon_{xy} \end{Bmatrix} \tag{10.5}$$

where c_{ij} $(c_{ji} = c_{ij})$ are the elasticity (material) constants for an orthotropic medium with the material principal directions coinciding with the coordinate axes (x, y, z) used to describe the problem. The c_{ij} can be expressed in terms of the engineering constants (E_1, E_2, ν_{12}, and G_{12}) for an orthotropic material, for plane stress and plane strain given below ($\nu_{12}E_2 = \nu_{21}E_1$):

Plane stress

$$c_{11} = \frac{E_1}{1-\nu_{12}\nu_{21}}, \qquad c_{22} = \frac{E_2}{1-\nu_{12}\nu_{21}}$$
$$c_{12} = \nu_{21}c_{11} = \nu_{12}c_{22}, \qquad c_{66} = G_{12} \tag{10.6}$$

Plane strain ($\nu_{23} = \nu_{12}$)

$$c_{11} = \frac{E_1(1-\nu_{12})}{1-\nu_{12}-2\nu_{12}\nu_{21}}, \qquad c_{22} = \frac{E_2(1-\nu_{12}\nu_{21})}{(1+\nu_{12})(1-\nu_{12}-2\nu_{12}\nu_{21})}$$
$$c_{12} = \frac{\nu_{12}E_2}{1-\nu_{12}-2\nu_{12}\nu_{21}}, \qquad c_{66} = G_{12} \tag{10.7}$$

For an isotropic material,

$$E_1 = E_2 = E, \qquad \nu_{12} = \nu_{21} = \nu, \qquad G = \frac{E}{2(1+\nu)} \tag{10.8}$$

BOUNDARY CONDITIONS

Natural

$$\left.\begin{aligned} t_x &\equiv \sigma_x n_x + \sigma_{xy} n_y = \hat{t}_x \\ t_y &\equiv \sigma_{xy} n_x + \sigma_y n_y = \hat{t}_y \end{aligned}\right\} \quad \text{on } \Gamma_2 \tag{10.9}$$

Essential

$$u = \hat{u}, \qquad v = \hat{v} \quad \text{on } \Gamma_1 \tag{10.10}$$

where (n_x, n_y) are the components (or direction cosines) of the unit normal vector $\hat{\mathbf{n}}$ on the boundary Γ, Γ_1 and Γ_2 are (disjoint) portions of the boundary (Γ_1 and Γ_2 do not overlap except for a small number of discrete points—singular points), $\hat{t}_x$ and $\hat{t}_y$ are specified boundary stresses (or tractions), and $\hat{u}$ and $\hat{v}$ are specified displacements. Only one element of each pair, (u, t_x) and (v, t_y), may be specified at a boundary point.

Equations (10.3) can be expressed in terms of just the displacements u and v by substituting (10.4) into (10.5), and the result into (10.3):

$$\boxed{\begin{aligned} -\frac{\partial}{\partial x}\left(c_{11}\frac{\partial u}{\partial x} + c_{12}\frac{\partial v}{\partial y}\right) - \frac{\partial}{\partial y}\left[c_{66}\left(\frac{\partial u}{\partial y} + \frac{\partial v}{\partial x}\right)\right] &= f_x - \rho\frac{\partial^2 u}{\partial t^2} \\ -\frac{\partial}{\partial x}\left[c_{66}\left(\frac{\partial u}{\partial y} + \frac{\partial v}{\partial x}\right)\right] - \frac{\partial}{\partial y}\left(c_{12}\frac{\partial u}{\partial x} + c_{22}\frac{\partial v}{\partial y}\right) &= f_y - \rho\frac{\partial^2 v}{\partial t^2} \end{aligned}} \tag{10.11}$$

the boundary stress components (or tractions) can also be expressed in terms of the displacements:

$$
\begin{aligned}
t_x &= \left(c_{11}\frac{\partial u}{\partial x} + c_{12}\frac{\partial v}{\partial y}\right)n_x + c_{66}\left(\frac{\partial u}{\partial y} + \frac{\partial v}{\partial x}\right)n_y \\
t_y &= c_{66}\left(\frac{\partial u}{\partial y} + \frac{\partial v}{\partial x}\right)n_x + \left(c_{12}\frac{\partial u}{\partial x} + c_{22}\frac{\partial v}{\partial y}\right)n_y
\end{aligned}
\tag{10.12}
$$

This completes the review of the governing equations of a plane elastic body undergoing small deformations (i.e., with strains small compared with unity) in the (x, y) plane. The material of the body will be assumed in the present study to be linearly orthotropic.

10.3 WEAK FORMULATIONS

10.3.1 Preliminary Comments

Here we study two different ways of constructing the weak form and associated finite element model of the plane elasticity equations (10.11) and (10.12). The first uses the principle of virtual displacements (or the total potential energy principle), expressed in terms of matrices relating displacements to strains, and strains to stresses. This approach is used in most finite element texts on solid mechanics. The second follows a procedure consistent with the previous sections and employs the weak formulation of (10.11) and (10.12) to construct the finite element model. Of course, both methods give, mathematically, the *same* finite element model, but differ in their algebraic forms.

10.3.2 Principle of Virtual Displacements in Matrix Form

First, we rewrite (10.3)–(10.5) in matrix form:

$$
\begin{bmatrix} \partial/\partial x & 0 & \partial/\partial y \\ 0 & \partial/\partial y & \partial/\partial x \end{bmatrix} \begin{Bmatrix} \sigma_x \\ \sigma_y \\ \sigma_{xy} \end{Bmatrix} + \begin{Bmatrix} f_x \\ f_y \end{Bmatrix} = \rho \begin{Bmatrix} \ddot{u} \\ \ddot{v} \end{Bmatrix} \tag{10.13a}
$$

or

$$
[T^*]\{\sigma\} + \{f\} = \rho \begin{Bmatrix} \ddot{u} \\ \ddot{v} \end{Bmatrix} \tag{10.13b}
$$

and

$$
\begin{Bmatrix} \epsilon_x \\ \epsilon_y \\ 2\epsilon_{xy} \end{Bmatrix} = \begin{bmatrix} \partial/\partial x & 0 \\ 0 & \partial/\partial y \\ \partial/\partial y & \partial/\partial x \end{bmatrix} \begin{Bmatrix} u \\ v \end{Bmatrix} \tag{10.14a}
$$

or

$$
\{\epsilon\} = [T]\begin{Bmatrix} u \\ v \end{Bmatrix} \tag{10.14b}
$$

and

$$\{\sigma\} = [C]\{\epsilon\} \tag{10.15}$$

where $[T]$ and $[T^*]$ are matrices of differential operators. Note that $[T]$ is the transpose of $[T^*]$:

$$[T] = [T^*]^T = \begin{bmatrix} \partial/\partial x & 0 \\ 0 & \partial/\partial y \\ \partial/\partial y & \partial/\partial x \end{bmatrix} \tag{10.16}$$

Next, we use the (dynamic version of the) principle of virtual displacements [see Reddy (1984)] applied to a plane elastic body:

$$0 = \int_{V^e} (\sigma_{ij}\,\delta\epsilon_{ij} + \rho\ddot{u}_i\,\delta u_i)\,dV - \oint_{V_e} f_i\,\delta u_i\,dV - \oint_{S^e} \hat{t}_i\,\delta u_i\,ds \tag{10.17}$$

where V^e denotes the volume of element e $[V^e = \Omega^e \times (-\frac{1}{2}h, \frac{1}{2}h)]$, S^e is the boundary of Ω^e, h_e is the thickness of the element, δ denotes the variational operator, σ_{ij} and ϵ_{ij} are the components of stress and strain tensors, respectively, and f_i and $\hat{t}_i$ are the components of the body force and boundary stress vectors, respectively:

$$\sigma_{11} = \sigma_x, \qquad \sigma_{12} = \sigma_{xy}, \qquad \sigma_{22} = \sigma_y, \qquad \text{etc.}$$
$$f_1 = f_x, \qquad f_2 = f_y, \qquad t_1 = t_x, \qquad t_2 = t_y$$

The first term in (10.17) corresponds to the virtual strain energy stored in the body, the second represents the virtual work done by the body forces, and the third represents the virtual work done by the surface tractions. For plane stress problems with thickness h_e, we assume that all quantities are independent of the thickness coordinate z. Hence,

$$\begin{aligned} 0 = h_e \int_{\Omega^e} (\sigma_x\,\delta\epsilon_x + \sigma_y\,\delta\epsilon_y + 2\sigma_{xy}\,\delta\epsilon_{xy} + \rho\ddot{u}\,\delta u + \rho\ddot{v}\,\delta v)\,dx\,dy \\ - h_e \int_{\Omega^e} (f_x\,\delta u + f_y\,\delta v)\,dx\,dy - h_e \oint_{\Gamma^e} (t_x\,\delta u + t_y\,\delta v)\,ds \end{aligned} \tag{10.18}$$

where f_x and f_y are now body forces per unit area, and t_x and t_y are boundary forces per unit length. When the stresses are expressed in terms of strains through (10.5), and strains in terms of displacements by (10.4), (10.18) takes the form of the principle of total minimum potential energy.

Equation (10.18) can be rewritten using the notation introduced in (10.13)–(10.16):

$$\begin{aligned} 0 = h_e \int_{\Omega^e} \left(\begin{Bmatrix} \delta\epsilon_x \\ \delta\epsilon_y \\ 2\,\delta\epsilon_{xy} \end{Bmatrix}^T \begin{Bmatrix} \sigma_x \\ \sigma_y \\ \sigma_{xy} \end{Bmatrix} + \rho \begin{Bmatrix} \delta u \\ \delta v \end{Bmatrix}^T \begin{Bmatrix} \ddot{u} \\ \ddot{v} \end{Bmatrix} \right) dx\,dy \\ - h_e \int_{\Omega^e} \begin{Bmatrix} \delta u \\ \delta v \end{Bmatrix}^T \begin{Bmatrix} f_x \\ f_y \end{Bmatrix} dx\,dy - h_e \oint_{\Gamma^e} \begin{Bmatrix} \delta u \\ \delta v \end{Bmatrix}^T \begin{Bmatrix} t_x \\ t_y \end{Bmatrix} ds \end{aligned} \tag{10.19}$$

10.3.3 Weak Form of the Governing Differential Equations

Here we present an alternative procedure for the weak form of the plane elasticity equations (10.11)–(10.12). The present approach does not assume knowledge of the principles of virtual displacements or the total minimum potential energy (10.17). It is based on the three-step procedure used throughout this book. We begin with the governing differential equations (10.11) of plane elasticity. We use the three-step procedure for each of the two differential equations, multiplying the first equation with a weight function w_1 and integrating by parts to trade the differentiation equally between the weight function and the dependent variables. We have

$$0 = h_e\int_{\Omega^e}\left[\frac{\partial w_1}{\partial x}\left(c_{11}\frac{\partial u}{\partial x}+c_{12}\frac{\partial v}{\partial y}\right)+\frac{\partial w_1}{\partial y}c_{66}\left(\frac{\partial u}{\partial y}+\frac{\partial v}{\partial x}\right)-w_1 f_x+\rho w_1\ddot{u}\right]dx\,dy$$
$$-h_e\oint_{\Gamma^e}w_1\left[\left(c_{11}\frac{\partial u}{\partial x}+c_{12}\frac{\partial v}{\partial y}\right)n_x+c_{66}\left(\frac{\partial u}{\partial y}+\frac{\partial v}{\partial x}\right)n_y\right]ds \tag{10.20a}$$

Similarly, for the second equation, we have

$$0 = h_e\int_{\Omega^e}\left[\frac{\partial w_2}{\partial x}c_{66}\left(\frac{\partial u}{\partial y}+\frac{\partial v}{\partial x}\right)+\frac{\partial w_2}{\partial y}\left(c_{12}\frac{\partial u}{\partial x}+c_{22}\frac{\partial v}{\partial y}\right)-w_2 f_y+\rho w_2\ddot{v}\right]dx\,dy$$
$$-h_e\oint_{\Gamma^e}w_2\left[c_{66}\left(\frac{\partial u}{\partial y}+\frac{\partial v}{\partial x}\right)n_x+\left(c_{12}\frac{\partial u}{\partial x}+c_{22}\frac{\partial v}{\partial y}\right)n_y\right]ds \tag{10.20b}$$

The last step of the development is to identify the primary and secondary variables of the formulation and rewrite the boundary integrals in terms of the secondary variables. Examination of the boundary integrals in (10.20) reveals that the expressions in the square brackets constitute the secondary variables. By comparing these expressions with those in (10.12), it follows that the boundary forces t_x and t_y are the secondary variables. The weight functions w_1 and w_2 are like the first variations of u and v, respectively. Thus, the final weak form is given by (10.20), with the boundary integrals replaced by

$$h_e\oint_{\Gamma^e}w_1 t_x\,ds, \qquad h_e\oint_{\Gamma^e}w_2 t_y\,ds \tag{10.21}$$

respectively.

This completes the development of the weak formulation of the plane elasticity equations (10.11) and (10.12). The alternative formulation in (10.20) is exactly the same as that in (10.19); one is in matrix form and the other is in the form of explicit expressions. Therefore, the finite element models developed using the weak forms (10.19) and (10.20) would be the same.

10.4 FINITE ELEMENT MODEL

10.4.1 Matrix Form of the Model

First we develop the finite element model of the plane elasticity equations using the matrix form in (10.19). An examination of the weak form (10.19) or

(10.20) reveals that (i) u and v are the primary variables, which must be carried as the primary nodal degrees of freedom; and (ii) only first derivatives of u and v with respect to x and y appear. Therefore, u and v must be approximated by the Lagrange family of interpolation functions, and at least bilinear (i.e., linear in x and y) interpolation should be used. The simplest elements that satisfy those requirements are the linear triangular and rectangular elements. Although u and v are independent of each other, they are the components of the displacement vector. Therefore, both components should be approximated using the same interpolation.

Let u and v be approximated over Ω^e by the finite element interpolations

$$u \approx \sum_{j=1}^{n} u_j^e \psi_j^e(x, y), \qquad v \approx \sum_{j=1}^{n} v_j^e \psi_j^e(x, y) \tag{10.22}$$

For the moment, we shall not restrict ψ_j^e to any specific element, so that the finite element formulation to be developed is valid for any admissible element. For example, for a linear triangular element ($n = 3$), there are two (u_j^e, v_j^e) degrees of freedom per node and a total of six nodal displacements per element (see Fig. 10.3*a*). For a linear quadrilateral element, there are a total of eight nodal displacements per element (see Fig. 10.3*b*). Since the first derivatives of ψ_i^e for a triangular element are element-wise-constant, all strains (ϵ_x, ϵ_y, ϵ_{xy}) computed for the triangular element are element-wise-constant. Therefore, the linear triangular element for plane elasticity is known as the *constant-strain triangular* (*CST*) *element.* For a quadrilateral element, the first derivatives of ψ_i^e are not constant: $\partial\psi_i^e/\partial\xi$ is linear in η and constant in ξ, and $\partial\psi_i^e/\partial\eta$ is linear in ξ and constant in η [see Barlow (1976, 1989)].

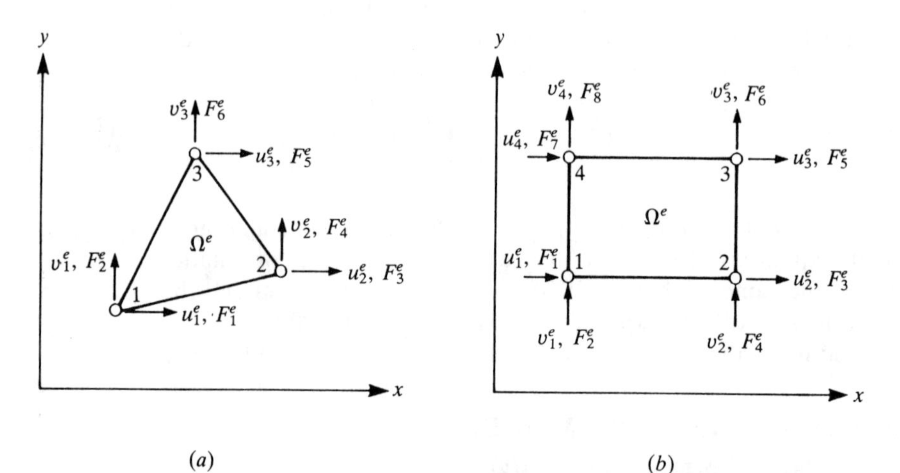

FIGURE 10.3
Linear triangular and rectangular elements for plane elasticity problems: (*a*) triangular element; (*b*) rectangular element.

The displacements and strains are (the element label e is omitted in the following)

$$\begin{Bmatrix} u \\ v \end{Bmatrix} = \begin{Bmatrix} \sum_{j=1}^{n} u_j \psi_j \\ \sum_{j=1}^{n} v_j \psi_j \end{Bmatrix} = \begin{bmatrix} \psi_1 & \psi_2 \dots \psi_n & 0 & 0 \dots 0 \\ 0 & 0 \dots 0 & \psi_1 & \psi_2 \dots \psi_n \end{bmatrix} \begin{Bmatrix} u_1 \\ u_2 \\ \vdots \\ u_n \\ v_1 \\ v_2 \\ \vdots \\ v_n \end{Bmatrix}$$

$$= \begin{bmatrix} \psi_1 & 0 & \psi_2 & 0 \dots \psi_n & 0 \\ 0 & \psi_1 & 0 & \psi_2 \dots 0 & \psi_n \end{bmatrix} \begin{Bmatrix} u_1 \\ v_1 \\ u_2 \\ v_2 \\ \vdots \\ u_n \\ v_n \end{Bmatrix} \equiv [\Psi]\{\Delta\} \tag{10.23}$$

$$\begin{Bmatrix} \ddot{u} \\ \ddot{v} \end{Bmatrix} = [\Psi]\{\ddot{\Delta}\} \tag{10.24}$$

and

$$\{\epsilon\} = [B]\{\Delta\}, \qquad \{\sigma\} = [C][B]\{\Delta\} \tag{10.25a}$$

where

$$[B] = [T][\Psi] \tag{10.25b}$$

and $[T]$ is the matrix of differential operators defined in (10.16). We have

$$\begin{Bmatrix} \delta u \\ \delta v \end{Bmatrix} = [\Psi]\{\delta\Delta\}, \qquad \{\delta\epsilon\} = [B]\{\delta\Delta\} \tag{10.26}$$

Substituting these expressions for the displacements and strains into (10.19), we obtain

$$\begin{aligned} 0 &= h_e \int_{\Omega^e} \{\delta\Delta\}^T([B]^T[C][B]\{\Delta\} + \rho[\Psi]^T[\Psi]\{\ddot{\Delta}\})\, dx\, dy \\ &\quad - h_e \int_{\Omega^e} \{\delta\Delta\}^T[\Psi]^T \begin{Bmatrix} f_x \\ f_y \end{Bmatrix} dx\, dy - h_e \oint_{\Gamma^e} \{\delta\Delta\}^T[\Psi]^T \begin{Bmatrix} t_x \\ t_y \end{Bmatrix} ds \\ &= \{\delta\Delta\}^T([K]\{\Delta\} + [M]\{\ddot{\Delta}\} - \{f\} - \{Q\}) \end{aligned} \tag{10.27}$$

Since this equation holds for any *arbitrary* variations $\{\delta\Delta\}$, it follows that the expression in parentheses should be identically zero, giving the result

$$\boxed{[M^e]\{\ddot{\Delta}^e\} + [K^e]\{\Delta^e\} = \{f^e\} + \{Q^e\}} \tag{10.28a}$$

where

$$\boxed{\begin{aligned} &[K^e] = h_e \int_{\Omega^e} [B^e]^T[C^e][B^e]\, dx\, dy, \quad [M^e] = \rho h_e \int_{\Omega^e} [\Psi^e]^T[\Psi^e]\, dx\, dy \\ &\{f^e\} = h_e \int_{\Omega^e} [\Psi^e]^T \begin{Bmatrix} f_x \\ f_y \end{Bmatrix} dx\, dy, \quad \{Q^e\} = h_e \oint_{\Gamma^e} [\Psi^e]^T \begin{Bmatrix} t_x \\ t_y \end{Bmatrix} ds \end{aligned}} \tag{10.28b}$$

The element mass matrix $[M^e]$ and stiffness matrix $[K^e]$ are of order $2n \times 2n$ and the element force vector $\{F^e\} = \{f^e\} + \{Q^e\}$ is of order $2n \times 1$.

10.4.2 Weak Form Model

Next we use the weak form (10.20) to construct the finite element model. Substituting (10.22) for u and v, and $w_1 = \psi_i^e$ and $w_2 = \psi_i^e$ [to obtain the ith algebraic equation associated with each of the weak statements in (10.20)], and writing the resulting algebraic equations in matrix form, we obtain

$$\begin{bmatrix} [M] & [0] \\ [0] & [M] \end{bmatrix} \begin{Bmatrix} \{\ddot{u}\} \\ \{\ddot{v}\} \end{Bmatrix} + \begin{bmatrix} [K^{11}] & [K^{12}] \\ [K^{12}]^T & [K^{22}] \end{bmatrix} \begin{Bmatrix} \{u\} \\ \{v\} \end{Bmatrix} = \begin{Bmatrix} \{F^1\} \\ \{F^2\} \end{Bmatrix} \tag{10.29a}$$

where

$$\begin{aligned} M_{ij} &= \int_{\Omega^e} c\psi_i\psi_j\, dx\, dy, \qquad c = \rho h \\ K_{ij}^{11} &= \int_{\Omega^e} h\left(c_{11}\frac{\partial\psi_i}{\partial x}\frac{\partial\psi_j}{\partial x} + c_{66}\frac{\partial\psi_i}{\partial y}\frac{\partial\psi_j}{\partial y}\right) dx\, dy \\ K_{ij}^{12} = K_{ji}^{21} &= \int_{\Omega^e} h\left(c_{12}\frac{\partial\psi_i}{\partial x}\frac{\partial\psi_j}{\partial y} + c_{66}\frac{\partial\psi_i}{\partial y}\frac{\partial\psi_j}{\partial x}\right) dx\, dy \\ K_{ij}^{22} &= \int_{\Omega^e} h\left(c_{66}\frac{\partial\psi_i}{\partial x}\frac{\partial\psi_j}{\partial x} + c_{22}\frac{\partial\psi_i}{\partial y}\frac{\partial\psi_j}{\partial y}\right) dx\, dy \\ F_i^1 &= \int_{\Omega^e} h\psi_i f_x\, dx\, dy + \oint_{\Gamma^e} h\psi_i t_x\, ds, \qquad F_i^2 = \int_{\Omega^e} h\psi_i f_y\, dx\, dy + \oint_{\Gamma^e} h\psi_i t_y\, ds \end{aligned} \tag{10.29b}$$

The coefficient matrix $[K^{12}]$, for example, corresponds to the coefficient of v in the first equation, i.e., the first superscript corresponds to the equation number and the second to the variable number. By expanding (10.28*a*), one can show that they yield the same equations as (10.29*a*).

10.4.3 Eigenvalue and Transient Problems

For natural vibration study of plane elastic bodies, (10.28*a*) or (10.29*a*) is reduced to an eigenvalue problem by replacing $\{\Delta\}$ by

$$\{\Delta\} = \{\bar{\Delta}\}e^{-i\omega t} \qquad (i = \sqrt{-1})$$

We obtain [cf. (8.170)]

$$\boxed{(-\omega^2[M^e] + [K^e])\{\bar{\Delta}^e\} = \{Q^e\}} \tag{10.30}$$

For transient analysis, following the discussions of Section 8.5.3, (10.28*a*) or (10.29*a*) can be reduced to a system of algebraic equations by using the Newmark integration scheme (8.171). We have [see Eq. (8.173)]

$$\boxed{[\hat{K}^e]_{s+1}\{\Delta^e\}_{s+1} = \{\hat{F}^e\}_{s,s+1}} \tag{10.31a}$$

where

$$\boxed{\begin{gathered} [\hat{K}^e]_{s+1} = [K^e]_{s+1} + a_3[M^e]_{s+1} \\ \{\hat{F}^e\}_{s,s+1} = \{F^e\}_{s+1} + [M^e]_{s+1}(a_3\{\Delta^e\}_s + a_4\{\dot{\Delta}^e\}_s + a_5\{\ddot{\Delta}^e\}_s) \\ a_3 = \frac{2}{\gamma(\Delta t)^2}, \qquad a_4 = \Delta t\, a_3, \qquad a_5 = \frac{1}{\gamma} - 1 \end{gathered}} \tag{10.31b}$$

For additional details, see Sections 6.2.3 and 8.5.3.

10.5 EVALUATION OF INTEGRALS

For the linear triangular (i.e., CST) element, ψ_i^e has the form

$$\psi_i^e = \frac{1}{2A_e}(\alpha_i^e + \beta_i^e x + \gamma_i^e y)$$

where A_e is the area of the triangular element, and $[B^e] = [T][\Psi^e]$ [see (10.23)–(10.25)] is constant. Since $[B^e]$ and $[C^e]$ are independent of x and y, the element stiffness matrix for the CST element is given by

$$[K^e] = h_e A_e [B^e]^T [C^e][B^e] \tag{10.32}$$

For the case in which the body force components f_x and f_y are element-wise-constant (say, equal to f_{x0}^e and f_{y0}^e, respectively), the force vector $\{f^e\}$ has the form

$$\{f^e\} = h_e \int_{\Omega^e} [\Psi^e]^T \begin{Bmatrix} f_{x0}^e \\ f_{y0}^e \end{Bmatrix} dx\, dy = \frac{A_e h_e}{3} \begin{Bmatrix} f_{x0}^e \\ f_{y0}^e \\ f_{x0}^e \\ f_{y0}^e \\ f_{x0}^e \\ f_{y0}^e \end{Bmatrix} \tag{10.33}$$

For a general quadrilateral element, it is not easy to compute the coefficients of the stiffness matrix by exact means. In such cases, we use the numerical integration discussed in Section 9.2. However, for a linear rectangular element of sides a and b, the element coefficient matrices in (8.35) and (8.44) can be used to obtain the stiffness matrix. The load vector for a linear rectangular element is given by

$$\{f^e\}_{8\times 1} = \frac{A_e h_e}{4} \begin{Bmatrix} f_{x0}^e \\ f_{y0}^e \\ f_{x0}^e \\ f_{y0}^e \\ \vdots \end{Bmatrix} \tag{10.34}$$

The vector $\{Q^e\}$ is computed only when the element Ω^e falls on the boundary of the domain on which tractions are specified (i.e., known). Computation of $\{Q^e\}$ involves the evaluation of line integrals (for any type of element), as explained in Section 8.2.6; see also Example 8.2. For plane elasticity problems, the surface tractions t_x and t_y take the place of q_n in single-variable problems [see (8.45)]. Often, in practice, it is convenient to express the surface tractions t_x and t_y in the element coordinates. In that case, $\{Q^e\}$ can be evaluated in the element coordinates and then transformed to the global coordinates for assembly. If $\{\bar{Q}^e\}$ denotes the element traction vector referred to the element coordinates then [see (4.58b)] the corresponding global vector is given by

$$\{Q^e\} = [R^e]^T \{\bar{Q}^e\} \tag{10.35a}$$

where

$$[R^e] = \begin{bmatrix} \cos\alpha & \sin\alpha & 0 & 0 & \\ -\sin\alpha & \cos\alpha & 0 & 0 & \\ 0 & 0 & \cos\alpha & \sin\alpha & \\ 0 & 0 & -\sin\alpha & \cos\alpha & \\ & & & & \ddots \end{bmatrix}_{2n\times 2n} \tag{10.35b}$$

and α is the angle between the global x axis and t_n.

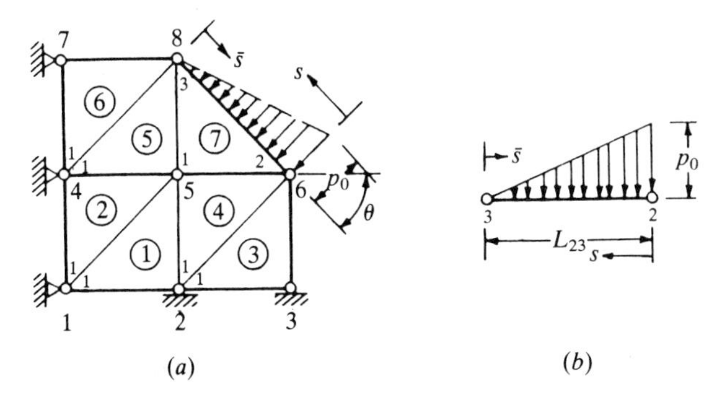

FIGURE 10.4
Computation of boundary force components for plane elasticity problems: (*a*) the constant strain triangle (CST); (*b*) a typical plane elasticity problem; (*c*) computation of the boundary forces.

As a specific example, consider the structure shown in Fig. 10.4(*a*). Side 2–3 of element 7 is subjected to a linearly varying normal force:

$$t_n \neq 0, \qquad t_s = 0$$

where the subscripts n and s refer to normal and tangential directions, respectively. We have for ($e = 7$)

$$\{\bar{Q}^e\} = \oint_{\Gamma^e} [\Psi^e]^T \begin{Bmatrix} t_n \\ t_s \end{Bmatrix} ds = \int_{\Gamma_{12}^e} [\Psi^e]^T \begin{Bmatrix} t_n \\ t_s \end{Bmatrix} ds + \int_{\Gamma_{23}^e} [\Psi^e]^T \begin{Bmatrix} t_n \\ 0 \end{Bmatrix} ds$$

$$+ \int_{\Gamma_{31}^e} [\Psi^e]^T \begin{Bmatrix} t_n \\ t_s \end{Bmatrix} ds \quad (10.36a)$$

The first and third integrals cannot be evaluated, because we do not know t_n and t_s on these sides of the element. However, by internal stress equilibrium, those portions cancel with like contributions from the neighboring elements (elements 4 and 5) in the assembled force vector of the structure. Thus, we must compute the integral over side 2–3 of the element. We have (for $e = 7$)

$$\{\bar{Q}^e\}_{2\text{-}3} = \int_0^{L_{23}} [\Psi^e]^T \begin{Bmatrix} t_n \\ 0 \end{Bmatrix} ds, \qquad t_n = -p_0\left(1 - \frac{s}{L_{23}}\right) \quad (10.36b)$$

where the minus sign in front of p_0 is added to account for the direction of the applied traction, which acts toward the body in the present case. The local coordinate system s used in the above expression is chosen along the side connecting node 2 to node 3, with its original at node 2 (see Fig. 10.4*b*). One is not restricted to this choice. If it is considered to be convenient to choose the local coordinate system $\bar{s}$, which is taken along side 3–2, with its origin at node 3 of element 7, we can write

$$\{\bar{Q}^e\}_{2\text{-}3} = \int_0^{L_{23}} [\Psi^e]^T \begin{Bmatrix} t_n \\ 0 \end{Bmatrix} d\bar{s}, \qquad t_n = -\frac{p_0\bar{s}}{L_{23}} \quad (10.36c)$$

where $[\Psi^e]$ is now expressed in terms of the local coordinate $\bar{s}$. We have

$$\{\bar{Q}^e\}_{2\text{-}3} = \int_0^{L_{23}} \begin{Bmatrix} 0 \\ 0 \\ \psi_2^7 t_n \\ 0 \\ \psi_3^7 t_n \\ 0 \end{Bmatrix} d\bar{s} = -\frac{L_{23} p_0}{6} \begin{Bmatrix} 0 \\ 0 \\ 2 \\ 0 \\ 1 \\ 0 \end{Bmatrix} \tag{10.37a}$$

The global components of this force vector are [set $\alpha = 90° - \theta$ in (10.35b)]

$$\{Q^e\}_{2\text{-}3} = -\frac{L_{23} p_0}{6} \begin{Bmatrix} 0 \\ 0 \\ 2\sin\theta \\ 2\cos\theta \\ \sin\theta \\ \cos\theta \end{Bmatrix} \tag{10.37b}$$

The same procedure applies to linear quadrilateral elements. In general, the loads due to specified boundary stresses can be computed using an appropriate local coordinate system and one-dimensional interpolation functions. When higher-order elements are involved, the same order one-dimensional interpolation functions must be used.

10.6 ASSEMBLY AND BOUNDARY AND INITIAL CONDITIONS

The assembly procedure for problems with many degrees of freedom is the same as that used when there is a single degree of freedom (see Section 8.2), except that the procedure should be applied to all degrees of freedom at each node. For example, consider the plane elastic structure shown in Fig. 10.4(a). There are eight nodes in the mesh; hence, the total size of the assembled stiffness matrix will be 16×16, and the force vector will be 16×1. The first two rows and columns of the global stiffness matrix correspond to global node 1, which has contributions from node 1 of elements 1 and 2. For instance, we have

$$K_{11} = K_{11}^1 + K_{11}^2, \quad K_{22} = K_{22}^1 + K_{22}^2, \quad K_{12} = K_{12}^1 + K_{12}^2, \quad K_{13} = K_{13}^1,$$
$$K_{33} = K_{33}^1 + K_{11}^3 + K_{11}^4, \quad K_{34} = K_{34}^1 + K_{12}^3 + K_{12}^4, \quad \text{etc.} \tag{10.38}$$

Note that K_{12}, for example, denotes the coupling stiffness coefficient between the first and second global degrees of freedom, both of which are at global node 1. Similar arguments apply for the assembly of the force vector.

With regard to the boundary conditions, the primary degrees of freedom are the displacements and the secondary degrees of freedom are the forces. At any node on the boundary of the domain of a plane elasticity problem, one has the following four distinct possibilities:

$$\begin{array}{ll} \text{Case 1} & u \text{ and } v \text{ are specified (and } t_x \text{ and } t_y \text{ are unknown)} \\ \text{Case 2} & u \text{ and } t_y \text{ are specified (and } t_x \text{ and } v \text{ are unknown)} \\ \text{Case 3} & t_x \text{ and } v \text{ are specified (and } u \text{ and } t_y \text{ are unknown)} \\ \text{Case 4} & t_x \text{ and } t_y \text{ are specified (and } u \text{ and } v \text{ are unknown)} \end{array} \tag{10.39}$$

In general, only one of the quantities of each of the pairs (u, t_x) and (v, t_y) must be specified at a point on the boundary.

For time-dependent problems, the initial displacement and velocity must be specified:

$$\begin{Bmatrix} u \\ v \end{Bmatrix} = \begin{Bmatrix} u_0 \\ v_0 \end{Bmatrix}, \quad \begin{Bmatrix} \dot{u} \\ \dot{v} \end{Bmatrix} = \begin{Bmatrix} \dot{u}_0 \\ \dot{v}_0 \end{Bmatrix} \tag{10.40}$$

10.7 EXAMPLES

Here we consider a couple of examples of plane elasticity problems to illustrate the load computation and imposition of boundary conditions. The stresses are evaluated at reduced Gauss points of the elements [see Barlow (1976, 1989)]. These examples are actually analyzed using the program FEM2DV2, which is discussed in Chapter 13.

Example 10.1. Consider a thin elastic plate subjected to a uniformly distributed edge load, as shown in Fig. 10.5(*a*). We wish to determine the equilibrium (i.e., static) solution. First we consider a two-element discretization of the plate by triangular elements, and then we perform all the algebra to obtain the nodal displacements.

The assembly of element matrices for elements with two degrees of freedom (DOF) is described in Section 10.6. For the finite element mesh at hand, the correspondence between the global and local nodes and stiffness is as follows:

Nodal correspondence		Stiffness correspondence	
Global (DOF)	Local (DOF)	Global	Local
1 (1, 2)	1 of element 1 (1, 2) 1 of element 2 (1, 2)	K_{11} K_{22} K_{12}	$K_{11}^1 + K_{11}^2$ $K_{22}^1 + K_{22}^2$ $K_{12}^1 + K_{12}^2$
2 (3, 4)	2 of element 1 (3, 4)	K_{33} K_{44} K_{34}	K_{33}^1 K_{44}^1 K_{34}^1
3 (5, 6)	3 of element 1 (5, 6) 2 of element 2 (3, 4)	K_{55} K_{66} K_{56}	$K_{55}^1 + K_{33}^2$ $K_{66}^1 + K_{44}^2$ $K_{56}^1 + K_{34}^2$
4 (7, 8)	3 of element 2 (5, 6)	K_{77} K_{88} K_{78}	K_{55}^2 K_{66}^2 K_{56}^2

(10.41)

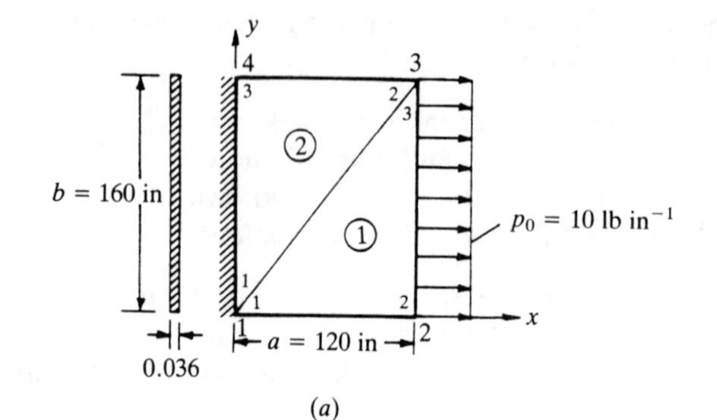

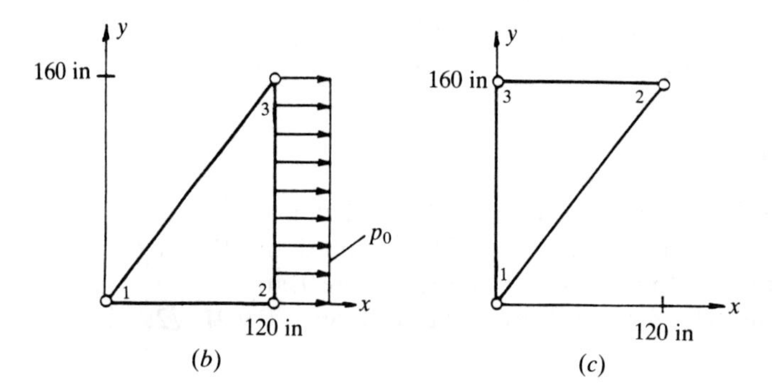

FIGURE 10.5
Geometry and finite element mesh of a plane elasticity problem by the CST elements: (*a*) a plane elasticity problem; (*b*) element 1; (*c*) element 2.

If two global nodes correspond to two (local) nodes of the same element then the corresponding stiffness coefficient is nonzero; otherwise it is zero.

The specified (primary) degrees of freedom are

$$U_1 = V_1 = U_4 = V_4 = 0 \tag{10.42}$$

The known secondary degrees of freedom (forces) are already included in $\{F^1\}$ and $\{F^2\}$. Equilibrium of forces requires that (note that p_0 is the force per unit length)

$$F_3^1 = \tfrac{1}{2}p_0 b, \quad F_4^1 = 0, \quad F_5^1 + F_3^2 = \tfrac{1}{2}p_0 b, \quad F_6^1 + F_4^2 = 0 \tag{10.43}$$

The first two rows and columns and the last two rows and columns of the assembled $[K]$ can be deleted (since the specified boundary conditions are homogeneous) to obtain the condensed form of the matrix equation:

$$\begin{bmatrix} K_{33}^1 & K_{34}^1 & K_{35}^1 & K_{36}^1 \\ K_{43}^1 & K_{44}^1 & K_{45}^1 & K_{46}^1 \\ K_{53}^1 & K_{54}^1 & K_{55}^1 + K_{33}^2 & K_{56}^1 + K_{34}^2 \\ K_{63}^1 & K_{64}^1 & K_{65}^1 + K_{43}^2 & K_{66}^1 + K_{44}^2 \end{bmatrix} \begin{Bmatrix} U_2 \\ V_2 \\ U_3 \\ V_3 \end{Bmatrix} = \begin{Bmatrix} \tfrac{1}{2}p_0 b \\ 0 \\ \tfrac{1}{2}p_0 b \\ 0 \end{Bmatrix} \tag{10.44}$$

or (using $a = 120$ in, $b = 160$ in, $h = 0.036$ in, $\nu = 0.25$, $E = 30 \times 10^6$ lb in^{-2}, and $p_0 =$

10 lb in^{-1})

$$10^4\begin{bmatrix} 93.0 & -36.0 & -16.2 & 14.4 \\ -36.0 & 72.0 & 21.6 & -43.2 \\ -16.2 & 21.6 & 93.0 & 0.0 \\ 14.4 & -43.2 & 0.0 & 72.0 \end{bmatrix} \begin{Bmatrix} U_2 \\ V_2 \\ U_3 \\ V_3 \end{Bmatrix} = \begin{Bmatrix} 800.0 \\ 0.0 \\ 800.0 \\ 0.0 \end{Bmatrix} \tag{10.45}$$

Inverting the matrix, we obtain

$$\begin{Bmatrix} U_2 \\ V_2 \\ U_3 \\ V_3 \end{Bmatrix} = \frac{10^{-6}}{3} \begin{bmatrix} 4.07 & 2.34 & 0.17 & 0.59 \\ 2.34 & 8.65 & -1.6 & 4.72 \\ 0.17 & -1.6 & 3.63 & -0.99 \\ 0.59 & 4.72 & -0.99 & 6.88 \end{bmatrix} \begin{Bmatrix} 800.0 \\ 0.0 \\ 8.00 \\ 0.0 \end{Bmatrix}$$

$$= 10^{-4} \begin{Bmatrix} 11.28 \\ 1.97 \\ 10.10 \\ -1.09 \end{Bmatrix} \text{in} \tag{10.46}$$

Note that the solution is not symmetric (i.e., $U_2 \neq U_3$, $V_2 \neq -V_3$) about the horizontal centerline. The exact solution should be symmetric about the centerline. This is due to the asymmetry in the finite element discretization.

Table 10.1 gives the finite element solutions (obtained using FEM2DV2) for the displacements at the points (120, 0) and (120, 160) of isotropic and orthotropic plates

TABLE 10.1
Finite element results for a thin plate (plane stress assumption) using various meshes of triangular and rectangular elements and material properties†

Mesh	Material	U_2 ($\times 10^{-4}$)	V_2 ($\times 10^{-4}$)	U_3 ($\times 10^{-4}$)	V_3 ($\times 10^{-4}$)
4 3 / 1 2	Isotropic: $E = 30 \times 10^6$ lb in^{-2} $\nu = 0.25$ $G = E/2(1+\nu)$	11.291 10.853	1.9637 2.3256	10.113 10.853	−1.08 −2.3256
4 3 / 1 2	Orthotropic: $E_1 = 31 \times 10^6$ lb in^{-2} $E_2 = 2.7 \times 10^6$ lb in^{-2} $G_{12} = 0.75 \times 10^6$ lb in^{-2} $\nu_{12} = 0.28$	10.767 10.728	1.6662 2.6758	10.650 10.728	−1.579 −2.6758
4 3 / 1 2	Orthotropic: same as above	10.821 10.778	2.157 2.002	10.821 10.778	−2.157 −2.002

For each mesh, the first row corresponds to triangular elements and the second row to one rectangular element.

TABLE 10.2
Deflections and stresses in an isotropic plate subjected to uniform edge load (Example 10.1)

Mesh		$u(120, 0)$ $(\times 10^{-4})$	$v(120, 0)$ $(\times 10^{-4})$	σ_x	σ_y	σ_{xy}
	1×1	11.291	1.9637	285.9	14.40	10.80
				(80, 53.33)†	(80, 53.33)	(80, 53.33)
Triangular	2×2	11.372	2.1745	279.7	69.36	23.20
				(80, 53.33)	(20, 53.33)	(40, 26.67)
	4×4	11.284	2.1255	280.0	69.59	35.93
				(80, 53.33)	(10, 26.67)	(20, 13.33)
	1×1	10.853	2.3256	277.8	25.84	0.0
				(60, 80)	(60, 80)	(60, 80)
Rectangular	2×2	11.078	2.0212	277.8	37.46	13.23
				(30, 40)	(30, 40)	(30, 40)
	4×4	11.150	2.0094	280.4	49.74	27.73
				(75, 60)	(15, 60)	(15, 20)

† Location of the stress.

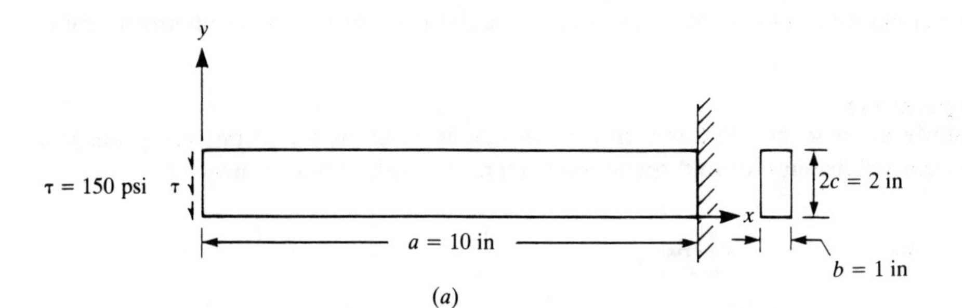

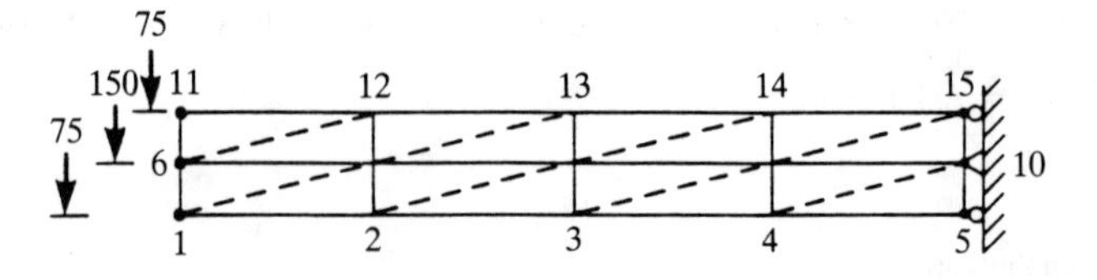

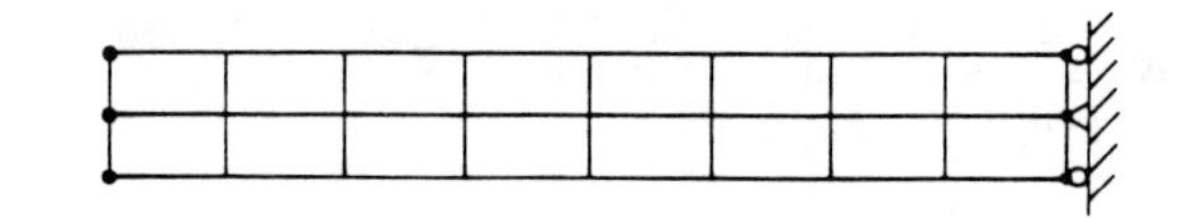

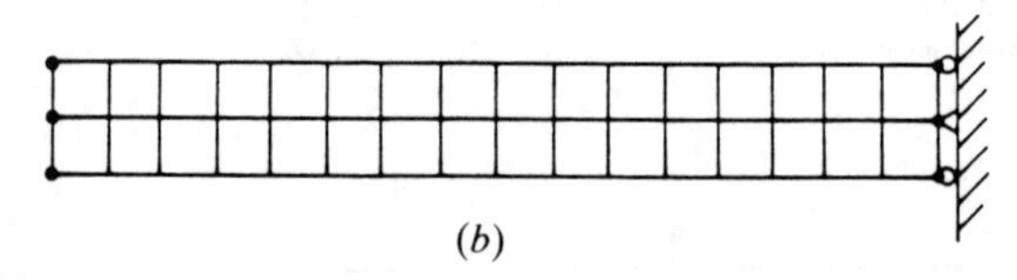

FIGURE 10.6
Finite-element meshes for an end-loaded cantilever beam.

for the meshes shown. Table 10.2 gives the deflections and stresses obtained with refined meshes of triangular and rectangular elements.

Example 10.2. Consider the cantilever beam ($E = 30 \times 10^6$ psi, $\nu = 0.25$, $a = 10$ in, $b = c = 1$ in) shown in Fig. 10.6(a). We wish to determine, using the elasticity equations, the maximum deflection and bending stress in the beam when it is subjected to a uniformly distributed shear stress $\tau = 150$ psi. The boundary conditions of the problem are

$$\begin{aligned} u(a, y) = 0, \quad & v(a, c) = 0 \\ t_x = t_y = 0 \quad & \text{at } y = 0, 2c \text{ for any } x \\ t_x = 0, \quad t_y = -\tau \quad & \text{at } x = 0 \text{ for any } y \\ t_y = 0 \quad & \text{at } x = a \text{ and for any } y \neq c \end{aligned} \tag{10.47}$$

We shall solve the problem using the plane stress assumption. The elastic coefficients c_{ij} for the plane stress case are defined (assuming that steel is isotropic) by

$$c_{11} = c_{12} = \frac{E}{1 - \nu^2}, \quad c_{12} = \frac{E\nu}{1 - \nu^2}, \quad c_{66} = \frac{E}{2(1 + \nu)} \; (= G) \tag{10.48}$$

Three different finite element meshes, increasingly refined, are shown in Fig. 10.6(b). These meshes are those consisting of linear rectangular elements. Equivalent triangular element meshes are obtained by joining node 1 to node 3 of each rectangular element (see the dashed lines). Equivalent meshes of nine-node quadratic Lagrange elements are obtained by considering a 2×2 mesh of linear Lagrange elements as a quadratic element.

For the finite element model, the boundary conditions on the primary and secondary variables, e.g., for the 15-node mesh, are given by

$$\begin{aligned} U_5 = U_{10} = V_{10} = U_{15} = 0.0 \\ F_1^y = -75.0, \quad F_6^y = -150.0, \quad F_{11}^y = -75.0 \end{aligned} \tag{10.49}$$

and all other forces are zero on the boundary.

Table 10.3 gives a comparison of the finite element solutions with the elasticity solutions for the tip deflection (i.e., the deflection at the center node of the left end) and bending stress σ_x, obtained using two-dimensional elasticity theory [see Reddy (1984)]. The linear triangular element mesh has the slowest convergence compared with the linear and quadratic rectangular elements.

The final example in this chapter deals with free vibration and transient analysis of a plane elasticity problem. We consider the cantilever beam of Example 10.2.

Example 10.3. Consider the cantilever beam shown in Fig. 10.6a. We wish to determine the natural frequencies and transient response using the plane elasticity elements. We use the finite element meshes of linear triangular and rectangular elements shown in Fig. 10.6(b) and their nodal equivalent meshes of quadratic elements to analyze the problem. Table 10.4 gives a comparison of the first 10 natural frequencies obtained with various meshes. The convergence of the natural frequencies with mesh refinement is clear.

TABLE 10.3
Comparison of the finite element solution with the elasticity solution for a cantilever beam subjected to a uniform shear load at the free end

	Tip deflection, $-v \times 10^{-2}$			Normal stress σ_x		
Number of nodes	**Linear triangles**	**Linear rectangles**	**Quadratic rectangles**	**Linear triangles**	**Linear rectangles**	**Quadratic rectangles**
15	0.16112	0.31335	0.50310	1209 (15)†	1196 (8.75, 1.5)‡	2196 (8.943, 1.577)
27	0.26623	0.43884	0.51288	2270 (31)	1793 (9.375, 1.5)	2439 (9.471, 1.577)
51	0.316459	0.48779	0.51374	2829 (63)	2056 (9.6875, 1.5)	2526 (9.736, 1.577)
Elasticity§		0.51875 (0.0, 1.0)		2876 (9.583, 1.667)	2180 (9.6875, 1.5)	2528 (9.736, 1.577)

† Element number.
‡ Quadrature points.
§ From Reddy (1984), p. 53: $v(0, 1) = -(PL^3/3EI)[1 + 3(1 + \nu)/L^2]$; $\sigma_x = Px(1 - y)/I$, $I = \frac{2}{3}$.

For transient analysis with the linear acceleration scheme ($\alpha = 0.5$, $\gamma = \frac{1}{3}$), the time step Δt is given by the inequality

$$\Delta t < \Delta t_{cr} = \left(\frac{12}{\lambda_{max}}\right)^{1/2} \tag{10.50}$$

For the 4×1 mesh of rectangular elements, for example, we have $\Delta t_{cr} = 1.617 \times 10^{-5}$. Figure 10.7 contains plots of the tip deflection $v(0, 1, t)$ versus time, as predicted using the 4×1 mesh of linear rectangular elements and the following two time integration schemes: (1) $\alpha = \gamma = \frac{1}{2}$ and (2) $\alpha = \frac{1}{2}$, $\gamma = \frac{1}{3}$. The time step used in these computations,

TABLE 10.4
Comparison of first ten frequencies of the cantilever beam of Example 10.3 as computed using various meshes of linear and quadratic triangular and rectangular elements

	Triangular elements				Rectangular elements			
ω	Linear element		Quadratic element		Linear element		Quadratic element	
	4×2	8×2	2×1	4×1	4×2	8×2	2×1	4×1
1	2,019.4	1,583.0	1,186.4	1,156.7	1,465.5	1,242.3	1,169.9	1,151.8
2	9,207.4	8,264.0	7,896.6	6,496.5	8,457.9	6,845.8	7,179.7	6,341.4
3	10,449.6	9,177.7	9,158.2	9,156.0	9,218.4	9,171.7	9,158.2	9,156.0
4	25,339.2	19,540.5	18,369.1	16,219.9	22,334.0	16,887.7	17,890.8	15,572.7
5	29,193.2	27,843.9	27,805.3	27,441.7	29,113.3	27,836.8	27,869.8	27,266.3
6	42,363.4	32,727.8	40,399.2	28,696.9	40,309.7	29,433.6	39,583.7	27,442.2
7	52,937.0	46,840.4	50,469.6	39,762.6	52,991.9	44,231.1	50,964.4	39,302.3
8	67,964.6	48,014.4	66,260.9	45,815.6	66,842.5	47,441.0	67,015.3	45,839.9
9	76,833.2	61,560.4	74,582.1	57,429.5	74,523.3	60,078.3	74,064.6	56,949.9
10	79,443.0	68,257.4	79,241.8	64,867.4	76,515.5	67,813.3	80,029.3	64,636.0

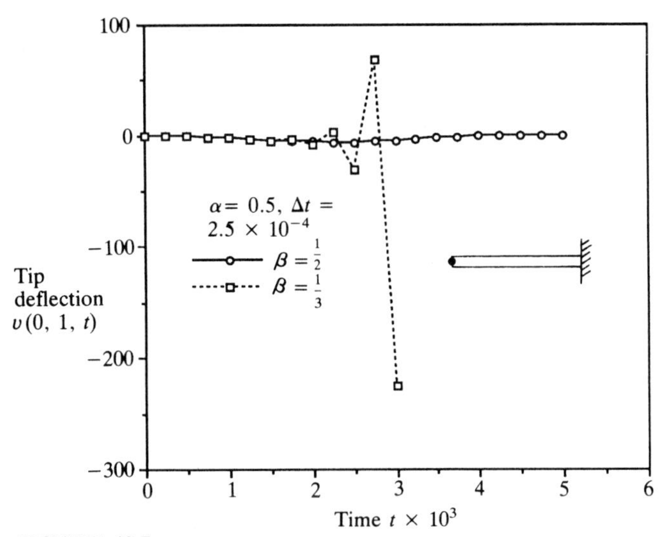

FIGURE 10.7
Stability of the finite element solutions predicted by two different time integration schemes (Example 10.3). The 4×1 mesh of linear rectangular elements is used.

$\Delta t = 2.5 \times 10^{-4} > \Delta t_{cr}$, makes the second scheme yield unstable solutions. It should be noted that if Δt is close to Δt_{cr} but $\Delta t > \Delta t_{cr}$ then the solution predicted by the linear acceleration scheme may be stable for the first several time steps, but it will eventually become unstable. Figures 10.8 and 10.9 show the relative accuracies of solutions predicted by triangular and rectangular elements. It should be noted that the linear triangular element yields a less accurate solution than the linear rectangular element.

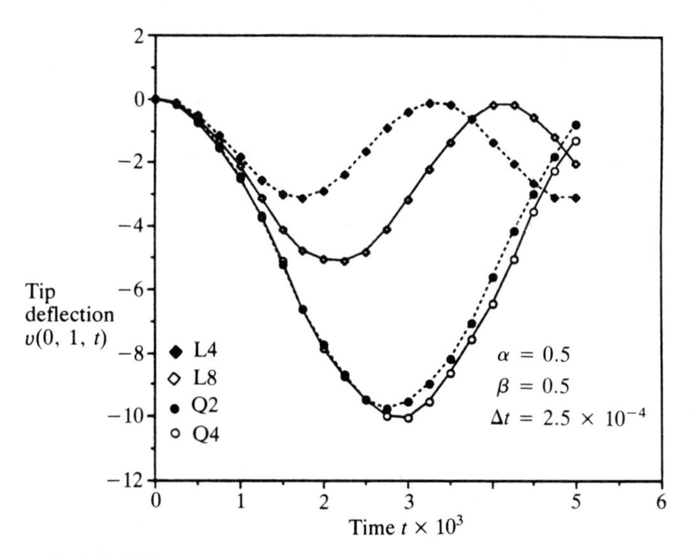

FIGURE 10.8
Tip deflection $v(0, 1, t)$ versus time, as predicted by various meshes of linear (L) and quadratic (Q) triangular elements: L4, 4×2 mesh of linear elements; L8, 8×2 mesh of linear elements; Q2, 2×1 mesh of quadratic elements; Q4, 4×1 mesh of quadratic elements.

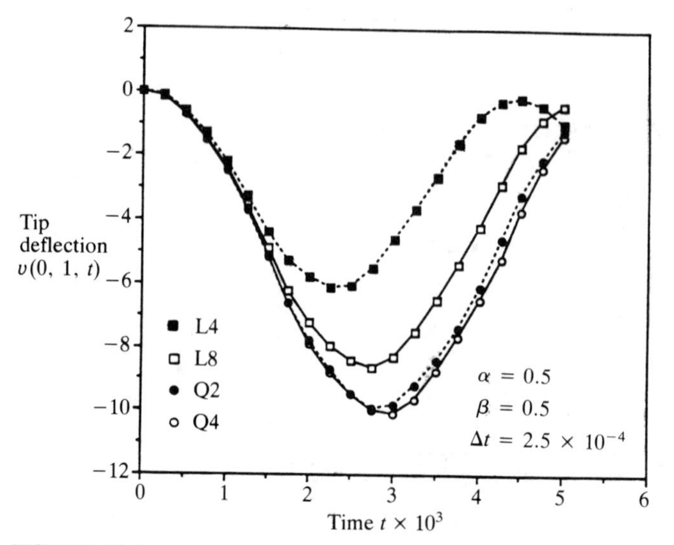

FIGURE 10.9
Tip deflection versus time, as predicted by various meshes of rectangular elements: L4, 4×2 mesh of linear elements; L8, 8×2 mesh of linear elements; Q2, 2×1 mesh of nine-node quadratic elements; Q4, 4×1 mesh of nine-node quadratic elements.

10.8 SUMMARY

In this chapter an introduction to the equations of plane elasticity (i.e., two-dimensional problems of elasticity) has been presented, and their finite element models have been formulated. The plane strain and plane stress problems, which differ only in the use of constitutive relations, have been discussed. The governing equations have been expressed in terms of the displacements, and their weak form and finite element model developed in two alternative ways: (i) the matrix formulation, which is standard in most finite element textbooks on solid mechanics; and (ii) the three-step procedure, which is used throughout the book. The eigenvalue and time-dependent problems of plane elasticity have also been discussed. Some numerical examples have been presented to illustrate the evaluation of element stiffness matrices and force vectors.

PROBLEMS

10.1–10.3. Compute the contribution of the surface forces to the global force degrees of freedom in the plane elasticity problems given in Figs. P10.1–P10.3.

10.4–10.6. Give the connectivity matrices and the specified primary degrees of freedom for the plane elasticity problems shown in Figs. P10.1–P10.3. Give only the first three rows of the connectivity matrix.

10.7. Consider a cantilevered beam of length 6 cm, width 2 cm, thickness 1 cm, and material properties $E = 3 \times 10^7\ \text{N cm}^{-2}$ and $\nu = 0.3$, and subjected to a bending moment of 600 N cm at the free end (see Fig. P.10.7). Replace the moment by

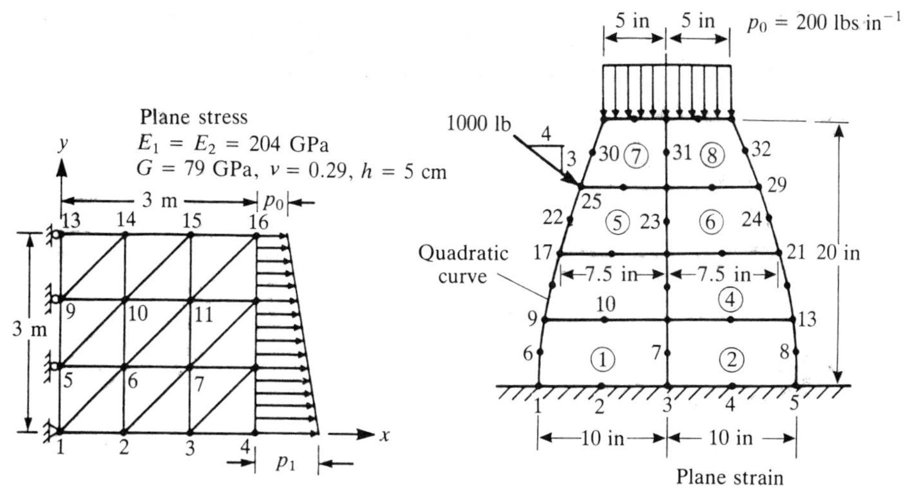

FIGURE P10.1

FIGURE P10.2

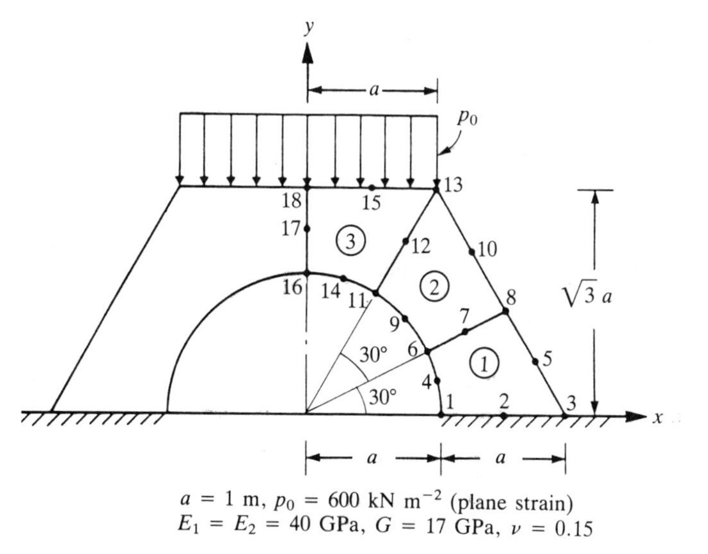

FIGURE P10.3

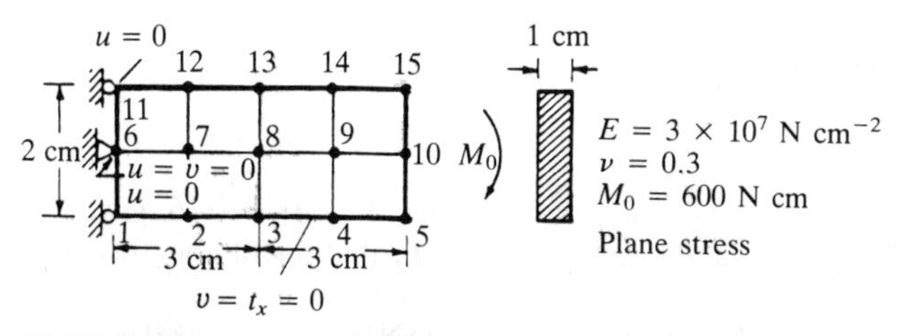

FIGURE P10.7

an equivalent distributed force at $x = 6$ cm, and model the domain by (a) a 4×2 mesh of linear rectangular elements and (b) quadratic rectangular elements. Identify the specified displacements and global forces.

Answer: $F_5^x = -300$ and $F_{15}^x = 300$.

10.8. Consider the ("transition") element shown in Fig. P10.8. Define the generalized displacement vector of the element by

$$\{u\} = \{u_1, v_1, \theta_1, u_2, v_2, u_3, v_3\}^T$$

and represent the displacement components u and v by

$$u = \psi_1 u_1 + \psi_2 u_2 + \psi_3 u_3 + \tfrac{1}{2} h \eta \psi_1 \theta_1, \qquad v = \psi_1 v_1 + \psi_2 v_2 + \psi_3 v_3$$

where ψ_1 is the interpolation function for the beam, and ψ_2 and ψ_3 are the interpolation functions for nodes 2 and 3:

$$\psi_1 = \tfrac{1}{2}(1 - \xi), \qquad \psi_2 = \tfrac{1}{4}(1 + \xi)(1 - \eta), \qquad \psi_3 = \tfrac{1}{4}(1 + \xi)(1 + \eta)$$

Derive the stiffness matrix for the element.

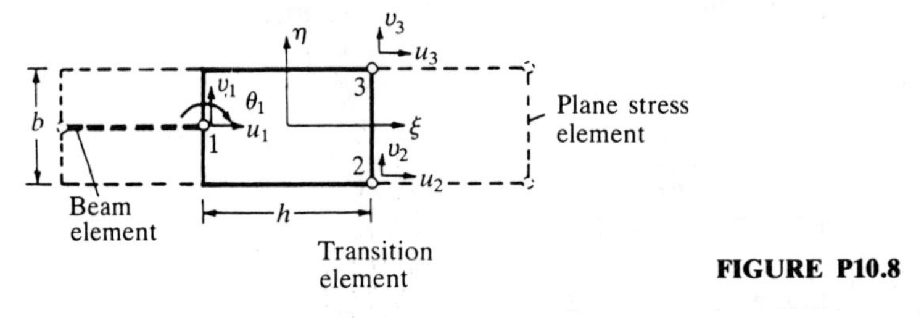

FIGURE P10.8

10.9. Consider the square, isotropic, elastic body of thickness h shown in Fig. P10.9. Suppose that the displacements are approximated by

$$u(x, y) = (1 - x) y u_1 + x(1 - y) u_2, \qquad v(x, y) = 0$$

Assuming that the body is in a state of plane stress, derive the 2×2 stiffness matrix for the unit square:

$$[K]\begin{Bmatrix} u_1 \\ u_2 \end{Bmatrix} = \begin{Bmatrix} F_1 \\ F_2 \end{Bmatrix}$$

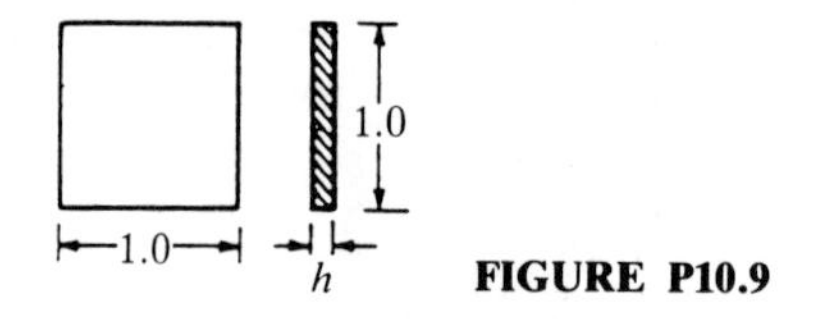

FIGURE P10.9

10.10–10.16. For the plane elasticity problems shown in Figs. P10.10–P10.16, give the boundary degrees of freedom and compute the contribution of the specified forces to the nodes.

Answer: (Problem 10.10) $F_7^y = 1000$ N, $F_{11}^y = 4000$ N, and $F_{18}^y = 1000$ N.

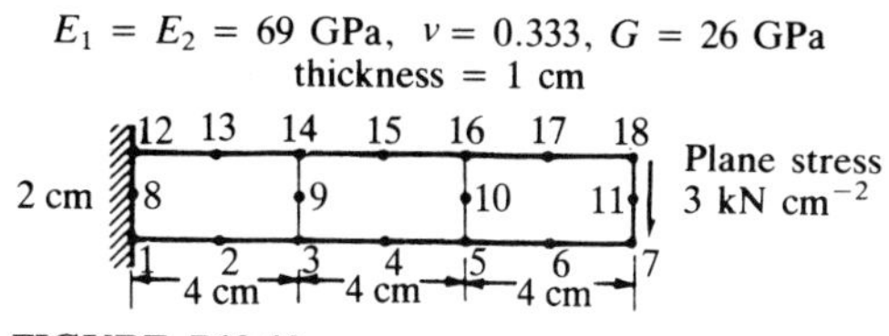

FIGURE P10.10

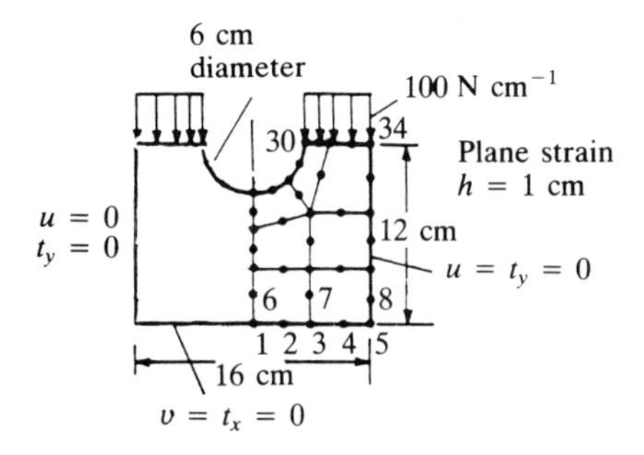

FIGURE P10.11

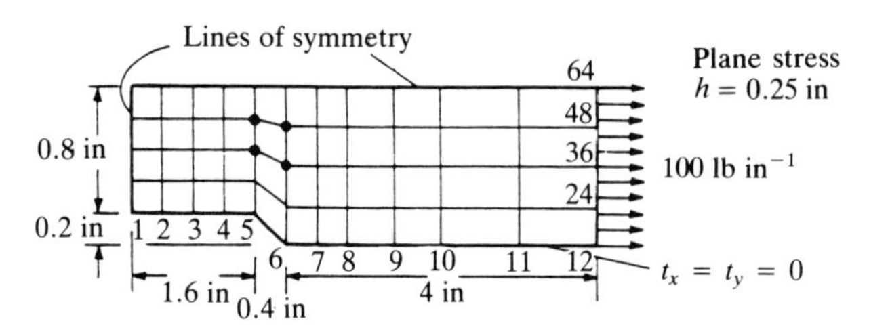

FIGURE P10.12

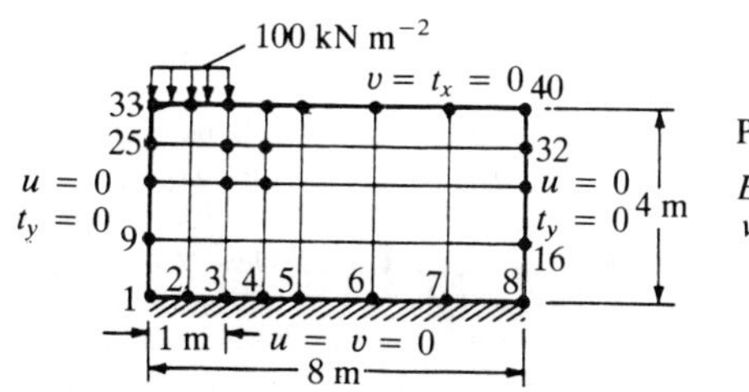

Plane strain

$E = 40$ GPa, $G = 17.4$ GPa
$\nu = 0.15$, thickness = 1 cm

FIGURE P10.13

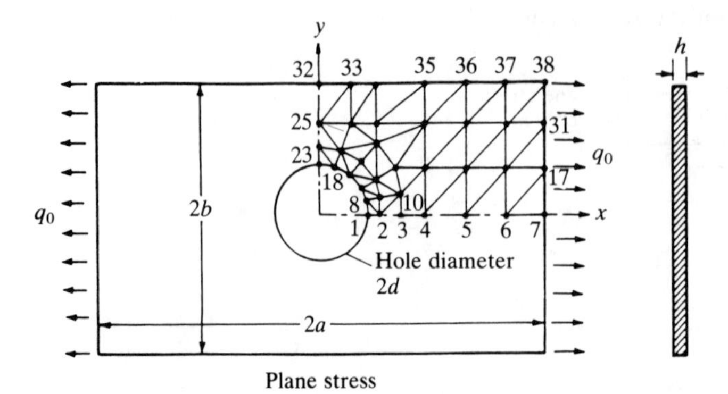

FIGURE P10.14

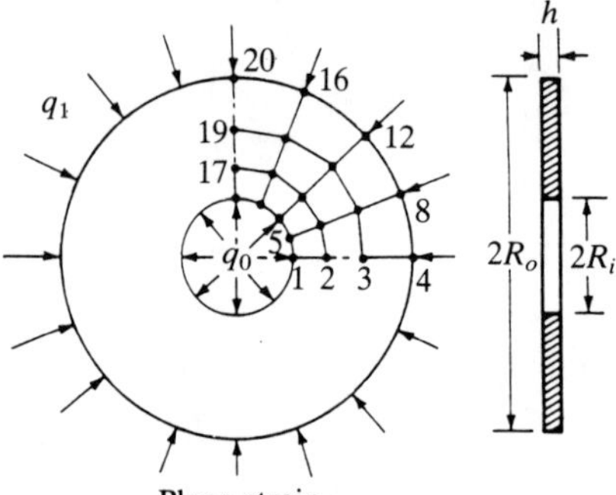

FIGURE P10.15

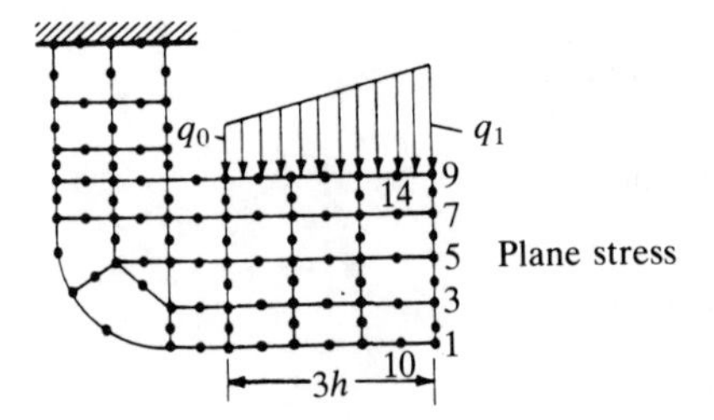

FIGURE P10.16

REFERENCES FOR ADDITIONAL READING

Barlow, J.: "Optimal Stress Locations in Finite Element Models," *International Journal for Numerical Methods in Engineering,* vol. 10, pp. 243–251, 1976.

———: "More on Optimal Stress Points—Reduced Integration, Element Distortions and Error Estimation," *International Journal for Numerical Methods in Engineering,* vol. 28, pp. 1487–1504, 1989.

Budynas, R. G.: *Advanced Strength and Applied Stress Analysis,* McGraw-Hill, New York, 1977.

Dym, C. L., and I. H. Shames: *Solid Mechanics: A Variational Approach,* McGraw-Hill, New York, 1973.

Ugural, A. C., and S. K. Fenster: *Advanced Strength and Applied Elasticity,* Elsevier, New York, 1975.

Volterra, E., and J. H. Gaines: *Advanced Strength of Materials,* Prentice-Hall, Englewood Cliffs, NJ, 1971.

Reddy, J. N.: *Energy and Variational Methods in Applied Mechanics,* John Wiley, New York, 1984.

——— and M. L. Rasmussen: *Advanced Engineering Analysis,* John Wiley, New York, 1982.

Rektorys, K.: *Variational Methods in Mathematics, Science and Engineering,* D. Reidel, Boston, MA, 1980.

———: *The Method of Discretization in Time,* D. Reidel, Boston, MA, 1982.

Timoshenko, S. P., and J. N. Goodier: *Theory of Elasticity,* 3d ed., McGraw-Hill, New York, 1970.

CHAPTER 11

FLOWS OF VISCOUS INCOMPRESSIBLE FLUIDS

11.1 PRELIMINARY COMMENTS

An introduction to fluids and their governing equations was given in Section 3.3.2. The finite element analysis of fluid flow problems that can be described as one-dimensional systems was also discussed there. Two-dimensional flows of inviscid incompressible fluids (i.e., potential flows) were considered in Section 8.4.2. Potential flow problems were cast in terms of either the stream function or the velocity potential, and the governing equation in each case was the Laplace equation (i.e., a second-order partial differential equation in a single variable; see Chapter 8).

In this section, we consider the finite element analysis of two-dimensional flows of viscous incompressible fluids. These problems are governed by a set of coupled partial differential equations in terms of the velocity components and pressure. When the speed of the flow is low, the nonlinear terms due to inertial effects can be neglected. Such a flow is called Stokes flow, and the resulting equations are termed the Stokes equations. Here we describe two different finite element models of Stokes flow. We begin with a review of the pertinent equations governing slow, laminar flows of viscous incompressible fluids (see Section 3.3.2).

11.2 GOVERNING EQUATIONS

Consider the slow flow of a viscous incompressible fluid in a closed domain Ω:

$$\frac{D}{Dt} \equiv \frac{\partial}{\partial t} + \mathbf{v} \cdot \boldsymbol{\nabla} \approx \frac{\partial}{\partial t} \qquad \text{(slow: inertial effects are negligible)}$$

$$\mu \neq 0 \qquad \text{(viscous)}$$

$$\frac{D\rho}{Dt} = 0 \quad (\rho = \text{constant}) \qquad \text{(incompressible)}$$

Suppose that the domain is very long in one direction—say, the z direction. Then, if there is no flow along this direction and the velocity components in the other two directions are independent of z, the flow can be approximated by a two-dimensional model. The governing equations are summarized below [see (3.102)–(3.106)]:

Conservation of linear momentum

$$\boxed{\begin{aligned} &\rho \frac{\partial u}{\partial t} - \frac{\partial}{\partial x}\left(2\mu \frac{\partial u}{\partial x}\right) - \frac{\partial}{\partial y}\left[\mu\left(\frac{\partial u}{\partial y} + \frac{\partial v}{\partial x}\right)\right] + \frac{\partial P}{\partial x} - f_x = 0 \\ &\rho \frac{\partial v}{\partial t} - \frac{\partial}{\partial x}\left[\mu\left(\frac{\partial u}{\partial y} + \frac{\partial v}{\partial x}\right)\right] - \frac{\partial}{\partial y}\left(2\mu \frac{\partial v}{\partial y}\right) + \frac{\partial P}{\partial y} - f_y = 0 \end{aligned}} \tag{11.1}$$

Conservation of mass

$$\boxed{\frac{\partial u}{\partial x} + \frac{\partial v}{\partial y} = 0} \tag{11.2}$$

Here u and v are the velocity components along the x and y directions, respectively, P is the pressure, f_x and f_y are the components of the body force, μ is the viscosity, and ρ is the density of the fluid. The boundary and initial conditions are of the form

$$\left.\begin{aligned} t_x &\equiv 2\mu \frac{\partial u}{\partial x} n_x + \mu\left(\frac{\partial u}{\partial y} + \frac{\partial v}{\partial x}\right) n_y - P n_x = \hat{t}_x \\ t_y &\equiv \mu\left(\frac{\partial u}{\partial y} + \frac{\partial v}{\partial x}\right) n_x + 2\mu \frac{\partial v}{\partial y} n_y - P n_y = \hat{t}_y \end{aligned}\right\} \text{ on } \Gamma_1 \tag{11.3a}$$

$$u = \hat{u}, \quad v = \hat{v} \qquad \text{on } \Gamma_2$$

$$u(x, y, 0) = u_0(x, y), \qquad v(x, y, 0) = v_0(x, y) \tag{11.3b}$$

The same four possibilities as given in (10.39) exist for the specification of the boundary conditions in fluid flow problems; i.e. only one element of each of the pairs (u, t_x) and (v, t_y) can be specified at any point on the boundary. The boundary stresses are

$$
\begin{aligned}
t_x &= \left(2\mu \frac{\partial u}{\partial x} - P\right)n_x + \mu\left(\frac{\partial u}{\partial y} + \frac{\partial v}{\partial x}\right)n_y \\
t_y &= \mu\left(\frac{\partial u}{\partial y} + \frac{\partial v}{\partial x}\right)n_x + \left(2\mu \frac{\partial v}{\partial y} - P\right)n_y
\end{aligned}
\tag{11.3c}
$$

There are three partial differential equations (11.1), (11.2) in three unknowns (u, v and P). It is possible to reduce the number of equations and unknowns by the introduction of the stream function ψ and/or the vorticity ζ:

$$
\frac{\partial \psi}{\partial x} = -v, \qquad \frac{\partial \psi}{\partial y} = u, \qquad \zeta = \frac{\partial u}{\partial y} - \frac{\partial v}{\partial x} = \nabla^2 \psi
\tag{11.4}
$$

If both functions are used, the three equations can be reduced to two second-order equations [because the continuity equation (11.2) is identically satisfied by the stream function]. If ζ is not used, the three equations can be reduced to a single fourth-order equation in ψ. Since these alternative formulations are not as physically clear as the original equations (especially for the imposition of boundary conditions), or require higher-order elements (for the fourth-order equation for ψ), we shall confine our attention here to the development of finite element models of (11.1) and (11.2) in terms of the primitive variables (u, v, and P).

In the present study, we shall consider two different finite element models of (11.1) and (11.2). The first is a natural and direct formulation in which the three equations in u, v, P are used in their original form. This formulation is known as the *velocity–pressure formulation.* The other is based on the interpretation that the continuity equation (11.2) is an additional relation among the velocity components (i.e., a constraint on u and v), and this constraint is satisfied in a least-squares (i.e., approximate) sense. This particular method of including the constraint in the formulation is known as the *penalty function method,* and the model is termed as the *penalty finite element model.* It is informative to note that the velocity–pressure formulation is the same as the Lagrange multiplier formulation, wherein the constraint is included by means of the Lagrange multiplier. The Lagrange multiplier turns out to be the negative of the pressure.

11.3 VELOCITY–PRESSURE FINITE ELEMENT MODEL

The weak forms of (11.1) and (11.2) over an element Ω^e can be obtained using the three-step procedure discussed in Chapter 8 and revisited in Chapter 10 for plane elasticity problems. We multiply the three equations in (11.1) and (11.2), with three different weight functions w_1, w_2, and w_3, and integrate over the element:

$$
\begin{aligned}
0 &= \int_{\Omega^e} w_1 \left\{ \rho \frac{\partial u}{\partial t} - \frac{\partial}{\partial x}\left(2\mu \frac{\partial u}{\partial x}\right) - \frac{\partial}{\partial y}\left[\mu\left(\frac{\partial u}{\partial y} + \frac{\partial v}{\partial x}\right)\right] + \frac{\partial P}{\partial x} - f_x \right\} dx\, dy \\
0 &= \int_{\Omega^e} w_2 \left\{ \rho \frac{\partial v}{\partial t} - \frac{\partial}{\partial x}\left[\mu\left(\frac{\partial u}{\partial y} + \frac{\partial v}{\partial x}\right)\right] - \frac{\partial}{\partial y}\left(2\mu \frac{\partial v}{\partial y}\right) + \frac{\partial P}{\partial y} - f_y \right\} dx\, dy \qquad (11.5) \\
0 &= \int_{\Omega^e} w_3 \left(\frac{\partial u}{\partial x} + \frac{\partial v}{\partial y}\right) dx\, dy
\end{aligned}
$$

where w_1, w_2, and w_3 are the weight functions. The latter can be interpreted physically as follows. Since the first equation is the momentum equation and $f_x\, dx\, dy$ denotes the force, w_1 must be like the x component of velocity (u), so that the product $f_x w_1$ gives the power. Similarly, w_2 must be like the y component of velocity (v). The third equation represents the volume change in an element of dimensions dx and dy. Therefore, w_3 must be like a force that causes the volume change. Volume changes occur under the action of hydrostatic pressure; hence, w_3 is like $-P$:

$$\boxed{w_1 \sim u, \qquad w_2 \sim v, \qquad w_3 \sim -P} \qquad (11.6)$$

This interpretation is useful in developing the finite element model, because w_1, for example, will be replaced by the ith interpolation function used in the approximation of u. Similarly, w_3 will be replaced by the ith interpolation function used in the approximation of P. When different interpolations are used for (u, v) and P, this interpretation becomes necessary.

The second step of the weak formulation for the present case needs some comments. The integration by parts to distribute differentiation equally among the variables (u, v, P) and weight functions (w_1, w_2, w_3) helps relax the continuity requirements on the finite element approximation functions used for u, v, and P. However, in any problem, such trading of differentiability is subject to the restriction that the resulting boundary expressions must be physically meaningful. Otherwise, the secondary variables of the formulation may not be the quantities the physical problem admits as boundary conditions. Examination of the boundary stress components t_x and t_y shows that the pressure term is a part of them [see (11.3c)]. For example, t_x is the x component of the *total* boundary stress, which is the sum of the viscous boundary stress, $2\mu(\partial u/\partial x)n_x + \mu(\partial u/\partial y + \partial v/\partial x)n_y$, and the hydrostatic boundary stress, $-Pn_x$. Therefore, it is necessary that each term, except the body force term, must be integrated by parts in the two momentum equations. By trading the differentiation from P onto w_1 and w_2 in the two momentum equations, we gain both physically meaningful natural boundary conditions and the symmetry of the finite element equations, as we shall see shortly. No integration by parts is used in the continuity equation, because no relaxation of continuity on u and v can be accomplished; further, the resulting boundary conditions are not physically justifiable.

Keeping the above comments in mind, we carry out the remaining two steps of the weak formulation. The result is

$$
\begin{aligned}
0 &= \int_{\Omega^e} \left[\rho w_1 \frac{\partial u}{\partial t} + 2\mu \frac{\partial w_1}{\partial x}\frac{\partial u}{\partial x} + \mu \frac{\partial w_1}{\partial y}\left(\frac{\partial u}{\partial y} + \frac{\partial v}{\partial x}\right) \right. \\
&\qquad \left. - \frac{\partial w_1}{\partial x} P - w_1 f_x \right] dx\, dy - \oint_{\Gamma^e} w_1 t_x\, ds \\
0 &= \int_{\Omega^e} \left[\rho w_2 \frac{\partial v}{\partial t} + \mu \frac{\partial w_2}{\partial x}\left(\frac{\partial u}{\partial y} + \frac{\partial v}{\partial x}\right) + 2\mu \frac{\partial w_2}{\partial y}\frac{\partial v}{\partial y} \right. \\
&\qquad \left. - \frac{\partial w_2}{\partial y} P - w_2 f_y \right] dx\, dy - \oint_{\Gamma^e} w_2 t_y\, ds \\
0 &= -\int_{\Omega^e} w_3 \left(\frac{\partial u}{\partial x} + \frac{\partial v}{\partial y}\right) dx\, dy
\end{aligned}
\tag{11.7}
$$

Note that there is no boundary integral involving w_3, because no integration by parts is used. This implies that P is *not* a primary variable; it is part of the secondary variables (t_x and t_y). This in turn requires that P not be made continuous across interelement boundaries. If P by itself is not specified in a problem (but t_x and t_y are specified) then P is arbitrarily set equal to a value at some node to determine the constant part of the pressure. Thus, P can be determined only to within an arbitrary constant. The minus sign in the third statement is inserted because $P \sim -w_3$, which makes the resulting finite element model symmetric.

An examination of the weak form reveals that u and v are the primary variables that should be made continuous at interelement boundaries, while P is a nodal variable that is not to be made continuous across the interelement boundaries. Therefore, the Lagrange family of finite elements can be used for (u, v, P). The weak form shows that the minimum continuity requirements on (u, v, P) are

$$(u, v) \text{ linear in } x \text{ and } y$$

$$P \text{ constant}$$

Thus, there are different continuity requirements on the interpolation of the velocity field and pressure:

$$
\boxed{u = \sum_{j=1}^{n} u_j \psi_j^e, \qquad v = \sum_{j=1}^{n} v_j \psi_j^e, \qquad P = \sum_{j=1}^{m} P_j \phi_j^e}
\tag{11.8}
$$

where ψ_j^e and ϕ_j^e are interpolation functions of different order ($m < n$; $m + 1 = n$). Substituting (11.8) into (11.7), we obtain the finite element model:

$$
\begin{bmatrix} [M] & [0] & [0] \\ [0] & [M] & [0] \\ [0] & [0] & [0] \end{bmatrix}
\begin{Bmatrix} \{\dot{u}\} \\ \{\dot{v}\} \\ \{\dot{P}\} \end{Bmatrix}
+
\begin{bmatrix} [K^{11}] & [K^{12}] & [K^{13}] \\ [K^{21}] & [K^{22}] & [K^{23}] \\ [K^{31}] & [K^{32}] & [K^{33}] \end{bmatrix}
\begin{Bmatrix} \{u\} \\ \{v\} \\ \{P\} \end{Bmatrix}
=
\begin{Bmatrix} \{F^1\} \\ \{F^2\} \\ \{0\} \end{Bmatrix}
\tag{11.9a}
$$

where $(i, j = 1, 2, \ldots, n; J = 1, 2, \ldots, m)$

$$M_{ij} = \int_{\Omega^e} \rho \psi_i \psi_j \, dx \, dy$$

$$[K^{11}] = 2\mu[S^{11}] + \mu[S^{22}], \qquad [K^{12}] = \mu[S^{12}]^T, \qquad [K^{21}] = [K^{12}]^T$$

$$[K^{22}] = \mu[S^{11}] + 2\mu[S^{22}], \qquad [K^{33}] = [0] \text{ (zero matrix)}$$

$$S_{ij}^{11} = \int_{\Omega^e} \frac{\partial \psi_i}{\partial x}\frac{\partial \psi_j}{\partial x} dx\, dy, \qquad S_{ij}^{12} = \int_{\Omega^e} \frac{\partial \psi_i}{\partial x}\frac{\partial \psi_j}{\partial y} dx\, dy, \qquad S_{ij}^{22} = \int_{\Omega^e} \frac{\partial \psi_i}{\partial y}\frac{\partial \psi_j}{\partial y} dx\, dy \tag{11.9b}$$

$$F_i^1 = \int_{\Omega^e} f_x \psi_i \, dx\, dy + \oint_{\Gamma^e} t_x \psi_i \, ds, \qquad F_i^2 = \int_{\Omega^e} f_y \psi_i \, dx\, dy + \oint_{\Gamma^e} t_y \, \psi_i \, ds$$

$$K_{iJ}^{13} = -\int_{\Omega^e} \frac{\partial \psi_i}{\partial x} \phi_J \, dx\, dy, \qquad K_{iJ}^{23} = -\int_{\Omega^e} \frac{\partial \psi_i}{\partial y} \phi_J \, dx\, dy$$

We note that $[K^{33}] = [0]$, because the continuity equation does not contain P. Therefore, the assembled equations will also have zeros in diagonal elements corresponding to the nodal values of P.

Both triangular and rectangular elements for the velocity–pressure model are shown in Fig. 11.1. Since the pressure does not appear at all nodes and it is not made continuous between elements (i.e., a discontinuous pressure approximation is used), the assembly procedure is applicable to only the velocity degrees of freedom. Thus, a finite element mesh with N global nodes will have a total of $2N + M$ unknowns, where N is the total number of nodes in the mesh and M is the number of pressure degrees of freedom.

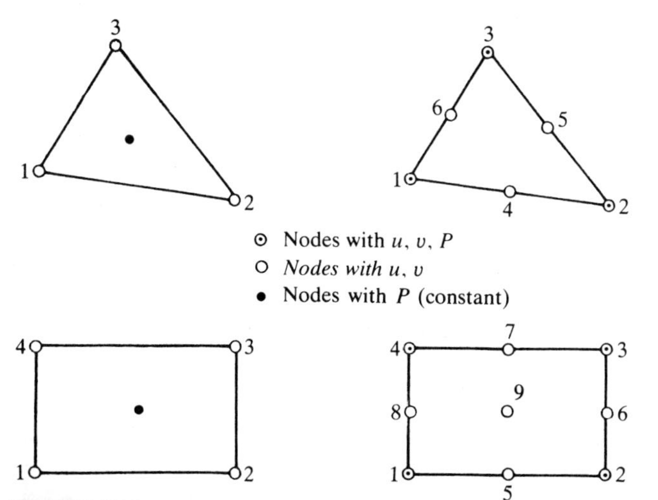

FIGURE 11.1
Linear and quadratic triangular and rectangular finite elements for the pressure–velocity model of incompressible fluid flow.

11.4 PENALTY–FINITE ELEMENT MODEL

11.4.1 Penalty Function Method

The penalty function method, like the Lagrange multiplier method, allows us to reformulate a problem with constraints as one without constraints. The basic idea of the method can be described by considering a constrained algebraic problem:

minimize the function $f(x, y)$ subject to the constraint $G(x, y) = 0$

In the Lagrange multiplier method, the problem is reformulated as one of determining the stationary (or critical) points of the modified function $F_L(x, y)$,

$$F_L(x, y) = f(x, y) + \lambda G(x, y) \tag{11.10}$$

subject to no constraints. Here λ denotes the *Lagrange multiplier*. The solution to the problem is obtained by setting partial derivatives of F_L with respect to x, y, and λ equal to zero:

$$\frac{\partial F_L}{\partial x} = 0, \qquad \frac{\partial F_L}{\partial y} = 0, \qquad \frac{\partial F_L}{\partial \lambda} = 0 \tag{11.11}$$

which gives three equations in the three unknowns (x, y, λ).

In the penalty function method, the problem is reformulated as one of finding the minimum of the modified function F_P,

$$F_P(x, y) = f(x, y) + \tfrac{1}{2}\gamma[G(x, y)]^2 \tag{11.12}$$

where γ is a preassigned weight parameter, called the *penalty parameter*. The factor $\frac{1}{2}$ in (11.12) is used for convenience: when F_P is differentiated with respect to its arguments, the factor will be cancelled by the power on $G(x, y)$. The solution to the modified problem is given by the following two equations:

$$\frac{\partial F_P}{\partial x} = 0, \qquad \frac{\partial F_P}{\partial y} = 0 \tag{11.13a}$$

The solution (x_γ, y_γ) of these will be a function of the penalty parameter γ. The larger the value of γ, the more exactly is the constraint satisfied (in a least-squares sense), and (x_γ, y_γ) approaches the actual solution (x, y) as $\gamma \to \infty$. An approximation to the Lagrange multiplier is computed from the equation (by comparing δF_L with δF_P)

$$\lambda_\gamma = \gamma G(x_\gamma, y_\gamma) \tag{11.13b}$$

We consider a specific example to illustrate the above ideas.

Example 11.1. Minimize the quadratic function

$$f(x, y) = 4x^2 - 3y^2 + 2xy + 6x - 3y + 5 \tag{11.14a}$$

subject to the constraint

$$G(x, y) \equiv 2x + 3y = 0 \tag{11.14b}$$

Geometrically, we see that the inflection point of the surface $f(x, y)$ that is on the line $2x + 3y = 0$ is the solution. We solve the problem using the Lagrange multiplier method and the penalty function method.

Lagrange multiplier method. The modified functional is

$$F_L(x, y) = f(x, y) + \lambda(2x + 3y) \tag{11.15}$$

and we have

$$\begin{aligned} \frac{\partial F_L}{\partial x} &= 8x + 2y + 6 + 2\lambda = 0 \\ \frac{\partial F_L}{\partial y} &= -6y + 2x - 3 + 3\lambda = 0 \\ \frac{\partial F_L}{\partial \lambda} &= 2x + 3y = 0 \end{aligned} \tag{11.16}$$

Solving the three algebraic equations, we obtain

$$x = -3, \qquad y = 2, \qquad \lambda = 7 \tag{11.17}$$

Penalty function method. The modified functional is

$$F_P(x, y) = f(x, y) + \tfrac{1}{2}\gamma(2x + 3y)^2 \tag{11.18}$$

and we have

$$\begin{aligned} \frac{\partial F_P}{\partial x} &= 8x + 2y + 6 + 2\gamma(2x + 3y) = 0 \\ \frac{\partial F_P}{\partial y} &= -6y + 2x - 3 + 3\gamma(2x + 3y) = 0 \end{aligned} \tag{11.19}$$

The solution of these equations is

$$x_\gamma = \frac{15 - 36\gamma}{-26 + 12\gamma}, \qquad y_\gamma = \frac{18 + 24\gamma}{-26 + 12\gamma} \tag{11.20a}$$

The Lagrange multiplier is given by

$$\lambda_\gamma = \gamma G(x_\gamma, y_\gamma) = \frac{84\gamma}{-26 + 12\gamma} \tag{11.20b}$$

Clearly, in the limit $\gamma \to \infty$, the penalty function solution approaches the exact solution:

$$\lim_{\gamma\to\infty} x_\gamma = -3, \qquad \lim_{\gamma\to\infty} y_\gamma = 2 \qquad \lim_{\gamma\to\infty} \lambda_\gamma = 7$$

An approximate solution (11.20) to the problem can be obtained, to within a desired accuracy, by selecting a finite value of the penalty parameter (see Table 11.1).

11.4.2 Formulation of the Flow Problem as a Constrained Problem

The continuum problem at hand, namely, the flow of a viscous incompressible fluid, can be stated as one of minimizing a functional subjected to a constraint. From the weak form in (11.7), we have the following linear and bilinear forms,

TABLE 11.1
Convergence of the penalty function solution with increasing penalty parameter

γ	x_γ	y_γ	λ_γ	$E_\gamma = 2x_\gamma + 3y_\gamma$
0	−0.5769	−0.6923	0.0000	−3.2308
1	1.5000	−3.0000	−6.0000	−6.0000
10	−3.6702	2.7447	8.9362	0.8936
100	−3.0537	2.0596	7.1550	0.0716
1,000	−3.0053	2.0058	7.0152	0.0068
10,000	−3.0005	2.0006	7.0015	0.0008
∞	−3.0000	2.0000	7.0000	0.0000

for the steady-state case, when the velocity field satisfies the constraint $\partial u/\partial x + \partial v/\partial y = 0$:

$$l(w_1, w_2) = \int_{\Omega^e} (f_x w_1 + f_y w_2)\,dx\,dy + \oint_{\Gamma^e} (t_x w_1 + t_y w_2)\,ds \tag{11.21a}$$

$$B((w_1, w_2), (u, v)) = \mu \int_{\Omega^e} \left[2\left(\frac{\partial w_1}{\partial x}\frac{\partial u}{\partial x} + \frac{\partial w_2}{\partial y}\frac{\partial v}{\partial y} \right) + \left(\frac{\partial w_1}{\partial y} + \frac{\partial w_2}{\partial x} \right)\left(\frac{\partial u}{\partial y} + \frac{\partial v}{\partial x} \right) \right] dx\,dy \tag{11.21b}$$

The pressure term does not appear in the bilinear form, and the third equation in (11.7) is not used, because the continuity equation is identically satisfied by the velocity field (u, v); i.e., (u, v) are subject to the constraint $\partial u/\partial x + \partial v/\partial y = 0$. From the discussion in Chapter 2 [see (2.40)–(2.44)], the quadratic functional is given by

$$\begin{aligned} I((u, v)) &= \tfrac{1}{2}B((u, v), (u, v)) - l((u, v)) \\ &= \mu \int_{\Omega^e} \left[\left(\frac{\partial u}{\partial x}\right)^2 + \left(\frac{\partial v}{\partial y}\right)^2 + \frac{1}{2}\left(\frac{\partial u}{\partial y} + \frac{\partial v}{\partial x} \right)^2 \right] dx\,dy \\ &\quad - \int_{\Omega^e} (f_x u + f_y v)\,dx\,dy - \oint_{\Gamma^e} (t_x u + t_y v)\,ds \end{aligned} \tag{11.22}$$

where the velocity components (u, v) satisfy the continuity equation (11.2).

Now we state that (11.1) and (11.2) governing the steady flow of viscous incompressible fluids are equivalent to minimizing $I(u, v)$ of (11.22) subject to the contsraint

$$G(u, v) \equiv \frac{\partial u}{\partial x} + \frac{\partial v}{\partial y} = 0 \tag{11.23}$$

The constrained problem can be reformulated as an unconstrained problem using the Lagrange multiplier or penalty function methods. These are discussed next.

11.4.3 Lagrange Multiplier Formulation

In the Lagrange multiplier method, the constrained problem is reformulated as one of finding the stationary points of the modified functional

$$I_L(u, v, \lambda) \equiv I(u, v) + \int_{\Omega^e} \lambda G(u, v)\, dx\, dy \tag{11.24}$$

The necessary condition for I_L to have a stationary point is that

$$\delta I_L = \delta_u I_L + \delta_v I_L + \delta_\lambda I_L = 0$$

or

$$\delta_u I_L = 0, \qquad \delta_v I_L = 0, \qquad \delta_\lambda I_L = 0 \tag{11.25}$$

where δ_u, δ_v, and δ_λ denote the partial variations (see Section 2.2.4) with respect to u, v, and λ, respectively. The three equations in (11.25) are exactly the same as the weak forms in (11.7), with $\lambda = -P$, $\delta u = w_1$, $\delta v = w_2$, and $\delta\lambda = -w_3$. Therefore, the velocity–pressure finite element model can also be termed the Lagrange multiplier finite element model.

11.4.4 Penalty Function Formulation

In the penalty function method, the constrained problem is reformulated as an unconstrained problem as follows: minimize the modified functional

$$I_P(u, v) \equiv I(u, v) + \tfrac{1}{2}\gamma_e \int_{\Omega^e} [G(u, v)]^2\, dx\, dy \tag{11.26}$$

where the penalty parameter γ_e can be chosen element-wise. The necessary conditions for the minimum of I_P are

$$\delta_u I_P = 0, \qquad \delta_v I_P = 0 \tag{11.27}$$

We have

$$\begin{aligned}
\delta_u I_P &= \delta_u I + \gamma_e \int_{\Omega^e} G(u, v)\, \delta_u G(u, v)\, dx\, dy \\
&= \int_{\Omega^e} \left[2\mu \frac{\partial \delta u}{\partial x}\frac{\partial u}{\partial x} + \mu \frac{\partial \delta u}{\partial y}\left(\frac{\partial u}{\partial y} + \frac{\partial v}{\partial x}\right)\right] dx\, dy - \int_{\Omega^e} \delta u\, f_x\, dx\, dy - \oint_{\Gamma^e} \delta u\, t_x\, ds \\
&\quad + \gamma_e \int_{\Omega^e} \frac{\partial \delta u}{\partial x}\left(\frac{\partial u}{\partial x} + \frac{\partial v}{\partial y}\right) dx\, dy = 0
\end{aligned} \tag{11.28a}$$

$$\begin{aligned}
\delta_v I_P &= \int_{\Omega^e} \left[2\mu \frac{\partial \delta v}{\partial y}\frac{\partial v}{\partial y} + \mu \frac{\partial \delta v}{\partial x}\left(\frac{\partial u}{\partial y} + \frac{\partial v}{\partial x}\right)\right] dx\, dy - \int_{\Omega^e} \delta v\, f_y\, dx\, dy - \oint_{\Gamma^e} \delta v\, t_y\, ds \\
&\quad + \gamma_e \int_{\Omega^e} \frac{\partial \delta v}{\partial y}\left(\frac{\partial u}{\partial x} + \frac{\partial v}{\partial y}\right) dx\, dy = 0
\end{aligned} \tag{11.28b}$$

These two statements provide the weak forms for the penalty finite element model. We note that the pressure, which is the Lagrange multiplier, does not appear explicitly in the weak form (11.28), although it is a part of the boundary stresses t_x and t_y. The time-derivative terms can be added to (11.28) without affecting the above discussion:

$$\int_{\Omega^e} \rho\, \delta u \frac{\partial u}{\partial t}\, dx\, dy, \qquad \int_{\Omega^e} \rho\, \delta v \frac{\partial v}{\partial t}\, dx\, dy \tag{11.29}$$

The penalty finite element model is obtained from (11.28*a,b*) with the inertia terms (11.29) added, by substituting (11.8) for the velocity field, and $\delta u = \psi_i$ and $\delta v = \psi_i$:

$$\begin{bmatrix} [M] & [0] \\ [0] & [M] \end{bmatrix} \begin{Bmatrix} \{\dot{u}\} \\ \{\dot{v}\} \end{Bmatrix} + \begin{bmatrix} [\bar{K}^{11}] & [\bar{K}^{12}] \\ [\bar{K}^{21}] & [\bar{K}^{22}] \end{bmatrix} \begin{Bmatrix} \{u\} \\ \{v\} \end{Bmatrix} = \begin{Bmatrix} \{F^1\} \\ \{F^2\} \end{Bmatrix} \tag{11.30}$$

where F_i^1 and F_i^2 are as defined in (11.9*b*), and

$$\begin{gathered} [\bar{K}^{11}] = [K^{11}] + \gamma[S^{11}], \qquad [\bar{K}^{12}] = [K^{12}] + \gamma[S^{12}], \\ [\bar{K}^{22}] = [K^{22}] + \gamma[S^{22}] \end{gathered} \tag{11.31}$$

The matrices $[K^{\alpha\beta}]$ and $[S^{\alpha\beta}]$ are defined in (11.9*b*).

For the unsteady case, (11.9*a*) or (11.30) are further approximated using a time approximation scheme. First, let us write (11.30) in the concise form

$$[M]\{\dot{\Delta}\} + [K]\{\Delta\} = \{F\} \tag{11.32}$$

where $\{\Delta\}$ denotes the vector of nodal velocities. Using the α family of approximation [see (8.162)–(8.164)], we reduce (11.32) to

$$[\hat{K}]_{s+1}\{\Delta\}_{s+1} = \{\hat{F}\}_{s,s+1} \tag{11.33a}$$

where

$$\begin{gathered} [\hat{K}]_{s+1} = [M] + a_1[K]_{s+1} \\ \{\hat{F}\}_{s+1} = (\alpha\{F\}_{s+1} + (1-\alpha)\{F\}_s)\, \Delta t_{s+1} + ([M] - a_2[K]_s)\{\Delta\}_s \\ a_1 = \alpha\, \Delta t, \qquad a_2 = (1-\alpha)\, \Delta t \end{gathered} \tag{11.33b}$$

The Lagrange multiplier (i.e., the negative of the pressure) for any $t > 0$ is computed from the equation

$$\boxed{\lambda_\gamma = -P_\gamma^e = \gamma G(u_\gamma, v_\gamma) = \gamma\left(\frac{\partial u_\gamma}{\partial x} + \frac{\partial v_\gamma}{\partial y}\right)} \tag{11.34}$$

where (u_γ, v_γ) is the finite element solution of (11.33).

11.4.5 Computational Aspects

Some comments are in order on the numerical evaluation of the integrals and pressure computation in the penalty model. Note that the penalty finite element model for the steady case is of the form

$$(\mu[K^1] + \gamma[K^2])\{\Delta\} = \{F\} \tag{11.35}$$

where $[K^1]$ is the contribution from the viscous terms and $[K^2]$ is from the penalty terms, which come from the incompressibility constraint (11.2). In theory, as we increase the value of γ, the mass conservation constraint is satisfied more exactly. However, in practice, for some large value of γ, the contribution from the viscous terms would be negligibly small compared with the penalty terms in the computer. Thus, in the limit as γ tends to a large number, we have

$$\lim_{\gamma\to\infty} \frac{1}{\gamma}\left((\mu[K^1]+\gamma[K^2])\{\Delta\}-\{F\}\right)=\{0\} \tag{11.36a}$$

or

$$[K^2]\{\Delta\}=\{0\} \quad \text{as } \gamma\to\infty \tag{11.36b}$$

Thus, if $[K^2]$ is a nonsingular (i.e., invertible) matrix, the solution of (11.36b) for large γ is trivial: $\{\Delta\}=\{0\}$. While this solution satisfies the continuity equation, it does not satisfy the momentum equations. In this case, the discrete problem (11.35) is said to be overconstrained or "locked." If $[K^2]$ is singular then the sum $\mu[K^1]+\gamma[K^2]$ is nonsingular (because $[K^1]$ is nonsingular), and a nontrivial solution to the problem can be obtained.

The numerical problem described above is eliminated by proper evaluation of the integrals in $[K^1]$ and $[K^2]$. It is found that if the coefficients of $[K^2]$ (i.e., the penalty terms) are evaluated using a numerical integration rule (see Section 9.2) of an order less than that required to integrate them exactly, the finite element equations (11.35) give acceptable solutions for the velocity field. This technique of under-integrating the penalty terms is known in the literature as *reduced integration.*

For a linear rectangular element, for example, the coefficients K_{ij}^1 are evaluated using 2×2 Gauss quadrature, and K_{ij}^2 are evaluated using 1×1 Gauss quadrature. One-point quadrature yields a singular $[K^2]$. Therefore, (11.36b) cannot be inverted, whereas $\mu_e[K^1]+\gamma_e[K^2]$ is nonsingular and can be inverted (after assembly and imposition of boundary conditions) to obtain a good finite element solution of the original problem.

For linear and quadratic triangular elements, it is found that a good solution for the velocity field is obtained without using reduced integration on penalty terms. For example, the four-point integration (i.e., full integration) of $[K^1]$ and $[K^2]$ for the linear triangular element gives exactly the same velocity field as when four-point (2×2) integration is used to evaluate $[K^1]$ and either three- or one-point integration is used for $[K^2]$. Similarly, seven-point integration of $[K^1]$ and $[K^2]$ for the quadratic triangular element gives the same velocity field as that obtained using seven-point integration of $[K^1]$ and three- or four-point integration of $[K^2]$. Since reduced integration saves computational time, it is used to evaluate $[K^2]$ in the present study.

Concerning the post-computation of pressure in the penalty model, in general, the pressure computed from (11.34) at the integration points is not always reliable and accurate. Various techniques have been proposed in the literature to obtain accurate pressure fields. It is found that for both triangular and rectangular elements, the pressure computed at the reduced Gauss

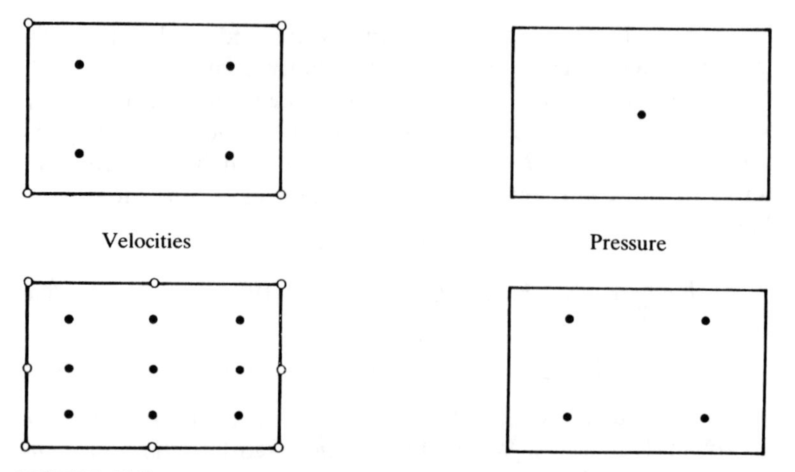

FIGURE 11.2
Finite elements for the approximation of velocities and the post-computation of the pressure (open circles denote nodes and dark circles integration points) in the penalty finite element model. A three-point evaluation of pressure is also suggested when velocities are approximated by quadratic polynomials.

quadrature points (the same integration points that are used to evaluate the penalty terms $[K^2]$) is better than that obtained with full integration points. Thus, pressure should be computed from (11.34) at the Gauss points of the reduced integration. The pressures predicted using the linear elements, especially for coarse meshes, are seldom acceptable. Quadratic elements are known to yield more reliable results. In general, triangular elements do not yield stable solutions for pressures. A mathematical study of the penalty elements (mostly rectangular ones) can be found in the work of Oden (1982).

Figure 11.2 shows linear and quadratic finite elements with integration points for the viscous and penalty terms. Other elements and their convergence and stability characteristics can be found in Oden and Carey (1983). Of these elements, only rectangular ones give stable solutions for pressures; and for uniform meshes and problems with smooth solutions, all elements give good velocity fields. The reader should be cautioned that pressures produced by the penalty elements can be unstable for some problems and coarse meshes, especially when triangular elements are used.

The choice of penalty parameter is largely dictated by the ratio of the magnitude of penalty terms to the viscous terms, the mesh, and the word length in the computer. Generally, a value of $\gamma = 10^4\mu$–$10^{12}\mu$ gives good results. It is found that the pressure is more sensitive to the value of γ than the velocity field, and triangular elements permit a smaller range of γ ($10^4\mu$–$10^8\mu$) than the rectangular elements for good pressures.

11.5 EXAMPLES

Example 11.2. Consider the flow of a viscous incompressible material squeezed between two long parallel plates (see Fig. 11.3a). When the length of the plates is very

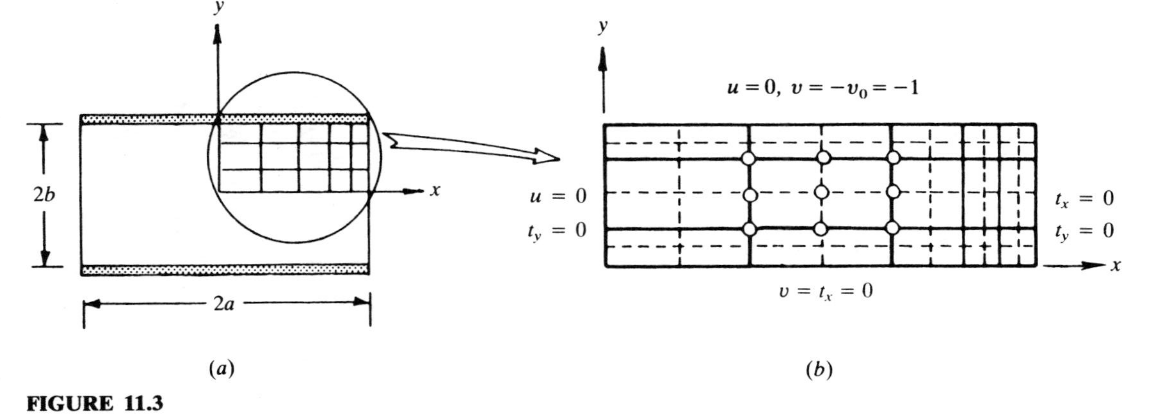

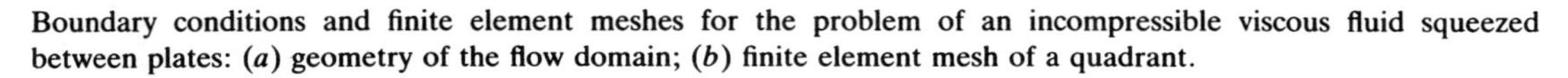

FIGURE 11.3
Boundary conditions and finite element meshes for the problem of an incompressible viscous fluid squeezed between plates: (*a*) geometry of the flow domain; (*b*) finite element mesh of a quadrant.

large compared with both their width and the distance between them, we have a case of plane flow (in the plane formed by the plate width and the distance between them). Assumining that a state of plane flow exists, we determine the velocity and pressure fields. Let V_0 be the velocity with which the two plates are moving toward each other (i.e., squeezing out the fluid), and let $2b$ and $2a$ denote, respectively, the distance between and the width of the plates (see Fig. 11.3a). An approximate analytical solution to this two-dimensional problem is provided by Nadai (1963):

$$
\begin{aligned}
u &= \frac{3V_0 x}{2b}\left(1 - \frac{y^2}{b^2}\right), \qquad v = -\frac{V_0 y}{2b}\left(3 - \frac{y^2}{b^2}\right)\\
P &= \frac{P_0}{a^2}(a^2 + y^2 - x^2), \qquad P_0 = \frac{3\mu V_0 a^2}{2b^3}\\
\sigma_x &\equiv 2\mu\frac{\partial u}{\partial x} - P = \frac{P_0}{a^2}(x^2 - 3y^2 - a^2 + 2b^2)\\
\sigma_y &\equiv 2\mu\frac{\partial v}{\partial y} - P = \frac{P_0}{a^2}(x^2 + y^2 - a^2 - 2b^2)\\
\tau_{xy} &\equiv \mu\left(\frac{\partial u}{\partial y} + \frac{\partial v}{\partial x}\right) = \frac{-2P_0 xy}{a^2}
\end{aligned}
\tag{11.37}
$$

Owing to the biaxial symmetry of the problem, it suffices to model only a quadrant of the domain. A 5×3 nonuniform mesh of nine-node quadratic elements is used in the velocity–pressure model, and a 10×6 mesh of four-node linear elements and a 5×3 mesh of quadratic elements is used in the penalty model (see Fig. 11.3b). The nonuniform mesh, with smaller elements near the free surface (i.e., at $x = a$), is used to accurately approximate the singularity in the shear stress at the point $(a, b) = (6, 2)$. The mesh used for the penalty model has exactly the same number of nodes as the mesh used for the velocity–pressure model. There are no specified

TABLE 11.2
Comparison of the horizontal velocity $u(x, 0)$ obtained using the penalty model† and the mixed (velocity–pressure) model with the analytical solution

	$\gamma = 1.0$		$\gamma = 100$		$\gamma = 10^8$		Mixed model	Mixed model	Analytical solution
x	4-node	9-node	4-node	9-node	4-node	9-node	8-node	9-node	
1	0.0303	0.0310	0.6563	0.6513	0.7576	0.7505	0.7496	0.7497	0.7500
2	0.0677	0.0691	1.3165	1.3062	1.5135	1.4992	1.5038	1.5031	1.5000
3	0.1213	0.1233	1.9911	1.9769	2.2756	2.2557	2.2563	2.2561	2.2500
4	0.2040	0.2061	2.6960	2.6730	3.0541	3.0238	3.0213	3.0203	3.0000
4.5	0.2611	0.2631	3.0718	3.0463	3.4648	3.4307	3.4331	3.4292	3.3750
5.0	0.3297	0.3310	3.4347	3.3956	3.8517	3.8029	3.8249	3.8165	3.7500
5.25	0.3674	0.3684	3.6120	3.5732	4.0441	3.9944	4.0074	3.9893	3.9375
5.5	0.4060	0.4064	3.7388	3.6874	4.1712	4.1085	4.1450	4.1204	4.1250
5.75	0.4438	0.4443	3.8316	3.7924	4.2654	4.2160	4.2188	4.2058	4.3125
6.0	0.4793	0.4797	3.8362	3.7862	4.2549	4.1937	4.2659	4.2364	4.5000

† The three-point Gauss rule for nonpenalty terms and the two-point Gauss rule for penalty terms are used for quadratic elements.

nonzero secondary variables in the problem. The velocities $u(x, 0)$ obtained using the two models are compared with the analytical solution in Table 11.2. The influence of the penalty parameter on the accuracy of the solution is clear from the results.

Next, a 12×8 mesh of linear elements and a 6×4 mesh of quadratic elements are used to evaluate the relative accuracies of the rectangular and triangular elements. The 12×8 mesh of linear triangular elements with full integration of $[K^1]$ and $[K^2]$ [see (11.35)] and selective integration (i.e., full integration of $[K^1]$ and reduced integration of $[K^2]$) both give the same results for the velocity field. However, in both cases, erroneous results for pressure and stresses are obtained. The 6×4 mesh of quadratic triangular elements with full and selective integrations gives the same velocity fields, while the stresses and pressure are predicted to be the same at the same quadrature points. Both the 12×8 mesh of linear rectangular elements and the 6×4 mesh of nine-node rectangular elements give good results for velocities, pressure, and stresses. Figure 11.4 shows plots of the velocity $u(x, y)$ for $x = 4$ and 6, and Fig. 11.5 shows plots of pressure $P(x, y)$, for $y = \text{constant}$, computed using the quadratic triangular and nine-node rectangular elements. The pressure in the penalty model was computed using (11.34) with the 2×2 Gauss rule for the quadratic rectangular element and the one-point formula for the quadratic triangular element. If the pressure in the penalty model were computed using the full quadrature rule for rectangular elements, we should obtain erroneous values. The linear triangular element with full as well as reduced integrations gives unstable pressures, while the quadratic triangular element with one- or two-point rules yields good results. In general, the same quadrature rule as that used for the evaluation of the penalty terms in the coefficient matrix must be used to evaluate the pressure, and one should avoid using the linear triangular element.

The next example deals with the finite element analysis of a lubrication problem.

Example 11.3 Flow of a viscous lubricant in a slider bearing. The slider (or slipper) bearing consists of a short sliding pad moving at a velocity $u = U_0$ relative to a stationary pad inclined at a small angle with respect to the stationary pad, and the small gap between the two pads is filled with a lubricant (see Fig. 11.6*a*). Since the ends of the bearing are generally open, the pressure P_0 there is atmospheric. If the upper pad is parallel to the base plate, the pressure everywhere in the gap must be atmospheric

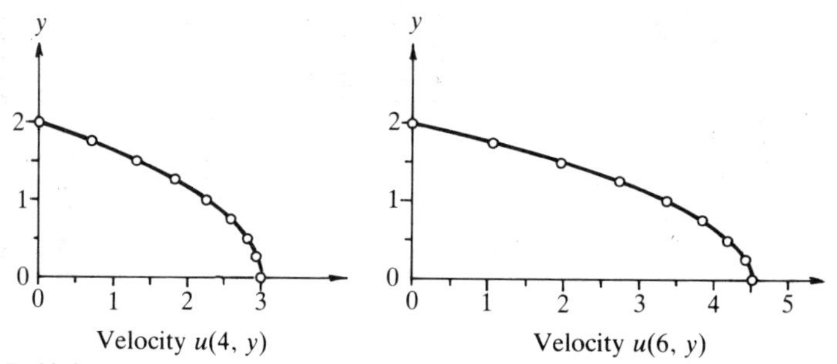

FIGURE 11.4
Horizontal velocity distributions at $x = 3$ and 6 for the problem of a viscous incompressible fluid squeezed between parallel plates (the solid lines represent the analytical solutions and the circles finite element results with all elements for the meshes used).

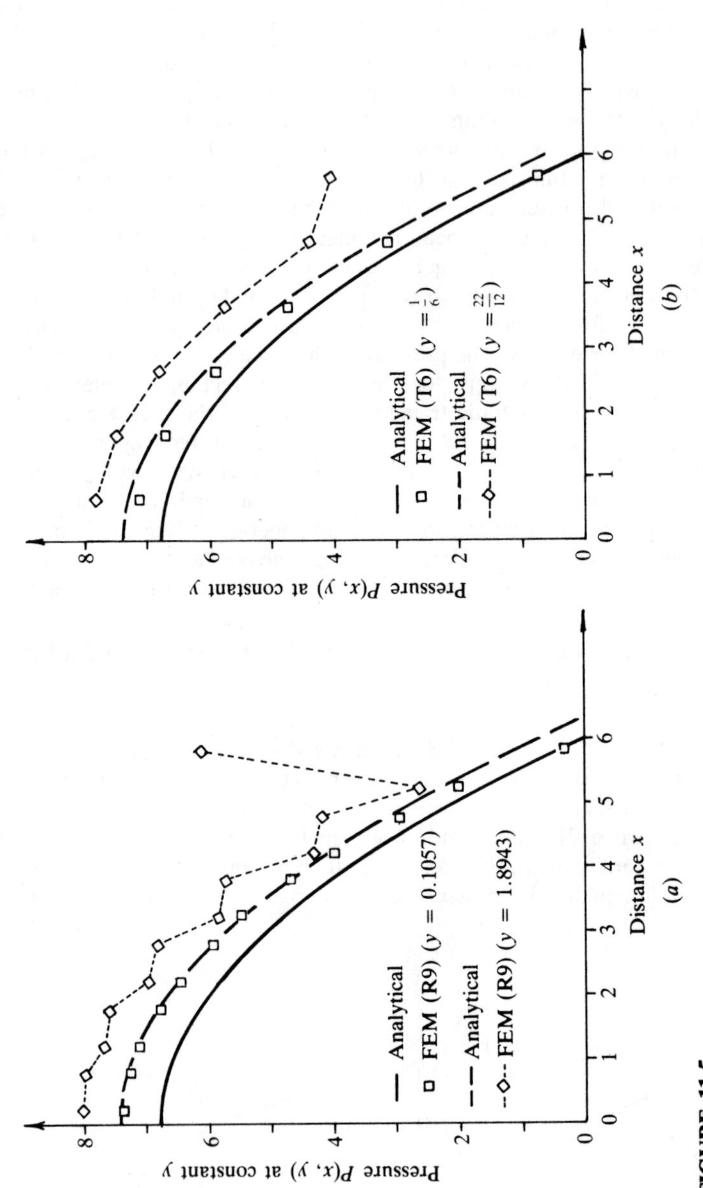

FIGURE 11.5
Comparison of the pressure variations computed using the finite element model and Nadai's analytical solution for a viscous fluid squeezed between two parallel plates (Example 11.2).

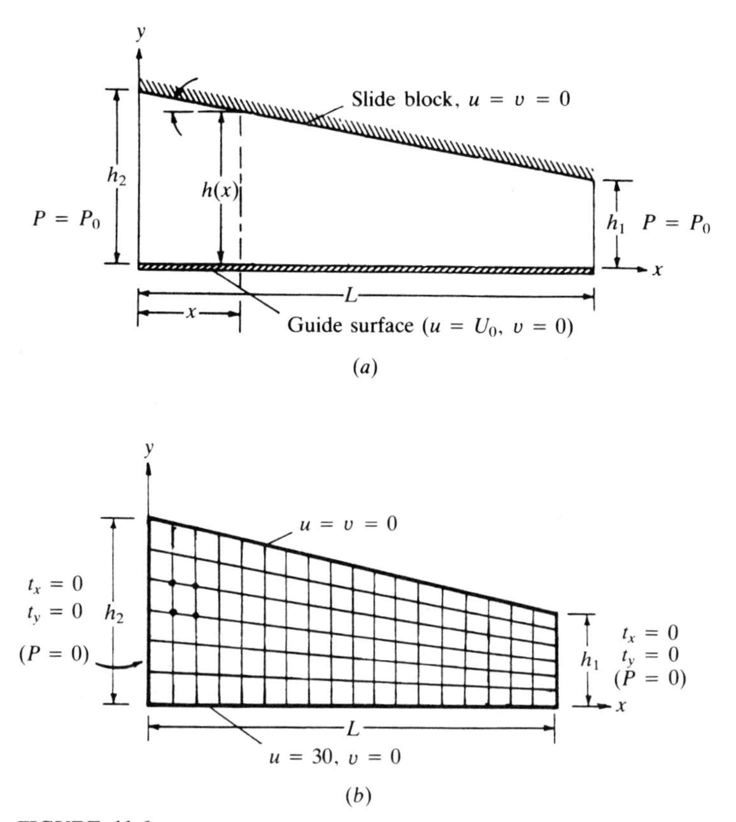

FIGURE 11.6
Finite element analysis of a slider bearing: (*a*) geometry and boundary conditions; (*b*) finite element mesh of linear quadrilateral elements.

(because dP/dx is a constant for flow between parallel plates), and the bearing cannot support any transverse load. If the upper pad is inclined to the base pad, a pressure distribution (in general, a function of x and y) is set up in the gap. For large values of U_0, the pressure generated can be of sufficient magnitude to support heavy loads normal to the base pad.

Analytical solution. Since the width of the gap and the angle of inclination are in general small, it can be assumed with good accuracy that the pressure is not a function of y. Assuming a two-dimensional flow and a small angle of inclination, and neglecting the normal stress gradient (in comparison with the shear stress gradient), the equations governing the motion of the lubricant between the pads can be written as

$$\frac{dP}{dx} = \mu \frac{\partial^2 u}{\partial y^2}, \qquad u = u(x, y), \qquad v = 0 \tag{11.38a}$$

with the boundary conditions

$$\begin{aligned} u &= U_0 \quad \text{at } y = 0 \\ u &= 0, \quad \text{for } y = h(x) \end{aligned} \tag{11.38b}$$

where

$$h(x) = h_2 + \frac{h_1 - h_2}{L} x \tag{11.38c}$$

The solution (u, P) of (11.38) is

$$u = \left(U_0 - \frac{1}{2\mu} h^2 \frac{dP}{dx} \frac{y}{h}\right)\left(1 - \frac{y}{h}\right), \qquad P = \frac{6\mu U_0 L}{h_2^2 - h_1^2} \frac{(h_2 - h)(h - h_1)}{h^2} \tag{11.39}$$

The shear stress is given by

$$\tau_{xy} = \mu \frac{\partial u}{\partial y} = \frac{dP}{dx}(y - \tfrac{1}{2}h) - \mu \frac{U_0}{h}, \tag{11.40}$$

In our computations, we choose

$$\begin{aligned} h_2 = 2h_1 = 8 \times 10^{-4}\,\text{ft}, \qquad L = 0.36\,\text{ft}, \\ \mu = 8 \times 10^{-4}\,\text{lbf s ft}^{-2}, \qquad U_0 = 30\,\text{ft s}^{-1} \end{aligned} \tag{11.41}$$

Finite element solution. It should be pointed out that the assumption concerning the pressure not being a function of y is not necessary in the finite element analysis. First, we use a mesh (mesh 1) of 18×6 linear quadrilateral elements to analyze the problem. The boundary conditions are given in Fig. 11.6(b). Figure 11.7 shows plots of the pressure distributions along the length of the bearing for various values of the penalty parameter. The converged pressure is slightly higher than the analytical solution.

To obtain more accurate solutions, a graded mesh (mesh 2) of 128 linear quadrilateral elements (153 nodes) (equivalently, 32 nine-node quadrilateral elements or 64 six-node triangular elements) is used. For this problem, the mesh of quadratic triangular elements gives unstable pressures (see Fig. 11.8), whereas both linear and quadratic elements give excellent results for velocities, pressure, and shear stress.

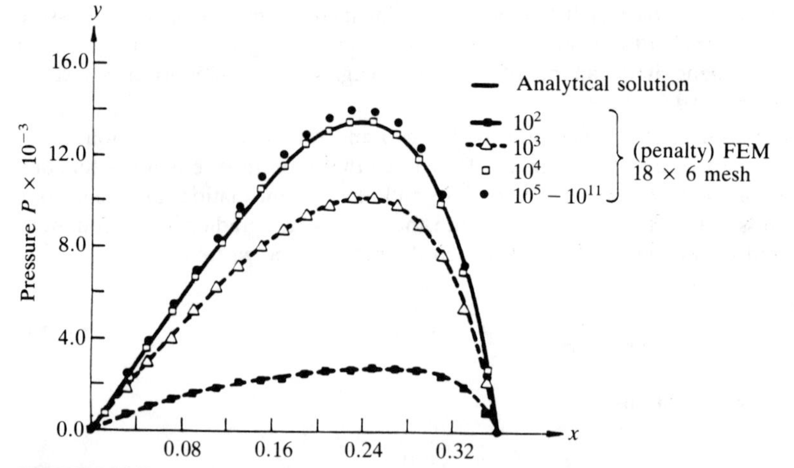

FIGURE 11.7
Influence of the penalty parameter on the variation of the pressure along the length of the slider bearing. A mesh of linear quadrilateral elements is used.

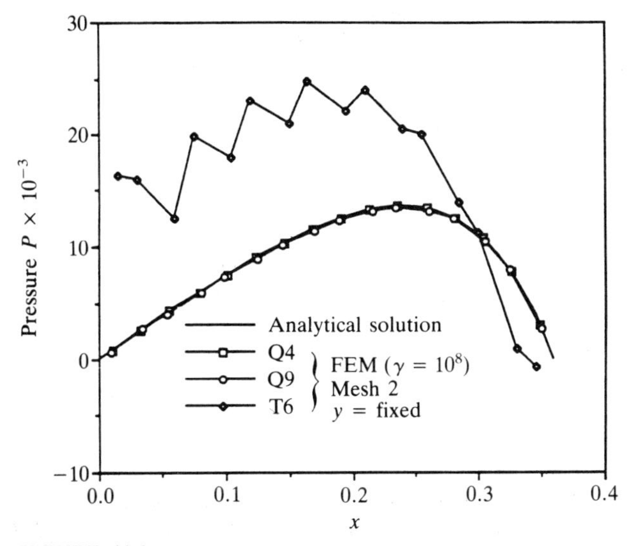

FIGURE 11.8
Variation of pressure $P(x, y_0)$ along the length of the bearing (y_0 is the y coordinate of the first integration point of the first element): Q4, four-node quadrilateral elements; Q9, nine-node quadrilateral elements; T6; six-node triangular element.

Figures 11.9 and 11.10 give a comparison of the finite element solutions (for $\gamma = 10^8$) with the analytical solutions for the horizontal velocity, pressure, and shear stress.

The last example of this section is devoted to the transient analysis of the problem of Example 11.2.

Example 11.4 Time-dependent analysis of fluid squeezed between plates. Consider the unsteady flow of a viscous fluid squeezed between two parallel plates (see Fig.

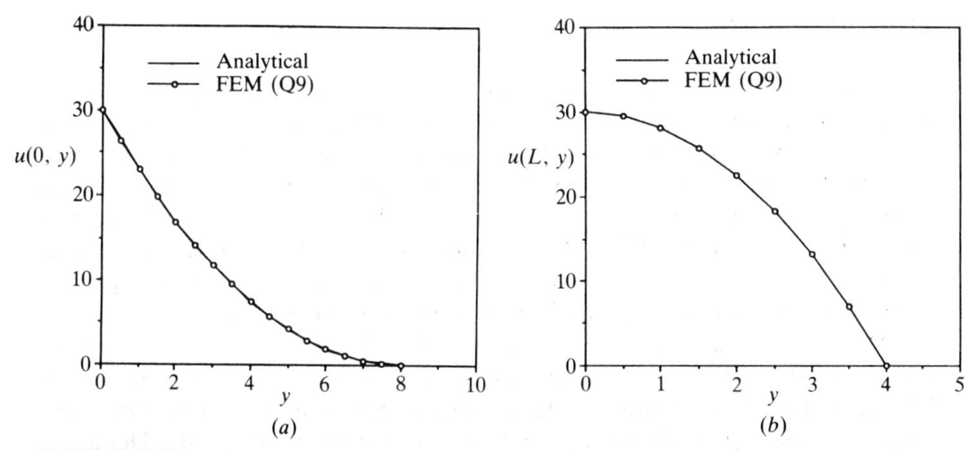

FIGURE 11.9
Comparison of the finite element solution $u(x_0, y)$ with the analytical solution of the slider bearing problem in Example 11.3 (the finite element solution is obtained using 32 nine-node rectangular elements).

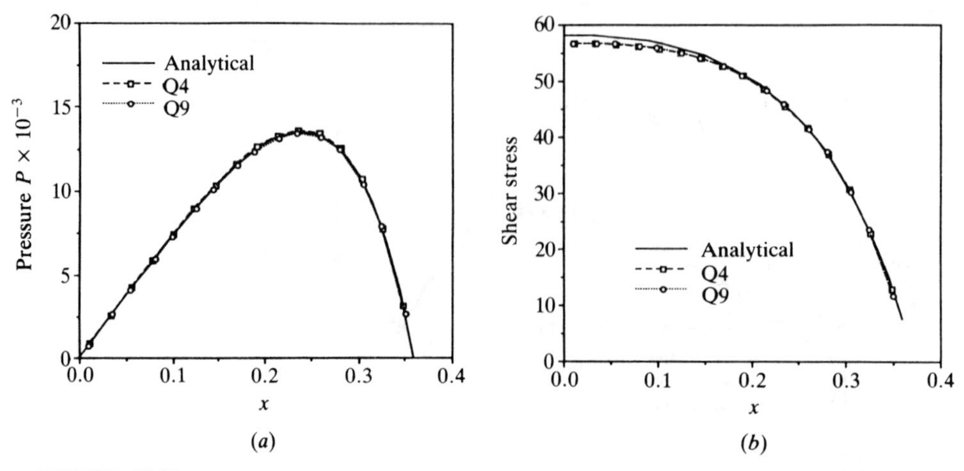

FIGURE 11.10
Comparison of the finite element solutions for pressure and shear stress with the analytical solution of the slider bearing problem of Example 11.3: Q4, four-node quadrilateral element; Q9, nine-node quadrilaterial element. The pressure and shear stress are computed at the Gauss points closest to the bottom wall.

11.3*a*). The flow is induced by the uniform motion of the plates toward each other. The boundary conditions of the model are the same as shown in Fig. 11.3(*b*). The initial boundary conditions are assumed to be zero.

We use the 6×4 mesh of nine-node quadratic elements, employed in Example 11.2, to model the problem. Figure 11.11(*a*) shows plots of the horizontal velocity $u(6, y)$ as a function of y for various times and for two different time steps, and Fig. 11.11(*b*) shows the evolution of the velocity $u(6, 0)$ with time. The transient solution becomes steady around $t = 1.5$ (for a difference of 10^{-3} between the solutions at two consecutive time steps).

11.6 SUMMARY

Finite element models of the equations governing two-dimensional flows of viscous incompressible fluids have been developed. Two different types of finite element models have been presented: (i) the velocity–pressure finite element model, with (u, v, P) as the primary nodal degrees of freedom; and (ii) the penalty finite element model, with (u, v) as the primary nodal degrees of freedom. In the penalty function method, the pressure is calculated from the velocity field in the post-computation. The coefficient matrix in the penalty finite element model is evaluated using mixed integration: full integration for the viscous terms and reduced integration for the penalty terms (i.e., terms associated with the incompressibility or divergence-free condition on the velocity field). Both triangular and rectangular elements have been discussed. In general, triangular elements do not yield accurate pressure fields. The linear and quadratic quadrilateral elements are more reliable for pressure as well as for velocity fields in the penalty finite element model.

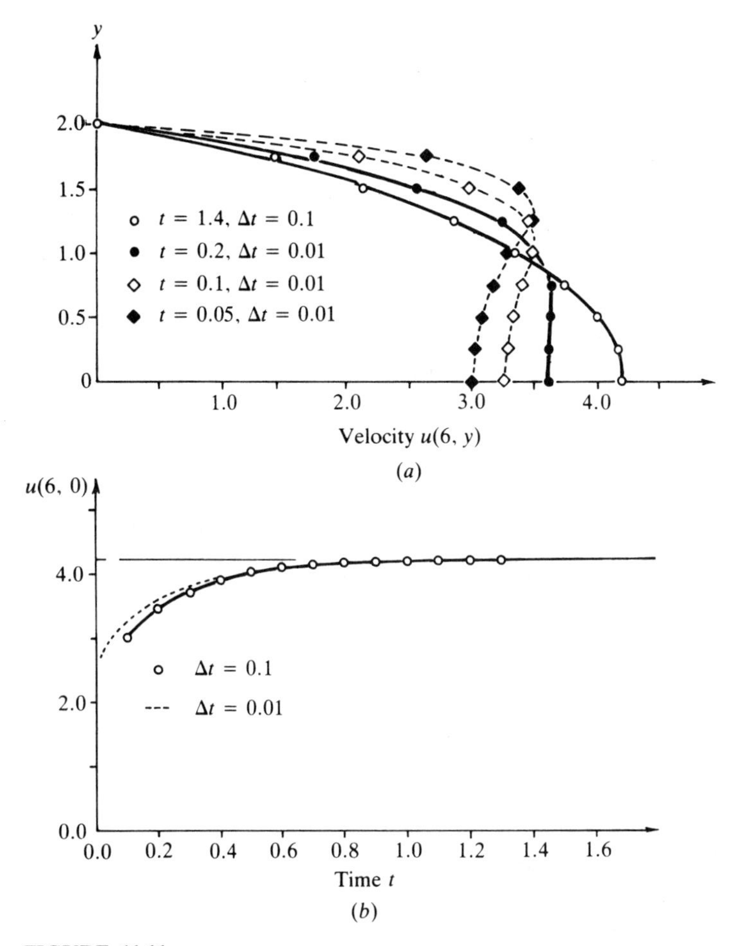

FIGURE 11.11
Transient analysis of a viscous fluid squeezed between parallel plates: (*a*) velocity $u(6, y)$ versus y; (*b*) velocity $u(6, 0)$ versus t.

PROBLEMS

11.1. Consider (11.1) and (11.2) in cylindrical coordinates (r, θ, z). For axisymmetric viscous incompressible flows (i.e., where the flow field is independent of θ), we have

$$\rho \frac{\partial u}{\partial t} = \frac{1}{r}\frac{\partial}{\partial r}(r\sigma_r) - \frac{\sigma_\theta}{r} + \frac{\partial \sigma_{rz}}{\partial z} + f_r \tag{i}$$

$$\rho \frac{\partial w}{\partial t} = \frac{1}{r}\frac{\partial}{\partial r}(r\sigma_{rz}) + \frac{\partial \sigma_z}{\partial z} + f_z \tag{ii}$$

$$\frac{1}{r}\frac{\partial}{\partial r}(ru) + \frac{\partial w}{\partial z} = 0 \tag{iii}$$

where

$$\sigma_r = -P + 2\mu \frac{\partial u}{\partial r}, \qquad \sigma_\theta = -P + 2\mu \frac{u}{r}, \qquad \sigma_z = -P + 2\mu \frac{\partial w}{\partial z}, \qquad \sigma_{rz} = \mu\left(\frac{\partial u}{\partial z} + \frac{\partial w}{\partial r}\right)$$

Develop the semidiscrete finite element model of the equations using the pressure–velocity formulation.

11.2. Develop the semidiscrete finite element model of the equations in Problem 11.1 using the penalty function formulation.

11.3. Write the fully discretized finite element equations of the finite element models in Problems 11.1 and 11.2. Use the α family of approximation.

11.4. The equations governing unsteady slow flow of viscous incompressible fluids in the (x, y) plane can be expressed in terms of the vorticity ζ and stream function ψ:

$$\rho \frac{\partial \zeta}{\partial t} - \mu \nabla^2 \zeta = 0, \qquad \zeta - \nabla^2 \psi = 0$$

Develop the semidiscrete finite element model of these equations. Discuss the meaning of the secondary variables. Use the α family of approximation to reduce the ordinary differential equations to algebraic equations.

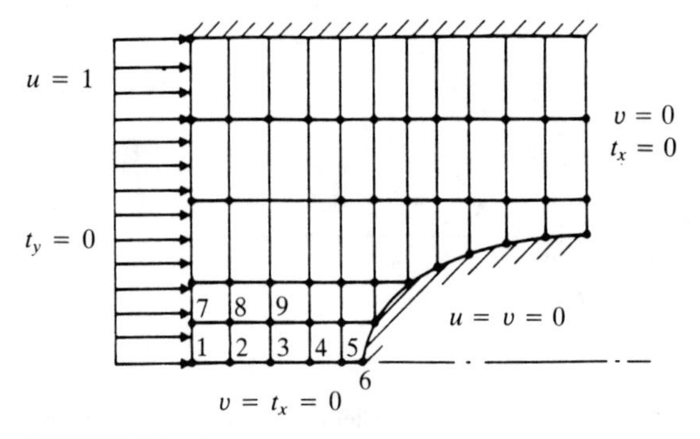

FIGURE P11.5

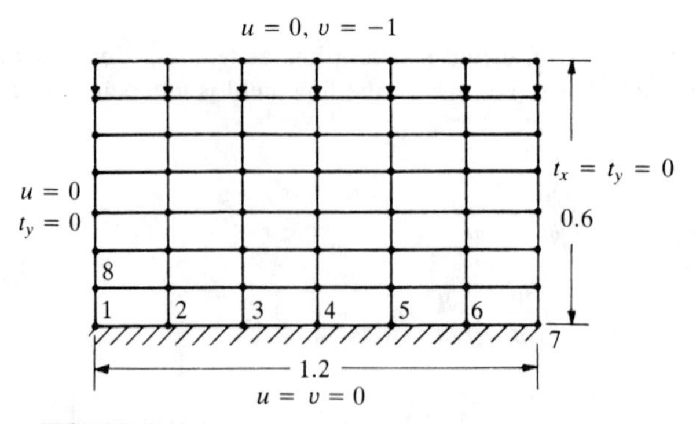

FIGURE P11.6

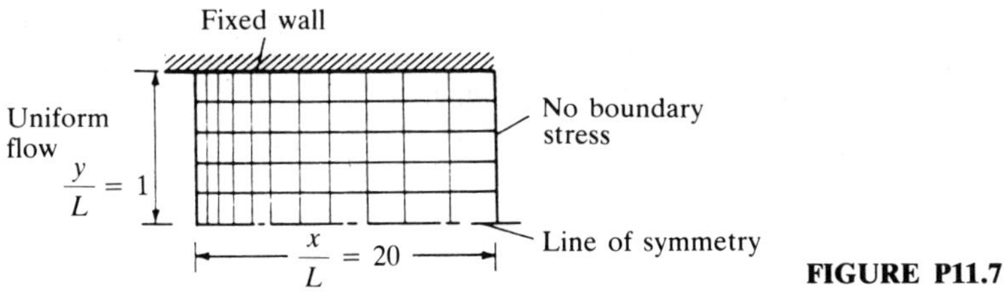

FIGURE P11.7

11.5–11.7. For the viscous flow problems given in Figs. P11.5–P11.7, give the specified primary and secondary degrees of freedom and their values.

11.8. Consider the flow of a viscous incompressible fluid in a square cavity (Fig. P11.8). The flow is induced by the movement of the top wall (or lid) with a constant velocity $u = 1.0$. For a 4×4 mesh of linear elements, identify the primary and secondary degrees of freedom.

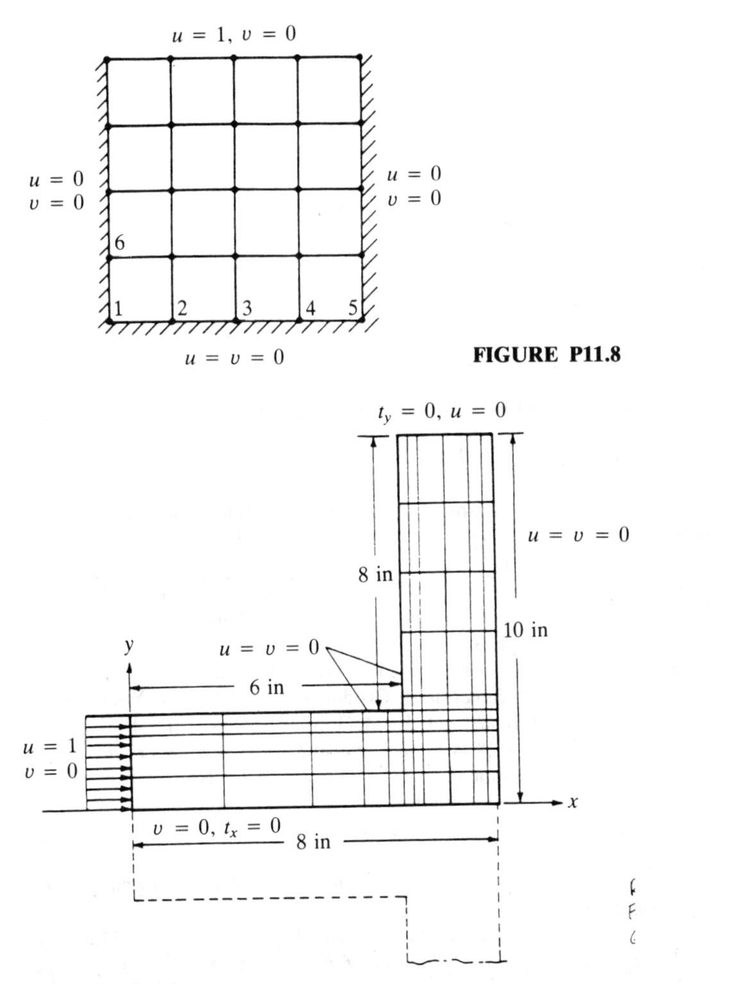

FIGURE P11.8

FIGURE P11.9

11.9. Consider the flow of a viscous incompressible fluid in a 90° plane tee. Using the symmetry and the mesh shown in Fig. P11.9, write the specified primary and secondary variables for the computational domain.

11.10. Repeat Problem 11.9 for the geometry shown in Fig. P11.10.

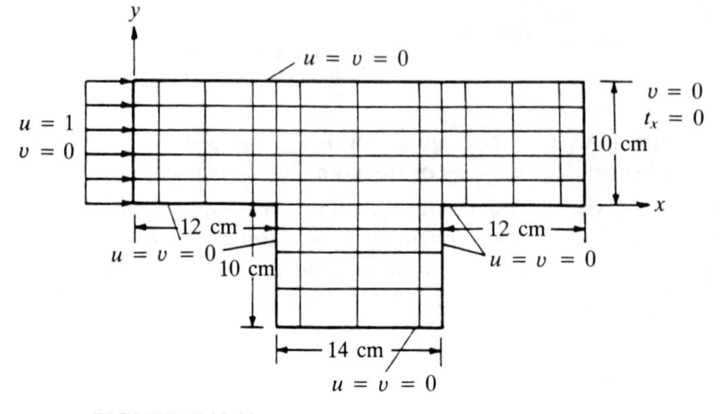

FIGURE P11.10

REFERENCES FOR ADDITIONAL READING

(For references on fluid mechanics, see References for Additional Reading in Chapter 8.)

Babuska, I.: "The Finite Element Method with Lagrange Multipliers," *Numerical Mathematics,* vol. 20, pp. 179–192, 1973*a*.

———: "The Finite Element Method with Penalty," *Mathematics of Computation,* vol. 27, pp. 221–228, 1973*b*.

Bird, R. B., W. E. Stewart, and E. N. Lightfoot: *Transport Phenomena,* John Wiley, New York, 1960.

Ciarlet, P. G.: *The Finite Element Method for Elliptic Problems,* North-Holland, Amsterdam, 1978.

Courant, R.: "Variational Methods for the Solution of Problems of Equilibrium and Vibrations," *Bulletin of the American Mathematical Society,* vol. 49, pp. 1–23, 1943.

Hestenes, M. R.: *Calculus of Variations and Optimal Control,* John Wiley, New York, 1966.

———: *Optimization Theory: The Finite Dimensional Case,* John Wiley, New York, 1975.

Hughes, J. T. R.: *The Finite Element Method (Linear, Static and Dynamic Finite Element Analysis),* Prentice-Hall, Englewood Cliffs, NJ, 1987).

Ladyzhenskaya, O. A.: *The Mathematical Theory of Viscous Incompressible Flow,* 2d ed. (translated from the Russian by R. A. Silverman), Gordon and Breach, New York, 1969.

Nadai, A.: *Theory of Flow and Fracture of Solids,* vol. II, McGraw-Hill, New York, 1963.

Oden, J. T.: "RIP methods for Stokesian flows," in *FInite Element Method in Flow Problems,* vol. IV, eds. R. H. Gallagher, O. C. Zienkiewicz, J. T. Oden, and D. Norrie, John Wiley, London, 1982.

——— and G. F. Carey: *Finite Elements, Mathematical Aspects,* vol. IV, Prentice-Hall, Englewood Cliffs, NJ, 1983.

Polyak, B. T.: "The Convergence Rate of the Penalty Function Method," *USSR Computational Mathematics and Mathematical Physics,* vol. 11, pp. 1–12, 1971).

Reddy, J. N.: "On the Accuracy and Existence of Solutions to Primitive Variable Models of Viscous Incompressible Fluids, *International Journal of Engineering Science,* vol. 16, pp. 921–929, 1978.

———: "On the Finite Element Method with Penalty for Incompressible Fluid Flow Problems," in *The Mathematics of Finite Elements and Applications* III, ed. J. R. Whiteman, Academic Press, New York, 1979*a*.

———: *Applied Functional Analysis and Variational Methods in Engineering,* McGraw-Hill, New York, 1986; Krieger, Melbourne, FL, 1991.

Schlichting, H.: *Boundary-Layer Theory* (translated by J. Kestin), 7th ed., McGraw-Hill, New York, 1979.

CHAPTER 12

BENDING OF ELASTIC PLATES

12.1 INTRODUCTION

The word "plate" refers to solid bodies that are bounded by two parallel planes whose lateral dimensions are large compared with the separation between them. Geometrically, plates are similar to the plane elastic bodies considered in Chapter 10; however, plates are subjected to transverse loads (i.e., loads perpendicular to the plane of the plate). A plate is a two-dimensional analog of a beam. Therefore, the deformation of a plate can be described by *classical plate theory* or by *first-order shear deformation theory*. Classical plate theory is an extension of the Euler–Bernoulli beam theory to plates, and is known as the *Kirchhoff plate theory*. First-order shear deformation theory is an extension of the Timoshenko beam theory, and is known as the *Hencky–Mindlin plate theory*. A review of these plate theories and other refined theories can be found in Reddy (1984, 1990*a, b*).

Classical plate theory is based on the assumptions that a straight line perpendicular to the plane of the plate is (1) inextensible, (2) remains straight, and (3) rotates such that it remains perpendicular to the tangent to the deformed surface (see Fig. 12.1*a*). These assumptions are equivalent to specifying

$$\epsilon_z = 0, \qquad \epsilon_{yz} = 0, \qquad \epsilon_{xz} = 0 \tag{12.1}$$

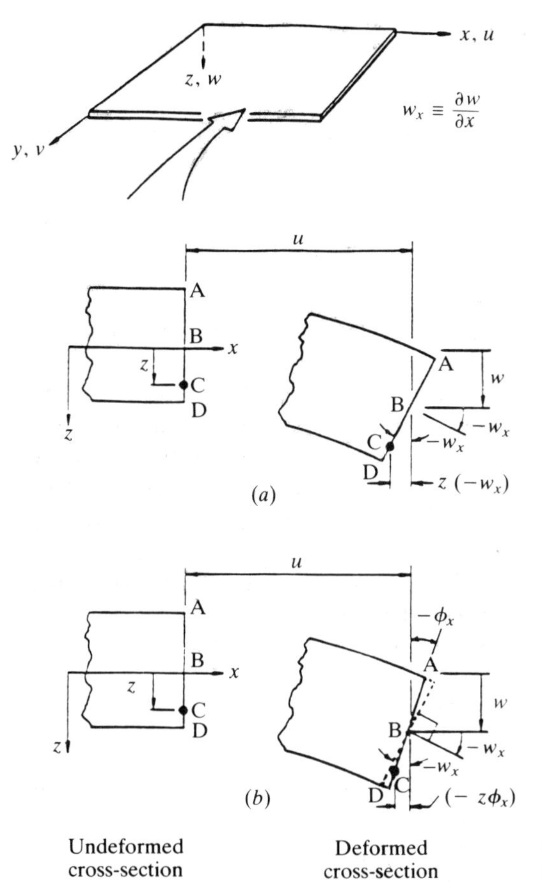

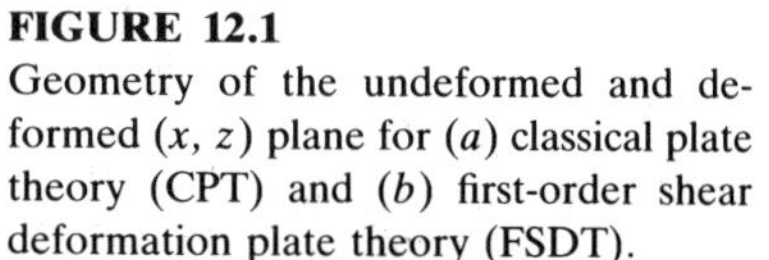

FIGURE 12.1
Geometry of the undeformed and deformed (x, z) plane for (a) classical plate theory (CPT) and (b) first-order shear deformation plate theory (FSDT).

In the first-order shear deformation theory, the normality condition, assumption (3), is removed (see Fig. 12.1b), resulting in a constant state of transverse shear strains through the thickness and zero transverse normal strain:

$$\epsilon_z = 0, \qquad \epsilon_{yz} = \epsilon_{yz}(x, y), \qquad \epsilon_{xz} = \epsilon_{xz}(x, y) \tag{12.2}$$

In this section, we briefly review the governing equations of these two theories and develop their finite element models. Let us consider a plate made up of several finite elements. The volume of a typical element is $V^e = \Omega^e \times (-\frac{1}{2}h_e, \frac{1}{2}h_e)$, where Ω^e denotes the midplane of the element and h_e is its thickness. The boundary of Ω^e is denoted by Γ^e. A plate of nonuniform thickness can be approximated by elements of uniform thickness. We assume that the midplanes of all elements are in the same plane Ω, the midplane of the plate.

12.2 CLASSICAL PLATE MODEL

12.2.1 Displacement Field

The classical theory of plates is based on the displacement field (see Fig. 12.1*a*)

$$\boxed{\begin{aligned} &u_1(x, y, z, t) = u - z\frac{\partial w}{\partial x}, \qquad u_2(x, y, z, t) = v - z\frac{\partial w}{\partial y} \\ &u_3(x, y, z, t) = w(x, y, t) \end{aligned}} \tag{12.3}$$

where (u_1, u_2, u_3) denote the displacements of the point (x, y, z) along the x, y and z directions, and (u, v, w) represent displacements of a point on the midplane $(x, y, 0)$ at time t. The equations of motion of the plate can be derived using the principle of virtual displacements [see Reddy (1984)], which also provides the weak form for the finite element model. For a linear theory based on infinitesimal strains and orthotropic materials, the in-plane displacements (u, v) are uncoupled from the transverse deflection. The in-plane displacements (u, v) are governed by the plane elasticity equations discussed in Chapter 10. Hence, we discuss only the equations governing the (bending) deflection w and the associated finite element model.

The linear strains due to the displacement w in (12.3) are

$$\begin{Bmatrix} \epsilon_x \\ \epsilon_y \\ 2\epsilon_{xy} \end{Bmatrix} = -z \begin{Bmatrix} \dfrac{\partial^2 w}{\partial x^2} \\ \dfrac{\partial^2 w}{\partial y^2} \\ 2\dfrac{\partial^2 w}{\partial x\, \partial y} \end{Bmatrix} \tag{12.4}$$

and $\epsilon_{xz} = 0$, $\epsilon_{yz} = 0$, and $\epsilon_z = 0$.

12.2.2 Virtual Work Statement

The principle of virtual displacements for the time-dependent case is

$$\begin{aligned} 0 = &\int_{V^e} \left[\rho z^2 \left(\frac{\partial \delta w}{\partial x} \frac{\partial^3 w}{\partial x\, \partial t^2} + \frac{\partial \delta w}{\partial y} \frac{\partial^3 w}{\partial y\, \partial t^2} \right) + \rho\, \delta w \frac{\partial^2 w}{\partial t^2} \right. \\ &\left. + \delta\epsilon_x\, \sigma_x + \delta\epsilon_y\, \sigma_y + 2\, \delta\epsilon_{xy}\, \sigma_{xy} \right] dV \\ &- \int_{\Omega^e} f\, \delta w\, dx\, dy - \oint_{\Gamma^e} \left(-M_n \frac{\partial \delta w}{\partial n} + V_n\, \delta w \right) ds \end{aligned} \tag{12.5}$$

Here the first three terms represent the virtual work done by the inertial forces in the three coordinate directions, while the remaining terms in the volume integral represent the virtual strain energy stored in the plate. The last two integrals, one defined on the midplane Ω^e and the other on the boundary Γ^e,

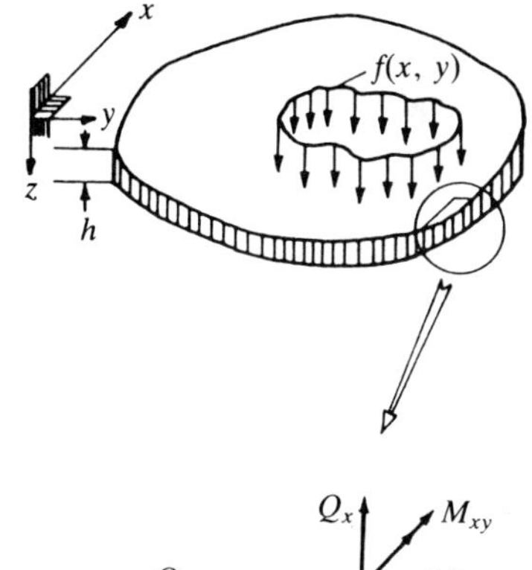

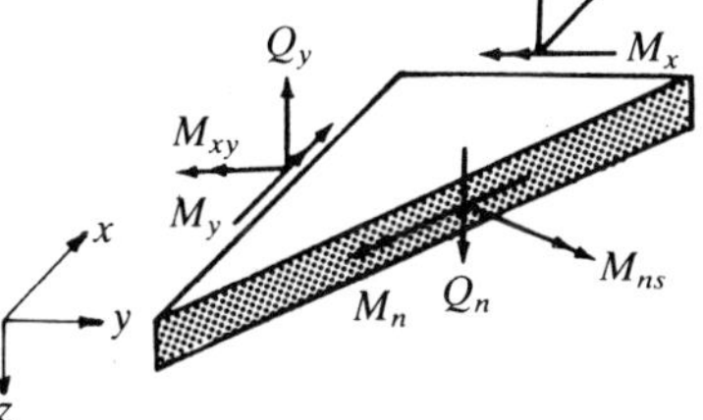

Use the right-hand-rule convention to determine the moment direction (shown with double arrow)

Boundary conditions for plates

Clamped	Simply supported 1	Simply supported 2	Free
$w=0$	$w=0$	$w=0$	$Q_n=0$
$\phi_s=0$	$M_{ns}=0$	$\phi_s=0$	$M_{ns}=0$
$\phi_n=0$	$M_n=0$	$M_n=0$	$M_n=0$

FIGURE 12.2
Geometry, moment, and shear force resultants, and various boundary conditions for a plate element.

denote the virtual work done by the transversely distributed load f and edge loads M_n and V_n (see Fig. 12.2). Since $V^e = \Omega^e \times (-\frac{1}{2}h_e, \frac{1}{2}h_e)$, and the displacements, strains, and stresses are separable into functions of x and y alone and functions of z,

$$F(x, y, z) = g(x, y)f(z) \tag{12.6a}$$

we can write the volume integral as

$$\int_{V^e} (\cdot)\, dV = \int_{\Omega^e} \int_{-h/2}^{h/2} (\cdot)\, dz\, dx\, dy \tag{12.6b}$$

Substitution of (12.3) and (12.4) for virtual displacements and strains into

(12.5) and integrating with respect to z, we obtain

$$0 = \int_{\Omega^e} \left(\delta w \, I_0 \frac{\partial^2 w}{\partial t^2} + \frac{\partial \delta w}{\partial x} I_2 \frac{\partial^3 w}{\partial x \, \partial t^2} + \frac{\partial \delta w}{\partial y} I_2 \frac{\partial^3 w}{\partial y \, \partial t^2} - M_x \frac{\partial^2 \delta w}{\partial x^2} - M_y \frac{\partial^2 \delta w}{\partial y^2} - 2M_{xy} \frac{\partial^2 \delta w}{\partial x \, \partial y} - f \, \delta w \right) dx \, dy - \oint_{\Gamma^e} \left(-M_n \frac{\partial \delta w}{\partial n} + V_n \, \delta w \right) ds \tag{12.7}$$

where (M_x, M_y, M_{xy}) are the bending moments (see Fig. 12.2),

$$M_x = \int_{-h/2}^{h/2} \sigma_x z \, dz, \qquad M_y = \int_{-h/2}^{h/2} \sigma_y z \, dz, \qquad M_{xy} = \int_{-h/2}^{h/2} \sigma_{xy} z \, dz \tag{12.8}$$

and I_0 and I_2 are the mass moments of inertia,

$$I_0 = \int_{-h/2}^{h/2} \rho \, dz = \rho h, \qquad I_2 = \int_{-h/2}^{h/2} \rho z^2 \, dz = \tfrac{1}{12} \rho h^3 \tag{12.9}$$

Note that M_n and M_{ns} denote the normal and twisting moments on an edge, and V_n is the shear force. It can be shown that, for a linear orthotropic material, the bending moments are related to the derivatives of the transverse deflection w:

$$\boxed{\begin{aligned} M_x &= -\left(D_{11} \frac{\partial^2 w}{\partial x^2} + D_{12} \frac{\partial^2 w}{\partial y^2} \right), \qquad M_y = -\left(D_{12} \frac{\partial^2 w}{\partial x^2} + D_{22} \frac{\partial^2 w}{\partial y^2} \right) \\ M_{xy} &= -2D_{66} \frac{\partial^2 w}{\partial x \, \partial y} \end{aligned}} \tag{12.10}$$

where D_{ij} are the plate rigidities,

$$\begin{aligned} D_{11} &= \frac{E_1 h^3}{12(1 - \nu_{12}\nu_{21})}, \qquad D_{22} = \frac{E_2 h^3}{12(1 - \nu_{12}\nu_{21})} \\ D_{12} &= \frac{\nu_{12} E_2 h^3}{12(1 - \nu_{12}\nu_{21})}, \qquad D_{66} = \tfrac{1}{12} G_{12} h^3 \end{aligned} \tag{12.11}$$

The boundary forces M_n and V_n can be related to M_x, M_y, and M_{xy} by

$$\begin{aligned} M_n &= M_x n_x^2 + M_y n_y^2 + 2M_{xy} n_x n_y, \qquad V_n = \hat{Q}_n + \frac{\partial M_{ns}}{\partial s} \\ \hat{Q}_n &= Q_x n_x + Q_y n_y + I_2 \left(\frac{\partial^3 w}{\partial x \, \partial t^2} n_x + \frac{\partial^3 w}{\partial y \, \partial t^2} n_y \right) \\ M_{ns} &= (M_y - M_x) n_x n_y + M_{xy}(n_x^2 - n_y^2) \\ Q_x &= \frac{\partial M_x}{\partial x} + \frac{\partial M_{xy}}{\partial y}, \qquad Q_y = \frac{\partial M_{xy}}{\partial x} + \frac{\partial M_y}{\partial y} \end{aligned} \tag{12.12}$$

where (n_x, n_y) are the direction cosines of the unit normal on the boundary Γ^e. Substituting (12.10) into (12.7) yields the weak form of the problem:

$$0 = \int_{\Omega^e} \left[\delta w\, I_0 \frac{\partial^2 w}{\partial t^2} + \frac{\partial \delta w}{\partial x} I_2 \frac{\partial^3 w}{\partial x\, \partial t^2} + \frac{\partial \delta w}{\partial y} I_2 \frac{\partial^3 w}{\partial y\, \partial t^2} \right.$$
$$+ \left(D_{11} \frac{\partial^2 w}{\partial x^2} + D_{12} \frac{\partial^2 w}{\partial y^2} \right) \frac{\partial^2 \delta w}{\partial x^2} + \left(D_{12} \frac{\partial^2 w}{\partial x^2} + D_{22} \frac{\partial^2 w}{\partial y^2} \right) \frac{\partial^2 \delta w}{\partial y^2}$$
$$\left. + 4D_{66} \frac{\partial^2 w}{\partial x\, \partial y} \frac{\partial^2 \delta w}{\partial x\, \partial y} - \delta w\, f \right] dx\, dy$$
$$- \oint_{\Gamma^e} \left(-M_n \frac{\partial \delta w}{\partial n} + V_n\, \delta w \right) ds \tag{12.13}$$

It is informative to note that the differential equation governing w is

$$\frac{\partial^2}{\partial x^2} \left(D_{11} \frac{\partial^2 w}{\partial x^2} + D_{12} \frac{\partial^2 w}{\partial y^2} \right) + \frac{\partial^2}{\partial y^2} \left(D_{12} \frac{\partial^2 w}{\partial x^2} + D_{22} \frac{\partial^2 w}{\partial y^2} \right)$$
$$+ 2 \frac{\partial^2}{\partial x\, \partial y} \left(2D_{66} \frac{\partial^2 w}{\partial x\, \partial y} \right) - f + I_0 \frac{\partial^2 w}{\partial t^2} - I_2 \frac{\partial^2}{\partial t^2} \left(\frac{\partial^2 w}{\partial x^2} + \frac{\partial^2 w}{\partial y^2} \right) = 0 \tag{12.14}$$

Note that the expressions in parentheses in the first three terms are the bending moments $-M_x$, $-M_y$, and $-M_{xy}$, respectively [see (12.10)].

Of course, we can construct the weak form of (12.14) using the three-step procedure. We obtain

$$0 = \int_{\Omega^e} \left[\frac{\partial^2 v}{\partial x^2} \left(D_{11} \frac{\partial^2 w}{\partial x^2} + D_{12} \frac{\partial^2 w}{\partial y^2} \right) + \frac{\partial^2 v}{\partial y^2} \left(D_{12} \frac{\partial^2 w}{\partial x^2} + D_{22} \frac{\partial^2 w}{\partial y^2} \right) \right.$$
$$\left. + 4D_{66} \frac{\partial^2 v}{\partial x\, \partial y} \frac{\partial^2 w}{\partial x\, \partial y} - vf + I_0 v \frac{\partial^2 w}{\partial t^2} + I_2 \left(\frac{\partial v}{\partial x} \frac{\partial^3 w}{\partial x\, \partial t^2} + \frac{\partial v}{\partial y} \frac{\partial^3 w}{\partial y\, \partial t^2} \right) \right] dx\, dy$$
$$- \oint_{\Gamma^e} v \left[\left(\frac{\partial M_x}{\partial x} + \frac{\partial M_{xy}}{\partial y} \right) n_x + \left(\frac{\partial M_{xy}}{\partial x} + \frac{\partial M_y}{\partial y} \right) n_y + I_2 \left(\frac{\partial^3 w}{\partial x\, \partial t^2} n_x + \frac{\partial^3 w}{\partial y\, \partial t^2} n_y \right) \right] ds$$
$$+ \oint_{\Gamma^e} \left[\frac{\partial v}{\partial x} (M_x n_x + M_{xy} n_y) + \frac{\partial v}{\partial y} (M_{xy} n_x + M_y n_y) \right] ds \tag{12.15}$$

where v denotes the weight function, which can be interpreted as the first variation, $\delta w = v$. Next, we convert the derivatives with respect to the rectangular coordinates (x, y) to those with respect to the local, normal and tangential, coordinates (n, s). We use the identities

$$\frac{\partial v}{\partial x} = n_x \frac{\partial v}{\partial n} - n_y \frac{\partial v}{\partial s}, \qquad \frac{\partial v}{\partial y} = n_x \frac{\partial v}{\partial s} + n_y \frac{\partial v}{\partial n} \tag{12.16}$$

in the boundary integrals of (12.15). The boundary integrals become

$$-\oint_{\Gamma^e} v\hat{Q}_n\,ds + \oint_{\Gamma^e}\left(\frac{\partial v}{\partial n}M_n + \frac{\partial v}{\partial s}M_{ns}\right)ds \tag{12.17}$$

where $\hat{Q}_n$, M_n, and M_{ns} are the quantities defined in (12.12). The second term in the second integral is integrated by parts to yield the final expression

$$-\oint_{\Gamma^e}\left[v\left(\hat{Q}_n + \frac{\partial M_{ns}}{\partial s}\right) - \frac{\partial v}{\partial n}M_n\right]ds \tag{12.18}$$

The expression in parentheses is denoted by V_n, and its specification is known as the *Kirchhoff free-edge condition*. We assumed that $[vM_{ns}]_s = 0$.

12.2.3 Finite Element Model

An examination of the boundary terms in the weak form (12.13) suggests that the essential boundary conditions involve specifying the transverse deflection w and the normal derivative of w, which constitute the primary variables of the problem (cf. the Euler–Bernoulli beam model). Hence, the finite element interpolation of w must be such that w, $\partial w/\partial n$, and $\partial w/\partial s$ are continuous across the interelement boundaries. Note that $\partial/\partial n$ and $\partial/\partial s$ are related to the global derivatives $\partial/\partial x$ and $\partial/\partial y$ by the relations [the inverses of those in (12.16)]

$$\frac{\partial}{\partial n} = n_x\frac{\partial}{\partial x} + n_y\frac{\partial}{\partial y}, \qquad \frac{\partial}{\partial s} = n_x\frac{\partial}{\partial y} - n_y\frac{\partial}{\partial x} \tag{12.19}$$

Thus, the primary variables at the nodes should be [this also follows directly from (12.15)]

$$w, \quad \frac{\partial w}{\partial x}, \quad \frac{\partial w}{\partial y}$$

Finite elements that require continuity of w and its first derivatives are called C^1 elements.

Suppose that w is interpolated by expressions of the form

$$\boxed{w = \sum_{j=1}^{n} \Delta_j \phi_j(x, y)} \tag{12.20}$$

where Δ_j denote the nodal values of w and its derivatives, and $\phi_j(x, y)$ are the Hermite interpolation functions. A rectangular element with four nodes, with $(w, \partial w/\partial x, \partial w/\partial y)$ at each node, requires the 12-term ($n = 12$) polynomial approximation of w to obtain expressions for ϕ_j:

$$w = a_1 + a_2x + a_3y + a_4xy + a_5x^2 + a_6y^2 + a_7x^2y + a_8xy^2 + a_9x^3 + a_{10}y^3 + a_{11}x^3y + a_{12}xy^3 \tag{12.21}$$

The polynomial is *not* a complete fourth-order polynomial; it is a complete third-order polynomial. For a three-node triangular element, the nine-term ($n = 9$) polynomial is selected to obtain ϕ_j:

$$w = a_1 + a_2 x + a_3 y + a_4 xy + a_5 x^2 + a_6 y^2 + a_7(x^2 y + xy^2) + a_8 x^3 + a_9 y^3 \tag{12.22}$$

This is an incomplete third-order polynomial, because $x^2 y$ and $y^2 x$ do not vary independently.

Some comments are in order on the interelement continuity of w and $\partial w/\partial n$ for the four-node rectangular element ($n = 12$) and three-node triangular element ($n = 9$). First let us consider the rectangular element. We note from (12.21) that w varies as a cubic along any line $x = \text{constant}$ or $y = \text{constant}$. Along a given side, there are two nodes and two values (w and its normal derivative) per node to define the cubic variation uniquely. Hence, w is uniquely defined along the element boundary and is continuous along interelement boundaries. The normal derivative, say, $\partial w/\partial x$ on a line $x = \text{constant}$, also varies as a cubic function of y along the side. Since only two values of $\partial w/\partial x$ are available on the line, the cubic variation cannot be uniquely defined, and the normal slope continuity is not satisfied. In addition, $\partial^2 w/\partial x\, \partial y$ is not single-valued at the corner points of the element. Elements that violate any of the continuity conditions are known as *nonconforming elements*. Thus the four-node rectangular element with w represented by (12.21) is a nonconforming element. We denote it as CPT(N). Despite this deficiency, the element is known to give good results. A similar discussion leads to the conclusion that the three-node triangular element is nonconforming. In addition, the triangular element is found to have convergence problems and singular behavior for certain meshes. The triangular plate bending elements will not be discussed in this study. The four-node element with (w, $\partial w/\partial x$, $\partial w/\partial y$, $\partial^2 w/\partial x\, \partial y$) as nodal degrees of freedom requires a complete quartic polynomial (obtained from the tensor products of the 1-D Hermite cubic polynomials). This element is a conforming element, and it is denoted as CPT(C). The conforming and nonconforming rectangular plate elements are shown in Fig. 12.3(a). The interpolation functions for the two rectangular elements are listed in Table 9.1. For further discussion on various plate bending elements, the reader is referred to Zienkiewicz and Taylor (1991) and references therein.

Substitution of (12.20) for w and $\delta w = \phi_i$ ($i = 1, 2, \ldots, n$) into (12.13) gives the finite element model

$$[M^e]\{\ddot{\Delta}^e\} + [K^e]\{\Delta^e\} = \{f^e\} + \{Q^e\} \tag{12.23a}$$

where

$$K_{ij}^e = \int_{\Omega^e} \left[D_{11} \frac{\partial^2 \phi_i}{\partial x^2} \frac{\partial^2 \phi_j}{\partial x^2} + D_{12} \left(\frac{\partial^2 \phi_i}{\partial x^2} \frac{\partial^2 \phi_j}{\partial y^2} + \frac{\partial^2 \phi_i}{\partial y^2} \frac{\partial^2 \phi_j}{\partial x^2} \right) \right.$$
$$\left. + D_{22} \frac{\partial^2 \phi_i}{\partial y^2} \frac{\partial^2 \phi_j}{\partial y^2} + 4D_{66} \frac{\partial^2 \phi_i}{\partial x\, \partial y} \frac{\partial^2 \phi_j}{\partial x\, \partial y} \right] dx\, dy \tag{12.23b}$$

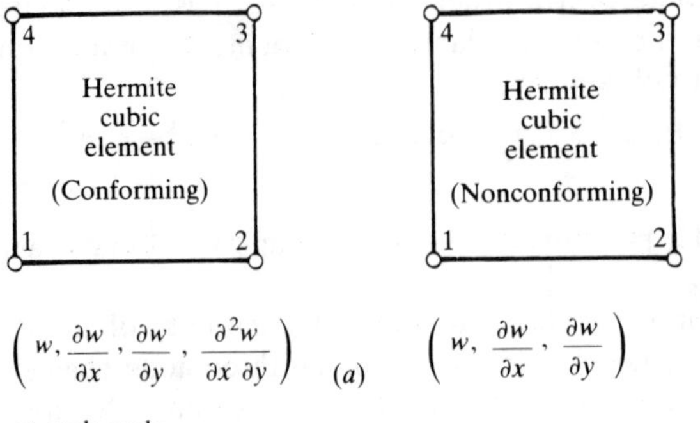

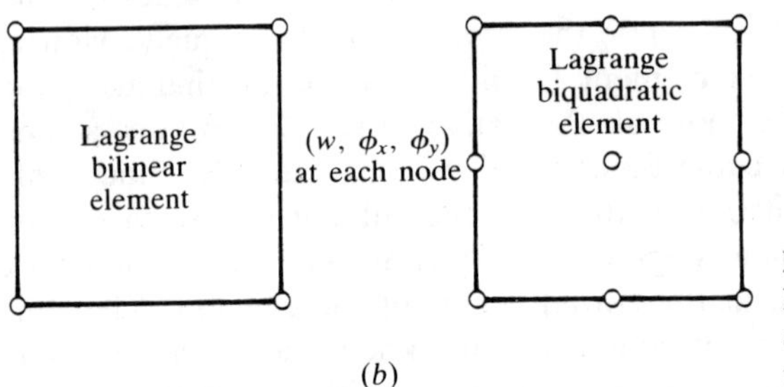

FIGURE 12.3
Finite elements for (*a*) displacement and (*b*) shear deformation models of plates.

$$M_{ij}^e = \int_{\Omega^e} \left[I_0 \phi_i \phi_j + I_2 \left(\frac{\partial \phi_i}{\partial x} \frac{\partial \phi_j}{\partial x} + \frac{\partial \phi_i}{\partial y} \frac{\partial \phi_j}{\partial y} \right) \right] dx\, dy$$

$$f_i^e = \int_{\Omega^e} f \phi_i \, dx\, dy, \qquad Q_i^e = \oint_{\Gamma^e} \left(-M_n \frac{\partial \phi_i}{\partial n} + V_n \phi_i \right) ds$$

The rotatory inertia, $I_2 = \frac{1}{12}\rho h^3$, where h is the plate thickness, is neglected in most books. This completes the development of the classical plate bending finite element.

12.3 SHEAR DEFORMABLE PLATE MODEL

12.3.1 Displacement Field

Of all the shear deformation plate theories available in the literature [see Reddy (1989, 1990*a*, *b*) for reviews], first-order shear deformation theory (FSDT) is the one most commonly used in the modeling of thick plates. The first-order theory is based on the displacement field (see Fig. 12.1*b*)

$$\boxed{u_1 = u + z\phi_x, \qquad u_2 = v + z\phi_y, \qquad u_3 = w} \tag{12.24}$$

where (u, v, w) are the displacements of a point $(x, y, 0)$, and ϕ_x and ϕ_y are the rotations of the transverse normal about the y and $-x$ axes, respectively. Once again, we develop the equations governing the bending deflections (w, ϕ_x, ϕ_y) only, because the in-plane displacements (u, v) are uncoupled from (w, ϕ_x, ϕ_y).

The bending strains associated with (12.24) are

$$\begin{Bmatrix} \epsilon_x \\ \epsilon_y \\ 2\epsilon_{xy} \\ 2\epsilon_{xz} \\ 2\epsilon_{yz} \end{Bmatrix} = \begin{Bmatrix} z\, \partial\phi_x/\partial x \\ z\, \partial\phi_y/\partial y \\ z(\partial\phi_x/\partial y + \partial\phi_y/\partial x) \\ \phi_x + \partial w/\partial x \\ \phi_y + \partial w/\partial y \end{Bmatrix} \tag{12.25}$$

Note that the transverse shear strains are nonzero and $\epsilon_z = 0$.

12.3.2 Virtual Work Statement

Substituting the displacement field (12.24) and strains (12.25) into the statement of the principle of virtual displacements, we obtain

$$\begin{aligned} 0 = & \int_{V^e} \left(\delta\phi_x\, \rho z^2 \frac{\partial^2 \phi_x}{\partial t^2} + \delta\phi_y\, \rho z^2 \frac{\partial^2 \phi_y}{\partial t^2} + \delta w\, \rho \frac{\partial^2 w}{\partial t^2} + \delta\epsilon_x\, \sigma_x + \delta\epsilon_y\, \sigma_y \right. \\ & \left. + 2\,\delta\epsilon_{xy}\, \sigma_{xy} + 2\,\delta\epsilon_{xz}\, \sigma_{xz} + 2\,\delta\epsilon_{yz}\, \sigma_{yz} \right) dV - \int_{\Omega^e} \delta w\, f\, dx\, dy \\ & - \oint_{\Gamma^e} (\delta\phi_n\, M_n + \delta\phi_s\, M_{ns} + \delta w\, Q_n)\, ds \\ = & \int_{\Omega^e} \left[I_0\, \delta w \frac{\partial^2 w}{\partial t^2} + I_2 \left(\delta\phi_x \frac{\partial^2 \phi_x}{\partial t^2} + \delta\phi_y \frac{\partial^2 \phi_y}{\partial t^2} \right) + M_x \frac{\partial \delta\phi_x}{\partial x} + M_y \frac{\partial \delta\phi_y}{\partial y} \right. \\ & + M_{xy} \left(\frac{\partial \delta\phi_x}{\partial y} + \frac{\partial \delta\phi_y}{\partial x} \right) + Q_x \left(\delta\phi_x + \frac{\partial \delta w}{\partial x} \right) \\ & \left. + Q_y \left(\delta\phi_y + \frac{\partial \delta w}{\partial y} \right) - \delta w\, f \right] dx\, dy \\ & - \oint_{\Gamma^e} (\delta\phi_n\, M_n + \delta\phi_s\, M_{ns} + \delta w\, Q_n)\, ds \end{aligned} \tag{12.26}$$

where M_n, M_{ns}, and Q_n are defined by (12.12), and

$$\begin{gathered} Q_x = \int_{-h/2}^{h/2} \sigma_{xz}\, dz = A_{55} \left(\phi_x + \frac{\partial w}{\partial x} \right), \quad Q_y = \int_{-h/2}^{h/2} \sigma_{yz}\, dz = A_{44} \left(\phi_y + \frac{\partial w}{\partial y} \right) \\ M_x = D_{11} \frac{\partial \phi_x}{\partial x} + D_{12} \frac{\partial \phi_y}{\partial y}, \quad M_y = D_{12} \frac{\partial \phi_x}{\partial x} + D_{22} \frac{\partial \phi_y}{\partial y} \\ M_{xy} = D_{66} \left(\frac{\partial \phi_x}{\partial y} + \frac{\partial \phi_y}{\partial x} \right) \end{gathered} \tag{12.27}$$

$$Q_n = Q_x n_x + Q_y n_y, \qquad \phi_n = \phi_x n_x + \phi_y n_y, \qquad \phi_s = \phi_y n_x - \phi_x n_y$$
$$A_{44} = kG_{23} h, \qquad A_{55} = kG_{13} h$$

Here k denotes the shear correction coefficient ($k = \frac{5}{6}$). This coefficient is introduced to account for the discrepancy between the distribution of transverse shear stresses of the first-order theory and the actual distribution (see Section 4.4 on the Timoshenko beam element). Equation (12.26) contains three weak forms for the three displacements (w, ϕ_x, ϕ_y):

$$0 = \int_{\Omega^e} \left(I_0\, \delta w \frac{\partial^2 w}{\partial t^2} + Q_x \frac{\partial \delta w}{\partial x} + Q_y \frac{\partial \delta w}{\partial y} - f\, \delta w \right) dx\, dy - \oint_{\Gamma^e} \delta w\, Q_n\, ds$$

$$0 = \int_{\Omega^e} \left(I_2\, \delta\phi_x \frac{\partial^2 \phi_x}{\partial t^2} + M_x \frac{\partial \delta\phi_x}{\partial x} + M_{xy} \frac{\partial \delta\phi_x}{\partial y} + Q_x\, \delta\phi_x \right) dx\, dy$$

$$- \oint_{\Gamma^e} \delta\phi_x\, (M_n n_x - M_{ns} n_y)\, ds \tag{12.28}$$

$$0 = \int_{\Omega^e} \left(I_2\, \delta\phi_y \frac{\partial^2 \phi_y}{\partial t^2} + M_{xy} \frac{\partial \delta\phi_y}{\partial x} + M_y \frac{\partial \delta\phi_y}{\partial y} + Q_y\, \delta\phi_y \right) dx\, dy$$

$$- \oint_{\Gamma^e} \delta\phi_y\, (M_n n_y + M_{ns} n_x)\, ds$$

The governing differential equations of first-order shear deformation theory are

$$\begin{aligned} \frac{\partial Q_x}{\partial x} + \frac{\partial Q_y}{\partial y} + f &= I_0 \frac{\partial^2 w}{\partial t^2} \\ \frac{\partial M_x}{\partial x} + \frac{\partial M_{xy}}{\partial y} - Q_x &= I_2 \frac{\partial^2 \phi_x}{\partial t^2} \\ \frac{\partial M_{xy}}{\partial x} + \frac{\partial M_y}{\partial y} - Q_y &= I_2 \frac{\partial^2 \phi_y}{\partial t^2} \end{aligned} \tag{12.29}$$

where (Q_x, Q_y, M_x, M_{xy}, M_y) are defined in (12.27). The three-step procedure of developing weak forms can be used to develop the weak forms of (12.29), which will be equivalent to those listed in (12.28). To see the equivalence, the following identities must be used:

$$\begin{aligned} M_n n_x - M_{ns} n_y &= M_x n_x + M_{xy} n_y \equiv \hat{M}_n \\ M_n n_y + M_{ns} n_x &= M_{xy} n_x + M_y n_y \equiv \hat{M}_{ns} \\ \phi_x = \phi_n n_x - \phi_s n_y, &\qquad \phi_y = \phi_n n_y + \phi_s n_x \end{aligned} \tag{12.30}$$

12.3.3 Finite Element Model

We note from the boundary integrals that the primary variables of the theory are (w, ϕ_x, ϕ_y) and the secondary variables are (Q_n, M_n, M_{ns}) [or a linear combination of Q_x, Q_y, M_x, M_y and M_{xy}; see (12.12)]. Therefore, the Lagrange interpolation of w, ϕ_x and ϕ_y is admissible for the first-order theory.

We assume finite element interpolation of w, ϕ_x, and ϕ_y in the form

$$w = \sum_{j=1}^{n} w_j \psi_j^1, \quad \phi_x = \sum_{j=1}^{m} S_j^x \psi_j^2, \quad \phi_y = \sum_{j=1}^{m} S_j^y \psi_j^2 \tag{12.31}$$

where ψ_j^1 and ψ_j^2 are interpolation functions used for w and (ϕ_x, ϕ_y), respectively. In general, ψ_j^1 and ψ_j^2 are polynomials of different degree. However, in the present study, we take $\psi_j^1 = \psi_j^2 \equiv \psi_j$. This choice, as discussed for the Timoshenko beam element, requires the use of reduced integration for the evaluation of stiffness coefficients associated with the transverse shear strains. Substituting (12.31) into (12.28), we obtain the finite element model

$$\begin{bmatrix} [M^{11}] & [0] & [0] \\ & [M^{22}] & [0] \\ \text{symmetric} & & [M^{33}] \end{bmatrix} \begin{Bmatrix} \{\ddot{w}\} \\ \{\ddot{S}^x\} \\ \{\ddot{S}^y\} \end{Bmatrix} + \begin{bmatrix} [K^{11}] & [K^{12}] & [K^{13}] \\ & [K^{22}] & [K^{23}] \\ \text{symmetric} & & [K^{33}] \end{bmatrix} \begin{Bmatrix} \{w\} \\ \{S^x\} \\ \{S^y\} \end{Bmatrix} = \begin{Bmatrix} \{F^1\} \\ \{F^2\} \\ \{F^3\} \end{Bmatrix} \tag{12.32a}$$

where

$$M_{ij}^{11} = I_0 M_{ij}, \quad M_{ij}^{22} = M_{ij}^{33} = I_2 M_{ij}, \quad M_{ij} = \int_{\Omega^e} \psi_i \psi_j \, dx\, dy$$

$$K_{ij}^{11} = \int_{\Omega^e} \left(A_{55} \frac{\partial \psi_i}{\partial x} \frac{\partial \psi_j}{\partial x} + A_{44} \frac{\partial \psi_i}{\partial y} \frac{\partial \psi_j}{\partial y} \right) dx\, dy$$

$$K_{ij}^{12} = \int_{\Omega^e} A_{55} \frac{\partial \psi_i}{\partial x} \psi_j \, dx\, dy, \quad K_{ij}^{13} = \int_{\Omega^e} A_{44} \frac{\partial \psi_i}{\partial y} \psi_j \, dx\, dy$$

$$K_{ij}^{22} = \int_{\Omega^e} \left(D_{11} \frac{\partial \psi_i}{\partial x} \frac{\partial \psi_j}{\partial x} + D_{66} \frac{\partial \psi_i}{\partial y} \frac{\partial \psi_j}{\partial y} + A_{55} \psi_i \psi_j \right) dx\, dy \tag{12.32b}$$

$$K_{ij}^{23} = \int_{\Omega^e} \left(D_{12} \frac{\partial \psi_i}{\partial x} \frac{\partial \psi_j}{\partial y} + D_{66} \frac{\partial \psi_i}{\partial y} \frac{\partial \psi_j}{\partial x} \right) dx\, dy$$

$$K_{ij}^{33} = \int_{\Omega^e} \left(D_{66} \frac{\partial \psi_i}{\partial x} \frac{\partial \psi_j}{\partial x} + D_{22} \frac{\partial \psi_i}{\partial y} \frac{\partial \psi_j}{\partial y} + A_{44} \psi_i \psi_j \right) dx\, dy$$

$$F_i^1 = \int_{\Omega^e} f \psi_i \, dx\, dy + \oint_{\Gamma^e} Q_n \psi_i \, dx, \quad F_i^2 = \oint_{\Gamma^e} \hat{M}_n \psi_i \, dx, \quad F_i^3 = \oint_{\Gamma^e} \hat{M}_{ns} \psi_i \, ds$$

The element stiffness and mass matrices in (12.32a) are of order $3n \times 3n$, where n is the number of nodes per element. When the four-node rectangular element is used, the element matrices are of order 12×12, while they are 27×27 for the nine-node element (see Fig. 12.3b).

12.3.4 Shear Locking and Reduced Integration

It should be noted that the inclusion of the transverse shear strains (i.e., terms involving A_{44} and A_{55}) in the equations presents computational difficulties when the side-to-thickness ratio of the plate is large (i.e., when the plate becomes thin). For thin plates, the transverse shear strains $2\epsilon_{xz} = \phi_x + \partial w/\partial x$ and $2\epsilon_{yz} = \phi_y + \partial w/\partial y$ are negligible, and consequently the element stiffness matrix becomes stiff and yields erroneous results for the generalized displacements (w_i, S_i^x, S_i^y). This phenomenon is known as *shear locking,* and it can be interpreted as being caused by the inclusion of the following constraints into the variational form [see Reddy (1979, 1980) and Averill and Reddy (1990, 1992)]:

$$\phi_x + \frac{\partial w}{\partial x} = 0, \qquad \phi_y + \frac{\partial w}{\partial y} = 0 \tag{12.33}$$

The energy due to transverse shear strains (i.e., the penalty term) in the total potential energy of the shear deformable plate is given by

$$\frac{1}{2}\int_{\Omega^e}\left[A_{44}\left(\phi_y + \frac{\partial w}{\partial y}\right)^2 + A_{55}\left(\phi_x + \frac{\partial w}{\partial x}\right)^2\right] dx\, dy \tag{12.34}$$

The locking observed in the displacement finite element model of the first-order theory is the result of the fact that the discrete form of (12.33) is not satisfied when the plate is very thin. Of course, when the plate is thick, the relations (12.33) do not have to be satisfied, and the locking does not occur (at least, it is not severe enough to give completely wrong results). However, for thin plates, the constraints (12.33) are valid but not satisfied in the numerical model, and we therefore face the same problem as in the Timoshenko beam model. Therefore, we use the same remedy as before: reduced integration to evaluate stiffness coefficients involving the transverse shear terms. For example, when a four-node rectangular element is used, the one-point Gauss rule should be used to evaluate the shear energy terms (i.e., terms involving A_{44} and A_{55}), while the two-point Gauss rule should be used for all other terms. When an eight- or nine-node rectangular element is used, the two- and three-point Gauss rules should be used to evaluate the shear and bending terms, respectively. For the triangular elements, we use one- and three-point integrations for transverse shear stiffnesses in the linear and quadratic elements, respectively. However, in the present study, we shall not consider triangular plate bending elements.

12.4 EIGENVALUE AND TIME-DEPENDENT PROBLEMS

Equations (12.23*a*) and (12.32*a*) can be reduced to appropriate forms depending on the type of analysis. For static analysis, we set the inertia term $\{\ddot{\Delta}\}$ equal to zero and solve the problem

$$[K]\{\Delta\} = \{F\} \tag{12.35}$$

For natural vibration problems, we replace the inertia term by

$$\{\ddot{\Delta}\} = -\omega^2\{\Delta\} \qquad (\text{or } \{\Delta\} = \{\Delta_0\}e^{-i\omega t})$$

Then (12.23*a*) or (12.32*a*) takes the form of an eigenvalue problem:

$$([K^e] - \omega^2[M^e])\{\Delta_0\} = \{0\} \tag{12.36}$$

For buckling analysis (i.e., to determine the value of the in-plane compressive force at which the plate buckles), we replace the mass matrix in (12.36) with the stability matrix $[G]$, and ω^2 with the buckling load λ. This expression comes from the in-plane force due to the nonlinear inplane strains (underlined):

$$\begin{gathered} \epsilon_x = \underline{\frac{1}{2}\left(\frac{\partial w}{\partial x}\right)^2} - z\frac{\partial^2 w}{\partial x^2}, \qquad \epsilon_y = \underline{\frac{1}{2}\left(\frac{\partial w}{\partial y}\right)^2} - z\frac{\partial^2 w}{\partial y^2}, \\ 2\epsilon_{xy} = \underline{\frac{\partial w}{\partial x}\frac{\partial w}{\partial y}} - 2z\frac{\partial^2 w}{\partial x\,\partial y} \end{gathered} \tag{12.37}$$

For classical plate theory, $[G]$ is given by

$$G_{ij} = \int_{\Omega^e}\left[\bar{N}_x\frac{\partial \phi_i}{\partial x}\frac{\partial \phi_j}{\partial x} + \bar{N}_y\frac{\partial \phi_i}{\partial y}\frac{\partial \phi_j}{\partial y} + \bar{N}_{xy}\left(\frac{\partial \phi_i}{\partial x}\frac{\partial \phi_j}{\partial y} + \frac{\partial \phi_i}{\partial y}\frac{\partial \phi_j}{\partial x}\right)\right] dx\,dy \tag{12.38}$$

where $\bar{N}_x$, $\bar{N}_y$, and $\bar{N}_{xy}$ are the applied in-plane force resultants. The eigenvalue λ represents the ratio of the actual buckling load to the applied in-plane forces:

$$\lambda = N_x/\bar{N}_x = N_y/\bar{N}_y = N_{xy}/\bar{N}_{xy} \tag{12.39}$$

For first-order shear deformation theory, only $[M^{11}]$ in (12.32*a*) is replaced by $[G]$ (with $\phi_i = \psi_i$), and $[M^{22}]$ and $[M^{33}]$ are set equal to zero. Hence, the stability matrix is not positive definite, requiring special eigenvalue solvers for the first-order theory.

To solve a time-dependent problem, one must approximate the time derivatives in (12.23*a*) or (12.32*a*) to obtain algebraic equations relating $\{\Delta\}$ at time $t + \Delta t$ to $\{\Delta\}$ at time t, where Δt is the time step. In the Newmark integration scheme, the vectors $\{\Delta\}$ and $\{\dot{\Delta}\}$ at time $t = (s+1)\,\Delta t$ are approximated by the expressions

$$\begin{aligned} \{\dot{\Delta}\}_{s+1} &= \{\dot{\Delta}\}_s + \big((1-\alpha)\{\ddot{\Delta}\}_s + \alpha\{\ddot{\Delta}\}_{s+1}\big)\,\Delta t \\ \{\Delta\}_{s+1} &= \{\Delta\}_s + \{\dot{\Delta}\}_s\,\Delta t + \tfrac{1}{2}\big((1-\gamma)\{\ddot{\Delta}\}_s + \gamma\{\ddot{\Delta}\}_{s+1}\big)(\Delta t)^2 \end{aligned} \tag{12.40}$$

where α and γ are parameters that control the accuracy and stability of the scheme, and the subscript s indicates that the vectors are evaluated at the sth time step (i.e., at time $t = s\,\Delta t$).

Rearranging (12.23*a*) and (12.40), or (12.32*a*) and (12.40), we obtian [see (8.171)–(8.174)]

$$[\hat{K}]_{s+1}\{\Delta\}_{s+1} = \{\hat{F}\}_{s,s+1} \tag{12.41a}$$

where $\{\Delta\}_{s+1}$ denotes the value of $\{\Delta\}$ at time $t = (s+1)\,\Delta t$, s being the time step number, and

$$[\hat{K}]_{s+1} = [K]_{s+1} + a_3[M]_{s+1}$$
$$\{\hat{F}\}_{s,s+1} = \{F\}_{s+1} + [M]_{s+1}(a_3\{\Delta\}_s + a_4\{\dot{\Delta}\}_s + a_5\{\ddot{\Delta}\}_s) \qquad (12.41b)$$
$$a_3 = \frac{2}{\gamma(\Delta t)^2}, \qquad a_4 = a_3\,\Delta t, \qquad a_5 = \frac{1}{\gamma} - 1$$

Once the solution $\{\Delta\}$ is known at time $t_{s+1} = (s+1)\,\Delta t$, the first and second derivatives of $\{\Delta\}$ (velocity and acceleration) at t_{s+1} can be computed from

$$\begin{aligned}\{\ddot{\Delta}\}_{s+1} &= a_3(\{\Delta\}_{s+1} - \{\Delta\}_s) - a_4\{\dot{\Delta}\}_s - a_5\{\ddot{\Delta}\}_s,\\ \{\dot{\Delta}\}_{s+1} &= \{\dot{\Delta}\}_s + a_2\{\ddot{\Delta}\}_s + a_1\{\ddot{\Delta}\}_{s+1}\end{aligned} \qquad (12.42)$$

where $a_1 = \alpha\,\Delta t$ and $a_2 = (1-\alpha)\,\Delta t$.

12.5 EXAMPLES

Here we consider several numerical examples of plate bending, involving plates with different boundary conditions. Figure 12.2 shows various types of boundary conditions on an edge of a plate element. When the edge is parallel to the x or y axis, the normal and tangential components of a variable become the y and x (or x and y) components, respectively, of the variable. For classical plate theory, ϕ_n and ϕ_s in Fig. 12.2 must be replaced with $-\partial w/\partial n$ and $-\partial w/\partial s$, respectively. In all cases, the stresses are computed at the reduced Gauss points [see Barlow (1976, 1989)].

The first example deals with the effect of reduced integration on the deflections and stresses of a simply supported plate as computed using the shear deformable plate element.

Example 12.1. Consider a simply supported, isotropic, square plate subjected to a uniformly distributed transverse load f_0. We shall solve the problem using the classical and shear deformable plate bending elements. Owing to the biaxial symmetry, we need model only a quadrant of the plate. The essential boundary conditions at simply supported edges ($x = \frac{1}{2}a$ and $y = \frac{1}{2}a$), for the first-order theory, are (see Fig. 12.4)

$$\begin{aligned} w = 0, \quad \phi_y = 0 \quad \text{at } x = \tfrac{1}{2}a\\ w = 0, \quad \phi_x = 0 \quad \text{at } y = \tfrac{1}{2}a\end{aligned} \qquad (12.43)$$

The essential (or geometric) boundary conditions along the symmetry lines ($x = 0$ and $y = 0$) are

$$\phi_x(0, y) = \phi_y(x, 0) = 0 \qquad (12.44)$$

The natural or force boundary conditions, which enter the finite element equations through the forces $\{F^i\}$, are

$$\begin{aligned} Q_n &= 0 \quad \text{along } x = 0 \text{ and } y = 0\\ M_x &= 0 \quad \text{along } y = 0 \text{ and } x = \tfrac{1}{2}a\\ M_y &= 0 \quad \text{along } x = 0 \text{ and } y = \tfrac{1}{2}a\end{aligned} \qquad (12.45)$$

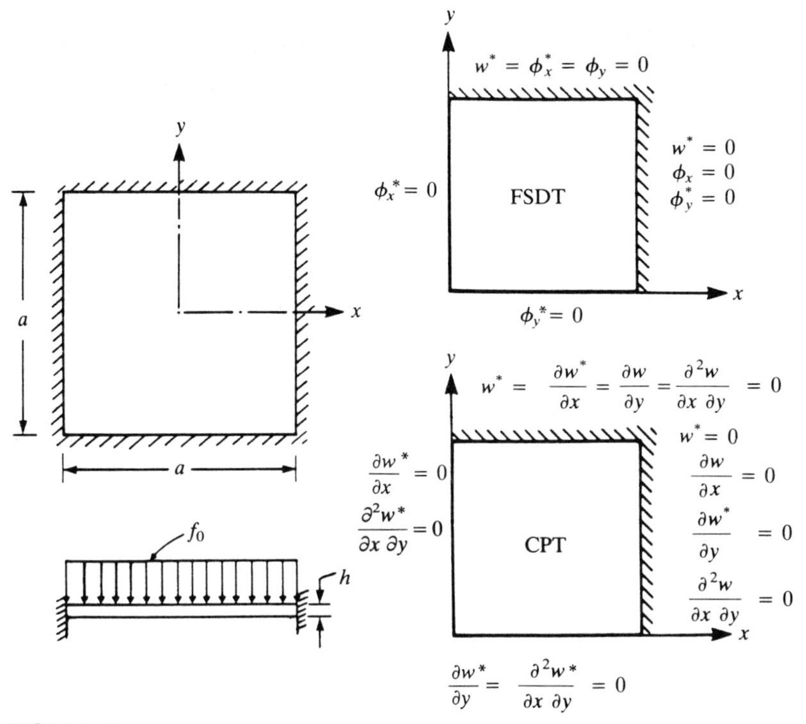

FIGURE 12.4
Geometry and coordinate system used for a square plate: CPT, classical plate theory; FSDT, first-order shear deformation theory. Boundary conditions for simply supported (quantities with an asterisk only) and clamped edges are shown for both theories. Boundary conditions along the lines of symmetry are also shown.

The essential boundary conditions for a quadrant, when using the classical plate element, are shown in Fig. 12.4.

For a linear (four-node) rectangular element of shear deformation theory, the contribution of a uniformly distributed load f_0 is given by

$$f_i = \int_0^{h_x} \int_0^{h_y} f_0 \psi_i \, dx \, dy = \tfrac{1}{4} h_x h_y f_0 \tag{12.46}$$

where h_x and h_y are the plane-form dimensions of the element. This contribution goes to the first, fourth, seventh, and tenth nodal degrees of freedom of the element (corresponding to w). For classical plate theory, the nodal forces are computed using the definition

$$f_i = \int_{\Omega^e} f_0 \phi_i \, dx \, dy \tag{12.47}$$

Except for the nodal forces given by (12.46) and (12.47), all other specified nodal forces of the problem are zero.

The effect of reduced integration, thickness, and mesh on the center deflection and stress is investigated, and the results are presented in Table 12.1. The nondimen-

TABLE 12.1
The effect of reduced integration, thickness, and mesh refinement on the center deflection and stress† of a simply supported, isotropic ($\nu = 0.25$), square plate under a uniform transverse load f_0 (Example 12.1)

		1×1 linear		2×2 linear		4×4 linear		2×2 quadratic		Exact‡	
a/h	Integration	$\bar{w}$	$\bar{\sigma}_x$	$\bar{w}$	$\bar{\sigma}_x$	$\bar{w}$	$\bar{\sigma}_x$	$\bar{w}$	$\bar{\sigma}_x$	$\bar{w}$	$\bar{\sigma}_x$
10	F	0.964	0.0182	2.474	0.1185	3.883	0.216	4.770	0.2899	4.791	0.2762
	M	3.950	0.0953	4.712	0.2350	4.773	0.2661	4.799	0.2715		
20	F	0.270	0.0053	0.957	0.0476	2.363	0.1375	4.570	0.2683	4.625	0.2762
	M	3.669	0.0954	4.524	0.2350	4.603	0.2660	4.633	0.2715		
40	F	0.0695	0.0014	0.279	0.0140	0.9443	0.0558	4.505	0.2699	4.584	0.2762
	M	3.599	0.0953	4.375	0.2349	4.560	0.2661	4.592	0.2714		
50	F	0.0045	0.0001	0.182	0.0092	0.6515	0.0386	4.496	0.2667	4.579	0.2762
	M	3.590	0.0953	4.472	0.2350	4.555	0.2660	4.587	0.2714		
100	F	0.011	0.0002	0.047	0.0024	0.182	0.0108	4.482	0.2664	4.572	0.2762
	M	3.579	0.0953	4.465	0.235	4.548	0.2661	4.580	0.2715		
CPT(N)		5.643	0.2599	4.857	0.2738	4.643	0.2758	—	—	4.570	0.2762
CPT(C)		4.638	0.2616	4.574	0.2714	4.570	0.2750	—	—	4.570	0.2762

† $\bar{w} = wEh^3 \times 10^2/f_0a^4$, $\bar{\sigma}_x = \sigma_x(A, A, \pm\frac{1}{2}h)h^2/f_0a^2$, $A = \frac{1}{4}a$ (1 × 1 linear), $\frac{1}{8}a$ (2 × 2 linear), $\frac{1}{16}a$ (4 × 4 linear), $0.05283a$ (2 × 2 quadratic).
‡ From Reddy (1984).

sionalized center displacement and stress are defined as

$$\bar{w} = \frac{w(0, 0)E_2h^3 \times 10^2}{f_0a^4}, \qquad \bar{\sigma}_x = \left(\frac{h}{a}\right)^2 \frac{1}{f_0}\,\sigma_x(A, A, \pm\tfrac{1}{2}h) \tag{12.48}$$

where A is the Gauss-point location with respect to the (x, y) system located at the center of the plate:

	2 × 2 Q4	4 × 4 Q4	2 × 2 Q9
A	$0.125a$	$0.0625a$	$0.05283a$

where Q4 denotes the four-node rectangular element and Q9 the nine-node rectangular element.

In Table 12.1, F denotes full integration for all terms, and M denotes mixed integration: full integration for bending terms and reduced integration for the shear terms. The following observations can be made from the results of Table 12.1:

1. The nine-node element gives virtually the same results for full (3 × 3 Gauss rule) and mixed (3 × 3 and 2 × 2 Gauss rules) integrations. However, the results obtained using the mixed integration are closest to the exact solution.
2. Full integration gives less accurate results than mixed integration, and the error increases with an increase in side-to-thickness ratio. This implies that mixed integration is essential for thin plates, especially when modeled by lower-order elements.

3. Full integration results in smaller errors for quadratic elements and refined meshes than for linear elements and/or coarse meshes.
4. The conforming plate finite element CPT(C) gives accurate results when compared with the nonconforming plate element CPT(N).

Table 12.2 gives the natural frequencies of the simply supported plate, obtained using various elements and meshes. Note that classical plate theory overpredicts the natural frequencies.

The next example deals with a clamped square plate under a distributed transverse load. The mixed integration rule is used in the evaluation of stiffness coefficients of the FSDT model.

Example 12.2. Consider an isotropic ($\nu = 0.3$) square plate under a uniform load of intensity f_0. We shall consider clamped boundary conditions (see Fig. 12.4). Note that, for the case of the conforming displacement model based on CPT, we must also specify the boundary conditions on the cross-derivative $\partial^2 w/\partial x\, \partial y$. Once again, we exploit the biaxial symmetry and model only a quadrant of the plate.

Table 12.3 gives the center deflection $\bar{w} = w(0, 0)D \times 10^2/f_0 a^4$ $[D = Eh^3/12(1 - \nu^2)]$ and center normal stress $\bar{\sigma}_x = \sigma_x(A, A) \times 10/f_0$ as obtained using uniform meshes of the two types of elements: CPT(N) and CPT(C). In both models, the point

TABLE 12.2
The first three symmetric vibrational frequencies ($\bar{\omega} \times 10^{-2}$) of an isotropic simply supported square plate obtained using classical and shear deformation theories ($a/h = 10$)

Theory	Mesh	$\bar{\omega}_{11}$	$\bar{\omega}_{13}$	$\bar{\omega}_{33}$
FSDT	1 × 1 Q4	0.0746	—	—
	2 × 2 Q4	0.0608	0.4473	0.4810
	4 × 4 Q4	0.0579	0.2913	0.4654
	1 × 1 Q9	0.0575	0.4030	0.5476
	2 × 2 Q9	0.0570	0.2651	0.4342
	Exact	0.0569	0.2552	0.4217
CPT†	1 × 1 Q4	0.0535	0.3118	0.3565
		(0.0597)	(0.2912)	(0.3360)
	2 × 2 Q4	0.0567	0.2762	0.4406
		(0.0584)	(0.4842)	(0.4842)
	4 × 4 Q4	0.0579	0.2792	0.4665
		(0.0584)	(0.2821)	(0.4900)
	Exact	0.0584	0.2829	0.4943

† Numbers without parentheses are computed using the noncomforming element and those in parentheses are computed using the conforming element $[\bar{\omega} = \omega(\rho/E)^{1/2}a^2/h]$.

TABLE 12.3
Comparison of the center deflection and normal stress of a clamped square plate under a uniformly distributed load as obtained using various finite element meshes (Example 12.2)

Mesh	Displacement model (CPT)	Displacement model (FSDT; $a/h = 10$) Linear	Quadratic
		Center deflection $\bar{w}$	
1×1	0.1943	0.0357	—
2×2	0.1265	0.1459	0.1757
4×4	0.1266	0.1495	0.1586
Exact	0.1266		
		Center stress $\bar{\sigma}_x$	
1×1	2.443	0.000	—
2×2	1.415	1.142	1.321
4×4	1.381	1.333	1.345

$(x, y) = (A, A)$ is defined as follows:

Element type		Location A 1×1	2×2 (1×1)	4×4 (2×2)	8×8 (4×4)
CPT model		0.05635a	0.02817a	0.01409a	0.03125a
FSDT model	linear	0.25a	0.125a	0.0625a	0.03125a
FSDT model	quadratic	—	0.1057a	0.0528a	0.02642a

The distance is measured from the center of the plate. The numbers in parentheses above denote the mesh of quadratic Q9 (i.e., nine-node) elements. Clearly, both finite element models exhibit good convergence characteristics. The difference between the CPT and FSDT is attributed to the inclusion of transverse shear strains in FSDT.

The next example deals with a simply supported orthotropic plate under a uniformly distributed transverse load.

Example 12.3. Here we consider an orthotropic plate with the following (graphite–epoxy) material properties ($\nu_{21} = \nu_{12}E_2/E_1$):

$$E_1 = 31.8 \times 10^6 \text{ psi}, \quad E_2 = 1.02 \times 10^6 \text{ psi}, \quad \nu_{12} = 0.31, \quad G_{12} = G_{23} = G_{13} = 0.96 \text{ psi} \tag{12.49}$$

The nondimensionalized center deflection $\bar{w}$ and normal stress $\bar{\sigma}_x$ obtained using the conforming CPT(C) element and the FSDT element are compared in Table 12.4.

TABLE 12.4
Comparison of the center deflection ($\bar{w} = wH \times 10^3/f_0a^4$) and normal stress ($\bar{\sigma}_x = \sigma_x \times 10\, h^2/f_0a^2$) of a graphite–epoxy, simply supported square plate under a uniform transverse load (Example 12.3)

Mesh	Displacement model (CPT)	Displacement model (FSDT; $a/h = 10$) Linear	Quadratic
	Center deflection $\bar{w}$		
2×2	0.9220	1.2545	1.2715
4×4	0.9224	1.2186	1.2147
8×8	0.9224	1.2152	1.2147
Exact†	0.9225	1.215	
	Center stress $\bar{\sigma}_x$		
2×2	7.678	6.277	7.192
4×4	7.616	7.256	7.399
8×8	7.600	7.449	7.478
Exact†	7.595	7.512	

† From Reddy (1984); $H = D_{12} + 2D_{66}$.

The final example deals with the transient response of an isotropic plate subjected to a sudden uniform patch loading.

Example 12.4. Consider an isotropic ($\nu = 0.3$, $\rho = 1.0$), simply supported, rectangular plate ($a/b = \sqrt{2}$, $h/b = 0.2$) under a suddenly applied uniformly distributed load on a square ($c/b = 0.4$) area at the center:

$$a/b = \sqrt{2}, \qquad \Delta t = 0.01$$

$$f = f_0(x, y)H(t), \qquad \text{where} \quad f_0(x, y) = \begin{cases} 1 & \text{for } 0 < x, y \leq 0.2 \\ 0 & \text{for } x, y > 0.2 \end{cases} \tag{12.50}$$

and $H(t)$ is the Heaviside unit step function. The geometry and boundary conditions are shown in Fig. 12.5. A nonuniform mesh of 4×4 nine-node shear deformation elements is used in a quadrant of the plate.

The center deflection and bending moments of the present linear analysis are compared with the analytical thick- and thin-plate solutions of Reismann and Lee (1969) in Fig. 12.6. We note significant difference between the solutions of the two theories. The present finite element solutions for the center deflection and bending moment are in excellent agreement with the thick-plate solution of Reismann and Lee. Since the bending moment in the finite element method is calculated at the Gauss points, it is not expected to match exactly with the analytical solution at the center of the plate.

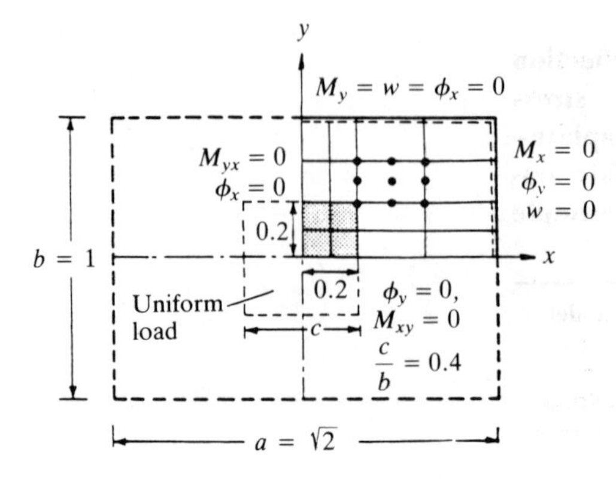

FIGURE 12.5
Domain, boundary conditions, and finite element mesh for the bending of a rectangular plate under a suddenly applied pulse loading at the central square area.

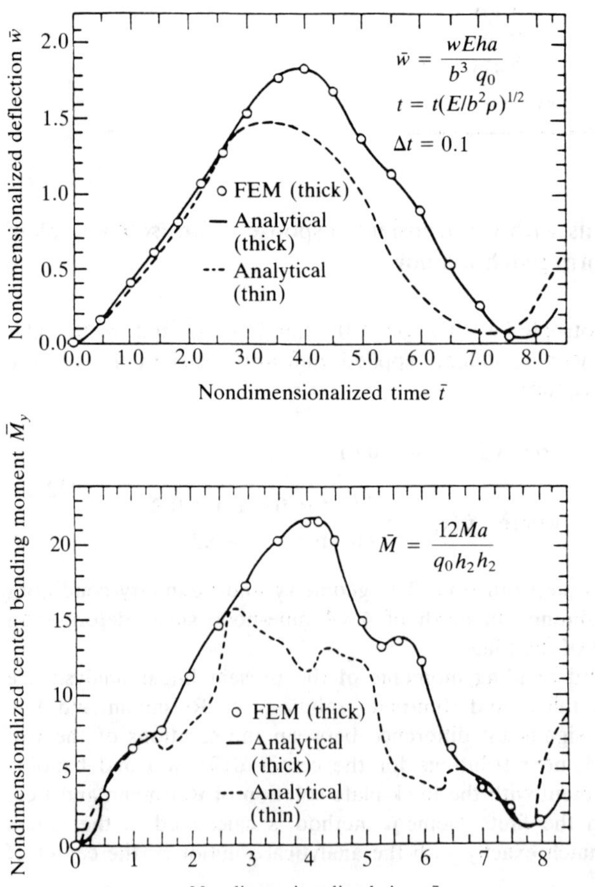

FIGURE 12.6
Comparison of the finite element solution with the analytical solution for a simply supported rectangular isotropic plate under a suddenly applied pulse loading at the central square area.

12.6 SUMMARY

Finite element models of the classical (i.e., Kirchhoff) and first-order shear deformation plate theories have been developed in this chapter. The classical plate theory requires C^1 continuity of the transverse deflection w (i.e., the deflection and its derivatives are continuous between elements), whereas first-order shear deformation theory requires C^0 continuity (i.e., only the variables are continuous between elements) of the generalized displacements (w, ϕ_x, ϕ_y). Triangular and rectangular elements with C^1 continuity have been discussed. Two four-node rectangular elements, one with $(w, \partial w/\partial x, \partial w/\partial y)$ and another with $(w, \partial w/\partial x, \partial w/\partial y, \partial^2 w/\partial x\, \partial y)$ as degrees of freedom, have been presented. The first does not satisfy continuity of the normal derivative along element sides, and is called a nonconforming element. The second is a conforming element. Linear and quadratic rectangular elements of first-order shear deformation theory have been developed. They require selective evaluation of the stiffness coefficients. The bending stiffness coefficients are evaluated using full integration and the transverse shear stiffness coefficients are evaluated using reduced integration to avoid the shear locking that occurs when these elements are applied to thin plates. Finite element models of vibration, stability, and static and dynamic response have been developed for the two theories.

PROBLEMS

12.1–12.9. For the plate bending problems (CPT and FSDT) given in Figs. P12.1–P12.9, give the specified primary and secondary degrees of freedom and their values for the 4×4 meshes shown. The dashed lines in the figures indicate simply supported boundary conditions: those of types "simply supported 1" and "simply supported 2" (see Fig. 12.2) are indicated by SS-1 and SS-2, respectively. Use E, ν, h, a, and b in formulating the data.

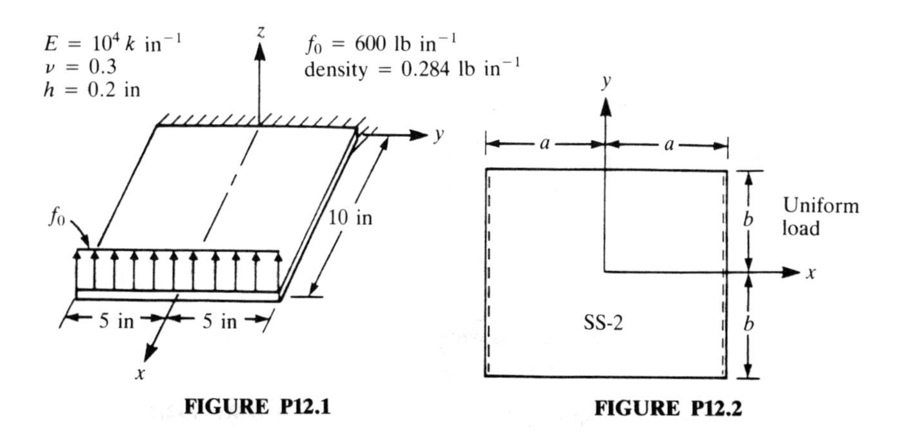

FIGURE P12.1

FIGURE P12.2

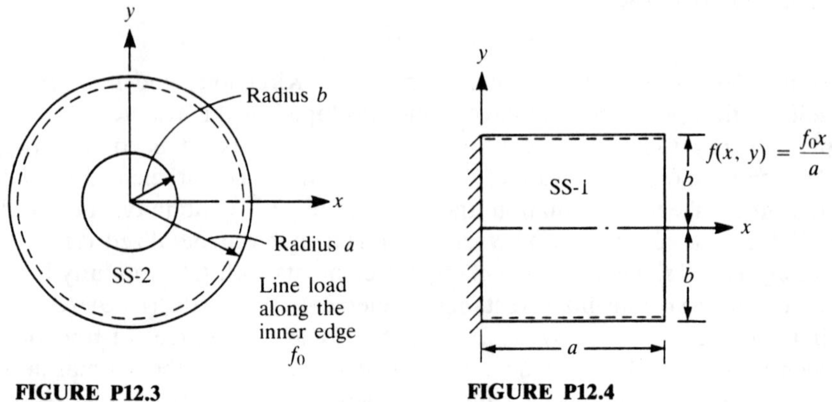

FIGURE P12.3

FIGURE P12.4

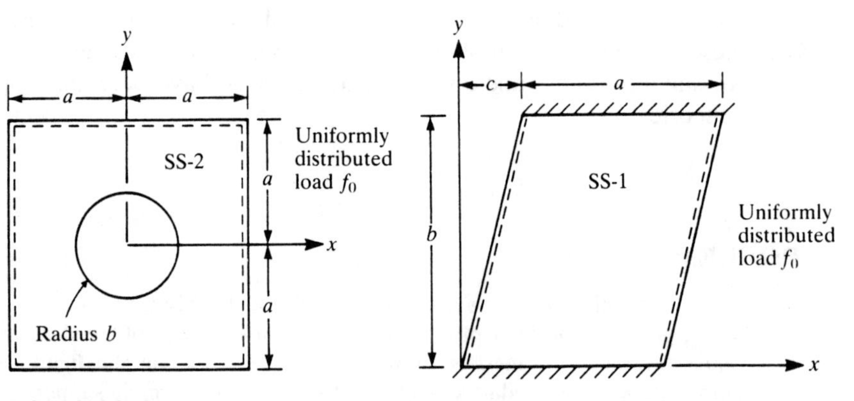

FIGURE P12.5

FIGURE P12.6

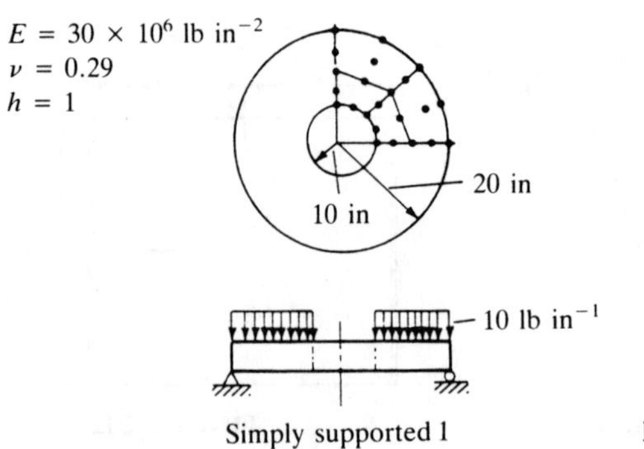

FIGURE P12.7

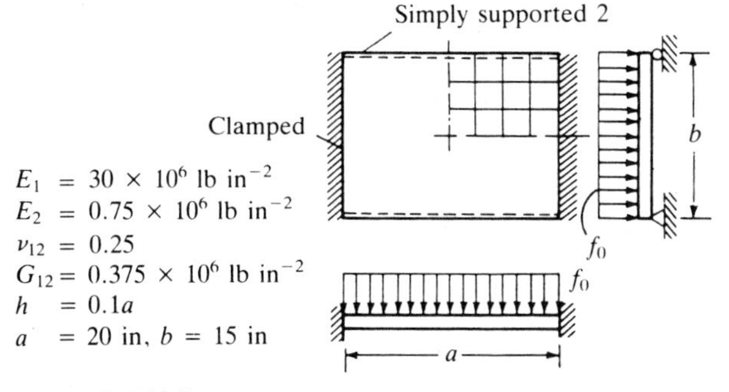

FIGURE P12.8

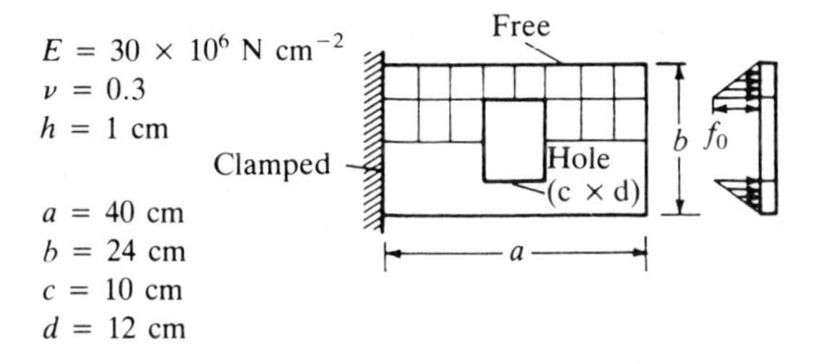

FIGURE P12.9

12.10. Give an algebraic form for the elements of the stiffness matrix (in a local coordinate system) for a thin plane elastic body subjected to both in-plane and transverse loads.

Hint: Combine the element matrices of the plane stress element with the plate bending element—analogous to the construction of a frame element from bar and beam elements.

12.11. Consider an isotropic annular plate ($E = 30 \times 10^6$ lb in^{-2}, $\nu = 0.29$) with the outer edges clamped and subjected to uniform loading (see Fig. P12.7). Formulate the necessary data for the problem.

12.12. Repeat Problem 12.11 for the case in which the plate is subjected to a load varying linearly with radial distance.

REFERENCES FOR ADDITIONAL READING

Averill, R. C., and J. N. Reddy: "On the Behavior of Plate Elements Based on the First-Order Shear Deformation Theory," *Engineering Computations,* vol. 7, pt. 1, pp. 57–74, 1990.

——— and ———: "An Assessment of Four-Noded Plate Finite Elements Based on a Generalized Third-Order Theory," *International Journal for Numerical Methods in Engineering,* vol. 33, pp. 1553–1572, 1992.

Barlow, J.: "Optimal Stress Locations in Finite Element Models," *International Journal for Numerical Methods in Engineering,* vol. 10, pp. 243–251, 1976.

———: "More on Optimal Stress Points—Reduced Integration Element Distortions and Error Estimation," *International Journal for Numerical Methods in Engineering,* vol. 28, pp. 1487–1504, 1989.

Dym, C. L., and I. H. Shames: *Solid Mechanics: A Variational Approach,* McGraw-Hill, New York, 1973.

Goudreau, G. L., and R. L. Taylor: "Evaluation of Numerical Integration Methods in Elastodynamics," *Journal of Computer Methods in Aplied Mechanics and Engineering,* vol. 2, pp. 69–97, 1973.

Hughes, T. J. R., and W. K. Liu: "Implicit–Explicit Finite Elements in Transient Analysis: Stability Theory, and Implementation and Numerical Examples," *Journal of Applied Mechanics,* vol. 45, pp. 371–378, 1978.

Jounson, D. E.: "A Proof of the Stability of the Hubolt Method," *Journal of the American Institute for Aeronautics and Astronautics,* vol. 4, pp. 1450–1451, 1966.

Krieg, R. D.: "Unconditional Stability in Numerical Time Integration Methods," *Journal of Applied Mechanics,* vol. 40, pp. 417–421, 1973.

Lax, P. D., and R. D. Richtmyer: "Survey of the Stability of Finite Difference Equations," *Communications in Pure and Applied Mathematics,* vol. 9, pp. 267–293, 1956.

Leech, J. W.: "Stability of Finite-Difference Equations for the Transient Response of a Flat Plate," *Journal of the American Institute for Aeronautics and Astronautics,* vol. 3, pp. 1172–1173 1965.

Levy, S., and W. D. Kroll: "Errors Introduced by Finite Space and Time Increments in Dynamic Response Computation," *Journal of Research, National Bureau of Standards,* vol. 51, pp. 57–68, 1953.

Nickell, R. E.: "On the Stability of Approximation Operators in Problems of Structural Dynamics," *International Journal for Solids and Structures,* vol. 7, pp. 301–319, 1971.

———: "Direct Integration in Structural Dynamics," *Journal of Engineering Mechanics Division, ASCE,* vol. 99, pp. 303–317, 1973.

Reddy, J. N.: "Simple Finite Elements with Relaxed Continuity for Nonlinear Analysis of Plates," in *Finite Element Methods in Engineering,* eds. A. P. Kabaila and V. A. Pulmano, pp. 265–281, The University of New South Wales, Sydney, 1979.

———: "A Penalty Plate Bending Element for the Analysis of Laminated Anisotropic Composite Plates," *International Journal for Numerical Methods in Engineering,* vol. 15, pp. 1187–1206, 1980.

———: "On the Solutions to Forced Motions of Rectangular Composite Plates," *Journal of Applied Mechanics,* vol. 49, pp. 403–408, 1982.

———: *Energy and Variational Methods in Applied Mechanics,* John Wiley, New York, 1984.

———: "A Review of Refined Theories of Laminated Composite Plates," *Shock and Vibration Digest,* vol. 22, pt. 7, pp. 3–17, 1990*a*.

———: "A General Non-Linear Third-Order Theory of Plates with Moderate Thickness," *Journal of Non-Linear Mechanics,* vol. 25, pp. 677–686, 1990*b*.

Reismann, H., and Y. Lee: "Forced Motions of Rectangular Plates," in *Developments in Theoretical and Applied Mechanics,* vol. 4, ed. D. Frederick, pp. 3–18, Pergamon Press, New York, 1969.

Szilard, R.: *Theory and Analysis of Plates,* Prentice-Hall, Englewood Cliffs, NJ, 1974.

Timoshenko, S., and S. Woinowsky-Krieger: *Theory of Plates and Shells,* 2d ed., McGraw–Hill, New York, 1959.

Tsay, C. S., and J. N. Reddy: "Bending, Stability, and Free Vibration of Thin Orthotropic Plates by Simplified Mixed Finite Elements," *Journal of Sound and Vibration,* vol. 59, pp. 307–311, 1978.

Tsui, T. Y., and P. Tong: "Stability of Transient Solution of Moderately Thick Plate by Finite Difference Method," *Journal of the American Institute for Aeronautics and Astronautics,* vol. 9, pp. 2062–2063, 1971.

Zienkiewicz, O. C., and R. L. Taylor: *The Finite Element Method,* vol. 1: *Linear Problems,* McGraw–Hill, New York, 1989.

——— and ———: *The Finite Element Method,* vol. 2: *Nonlinear Problems,* McGraw–Hill, New York, 1991.

———, ——— and J. M. Too: "Reduced Integration in General Analysis of Plates and Shells," *International Journal for Numerical Methods in Engineering,* vol. 3, pp. 275–290, 1972.

CHAPTER 13

COMPUTER IMPLEMENTATION

13.1 INTRODUCTION

In Chapter 7, we discussed some basic ideas concerning the development of a typical finite element program, and the use of FEM1DV2 in the solution of one-dimensional problems was illustrated via many example problems. Specific details of various logical units of a finite element program for one-dimensional problems were given. Most of the ideas presented there are also valid for two-dimensional problems. The imposition of the boundary conditions and the solution of the equations remain the same as in one-dimensional problems. Here we focus attention on the computer implementation of two-dimensional elements. The use of a model program FEM2DV2 (a revised and combined version of the FEM2D and PLATE programs in the first edition of the book) is discussed. The program FEM2DV2 contains linear and quadratic triangular and rectangular elements, and it can be used for the solution of heat conduction and convection problems, laminar flows of viscous incompressible fluids using the penalty function formulation, plane elasticity problems, and plate bending problems using classical and shear deformation theories. A flow chart of FEM2DV2 is given in Fig. 13.1.

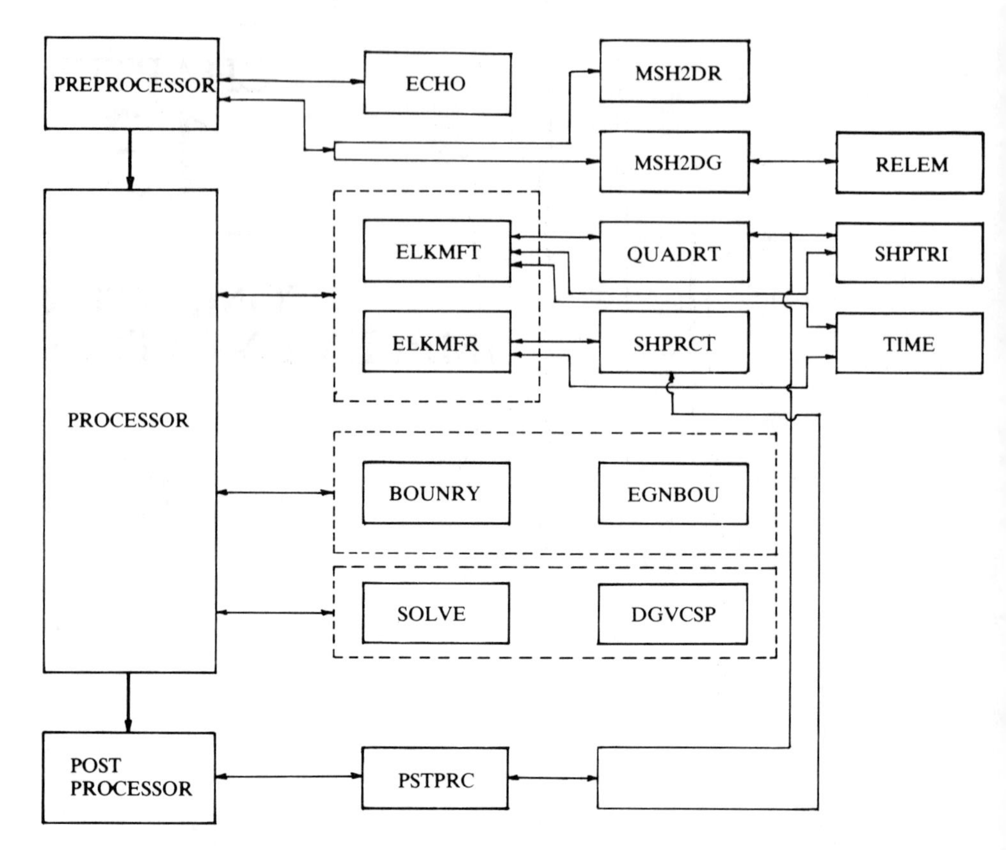

FIGURE 13.1

In two dimensions, the element calculations are more involved than in one dimension, owing to the following considerations:

1. Various geometric shapes of elements.
2. Single as well as multivariable problems.
3. Integrations performed over areas as opposed to along lines (for one-dimensional elements).
4. Mixed-order integrations used in certain formulations (shear-deformable plates and penalty function formulations of viscous incompressible fluids).

13.2 PREPROCESSOR

In the preprocessor unit, the program MSH2DR is used to generate triangular- and rectangular-element meshes of rectangular domains. The subroutine requires minimal input, but is not general enough to generate finite element meshes of arbitrary domains. The subroutine MSH2DG is more general, and can be used to generate meshes for nonrectangular domains. Of course, one

BOUNRY Subroutine to impose specified boundary conditions on the primary and secondary variables when a steady-state or transient analysis is carried out.

DGVCSP An IBM system-supplied subroutine to solve an eigenvalue problem of the type $[K]\{U\} = L(M)\{U\}$, where $[K]$ and $[M]$ are positive definite matrices.

ECHO Subroutine to echo the input data to the program.

EGNBOU Subroutine to impose specified boundary conditions on the primary variables when an eigenvalue problem is analyzed.

ELKMFR Subroutine to compute the element $[K]$, $[M]$, and $\{F\}$ matrices and vectors for the rectangular elements.

ELKMFT Subroutine to compute the element $[K]$, $[M]$, and $\{F\}$ matrices and vectors for the triangular elements.

MSH2DG Subroutine to generate mesh for general domains.

MSH2DR Subroutine to generate mesh for rectangular domains only.

PSTPRC Subroutine to post-complete the solution, gradient of solution, and stresses.

QUADRT Subroutine to generate the quadrature points and weights for the triangular elements.

RELEM Subroutine called in MSH2DG to generate element data.

SHPRCT Subroutine to compute the shape functions for linear and quadratic (eight- and nine-node) rectangular elements.

SHPTRI Subroutine to compute the shape functions for linear and quadratic triangular elements.

SOLVE Subroutine to solve a banded, symmetric system of algebraic equations.

TIME Subroutine to compute the equivalent coefficient matrices and columen vectors for parabolic and hyperbolic equations when time-dependent analysis is carried out.

FIGURE 13.1 (Continued)

can use any other mesh generation program in place of MSH2DR or MSH2DG. The subroutines MSH2DR and MSH2DG generate the connectivity matrix (array NOD) and the global coordinates of the nodes (array GLXY). When the mesh generators cannot be used, the mesh information should be read in.

13.3 ELEMENT COMPUTATIONS: PROCESSOR

Element calculations for linear and quadratic triangular (ELKMFT) and quadrilateral (ELKMFR) elements can be carried out according to the developments presented in Chapters 8 and 9. The principal steps involved are as follows:

1. Development of a subroutine for the evaluation of the interpolation functions and their derivatives with respect to the global coordinates [see (9.42)–(9.45)].

2. Numerical integration of the coefficients of the element matrices using numerical quadrature formulas [see (9.54) and (9.67)].
3. Setting up of the element matrices required for the class of problems being solved (e.g., static, transient, and eigenvalue problems).

The subroutines SHPTRI and SHPRCT (called in a do-loop on the number of quadrature points) contain the expressions of the interpolation functions and their derivatives for various-order triangular (TRI) and rectangular (RCT) elements, respectively. The derivatives of the interpolation functions with respect to global coordinates [(9.42)] are also computed in these subroutines. The Fortran implementation of (9.42)–(9.45) is summarized below.

```
C     ________________________________________________________________
C
C     DEFINITION OF THE VARIABLES
C
C     SF(I)........Interpolation function for node I of the element
C     DSF(J,I).....Derivative of SF(I) with respect to XI if J=1 and
C                  and ETA if J=2
C     GDSF(J,I)....Derivative of SF(I) with respect to  X if J=1 and
C                  and  Y  if J=2
C     ELXY(I,J)....Global coordinates of the element Ith node:
C                  J=1: the x coordinate; J=2: the y coordinate
C     GJ(I,J)......Jacobian matrix
C     GJINV(I,J)...Inverse of the jacobian matrix
C     DET..........Determinant of the Jacobian matrix
C     ________________________________________________________________
C
C     Given the interpolation functions {SF} and their derivatives with
C     respect to the natural coordinates  [DSF], compute  the  Jacobian
C     matrix [GJ] [see Eqn. (9.44b)] and its inverse [GJINV]:
C
      DO 40 I = 1,2
      DO 40 J = 1,2
      GJ(I,J)   = 0.0
      DO 40 K = 1,NPE
   40 GJ(I,J)   = GJ(I,J) + DSF(I,K)*ELXY(K,J)
C
      DET = GJ(1,1)*GJ(2,2)-GJ(1,2)*GJ(2,1)
      GJINV(1,1) = GJ(2,2)/DET
      GJINV(2,2) = GJ(1,1)/DET
      GJINV(1,2) = -GJ(1,2)/DET
      GJINV(2,1) = -GJ(2,1)/DET
C
C     Compute the first derivatives of the interpolation functions with
C     respect to the global coordinates [GDSF] [see Eqn. (9.42)]:
C
          DO 50 I   = 1,2
          DO 50 J   = 1,NPE
          GDSF(I,J) = 0.0
          DO 50 K   = 1, 2
   50     GDSF(I,J) = GDSF(I,J) + GJINV(I,K)*DSF(K,J)
C     ________________________________________________________________
```

The two-dimensional problems of interest to us here require the evaluation of element matrices that involve products of interpolation functions and their derivatives with respect to the global coordinates. Since the integrals are evaluated numerically, the integrands must be evaluated at the quadrature points and summed over the number of integration points. Thus, evaluation of the interpolation functions and their derivatives must be carried out inside the do-loops. To fix ideas, for rectangular elements, we define

SF(I) — interpolation function ψ_I of the Ith node of an element

GDSF(K, I) — global derivative with respect to x_K (i.e., derivative with respect to the global coordinate x_K) of interpolation function ψ_I [GDSF(K, I) $= \partial\psi_I/\partial x_K$, $x_1 = x$, $x_2 = y$]

CONST — product of the Jacobian (i.e., determinant of $[\mathcal{J}]$) with the weights corresponding to the Gauss integration point (ξ_{NI}, η_{NJ})
= DET * GAUSWT(NI, NGP) * GAUSWT(NJ, NGP)

Then S_{ij}^{00}, S_{ij}^{11}, . . . of (8.35),

$$S_{ij}^{\alpha\beta} = \int_{\Omega^e} \frac{\partial\psi_i}{\partial x_\alpha}\frac{\partial\psi_j}{\partial x_\beta}\,dx\,dy \tag{13.1}$$

are given in Fortran form by

$$\begin{aligned}
&\text{S00(I, J)} = \text{S00(I, J)} + \text{SF(I)} * \text{SF(J)} * \text{CONST}\\
&\text{S11(I, J)} = \text{S11(I, J)} + \text{GDSF(1, I)} * \text{GDSF(1, J)} * \text{CONST}\\
&\text{S12(I, J)} = \text{S12(I, J)} + \text{GDSF(1, I)} * \text{GDSF(2, J)} * \text{CONST}\\
&\text{S22(I, J)} = \text{S22(I, J)} + \text{GDSF(2, I)} * \text{GDSF(2, J)} * \text{CONST}
\end{aligned} \tag{13.2}$$

The summed values of S00(I, J), S11(I, J), . . . represent the numerical values of the integral coefficients in (13.1). The Fortran statements listed below summarize the discussion:

```
C     ________________________________________________________________
C
C     NUMERICAL EVALUATION OF ELEMENT COEFFICIENT MATRICES IN EQ.(13.1)
C        FOR QUADRILATERAL ELEMENTS (WITH ISOPARAMETRIC FORMULATION)
C
C      NPE = Number of nodes per elements
C      IPDF= Number of integration points (i.e., Gauss points)
C
C     ________________________________________________________________
      DIMENSION  GAUSPT(5,5), GAUSWT(5,5), SF(9), GDSF(2,9), ELXY(9,2),
     1           S00(9,9), S11(9,9), S12(9,9), S21(9,9), S22(9,9)
C
      DATA GAUSPT/5*0.0D0, -0.57735027D0, 0.57735027D0, 3*0.0D0,
     2  -0.77459667D0, 0.0D0, 0.77459667D0, 2*0.0D0, -0.86113631D0,
     3  -0.33998104D0, 0.33998104D0, 0.86113631D0, 0.0D0, -0.90617984D0,
     4  -0.53846931D0, 0.0D0, 0.53846931D0, 0.90617984D0/
C
      DATA GAUSWT/2.0D0, 4*0.0D0, 2*1.0D0, 3*0.0D0, 0.55555555D0,
     2   0.88888888D0, 0.55555555D0, 2*0.0D0, 0.34785485D0,
     3 2*0.65214515D0, 0.34785485D0, 0.0D0, 0.23692688D0,
     4   0.47862867D0, 0.56888888D0, 0.47862867D0, 0.23692688D0/
C
C     Initialize the arrays
C
      DO 120 I = 1,NPE
         DO 120 J = 1,NPE
            S00(I,J)= 0.0
            S11(I,J)= 0.0
```

```
            S12(I,J)= 0.0
            S21(I,J)= 0.0
            S22(I,J)= 0.0
  120 CONTINUE
C
C     DO-loops on numerical (Gauss) integration begin here:____________
C
      DO 200 NI = 1,IPDF
      DO 200 NJ = 1,IPDF
      XI  = GAUSPT(NI,IPDF)
      ETA = GAUSPT(NJ,IPDF)
C
C     SHPRCT (SHaPe functions for ReCTangular elements) is called here
C     to compute {SF} and [GDSF]:
C
      CALL SHPRCT (NPE,XI,ETA,ELXY,DET,SF,GDSF)
      CNST = DET*GAUSWT(NI,IPDF)*GAUSWT(NJ,IPDF)
C
C     Compute Sij(I,J) of Eqn. (13.1):
C
      DO 180 I=1,NPE
         DO 180 J=1,NPE
            S00(I,J) = S00(I,J) + SF(I)*SF(J)*CNST
            S11(I,J) = S11(I,J) + GDSF(1,I)*GDSF(1,J)*CNST
            S12(I,J) = S12(I,J) + GDSF(1,I)*GDSF(2,J)*CNST
            S21(I,J) = S21(I,J) + GDSF(2,I)*GDSF(1,J)*CNST
            S22(I,J) = S22(I,J) + GDSF(2,I)*GDSF(2,J)*CNST
  180    CONTINUE
  200 CONTINUE
C     _____________________________________________________________________
```

To set up the element coefficient matrices of a given problem, we make use of the element matrices defined above. As an example, consider the problem described by (8.1). The element coefficient matrix and the column vectors for the problem are given by (8.14*b*). The element matrix K_{ij} [ELK(I, J)] can be expressed in terms of S_{ij}^{00}, S_{ij}^{11}, . . . by

$$\begin{aligned}\text{ELK(I, J)} &= \text{A00} * \text{S00(I, J)} + \text{A11} * \text{S11(I, J)} + \text{A12} * \text{S12(I, J)} \\ &\quad + \text{A21} * \text{S12(J, I)} + \text{A22} * \text{S22(I, J)}\end{aligned}$$

where $a_{00} = \text{A00}$, $a_{11} = \text{A11}$, $a_{12} = \text{A12}$, $a_{21} = \text{A21}$, and $a_{22} = \text{A22}$ are the constant coefficients of the differential equation (8.1).

In multivariable problems, the element matrices are themselves defined in terms of submatrices, as was the case for plane elasticity, fluid flow, and plate bending. In such cases, the nodal degrees of freedom should be renumbered to reduce the half-bandwidth of the assembled coefficient matrix. For example, consider the element equations (10.29*a*) associated with plane elasticity problems. The element nodal variables Δ_i are given (say, for a linear rectangular element) by

$$\begin{Bmatrix} \Delta_1 \\ \Delta_2 \\ \Delta_3 \\ \\ \vdots \\ \\ \Delta_8 \end{Bmatrix} = \begin{Bmatrix} u_1 \\ u_2 \\ u_3 \\ u_4 \\ v_1 \\ v_2 \\ v_3 \\ v_4 \end{Bmatrix} \tag{13.3}$$

Thus, at any node, the difference between the label number of the first degree of freedom and that of the second degree of freedom is 4 (in a general case, the difference is n, where n is the number of nodes per element). This difference contributes to an increase in the half-bandwidth of the assembled coefficient matrix. To remedy this situation, we reorder the element nodal degrees of freedom as follows:

$$\begin{Bmatrix} \Delta_1 \\ \Delta_2 \\ \Delta_3 \\ \Delta_4 \\ \vdots \\ \Delta_{2n-1} \\ \Delta_{2n} \end{Bmatrix} = \begin{Bmatrix} u_1 \\ v_1 \\ u_2 \\ v_2 \\ \vdots \\ u_n \\ v_n \end{Bmatrix} \tag{13.4}$$

In reordering the nodal degrees of freedom, we must retain the symmetry, if one is present, of the system of algbebraic equations. This is accomplished by renumbering the equations in the same way as the nodal degrees of freedom. To illustrate how this can be done, we consider a set of four equations in four unknowns:

$$\begin{aligned} S_{11}^{11}u_1 + S_{12}^{11}u_2 + S_{11}^{12}v_1 + S_{12}^{12}v_2 &= F_1^1 \\ S_{21}^{11}u_1 + S_{22}^{11}u_2 + S_{21}^{12}v_1 + S_{22}^{12}v_2 &= F_2^1 \\ S_{11}^{21}u_1 + S_{12}^{21}u_2 + S_{11}^{22}v_1 + S_{12}^{22}v_2 &= F_1^2 \\ S_{21}^{21}u_1 + S_{22}^{21}u_2 + S_{21}^{22}v_1 + S_{22}^{22}v_2 &= F_2^2 \end{aligned} \tag{13.5a}$$

or, in matrix form,

$$\begin{bmatrix} [S^{11}] & [S^{12}] \\ [S^{21}] & [S^{22}] \end{bmatrix} \begin{Bmatrix} \{u\} \\ \{v\} \end{Bmatrix} = \begin{Bmatrix} \{F^1\} \\ \{F^2\} \end{Bmatrix} \tag{13.5b}$$

Now letting

$$\Delta_1 = u_1, \qquad \Delta_2 = v_1, \qquad \Delta_3 = u_2, \qquad \Delta_4 = v_2 \tag{13.6}$$

(i.e., the third nodal variable is renamed as the second, and vice versa) and rearranging (13.5) (i.e., the third equation becomes the second equation, and vice versa), we obtain

$$\begin{aligned} S_{11}^{11}\Delta_1 + S_{11}^{12}\Delta_2 + S_{12}^{11}\Delta_3 + S_{12}^{12}\Delta_4 &= F_1^1 \\ S_{11}^{21}\Delta_1 + S_{11}^{22}\Delta_2 + S_{12}^{21}\Delta_3 + S_{12}^{22}\Delta_4 &= F_1^2 \\ S_{21}^{11}\Delta_1 + S_{21}^{12}\Delta_2 + S_{22}^{11}\Delta_3 + S_{22}^{12}\Delta_4 &= F_2^1 \\ S_{21}^{21}\Delta_1 + S_{21}^{22}\Delta_2 + S_{22}^{21}\Delta_3 + S_{22}^{22}\Delta_4 &= F_2^2 \end{aligned} \tag{13.7a}$$

or, in matrix form,

$$[S]\{\Delta\} = \{F\} \tag{13.7b}$$

where

$$\begin{gathered} S_{ij} = S_{\alpha\beta}^{11}, \quad S_{i,j+1} = S_{\alpha\beta}^{12}, \quad S_{i+1,j} = S_{\alpha\beta}^{21}, \quad S_{i+1,j+1} = S_{\alpha\beta}^{22} \\ F_i = F_\alpha^1, \quad F_{i+1} = F_\alpha^2 \\ i = 2\alpha - 1, \quad j = 2\beta - 1, \quad \alpha, \beta = 1, 2 \end{gathered} \tag{13.7c}$$

The above discussion applies to any number of degrees of freedom per node (NDF). Computer implementation of (13.7) is straightforward. An example of this procedure is given by the following Fortran statements, which correspond to (10.29):

```
C
C     ____________________________________________________________________
C
C         REARRANGING THE ELEMENT EQUATIONS OF MULTIVARIABLE PROBLEMS
C             (Illustrated using the plane elasticity FE model)
C
C         NDF........ = Number of degrees of freedom per node
C         NPE........ = Number of nodes per element
C         CMAT(I,J).. = Matrix of elastic coefficients
C         ELK(I,J)... = Element stiffness matrix coefficients
C
C     ____________________________________________________________________
C
C     Dimension the arrays. For example (NN = NPE*NDF):
C                        ELK(NN,NN),CMAT(3,3)
C
C     Compute the coefficients  ELK11(I,J),  ELK12(I,J),  ELK21(I,J) and
C     ELK22(I,J) of Eqn. (10.29) using  numerical  integration.  We have
C     (the following statements go inside the loops on Gauss quadrature;
C     see the fortran statements for the evaluation of Sij):
C
      DO 140 I=1,NPE
         DO 120 J=1,NPE
            ELK11(I,J) = ELK11(I,J) + CMAT(1,1)*S11 + CMAT(3,3)*S22
            ELK12(I,J) = ELK12(I,J) + CMAT(1,2)*S12 + CMAT(3,3)*S21
            ELK21(I,J) = ELK21(I,J) + CMAT(1,2)*S21 + CMAT(3,3)*S12
            ELK22(I,J) = ELK22(I,J) + CMAT(3,3)*S11 + CMAT(2,2)*S22
  120    CONTINUE
  140 CONTINUE
C
C     We are now ready to rearrange the coefficients
C
      II=1
      DO 180 I=1,NN
         JJ=1
         DO 160 J=1,NN
            ELK(II,JJ)       = ELK11(I,J)
            ELK(II,JJ+1)     = ELK12(I,J)
            ELK(II+1,JJ)     = ELK21(I,J)
            ELK(II+1,JJ+1)   = ELK22(I,J)
  160    JJ = NDF*J+1
  180 II = NDF*I+1
C     ____________________________________________________________________
```

13.4 APPLICATIONS OF THE COMPUTER PROGRAM FEM2DV2

13.4.1 Introduction

The computer program FEM2DV2 (see Appendix 2 for the source listing) is developed to solve the following types of problems:

1. Single-variable problems, including convective-type boundary conditions for heat transfer problems,

$$c_t\left(\frac{\partial u}{\partial t}+\frac{\partial^2 u}{\partial t^2}\right)-\frac{\partial}{\partial x}\left(a_x\frac{\partial u}{\partial x}\right)-\frac{\partial}{\partial y}\left(a_y\frac{\partial u}{\partial y}\right)+a_0u=f \tag{13.8a}$$

with

$$\begin{gathered} c_t=c_0+c_xx+c_yy, \qquad a_x=a_{10}+a_{1x}x+a_{1y}y, \\ a_y=a_{20}+a_{2x}x+a_{2y}y, \qquad f_0=f_0+f_xx+f_yy \\ a_0=\text{constant} \end{gathered} \tag{13.8b}$$

2. The plane elasticity problems of Chapter 10 [(10.11)]
3. Viscous incompressible fluid flows using the penalty function formulation of Chapter 11 [(11.1) with P replaced by (11.34)]
4. Plate problems using classical and shear deformation theories (only with rectangular elements) of Chapter 12 [(12.29)].

The first category of problems is quite general and includes, as special cases, many other field problems. As a special case, axisymmetric problems can be analyzed. The last three categories are specialized to linear elasticity, linear (i.e., Stokes) viscous incompressible fluid flow, and linear plate bending.

The type of gradient of the solution computed (in subroutine PSTPRC) for single-variable problems differs for different physical problems. For heat transfer problems, we have

$$q_x=-a_x\frac{\partial u}{\partial x}, \qquad q_y=-a_y\frac{\partial u}{\partial y} \tag{13.9}$$

The same definition applies to the velocity potential formulation of inviscid fluid flows (to calculate velocity components). In the stream function formulation, the velocity components (u, v) are defined by

$$u=q_y=a_y\frac{\partial u}{\partial y}, \qquad v=q_x=-a_x\frac{\partial u}{\partial x} \tag{13.10}$$

The (total) stresses for multivariable problems are computed using the constitutive equations, with the strains (or strain rates for fluid flow problems) computed at the reduced Gauss points using the strain–displacement relations. The spatial variation of the derivatives of the solution is dependent on the element type.

For heat transfer problems (i.e., ITYPE = 0), the variable ICONV is used to indicate the presence (ICONV = 1) or absence (ICONV = 0) of convective boundaries. When convective boundaries are involved (i.e., ICONV = 1), the elements whose boundaries coincide with such a boundary will have additional contributions to their coefficient matrices [see (8.99)–(8.103)]. The array IBN is used to store elements that have convective

TABLE 13.1
Description of the input variables to the program FEM2DV2

```
* Data Card 1:

      TITLE  - Title of the problem being solved (80 characters)

* Data Card 2:

      ITYPE  - Problem type:
               ITYPE = 0, Single variable problems
               ITYPE = 1, Viscous incompressible flow problems
               ITYPE = 2, Plane elasticity problems
               ITYPE = 3, Plate bending problems by the FSDT
               ITYPE = 4, Plate bending problems by the CPT(N)
               ITYPE = 5, Plate bending problems by the CPT(C)

      IGRAD  - Indicator for computing the gradient of the solution
               or stresses in the postprocessor:
               IGRAD = 0, No postprocessing is required
               IGRAD > 0, Postprocessing is required

               When ITYPE=0 and IGRAD=1, the gradient is computed
               as in Eq. (13.9);  for ITYPE=0 and IGRAD > 1,  the
               gradient is computed as in Eq (13.10).

      ITEM   - Indicator for dynamic analysis:
               ITEM  = 0, Static analysis is required
               ITEM  > 0, Either eigenvalue or transient analysis
                          is required:

                          ITEM = 1, Parabolic equation is analyzed
                          ITEM = 2, Hyperbolic equation is analyzed

      NEIGN  - Indicator for eigenvalue analysis:
               NEIGN = 0, Static or transient analysis is required
               NEIGN > 0, Eigenvalue analysis is required:
                          NEIGN= 1, Frequency/vibration analysis
                          NEIGN> 1, Stability analysis of plates

* Data Card 3:  SKIP the card if NEIGN = 0

      NVALU  - Number of eigenvalues to be printed
      NVCTR  - Indicator for printing eigenvectors:
               NVCTR = 0, Do not print eigenvectors
               NVCTR > 0, Print eigenvectors

* Data Card 4:

      IELTYP - Element type used in the analysis:
               IELTYP= 0, Triangular elements
               IELTYP> 0, Quadrilateral elements

      NPE    - Nodes per element:
               NPE   = 3, Linear triangle (IELTYP=0)
               NPE   = 4, Linear quadrilateral (IELTYP>0)
               NPE   = 6, Quadratic triangle (IELTYP=0)
               NPE   = 8 or 9, Quadratic quadrilateral (IELTYP>0)
```

TABLE 13.1 (Continued)

```
        MESH    - Indicator for mesh generation by the program:
                  MESH  = 0, Mesh is not generated by the program
                  MESH  = 1, Mesh is generated by the program for
                             rectangular domains (by MSH2DR)
                  MESH  > 1, Mesh is generated by the program for
                             general domains (by MSH2DG)

        NPRNT   - Indicator for printing certain output:
                  NPRNT = 0, Not print array NOD, element matrices,
                             or global matrices
                  NPRNT = 1, Print array NOD and Element 1 matrices:
                             [ELK] and {ELF}
                  NPRNT = 2, Print array NOD and assembled matrices,
                             [GLK] and {GLF}
                  NPRNT > 2, Combination of NPRNT=1 and NPRNT=2

  * Data Card 5:  SKIP the card if MESH.EQ.1

        NEM     - Number of elements in the mesh when the user inputs
                  the mesh or the mesh is generated by MSH2DG

        NNM     - Number of  nodes   in the mesh when the user inputs
                  the mesh or the mesh is generated by MSH2DG

  * Data Card 6:  SKIP the card if  MESH.NE.0; otherwise, the card is
                  read in a loop on the number of elements (N=1, NEM)

        NOD(N,I)-Connectivity for the N-th element (I=1,NPE)

  * Data Card 7:  SKIP the card if  MESH.NE.0

        GLXY(I,J)-Global x and y coordinates of I-th global node in
                  the mesh (J=1, x-coordinate, J=2, y-coordinate)
                  Loops on I and J are: ((J=1,2), I=1,NNM); the NNM
                  pairs of (x,y)--coordinates are read sequentially

  _____ The next FOUR data cards are read in subroutine  MSH2DG ______
  * * * * *  SKIP Cards 8, 9, 10  and  11  unless MESH.GT.1  * * * * *

  * Data Card 8:

        NRECL   - Number of line records to be read in the mesh

  * Data Card 9: Read the following variables NRECL times

        NOD1    - First global node number of the line segment
        NODL    - Last global node number of the line segment
        NODINC  - Node increment on the line
        X1      - The global x-coordinate of NOD1
        Y1      - The global y-coordinate of NOD1
        XL      - The global x-coordinate of NODL
        YL      - The global y-coordinate of NODL
        RATIO   - The ratio of the first element length to the last
                  element length

  * Data Card 10:

        NRECEL  - Number of rows of elements to be read in the mesh
```

TABLE 13.1 (Continued)

```
* Data Card 11: Read the following variables NRECEL times

        NEL1    - First element number of the row
        NELL    - Last element  number of the row
        IELINC  - Increment of element number in the row
        NODINC  - Increment of global node number in the row
        NPE     - Number of nodes in each element
        NODE(I) - Connectivity array of the first element in the row
                  (I=1,NPE)

* Data Card 12:  SKIP the card if  MESH.NE.1

        NX      - Number of elements in the x-direction
        NY      - Number of elements in the y-direction

* Data Card 13:  SKIP the card if  MESH.NE.1

        X0      - The x-coordinate of global node 1
        DX(I)   - The x-dimension of the I-th element  (I=1,NX)

* Data Card 14:  SKIP the card if  MESH.NE.1

        Y0      - The y-coordinate of global node 1
        DY(I)   - The y-dimension of the I-th element  (I=1,NY)

* Data Card 15:

        NSPV    - The number of specified primary variables

* Data Card 16:  SKIP the card if NSPV.EQ.0

        ISPV(I,J)-Node number and LOCAL degree of freedom number of
                  the I-th specified primary variable:
                  ISPV(I,1)=Node number; ISPV(I,2)=Local DOF number

                  The loops on I and J are: ((J=1,2),I=1,NSPV)

* Data Card 17:  SKIP the card if NSPV.EQ.0  or  NEIGN.NE.0

        VSPV(I) - Specified value of the I-th primary variable
                  (I=1,NSPV)

* Data Card 18:  SKIP the card if NEIGN.NE.0

        NSSV    - Number of (nonzero) specified secondary variables

* Data Card 19:  SKIP the card if NSSV.EQ.0 or NEIGN.NE.0

        ISSV(I,J)-Node number and LOCAL degree of freedom number of
                  the I-th specified secondary variable:
                  ISSV(I,1)=Node number; ISSV(I,2)=Local DOF number

                  The loops on I and J are: ((J=1,2),I=1,NSSV)

* Data Card 20:  SKIP the card if NSSV.EQ.0  or  NEIGN.NE.0

        VSSV(I) - Specified value of the I-th secondary variable
                  (I=1,NSSV)
```

TABLE 13.1 (Continued)

```
---------------------------------------------------------------------

Data Cards 21 through 27 are for SINGLE VARIABLE PROBLEMS (ITYPE=0)

* Data Card 21:  SKIP the card if ITYPE.NE.0

        A10     |
        A1X     |   Coefficients of the differential equation
        A1Y     |   a11 = A10 +A1X*X + A1Y*Y

* Data Card 22:  SKIP the card if ITYPE.NE.0

        A20     |
        A2X     |   Coefficients of the differential equation
        A2Y     |   a22 = A20 +A2X*X + A2Y*Y

* Data Card 23:  SKIP the card if ITYPE.NE.0

        A00       -  Coefficient of the differential equation

* Data Card 24:  SKIP the card if ITYPE.NE.0

        ICONV     -  Indicator for convection boundary conditions:
                     ICONV = 0, No convection boundary conditions
                     ICONV > 0, Convection boundary conditions present

* Data Card 25:  SKIP the card if ITYPE.NE.0 or ICONV.EQ.0

        NBE       -  Number elements with convection

* Data Card 26:  SKIP the card if ITYPE.NE.0 or ICONV.EQ.0
                 The following cards are read for each I, I=1,NBE

        IBN(I)    -  I-th element number with convection
        BETA(I)   -  Film coefficient for convection on I-th element
        TINF(I)   -  Ambient temperature of the I-th element

* Data Card 27:  SKIP the card if ITYPE.NE.0 or ICONV.EQ.0

        INOD(I,J)-   Local node numbers of the side with convection
                     (J=1,2; for quadratic elements, give end nodes)
                     Loops on I and J are: ((J=1,2), I=1,NBE)

---------------------------------------------------------------------

_____ Data Card 28 is for VISCOUS FLUID FLOWS (ITYPE = 1) only ____

* Data Card 28:  SKIP the card if ITYPE.NE.1

        AMU    - Viscosity of the fluid
        PENLTY - Value of the penalty parameter

---------------------------------------------------------------------

____ Data Cards 29 & 30 are for PLANE ELASTICITY (ITYPE=2) only ___

* Data Card 29:  SKIP the card if ITYPE.NE.2

        LNSTRS - Flag for PLANE STRESS or PLANE STRAIN problems:
                 LNSTRS=0, Plane strain elastic problems
```

TABLE 13.1 (Continued)

```
                    LNSTRS>0, Plane stress elastic problems

 * Data Card 30:    SKIP the card if ITYPE.NE.2

          E1      - Young's moduli along the global x-axis
          E2      - Young's moduli along the global y-axis
          ANU12   - Poisson's ratio in the xy-plane
          G12     - Shear modulus in the xy-plane
          THKNS   - Thickness of the plane elastic body analyzed

 ------------------------------------------------------------------

 __ Data Card 31 is for PLATE BENDING PROBLEMS (ITYPE=3 to 5) only _

 * Data Card 31:    SKIP the card unless ITYPE.GE.3

          E1      - Young's moduli along the global x-axis
          E2      - Young's moduli along the global y-axis
          ANU12   - Poisson's ratio in the xy-plane
          G12     - Shear modulus in the xy-plane
          G13     - Shear modulus in the xz-plane
          G23     - Shear modulus in the yz-plane
          THKNS   - Thickness of the plate analyzed

 ------------------------------------------------------------------

 __________ Remaining data cards are for ALL problem types __________

 * Data Card 32:    SKIP the card if NEIGN.NE.0

          F0      |
          FX      |  Coefficients to define the source term:
          FY      |   f = F0 + FX*x + FY*y

 ------------------------------------------------------------------

 __________ Cards 33 thru 37 are for DYNAMIC ANALYSIS only __________

 * Data Card 33:    SKIP the card if ITEM.EQ.0

          C0      |
          CX      |  Coefficients defining the temporal parts of the
          CY      |  differential equations:

                     CT = C0 + CX*x + CY*y, when ITYPE = 0 or 1

                     CT = (C0 + CX*x + CY*y)*THKNS, when ITYPE = 2

                     I0 = C0*THKNS                 |
                     I2 = C0*(THKNS**3)/12         | when ITYPE=3 to 5
                     and CX and CY are not used    | and NEIGN.LE.1

                     C0, CX, and CY denote the buckling parameters
                     when ITYPE.EQ.3 and NEIGN.GT.1

 * Data Card 34:    SKIP the card if ITEM.EQ.0 or NEIGN.NE.0

          NTIME   - Number of time steps for the transient solution
          NSTP    - Time step number at which the source is removed
          INTVL   - Time step interval at which to print the solution
```

TABLE 13.1 (Continued)

```
          INTIAL - Indicator for nature of initial conditions:

                   INTIAL=0, Zero initial conditions are used
                   INTIAL>0, Non-zero initial conditions are used

 * Data Card 35:   SKIP the card if ITEM.EQ.0 or NEIGN.NE.0

          DT     - Time step used for the transient solution

          ALFA   - Parameter in the alfa-family of time approximation
                   used for parabolic equations:

                   ALFA=0,   The forward difference scheme (C.S.)@
                   ALFA=0.5, The Crank-Nicolson scheme (stable)
                   ALFA=2/3, The Galerkin scheme (stable)
                   ALFA=1,   The backward difference scheme (stable)

                  @C.S.=Conditionally Stable; For all schemes with
                   ALFA < 0.5, the time step DT is restricted to:
                   DT < 2/[MAXEGN*(1-2*ALFA)], where MAXEGN is the
                   maximum eigenvalue of the discrete problem.

          GAMA   - Parameter in the Newmark time integration scheme
                   used for hyperbolic equations:
                   GAMA=0.5,  Constant-average acceleration (stable)
                   GAMA=1/3,  Linear acceleration scheme (C.S.)@
                   GAMA=0.0,  The central difference scheme (C.S.)

                   ALFA = 0.5 for all schemes; For schemes for which
                   ALFA.LE.0.5 and GAMA < ALFA, DT is restricted to:
                   DT < 2/SQRT[MAXEGN*(ALFA-GAMA)], MAXEGN being the
                   maximum eigenvalue of the discrete system.

          EPSLN  - A small parameter to check if the solution has
                   reached a steady state

 * Data Card 36:   SKIP the card if ITEM or INTIAL.EQ.0, or NEIGN.NE.0

          GLU(I) - Vector of initial values of the primary variables
                   (I=1,NEQ, NEQ=Number of nodal values in the mesh)

 * Data Card 37:   SKIP if ITEM.LE.1, NEIGN.NE.0, or INTIAL.EQ.0

          GLV(I) - Vector of initial values of the first derivative
                   of the primary variables (velocity)
                   (I=1,NEQ, NEQ=Number of nodal values in the mesh)
```

boundaries, and the array INOD is used to store the pairs of element local nodes (of elements in array IBN) that are on the convective boundary (to specify the side of the element on the convective boundary). If an element has more than one of its sides on the convective boundary, it should be repeated as many times as the number of its sides on the convective boundary.

A complete description of the input variables of the program FEM2DV2 is given in Table 13.1, which contains the values of the key variables for the four classes (ITYPE = 0, 1, 2, . . . , 5) of problems. In the next section, the application of the program FEM2DV2 is illustrated via several examples.

13.4.2 Description of Mesh Generators

In this section, the input data to FEM2DV2 for several example problems is given. The example problems are selected from those discussed in earlier sections of this chapter. A major limitation of the program lies in the mesh generation [i.e., the computation of arrays NOD(I, J) and GLXY(I, J) for arbitrary domains]. For such problems, the user is required to input the mesh information, which can be a tedious job if many elements are used. Of course, the program can be modified to accept any other mesh generation subroutines.

First, let us consider MSH2DR. The program is restricted to rectangular domains with sides parallel to the global x and y axes. The subroutine requires the following input data:

NX	number of elements in the x direction
NY	number of elements in the y direction
(X_0, Y_0)	global coordinates of global node 1, which is located at the lower left corner of the rectangular domain (see Fig. 13.2)
DX(I)	the array of element lengths along the x direction
DY(I)	the array of element lengths along the y direction

The node and element numbering schemes for triangular and rectangular element meshes generated by MSH2DR are shown in Fig. 13.2.

Next, we consider MSH2DG, which is relatively more general than MSH2DR. The program, based on the straight line generation logic used by Akay *et al.* (1987), requires the user to sketch a desired mesh with certain regularity of node and element numbering. It exploits the regularity to generate the mesh. The program MSH2DG requires the following input (except for NEM and NNM, all other variables are read from the subroutine):

NEM	number of total elements in the mesh
NNM	number of total nodes in the mesh
NRECL	number of line-segment records

For each line segment, read the following variables:

NOD1	first node number on the line segment
NODL	last node number on the line segment
NODINC	increment between two consecutive nodes on the line
(X1, Y1)	global coordinates of the first node, NOD1 on the line
(XN, YN)	global coordinates of the last node, NODL on the line
RATIO	ratio of the first element length to the last element length

Similar information on the elements is read:

NRECEL	number of rows of elements

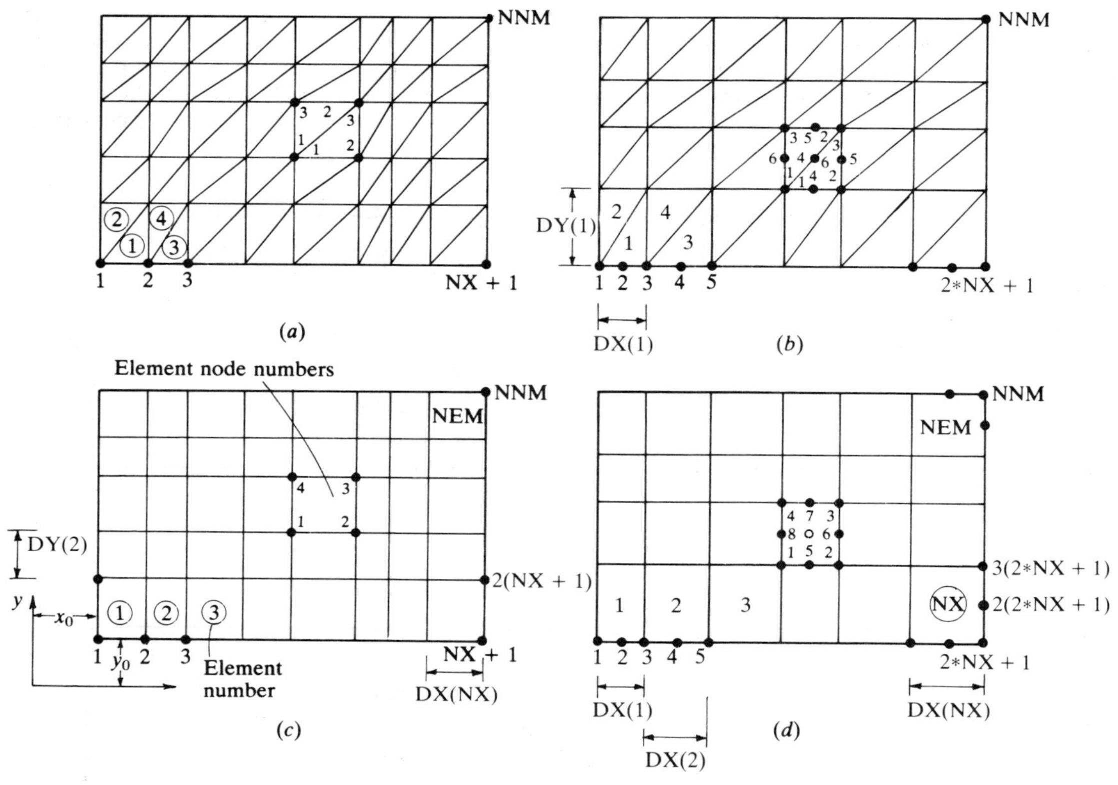

FIGURE 13.2
Global node numbering and element numbering system used in subroutine MSH2DR for the generation of meshes of (a) linear triangular elements [NNM = (NX + 1) * (NY + 1)]; (b) quadratic triangular elements [NNM = (2 * NX + 1) * (2 * NY + 1)]; (c) linear rectangular elements [NNM = (NX + 1) * (NY + 1)]; and (d) quadratic rectangular elements. Either the nine- or the eight-node rectangular elements can be used.

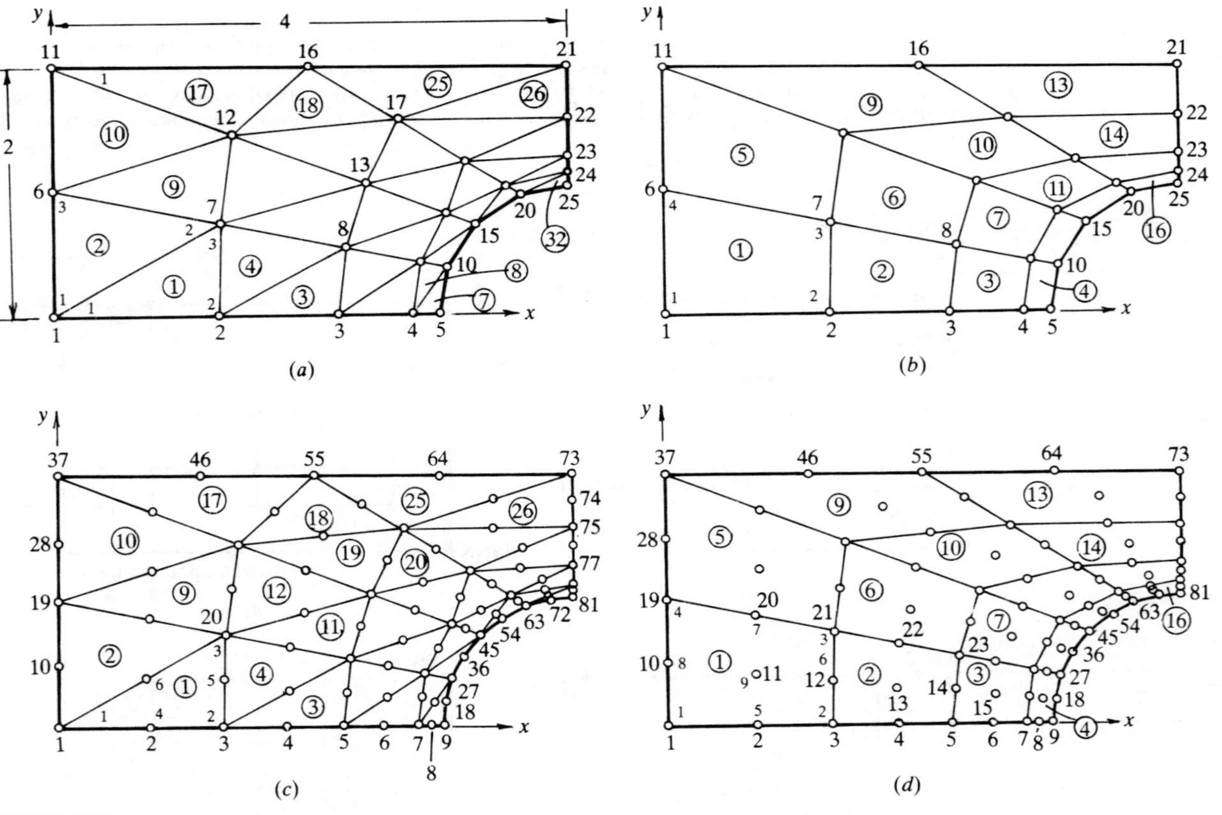

FIGURE 13.3
Typical examples of mesh generation using subroutine MSH2DG: (*a*) mesh of linear triangular elements; (*b*) mesh cf linear quadrilateral elements; (*c*) mesh of quadratic triangles; (*d*) mesh of nine-node quadratic quadrilateral elements (eight-node quadrilaterals can also be generated in the same way). Global node numbers, element numbers, and a typical element node numbering scheme are shown.

For each row of elements, read the following variables:

NEL1	first element number in the row
NELL	last element number in the row
NODINC	increment between respective nodes of consecutive elements in the row
NPE	number of nodes per element
NODE(I)(I = 1, NPE)	nodal connectivity for element NEL1

The type of data being read in MSH2DG should give some indication of the restrictions of the program. The node and element numbering should be regular along the lines and rows being read. Figures 13.3(*a–d*) show typical examples of meshes of linear and quadratic triangular and quadrilateral elements. For each of these meshes, the input data required for MSH2DG is listed in Table 13.2.

13.4.3 Applications (Illustrative Examples)

Example 13.1 Solution of the Poisson equation of Example 8.1. We consider the Poisson equation $-\nabla^2 u = 1$ in a unit square. We represent the domain by linear triangles or rectangles (see Fig. 13.4).

Triangular elements. We use the 2 × 2 mesh shown in Fig. 13.4(*a*). We have

$$\text{ITYPE} = 0, \quad \text{IGRAD} = 1, \quad \text{ITEM} = 0, \quad \text{NEIGN} = 0$$

$$\text{IELTYP} = 0, \quad \text{NPE} = 3, \quad \text{MESH} = 1, \quad \text{NPRNT} = 0$$

Note that we chose to generate the mesh using subroutine MSH2DR. Therefore, we must specify the number of subdivisions and their lengths along each direction:

$$\text{NX} = 2, \quad \text{NY} = 2, \quad \text{X0} = 0.0, \quad \text{Y0} = 0.0$$

$$\text{DX(1)} = 0.5, \quad \text{DX(2)} = 0.5, \quad \text{DY(1)} = 0.5, \quad \text{DY(2)} = 0.5$$

The number of specified primary variables (NSPV), the node numbers and the specified local degree of freedom (ISPV), and their specified values (VSPV) for the problem are

$$\text{NSPV} = 5, \quad \text{ISPV(I, J)} = (3, 1; 6, 1; 7, 1; 8, 1; 9, 1), \quad \text{VSPV(I)} = (0.0, 0.0, 0.0, 0.0, 0.0)$$

There are no specified secondary variables: NSSV = 0.

The coefficients a_x and a_y of the differential equation are unity, $a_0 = 0$, the source term f is unity, and there is no convection:

$$\text{A10} = 1.0, \quad \text{A1X} = 0.0, \quad \text{A1Y} = 0.0, \quad \text{A20} = 1.0, \quad \text{A2X} = 0.0, \quad \text{A2Y} = 0.0$$

$$\text{A00} = 0.0, \quad \text{ICONV} = 0, \quad \text{F0} = 1.0, \quad \text{FX} = 0.0, \quad \text{FY} = 0.0$$

Rectangular elements. For the 2 × 2 (four-element) mesh of rectangular elements (see Fig. 13.4*b*), the data input to the program differs only in the specification of the element type and the number of nodes per element:

$$\text{IELTYP} = 1, \quad \text{NPE} = 4$$

An edited output of the program for the problem is presented in Table 13.3.

TABLE 13.2
The input data to Program FEM2DV2 to generate the finite element meshes shown in Figs. 13.3(*a–d*)

```
(a) Mesh of linear triangles (see Fig.13.3a):

 32  25                                               NEM, NNM

  5                                                   NRECL
  1   5  1  0.0  0.0  3.0       0.0       6.0    |
  6  10  1  0.0  1.0  3.07612   0.38268   6.0    |    NOD1,NODL,NODINC,
 11  15  1  0.0  2.0  3.29289   0.7071    6.0    |    X1,Y1,XL,YL,RATIO
 16  20  1  2.0  2.0  3.61732   0.92388   6.0    |    for each line segment
 21  25  1  4.0  2.0  4.0       1.0       6.0    |

  8                                                   NRECEL
  1   7  2  1  3    1  2   7                     |
  2   8  2  1  3    1  7   6                     |
  9  15  2  1  3    6  7  12                     |    NEL1,NELL,IELINC,NODINC,
 10  16  2  1  3    6 12  11                     |    NPE, NOD(I,J)
 17  23  2  1  3   11 12  16                     |    for each line
 18  24  2  1  3   12 17  16                     |
 25  31  2  1  3   16 17  21                     |
 26  32  2  1  3   17 22  21                     |

(b) Mesh of four-node quadrilaterals (see Fig. 13.3b):

 16  25                                               NEM, NNM

  5                                                   NRECL
  1   5  1  0.0  0.0  3.0       0.0       6.0    |
  6  10  1  0.0  1.0  3.07612   0.38268   6.0    |    NOD1,NODL,NODINC,
 11  15  1  0.0  2.0  3.29289   0.7071    6.0    |    X1,Y1,XL,YL,RATIO
 16  20  1  2.0  2.0  3.61732   0.92388   6.0    |    for each line segment
 21  25  1  4.0  2.0  4.0       1.0       6.0    |

  4                                                   NRECEL
  1   4  1  1  4    1  2   7   6                 |
  5   8  1  1  4    6  7  12  11                 |    NEL1,NELL,IELINC,NODINC,
  9  12  1  1  4   11 12  17  16                 |    NPE,NOD(I,J)
 13  16  1  1  4   16 17  22  21                 |
```

Example 13.2 Heat transfer with convection boundary conditions. Consider a square region of 1 m × 1 m. The left side of the region (i.e., $x = 0$) is maintained at 100°C, while the boundary $y = 1$ m is maintained at 500°C. The boundaries $x = 1$ m and $y = 0$ are exposed to an ambient temperature of 100°C, and the film coefficient $\beta = 10\ \mathrm{W\,m^{-2}\,{}^\circ C^{-1}}$. There is no internal heat generation ($f = 0$). The conductivity is taken to be $k_x = k_y = 12.5\ \mathrm{W\,m^{-1}\,{}^\circ C^{-1}}$.

The input variables associated with convective boundary conditions are

$$\text{ICONV} = 1, \quad \text{NBE} = 16 \quad \text{(for an } 8 \times 8 \text{ mesh of linear rectangular elements)}$$

$$[\text{IBN(I), BETA(I), TINF(I)}] = [1, 10.0, 100.0; 2, 10.0, 100.0; \ldots]$$

$$[\text{INOD(I, J)}] = [1, 2; 1, 2; \ldots]$$

$$\text{A10} = 12.5, \quad \text{A1X} = 0.0, \quad \text{A1Y} = 0.0, \quad \text{A20} = 12.5,$$

$$\text{A2X} = 0.0, \quad \text{A2Y} = 0.0, \quad \text{A00} = 0.0$$

Table 13.4 gives the input data for the 8 × 8 mesh of linear rectangular elements. Figures 13.5 and 13.6 show plots of temperature variations and heat flow

$$q_x = -a_x \frac{\partial T}{\partial x}, \qquad q_y = -a_y \frac{\partial T}{\partial y}$$

TABLE 13.2 (Continued)

(c) Mesh of quadratic triangles (see Fig. 13.3c):

```
32  81                                               NEM, NNM

 9                                                   NRECL
 1   9  1  0.0  0.0  3.0      0.0      6.0
10  18  1  0.0  0.5  3.01921  0.19509  6.0
19  27  1  0.0  1.0  3.07612  0.38268  6.0
28  36  1  0.0  1.5  3.16853  0.55557  6.0          NOD1,NODL,NODINC,
37  45  1  0.0  2.0  3.29289  0.7071   6.0          X1,Y1,XL,YL,RATIO
46  54  1  1.0  2.0  3.44443  0.83147  6.0          for each line segment
55  63  1  2.0  2.0  3.61732  0.92388  6.0
64  72  1  3.0  2.0  3.80491  0.98078  6.0
73  81  1  4.0  2.0  4.0      1.0      6.0

 8                                                   NRECEL
 1   7  2  2  6    1  3 21  2 12 11
 2   8  2  2  6    1 21 19 11 20 10
 9  15  2  2  6   19 21 39 20 30 29
10  16  2  2  6   19 39 37 29 38 28                 NEL1,NELL,IELINC,NODINC,
17  23  2  2  6   37 39 55 38 47 46                 NPE,NOD(I,J)
18  24  2  2  6   39 57 55 48 56 47
25  31  2  2  6   55 57 73 56 65 64
26  32  2  2  6   57 75 73 66 74 65
```

(d) Mesh of quadratic (nine-node) quadrilaterals (see Fig. 13.3d):

```
16  81                                               NEM, NNM

 9                                                   NRECL
 1   9  1  0.0  0.0  3.0      0.0      6.0
10  18  1  0.0  0.5  3.01921  0.19509  6.0
19  27  1  0.0  1.0  3.07612  0.38268  6.0
28  36  1  0.0  1.5  3.16853  0.55557  6.0          NOD1,NODL,NODINC,
37  45  1  0.0  2.0  3.29289  0.7071   6.0          X1,Y1,XL,YL,RATIO
46  54  1  1.0  2.0  3.44443  0.83147  6.0          for each line segment
55  63  1  2.0  2.0  3.61732  0.92388  6.0
64  72  1  3.0  2.0  3.80491  0.98078  6.0
73  81  1  4.0  2.0  4.0      1.0      6.0

 4                                                   NRECEL
 1   4  1  2  9    1  3 21 19  2 12 20 10 11
 5   8  1  2  9   19 21 39 37 20 30 38 28 29        NEL1,NELL,IELINC,NODINC,
 9  12  1  2  9   37 39 57 55 38 48 56 46 47        NPE,NOD(I,J)
13  16  1  2  9   55 57 75 73 56 66 74 64 65
```

along the boundaries. Note that q_x is linear in x and constant in y, and q_y is linear in y and constant in x (for constant a_x and a_y). A nonuniform mesh with smaller elements in the high-gradient region gives more accurate results:

$$\{\mathrm{DY(I)}\} = \{\mathrm{DX(I)}\} = (0.25, 0.125, 0.125, 0.125, 0.125, 0.125, 0.0625, 0.0625)$$

The program can also be used to analyze an axisymmetric problem. For example, consider a finite cylinder of radius $R_0 = 1$ m and length $L = 1$ m. The bottom and top of the cylinder are maintained at $T_0 = 100°C$, while the surface is exposed to an ambient temperature $T_\infty = 100°\mathrm{C}$ ($\beta = 10\ \mathrm{W\,m^{-2}\,°C^{-1}}$). For this case, the governing differential equation is (8.104). The coefficients A10, A1X, A1Y, A20, A2X, A2Y, and A00 are

$$\mathrm{A10} = 0.0, \quad \mathrm{A1X} = 2\pi k_r, \quad \mathrm{A1Y} = 0.0$$

$$\mathrm{A20} = 0.0, \quad \mathrm{A2X} = 2\pi k_z, \quad \mathrm{A2Y} = 0.0$$

$$\mathrm{A00} = 0.0$$

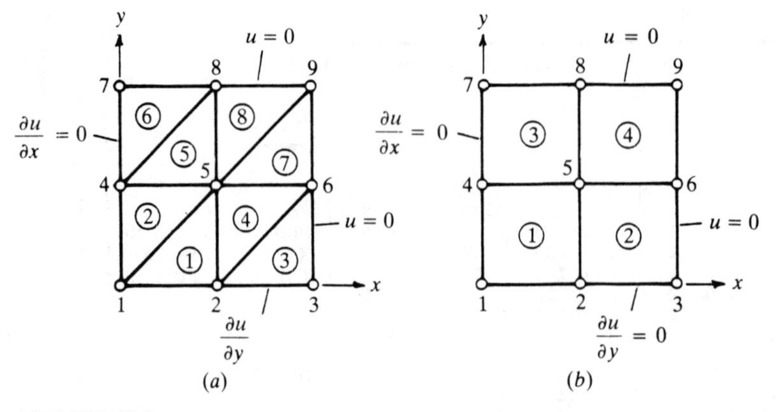

FIGURE 13.4
Finite element meshes of linear triangular and rectangular elements used in Example 13.1: (*a*) 2×2 mesh of triangles; (*b*) 2×2 mesh of rectangles.

The uniform internal heat generation f_0 (if not zero) is entered as

$$\mathrm{F0} = 0.0, \qquad \mathrm{FX} = 2\pi f_0, \qquad \mathrm{FY} = 0.0$$

Example 13.3 Flow about a circular cylinder (Example 8.7). Consider the flow of an inviscid fluid around a cylinder. We shall use the stream function and velocity potential formulations. Since the domain is not rectangular, we should use MSH2DG (i.e., MESH = 2). We consider the mesh of 25 nodes and 32 triangular elements shown in Fig. 13.7(*a*). We have ITYPE = 0 and IGRAD = 1 in the velocity potential formulation, and IGRAD = 2 in the stream function formulation, ITEM = 0, NEIGN = 0 and IELTYP = 0 for triangles, and IELTYP = 1 for quadrilaterals, MESH = 2, NPRNT = 0. The input for MSH2DG is given in Table 13.5 for triangular and quadrilateral elements.

In the stream function formulation, we have NSPV = 13 and NSSV = 0; and in the velocity potential formulation, we have NSPV = 5 and NSSV = 3. The coefficients are

$$\mathrm{A10} = 1.0, \quad \mathrm{A20} = 1.0, \quad \mathrm{A1X} = 0.0, \quad \mathrm{A1Y} = 0.0, \quad \mathrm{A2X} = 0.0, \quad \mathrm{A2Y} = 0.0$$

$$\mathrm{A00} = 0.0, \quad \mathrm{F0} = 0.0, \quad \mathrm{FX} = 0.0, \quad \mathrm{FY} = 0.0$$

The partial input of the problem is given in Table 13.6.

Example 13.4 Eigenvalue and transient analysis (Examples 8.9 and 8.10). We consider the eigenvalue and transient problems discussed in Examples 8.9 and 8.10. The governing differential equation is [see (8.165)]

$$\frac{\partial u}{\partial t} - \left(\frac{\partial^2 u}{\partial x^2} + \frac{\partial^2 u}{\partial y^2}\right) = 1$$

For eigenvalue analysis, we set ITEM = 1 and NEIGN = 1; for the transient analysis, we set ITEM = 1 (parabolic equation) and NEIGN = 0. In addition, we must

TABLE 13.3
Output from FEM2DV2 for the problem in Example 13.1

```
     *** ECHO OF THE INPUT DATA STARTS ***

  Example 8.1: Solution of the Poisson equation on a square domain
    0   1   0   0                                    ITYPE,IGRAD,ITEM,NEIGN
    0   3   1   0                                    IELTYP,NPE,MESH,NPRNT

    2   2                                            NX,NY
    0.0  0.5  0.5                                    X0,DX(I)
    0.0  0.5  0.5                                    Y0,DY(I)

    5                                                NSPV
    3 1     6 1     7 1     8 1     9 1              ISPV(I,J)
    0.0     0.0     0.0     0.0     0.0              VSPV(I)

    0                                                NSSV

    1.0  0.0  0.0                                    A10, A1X, A1Y
    1.0  0.0  0.0                                    A20, A2X, A2Y
    0.0                                              A00

    0                                                ICONV

    1.0  0.0  0.0                                    F0,  FX,  FY

     **** ECHO OF THE INPUT DATA ENDS ****

  Example 8.1: Solution of the Poisson equation on a square domain
  ____________________________________________________________________________

           OUTPUT   FROM   PROGRAM   *FEM2DV2*    BY   J. N. REDDY
  ____________________________________________________________________________


                  ANALYSIS   OF   A  POISSON/LAPLACE  EQUATION

     COEFFICIENTS OF THE DIFFERENTIAL EQUATION:

        Coefficient, A10 ........................=  0.1000E+01
        Coefficient, A1X ........................=  0.0000E+00
        Coefficient, A1Y ........................=  0.0000E+00
        Coefficient, A20 ........................=  0.1000E+01
        Coefficient, A2X ........................=  0.0000E+00
        Coefficient, A2Y ........................=  0.0000E+00
        Coefficient, A00 ........................=  0.0000E+00


     CONTINUOUS SOURCE COEFFICIENTS:

        Coefficient, F0  ........................=  0.1000E+01
        Coefficient, FX  ........................=  0.0000E+00
        Coefficient, FY .........................=  0.0000E+00


        ******* A STEADY-STATE PROBLEM is analyzed *******
           *** A mesh of   TRIANGLES    is chosen by user ***

       FINITE ELEMENT MESH INFORMATION:

           Element type: 0 = Triangle; >0 = Quad.)..=   0
           Number of nodes per element, NPE ........=   3
           No. of primary deg. of freedom/node, NDF =   1
           Number of elements in the mesh, NEM .....=   8
           Number of nodes in the mesh, NNM ........=   9
           Number of equations to be solved, NEQ ...=   9
           Half bandwidth of the matrix GLK, NHBW ..=   5
```

TABLE 13.3 (Continued)

```
          Mesh subdivisions, NX and NY ...........=   2   2

          No. of specified PRIMARY variables, NSPV =   5
```

Node	x-coord.	y-coord.	Speci. primary & secondary variables (0, unspecified; >0, specified) Primary DOF	Secondary DOF
1	0.0000E+00	0.0000E+00	0	0
2	0.5000E+00	0.0000E+00	0	0
3	0.1000E+01	0.0000E+00	1	0
4	0.0000E+00	0.5000E+00	0	0
5	0.5000E+00	0.5000E+00	0	0
6	0.1000E+01	0.5000E+00	1	0
7	0.0000E+00	0.1000E+01	1	0
8	0.5000E+00	0.1000E+01	1	0
9	0.1000E+01	0.1000E+01	1	0

```
NUMERICAL INTEGRATION DATA:

   Full Integration polynomial degree, IPDF =   3
   Number of full integration points,  NIPF =   4
   Reduced Integration polynomial deg.,IPDR =   1
   No.  of reduced integration points, NIPR =   1
   Integ. poly. deg. for stress comp., ISTR =   1
   No. of integ. pts. for stress comp.,NSTR =   1

S O L U T I O N :
```

Node	x-coord.	y-coord.	Primary DOF
1	0.00000E+00	0.00000E+00	0.31250E+00
2	0.50000E+00	0.00000E+00	0.22917E+00
3	0.10000E+01	0.00000E+00	0.00000E+00
4	0.00000E+00	0.50000E+00	0.22917E+00
5	0.50000E+00	0.50000E+00	0.17708E+00
6	0.10000E+01	0.50000E+00	0.00000E+00
7	0.00000E+00	0.10000E+01	0.00000E+00
8	0.50000E+00	0.10000E+01	0.00000E+00
9	0.10000E+01	0.10000E+01	0.00000E+00

The orientation of gradient vector is measured from the positive x-axis

x-coord.	y-coord.	-a11(du/dx)	-a22(du/dy)	Flux Mgntd	Orientation
0.3333E+00	0.1667E+00	0.1667E+00	0.1042E+00	0.1965E+00	32.01
0.1667E+00	0.3333E+00	0.1042E+00	0.1667E+00	0.1965E+00	57.99
0.8333E+00	0.1667E+00	0.4583E+00	0.0000E+00	0.4583E+00	0.00
0.6667E+00	0.3333E+00	0.3542E+00	0.1042E+00	0.3692E+00	16.39
0.3333E+00	0.6667E+00	0.1042E+00	0.3542E+00	0.3692E+00	73.61
0.1667E+00	0.8333E+00	0.0000E+00	0.4583E+00	0.4583E+00	90.00
0.8333E+00	0.6667E+00	0.3542E+00	0.0000E+00	0.3542E+00	0.00
0.6667E+00	0.8333E+00	0.0000E+00	0.3542E+00	0.3542E+00	90.00

TABLE 13.4
Input data for the heat transfer problem in Example 13.2

```
Example 13.2: Convective heat transfer in a square region
  0   1   0   0                                         ITYPE,IGRAD,ITEM,NEIGN
  1   4   1   0                                         IELTYP,NPE,MESH,NPRNT

  8   8                                                                  NX,NY
  0.0  0.125  0.125  0.125  0.125  0.125  0.125  0.125  0.125  X0,DX(I)
  0.0  0.125  0.125  0.125  0.125  0.125  0.125  0.125  0.125  Y0,DY(I)

 17                                                     NSPV
 1 1   10 1   19 1   28 1   37 1   46 1   55 1   64 1    73 1   74 1   75 1
 76 1  77 1   78 1   79 1   80 1   81 1                 ISPV(I,J)
 100.0 100.0 100.0 100.0 100.0 100.0 100.0 100.0  500.0 500.0 500.0
 500.0 500.0 500.0 500.0 500.0 500.0                    VSPV(I)

  0                                                     NSSV

  10.0   0.0   0.0                                      A10, A1X, A1Y
  10.0   0.0   0.0                                      A20, A2X, A2Y
   0.0                                                  A00

  1                                                     ICONV
 16                                                     NBE
  1 10.0 100.0   2 10.0 100.0   3 10.0 100.0   4 10.0 100.0
  5 10.0 100.0   6 10.0 100.0   7 10.0 100.0   8 10.0 100.0
  8 10.0 100.0  16 10.0 100.0  24 10.0 100.0  32 10.0 100.0
 40 10.0 100.0  48 10.0 100.0  54 10.0 100.0  62 10.0 100.0   IBN,BETA,...
  1 2     1 2     1 2     1 2     1 2     1 2     1 2     1 2
  2 3     2 3     2 3     2 3     2 3     2 3     2 3     2 3     INOD(I,J)

  0.0      0.0   0.0                                    F0,  FX,  FY
```

input the following parameters:

$$\text{NTIME} = 20, \quad \text{NSTP} = 21 \ (>\text{NTIME}), \quad \text{INTVL} = 1$$

$$\text{INTIAL} = 0, \quad \text{DT} = 0.05, \quad \text{ALFA} = 0.5, \quad \text{GAMA} = 0.5 \ \text{(not used)}$$

$$\text{C0} = 1.0, \quad \text{CX} = 0.0, \quad \text{CY} = 0.0$$

The parameter NSTP allows removal of the source (i.e., f) at a given time step. For example, if NSTP = 5 then, at the fifth time step and at each subsequent time step, f will be set equal to zero.

Table 13.7 gives the input files for the two problems. For a discussion of the results, see Examples 8.9 and 8.10.

The problems in Examples 8.11 and 8.12 can be analyzed using FEM2DV2, with minor changes to the input data in Table 13.7. The data for the hyperbolic problem in Examples 8.11 and 8.12 is given in Table 13.8.

Example 13.5 Deformation of a cantilever plate (Example 10.2). This is a plane elasticity problem, with plane stress assumption (i.e., LNSTRS = 1). Here we consider both triangular and rectangular element meshes (see Fig. 13.7). MSH2DR can be used to generate the meshes (i.e., MESH = 1). For the 8×2 mesh of linear elements (or the 4×1 mesh of quadratic elements), we have the following input parameters:

$$\text{ITYPE} = 2, \quad \text{IGRAD} = 1 \quad (\text{or} > 0), \quad \text{ITEM} = 0$$

for static analysis and ITEM = 2 for dynamic analysis.

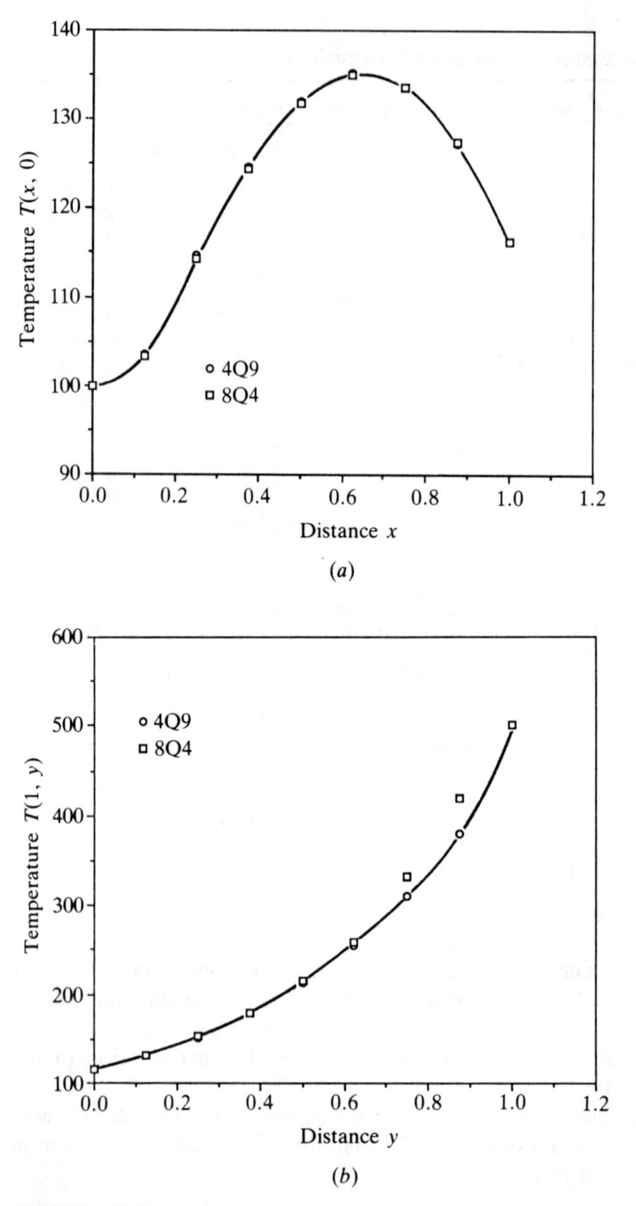

FIGURE 13.5
Temperature variations along the boundaries $y = 0$ and $x = 1$ for the convective heat transfer problem of Example 13.2: 4Q9, 4×4 mesh of nine-node quadrilateral elements; 8Q4, 8×8 mesh of four-node quadrilateral elements.

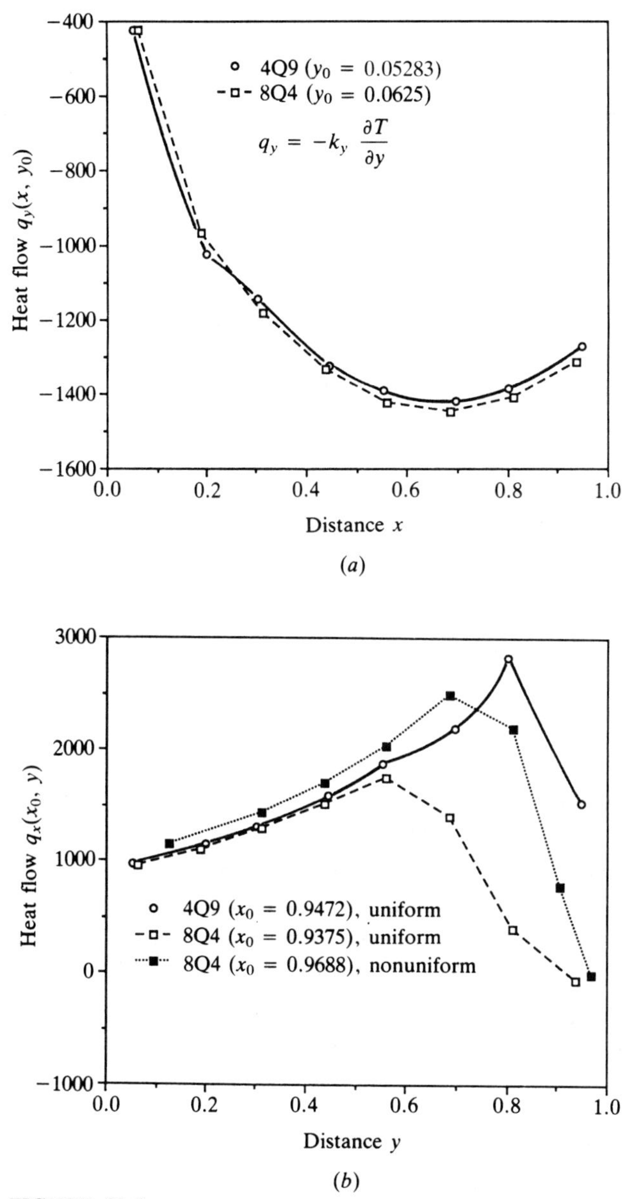

FIGURE 13.6
Variations of the heat flow along the horizontal ($x = 0.9472$ and 0.9375) and vertical ($y = 0.05283$ and 0.0625) boundaries (at the Gauss points nearest to the boundaries). See Fig. 13.5 for the explanation of the notation 4Q9 and 8Q4.

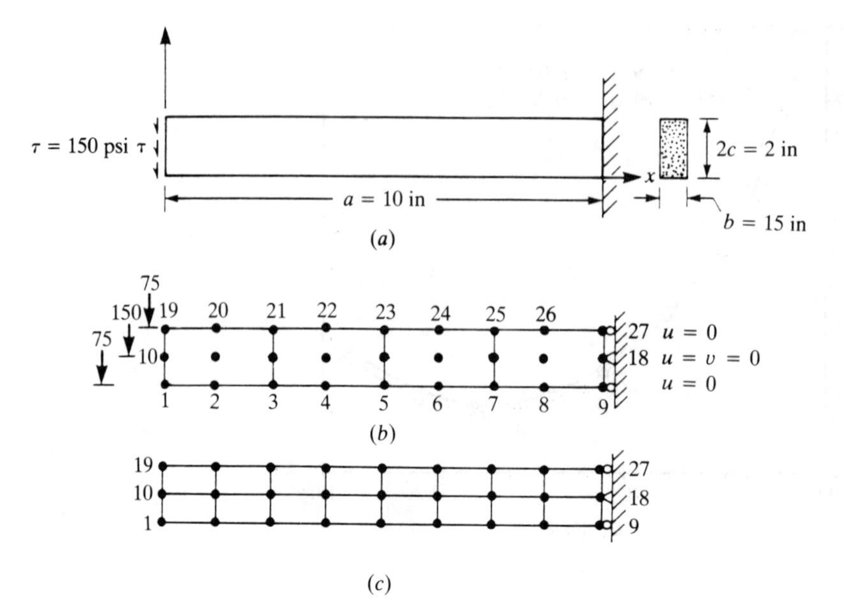

FIGURE 13.7
Bending of a cantilever plate using the elasticity equations: (*a*) geometry and loading; (*b*) mesh of nine-node quadratic elements (4 × 1); (*c*) mesh of linear rectangular elements (8 × 2).

There are four specified primary variables and three nonzero specified forces for the mesh:

$$\text{NSPV} = 4, \quad \text{ISPV(I, J)} = (9, 1; 18, 1; 18, 2; 27, 1)$$
$$\text{VSPV(I)} = (0.0, 0.0, 0.0, 0.0)$$
$$\text{NSSV} = 3, \quad \text{ISSV(I, J)} = (1, 2; 10, 2; 19, 2), \quad \text{VSSV(I)} = (-75.0, -150.0, -75.0)$$

Table 13.9 gives the data sets for natural vibration and transient analysis for the 4 × 1 mesh of nine-node quadratic elements. The data for the static case follows easily from the transient case. The results were discussed in Examples 10.2 and 10.3.

Example 13.6 Viscous fluid squeezed between two parallel plates (Example 11.4). We set up the data for a 12 × 8 uniform mesh of linear rectangular elements and the equivalent 6 × 4 mesh of nine-node elements. Most of the data are exactly the same for both meshes. We have

$$\text{ITYPE} = 1, \quad \text{IGRAD} = 1, \quad \text{ITEM} = 0 \text{ (for static analysis)}$$
$$\text{ITEM} = 1 \text{ (for transient analysis)}, \quad \text{NEIGN} = 0$$

Table 13.10 gives the complete input data for the 6 × 4 mesh of nine-node elements for the transient case.

Example 13.7 Bending of a simply supported isotropic plate under uniform load (Example 12.1). The 4 × 4 mesh of Hermite elements and 2 × 2 mesh of nine-node elements are used to model the square plate. Only rectangular elements are allowed (IELTYP = 1).

Classical plate model (ITYPE = 4 or 5)

$$\text{ITYPE} = 5 \text{ (conforming element)}, \quad \text{IGRAD} = 1$$

TABLE 13.5
Input data required for mesh generation by MSH2DG for the problem of inviscid flow around a circular cylinder

```
(a) Mesh of linear triangles (see Fig.13.3a):

 32  25                                                  NEM, NNM

  5                                                      NRECL
  1   5  1  0.0  0.0  3.0      0.0      6.0  |
  6  10  1  0.0  1.0  3.07612  0.38268  6.0  |  NOD1,NODL,NODINC,
 11  15  1  0.0  2.0  3.29289  0.7071   6.0  |  X1,Y1,XL,YL,RATIO
 16  20  1  2.0  2.0  3.61732  0.92388  6.0  |  for each line segment
 21  25  1  4.0  2.0  4.0      1.0      6.0  |

  8                                              NRECEL
  1   7  2  1  3   1  2   7                  |
  2   8  2  1  3   1  7   6                  |
  9  15  2  1  3   6  7  12                  |  NEL1,NELL,IELINC,NODINC,
 10  16  2  1  3   6 12  11                  |  NPE, NOD(I,J)
 17  23  2  1  3  11 12  16                  |  for each line
 18  24  2  1  3  12 17  16                  |
 25  31  2  1  3  16 17  21                  |
 26  32  2  1  3  17 22  21                  |

(b) Mesh of four-node quadrilaterals (see Fig. 13.3b):

 16  25                                                  NEM, NNM

  5                                                      NRECL
  1   5  1  0.0  0.0  3.0      0.0      6.0  |
  6  10  1  0.0  1.0  3.07612  0.38268  6.0  |  NOD1,NODL,NODINC,
 11  15  1  0.0  2.0  3.29289  0.7071   6.0  |  X1,Y1,XL,YL,RATIO
 16  20  1  2.0  2.0  3.61732  0.92388  6.0  |  for each line segment
 21  25  1  4.0  2.0  4.0      1.0      6.0  |

  4                                              NRECEL
  1   4  1  1  4   1  2   7   6              |
  5   8  1  1  4   6  7  12  11              |  NEL1,NELL,IELINC,NODINC,
  9  12  1  1  4  11 12  17  16              |  NPE,NOD(I,J)
 13  16  1  1  4  16 17  22  21              |
```

Shear deformation plate model

$$\text{ITYPE} = 3, \qquad \text{IGRAD} = 1$$

The classical plate model has three degrees of freedom per node when the nonconforming element is used (i.e., ITYPE = 4), and four degrees of freedom per node for the conforming element (i.e., ITYPE = 5). This must be taken into consideration in inputting boundary conditions.

Table 13.11 gives the input data for classical and shear deformation plate models. For the nonconforming element, the data is the same as that used for the shear deformation model, except that ITYPE = 4.

For free vibration (NEIGN = 1) and stability analysis (NEIGN = 2), most of the data remain the same. For transient analysis, the time step used for unconditionally stable schemes is arbitrary, but should be small enough to give a complete response curve.

TABLE 13.6
Input data for the problem in Example 13.3 (for triangles)†

```
Example 8.7: Flow around a circular cylinder (Streamfunction)
  0   2   0   0                                   ITYPE,IGRAD,ITEM,NEIGN
  0   3   2   0                                   IELTYP,NPE,MESH,NPRNT
 32  25                                           NEM, NNM

<--------- input data for MSH2DG from Table 13.5 goes here --------->

 13                                                           NSPV
  1 1    2 1    3 1    4 1    5 1   10 1   15 1   20 1   25 1
  6 1   11 1   16 1   21 1                                    ISPV(I,J)
  0.0    0.0    0.0    0.0    0.0    0.0    0.0    0.0    0.0
  1.0    2.0    2.0    2.0                                    VSPV(I)

  0                                                   NSSV

  1.0   0.0   0.0                                     A10, A1X, A1Y
  1.0   0.0   0.0                                     A20, A2X, A2Y
  0.0                                                 A00

  0                                                   ICONV

  0.0   0.0   0.0                                     F0,  FX,  FY

Example 8.7: Flow around a circular cylinder (velocity potential)
  0   1   0   0                                   ITYPE,IGRAD,ITEM,NEIGN
  0   3   2   0                                   IEL, NPE, MESH, NPRNT

 32  25                                           NEM, NNM

<--------- input data for MSH2DG from Table 13.5 goes here --------->

  5                                               NSPV
 21 1   22 1   23 1   24 1   25 1                 ISPV(I,J)
  0.0    0.0    0.0    0.0    0.0                 VSPV(I)

  3                                               NSSV
  1 1    6 1   11 1                               ISSV(I,J)
  0.5    1.0    0.5                               VSSV(I)

  1.0   0.0   0.0                                 A10, A1X, A1Y
  1.0   0.0   0.0                                 A20, A2X, A2Y
  0.0                                             A00

  0                                               ICONV

  0.0   0.0   0.0                                 F0,  FX,  FY
```

† For the nodal-equivalent mesh of quadrilateral elements, the input data differs only in specifying IELTYP = 1 and NEM = 16, and using the quadrilateral element data for MSH2DG from Table 13.5.

Example 13.8 Dynamic response of a clamped circular plate under suddenly applied uniform transverse loading. Consider a clamped circular plate of radius $R = 100$ in, thickness $h = 20$ in, modulus $E = 100 \text{ lb in}^{-2}$, Poisson ratio $\nu = 0.3$, and density $\rho = 10 \text{ lb s}^2 \text{ in}^{-4}$, subjected to a suddenly applied (i.e., $w = \phi_x = \phi_y = \dot{w} = \dot{\phi}_x = \dot{\phi}_y = 0$ at time $t = 0$) uniform loading of intensity $f_0 = 1 \text{ lb in}^{-2}$. We analyze the problem by modeling one quadrant (using the symmetry) of the plate by five nine-node elements (see Fig. 13.8*a*) and $\Delta t = 2.5$ s. The input data is presented in Table 13.12.

Plots of the center deflection and stress versus time are shown in Fig. 13.8(*b*).

TABLE 13.7
Input data for the eigenvalue and transient analysis of a parabolic equation over a square region ($a = b = 2$)†

```
Example 8.9: Eigenvalue analysis of a parabolic equation on a square
  0   0   1   1                                    ITYPE,IGRAD,ITEM,NEIGN
 10   0                                            NVALU, NVCTR

  1   4   1   0                                    IELTYP,NPE,MESH,NPRNT

  4   4                                            NX,NY
  0.0  0.25   0.25   0.25   0.25                   X0,DX(I)
  0.0  0.25   0.25   0.25   0.25                   Y0,DY(I)

  9                                                                 NSPV
  5 1   10 1   15 1   20 1   21 1   22 1   23 1   24 1  25 1        ISPV

  1.0   0.0  0.0                                   A10, A1X, A1Y
  1.0   0.0  0.0                                   A20, A2X, A2Y
  0.0                                              A00

  0                                                ICONV

   1.0  0.0  0.0                                   C0,  CX,  CY

Example 8.10: Transient analysis of a parabolic equation on a square
  0   0   1   0                                    ITYPE,IGRAD,ITEM,NEIGN
  1   4   1   0                                    IELTYP,NPE,MESH,NPRNT

  4   4                                            NX,NY
  0.0  0.25   0.25   0.25   0.25                   X0,DX(I)
  0.0  0.25   0.25   0.25   0.25                   Y0,DY(I)

  9                                                                 NSPV
  5 1   10 1   15 1   20 1   21 1   22 1   23 1   24 1  25 1        ISPV
  0.0   0.0    0.0    0.0    0.0    0.0    0.0    0.0   0.0         VSPV

  0                                                                 NSSV

  1.0   0.0  0.0                                   A10, A1X, A1Y
  1.0   0.0  0.0                                   A20, A2X, A2Y
  0.0                                              A00

  0                                                ICONV

   1.0  0.0  0.0                                   F0,  FX,  FY

   1.0  0.0  0.0                                   C0,  CX,  CY

  20    21    1    0                               NTIME,NSTP,INTVL,INTIAL
  0.05  0.5   0.5  1.0E-3                          DT,ALFA,GAMA,EPSLN
```

† Only a quadrant is modeled in the transient as well as the eigenvalue analysis (hence, only symmetric eigenvalues can be obtained).

13.5 SUMMARY

A description of finite element computer program FEM2DV2 and its application to problems discussed in Chapters 8–12 have been presented. The program can be used to analyze two-dimensional field problems and problems of plane elasticity, 2-D flows of viscous incompressible fluids, and plate bending. It allows static, eigenvalue, and time-dependent analyses. Linear and quadratic, and triangular and rectangular elements can be used. The mesh

TABLE 13.8
Input data for the eigenvalue and transient analysis of an hyperbolic equation over a rectangular region ($a = 4$, $b = 2$)†

```
Example 8.11: Natural vibration of a rectangular membrane
  0   0   2   1                                                 ITYPE,IGRAD,ITEM,NEIGN
  9   0                                                         NVALU, NVCTR
  1   4   1   0                                                 IELTYP,NPE,MESH,NPRNT
  4   4                                                         NX,NY
  0.0  1.0    1.0    1.0    1.0                                 X0,DX(I)
  0.0  0.5    0.5    0.5    0.5                                 Y0,DY(I)
  16                                                                             NSPV
   1  1   2 1    3 1    4 1    5 1    6 1   10 1   11 1
  15  1  16 1   20 1   21 1   22 1   23 1   24 1   25 1                          ISPV
  12.5  0.0  0.0                                                A10, A1X, A1Y
  12.5  0.0  0.0                                                A20, A2X, A2Y
  0.0                                                           A00
  0                                                             ICONV
  2.5  0.0  0.0                                                 C0,  CX,  CY

Example 8.12: Transient analysis of an hyperbolic equation (membrane)
  0   0   2   0                                                 ITYPE,IGRAD,ITEM,NEIGN
  1   4   1   0                                                 IELTYP,NPE,MESH,NPRNT
  4   4                                                         NX,NY
  0.0  0.5    0.5    0.5    0.5                                 X0,DX(I)
  0.0  0.25   0.25   0.25   0.25                                Y0,DY(I)
  9                                                                              NSPV
  5 1  10 1   15 1   20 1   21 1   22 1   23 1   24 1  25 1                      ISPV
  0.0   0.0    0.0    0.0    0.0    0.0    0.0    0.0    0.0                     VSPV
  0                                                                              NSSV
  12.5  0.0  0.0                                                A10, A1X, A1Y
  12.5  0.0  0.0                                                A20, A2X, A2Y
  0.0                                                           A00
  0                                                             ICONV
  0.0  0.0  0.0                                                 F0,  FX,  FY
  2.5  0.0  0.0                                                 C0,  CX,  CY
  20     21    1    1                                           NTIME,NSTP,INTVL,INTIAL
  0.025 0.5  0.5  1.0E-3                                        DT,ALFA,GAMA,EPSLN
  0.400    0.375    0.300    0.175    0.0
  0.375    0.35156  0.28125  0.16406  0.0
  0.300    0.28125  0.225    0.13125  0.0
  0.175    0.16406  0.13125  0.076563 0.0
  0.0      0.0      0.0      0.0      0.0                       Initial cond., GLU(I)
  0.0      0.0      0.0      0.0      0.0
  0.0      0.0      0.0      0.0      0.0
  0.0      0.0      0.0      0.0      0.0
  0.0      0.0      0.0      0.0      0.0
  0.0      0.0      0.0      0.0      0.0                       Initial cond., GLV(I)
```

† Only a quadrant is modeled in the transient analysis, and the whole domain is modeled in the eigenvalue analysis.

generators are fairly general for generating finite element meshes for geometrically complex domains. The program FEM2DV2 is a true reflection of the theory presented in Chapters 8–12, and it can be extended to analyze other field problems with appropriate modifications.

TABLE 13.9
Input data for the dynamic analysis of a cantilever beam

```
Example 10.5: Natural vibration of a cantilever beam  by elasticity
  2   0   2   1                                       ITYPE,IGRAD,ITEM,NEIGN
  10  0                                               NVALU,NVCTR
  1   9   1   1                                       IELTYP,NPE,MESH,NPRNT

  4   1                                               NX, NY
  0.0  2.5    2.5    2.5    2.5                       X0, DX(I)
  0.0  2.0                                            Y0, DY(I)

  4                                                   NSPV
  9  1  18 1   18 2   27 1                            ISPV

  1                                                   LNSTRS

  30.0E06  30.0E06  0.25  12.0E06  1.0                E1,E2,ANU12,G12,THKNS

  8.8255E-03  0.0   0.0                               C0, CX, CY
Example 10.5: Transient analysis of a cantilever beam by elasticity
  2   1   2   0                                       ITYPE,IGRAD,ITEM,NEIGN
  1   9   1   0                                       IELTYP,NPE,MESH,NPRNT

  4   1                                               NX, NY
  0.0  2.5    2.5    2.5    2.5                       X0,DX(I)
  0.0  2.0                                            Y0,DY(I)

  4                                                   NSPV
  9  1  18 1   18 2   27 1                            ISPV
  0.0   0.0    0.0    0.0                             VSPV

  3                                                   NSSV
  1  2    10  2  19  2                                ISSV
 -75.0  -150.0  -75.0                                 VSSV

  1                                                   LNSTRS

  30.0E06  30.0E06  0.25  12.0E06  1.0                E1,E2,ANU12,G12,THKNS

   0.0          0.0  0.0                              F0, FX, FY

   8.8255E-03  0.0  0.0                               C0, CX, CY

   20   21   1    0                                   NTIME,NSTP,INTVL,INTIAL

  0.25E-03      0.5  0.5  1.0E-3                      DT,ALFA,GAMA,EPSLN
```

TABLE 13.10
Input data for the viscous flow problem in Example 13.6

```
Example 11.5: Transient analysis of fluid squeezed between plates
  1   1   1   0                                     ITYPE,IGRAD,ITEM,NEIGN
  1   9   1   0                                     IEL, NPE, MESH, NPRNT

  6   4                                             NX, NY
  0.0  1.0  1.0  1.0  1.0  1.0  1.0                 X0, DX(I)
  0.0  0.5  0.5  0.5  0.5                           Y0, DY(I)

  47                                                NSPV
   1 1    1 2    2  2    3  2    4  2    5  2    6  2  7  2   8  2   9  2
  10 2   11 2   12  2   13  2   14  1   27  1   40  1  53  1  66  1  79  1
  92 1  105 1  105  2  106  1  106  2  107  1  107  2 108  1 108  2 109  1
 109 2  110 1  110  2  111  1  111  2  112  1  112  2 113  1 113  2 114  1
 114 2  115 1  115  2  116  1  116  2  117  1  117  2 ISPV(I,J)

  0.0   0.0    0.0    0.0    0.0    0.0    0.0   0.0   0.0    0.0
  0.0   0.0    0.0    0.0    0.0    0.0    0.0   0.0   0.0    0.0
  0.0   0.0   -1.0    0.0   -1.0    0.0   -1.0   0.0  -1.0    0.0
 -1.0   0.0   -1.0    0.0   -1.0    0.0   -1.0   0.0  -1.0    0.0
 -1.0   0.0   -1.0    0.0   -1.0    0.0   -1.0  VSPV(I)

  0                                                 NSSV

  1.0   1.0E8                                       AMU, PENLTY

  0.0   0.0   0.0                                   F0, FX, FY

  1.0   0.0   0.0                                   C0, CX, CY

  20    50    1      0                              NTIME,NSTP,INTVL,INTIAL
  0.1   0.5   0.25   1.0D-3                         DT, ALFA, GAMA, EPSLN
```

TABLE 13.11
Input data for the bending of a simply supported plate

```
Example 12.1: Bending of a simply supported isotropic plate--FSDT
  3   1   0   0                                   ITYPE,IGRAD,ITEM,NEIGN
  1   4   1   0                                   IEL, NPE, MESH, NPRNT

  4   4                                           NX, NY
  0.0  1.25   1.25   1.25   1.25                  X0, DX(I)
  0.0  1.25   1.25   1.25   1.25                  Y0, DY(I)

 27                                                                NSPV
  1  2   1  3   2  3   3  3   4  3   5  1   5  3   6  2   10 1
 10  3  11  2  15  1  15  3  16  2  20  1  20  3  21  1   21 2
 22  1  22  2  23  1  23  2  24  1  24  2  25  1  25  2   25 3     ISPV

 0.0    0.0    0.0    0.0    0.0    0.0    0.0    0.0    0.0
 0.0    0.0    0.0    0.0    0.0    0.0    0.0    0.0    0.0
 0.0    0.0    0.0    0.0    0.0    0.0    0.0    0.0    0.0    VSPV

 0                                                NSSV

 1.0E7  1.0E7  0.25  0.4E7 0.4E7 0.4E7   1.0     E1, E2, ANU12, G12,...

 1.0    0.0    0.0                                F0, FX, FY

Example 12.1: Bending of a simply supported isotropic plate--CPT(C)
  5   1   0   0                                   ITYPE,IGRAD,ITEM,NEIGN
  1   4   1   0                                   IEL, NPE, MESH, NPRNT

  1   1                                           NX, NY
  0.0  5.0                                        X0, DX(I)
  0.0  5.0                                        Y0, DY(I)

 12                                               NSPV
 1  2   1  3   2  1   2  3   3  1   3  2   4  1   4  2   4  3
 1  4   2  4   3  4                               ISPV(I,J)

 0.0    0.0    0.0    0.0    0.0    0.0    0.0    0.0    0.0
 0.0    0.0    0.0                                VSPV(I)

 0                                                NSSV

 1.0E7  1.0E7  0.25  0.4E7  0.4E7  0.4E7  1.0   E1,E2,ANU12,G12,...

 1.0    0.0    0.0                                F0, FX, FY
```

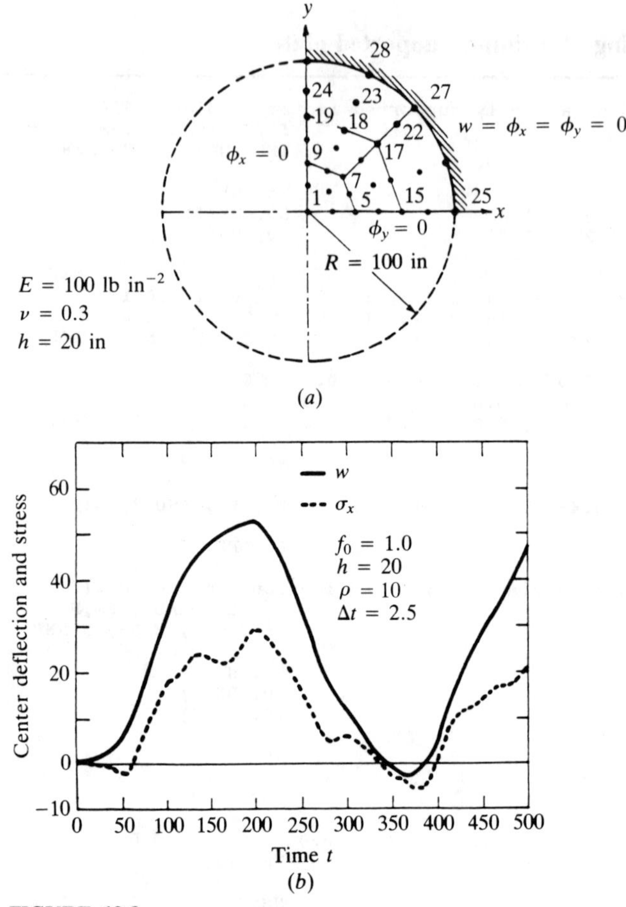

FIGURE 13.8
Clamped circular plate under a suddenly applied uniform load: (*a*) geometry and boundary conditions; (*b*) center deflection and stress versus time.

TABLE 13.12
Input data for the transient analysis of a circular plate

```
Example 13.8: Transient response of a clamped circular plate--FSDT
  3   1   2   0                                          ITYPE,IGRAD,ITEM,NEIGN
  1   9   0   0                                          IEL, NPE, MESH, NPRNT

  5  29                                                  NEM, NNM

  1   5   7   9   2   6   8   4   3                      NOD(1,I), I=1,NPE
  5  15  17   7  10  16  12   6  11                      NOD(2,I), I=1,NPE
  9   7  17  19   8  12  18  14  13                      NOD(3,I), I=1,NPE
 15  25  27  17  20  26  22  16  21                      NOD(4,I), I=1,NPE
 19  17  27  29  18  22  28  24  23                      NOD(5,I), I=1,NPE

 0.0000  0.0000   16.5000  0.0000     11.6673  11.6673
 0.0000  16.5     33.0     0.0        30.488   12.6286
 23.3345 23.3345  12.6286  30.488     0.0      33.0
 49.5    0.0      45.732   18.9428    35.0     35.0
 18.9428 45.732   0.0      49.5       66.0     0.0
 60.976  25.2571  46.669   46.669     25.2571  60.976
 0.0     66.0     83.0     0.0        76.682   31.7627
 58.6898 58.6898  31.7627  76.682     0.0      83.0
 100.0   0.0      92.388   38.2683    70.7107  70.7107
 38.2683 92.388   0.0      100.0                            GLXY(I,J)

 27                                                                NSPV
  1  2   1  3   2  3   4  2   5  3   9  2  10  3  14  2  15 3
 19  2  20  3  24  2  25  1  25  2  25  3  26  1  26  2  26 3
 27  1  27  2  27  3  28  1  28  2  28  3  29  1  29  2  29 3     ISPV

 0.0    0.0    0.0    0.0    0.0    0.0    0.0    0.0    0.0
 0.0    0.0    0.0    0.0    0.0    0.0    0.0    0.0    0.0
 0.0    0.0    0.0    0.0    0.0    0.0    0.0    0.0    0.0     VSPV

 0                                                       NSSV

 1.0E2  1.0E2  0.30  0.3845E2  0.3845E2  0.3845E2  20.0  E1,E2,...

 1.0    0.0    0.0                                       F0, FX, FY

 10.0   0.0    0.0                                       C0, CX, CY

 40     41     10    0                                   NTIME,NSTP,INTVL,INTIAL
 2.5    0.5    0.5   1.0E-03                             DT, ALFA, GAMA, EPSLN
```

PROBLEMS

Note that all of the problems must be analyzed using FEM2DV2. The results obtained from the program should be evaluated for their accuracy in the light of analytical solutions for qualitative understanding of the solution of the problem. New problems can be generated from those given here by changing the problem data, mesh, type of element, etc. For time-dependent problems, the time step and number of time steps should be chosen such that the solution pattern is established or a steady state is reached. When specific material properties are not given, use values such that the solution can be interpreted as the nondimensional solution of the problem.

Field problems (Chapter 8)

13.1. Investigate the convergence of the solution to Problem 8.26 using 2×2, 4×4, and 8×8 meshes of linear triangular elements, and compare the results (in graphical or tabular form) with the analytical solution.

13.2. Repeat Problem 13.1 with rectangular elements.

13.3. Repeat Problem 13.1 for the case $u_0 = 1$ (see Problem 8.27 for the analytical solution).

13.4. Repeat Problem 13.3 with rectangular elements.

13.5. Investigate the convergence of the solution to Problem 8.31 using 2×2, 4×4, and 8×8 meshes of linear triangular elements and equivalent meshes of quadratic triangular elements.

13.6. Repeat Problem 13.5 using rectangular elements.

13.7. Analyze the axisymmetric problem in Problem 8.36 using 4×1 and 8×1 linear rectangular elements, and compare the solution with the exact solution.

13.8. Analyze the axisymmetric problem in Problem 8.37 using 4×4 and 8×8 meshes of linear rectangular elements.

13.9. Analyze Problem 8.26 for eigenvalues (take $c = 1.0$), using a 2×4 uniform mesh of triangular elements. Calculate the critical time step for a parabolic equation.

13.10. Analyze Problem 8.26 using a 2×4 mesh of triangles for transient response. Assume zero initial conditions. Use $\alpha = 0.5$ and $\Delta t = 0.01$. Investigate the stability of the solution when $\alpha = 0.0$ and $\Delta t = 0.005$. The number of time steps should be such that the solution reaches its peak value or a steady state.

13.11. Analyze Problem 8.31 for transient response (take $c = 1.0$) using a 4×4 mesh of linear rectangular elements and a 2×2 mesh of nine-node quadratic rectangular elements. Assume zero initial conditions. Investigate the stability and accuracy of the Crank–Nicolson scheme ($\alpha = 0.5$) and the forward difference scheme ($\alpha = 0$).

13.12. Repeat Problem 13.11 for the axisymmetric problem in Fig. P8.37. Assume zero initial conditions.

Heat transfer (Chapter 8)

13.13. Analyze the heat transfer problem in Problem 8.38 using an 8×16 mesh of linear triangular elements and an equivalent mesh of linear rectangular elements.

13.14. Analyze the heat transfer problem in Fig. P8.39.

13.15. Analyze Problem 8.40 for nodal temperatures and heat flow across the boundaries. Use the following data: $k = 30\ \text{W m}^{-1}\,°\text{C}^{-1}$, $\beta = 60\ \text{W m}^{-2}\,°\text{C}^{-1}$, $T_\infty = 0°\text{C}$, $T_0 = 100°\text{C}$, $q_0 = 200,000\ \text{W m}^{-2}$, $f_0 = 10^7\ \text{W m}^{-3}$, and $a = 1$ cm.

13.16. Repeat Problem 13.15 with an equivalent mesh of triangular elements.

13.17. Analyze Problem 8.46 for nodal temperature and heat flows across the boundary. Take $k = 5\ \text{W m}^{-1}\,°\text{C}^{-1}$.

13.18. Repeat Problem 13.17 using equivalent meshes of nine-node rectangular elements.

13.19. Consider heat transfer in a rectangular domain with a central heated circular cylinder (see Fig. P13.19 for the geometry). Analyze the problem using the mesh of linear quadrilateral elements shown in Fig. 13.3(b).

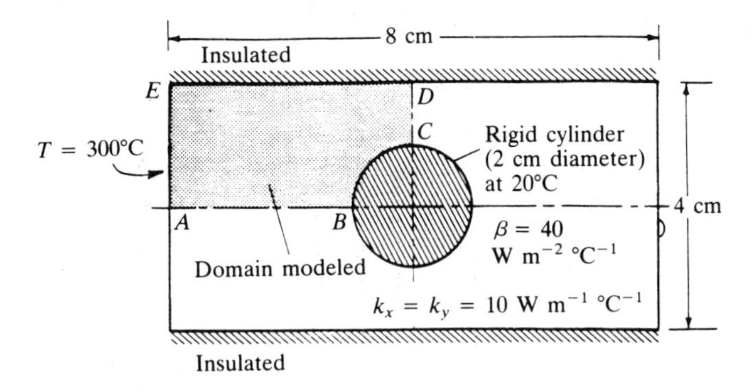

FIGURE P13.19

13.20. Repeat Problem 13.19 with the mesh of quadratic quadrilateral elements shown in Fig. 13.3(d).

13.21. Analyze the heat transfer problem in Fig. P8.42 with (a) 2×2 and (b) 4×4 meshes of linear rectangular elements.

13.22. Repeat Problem 13.21 with triangular elements.

13.23. Analyze the problem in Fig. P8.43 with (a) 3×3, and (b) 6×6 meshes of linear rectangular elements. Take $k = 10\ \text{W m}^{-1}\,°\text{C}^{-1}$.

13.24. Repeat Problem 13.23 with linear triangular elements.

13.25. Analyze the heat transfer problem in Fig. P8.45 with a 4×4 mesh of linear rectangular elements and an equivalent mesh of quadratic (nine-node) elements. Take $a = 1$ cm, $T_0 = 100°\text{C}$, and $k = 3\ \text{W m}^{-1}\,°\text{C}^{-1}$.

13.26. Determine the transient response of the problem in Fig. P8.43 using a 6×6 mesh of rectangular elements. Use the Crank–Nicolson scheme.

13.27. Analyze the problem in Fig. P8.46 for transient response using (a) $\alpha = 0$ and (b) $\alpha = 0.5$.

13.28. Analyze the axisymmetric problem in Fig. P8.36 using the Crank–Nicolson method. Use an 8×1 mesh of linear rectangular elements.

Ground water and inviscid flows (Chapter 8)

13.29. Analyze the groundwater flow in Problem 8.50 using a 16×8 mesh of linear rectangular elements, and homogeneous properties $a_{11} = a_{22} = 1\ \text{m day}^{-1}$.

13.30. Repeat Problem 13.29 with a 16×18 mesh of linear triangular elements.

13.31. Analyze Problem 8.49 with an 8×4 mesh of linear triangular elements.

13.32. Repeat Problem 8.49 with an equivalent mesh of linear quadrilateral elements.

13.33. Repeat Problem 8.49 with an equivalent mesh of quadratic triangular elements.

13.34. Repeat Problem 8.49 with an equivalent mesh of quadratic (nine-node) quadrilateral elements.

13.35. Analyze Problem 8.48 using linear rectangular elements and an equivalent mesh of quadratic rectangular elements. Use an 8×6 mesh in the first rectangle and an 8×4 mesh in the second. The meshes should be refined in the horizontal direction to have smaller elements around the sheet pile.

13.36. Analyze the flow around cylinder of elliptical cross-section (see Fig. P8.52). Use the symmetry and an appropriate mesh of linear triangular elements. Use the stream function approach.

13.37. Repeat Problem 13.36 using the velocity potential formulation.

13.38. Repeat Problem 13.36 with quadrilateral elements.

13.39. Repeat Problem 13.36 using the velocity potential formulation and quadrilateral elements.

13.40. Analyze the problem in Fig. P8.53 using (*a*) the stream function and (*b*) the velocity potential formulations. Use an 8×8 mesh in the first rectangle and a 4×4 mesh in the second. Use a nonuniform mesh of linear rectangular elements such that smaller elements are used near the obstruction.

Solid mechanics (Chapter 8)

13.41. Analyze the torsion of a member of circular cross-section (see Fig. P8.54) for the state of shear stress distribution. Investigate the accuracy with mesh refinements (by subdividing the mesh in Fig. P8.54 with horizontal and vertical lines).

13.42. Analyze the torsion problem in Fig. P8.56.

13.43. Analyze the hollow-cross-section torsion problem in Fig. P8.57 using a mesh of (*a*) linear triangular elements in an octant and (*b*) linear rectangular elements in a quadrant. The meshes should be node-wise-equivalent.

13.44. Analyze the rectangular membrane problem in Fig. P8.59 with 4×4 and 8×8 meshes of linear rectangular elements in the computational domain. Take $a_{11} = a_{22} = 1$.

13.45. Repeat Problem 13.44 with triangular elements.

13.46. Repeat Problem 13.45 with equivalent meshes of quadratic elements.

13.47. Investigate the convergence of the solution of the circular membrane problem in Fig. P8.60 by subdividing each element into three elements. Compare the nodal values of the deflection obtained with the two meshes.

13.48. Determine the eigenvalues of the rectangular membrane in Fig. P8.59 using a 4×4 mesh in the half-domain. Use $c = 1.0$.

13.49. Analyze Problem 8.59 (with a 4×4 mesh in the half-domain) for transient response. Assume zero initial conditions and (*a*) $\alpha = \frac{1}{2}$, $\gamma = \frac{1}{2}$ and (*b*) $\alpha = \frac{1}{2}$, $\gamma = \frac{1}{3}$. Use $\Delta t \approx \Delta t_{cr}$ and a sufficient number of time steps to plot the results graphically.

13.50. Determine the eigenvalues of the circular membrane problem in Fig. P8.60 with a refined mesh obtained from that shown in the figure by subdividing the elements.

13.51. Determine the transient response of the problem in Fig. P8.60 (see Problem 13.50). Assume zero initial conditions, $c = 1$ and $f_0 = 1$. Use $\alpha = \gamma = 0.5$, $\Delta t = 0.05$, and plot the center deflection versus time t for $t = 0$–2.4.

Plane elasticity (Chapter 10)

13.52. Analyze the plane elasticity problem in Fig. P10.7 using (*a*) 4×2 and (*b*) 8×4 meshes of linear rectangular elements and equivalent meshes of quadratic (nine-node) elements. Evaluate the results (i.e., displacements and stresses) qualitatively. Use the plane stress assumption.

13.53. Repeat Problem 13.52 using triangular elements.

13.54–13.63. Analyze the plane elasticity problems shown in Figs. P10.1–10.3 and P10.10–P10.16 using suitable meshes of triangular or rectangular elements (the instructor should specify the element type and mesh).

13.64. Analyze the plane elasticity problem in Fig. P10.7 for natural frequencies. Use an 8×4 mesh of linear rectangular elements and an equivalent mesh of quadratic elements. Use a density of 0.0088.

13.65. Repeat Problem 13.64 with triangular elements.

13.66. Analyze Problem 13.64 for its transient response. Use $\alpha = \frac{1}{2}$, $\gamma = \frac{1}{2}$, and $\Delta t = 10^{-5}$. Assume zero initial conditions.

13.67–13.70. Analyze Problems 13.54–13.57 for their transient response. Use $\alpha = \frac{1}{2}$, $\gamma = \frac{1}{2}$, and $\Delta t = \Delta t_{cr}$.

Viscous incompressible fluids (Chapter 11)

13.71. Analyze the viscous flow problem in Problem 11.8 using an 8×8 mesh of linear rectangular elements. Plot the horizontal velocity $u(0.5, y)$ versus y, and the pressure along the top surface of the cavity. Investigate the effect of the penalty parameter on the solution (see Fig. P11.8).

13.72. Repeat Problem 13.71 with nine-node quadratic elements.

13.73. Repeat Problem 13.71 with eight-node quadratic elements.

13.74. Analyze the slider bearing problem of Example 11.4 to investigate the effect of the penalty parameter on the velocity and pressure fields. Use an 8×8 mesh of linear rectangular elements.

13.75. Repeat Problem 13.74 with nine-node quadratic elements.

13.76. Repeat Problem 13.74 with eight-node quadratic elements.

13.77. Analyze the problem of a viscous incompressible fluid being squeezed through a 4:1 contraction, as shown in Fig. P13.77. Take $L_1 = 10$, $L = 6$, $R_1 = 4$, and $R_2 = 1$, and linear quadrilateral elements. The inlet velocity $u(y)$ is the fully developed solution of the flow between parallel plates. Plot the velocity $u(x, y)$ and pressure along the horizontal centerline.

13.78. Analyze the cavity problem in Problem 13.71 for its transient solution. Use $\rho = 1.0$, zero initial conditions, penalty parameter $\gamma = 10^8$, time parameter $\alpha = 0.5$, and a time step of 0.005 to capture the evolution of $u(0.5, y)$ with time.

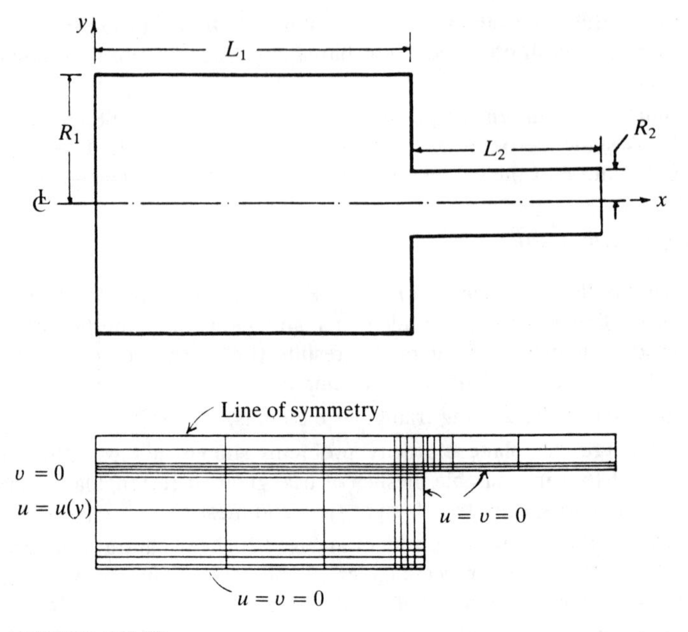

FIGURE P13.77

13.79. Analyze the slider bearing problem of Example 11.4 for its transient solution. Use $\rho = 20$, zero initial conditions, $\gamma = 10^8$, $\alpha = 0.5$, and a time step of 10^{-4}.

Plate bending (Chapter 12)

13.80. Analyze the plate problem in Fig. P12.1 using 2×1, 4×2 and 8×4 meshes of CPT(N) elements in the half-plate, and compare the maximum deflections and stresses.

13.81. Repeat Problem 13.80 with a CPT(C) element.

13.82. Repeat Problem 13.80 with an 8×4 mesh of linear rectangular elements and a 4×2 mesh of nine-node quadratic elements (first-order plate theory).

13.83. Analyze the plate problem in Fig. P12.4 using a 4×4 mesh of CPT(N) elements.

13.84. Repeat Problem 13.83 with CPT(C) elements.

13.85. Repeat Problem 13.83 with a 4×4 mesh of nine-node quadratic elements.

13.86–13.88. Analyze the plate bending problems shown in Figs. P12.7–P12.9 with the CPT(C) elements. Use the mesh shown.

13.89–13.91. Repeat Problems 13.86–13.88 using shear deformation theory. Use linear finite elements.

13.92. Analyze the annular plate in Fig. P12.3 using a four-element mesh (see Fig. P12.7) of CPT(C) elements.

13.93. Repeat Problem 13.92 with four nine-node FSDT elements.

13.94. Analyze the rhombic plate in Fig. P12.6 with a 4×4 mesh of nine-node (FSDT) elements.

13.95. Analyze the plate problem in Fig. P12.1 for its transient response. Use CPT(C) elements and a value of 0.1 for the density.

13.96. Analyze the plate problem in Fig. P12.1 for its transient response using the 4×2 mesh of nine-node quadratic elements (see Problem 13.82) and a value of 0.1 for the density.

13.97. Determine the transient response of the plate in Problem 13.83 with the CPT(C) element.

13.98. Determine the transient response of the plate in Problem 13.83 with the nine-node quadratic element.

13.99. Determine the transient response of the annular plate in Fig. P12.7 using FSDT elements.

REFERENCES FOR ADDITIONAL READING

(See also Chapters 3, 4, 6–12)

Fluid mechanics

Eskinazi, S.: *Principles of Fluid Mechanics,* Allyn and Bacon, Boston 1962.
Bird, R. B., W. E. Stewart, and E. N. Lightfoot: *Transport Phenomena,* John Wiley, New York, 1960.
Duncan, W. J., A. S. Thom, and A. D. Young: *Mechanics of Fluids,* 2d ed., Elsevier, New York, 1970.
Harr, M. E.: *Ground Water and Seepage,* McGraw-Hill, New York, 1962.
Nadai, A.: *Theory of Flow and Fracture of Solids,* vol. II, McGraw-Hill, New York, 1963.
Schlichting, H.: *Boundary-Layer Theory* (translated by J. Kestin), 7th ed., McGraw-Hill, New York, 1979.
Shames, I. H.: *Mechanics of Fluids,* McGraw-Hill, New York, 1962.
Vallentine, H. R.: *Applied Hydrodynamics,* Butterworths, London, 1959.
Verrujit, A.: *Theory of Groundwater Flow,* Gordon and Breach, New York, 1970.

Heat transfer

Carslaw, H. S., and J. C. Jaeger: *Conduction of Heat in Solids,* Clarendon Press, Oxford, 1959.
Holoman, J. P.: *Heat Transfer,* 6th ed., McGraw-Hill, New York, 1986.
Donea, J.: "On the Accuracy of Finite Element Solutions to the Transient Heat-Conduction Equation," *International Journal for Numerical Methods in Engineering,* vol. 8, pp. 103–110, 1974.
Kreith, F.: *Principles of Heat Transfer,* 3d ed., Harper & Row, New York, 1973.
Moore, A. H., B. Kaplan, and D. B. Mitchell: "A Comparison of Crandall and Crank–Nicolson Methods for Solving a Transient Heat Conduction Problem," *International Journal for Numerical Methods in Engineering,* vol. 9, pp. 938–943, 1975.
Myers, G. E.: *Analytical Methods in Conduction Heat Transfer,* McGraw-Hill, New York, 1972.
Özisik, M. N.: *Heat Transfer: A Basic Approach,* McGraw-Hill, New York, 1985.
Wilson, E. L., and R. E. Nickell: "Application of the Finite Element Method to Heat Conduction Analysis," *Journal for Nuclear Engineering and Design,* vol. 4, pp. 276–286, 1966.
Wood, W. L., and R. W. Lewis: "A Comparison of Time Marching Schemes for the Transient Heat Conduction Equation," *International Journal for Numerical Methods in Engineering,* vol. 9, pp. 679–690, 1975.

Plane elasticity

Budynas, R. G.: *Advanced Strength and Applied Stress Analysis,* McGraw-Hill, New York, 1977.
Harris, C. M., and C. E. Crede: *Shock and Vibration Handbook,* vol. 1, McGraw-Hill, New York, 1961.

Ugural, A. C., and S. K. Fenster, *Advanced Strength and Applied Elasticity,* Elsevier, New York, 1975.

Volterra, E., and J. H. Gaines: *Advanced Strength of Materials,* Prentice-Hall, Englewood Cliffs, NJ, 1971.

Plate bending

Dym, C. L., and I. H. Shames: *Solid Mechanics: A Variational Approach,* McGraw-Hill, New York, 1973.

Reddy, J. N.: "A Penalty-Plate Bending Element for the Analysis of Laminated Anisotropic Composite Plates," *International Journal for Numerical Methods in Engineering,* vol. 15, pp. 1187–1206, 1980.

———: "On the Solutions to Forced Motions of Rectangular Composite Plates," *Journal of Applied Mechanics,* vol. 49, pp. 403–408, 1982.

———: *Energy and Variational Methods in Applied Mechanics,* John Wiley, New York, 1984.

Szilard, R.: *Theory and Analysis of Plates,* Prentice-Hall, Englewood Cliffs, NJ, 1974.

Timoshenko, S., and S. Woinowsky-Krieger: *Theory of Plates and Shells,* 2d ed., McGraw-Hill, New York, 1959.

Tsay, C. S., and J. N. Reddy: "Bending, Stability, and Free Vibration of Thin Orthotropic Plates by Simplified Mixed Finite Elements," *Journal of Sound and Vibration,* vol. 59, pp. 307–311, 1978.

Computer Implementation

Akay, H. U., P. G. Willhite, and H. Didandeh: "UCODE2, Version 87.1," Computational Fluid Dynamics Laboratory, Department of Mechanical Engineering, Purdue University at Indianapolis, Indianapolis, 1987.

Bathe, K.-J.: *Finite Element Procedures in Engineering Analysis,* Prentice-Hall, Englewood Cliffs, NJ, 1982.

Burnett, D. S.: *Finite Element Analysis,* Addison-Wesley, Reading, MA, 1987.

Hughes, T. J. R.: *The Finite Element Method,* Prentice-Hall, Englewood Cliffs, NJ, 1987.

Zienkiewicz, O. C., and R. L. Taylor: *The Finite Element Method,* vol. 1, McGraw-Hill, New York, 1989.

PART 4

ADVANCED TOPICS

CHAPTER 14

WEIGHTED-RESIDUAL FINITE ELEMENT MODELS, AND FINITE ELEMENT MODELS OF NONLINEAR AND THREE-DIMENSIONAL PROBLEMS

14.1 INTRODUCTION

The introduction presented in the preceding chapters is sufficient to provide the essential ideas for the development of finite element models and the associated computer programs for most linear boundary and initial value problems, and eigenvalue problems in one and two dimensions. However, there are many additional ideas that deserve some comment. Here we discuss some immediate extensions of the present study to alternative formulations, three-dimensional problems, and nonlinear problems. The discussions are only meant to give some idea of the applicability of the finite element method to these advanced topics, and therefore details are not included here. The topics discussed in this chapter are as follows:

1. Alternative formulations
2. Nonlinear problems
3. Three-dimensional problems

The reader interested in a detailed treatment of any of these topics should consult the books listed in the References at the end of the chapter.

14.2 ALTERNATIVE FORMULATIONS

14.2.1 Introductory Comments

The finite element formulations presented in Chapters 3–12 were based on the weak formulation of the governing differential equations. These models can be termed the *Ritz finite element models* or *weak form finite element models*. In most cases, especially when the governing equations contain derivatives of even order, weak forms can be developed using the three-step procedure. In some cases, the governing equations must be recast in an alternative form that facilitates the use of lower-order interpolations and/or yields better accuracies for the secondary variables, which, as the reader knows, are discontinuous at the interelement boundaries. Here we discuss a couple of such formulations and demonstrate the use of the weighted-residual methods of Section 2.4.3 in the finite element modeling of the equations. More specifically, we shall study the following formulations:

1. Weighted-residual formulations
2. A mixed formulation of the Euler–Bernoulli beam theory

It should be noted that these specific equations have been chosen to illustrate the basic ideas behind the two formulations—the ideas are applicable to other equations, and to two- and three-dimensional problems.

14.2.2 Weighted-Residual Finite Element Models

Here we describe finite element models based on weighted-residual methods. Recall from the preceding chapters that the finite element models presented are based on a weak formulation. In this formulation, integration by parts is used to include the natural boundary conditions in the integral form. Weighted-residual methods are based on a weighted-integral form of a given differential equation. They are the natural and only choice for first-order equations, which do not have a weak formulation. For second- and higher-order equations, we have a choice between the weak formulation and the weighted residual formulation. We begin with the description of weighted-residual finite element models of a first-order equation in one dimension.

1-D FIRST-ORDER EQUATIONS. Consider the first-order equation

$$a\frac{du}{dx} + cu = f \quad \text{for } 0 < x < L \tag{14.1}$$

where a, c, and f are given functions of x. An example of a situation in which the above equation arises is given by Newton's law of cooling of a body with temperature u in an environment at temperature u_0:

$$\frac{du}{dt} + k(u - u_0) = 0$$

which is the same as (14.1) with $x = t$, $c = k$, $a = 1$, and $f = ku_0$.

Since the equation under consideration is of first order, there is no advantage in transferring the derivative to the weight function. Hence, the weighted-integral form to be used here naturally falls under the weighted-residual method. Over an element, we have

$$0 = \int_{x_A}^{x_B} w\left(a\frac{du}{dx} + cu - f\right) dx \tag{14.2}$$

Note that there are no "flux" terms in the equation. When u is approximated by an n-parameter approximation, we need n algebraic equations to solve for the parameters. These n equations are provided by n different (i.e., linearly independent) choices of the weight function w. Different choices of w dictate different methods.

Suppose that u is approximated as

$$u \approx U^e = \sum_{j=1}^{n} u_j^e \psi_j^e(x) \tag{14.3}$$

where u_j^e denote the nodal values of the variable U^e, and $\psi_j^e(x)$ are the interpolation functions. An examination of the weighted-integral statement (14.2) shows that ψ_j should be once-differentiable with respect to x. Hence, the minimum continuity on ψ_j is that they be linear. Consequently, the Lagrange family of interpolation functions are admissible.

The only requirements on the weight function w are (i) that it be integrable in the sense of (14.2), and (ii) that it belong to a linearly independent set $\{w_i\}_{i=1}^n$. The ith algebraic equation is obtained by replacing w in (14.2) with w_i. Different choices for the set $\{w_i\}$ have been suggested, and the resulting models bear the names of the original scientists who suggested them. The best-known choices for w_i are

1. The Petrov–Galerkin method: $w_i = \phi_i$ and $\phi_i \neq \psi_i$
2. The Bubnov–Galerkin method: $w_i = \psi_i$
3. The collocation method: $w_i = \delta(x - x_i)$, the Dirac delta function
4. The subdomain method: $w_i = \delta_{ij}$ in the ith subdomain
5. The least-squares method: $w_i = A(\psi_i)$, where A is the operator in the differential equation $Au = f$

(14.4)

The finite element models of (14.1) based on these methods are discussed next.

The Petrov–Galerkin model. Substitute (14.3) for u and $w = \phi_i$ into (14.2) to obtain

$$0 = \sum_{j=1}^{n} K_{ij}u_j - f_i \quad \text{or} \quad [K^e]\{u^e\} = \{f^e\} \tag{14.5a}$$

where

$$K_{ij}^e = \int_{x_A}^{x_B} \left(a\phi_i^e \frac{d\psi_j^e}{dx} + c\phi_i^e\psi_j^e\right) dx, \quad f_i^e = \int_{x_A}^{x_B} \phi_i^e f\, dx \tag{14.5b}$$

The Bubnov–Galerkin model. Here we substitute $w = \psi_i^e$ into (14.2) to obtain the finite element model

$$[K^e]\{u^e\} = \{f^e\} \tag{14.6a}$$

where

$$K_{ij}^e = \int_{x_A}^{x_B} \left(a\psi_i^e \frac{d\psi_j^e}{dx} + c\psi_i^e\psi_j^e \right) dx, \qquad f_i^e = \int_{x_A}^{x_B} f\psi_i^e \, dx \tag{14.6b}$$

Collocation model. In this case, we take $w = \delta(x - x_i)$, the Dirac delta function. The collocation points x_i can be chosen arbitrarily, usually as the quadrature points in $\Omega^e = [x_A, x_B]$. We have

$$K_{ij}^e = a(x_i)\frac{d\psi_j^e}{dx}(x_i) + c(x_i)\psi_j^e(x_i), \qquad f_i^e = f(x_i) \tag{14.7}$$

Subdomain model. In this model, the element domain Ω^e is subdivided further into n subdomains. The weight function for the ith subdomain Ω_i^e is unity over the subdomain and zero outside it:

$$w(x) = \begin{cases} 1 & \text{for } x \in \Omega_i^e \\ 0 & \text{for } x \notin \Omega_i^e \end{cases} \tag{14.8}$$

We obtain

$$K_{ij}^e = \int_{\Omega_i^e} \left(a\frac{d\psi_j^e}{dx} + c\psi_j^e \right) dx, \qquad f_i^e = \int_{\Omega_i^e} f \, dx \tag{14.9}$$

Least-squares model. The weight function in this case is $w \equiv A(\psi_i) = a\,d\psi_i^e/dx + c\psi_i^e$. Consequently, we have

$$K_{ij}^e = \int_{x_A}^{x_B} \left(a\frac{d\psi_i^e}{dx} + c\psi_i^e \right)\left(a\frac{d\psi_j^e}{dx} + c\psi_j^e \right) dx \tag{14.10a}$$

$$f_i^e = \int_{x_A}^{x_B} \left(a\frac{d\psi_i^e}{dx} + c\psi_i^e \right) f \, dx \tag{14.10b}$$

Note that only the least-squares method gives a symmetric coefficient matrix for a single first-order equation. It is possible to obtain a symmetric coefficient matrix in the Galerkin model of a set of first-order equations (see Problem 14.1). Next we consider an example.

Example 14.1. Let us solve (14.1) for the data $a = 1$, $c = 2$, $f = 1$, and $L = 1$, and the boundary condition $u(0) = 1$. We shall use two linear elements in the domain.

For the Bubnov–Galerkin model, we have

$$[K^e] = \frac{1}{2}\begin{bmatrix} -1 & 1 \\ -1 & 1 \end{bmatrix} + \frac{h_e}{3}\begin{bmatrix} 2 & 1 \\ 1 & 2 \end{bmatrix}, \qquad \{f^e\} = \frac{h_e}{2}\begin{Bmatrix} 1 \\ 1 \end{Bmatrix} \tag{14.11}$$

The assembled equations are given by

$$\left(\frac{1}{2}\begin{bmatrix} -1 & 1 & 0 \\ -1 & 0 & 1 \\ 0 & -1 & 1 \end{bmatrix} + \frac{1}{6}\begin{bmatrix} 2 & 1 & 0 \\ 1 & 4 & 1 \\ 0 & 1 & 2 \end{bmatrix}\right)\begin{Bmatrix} U_1 \\ U_2 \\ U_3 \end{Bmatrix} = \frac{1}{4}\begin{Bmatrix} 1 \\ 2 \\ 1 \end{Bmatrix} \tag{14.12}$$

Using the boundary condition $U_1 = 1$, we obtain from the last two equations

$$\frac{1}{6}\begin{bmatrix} 4 & 4 \\ -2 & 5 \end{bmatrix}\begin{Bmatrix} U_2 \\ U_3 \end{Bmatrix} = \frac{1}{6}\begin{Bmatrix} 5 \\ 1.5 \end{Bmatrix} \tag{14.13}$$

and their solution is $U_2 = 0.6786$ and $U_3 = 0.5714$. The exact solution is

$$u(x) = u(0)e^{-(c/a)x} - \frac{f}{c}(1 - e^{-(c/a)x}) \tag{14.14}$$

The exact values of u at $x = \frac{1}{2}$ and 1 are 0.6389 and 0.5677, respectively.

For the collocation model, we choose $x_1 = \frac{1}{3}h_e$ and $x_2 = \frac{2}{3}h_e$, and obtain

$$[K^e] = \frac{1}{h_e}\begin{bmatrix} -1 & 1 \\ -1 & 1 \end{bmatrix} + \frac{2}{3}\begin{bmatrix} 2 & 1 \\ 1 & 2 \end{bmatrix}, \qquad \{f^e\} = \begin{Bmatrix} 1 \\ 1 \end{Bmatrix} \tag{14.15}$$

Clearly, the element equations obtained using the collocation points $x_1 = \frac{1}{3}h_e$ and $x_2 = \frac{2}{3}h_e$ are the same as those obtained in the Bubnov–Galerkin method (actually a multiple of $\frac{1}{2}h_e$). Hence, we obtain the same solution as in the Bubnov–Galerkin method.

In the case of the subdomain model, we use $\Omega_1^e = (0, \frac{1}{2}h_e)$ and $\Omega_2^e = (\frac{1}{2}h_e, h_e)$. For this choice, we obtain

$$[K^e] = \frac{1}{2}\begin{bmatrix} -1 & 1 \\ -1 & 1 \end{bmatrix} + \frac{h_e}{4}\begin{bmatrix} 3 & 1 \\ 1 & 3 \end{bmatrix}, \qquad \{f^e\} = \frac{3h_e}{8}\begin{Bmatrix} 1 \\ 1 \end{Bmatrix} \tag{14.16}$$

The assembled equations become

$$\frac{1}{8}\begin{bmatrix} -1 & 5 & 0 \\ -3 & 6 & 5 \\ 0 & -3 & 7 \end{bmatrix}\begin{Bmatrix} U_1 \\ U_2 \\ U_3 \end{Bmatrix} = \frac{3}{16}\begin{Bmatrix} 1 \\ 2 \\ 1 \end{Bmatrix} \tag{14.17}$$

The solution, from the last two equations, is $U_2 = 0.6053$ and $U_3 = 0.4737$.

Last, we consider the least-squares model. We have

$$[K^e] = \frac{1}{h_e}\begin{bmatrix} 1 & -1 \\ -1 & 1 \end{bmatrix} + \frac{4h_e}{6}\begin{bmatrix} 2 & 1 \\ 1 & 2 \end{bmatrix} + \begin{bmatrix} -2 & 0 \\ 0 & 2 \end{bmatrix} \tag{14.18a}$$

$$\{f^e\} = \begin{Bmatrix} -1 \\ 1 \end{Bmatrix} + h_e\begin{Bmatrix} 1 \\ 1 \end{Bmatrix} \tag{14.18b}$$

The assembled equations are

$$\frac{1}{3}\begin{bmatrix} -2.5 & -0.5 & 0.0 \\ -0.5 & 7.0 & -0.5 \\ 0.0 & -0.5 & 9.5 \end{bmatrix}\begin{Bmatrix} U_1 \\ U_2 \\ U_3 \end{Bmatrix} = \begin{Bmatrix} -0.5 \\ 1.0 \\ 1.5 \end{Bmatrix} \tag{14.19}$$

TABLE 14.1
Comparison of the numerical solution of (14.1) with the exact solution (Example 14.1)

Nodal values	Bubnov–Galerkin	Collocation	Subdomain	Least-squares	Exact
$u(0.5)$	0.6786	0.6786	0.6053	0.6793	0.6839
$u(1.0)$	0.5714	0.5714	0.4737	0.5094	0.5677

The solution of these equations is

$$U_2 = 0.6793, \qquad U_3 = 0.5094$$

Table 14.1 gives a summary of the solutions obtained using various models.

1-D SECOND-ORDER EQUATIONS. Use of weighted-residual finite element models in the solution of second-order equations is more involved. We describe the finite element models for the model equation

$$-\frac{d}{dx}\left(a\frac{du}{dx}\right) = f \tag{14.20}$$

For the weak formulation of this equation, the approximation functions ψ_i^e must be continuous, as required by the weak form, and they are also required to be the interpolants of u only. The weak form includes the natural boundary conditions associated with the equation, and therefore interpolation of only u (not its derivatives) is required.

In the weighted-residual methods, we use the weighted-integral form of the differential equation without weakening the differentiability on u. Therefore, the continuity conditions on the interpolation functions used in the weighted-residual methods are dictated by the order of the differential equation. For example, the second-order differential equation (14.20) requires the approximation functions to be twice-differentiable with respect to x. In addition, the approximation solution must satisfy the end conditions on the primary and secondary variables (identified with the help of the weak formulation) of the problem. This amounts, for second-order equations, to using approximation functions that make the primary and secondary variables continuous at the nodes connecting two elements (i.e., interelement nodes). For the second-order equation under consideration, the natural boundary condition involves specifying the secondary variable $a\,du/dx$ at the element boundaries. Therefore, the interpolation functions must be selected such that u and $a\,du/dx$ are continuous across an interface between elements. This in turn implies, if a is continuous, that du/dx is continuous throughout the domain $\Omega = (0, L)$. Hence, a C^1 approximation (i.e., Hermite interpolation) of u is required.

Because u and du/dx are required to be continuous across an interface between elements, and a typical element (in one dimension) has two such interfaces, the polynomial approximation of u must involve four parameters, i.e., it must be a cubic polynomial. Thus, the finite element is a line element with two nodes and two degrees of freedom, u and du/dx, at each node. The element is different from the Lagrange cubic element, which has four nodes with one degree of freedom per node. The Lagrange interpolation functions (of any order) do not satisfy the continuity of du/dx across element interfaces, and therefore do not belong to $C^1(0, L)$. The two-node element, with continuous u and du/dx at element interfaces, is the Hermite cubic element developed in Section 4.2 for the Euler–Bernoulli beam element. We have

$$U^e(x) = \sum_{j=1}^{4} u_j^e \psi_j^e(x) \tag{14.21}$$

where u_1^e and u_3^e are the nodal values of U^e, u_2^e and u_4^e are the nodal values of dU^e/dx at the two nodes, and ψ_j^e are the Hermite cubic interpolation functions in (4.10). We consider various weighted-residual finite element models of (14.20).

The weighted-residual form of (14.20) over an element $\Omega^e = (x_A, x_B)$ is

$$0 = \int_{x_A}^{x_B} w\left[-\frac{d}{dx}\left(a\frac{du}{dx}\right) - f\right] dx \tag{14.22}$$

Substituting ϕ_i^e for the weight function w and (14.21) for u, we obtain

$$0 = \sum_{j=1}^{4} K_{ij}^e u_j^e - f_i^e, \quad \text{or} \quad [K^e]\{u^e\} = \{f^e\} \tag{14.23a}$$

where

$$K_{ij}^e = \int_{x_A}^{x_B} \left[-\phi_i^e \frac{d}{dx}\left(a\frac{d\psi_j^e}{dx}\right)\right] dx, \qquad f_i^e = \int_{x_A}^{x_B} \phi_i^e f\, dx \tag{14.23b}$$

Equation (14.23a) is the Petrov–Galerkin model of (14.20) when $\phi_i^e \neq \psi_i^e$.

For different choices of ϕ_i^e in (14.23b), we obtain different finite element models. These are presented below.

The Bubnov–Galerkin model. For $\phi_i^e = \psi_i^e$, (14.23b) becomes

$$K_{ij}^e = \int_{x_A}^{x_B} -\psi_i^e \frac{d}{dx}\left(a\frac{d\psi_j^e}{dx}\right) dx, \qquad f_i^e = \int_{x_A}^{x_B} \psi_i^e f\, dx \tag{14.24}$$

Least-squares model. For $\phi_i^e = A(\psi_i^e) \equiv -(d/dx)(a\, d\psi_i^e/dx)$, we have

$$K_{ij}^e = \int_{x_B}^{x_A} \frac{d}{dx}\left(a\frac{d\psi_i^e}{dx}\right)\frac{d}{dx}\left(a\frac{d\psi_j^e}{dx}\right) dx, \qquad f_i^e = -\int_{x_A}^{x_B} \frac{d}{dx}\left(a\frac{d\psi_i^e}{dx}\right) f\, dx \tag{14.25}$$

Collocation model. For $\phi_i^e = \delta(x - x_i)$, (14.23$b$) takes the form

$$K_{ij}^e = -\left\{\frac{d}{dx}\left[a(x)\frac{d\psi_j^e}{dx}(x)\right]\right\}_{x=x_i}, \qquad f_i^e = f(x_i) \tag{14.26}$$

where x_i are the collocation points.

While the Bubnov–Galerkin and least-squares models have the same form as the Ritz model, i.e., they are defined by integral expressions, the collocation model does not. In the latter, one simply evaluates the coefficient matrices and column vector at the collocation points, instead of evaluating the integral expressions. The number of collocation points should be equal to the number of unknowns, after imposition of the boundary conditions of the problem. For second-order equations, we have two boundary conditions and $2(N+1)$ nodal degrees of freedom for an N-element model. Hence, a total of $2N$ collocation points, two per element, are needed. Also, note that the coefficient matrix in (14.26) is of order 2×4 ($i=1,2$), and there is no overlap of element matrices because there is no summation of equations over the number of elements. However, the continuity conditions on nodal variables are imposed in all other models.

Example 14.2. Consider the boundary value problem

$$-\frac{d}{dx}\left[(1+x)\frac{du}{dx}\right]=0 \quad \text{for } 0<x<1 \tag{14.27a}$$

$$u(0)=0, \qquad \left[(1+x)\frac{du}{dx}\right]\bigg|_{x=1}=1 \tag{14.27b}$$

The exact solution of this problem is

$$u=\ln(1+x) \tag{14.28}$$

We wish to solve the problem using various weighted-residual finite element models.

Consider a two-element discretization of the problem. There are three nodes and six degrees of freedom. The known degrees of freedom for the weighted-residual models are

$$U_1=0, \qquad U_6=0.5 \tag{14.29}$$

whereas, in the four-element Ritz model (i.e., the weak form finite element model) with linear elements, they are

$$U_1=0, \qquad U_5=1.0 \tag{14.30}$$

For the collocation model, the two-point Gauss quadrature points are used as the collocation points. The two- and four-element finite element solutions obtained from three weighted-residual finite element models, all using the Hermite cubics, are compared with the exact solution and the solution of the weak form finite element model in Table 14.2. All models give accurate results.

2-D SECOND-ORDER EQUATIONS. Here we describe the Bubnov–Galerkin, least-squares, and collocation finite element models of a second-order equation in two dimensions. We consider the Poisson equation ($Au=f$)

$$-\frac{\partial}{\partial x}\left(a_{11}\frac{\partial u}{\partial x}\right)-\frac{\partial}{\partial y}\left(a_{22}\frac{\partial u}{\partial y}\right)=f \quad \text{in } \Omega \tag{14.31}$$

TABLE 14.2
Comparison of weighted-residual finite-element solutions† with the exact solution of the problem in (14.27)

x		Exact	Collocation	Bubnov–Galerkin	Least-squares	Ritz
0.00	u	0.00000	0.00000	0.00000	0.00000	0.00000
	u'	1.00000	0.99390	0.99604	0.99902	0.99612
			0.99967	0.99930	0.99993	0.99930
0.25	u	0.22314	—	—	—	—
			0.22326	0.22313	0.22315	0.22313
	u'	0.80000	—	—	—	—
			0.80028	0.80004	0.79997	0.80004
0.50	u	0.40547	0.40487	0.40537	0.40554	0.40538
			0.40562	0.40546	0.40547	0.40546
	u'	0.66667	0.66299	0.66707	0.66645	0.66728
			0.66646	0.66673	0.66665	0.66674
0.75	u	0.55962	—	—	—	—
			0.55975	0.55961	0.55962	0.55961
	u'	0.57143	—	—	—	—
			0.57123	0.57146	0.57142	0.57148
1.00	u	0.69315	0.69202	0.69309	0.69324	0.69315
			0.69325	0.69314	0.69315	0.69315
	u'	0.50000	0.50000	0.50000	0.50000	0.50102
			0.50000	0.50000	0.50000	0.50010

† The first line in each case corresponds to two elements and the second line to four.

In the weighted-residual finite element model, we seek an approximate solution U^e of u in Ω^e in the form

$$U^e = \sum_{j=1}^{16} u_j^e \psi_j^e(x, y) \tag{14.32}$$

where $\psi_j^e(x, y)$ are the conforming Hermite interpolation functions of the four-node rectangular element (see Table 9.1), and u_1^e, u_5^e, u_9^e, and u_{13}^e are the nodal values of U^e at the four nodes of the element, u_2^e, u_6^e, u_{10}^e, and u_{14}^e are the nodal values of $\partial U^e/\partial x$ at the four nodes, and so on. The weighted-residual statement of (14.31) over an element is given by

$$0 = \int_{\Omega^e} w\left[-\frac{\partial}{\partial x}\left(a_{11}\frac{\partial u}{\partial x}\right) - \frac{\partial}{\partial y}\left(a_{22}\frac{\partial u}{\partial y}\right) - f\right] dx\, dy \tag{14.33}$$

Substituting $w = \phi_i^e$ and replacing u by (14.32), we obtain the usual form of the element equations

$$[K^e]\{u^e\} = \{f^e\} \tag{14.34}$$

The specific forms of $[K^e]$ and $\{f^e\}$ are defined below for various models.

The Bubnov–Galerkin model. For $w = \psi_i^e$, the coefficient matrix $[K^e]$ and column vector $\{f^e\}$ are defined by

$$K_{ij}^e = -\int_{\Omega^e} \psi_i^e \left[\frac{\partial}{\partial x}\left(a_{11}\frac{\partial \psi_j^e}{\partial x}\right) + \frac{\partial}{\partial y}\left(a_{22}\frac{\partial \psi_j^e}{\partial y}\right)\right] dx\, dy \tag{14.35a}$$

$$f_i^e = \int_{\Omega^e} \psi_i^e f\, dx\, dy \tag{14.35b}$$

The least-squares model. For $w = A(\psi_i^e)$, where A is the differential operator in (14.31), the coefficients of $[K^e]$ and $\{f^e\}$ are defined by

$$K_{ij}^e = \int_{\Omega^e} \left[\frac{\partial}{\partial x}\left(a_{11}\frac{\partial \psi_i^e}{\partial x}\right) + \frac{\partial}{\partial y}\left(a_{22}\frac{\partial \psi_i^e}{\partial y}\right)\right]\left[\frac{\partial}{\partial x}\left(a_{11}\frac{\partial \psi_j^e}{\partial x}\right) + \frac{\partial}{\partial y}\left(a_{22}\frac{\partial \psi_j^e}{\partial y}\right)\right] dx\, dy \tag{14.36a}$$

$$f_i^e = -\int_{\Omega^e} \left[\frac{\partial}{\partial x}\left(a_{11}\frac{\partial \psi_i^e}{\partial x}\right) + \frac{\partial}{\partial y}\left(a_{22}\frac{\partial \psi_i^e}{\partial y}\right)\right] f\, dx\, dy \tag{14.36b}$$

The collocation model. In this model, we select four collocation points per element and satisfy (14.31) exactly at those four points of each element. For the best results, the Gauss quadrature points are selected. We have

$$K_{ij}^e = \left[-\frac{\partial}{\partial x}\left(a_{11}^e\frac{\partial \psi_j^e}{\partial x}\right) - \frac{\partial}{\partial y}\left(a_{22}^e\frac{\partial \psi_j^e}{\partial y}\right)\right]\Bigg|_{(x,y)=(x_i,y_i)} \tag{14.37a}$$

$$f_i^e = f_e(x_i, y_i) \tag{14.37b}$$

for $i = 1, \ldots, 4$ and $j = 1, 2, \ldots, 16$. Note that the coefficient matrix is rectangular (4×16) and that each coefficient of the global matrix and column vector has a contribution from no more than one element. After imposing the boundary conditions of the problem, the number of linearly independent equations (which is equal to four times the number of elements in the finite element mesh) will be equal to the number of unknown nodal degrees of freedom.

We now consider an application of the weighted-residual models described in (14.35)–(14.37).

Example 14.3. Consider the Dirichlet problem for the Poisson equation [cf. (8.150)]

$$\begin{aligned} -\nabla^2 u &= 2 \quad \text{in } \Omega \\ u &= 0 \quad \text{on } \Gamma \end{aligned} \tag{14.38}$$

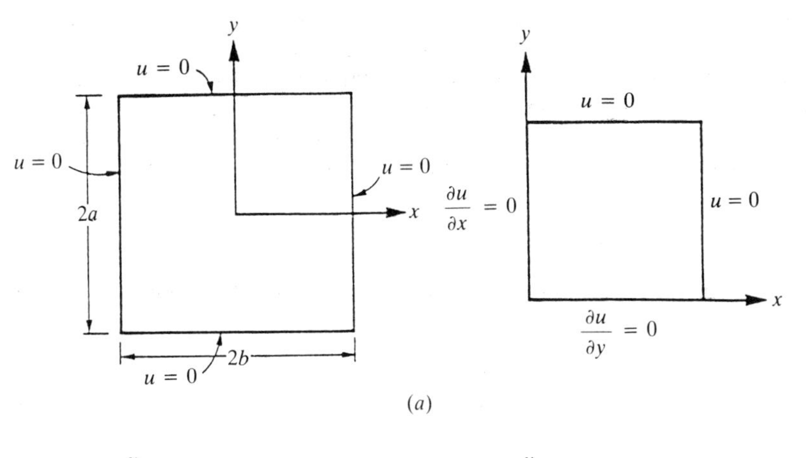

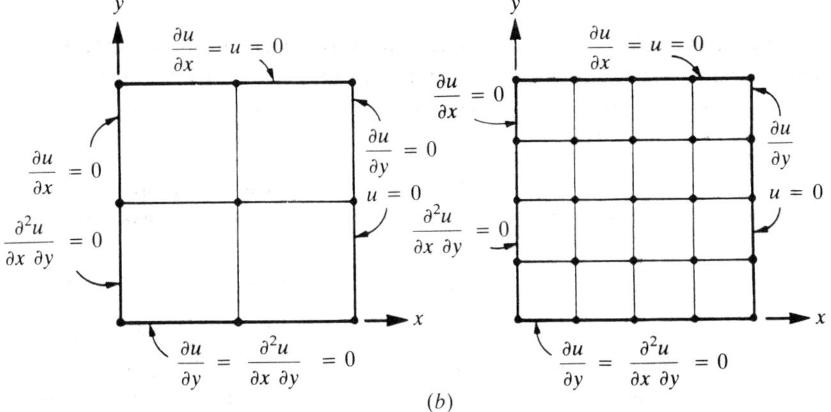

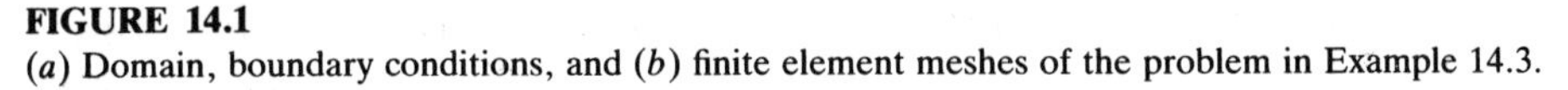

FIGURE 14.1
(*a*) Domain, boundary conditions, and (*b*) finite element meshes of the problem in Example 14.3.

where Ω is a square region (see Fig. 14.1*a*). The exact solution of this problem is given by (8.152*a*). Exploiting the biaxial symmetry, we model only a quadrant, say the region bounded by the positive axes.

Two different uniform meshes (see Fig. 14.1*b*) of Hermite rectangular elements are used to solve the problem, employing various finite element models. Since the element has u, $\partial u/\partial x$, $\partial u/\partial y$, and $\partial^2 u/\partial x\, \partial y$ as nodal degrees of freedom, we must impose all known boundary values of these quantities (which are not readily known a priori). This can be considered as a drawback of the weighted-residual finite element models (especially, the collocation finite-element model) compared with the Ritz finite element model of a second-order equation, where the only boundary conditions are on u and $\partial u/\partial n$. For the problem at hand, we impose the following boundary conditions: $u = 0$ on the lines $x = 1$ and $y = 1$, $\partial u/\partial x = 0$ on the line $x = 0$, and $\partial u/\partial y = 0$ on the line $y = 0$. The boundary conditions for the weighted residual models are indicated in Fig. 14.1(*b*). For the 2×2 mesh, the number of known boundary conditions is 20, whereas the number of total nodal variables is 36. Thus, for the collocation model, we have 16 unknowns and 16 equations, 4 from each element. Similarly, for the 4×4

mesh, we have 36 boundary conditions among 100 nodal variables, requiring 64 collocation equations, which are provided by the 16 elements.

The finite element solutions obtained from the three finite element models for the two meshes are compared with the exact solution (8.152*a*) in Table 14.3. The collocation finite element solution is relatively more accurate than the other two solutions. The numerical convergence of all three models is apparent from the results.

14.2.3 Mixed Formulations

The equation governing bending of beams according to the Euler–Bernoulli beam theory is

$$\frac{d^2}{dx^2}\left(EI\frac{d^2w}{dx^2}\right)-f=0 \tag{14.39}$$

TABLE 14.3
Comparison of the various finite element solutions† with the exact solution of the Dirichlet problem for the Poisson equation (14.38) (Example 14.3)

		$u(x, y)$				$\partial u/\partial x$			
x	y	**Exact**	**Bubnov–Galerkin‡**	**Least-squares**	**Col-location**	**Exact**	**Bubnov–Galerkin‡**	**Least-squares**	**Col-location**
0.0	0.0	0.58936	0.58903 0.58935	0.58902 0.58935	0.58932 0.58937	0.00000	0.00000	0.00000	0.00000
0.25	0.0	0.55776	0.55774	0.55774	0.55776	0.25538	0.25568	0.25568	0.25570
0.50	0.0	0.45868	0.45837 0.45866	0.45846 0.45886	0.45862 0.45868	0.54549	0.54436 0.54541	0.54540 0.54543	0.54548 0.54555
0.75	0.0	0.27945	0.27944	0.27945	0.27946	0.90265	0.90176	0.90184	0.90192
1.00	0.00	0.00000	0.00000	0.00000	0.00000	1.33490	1.34628 1.35040	1.34912 1.35064	1.35068 1.35064
0.25	0.25	0.52830	0.52827	0.52827	0.52829	0.23827	0.23680	0.23859	0.23859
0.50	0.25	0.43560	0.43557	0.43558	0.43559	0.51168	0.51162	0.51162	0.51166
0.75	0.25	0.26665	0.26664	0.26665	0.26665	0.85513	0.85416	0.85432	0.85440
1.00	0.25	0.00000	0.00000	0.00000	0.00000	1.28190	1.29712	1.29744	1.29752
0.50	0.50	0.36230	0.36192 0.36226	0.36197 0.36226	0.36225 0.36228	0.40785	0.40744 0.40789	0.40692 0.40783	0.40724 0.40782
0.75	0.50	0.22548	0.22544	0.22545	0.22547	0.70416	0.70317	0.70314	0.70337
1.00	0.50	0.00000	0.00000	0.00000	0.00000	1.1102	1.11576 1.12448	1.1182 1.1252	1.12268 1.12552
0.75	0.75	0.14564	0.14557	0.14557	0.14563	0.42422	0.42355	0.42309	0.42315

† The first line in each case corresponds to the solution obtained using a 2 × 2 mesh and the second to a 4 × 4 mesh.

‡ The Ritz finite-element solution coincides with the Galerkin solution for the same choice of the Hermite interpolation.

The weak form finite element model of this equation requires Hermite cubic interpolation of the transverse deflection w. On the other hand, a weighted-integral finite element model of (14.39) requires the approximation W^e of w to have a nonzero fourth-order derivative, and the element should include w, dw/dx, $EI\, d^2w/dx^2$, and $(d/dx)(EI\, d^2w/dx^2)$ as the nodal degrees of freedom (i.e., seventh-degree Hermite polynomials). To reduce the differentiability requirements on w in the weak form and include the bending moment (or stress) as a nodal degree of freedom, (14.39) can be decomposed into a pair of lower-order equations:

$$\frac{d^2M}{dx^2} - f = 0, \qquad \frac{d^2w}{dx^2} - \frac{M}{b} = 0 \quad (b \neq 0) \tag{14.40}$$

The assumption that the function $b = EI \neq 0$ always holds in practice because neither the modulus of elasticity nor the moment of inertia is zero; in fact, we always have $b > 0$. We can develop either a weak form finite element model or a weighted-integral finite element model of (14.40). Here we discuss the weak form finite element model.

WEAK FORM. The weak form associated with the pair (14.40) over a typical 1-D element (a two-node line element) is given by

$$\begin{aligned} 0 &= \int_{x_A}^{x_B} u\left(\frac{d^2M}{dx^2} - f\right) dx \\ &= -\int_{x_A}^{x_B} \left(\frac{du}{dx}\frac{dM}{dx} + uf\right) dx - u(x_A)\left(\frac{dM}{dx}\right)\bigg|_{x_A} + u(x_B)\left(\frac{dM}{dx}\right)\bigg|_{x_B} \end{aligned} \tag{14.41a}$$

$$\begin{aligned} 0 &= \int_{x_A}^{x_B} v\left(\frac{d^2w}{dx^2} - \frac{M}{b}\right) dx \\ &= -\int_{x_A}^{x_B} \left(\frac{dv}{dx}\frac{dw}{dx} + \frac{vM}{b}\right) dx - v(x_A)\left(\frac{dw}{dx}\right)\bigg|_{x_A} + v(x_B)\left(\frac{dw}{dx}\right)\bigg|_{x_B} \end{aligned} \tag{14.41b}$$

where u and v are the weight functions, which can be viewed as the variations of w and M, respectively ($u \sim \delta w$ and $v \sim \delta M$). The boundary terms in (14.41) indicate that the specifications of w and M constitute the essential boundary conditions, and the specifications of the derivatives dw/dx and dM/dx constitute natural boundary conditions of the problem:

Essential boundary condition

$$\text{specify } w \text{ and } M \text{ (primary variables)} \tag{14.42a}$$

Natural boundary conditions

$$\text{specify } \frac{dw}{dx} \text{ and } \frac{dM}{dx} \text{ (secondary variables)} \tag{14.42b}$$

Note that the natural boundary condition on the bending moment M in the conventional formulation (i.e., the Euler–Bernoulli beam element) becomes

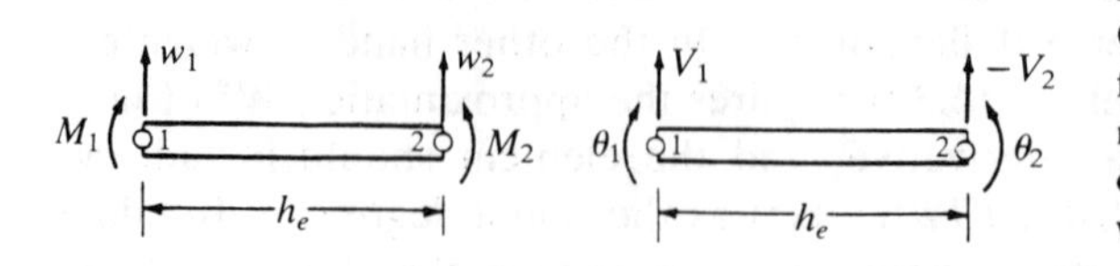

FIGURE 14.2
Generalized displacements and forces for the mixed finite element formulation of the fourth-order equation (14.39) [or the weak form model of (14.40)].

the essential boundary condition in the present formulation. Since both the displacement variable and force variables are used as the nodal degrees of freedom, the finite element model of (14.42) is called a mixed finite element model.

FINITE ELEMENT MODEL. The finite element model of (14.41) is obtained by substituting finite element approximations of the form (see Fig. 14.2 for the nodal variables)

$$w = \sum_{j=1}^{m} w_j \psi_j, \qquad M = \sum_{j=1}^{n} M_j \phi_j \tag{14.43}$$

into the weak forms (14.41). Note that, in general, w and M can be interpolated by different sets of interpolation functions ψ_j and ϕ_j. The finite element equations can be expressed conveniently in the form

$$\begin{bmatrix} [K^{11}] & [K^{12}] \\ [K^{12}]^T & [K^{22}] \end{bmatrix} \begin{Bmatrix} \{w\} \\ \{M\} \end{Bmatrix} = \begin{Bmatrix} \{F^1\} \\ \{F^2\} \end{Bmatrix} \tag{14.44a}$$

where the matrix coefficients $K_{ij}^{\alpha\beta}$ of the matrix $[K^{\alpha\beta}]$ (α, $\beta = 1, 2$) are given by

$$K_{ij}^{11} = 0, \qquad K_{ij}^{12} = \int_{x_A}^{x_B} \frac{d\psi_i}{dx}\frac{d\phi_j}{dx}\,dx = K_{ji}^{21}, \qquad K_{ij}^{22} = \int_{x_A}^{x_B} \frac{1}{b}\,\phi_i\phi_j\,dx$$

$$F_i^1 = \int_{x_A}^{x_B} (-\psi_i f)\,dx - \psi_i(x_A)V_1 - \psi_i(x_B)V_2, \qquad F_i^2 = \phi_i(x_A)\theta_1 + \phi_i(x_B)\theta_2 \tag{14.44b}$$

$$V_1 \equiv \left(\frac{dM}{dx}\right)\bigg|_{x_A}, \quad V_2 \equiv \left(-\frac{dM}{dx}\right)\bigg|_{x_B}, \quad \theta_1 \equiv \left(-\frac{dw}{dx}\right)\bigg|_{x_A}, \quad \theta_2 \equiv \left(\frac{dw}{dx}\right)\bigg|_{x_B}$$

The quantities V_i and θ_i are the secondary variables (shear force and rotation, respectively) associated with w and M at the nodes. It should be pointed out that (14.41) are used in a specific order that gives a symmetric coefficient matrix in (14.44a).

In order to compute the matrix coefficients in (14.44b), we must choose appropriate functions for ψ_i and ϕ_i. For the sake of simplicity, we take $\psi_i = \phi_i$ and ψ_i to be the linear interpolation functions. We obtain ($m = n = 2$), for constant b and f,

$$\begin{aligned} [K^{12}] &= \frac{1}{h_e}\begin{bmatrix} 1 & -1 \\ -1 & 1 \end{bmatrix}, \qquad [K^{22}] = \frac{h_e}{6b_e}\begin{bmatrix} 2 & 1 \\ 1 & 2 \end{bmatrix} \\ \{F^1\} &= \frac{-f_e h_e}{2}\begin{Bmatrix} 1 \\ 1 \end{Bmatrix} - \begin{Bmatrix} V_1 \\ V_2 \end{Bmatrix}, \qquad \{F^2\} = \begin{Bmatrix} \theta_1 \\ \theta_2 \end{Bmatrix} \end{aligned} \tag{14.45}$$

Rearranging the primary variables, the element equations (14.44a) can be expressed as

$$\frac{1}{h_e}\begin{bmatrix} 0 & 1 & 0 & -1 \\ 1 & 2\alpha_e & -1 & \alpha_e \\ 0 & -1 & 0 & 1 \\ -1 & \alpha_e & 1 & 2\alpha_e \end{bmatrix}\begin{Bmatrix} w_1 \\ M_1 \\ w_2 \\ M_2 \end{Bmatrix} = -\frac{f_e h_e}{2}\begin{Bmatrix} 1 \\ 0 \\ 1 \\ 0 \end{Bmatrix} + \begin{Bmatrix} -V_1 \\ \theta_1 \\ -V_2 \\ \theta_2 \end{Bmatrix} \tag{14.46}$$

where $\alpha_e = h_e^2/6b_e$.

The usual assembly procedure, which assumes interelement continuity of the primary variables, would result in a continuous bending moment field. This feature is not desirable when the problem involves specified concentrated (i.e., point) moments. Therefore, in general, the moment degrees of freedom should be either eliminated at the element level or not made continuous during the assembly of element equations (14.46). The following example illustrates this point.

Example 14.4. Consider the clamped, simply supported beam shown in Fig. 14.3. We wish to determine its deflection and bending moment using the Euler–Bernoulli beam theory, i.e., (14.40). We take $b = EI =$ constant and $f = 0$. The boundary conditions are

$$w(0) = 0, \qquad \left(EI\frac{d^2w}{dx^2}\right)\bigg|_{x=h^-}^{x=h^+} = -M_0, \qquad w(2h) = \theta(2h) = 0 \tag{14.47}$$

We wish to analyze the problem using the mixed finite element in (14.44).

For simplicity, we shall use the minimum number of elements that allows us to impose the specified generalized displacements and forces. Thus, we have two elements and three nodes in the mesh. The assembled stiffness matrix is given by

$$\frac{1}{h}\begin{bmatrix} 0 & 1 & 0 & -1 & 0 & 0 \\ 1 & 2\alpha & -1 & \alpha & 0 & 0 \\ 0 & -1 & 0 & 2 & 0 & -1 \\ -1 & \alpha & 2 & 4\alpha & -1 & \alpha \\ 0 & 0 & 0 & -1 & 0 & 1 \\ 0 & 0 & -1 & \alpha & 1 & 2\alpha \end{bmatrix}\begin{Bmatrix} U_1 \\ U_2 \\ U_3 \\ U_4 \\ U_5 \\ U_6 \end{Bmatrix} = \begin{Bmatrix} -V_1^1 \\ \theta_1^1 \\ -V_2^1 - V_1^2 \\ \theta_2^1 + \theta_1^2 \\ -V_2^2 \\ \theta_2^2 \end{Bmatrix} \tag{14.48}$$

The boundary conditions and equilibrium of secondary variables require

$$U_1 = U_2 = U_5 = 0, \qquad -V_2^1 - V_1^2 = 0, \qquad \theta_2^1 + \theta_1^2 = 0, \qquad \theta_2^2 = 0 \tag{14.49}$$

Equations (14.48) become

$$\frac{1}{h}\begin{bmatrix} 0 & 2 & -1 \\ 2 & 4\alpha & \alpha \\ -1 & \alpha & 2\alpha \end{bmatrix}\begin{Bmatrix} U_3 \\ U_4 \\ U_6 \end{Bmatrix} = \begin{Bmatrix} 0 \\ 0 \\ 0 \end{Bmatrix} \tag{14.50}$$

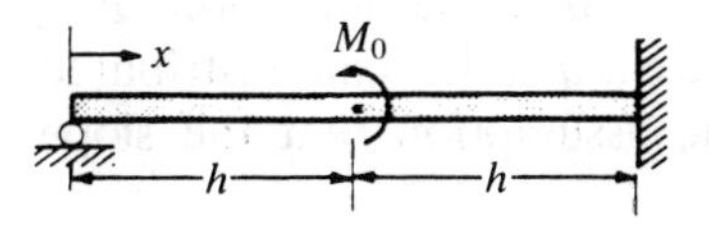

FIGURE 14.3
Clamped, simply supported beam considered in Example 14.4.

If we use the additional boundary condition $U_4 = M_0$, we obtain an inconsistent set of equations. This is the result of the assumption

$$M_2^1 = M_1^2 = U_4 \tag{14.51}$$

used in the assembly. In the present problem, the moment is discontinuous and therefore (14.51) does not hold. The correct condition is

$$M_2^1 - M_1^2 = M_0 \tag{14.52}$$

Suppose that $U_4 = M_1^2 = M_2^1 - M_0$. Then the equations of element 2 become

$$\frac{1}{h}\begin{bmatrix} 0 & 1 & 0 & -1 \\ 1 & 2\alpha & -1 & \alpha \\ 0 & -1 & 0 & 1 \\ -1 & \alpha & 1 & 2\alpha \end{bmatrix} \begin{Bmatrix} w_1^2 \\ M_2^1 - M_0 \\ w_2^2 \\ M_2^2 \end{Bmatrix} = \begin{Bmatrix} -V_1^2 \\ \theta_1^2 \\ -V_2^2 \\ \theta_2^2 \end{Bmatrix} \tag{14.53}$$

The assembled equations for the unknowns, after using the boundary condition (14.52), become

$$\frac{1}{h}\begin{bmatrix} 0 & 2 & -1 \\ 2 & 4\alpha & \alpha \\ -1 & \alpha & 2\alpha \end{bmatrix} \begin{Bmatrix} U_3 \\ M_2^1 \\ U_6 \end{Bmatrix} = \frac{M_0}{h} \begin{Bmatrix} 1 \\ 2\alpha \\ \alpha \end{Bmatrix} \tag{14.54}$$

where the right-hand side comes from the second element when M_1^2 is replaced by $M_1^2 = M_2^1 - M_0$. The solution is

$$U_3 = -\frac{M_0 h^2}{32EI}, \quad M_2^1 = \frac{9M_0}{16}, \quad U_6 = \frac{M_0}{8}, \quad V_1^1 = \frac{9M_0}{16h}, \quad V_2^1 = -\frac{9M_0}{16h} \tag{14.55}$$

Recalling the sign convention for the bending moments in the mixed element (see Fig. 14.2), we note that the solution (14.55) coincides with that given by the conventional model. The reader should verify this statement.

14.3 NONLINEAR PROBLEMS

14.3.1 General Comments

Many engineering problems are described by nonlinear differential equations. Under certain simplifying assumptions, these problems can be described by linear differential equations. For example, the equations governing the large-deflection bending of elastic beams are

$$\begin{aligned} -\frac{d}{dx}\left\{EA\left[\frac{du}{dx} + \frac{1}{2}\left(\frac{dw}{dx}\right)^2\right]\right\} - q = 0 \\ \frac{d^2}{dx^2}\left(EI\frac{d^2w}{dx^2}\right) - \frac{d}{dx}\left\{EA\frac{dw}{dx}\left[\frac{du}{dx} + \frac{1}{2}\left(\frac{dw}{dx}\right)^2\right]\right\} - f = 0 \end{aligned} \tag{14.56}$$

where u is the longitudinal displacement, w is the transverse deflection, E is the modulus of elasticity, A is the cross-sectional area, q is the axial distributed load, and f is the transverse loading. Under the assumption that the slope

dw/dx is small compared with unity [i.e., $(dw/dx)\ (du/dx)$, $(dw/dx)^2 \approx 0$], (14.56) become uncoupled and reduce to (3.1) and (14.39), respectively. However, when the slope dw/dx is not too small, we must solve the coupled set of nonlinear equations (14.56).

Another example of a nonlinear problem is provided by the equation governing the flow of a viscous incompressible fluid. When the convective effects are larger than the viscous effects, (11.1) should be modified to include the convective terms, that is, the Navier–Stokes equation should be solved:

$$\boxed{\begin{aligned} &\rho\left(u\frac{\partial u}{\partial x}+v\frac{\partial u}{\partial y}\right)-\mu\left[2\frac{\partial^2 u}{\partial x^2}+\frac{\partial}{\partial y}\left(\frac{\partial u}{\partial y}+\frac{\partial v}{\partial x}\right)\right]+\frac{\partial P}{\partial x}=f_x \\ &\rho\left(u\frac{\partial v}{\partial x}+v\frac{\partial v}{\partial y}\right)-\mu\left[\frac{\partial}{\partial x}\left(\frac{\partial u}{\partial y}+\frac{\partial v}{\partial x}\right)+2\frac{\partial^2 v}{\partial y^2}\right]+\frac{\partial P}{\partial y}=f_y \\ &\frac{\partial u}{\partial x}+\frac{\partial v}{\partial y}=0 \end{aligned}} \tag{14.57}$$

where ρ is the density of the fluid, and all other symbols have the same meaning as in Chapter 11.

The finite element formulation of nonlinear problems proceeds in much the same way as for linear problems. The main difference lies in the solution of the finite element algebraic equations. Here we describe some details of the formulation, and comment about the solution procedure using equations of large-deflection bending of beams and the Navier–Stokes equations.

14.3.2 Large-Deflection Bending of (Euler–Bernoulli) Beams

Here we use the weak forms of (14.56) to develop the finite element model. These weak forms over an element (x_A, x_B) are

$$0=\int_{x_A}^{x_B}\left\{EA\frac{dv_1}{dx}\left[\frac{du}{dx}+\frac{1}{2}\left(\frac{dw}{dx}\right)^2\right]-v_1 q\right\}dx-Q_1^e v_1(x_A)-Q_4^e v_1(x_B) \tag{14.58a}$$

where v_1 is the weight function ($v_1 \sim \delta u$) and

$$Q_1^e=-\left\{EA\left[\frac{du}{dx}+\frac{1}{2}\left(\frac{dw}{dx}\right)^2\right]\right\}\bigg|_{x_A},\qquad Q_4^e=\left\{EA\left[\frac{du}{dx}+\frac{1}{2}\left(\frac{dw}{dx}\right)^2\right]\right\}\bigg|_{x_B} \tag{14.58b}$$

Similarly,

$$\begin{aligned}0=\int_{x_A}^{x_B}&\left\{EI\frac{d^2v_2}{dx^2}\frac{d^2w}{dx^2}+EA\frac{dv_2}{dx}\frac{dw}{dx}\left[\frac{du}{dx}+\frac{1}{2}\left(\frac{dw}{dx}\right)^2\right]-v_2 f\right\}\\ &-Q_2^e v_2(x_A)-Q_3^e\left(-\frac{dv_2}{dx}\right)\bigg|_{x_A}-Q_5^e v_2(x_B)-Q_6^e\left(-\frac{dv_2}{dx}\right)\bigg|_{x_B}\end{aligned} \tag{14.59a}$$

where v_2 is the weight function ($v_2 \sim \delta w$), and

$$Q_2^e = \left\{\frac{d}{dx}\left(EI\frac{d^2w}{dx^2}\right) - EA\frac{dw}{dx}\left[\frac{du}{dx} + \frac{1}{2}\left(\frac{dw}{dx}\right)^2\right]\right\}\bigg|_{x_A}, \quad Q_3^e = \left(EI\frac{d^2w}{dx^2}\right)\bigg|_{x_A}$$

$$Q_5^e = -\left\{\frac{d}{dx}\left(EI\frac{d^2w}{dx^2}\right) - EA\frac{dw}{dx}\left[\frac{du}{dx} + \frac{1}{2}\left(\frac{dw}{dx}\right)^2\right]\right\}\bigg|_{x_B}, \quad Q_6^e = -\left(EI\frac{d^2w}{dx}\right)\bigg|_{x_B} \tag{14.59b}$$

The primary variables of the formulation (as in the frame element formulation of the Euler–Bernoulli beam theory) are

$$u, \quad w, \quad -\frac{dw}{dx}$$

From the discussions presented in Chapters 3 and 4, it is clear that we must use a Lagrange interpolation of u and a Hermite interpolation of w:

$$u = \sum_{j=1}^{n} u_j\psi_j(x), \qquad w = \sum_{j=1}^{m} s_j\phi_j(x) \tag{14.60}$$

where ψ_j are the Lagrange interpolation functions of degree $n-1$, and ϕ_j are the Hermite interpolation functions of degree $m-1$. For $n=2$ and $m=4$, the elements used for u and w contain the same number of nodes, which is convenient in the computer implementation of the model (see Fig. 14.4).

Substituting the approximations (14.60) for u and w, and $v_1 = \psi_i^e$ and $v_2 = \phi_i^e$, into (14.58a) and (14.59a), we obtain the finite element model

$$\begin{bmatrix} [K^{11}] & [K^{12}] \\ [K^{21}] & [K^{22}] \end{bmatrix} \begin{Bmatrix} \{u\} \\ \{s\} \end{Bmatrix} = \begin{Bmatrix} \{F^1\} \\ \{F^2\} \end{Bmatrix} \tag{14.61a}$$

where

$$K_{ij}^{11} = \int_{x_A}^{x_B} EA\frac{d\psi_i}{dx}\frac{d\psi_j}{dx}\,dx, \quad f_i^1 = \int_{x_A}^{x_B} \psi_i q\,dx + Q_{3i-2} \quad (i, j = 1, 2, \ldots, n)$$

$$K_{ij}^{12} = \int_{x_A}^{x_B} \tfrac{1}{2}EA\frac{dw}{dx}\frac{d\psi_i}{dx}\frac{d\phi_j}{dx}\,dx, \quad K_{ji}^{21} = \int_{x_A}^{x_B} EA\frac{dw}{dx}\frac{d\phi_j}{dx}\frac{d\psi_i}{dx}\,dx$$

$$(i = 1, 2, \ldots, n; j = 1, 2, \ldots, m)$$

$$K_{ij}^{22} = \int_{x_A}^{x_B} EI\frac{d^2\phi_i}{dx^2}\frac{d^2\phi_j}{dx^2}\,dx + \int_{x_A}^{x_B} \tfrac{1}{2}EA\left(\frac{dw}{dx}\right)^2\frac{d\phi_i}{dx}\frac{d\phi_j}{dx}\,dx \tag{14.61b}$$

$$(i, j = 1, 2, \ldots, m)$$

$$F_i^2 = \int_{x_A}^{x_B} \phi_i f\,dx + Q_{i+I}, \quad \begin{cases} I = 1 & \text{if } i = 1 \text{ or } 2 \\ I = 2 & \text{if } i = 3 \text{ or } 4 \end{cases}$$

This completes the finite element model of the nonlinear bending of beams.

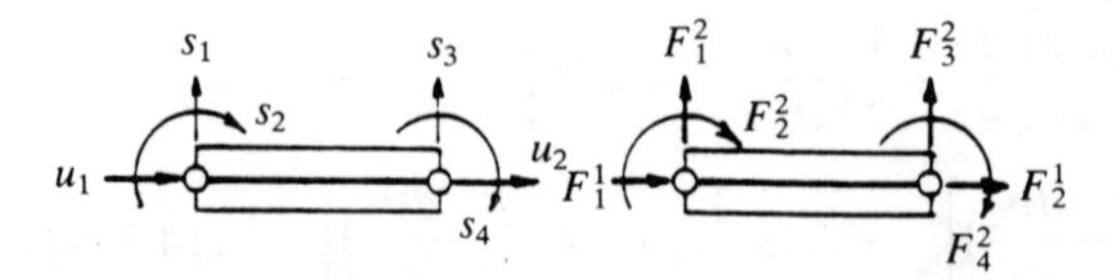

FIGURE 14.4
Nonlinear beam bending element based on linear interpolation of the axial displacement u and Hermite cubic interpolation of the transverse deflection.

14.3.3 Solution Methods for Nonlinear Algebraic Equations

Note that the element stiffness matrix in (14.61*a*) is nonlinear and asymmetric. Therefore, the assembled equations will be nonlinear and asymmetric. The assembled nonlinear equations must be solved, after imposing boundary conditions, using an iterative method, which seeks an approximate solution to the algebraic equations by linearization. The *direct iteration method,* also known as the *Picard method,* is based on the scheme

$$\boxed{[K(\{\Delta\}^r)]\{\Delta\}^{r+1} = \{F\}} \tag{14.62}$$

where $\{\Delta\}^r$ denotes the solution at the rth iteration. Thus, in the direct iteration method, the coefficients K_{ij} (and hence $K_{ij}^{\alpha\beta}$) are evaluated using the solution $\{\Delta\}^r$ from the previous iteration, and the solution at the $(r+1)$th iteration is obtained by solving (14.62):

$$\{\Delta\}^{r+1} = [K(\{\Delta\}^r)]^{-1}\{F\} \tag{14.63}$$

At the beginning of the iteration (i.e., $r = 0$), we assume a solution $\{\Delta\}^0$, based on our qualitative understanding of the solution behavior. For example, $\{\Delta\}^0 = \{0\}$ for large-deflection bending would reduce the nonlinear stiffness matrix to a linear one, and (14.63) would yield the linear solution of the problem at the end of the first iteration, $\{\Delta\}^1$. The iteration is continued [i.e., (14.63) is solved in each iteration] until the difference between $\{\Delta\}^r$ and $\{\Delta\}^{r+1}$ reduces to a preselected error tolerance. The *error criterion* is of the form (other criteria can also be used)

$$\boxed{\left(\frac{\sum_{I=1}^{N} |\Delta_I^{r+1} - \Delta_I^r|^2}{\sum_{I=1}^{N} |\Delta_I^{r+1}|^2}\right)^{1/2} < \epsilon \quad (\text{say}, 10^{-3}).} \tag{14.64}$$

where N is the total number of primary unknowns (i.e., generalized displacements) in the finite element mesh.

The other iterative method is the *Newton–Raphson method,* which is based on the Taylor series expansion of the algebraic equations (14.62) about the known solution $\{\Delta\}^r$. To describe the method, we rewrite (14.62) in the form

$$\{R\} \equiv [K]\{\Delta\} - \{F\} = 0 \tag{14.65}$$

where $\{R\}$ denotes the residual. Expanding $\{R\}$ about $\{\Delta\}^r$, we obtain

$$\{0\} = \{R\} = \{R\}^r + \left(\frac{\partial\{R\}}{\partial\{\Delta\}}\right)^r (\{\Delta\}^{r+1} - \{\Delta\}^r)$$

$$+ \frac{1}{2!}\left(\frac{\partial^2\{R\}}{\partial\{\Delta\}^2}\right)^r (\{\Delta\}^{r+1} - \{\Delta\}^r)^2 + \ldots \tag{14.66a}$$

or

$$0 \approx \{R\}^r + [K^T]^r\{\delta\Delta\} + O(\{\delta\Delta\})^2 \tag{14.66b}$$

where $[K^T]$ is the *tangent (stiffness) matrix*,

$$\boxed{\begin{aligned}&[K^T]^r \equiv \left[\frac{\partial\{R\}}{\partial\{\Delta\}}\right] \quad \text{evaluated at} \quad \{\Delta\} = \{\Delta\}^r\\ &\{\delta\Delta\} = \{\Delta\}^{r+1} - \{\Delta\}^r \quad \text{(incremental solution)}\end{aligned}} \tag{14.66c}$$

For structural problems with variational principles, it can be shown that $[K^T]$ is symmetric even if $[K]$ is not. From (14.66*b*), we have

$$\boxed{\{\delta\Delta\} = -[K^T]^{-1}\{R\}^r = [K^T(\{\Delta\}^r)]^{-1}(\{F\} - [K(\{\Delta\}^r)]\{\Delta\}^r)} \tag{14.67a}$$

and the total solution at the $(r+1)$th iteration is given by

$$\{\Delta\}^{r+1} = \{\Delta\}^r + \{\delta\Delta\} \tag{14.67b}$$

The iteration in (14.66*a*) is continued until the convergence criteria in (14.64) is satisfied or the residual $\{R\}$ [measured in the same way as the solution error in (14.64)] in (14.65) is less than a certain preselected value. For additional details on iterative methods, the References at the end of the chapter.

14.3.4 The 2-D Navier–Stokes Equations

The weak formulation of (14.57) over an element is given by (see Section 11.3 for details)

$$\begin{aligned}&\int_{\Omega^e}\left\{w_1\rho\left(u\frac{\partial u}{\partial x} + v\frac{\partial u}{\partial y}\right) + [\cdots]\right\}dx\,dy - \oint_{\Gamma^e} w_1 t_x\,ds = 0\\ &\int_{\Omega^e}\left\{w_2\rho\left(u\frac{\partial v}{\partial x} + v\frac{\partial v}{\partial y}\right) + [\cdots]\right\}dx\,dy - \oint_{\Gamma^e} w_2 t_y\,ds = 0\\ &\int_{\Omega^e} w_3\left(\frac{\partial u}{\partial x} + \frac{\partial v}{\partial y}\right)dx\,dy = 0\end{aligned} \tag{14.68}$$

where $[\cdots]$ denotes the expressions (omitting the time-derivative terms) in the square brackets of (11.7). The finite element model of these equations is given by

$$\begin{bmatrix}[\bar{K}^{11}] & [K^{12}] & [K^{13}]\\ [K^{12}]^T & [\bar{K}^{22}] & [K^{23}]\\ [K^{13}]^T & [K^{23}]^T & [0]\end{bmatrix}\begin{Bmatrix}\{u\}\\ \{v\}\\ \{P\}\end{Bmatrix} = \begin{Bmatrix}\{F^1\}\\ \{F^2\}\\ \{0\}\end{Bmatrix} \tag{14.69}$$

where $[K^{\alpha\beta}]$ and $\{F^{\alpha}\}$ $(\alpha, \beta = 1, 2, 3)$ are as defined in (11.9*b*), and

$$\bar{K}_{ij}^{11} = K_{ij}^{11} + \rho \int_{\Omega^e} \psi_i\left(\bar{u}\frac{\partial \psi_j}{\partial x} + \bar{v}\frac{\partial \psi_j}{\partial y}\right) dx\, dy$$

$$\bar{K}_{ij}^{22} = K_{ij}^{22} + \rho \int_{\Omega^e} \psi_i\left(\bar{u}\frac{\partial \psi_i}{\partial x} + \bar{v}\frac{\partial \psi_j}{\partial y}\right) dx\, dy \tag{14.70}$$

$$\bar{u} = \sum_{i=1}^{r} \bar{u}_i\psi_i, \qquad \bar{v} = \sum_{i=1}^{r} \bar{v}_i\psi_i$$

where $\bar{u}$ and $\bar{v}$ are velocity components that are assumed to be known.

Clearly, the element coefficient matrix, and hence the global coefficient matrix, are unsymmetric, and they depend on the velocity field, which is not known a priori. Therefore, an iterative solution procedure, such as the direct iteration method, is required. At the beginning of the first iteration, the velocity field can be set equal to zero to obtain the linear solution. In the second iteration, the coefficient matrices are evaluated using the velocity field obtained in the first iteration, and the assembled equations are solved again for the nodal velocities. This procedure is repeated until the velocity field obtained at the end of two consecutive iterations differs by a small preassigned number. The convergence criterion can be expressed as

$$\frac{\left[\sum_{i=1}^{N} \left(|U_i^{(r)} - U_i^{(r+1)}|^2 + |V_i^{(r)} - V_i^{(r+1)}|^2\right)\right]^{1/2}}{\left[\sum_{i=1}^{N} \left(|U_i^{(r+1)}|^2 + |V_i^{(r+1)}|^2\right)\right]^{1/2}} \leqslant \epsilon \tag{14.71}$$

where (U_i^r, V_i^r) denote the velocities at node i at the rth iteration. We note that the tangent matrix for the Navier–Stokes equations is not symmetric (because no variational principle exists for this case). The iterative solution methods discussed in Section 14.3.3 are also applicable here.

14.4 THREE-DIMENSIONAL PROBLEMS

Most of the basic ideas covered in Chapters 3 and 8 for one- and two-dimensional problems can be extended to three-dimensional problems. For the sake of completeness, we discuss here the finite element formulation of the Poisson equation in three dimensions, and describe some of the three-dimensional elements.

Consider the Poisson equation

$$-\frac{\partial}{\partial x}\left(k_1\frac{\partial u}{\partial x}\right) - \frac{\partial}{\partial y}\left(k_2\frac{\partial u}{\partial y}\right) - \frac{\partial}{\partial z}\left(k_3\frac{\partial u}{\partial z}\right) = f \quad \text{in } \Omega \tag{14.72a}$$

$$u = \hat{u} \quad \text{on } \Gamma_1, \qquad k_1\frac{\partial u}{\partial x}n_x + k_2\frac{\partial u}{\partial y}n_y + k_3\frac{\partial u}{\partial z}n_z = \hat{q} \quad \text{on } \Gamma_2 \tag{14.72b}$$

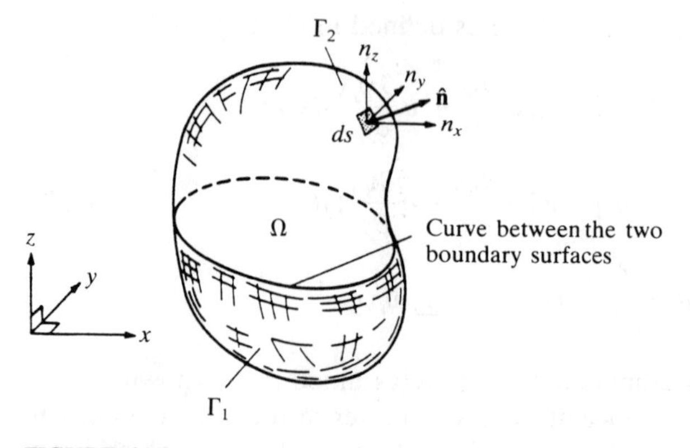

FIGURE 14.5
Three-dimensional domain showing a surface element ds, unit normal and its components $\hat{\mathbf{n}} = (n_x, n_y, n_z)$, and two portions of the boundary Γ.

where $k_i = k_i(x, y, z)$ and $f = f(x, y, z)$ are given functions of position in a three-dimensional domain Ω, and $\hat{u}$ and $\hat{q}$ are specified functions of position on the portions Γ_1 and Γ_2, respectively, of the surface Γ of the domain (see Fig. 14.5). Suppose that the domain Ω is discretized by some three-dimensional elements Ω^e, such as tetrahedral and prism elements (which are the three-dimensional extensions of the triangular and rectangular elements).

The variational formulation of (14.72) over an element Ω^e is given by

$$0 = \int_{\Omega^e} v\left[-\frac{\partial}{\partial x}\left(k_1\frac{\partial u}{\partial x}\right) - \frac{\partial}{\partial y}\left(k_2\frac{\partial u}{\partial y}\right) - \frac{\partial}{\partial z}\left(k_3\frac{\partial u}{\partial z}\right) - f\right] dx\,dy\,dz$$

$$= \int_{\Omega^e}\left[k_1\frac{\partial v}{\partial x}\frac{\partial u}{\partial x} + k_2\frac{\partial v}{\partial y}\frac{\partial u}{\partial y} + k_3\frac{\partial v}{\partial z}\frac{\partial u}{\partial z} - vf\right] dx\,dy\,dz - \oint_{\Gamma^e} vq_n\,ds \qquad (14.73a)$$

where

$$q_n \equiv k_1\frac{\partial u}{\partial x}n_x + k_2\frac{\partial u}{\partial y}n_y + k_3\frac{\partial u}{\partial z}n_z \qquad (14.73b)$$

Clearly, the primary variable is u and the secondary variable is q_n.

Assuming a finite element interpolation of the form

$$u = \sum_{j=1}^{n} u_j\psi_j^e(x, y, z) \qquad (14.74)$$

over the element Ω^e, and substituting $v = \psi_i$ and (14.74) into (14.73a), we obtain

$$[K^e]\{u^e\} = \{f^e\} + \{Q^e\} \tag{14.75a}$$

where

$$K_{ij}^e = \int_{\Omega^e} \left(k_1 \frac{\partial \psi_i}{\partial x}\frac{\partial \psi_j}{\partial x} + k_2 \frac{\partial \psi_i}{\partial y}\frac{\partial \psi_j}{\partial y} + k_3 \frac{\partial \psi_i}{\partial z}\frac{\partial \psi_j}{\partial z} \right) dx\,dy\,dz$$
$$f_i^e = \int_{\Omega^e} f\psi_i \, dx\,dy\,dz, \qquad Q_i^e = \oint_{\Gamma^e} q_n \psi_i \, ds \tag{14.75b}$$

The interpolation functions ψ_i^e have the same general properties as for two-dimensional elements:

$$\sum_{i=1}^{n} \psi_i^e(x, y, z) = 1, \qquad \psi_i^e(x_j, y_j, z_j) = \delta_{ij} \tag{14.76}$$

The assembly of equations, the imposition of boundary conditions, and the solution of the equations are completely analogous to those described in Chapter 8 for the two-dimensional problems. Next, we comment on the geometry of two linear 3-D elements and the element calculations.

The element matrices in (14.75*b*) require the use of interpolation functions that are at least linear in x, y, and z. Here we consider two linear elements: the tetrahedral element and the prism (or brick) element. These elements are described by approximations of the form

$$u(x, y, z) = a_0 + a_1 x + a_2 y + a_3 z \qquad \text{(four-node tetrahedral element)}$$
$$u(x, y, z) = a_0 + a_1 x + a_2 y + a_3 z + a_4 yz + a_5 xz + a_6 xy + a_7 xyz \qquad \text{(eight-node prism element)} \tag{14.77}$$

The interpolation functions can be determined as described in Chapters 8 and 9 for two-dimensional elements.

If the element matrices are to be evaluated numerically, the isoparametric element concept can be used. The geometry of the elements can be described by the transformation equations

$$x = \sum_{i=1}^{n} x_i \hat{\psi}_i(\xi, \eta, \zeta), \qquad y = \sum_{i=1}^{n} y_i \hat{\psi}_i(\xi, \eta, \zeta), \qquad z = \sum_{i=1}^{n} z_i \hat{\psi}_i(\xi, \eta, \zeta) \tag{14.78}$$

Under these transformations, the master tetrahedral and prism elements transform to arbitrary tetrahedral and hexahedral elements, as shown in Fig. 14.6. The definition of the Jacobian matrix and the numerical quadrature rules described in Chapter 9 can easily be extended to the three-dimensional case.

14.5 SUMMARY

Three advanced topics have been discussed briefly in this chapter: (i) weighted-residual and mixed finite element formulations of differential equa-

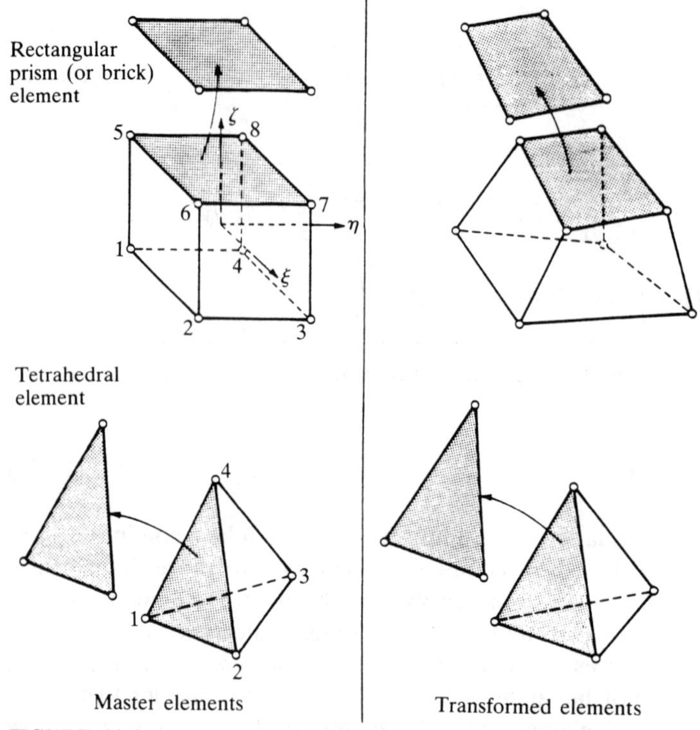

FIGURE 14.6
Linear three-dimensional elements: tetrahedral element and prism element (whose surfaces are two-dimensional triangular and rectangular elements, respectively).

tions; (ii) finite element models of nonlinear equations; and (iii) finite element models of three-dimensional problems. The weighted-residual models discussed include the Petrov–Galerkin model, Bubnov–Galerkin model, collocation model, subdomain model, and least-squares model. First- and second-order differential equations have been considered. Nonlinear finite element models of the Euler–Bernoulli beam theory and the Navier–Stokes equations governing 2-D viscous incompressible flows have been developed. Two iterative schemes, Picard and Newton–Raphson, for solving nonlinear algebraic equations have been discussed. Finally, the finite element formulation of the Poisson equation governing 3-D field problems has been developed. Each of these topics deserves detailed discussion. However, they are not within the scope of the present study. For further study, the reader should consult the books listed in the References.

PROBLEMS

14.1. Consider the second-order equation

$$-\frac{d}{dx}\left(a\frac{du}{dx}\right)=f \tag{i}$$

and rewrite it as a pair of first-order equations,

$$-\frac{du}{dx}+\frac{P}{a}=0, \qquad -\frac{dP}{dx}-f=0 \tag{ii}$$

Construct the weighted-residual finite element model of the equations, and specialize it to the Galerkin model. Assume interpolation in the form

$$u=\sum_{j=1}^{m} u_j\psi_j(x), \qquad P=\sum_{j=1}^{n} P_j\phi_j(x) \tag{iii}$$

and use the equations in (ii) in a sequence that yields symmetric element equations:

$$\begin{bmatrix} [K^{11}] & [K^{12}] \\ [K^{12}]^T & [K^{22}] \end{bmatrix} \begin{Bmatrix} \{u\} \\ \{P\} \end{Bmatrix} = \begin{Bmatrix} \{F^1\} \\ \{F^2\} \end{Bmatrix} \tag{iv}$$

The model can also be called a mixed model because (u, P) are of different kinds.

14.2. Evaluate the coefficient matrices $[K^{\alpha\beta}]$ in Problem 14.1 for $a=\text{constant}$ and column vectors $\{F^\alpha\}$ for $f=\text{constant}$. Assume that $\psi_i=\phi_i$ are the linear interpolation functions. Eliminate $\{P\}$ from the two sets of equations (iv) to obtain an equation of the form

$$[K]\{u\}=\{F\}$$

Compare the coefficient matrix $[K]$ and $\{F\}$ with those obtained with the weak form finite element model of (i). What conclusions can you draw?

14.3. Develop the least-squares finite element model of (ii) in Problem 14.1, and compute element coefficient matrices and vectors when $\psi_i=\phi_i$ are the linear interpolation functions.

14.4. Solve the problem in Example 3.1, Set 3 boundary conditions, using two linear elements of the least-squares model developed in Problem 14.3. Compare the results with the exact solution and those of the weak form finite element model.

14.5. For a cantilevered beam of length L and flexural rigidity EI, subjected to a concentrated moment M_0 at $x=\frac{1}{2}L$, obtain the mixed finite element solution and compare it with the exact solution. Do you see any discrepancy in the solutions?

14.6. Consider the pair of equations

$$\nabla u-\mathbf{q}/k=0, \quad \nabla\cdot\mathbf{q}+f=0 \quad \text{in } \Omega$$

where u and $\mathbf{q}$ are the dependent variables, and k and f are given functions of position (x, y) in a two-dimensional domain Ω. Derive the finite element formulation of the equations in the form

$$\begin{bmatrix} [K^{11}] & [K^{12}] & [K^{13}] \\ & [K^{22}] & [K^{23}] \\ \text{symmetric} & & [K^{33}] \end{bmatrix} \begin{Bmatrix} \{u\} \\ \{q^1\} \\ \{q^2\} \end{Bmatrix} = \begin{Bmatrix} \{F^1\} \\ \{F^2\} \\ \{F^3\} \end{Bmatrix}$$

Caution: Do not eliminate the variable u from the given equations.

14.7. Compute the element coefficient matrices $[K^{\alpha\beta}]$ and vectors $\{F^\alpha\}$ of Problem 14.6 using linear triangular elements for all variables. Assume that k is constant.

14.8. Repeat Problem 14.7 with linear rectangular elements.

14.9. Consider the following equations governing a thin, elastic, isotropic plate:

$$-S(M_1 - \nu M_2) - \frac{\partial^2 w}{\partial x^2} = 0, \qquad -S(M_2 - \nu M_1) - \frac{\partial^2 w}{\partial y^2} = 0$$

$$\frac{2}{(1+\nu)S} \frac{\partial^4 w}{\partial x^2\, \partial y^2} - \frac{\partial^2 M_1}{\partial x^2} - \frac{\partial^2 M_2}{\partial y^2} - f = 0$$

Where M_1 and M_2 are the bending moments, w is the transverse deflection, q is the distributed load, ν is the Poisson ratio, and S is a constant. Give
(*a*) the weak form of the equations, and
(*b*) the finite element model (M_1, M_2, and w are the dependent unknowns) in the form

$$\begin{bmatrix} [K^{11}] & [K^{12}] & [K^{13}] \\ & [K^{22}] & [K^{23}] \\ \text{symmetric} & & [K^{33}] \end{bmatrix} \begin{Bmatrix} \{w\} \\ \{M_1\} \\ \{M_2\} \end{Bmatrix} = \begin{Bmatrix} \{F^1\} \\ \{F^2\} \\ \{F^3\} \end{Bmatrix}$$

14.10. Use the interpolation

$$w = \sum_{j=1}^{4} w_j \psi_j, \qquad M_1 = \sum_{j=1}^{2} m_1^j \phi_j^1, \qquad M_2 = \sum_{j=1}^{2} m_2^j \phi_j^2$$

with

$$\psi_1 = (1-x)(1-y), \qquad \psi_2 = x(1-y), \qquad \psi_3 = xy, \qquad \psi_4 = (1-x)y$$

$$\phi_1^1 = 1 - x, \qquad \phi_2^1 = x, \qquad \phi_1^2 = 1 - y, \qquad \phi_2^2 = y$$

for a rectangular element with sides a and b to evaluate the matrices $[K^{\alpha\beta}]$ ($\alpha, \beta = 1, 2, 3$) in Problem 14.9.

14.11. Repeat Problem 14.10 for the case in which $\phi_i^1 = \phi_i^2 = \psi_i$.

14.12. Evaluate the element matrices in (14.61*b*) by assuming that the nonlinear parts in the element coefficients are element-wise-constant.

14.13. Give the finite element formulation of the following nonlinear equation over an element (x_A, x_B):

$$-\frac{d}{dx}\left(u \frac{du}{dx}\right) + 1 = 0 \quad \text{for } 0 < x < 1$$

$$\left(\frac{du}{dx}\right)\bigg|_{x=0} = 0, \qquad u(1) = \sqrt{2}$$

14.14. Compute the tangent coefficient matrix for the nonlinear problems in Problem 14.13. What restriction(s) should be placed on the initial guess vector?

14.15. Compute the tangent stiffness matrix $[K^T]$ in (14.66*c*) for the Euler–Bernoulli beam element in (14.61*b*).

14.16. Develop the nonlinear finite element model of the Timoshenko beam theory. Equations (14.56) are valid for this case, with the following changes. In place of $(d^2/dx^2)(b\, d^2w/dx^2)$ use $-(d/dx)(b\, d\Psi/dx) + GAk(dw/dx + \Psi)$ and add the following additional equation for w:

$$-\frac{d}{dx}\left[GAk\left(\frac{dw}{dx} + \Psi\right)\right] = f$$

See Section 4.4 for additional details.

14.17. Compute the tangent stiffness matrix for the Timoshenko beam element in Problem 14.16.

14.18. (*Natural convection in flow between heated vertical plates*) Consider the flow of a viscous incompressible fluid in the presence of a temperature gradient between two stationary long vertical plates. Assuming zero pressure gradient between the plates, we can write $u = u(y)$, $v = 0$, $T = T(y)$, and

$$0 = \rho\beta g(T - T_m) + \mu\frac{d^2u}{dy^2}, \qquad 0 = k\frac{d^2T}{dy^2} + \mu\left(\frac{du}{dy}\right)^2$$

where $T_m = \frac{1}{2}(T_0 + T_1)$ is the mean temperature of the two plates, g the gravitational acceleration, ρ the density, β the coefficient of thermal expansion, μ the viscosity, and k the thermal conductivity of the fluid. Give a finite element formulation of the equations and discuss the solution strategy for the computational scheme.

14.19. The equations given in Problem 14.18 are a special case of the more general Navier–Stokes equations (which are nonlinear and coupled) for two-dimensional flow:

(*a*) *Mass continuity*

$$\frac{\partial u}{\partial x} + \frac{\partial v}{\partial y} = 0$$

(*b*) *Conservation of momentum*

$$\rho\left(u\frac{\partial u}{\partial x} + v\frac{\partial u}{\partial y}\right) = -\frac{\partial P}{\partial x} + 2\mu\frac{\partial^2 u}{\partial x^2} + \mu\frac{\partial}{\partial y}\left(\frac{\partial u}{\partial y} + \frac{\partial v}{\partial x}\right) + \rho g\beta(T - T_m)$$

$$\rho\left(u\frac{\partial v}{\partial x} + v\frac{\partial v}{\partial y}\right) = -\frac{\partial P}{\partial y} + \mu\frac{\partial}{\partial x}\left(\frac{\partial u}{\partial y} + \frac{\partial v}{\partial x}\right) + 2\mu\frac{\partial^2 v}{\partial y^2}$$

(*c*) Energy equation

$$\rho c\left(u\frac{\partial T}{\partial x} + v\frac{\partial T}{\partial y}\right) = k\left(\frac{\partial^2 T}{\partial x^2} + \frac{\partial^2 T}{\partial y^2}\right)$$

where c is the specific heat (the x coordinate is taken vertically downward). Construct the finite element model of the equations, and discuss the computational strategy through a flow chart.

14.20. Derive the interpolation functions ψ_1, ψ_5, and ψ_8 for the eight-node prism element using the alternative procedure described in Section 8.2.

14.21. Give the finite element formulation of the heat conduction equation in three dimensions with convective boundary condition. Use a rectangular cartesian system.

14.22. Give the penalty finite element model of the three-dimensional flow equations in cartesian coordinates.

14.23. Repeat Problem 14.21 in cylindrical coordinates.

14.24. Repeat Problem 14.22 in cylindrical coordinates.

14.25. Evaluate the source vector components f_i^e and coefficients K_{ij}^e over a master prism element when f is a constant, f_0, and $k_1 = k_2 = k_3 =$ constant in (14.75*b*).

REFERENCES FOR ADDITIONAL READING

Baker, A. J.: *Finite Element Computational Fluid Mechanics,* McGraw-Hill, New York/Hemisphere, Washington, DC, 1983.

Bathe, K. J.: *Finite Elements in Engineering Analysis,* Prentice-Hall, Englewood Cliffs, NJ, 1982.

Carey, G. F., and J. T. Oden: *Finite Elements: A Second Course,* Prentice-Hall, Englewood Cliffs, NJ, 1983*a*.

——— and ———: *Finite Elements: Computational Aspects,* Prentice-Hall, Englewood Cliffs, NJ, 1983*b*.

Chandrupatla, T. R., and A. D. Belegundu: *Introduction to Finite Elements in Engineering*, Prentice-Hall, Englewood Cliffs, NJ, 1991.

Cook, R. D., S. D. Malkus, and M. E. Plesha: *Concepts and Applications of Finite Analysis,* 3d ed., John Wiley, New York, 1989.

Desai, C. S., and J. F. Abel: *Introduction to the Finite Element Method,* Van Nostrand-Reinhold, New York, 1972.

Gallagher, R. H.: *Finite Element Analysis: Fundamentals,* Prentice-Hall, Englewood Cliffs, NJ, 1975.

Huebner, K. H., and E. A. Thornton: *The Finite Element Method for Engineers,* 2d ed., John Wiley, New York, 1982.

Hughes, T. J. R.: *The Finite Element Method,* Prentice-Hall, Englewood Cliffs, NJ, 1986.

Irons, B., and S. Ahmad: *Techniques of Finite Elements,* Ellis Horwood, Chichester, Sussex/Halsted Press, New York, 1980.

———: "Quadrature Rules for Brick-Based Finite Elements," *International Journal for Numerical Methods in Engineering,* vol. 3, pp. 293–294, 1971.

Oden, J. T.: *Finite Elements of Nonlinear Continua,* McGraw-Hill, New York, 1972.

——— and G. F. Carey: *Finite Elements: Special Problems in Solid Mechanics,* vol. V, Prentice-Hall, Englewood Cliffs, NJ, 1984.

Rao, S. S.: *The Finite Element Method in Engineering,* Pergamon Press, New York, 1982.

Reddy, J. N.: *Energy and Variational Methods in Applied Mechanics,* John Wiley, New York, 1984.

———: *Applied Functional Analysis and Variational Methods in Engineering,* McGraw-Hill, New York, 1986.

Zienkiewicz, O. C., and R. L. Taylor: *The Finite Element Method,* vols. 1 and 2, McGraw-Hill, New York, 1989 and 1991.

APPENDICES

APPENDIX 1

COMPUTER PROGRAM FEM1DV2

```
C       Program Name: FEM1DV2                            Length: 2194 lines F1D00010
C       * * * * * * * * * * * * * * * * * * * * * * * * * * * * * * * * * * *F1D00020
C       *                       Program  FEM1DV2                            *F1D00030
C       *     (AN IN-CORE FINITE ELEMENT ANALYSIS COMPUTER PROGRAM)         *F1D00040
C       * * * * * * * * * * * * * * * * * * * * * * * * * * * * * * * * * * *F1D00050
C        ___________________________________________________________________ F1D00060
C       |                                                                   |F1D00070
C       |   This is a finite element computer program for the analysis      |F1D00080
C       |   of following three model equations and others:                  |F1D00090
C       |                                                                   |F1D00100
C       |   1. Heat transfer, fluid mechanics, bars, and cables:            |F1D00110
C       |                                                                   |F1D00120
C       |             CT.u* + CT.u** - (AX.u')' + CX.u = FX                 |F1D00130
C       |                                                                   |F1D00140
C       |   2. The Timoshenko beam and circular plate theory:               |F1D00150
C       |                                                                   |F1D00160
C       |              CT0.w** - [AX.(w' + s)]' + CX.w = FX                 |F1D00170
C       |              CT1.s** - (BX.s')' + AX.(w' + s) = 0                 |F1D00180
C       |                                                                   |F1D00190
C       |   3. The Euler-Bernoulli beam and circular plate theory:          |F1D00200
C       |                                                                   |F1D00210
C       |                CT.w** + (BX.w'')'' + CX.w = FX                    |F1D00220
C       |                                                                   |F1D00230
C       |  In the above equations  (')  and (*) denote  differentiations    |F1D00240
C       |  with respect to space x and time t,  and  AX, BX, CX, CT, and    |F1D00250
C       |  FX are functions of x only:                                      |F1D00260
C       |                                                                   |F1D00270
C       |    AX = AX0 + AX1.X,  BX = BX0 + BX1.X,  CX = CX0 + CX1.X         |F1D00280
C       |         CT = CT0 + CT1.X,  FX = FX0 + FX1.X + FX2.X.X             |F1D00290
C       |                                                                   |F1D00300
C       |    In addition to the three model equations, other equations      |F1D00310
C       |  (for example, disks, trusses, and frames) can be analyzed by     |F1D00320
C       |  the program.                                                     |F1D00330
C       |___________________________________________________________________|F1D00340
```

```
C                                                                        F1D00350
C     ___________________________________________________________________F1D00360
C     .                                                                 .F1D00370
C     .              KEY VARIABLES USED IN THE PROGRAM                  .F1D00380
C     .      (see Table 7.4 for a description of other variables)       .F1D00390
C     .                                                                 .F1D00400
C     . NDF....... Number of degrees of freedom per node                .F1D00410
C     . NEQ....... Number of equations in the model (before B. C.)      .F1D00420
C     . NGP....... Number of Gauss points used in the evaluation of     .F1D00430
C     .            the element coeffcients, [ELK], {ELF}, [ELM]         .F1D00440
C     . NHBW...... Half bandwidth of global coefficient matrix [GLK]    .F1D00450
C     . NN ....... Number of total degrees of freedom in the element    .F1D00460
C     . NPE....... Number of nodes per element                          .F1D00470
C     ___________________________________________________________________F1D00480
C     .                                                                 .F1D00490
C     .         DIMENSIONS OF VARIOUS ARRAYS IN THE PROGRAM             .F1D00500
C     .                                                                 .F1D00510
C     . Values of MXELM,MXNOD, etc. in the PARAMETER statement should   .F1D00520
C     .      be changed to meet the requirements of problem:            .F1D00530
C     .                                                                 .F1D00540
C     . MXELM..... Maximum number of elements in the mesh:              .F1D00550
C     . MXEBC..... Maximum number of speci. primary deg. of freedom     .F1D00560
C     . MXMBC..... Maximum number of speci. mixed boundary conditions   .F1D00570
C     . MXNBC..... Maximum number of speci. secondary deg. of freedom   .F1D00580
C     . MXNEQ..... Maximum number of equations in the FE model          .F1D00590
C     . MXNOD..... Maximum number of nodes in the mesh                  .F1D00600
C     .                                                                 .F1D00610
C     . NOTE:  The following dimension statement in subroutine JACOBI   .F1D00620
C     .        should be modified when MXNEQ is greater than 50:        .F1D00630
C     .          DIMENSION  V(50,50),VT(50,50),W(50,50),IH(50)          .F1D00640
C     .        The value of MXNEQ should be used in place of '50'       .F1D00650
C     ___________________________________________________________________F1D00660
C     .                                                                 .F1D00670
C     .               SUBROUTINES USED IN THE PROGRAM                   .F1D00680
C     .                                                                 .F1D00690
C     .  ASSMBL, AXLBX, BONDRY COEFNT, ECHO, JACOBI, MESH1D, MTRXML,     .F1D00700
C     .    PSTPRC, REACTN, SHP1D, SOLVER, TMSFRC, TMSTRS, TRSFRM        .F1D00710
C                                                                        F1D00720
C     ___________________________________________________________________F1D00730
      IMPLICIT REAL*8(A-H,O-Z)                                          F1D00740
      PARAMETER  (MXELM=20,MXNEQ=50,MXEBC=10,MXNBC=10,MXMBC=10,MXNOD=25)F1D00750
      DIMENSION  DCAX(MXELM,2),DCBX(MXELM,2),DCCX(MXELM,2),DCFX(MXELM,3)F1D00760
      DIMENSION  GU0(MXNEQ),GU1(MXNEQ),GU2(MXNEQ),GPU(MXNEQ),DX(MXNOD)  F1D00770
      DIMENSION  IBDY(MXEBC),ISPV(MXEBC,2),ISSV(MXNBC,2),INBC(MXMBC,2)  F1D00780
      DIMENSION  GLM(MXNEQ,MXNEQ),GLF(MXNEQ),GLX(MXNOD),NOD(MXELM,4)    F1D00790
      DIMENSION  CS(MXELM),SN(MXELM),CNT(MXELM),SNT(MXELM),XB(MXELM)    F1D00800
      DIMENSION  EGNVAL(MXNEQ),EGNVEC(MXNEQ,MXNEQ),GLK(MXNEQ,MXNEQ)     F1D00810
      DIMENSION  PR(MXELM),SE(MXELM),SL(MXELM),SA(MXELM),SI(MXELM)      F1D00820
      DIMENSION  HF(MXELM),VF(MXELM),PF(MXELM),F3(MXELM),TITLE(20)      F1D00830
      DIMENSION  UREF(MXMBC),VSPV(MXEBC),VSSV(MXNBC),VNBC(MXMBC)        F1D00840
      COMMON/STF1/ELK(9,9),ELM(9,9),ELF(9),ELX(4),ELU(9),ELV(9),ELA(9)  F1D00850
      COMMON/STF2/A1,A2,A3,A4,A5,AX0,AX1,BX0,BX1,CX0,CX1,CT0,CT1,FX0,   F1D00860
     *            FX1,FX2                                               F1D00870
      COMMON/IO/IN,IT                                                   F1D00880
C                                                                       F1D00890
C              _____________________________________________            F1D00900
C             |      P R E P R O C E S S O R   U N I T      |           F1D00910
C             |_____________________________________________|           F1D00920
C                                                                       F1D00930
      IN=5                                                              F1D00940
      IT=6                                                              F1D00950
      NT=0                                                              F1D00960
      NSSV=0                                                            F1D00970
      JVEC=1                                                            F1D00980
      TIME=0.0D0                                                        F1D00990
      TOLRNS=1.0D-06                                                    F1D01000
      CALL ECHO(IN,IT)                                                  F1D01010
C                                                                       F1D01020
      READ(IN,300) TITLE                                                F1D01030
      READ(IN,*) MODEL,NTYPE,ITEM                                       F1D01040
      READ(IN,*) IELEM,NEM                                              F1D01050
      READ(IN,*) ICONT,NPRNT                                            F1D01060
C                                                                       F1D01070
      IF(MODEL.GE.3)THEN                                                F1D01080
         NPE=2                                                          F1D01090
```

```
         IF(MODEL.EQ.4 .AND. NTYPE.GE.1)THEN                              F1D01100
             NDF=3                                                        F1D01110
         ELSE                                                             F1D01120
             NDF=2                                                        F1D01130
         ENDIF                                                            F1D01140
         IF(MODEL.EQ.4 .AND. NTYPE.EQ.2)THEN                              F1D01150
             IELEM=1                                                      F1D01160
         ELSE                                                             F1D01170
             IELEM=0                                                      F1D01180
         ENDIF                                                            F1D01190
      ELSE                                                                F1D01200
         IF(MODEL.EQ.2)THEN                                               F1D01210
             NDF=2                                                        F1D01220
             IF(NTYPE.GT.1)IELEM=1                                        F1D01230
         ELSE                                                             F1D01240
             NDF=1                                                        F1D01250
         ENDIF                                                            F1D01260
         NPE=IELEM+1                                                      F1D01270
      ENDIF                                                               F1D01280
C                                                                         F1D01290
      NNM = NEM*(NPE-1)+1                                                 F1D01300
      NN  = NPE*NDF                                                       F1D01310
      NEM1=NEM + 1                                                        F1D01320
      IF(MODEL.NE.4)THEN                                                  F1D01330
C                                                                         F1D01340
C     Data input for BAR-LIKE and BEAM problems (MODEL=1,2, AND 3)        F1D01350
C                                                                         F1D01360
          IF(ICONT.NE.0)THEN                                              F1D01370
              READ(IN,*) (DX(I), I=1,NEM1)                                F1D01380
              CALL MESH1D(NEM,NNM,NPE,NOD,MXELM,MXNOD,DX,GLX)             F1D01390
              READ(IN,*) AX0,AX1                                          F1D01400
              READ(IN,*) BX0,BX1                                          F1D01410
              READ(IN,*) CX0,CX1                                          F1D01420
              IF(ITEM.NE.3)THEN                                           F1D01430
                  READ(IN,*) FX0,FX1,FX2                                  F1D01440
              ENDIF                                                       F1D01450
          ELSE                                                            F1D01460
C                                                                         F1D01470
C     Read GLX, NOD, and element-wise continuous coefficients [DC.X]      F1D01480
C                                                                         F1D01490
              DO 10 N=1,NEM                                               F1D01500
              II=(N-1)*(NPE-1)+1                                          F1D01510
              JJ=N*(NPE-1)+1                                              F1D01520
              READ(IN,*) (GLX(K),K=II,JJ)                                 F1D01530
              READ(IN,*) (NOD(N,I),I=1,NPE)                               F1D01540
              READ(IN,*) (DCAX(N,I),I=1,2)                                F1D01550
              READ(IN,*) (DCBX(N,I),I=1,2)                                F1D01560
              READ(IN,*) (DCCX(N,I),I=1,2)                                F1D01570
   10         READ(IN,*) (DCFX(N,I),I=1,3)                                F1D01580
          ENDIF                                                           F1D01590
C                                                                         F1D01600
      ELSE                                                                F1D01610
C                                                                         F1D01620
C     Input data for plane TRUSS or FRAME structures (MODEL=4)            F1D01630
C                                                                         F1D01640
          READ(IN,*)NNM                                                   F1D01650
          IF(NTYPE.NE.0)THEN                                              F1D01660
              DO 20 N=1,NEM                                               F1D01670
              READ(IN,*) PR(N),SE(N),SL(N),SA(N),SI(N),CS(N),SN(N)        F1D01680
              READ(IN,*) HF(N),VF(N),PF(N),XB(N),CNT(N),SNT(N)            F1D01690
   20         READ(IN,*) (NOD(N,I),I=1,2)                                 F1D01700
          ELSE                                                            F1D01710
              DO 30 N=1,NEM                                               F1D01720
              READ(IN,*) SE(N),SL(N),SA(N),CS(N),SN(N),HF(N)              F1D01730
   30         READ(IN,*) (NOD(N,I),I=1,2)                                 F1D01740
          ENDIF                                                           F1D01750
      ENDIF                                                               F1D01760
      NEQ=NNM*NDF                                                         F1D01770
C                                                                         F1D01780
C     Read data on BOUNDARY CONDITIONS of three kinds: Dirichlet (PV)     F1D01790
C             Neumann (SV), and Newton's (MIXED) types                    F1D01800
C                                                                         F1D01810
      READ(IN,*) NSPV                                                     F1D01820
      IF(NSPV.NE.0)THEN                                                   F1D01830
          DO 40 NB=1,NSPV                                                 F1D01840
```

```
         IF(ITEM.GT.2)THEN                                          F1D01850
             READ(IN,*) (ISPV(NB,J),J=1,2)                          F1D01860
         ELSE                                                       F1D01870
             READ(IN,*) (ISPV(NB,J),J=1,2),VSPV(NB)                 F1D01880
         ENDIF                                                      F1D01890
   40    CONTINUE                                                   F1D01900
      ENDIF                                                         F1D01910
C                                                                   F1D01920
      IF(ITEM.LE.2)THEN                                             F1D01930
         READ(IN,*) NSSV                                            F1D01940
         IF(NSSV.NE.0)THEN                                          F1D01950
             DO 50 IB=1,NSSV                                        F1D01960
   50        READ(IN,*) (ISSV(IB,J),J=1,2),VSSV(IB)                 F1D01970
         ENDIF                                                      F1D01980
      ENDIF                                                         F1D01990
C                                                                   F1D02000
      READ(IN,*) NNBC                                               F1D02010
      IF(NNBC.NE.0)THEN                                             F1D02020
         DO 60 I=1, NNBC                                            F1D02030
   60    READ(IN,*) (INBC(I,J),J=1,2),VNBC(I),UREF(I)               F1D02040
      ENDIF                                                         F1D02050
C                                                                   F1D02060
      IF(ITEM .NE. 0)THEN                                           F1D02070
C                                                                   F1D02080
C     Input data here for TIME-DEPENDENT problems                   F1D02090
C                                                                   F1D02100
          READ(IN,*) CT0,CT1                                        F1D02110
          IF(ITEM.LE.2)THEN                                         F1D02120
              READ(IN,*) DT,ALFA,GAMA                               F1D02130
              READ(IN,*) INCOND,NTIME,INTVL                         F1D02140
              A1=ALFA*DT                                            F1D02150
              A2=(1.0-ALFA)*DT                                      F1D02160
              IF(INCOND.NE.0)THEN                                   F1D02170
                  READ(IN,*) (GU0(I),I=1,NEQ)                       F1D02180
              ELSE                                                  F1D02190
                  DO 70 I=1,NEQ                                     F1D02200
   70             GU0(I)=0.0                                        F1D02210
              ENDIF                                                 F1D02220
              IF(ITEM.EQ.2)THEN                                     F1D02230
                  A3=2.0/GAMA/(DT*DT)                               F1D02240
                  A4=A3*DT                                          F1D02250
                  A5=1.0/GAMA-1.0                                   F1D02260
                  IF(INCOND.NE.0)THEN                               F1D02270
                      READ(IN,*) (GU1(I),I=1,NEQ)                   F1D02280
                  ELSE                                              F1D02290
                      DO 80 I=1,NEQ                                 F1D02300
                      GU1(I)=0.0                                    F1D02310
   80                 GU2(I)=0.0                                    F1D02320
                  ENDIF                                             F1D02330
              ENDIF                                                 F1D02340
          ENDIF                                                     F1D02350
       ENDIF                                                        F1D02360
C                                                                   F1D02370
C     ----------------------------------------------------------    F1D02380
C      E N D    O F    T H E    I N P U T    D A T A               F1D02390
C     ----------------------------------------------------------    F1D02400
C     Compute the half BANDWIDTH of the coefficient matrix GLK      F1D02410
C                                                                   F1D02420
      NHBW=0.0                                                      F1D02430
      DO 90 N=1,NEM                                                 F1D02440
      DO 90 I=1,NPE                                                 F1D02450
      DO 90 J=1,NPE                                                 F1D02460
      NW=(IABS(NOD(N,I)-NOD(N,J))+1)*NDF                            F1D02470
   90 IF(NHBW.LT.NW) NHBW=NW                                        F1D02480
C                                                                   F1D02490
C     ----------------------------------------------------------    F1D02500
C               P R I N T    T H E    I N P U T    D A T A          F1D02510
C     ----------------------------------------------------------    F1D02520
C                                                                   F1D02530
      WRITE(IT,530)                                                 F1D02540
      WRITE(IT,310)                                                 F1D02550
      WRITE(IT,530)                                                 F1D02560
      WRITE(IT,300) TITLE                                           F1D02570
      WRITE(IT,320) MODEL,NTYPE                                     F1D02580
      WRITE(IT,350) IELEM,NDF,NEM,NEQ,NSPV,NSSV,NNBC                F1D02590
```

```
C                                                                      F1D02600
      IF(ITEM.NE.0)THEN                                                F1D02610
          IF(ITEM.LE.2)THEN                                            F1D02620
              WRITE(IT,330)                                            F1D02630
              WRITE(IT,390) CT0,CT1,ALFA,GAMA,DT,NTIME,INTVL           F1D02640
              IF(INCOND.NE.0)THEN                                      F1D02650
                  WRITE(IT,370)                                        F1D02660
                  WRITE(IT,540) (GU0(I),I=1,NEQ)                       F1D02670
                  IF(ITEM.EQ.2)THEN                                    F1D02680
                      WRITE(IT,380)                                    F1D02690
                      WRITE(IT,540) (GU1(I),I=1,NEQ)                   F1D02700
                  ENDIF                                                F1D02710
              ENDIF                                                    F1D02720
          ELSE                                                         F1D02730
              WRITE(IT,340)                                            F1D02740
              WRITE(IT,400) CT0,CT1                                    F1D02750
          ENDIF                                                        F1D02760
      ENDIF                                                            F1D02770
C                                                                      F1D02780
      IF(NSPV.NE.0)THEN                                                F1D02790
          WRITE(IT,480)                                                F1D02800
          DO 100 IB=1,NSPV                                             F1D02810
          IF(ITEM.LE.2)THEN                                            F1D02820
              WRITE(IT,490) (ISPV(IB,J),J=1,2),VSPV(IB)                F1D02830
          ELSE                                                         F1D02840
              WRITE(IT,490) (ISPV(IB,J),J=1,2)                         F1D02850
          ENDIF                                                        F1D02860
  100     CONTINUE                                                     F1D02870
      ENDIF                                                            F1D02880
C                                                                      F1D02890
      IF(NSSV.NE.0)THEN                                                F1D02900
          WRITE(IT,500)                                                F1D02910
          DO 110 IB=1,NSSV                                             F1D02920
  110     WRITE(IT,490) (ISSV(IB,J),J=1,2),VSSV(IB)                    F1D02930
      ENDIF                                                            F1D02940
C                                                                      F1D02950
      IF(NNBC.NE.0)THEN                                                F1D02960
          WRITE(IT,510)                                                F1D02970
          DO 120 I=1,NNBC                                              F1D02980
  120     WRITE(IT,490) (INBC(I,J),J=1,2),VNBC(I),UREF(I)              F1D02990
      ENDIF                                                            F1D03000
C                                                                      F1D03010
      IF(MODEL.NE.4)THEN                                               F1D03020
          IF(ICONT.EQ.1)THEN                                           F1D03030
              WRITE(IT,410)                                            F1D03040
              WRITE(IT,540) (GLX(I),I=1,NNM)                           F1D03050
              WRITE(IT,420)                                            F1D03060
              IF(MODEL.NE.3)THEN                                       F1D03070
                  WRITE(IT,440) AX0,AX1,BX0,BX1,CX0,CX1,FX0,FX1,FX2    F1D03080
              ELSE                                                     F1D03090
                  WRITE(IT,445) AX0,AX1,BX0,BX1,CX0,CX1                F1D03100
              ENDIF                                                    F1D03110
          ELSE                                                         F1D03120
              DO 130 N=1,NEM                                           F1D03130
              II=(N-1)*(NPE-1)+1                                       F1D03140
              JJ=N*(NPE-1)+1                                           F1D03150
              WRITE(IT,430) N,GLX(II),GLX(JJ)                          F1D03160
  130         WRITE(IT,440) (DCAX(N,I),I=1,2),(DCBX(N,I),I=1,2),       F1D03170
     *                      (DCCX(N,I),I=1,2),(DCFX(N,I),I=1,3)        F1D03180
          ENDIF                                                        F1D03190
      ELSE                                                             F1D03200
          DO 140 N=1,NEM                                               F1D03210
          WRITE(IT,460) N                                              F1D03220
          IF(NTYPE.NE.0)THEN                                           F1D03230
              WRITE(IT,450) PR(N),SE(N),SL(N),SA(N),SI(N),CS(N),SN(N), F1D03240
     *                      HF(N),VF(N),PF(N),XB(N),CNT(N),SNT(N),     F1D03250
     *                      (NOD(N,I),I=1,2)                           F1D03260
          ELSE                                                         F1D03270
              WRITE(IT,470) SE(N),SL(N),SA(N),CS(N),SN(N),HF(N),       F1D03280
     *                      (NOD(N,I),I=1,2)                           F1D03290
          ENDIF                                                        F1D03300
  140     CONTINUE                                                     F1D03310
      ENDIF                                                            F1D03320
C                                                                      F1D03330
```

```
C             |-----------------------------------------------------|            F1D03340
C             |           P R O C E S S O R   U N I T               |            F1D03350
C             |_____________________________________________________|            F1D03360
C                                                                                F1D03370
C     TIME MARCHING scheme begins here. For  ITEM=2, initial conditions F1D03380
C     on second derivatives of the solution are computed in the program F1D03390
C                                                                         F1D03400
      IF(ITEM.NE.0)THEN                                                   F1D03410
          IF(ITEM.EQ.1)THEN                                               F1D03420
              NT=NT+1                                                     F1D03430
              TIME=TIME+DT                                                F1D03440
          ENDIF                                                           F1D03450
      ENDIF                                                               F1D03460
C                                                                         F1D03470
      IF(ITEM.GT.2)NHBW=NEQ                                               F1D03480
C                                                                         F1D03490
C     Initialize global matrices and vectors                              F1D03500
C                                                                         F1D03510
  150 DO 160 I=1,NEQ                                                      F1D03520
      GLF(I)=0.0                                                          F1D03530
      DO 160 J=1,NHBW                                                     F1D03540
      IF(ITEM.GT.2)THEN                                                   F1D03550
          GLM(I,J)=0.0                                                    F1D03560
      ENDIF                                                               F1D03570
  160 GLK(I,J)=0.0                                                        F1D03580
C                                                                         F1D03590
C     Do-loop for ELEMENT CALCULATIONS and ASSEMBLY                       F1D03600
C                                                                         F1D03610
      DO 200 NE = 1, NEM                                                  F1D03620
      IF(MODEL.NE.4)THEN                                                  F1D03630
          IF(ICONT.NE.1) THEN                                             F1D03640
              AX0=DCAX(NE,1)                                              F1D03650
              AX1=DCAX(NE,2)                                              F1D03660
              BX0=DCBX(NE,1)                                              F1D03670
              BX1=DCBX(NE,2)                                              F1D03680
              CX0=DCCX(NE,1)                                              F1D03690
              CX1=DCCX(NE,2)                                              F1D03700
              FX0=DCFX(NE,1)                                              F1D03710
              FX1=DCFX(NE,2)                                              F1D03720
              FX2=DCFX(NE,3)                                              F1D03730
          ENDIF                                                           F1D03740
C                                                                         F1D03750
          L=0                                                             F1D03760
          DO 180 I=1,NPE                                                  F1D03770
          NI=NOD(NE,I)                                                    F1D03780
          ELX(I)=GLX(NI)                                                  F1D03790
          IF(ITEM.EQ.1 .OR. ITEM.EQ.2)THEN                                F1D03800
              LI=(NI-1)*NDF                                               F1D03810
              DO 170 J=1,NDF                                              F1D03820
              LI=LI+1                                                     F1D03830
              L=L+1                                                       F1D03840
              ELU(L)=GU0(LI)                                              F1D03850
              IF(ITEM.EQ.2 .AND. NT.GT.0)THEN                             F1D03860
                  ELV(L)=GU1(LI)                                          F1D03870
                  ELA(L)=GU2(LI)                                          F1D03880
              ENDIF                                                       F1D03890
  170         CONTINUE                                                    F1D03900
          ENDIF                                                           F1D03910
  180     CONTINUE                                                        F1D03920
C                                                                         F1D03930
          CALL COEFNT(IELEM,ITEM,MODEL,NDF,NPE,TIME,NTYPE,NE,F3,MXELM)    F1D03940
      ELSE                                                                F1D03950
          CALL TRSFRM (MXELM,MXNEQ,NE,NTYPE,PR,SE,SL,SA,SI,CS,SN,         F1D03960
     *                 CNT,SNT,HF,VF,PF,XB)                               F1D03970
      ENDIF                                                               F1D03980
C                                                                         F1D03990
      IF(NPRNT .NE.0)THEN                                                 F1D04000
          IF(NPRNT .LE.2)THEN                                             F1D04010
              IF(NE.EQ.1 .AND. NT.LE.1)THEN                               F1D04020
                  WRITE(IT,550)                                           F1D04030
                  DO 190 I=1,NN                                           F1D04040
  190             WRITE(IT,540) (ELK(I,J),J=1,NN)                         F1D04050
                  IF(ITEM.GT.2)THEN                                       F1D04060
                      WRITE(IT,360)                                       F1D04070
                      DO 195 I=1,NN                                       F1D04080
```

```
  195                WRITE(IT,540) (ELM(I,J),J=1,NN)                      F1D04090
                 ELSE                                                      F1D04100
                     WRITE(IT,560)                                         F1D04110
                     WRITE(IT,540) (ELF(I),I=1,NN)                         F1D04120
                 ENDIF                                                     F1D04130
             ENDIF                                                         F1D04140
         ENDIF                                                             F1D04150
      ENDIF                                                                F1D04160
C                                                                          F1D04170
C     Assemble element matrices                                            F1D04180
C                                                                          F1D04190
      CALL ASSMBL(NOD,MXELM,MXNEQ,NDF,NPE,NE,ITEM,GLK,GLM,GLF)             F1D04200
C                                                                          F1D04210
  200 CONTINUE                                                             F1D04220
      IF(NPRNT.EQ.2)THEN                                                   F1D04230
C                                                                          F1D04240
C     Print assembled coefficient matrices if required                     F1D04250
C                                                                          F1D04260
          WRITE(IT,570)                                                    F1D04270
          DO 210 I=1,NEQ                                                   F1D04280
  210     WRITE(IT,540) (GLK(I,J),J=1,NHBW)                                F1D04290
          IF(ITEM.GT.2)THEN                                                F1D04300
              WRITE(IT,575)                                                F1D04310
              DO 215 I=1,NEQ                                               F1D04320
  215         WRITE(IT,540) (GLM(I,J),J=1,NHBW)                            F1D04330
          ELSE                                                             F1D04340
              WRITE(IT,580)                                                F1D04350
              WRITE(IT,540) (GLF(I),I=1,NEQ)                               F1D04360
          ENDIF                                                            F1D04370
      ENDIF                                                                F1D04380
C                                                                          F1D04390
C     Call subroutine BONDRY to impose essential, natural and Newton's     F1D04400
C     type boundary conditions on the primary and secondary variables.     F1D04410
C                                                                          F1D04420
      CALL BONDRY(NEQ,NEQR,NHBW,NSPV,NSSV,NNBC,NDF,DT,ITEM,ALFA,IBDY,      F1D04430
     *            ISPV,ISSV,INBC,UREF,VSPV,VSSV,VNBC,GLK,GLM,GLF,GU0,      F1D04440
     *            MXEBC,MXNBC,MXMBC,MXNEQ)                                 F1D04450
C                                                                          F1D04460
      IF(ITEM.GT.2)THEN                                                    F1D04470
C                                                                          F1D04480
C     Call subroutine AXLBX to solve for the eigenvalues and eigenvectorF1D04490
C                                                                          F1D04500
          CALL AXLBX(NEQR,GLK,GLM,EGNVAL,EGNVEC,JVEC,NROT,MXNEQ)           F1D04510
C                                                                          F1D04520
          WRITE(IT,690) NROT                                               F1D04530
          DO 230 NVEC=1,NEQR                                               F1D04540
          FRQNCY=DSQRT(EGNVAL(NVEC))                                       F1D04550
          WRITE(IT,700)NVEC,EGNVAL(NVEC),FRQNCY                            F1D04560
  230     WRITE(IT,540)(EGNVEC(I,NVEC),I=1,NEQR)                           F1D04570
          STOP                                                             F1D04580
      ENDIF                                                                F1D04590
C                                                                          F1D04600
      IRES = 0                                                             F1D04610
C                                                                          F1D04620
C     Call subroutine SOLVER to solve the finite-element equations         F1D04630
C                                                                          F1D04640
      CALL SOLVER(MXNEQ,MXNEQ,NEQ,NHBW,GLK,GLF,IRES)                       F1D04650
C                                                                          F1D04660
      IF(ITEM.EQ.0)THEN                                                    F1D04670
          WRITE(IT,590)                                                    F1D04680
          WRITE(IT,540) (GLF(NI),NI=1,NEQ)                                 F1D04690
      ELSE                                                                 F1D04700
          IF(NT.EQ.0)THEN                                                  F1D04710
              DO 240 I=1,NEQ                                               F1D04720
  240         GU2(I)=GLF(I)                                                F1D04730
              NT=NT+1                                                      F1D04740
              TIME=TIME+DT                                                 F1D04750
              GOTO 150                                                     F1D04760
          ENDIF                                                            F1D04770
C                                                                          F1D04780
C     Compute and print current values of GU0, GU1, and GU2                F1D04790
C                                                                          F1D04800
          DO 250 I=1,NEQ                                                   F1D04810
          IF(ITEM.EQ.2)THEN                                                F1D04820
              ACCLRN=A3*(GLF(I)-GU0(I))-A4*GU1(I)-A5*GU2(I)                F1D04830
```

```
            GU1(I)=GU1(I)+A2*GU2(I)+A1*ACCLRN                           F1D04840
            GU2(I)=ACCLRN                                               F1D04850
            GPU(I)=GU0(I)                                               F1D04860
          ELSE                                                          F1D04870
            GPU(I)=GU0(I)                                               F1D04880
          ENDIF                                                         F1D04890
  250     GU0(I)=GLF(I)                                                 F1D04900
C                                                                       F1D04910
          DIFF=0.0                                                      F1D04920
          SOLN=0.0                                                      F1D04930
          DO 260 I=1,NEQ                                                F1D04940
          SOLN=SOLN+GU0(I)*GU0(I)                                       F1D04950
  260     DIFF=DIFF+(GLF(I)-GPU(I))**2                                  F1D04960
          PRCNT=DIFF/SOLN                                               F1D04970
          IF(PRCNT.LE.TOLRNS)THEN                                       F1D04980
              STOP                                                      F1D04990
          ELSE                                                          F1D05000
              IF(INTVL.LE.0)INTVL=1                                     F1D05010
              NTEN=(NT/INTVL)*INTVL                                     F1D05020
              IF(NTEN.EQ.NT)THEN                                        F1D05030
                  WRITE(IT,600) TIME, NT                                F1D05040
                  WRITE(IT,590)                                         F1D05050
                  WRITE(IT,540) (GU0(I),I=1,NEQ)                        F1D05060
                  IF(ITEM.NE.1) THEN                                    F1D05070
                      WRITE(IT,540) (GU1(I),I=1,NEQ)                    F1D05080
                      WRITE(IT,540) (GU2(I),I=1,NEQ)                    F1D05090
                  ENDIF                                                 F1D05100
                  NT=NT+1                                               F1D05110
                  TIME=TIME+DT                                          F1D05120
              ELSE                                                      F1D05130
                  NT=NT+1                                               F1D05140
                  TIME=TIME+DT                                          F1D05150
                  GOTO 150                                              F1D05160
              ENDIF                                                     F1D05170
          ENDIF                                                         F1D05180
      ENDIF                                                             F1D05190
C                                                                       F1D05200
C          |-------------------------------------------------------|    F1D05210
C          |          P O S T - P R O C E S S O R   U N I T        |    F1D05220
C          |-------------------------------------------------------|    F1D05230
C                                                                       F1D05240
      IF(NPRNT.LE.1)THEN                                                F1D05250
          IF(MODEL.EQ.1)THEN                                            F1D05260
              WRITE(IT,530)                                             F1D05270
          ELSE                                                          F1D05280
              IF(MODEL.EQ.4)THEN                                        F1D05290
                  WRITE(IT,630)                                         F1D05300
              ENDIF                                                     F1D05310
              WRITE(IT,520)                                             F1D05320
          ENDIF                                                         F1D05330
C                                                                       F1D05340
          IF(MODEL.EQ.1)THEN                                            F1D05350
              IF(NTYPE.EQ.0)THEN                                        F1D05360
                  WRITE(IT,610)                                         F1D05370
              ELSE                                                      F1D05380
                  WRITE(IT,620)                                         F1D05390
              ENDIF                                                     F1D05400
          ENDIF                                                         F1D05410
C                                                                       F1D05420
          IF(MODEL.EQ.2 .OR. MODEL.EQ.3)THEN                            F1D05430
              IF(NTYPE.EQ.0)THEN                                        F1D05440
                  WRITE(IT,650)                                         F1D05450
              ELSE                                                      F1D05460
                  WRITE(IT,660)                                         F1D05470
              ENDIF                                                     F1D05480
          ENDIF                                                         F1D05490
C                                                                       F1D05500
          IF(MODEL.EQ.4)THEN                                            F1D05510
              IF(NTYPE.EQ.0)THEN                                        F1D05520
                  WRITE(IT,680)                                         F1D05530
              ELSE                                                      F1D05540
                  WRITE(IT,670)                                         F1D05550
              ENDIF                                                     F1D05560
          ENDIF                                                         F1D05570
```

```
C                                                                       F1D05580
        IF(MODEL.EQ.1)THEN                                              F1D05590
           WRITE(IT,530)                                                F1D05600
        ELSE                                                            F1D05610
           WRITE(IT,520)                                                F1D05620
        ENDIF                                                           F1D05630
C                                                                       F1D05640
        IF(MODEL.LE.3)THEN                                              F1D05650
C                                                                       F1D05660
           CALL PSTPRC(F3,GLF,GLX,NOD,AX0,IELEM,NPE,MODEL,NTYPE,ITEM,F1D05670
     *                 MXELM,MXNEQ,MXNOD,NEM,NDF)                       F1D05680
        ELSE                                                            F1D05690
           CALL REACTN(MXELM,MXNEQ,NDF,NEM,NOD,NPE,NTYPE,PR,GLF,        F1D05700
     *                 SE,SL,SA,SI,CS,SN,CNT,SNT,HF,VF,PF,XB)           F1D05710
        ENDIF                                                           F1D05720
C                                                                       F1D05730
        IF(MODEL.EQ.1)THEN                                              F1D05740
           WRITE(IT,530)                                                F1D05750
        ELSE                                                            F1D05760
           WRITE(IT,520)                                                F1D05770
        ENDIF                                                           F1D05780
      ENDIF                                                             F1D05790
C                                                                       F1D05800
      IF(ITEM.EQ.0)STOP                                                 F1D05810
      IF(NT.LT.NTIME)THEN                                               F1D05820
         IF(PRCNT.GT.TOLRNS)THEN                                        F1D05830
            GOTO 150                                                    F1D05840
         ENDIF                                                          F1D05850
      ELSE                                                              F1D05860
         WRITE(IT,710)                                                  F1D05870
      ENDIF                                                             F1D05880
C                                                                       F1D05890
C     --------------------------------------------------------------    F1D05900
C                       F   O   R   M   A   T   S                       F1D05910
C     --------------------------------------------------------------    F1D05920
C                                                                       F1D05930
  300 FORMAT(20A4)                                                      F1D05940
  310 FORMAT(4X,'OUTPUT FROM PROGRAM   FEM1DV2  BY  J. N. REDDY')       F1D05950
  320 FORMAT(/,4X,'*** ANALYSIS OF MODEL',I2,', AND TYPE',I2,           F1D05960
     *       ' PROBLEM ***',/,15X,'(see the code below)',/,             F1D05970
     *       /,4X,'MODEL=1,NTYPE=0: A problem described by MODEL EQ. 1',F1D05980
     *       /,4X,'MODEL=1,NTYPE=1: A circular DISK (PLANE STRESS) ',   F1D05990
     *       /,4X,'MODEL=1,NTYPE>1: A circular DISK (PLANE STRAIN) ',   F1D06000
     *       /,4X,'MODEL=2,NTYPE=0: A Timoshenko BEAM (RIE) problem',   F1D06010
     *       /,4X,'MODEL=2,NTYPE=1: A Timoshenko PLATE (RIE) problem',  F1D06020
     *       /,4X,'MODEL=2,NTYPE=2: A Timoshenko BEAM (CIE) problem',   F1D06030
     *       /,4X,'MODEL=2,NTYPE>2: A Timoshenko PLATE (CIE) problem',  F1D06040
     *       /,4X,'MODEL=3,NTYPE=0: A Euler-Bernoulli BEAM problem',    F1D06050
     *       /,4X,'MODEL=3,NTYPE>0: A Euler-Bernoulli Circular plate',  F1D06060
     *       /,4X,'MODEL=4,NTYPE=0: A plane TRUSS problem',             F1D06070
     *       /,4X,'MODEL=4,NTYPE=1: A Euler-Bernoulli FRAME problem',   F1D06080
     *       /,4X,'MODEL=4,NTYPE=2: A Timoshenko (CIE) FRAME problem',/)F1D06090
  330 FORMAT(/,4X,'TIME-DEPENDENT (TRANSIENT) ANALYSIS ',/)             F1D06100
  340 FORMAT(/,4X,'E I G E N V A L U E   A N A L Y S I S',/)            F1D06110
  350 FORMAT(/,10X, 'Element type (0, Hermite,>0, Lagrange)..=',I4,/,   F1D06120
     *         10X, 'No. of deg. of freedom per node, NDF....=',I4,/,   F1D06130
     *         10X, 'No. of elements in the mesh, NEM........=',I4,/,   F1D06140
     *         10X, 'No. of total DOF in the model, NEQ......=',I4,/,   F1D06150
     *         10X, 'No. of specified primary DOF, NSPV......=',I4,/,   F1D06160
     *         10X, 'No. of specified secondary DOF, NSSV....=',I4,/,   F1D06170
     *         10X, 'No. of specified Newton B. C.: NNBC.....=',I4)     F1D06180
  360 FORMAT(/,3X,'Element coefficient matrix, [ELM]:',/)               F1D06190
  370 FORMAT(/,3X, 'Initial conditions on the primary variables:',/)    F1D06200
  380 FORMAT(/,3X, 'Initial cond. on time der. of primary variables:',/)F1D06210
  390 FORMAT(/,10X,'Coefficient, CT0.......................=',E12.4,/, F1D06220
     *         10X,'Coefficient, CT1.......................=',E12.4,/, F1D06230
     *         10X,'Parameter, ALFA........................=',E12.4,/, F1D06240
     *         10X,'Parameter, GAMA........................=',E12.4,/, F1D06250
     *         10X,'Time increment, DT.....................=',E12.4,/, F1D06260
     *         10X,'No. of time steps, NTIME...............=',I4,/,     F1D06270
     *         10X,'Time-step interval to print soln., INTVL=',I4,/)    F1D06280
  400 FORMAT(/,10X,'Coefficient, CT0.......................=',E12.4,/, F1D06290
     *         10X,'Coefficient, CT1.......................=',E12.4,/)  F1D06300
  410 FORMAT(/,3X,'Global coordinates of the nodes, {GLX}:',/)          F1D06310
```

```
420 FORMAT(/,3X,'Coefficients of the differential equation:',/)        F1D06320
430 FORMAT(/,5X,'Properties of Element =',I3,//,                        F1D06330
   *       10X,'XA  =',E12.4,5X,'XB  =',E12.4)                          F1D06340
440 FORMAT( 10X,'AX0 =',E12.4,5X,'AX1 =',E12.4,/,                       F1D06350
   *        10X,'BX0 =',E12.4,5X,'BX1 =',E12.4,/,                       F1D06360
   *        10X,'CX0 =',E12.4,5X,'CX1 =',E12.4,/,                       F1D06370
   *        10X,'FX0 =',E12.4,5X,'FX1 =',E12.4,5X,'FX2 =',E12.4,/)     F1D06380
445 FORMAT( 10X,'AX0 =',E12.4,5X,'AX1 =',E12.4,/,                       F1D06390
   *        10X,'BX0 =',E12.4,5X,'BX1 =',E12.4,/,                       F1D06400
   *        10X,'CX0 =',E12.4,5X,'CX1 =',E12.4,/)                       F1D06410
450 FORMAT(10X,'The poisson ratio,           PR........ =',E12.4,/,     F1D06420
   *      10X,'Modulus of elasticity,       SE........ =',E12.4,/,      F1D06430
   *      10X,'Length of the element,       SL........ =',E12.4,/,      F1D06440
   *      10X,'Area of cross section,       SA........ =',E12.4,/,      F1D06450
   *      10X,'Moment of inertia,           SI........ =',E12.4,/,      F1D06460
   *      10X,'Cosine of orientation,       CN........ =',E12.4,/,      F1D06470
   *      10X,'Sine of orientation,         SN........ =',E12.4,/,      F1D06480
   *      10X,'Axial body force (constant), HF........ =',E12.4,/,      F1D06490
   *      10X,'Transverse body force (cnst),VF........ =',E12.4,/,      F1D06500
   *      10X,'Internal point force,        PF........ =',E12.4,/,      F1D06510
   *      10X,'Location of PF from node 1,  XB........ =',E12.4,/,      F1D06520
   *      10X,'Orientation of PF: cosine,   CST....... =',E12.4,/,      F1D06530
   *      10X,'Orientation of PF: sine,     SNT....... =',E12.4,/,      F1D06540
   *      10X,'Nodal connectivity:          NOD(I,J).. =',2I6,/)        F1D06550
460 FORMAT(//,3X,'Element No. =', I3,/)                                 F1D06560
470 FORMAT(10X,'Modulus of elasticity,       SE........ =',E12.4,/,     F1D06570
   *      10X,'Length of the element,       SL........ =',E12.4,/,      F1D06580
   *      10X,'Area of cross section,       SA........ =',E12.4,/,      F1D06590
   *      10X,'Cosine of orientation,       CN........ =',E12.4,/,      F1D06600
   *      10X,'Sine of orientation,         SN........ =',E12.4,/,      F1D06610
   *      10X,'Axial body force (constant), HF........ =',E12.4,/,      F1D06620
   *      10X,'Nodal connectivity:          NOD(I,J).. =',2I6,/)        F1D06630
480 FORMAT(/,3X, 'Boundary information on primary variables:',/)        F1D06640
490 FORMAT(5X,2I5,2E15.5)                                               F1D06650
500 FORMAT(/,3X, 'Boundary information on secondary variables:',/)      F1D06660
510 FORMAT(/,3X, 'Boundary information on mixed boundary cond.:',/)     F1D06670
520 FORMAT(2X,78('_'),/)                                                F1D06680
530 FORMAT(2X,55('_'),/)                                                F1D06690
540 FORMAT(2X,5E13.5)                                                   F1D06700
550 FORMAT(/,3X,'Element coefficient matrix, [ELK]:',/)                 F1D06710
560 FORMAT(/,3X,'Element source vector, {ELF}:',/)                      F1D06720
570 FORMAT(/,3X,'Global coefficient matrix, [GLK]:',/)                  F1D06730
575 FORMAT(/,3X,'Global coefficient matrix, [GLM]:',/)                  F1D06740
580 FORMAT(/,3X,'Global source vector, {GLF}:',/)                       F1D06750
590 FORMAT(/,1X,'SOLUTION (values of PVs) at the NODES: ',/)            F1D06760
600 FORMAT(/,1X,'TIME =',E12.4,5X,'Time step number =',I3,/)            F1D06770
610 FORMAT(7X,' X ',5X,'P. Variable',2X,'S. Variable')                  F1D06780
620 FORMAT(7X,' X ',5X, 'Displacemnt',2X,'Radial Strs',2X,              F1D06790
   *    'Hoop Stress')                                                  F1D06800
630 FORMAT(/,9X,'Generalized forces in the element coordinates',/,      F1D06810
   * 5X,'(second line gives the results in the global coordinates)')   F1D06820
650 FORMAT(7X,'  X ',6X, 'Deflect.',5X,'Rotation',5X,'B. Moment',       F1D06830
   *    3X,'Shear Force')                                               F1D06840
660 FORMAT(7X,'  X ',6X, 'Deflect.',5X,'Rotation',4X,'Moment, Mr',      F1D06850
   *    3X,'Moment, Mt',3X,'Shear Force')                               F1D06860
670 FORMAT(3X,  'Ele  Force, H1   Force, V1  Moment, M1  Force, H2      F1D06870
   *Force, V2  Moment, M2')                                             F1D06880
680 FORMAT(3X,  'Ele  Force, H1   Force, V1   Force, H2  Force, V2')    F1D06890
690 FORMAT(/,5X,'Number of rotations taken in JACOBI =',I2,/)           F1D06900
700 FORMAT(/,5X,'EIGENVALUE(',I2,') = ',E14.6,2X,'SQRT(EGNVAL) = ',     F1D06910
   *    E13.5,/,5X,'EIGENVECTOR:')                                      F1D06920
710 FORMAT(/,5X,'***** Number of time steps exceeded NTIME *****',/)    F1D06930
    STOP                                                                F1D06940
    END                                                                 F1D06950
```

```
      SUBROUTINE ASSMBL(NOD,MXELM,MXNEQ,NDF,NPE,NE,ITEM,GLK,GLM,GLF)    F1D06960
C     ________________________________________________________________ F1D06970
C                                                                       F1D06980
C       The subroutine calls subroutines to compute element matrices    F1D06990
C            and assembles them in a upper banded matrix form           F1D07000
C                                                                       F1D07010
C        {ELF}.... Element source vector, {F}                           F1D07020
C        {ELK}.... Element coefficient matrix, [K]                      F1D07030
C        {ELM}.... Element coefficient matrix, [M]                      F1D07040
C        [NOD].... Connectivity matrix                                  F1D07050
C     ________________________________________________________________ F1D07060
C                                                                       F1D07070
      IMPLICIT REAL*8 (A-H,O-Z)                                         F1D07080
      DIMENSION  GLK(MXNEQ,MXNEQ),GLM(MXNEQ,MXNEQ),GLF(MXNEQ),          F1D07090
     *           NOD(MXELM,4)                                           F1D07100
      COMMON/STF1/ELK(9,9),ELM(9,9),ELF(9),ELX(4),ELU(9),ELV(9),ELA(9) F1D07110
      IF(ITEM.LE.2)THEN                                                 F1D07120
C                                                                       F1D07130
C     Assemble element coefficient matrix ELK and source vector ELF     F1D07140
C                                                                       F1D07150
          DO 50 I = 1, NPE                                              F1D07160
          NR = (NOD(NE,I) - 1)*NDF                                      F1D07170
          DO 40 II = 1, NDF                                             F1D07180
          NR = NR + 1                                                   F1D07190
          L = (I-1)*NDF + II                                            F1D07200
          GLF(NR) = GLF(NR) + ELF(L)                                    F1D07210
          DO 30 J = 1, NPE                                              F1D07220
          NCL = (NOD(NE,J)-1)*NDF                                       F1D07230
          DO 20 JJ = 1, NDF                                             F1D07240
          M = (J-1)*NDF + JJ                                            F1D07250
          NC = NCL-NR+JJ+1                                              F1D07260
          IF(NC)20,20,10                                                F1D07270
   10     GLK(NR,NC) = GLK(NR,NC) + ELK(L,M)                            F1D07280
   20     CONTINUE                                                      F1D07290
   30     CONTINUE                                                      F1D07300
   40     CONTINUE                                                      F1D07310
   50     CONTINUE                                                      F1D07320
      ELSE                                                              F1D07330
C                                                                       F1D07340
C                                                                       F1D07350
C     ASSEMBLE ELEMENT MATRICES INTO FULL GLOBAL MATRICES               F1D07360
C                                                                       F1D07370
          DO 100 I=1,NPE                                                F1D07380
          NR=(NOD(NE,I)-1)*NDF                                          F1D07390
          DO 90 II=1,NDF                                                F1D07400
          NR=NR+1                                                       F1D07410
          L=(I-1)*NDF+II                                                F1D07420
          DO 80 J=1,NPE                                                 F1D07430
          NC=(NOD(NE,J)-1)*NDF                                          F1D07440
          DO 70 JJ=1,NDF                                                F1D07450
          M=(J-1)*NDF+JJ                                                F1D07460
          NC=NC+1                                                       F1D07470
          GLK(NR,NC)=GLK(NR,NC)+ELK(L,M)                                F1D07480
   60     GLM(NR,NC)=GLM(NR,NC)+ELM(L,M)                                F1D07490
   70     CONTINUE                                                      F1D07500
   80     CONTINUE                                                      F1D07510
   90     CONTINUE                                                      F1D07520
  100     CONTINUE                                                      F1D07530
      ENDIF                                                             F1D07540
      RETURN                                                            F1D07550
      END                                                               F1D07560

      SUBROUTINE AXLBX(N,A,B,XX,X,NEGN,NR,MXNEQ)                        F1D07570
C     ________________________________________________________________ F1D07580
C                                                                       F1D07590
C       Subroutine to solve the EIGENVALUE PROBLEM:                     F1D07600
C                                                                       F1D07610
C                          [A]{X} = Lambda.[B]{X}                       F1D07620
C                                                                       F1D07630
C       The program can be used only for positive-definite [B] matrix  F1D07640
C        The dimensions of V, VT, W, and IH should be equal to MXNEQ    F1D07650
C     ________________________________________________________________ F1D07660
C                                                                       F1D07670
      IMPLICIT REAL*8 (A-H,O-Z)                                         F1D07680
```

```
      DIMENSION  A(MXNEQ,MXNEQ),B(MXNEQ,MXNEQ),XX(MXNEQ),X(MXNEQ,MXNEQ) F1D07690
      DIMENSION  V(50,50),VT(50,50),W(50,50),IH(50)                      F1D07700
C                                                                        F1D07710
C     Call JACOBI to diagonalize [B]                                     F1D07720
C                                                                        F1D07730
      CALL JACOBI (N,B,NEGN,NR,V,XX,IH,MXNEQ)                            F1D07740
C                                                                        F1D07750
C     Make diagonalized [B] symmetric                                    F1D07760
C                                                                        F1D07770
      DO 10 I=1,N                                                        F1D07780
      DO 10 J=1,N                                                        F1D07790
   10 B(J,I)=B(I,J)                                                      F1D07800
C                                                                        F1D07810
C     Check (to make sure) that [B] is positive-definite                 F1D07820
C                                                                        F1D07830
      DO 30 I=1,N                                                        F1D07840
      IF (B(I,I))20,30,30                                                F1D07850
   20 WRITE(6,80)                                                        F1D07860
      STOP                                                               F1D07870
   30 CONTINUE                                                           F1D07880
C                                                                        F1D07890
C     The eigenvectors of [B] are stored in array V(I,J)                 F1D07900
C     Form the transpose of [V] as [VT]                                  F1D07910
C                                                                        F1D07920
      DO 40 I=1,N                                                        F1D07930
      DO 40 J=1,N                                                        F1D07940
   40 VT(I,J)=V(J,I)                                                     F1D07950
C                                                                        F1D07960
C     Find the product [F]=[VT][A][V] and store in [A] to save storage   F1D07970
C                                                                        F1D07980
      CALL MTRXML (VT,N,N,A,N,W)                                         F1D07990
      CALL MTRXML (W,N,N,V,N,A)                                          F1D08000
C                                                                        F1D08010
C     Get [GI] from diagonalized [B], but store it in [B]                F1D08020
C                                                                        F1D08030
      DO 50 I=1,N                                                        F1D08040
   50 B(I,I)=1.0/DSQRT(B(I,I))                                           F1D08050
C                                                                        F1D08060
C     Find the product [Q]=[GI][F][GI]=[B][A][B] and store in [A]        F1D08070
C                                                                        F1D08080
      CALL MTRXML (B,N,N,A,N,W)                                          F1D08090
      CALL MTRXML (W,N,N,B,N,A)                                          F1D08100
C                                                                        F1D08110
C     We now have the form [Q]{Z}=Lamda{Z}. Diagonalize [Q] to obtain    F1D08120
C     the eigenvalues by calling JACOBI.                                 F1D08130
C                                                                        F1D08140
      CALL JACOBI (N,A,NEGN,NR,VT,XX,IH,MXNEQ)                           F1D08150
C                                                                        F1D08160
C     The eigenvalues are returned as diag [A].                          F1D08170
C                                                                        F1D08180
      DO 60 J=1,N                                                        F1D08190
   60 XX(J)=A(J,J)                                                       F1D08200
C                                                                        F1D08210
C     The eigenvectors are computed from the relation,                   F1D08220
C                     {X}=[V][GI]{Z}=[V][B][VT]                          F1D08230
C     since {Z} is stored in [VT].                                       F1D08240
C                                                                        F1D08250
      CALL MTRXML (V,N,N,B,N,W)                                          F1D08260
      CALL MTRXML (W,N,N,VT,N,X)                                         F1D08270
C                                                                        F1D08280
   80 FORMAT(/'*** Matrix [GLM] is NOT positive-definite ***')           F1D08290
      RETURN                                                             F1D08300
      END                                                                F1D08310

      SUBROUTINE BONDRY(NEQ,NEQR,NHBW,NSPV,NSSV,NNBC,NDF,DT,ITEM,ALFA,   F1D08320
     *                  IBDY,ISPV,ISSV,INBC,UREF,VSPV,VSSV,VNBC,         F1D08330
     *                  GLK,GLM,GLF,GU0,MXEBC,MXNBC,MXMBC,MXNEQ)         F1D08340
C     _________________________________________________________________F1D08350
C                                                                        F1D08360
C     The subroutine is used to implement specified boundary conditions F1D08370
C           on the assembled system of finite element equations         F1D08380
C     _________________________________________________________________F1D08390
C                                                                        F1D08400
      IMPLICIT REAL*8 (A-H,O-Z)                                          F1D08410
```

```
      DIMENSION  ISPV(MXEBC,2),ISSV(MXNBC,2),INBC(MXMBC,2),IBDY(MXEBC)   F1D08420
      DIMENSION  UREF(MXMBC),VSPV(MXEBC),VSSV(MXNBC),VNBC(MXMBC)         F1D08430
      DIMENSION  GLK(MXNEQ,MXNEQ),GLM(MXNEQ,MXNEQ),GLF(MXNEQ),GU0(MXNEQ)F1D08440
C                                                                        F1D08450
C     Impose boundary conditions for STATIC and TIME-DEPENDENT problems F1D08460
C                                                                        F1D08470
      IF(ITEM.LE.2)THEN                                                  F1D08480
C                                                                        F1D08490
C     Include specified PRIMARY degrees of freedom                       F1D08500
C                                                                        F1D08510
         IF(NSPV.NE.0)THEN                                               F1D08520
            DO 30 NB = 1,NSPV                                            F1D08530
            IE=(ISPV(NB,1)-1)*NDF+ISPV(NB,2)                             F1D08540
            IT=NHBW-1                                                    F1D08550
            I=IE-NHBW                                                    F1D08560
            DO 10 II=1,IT                                                F1D08570
            I=I+1                                                        F1D08580
            IF(I .GE. 1)THEN                                             F1D08590
               J=IE-I+1                                                  F1D08600
               GLF(I)=GLF(I)-GLK(I,J)*VSPV(NB)                           F1D08610
               GLK(I,J)=0.0                                              F1D08620
            ENDIF                                                        F1D08630
   10       CONTINUE                                                     F1D08640
            GLK(IE,1)=1.0                                                F1D08650
            GLF(IE)=VSPV(NB)                                             F1D08660
            I=IE                                                         F1D08670
            DO 20 II=2,NHBW                                              F1D08680
            I=I+1                                                        F1D08690
            IF(I .LE. NEQ)THEN                                           F1D08700
               GLF(I)=GLF(I)-GLK(IE,II)*VSPV(NB)                         F1D08710
               GLK(IE,II)=0.0                                            F1D08720
            ENDIF                                                        F1D08730
   20       CONTINUE                                                     F1D08740
   30       CONTINUE                                                     F1D08750
         ENDIF                                                           F1D08760
C                                                                        F1D08770
         IF(NSSV.NE.0)THEN                                               F1D08780
C                                                                        F1D08790
C     Include specified SECONDARY degrees of freedom                     F1D08800
C                                                                        F1D08810
            DO 40 NF = 1,NSSV                                            F1D08820
            NB=(ISSV(NF,1)-1)*NDF+ISSV(NF,2)                             F1D08830
            IF(ITEM.EQ.1)GLF(NB)=GLF(NB)+VSSV(NF)*DT                     F1D08840
   40       IF(ITEM.NE.1)GLF(NB)=GLF(NB)+VSSV(NF)                        F1D08850
         ENDIF                                                           F1D08860
C                                                                        F1D08870
         IF(NNBC.NE.0)THEN                                               F1D08880
C                                                                        F1D08890
C     Include specified MIXED boundary conditions                        F1D08900
C                                                                        F1D08910
            DO 50 IC=1,NNBC                                              F1D08920
            NC=(INBC(IC,1)-1)*NDF+INBC(IC,2)                             F1D08930
            IF(ITEM.EQ.1)THEN                                            F1D08940
               GLK(NC,1)=GLK(NC,1)+ALFA*DT*VNBC(IC)                      F1D08950
               GLF(NC)=GLF(NC)+DT*VNBC(IC)*(UREF(IC)                     F1D08960
     *                           -(1.0-ALFA)*GU0(NC))                    F1D08970
            ELSE                                                         F1D08980
               GLK(NC,1)=GLK(NC,1)+VNBC(IC)                              F1D08990
               GLF(NC)=GLF(NC)+VNBC(IC)*UREF(IC)                         F1D09000
            ENDIF                                                        F1D09010
   50       CONTINUE                                                     F1D09020
         ENDIF                                                           F1D09030
      ELSE                                                               F1D09040
         IF(NNBC.NE.0)THEN                                               F1D09050
C                                                                        F1D09060
C     Include specified MIXED boundary conditions                        F1D09070
C                                                                        F1D09080
            DO 70 IC=1,NNBC                                              F1D09090
            NC=(INBC(IC,1)-1)*NDF+INBC(IC,2)                             F1D09100
            GLK(NC,NC)=GLK(NC,NC)+VNBC(IC)                               F1D09110
   70       CONTINUE                                                     F1D09120
         ENDIF                                                           F1D09130
C                                                                        F1D09140
C        Impose boundary conditions for EIGENVALUE problems              F1D09150
C                                                                        F1D09160
```

```
          IF(NSPV.NE.0)THEN                                             F1D09170
              DO 80 IB=1,NSPV                                           F1D09180
   80         IBDY(IB)=(ISPV(IB,1)-1)*NDF+ISPV(IB,2)                    F1D09190
              DO 120 I=1,NSPV                                           F1D09200
              IMAX=IBDY(I)                                              F1D09210
              DO 110 J=I,NSPV                                           F1D09220
              IF(IBDY(J).GE.IMAX)THEN                                   F1D09230
                  IMAX=IBDY(J)                                          F1D09240
                  IKEPT=J                                               F1D09250
              ENDIF                                                     F1D09260
  110         CONTINUE                                                  F1D09270
              IBDY(IKEPT)=IBDY(I)                                       F1D09280
              IBDY(I)=IMAX                                              F1D09290
  120         CONTINUE                                                  F1D09300
              NEQR = NEQ                                                F1D09310
              DO 180 I=1,NSPV                                           F1D09320
              IB=IBDY(I)                                                F1D09330
              IF(IB .LT. NEQR)THEN                                      F1D09340
                  NEQR1=NEQR-1                                          F1D09350
                  DO 160 II=IB,NEQR1                                    F1D09360
                  DO 140 JJ=1,NEQR                                      F1D09370
                  GLM(II,JJ)=GLM(II+1,JJ)                               F1D09380
  140             GLK(II,JJ)=GLK(II+1,JJ)                               F1D09390
                  DO 150 JJ=1,NEQR                                      F1D09400
                  GLM(JJ,II)=GLM(JJ,II+1)                               F1D09410
  150             GLK(JJ,II)=GLK(JJ,II+1)                               F1D09420
  160             CONTINUE                                              F1D09430
              ENDIF                                                     F1D09440
              NEQR=NEQR-1                                               F1D09450
  180         CONTINUE                                                  F1D09460
          ENDIF                                                         F1D09470
      ENDIF                                                             F1D09480
      RETURN                                                            F1D09490
      END                                                               F1D09500

      SUBROUTINE COEFNT(IELEM,ITEM,MODEL,NDF,NPE,TIME,NTYPE,NE,F3,MXELM)F1D09510
C     __________________________________________________________________F1D09520
C                                                                       F1D09530
C     The subroutine computes the coefficient matrices and source vectorF1D09540
C          for the model problem described by Eqn. (1) (see MAIN)       F1D09550
C                                                                       F1D09560
C        X.........  Global (i.e., problem) coordinate                  F1D09570
C        XI .......  Local (i.e., element) coordinate                   F1D09580
C        H.........  Element length                                     F1D09590
C        {SF}......  Element interpolation (or shape) functions         F1D09600
C        {GDSF}....  First derivative of Sf w.r.t. X                    F1D09610
C        {GDDSF}...  Second derivative of SF w.r.t. X                   F1D09620
C        GJ........  Jacobian of the transformation                     F1D09630
C        [GAUSPT]..  4x4 matrix of Gauss points: N-th column correspondsF1D09640
C                    to the N-point Guass rule                          F1D09650
C        [GAUSWT]..  4x4 matrix of Gauss weights (see the comment above)F1D09660
C        [A],[B],..  Element matrices needed to compute ELK             F1D09670
C        [ELK].....  Element coefficient matrix [K]                     F1D09680
C        [ELM].....  Element 'mass' matrix [M]                          F1D09690
C     __________________________________________________________________F1D09700
C                                                                       F1D09710
      IMPLICIT REAL*8(A-H,O-Z)                                          F1D09720
      COMMON/STF1/ELK(9,9),ELM(9,9),ELF(9),ELX(4),ELU(9),ELV(9),ELA(9)  F1D09730
      COMMON/STF2/A1,A2,A3,A4,A5,AX0,AX1,BX0,BX1,CX0,CX1,CT0,CT1,FX0,   F1D09740
     *            FX1,FX2                                               F1D09750
      COMMON/SHP/SF(4),GDSF(4),GDDSF(4),GJ                              F1D09760
      DIMENSION GAUSPT(5,5),GAUSWT(5,5),F3(MXELM)                       F1D09770
C                                                                       F1D09780
      DATA GAUSPT/5*0.0D0,-.57735027D0,.57735027D0,3*0.0D0,-.77459667D0,F1D09790
     * 0.0D0,.77459667D0,2*0.0D0,-.86113631D0,-.33998104D0,.33998104D0, F1D09800
     *.86113631D0,0.0D0,-.906180D0,-.538469D0,0.0D0,.538469D0,.906180D0/F1D09810
C                                                                       F1D09820
      DATA GAUSWT/2.0D0,4*0.0D0,2*1.0D0,3*0.0D0,.55555555D0,.88888888D0,F1D09830
     * 0.55555555D0,2*0.0D0,.34785485D0,2*.65214515D0,.34785485D0,0.0D0,F1D09840
     * 0.236927D0,.478629D0,.568889D0,.478629D0,.236927D0/              F1D09850
C                                                                       F1D09860
      NN=NDF*NPE                                                        F1D09870
      H = ELX(NPE) - ELX(1)                                             F1D09880
```

```
      IF(IELEM .EQ. 0)THEN                                           F1D09890
          NGP=4                                                      F1D09900
      ELSE                                                           F1D09910
          NGP = IELEM+1                                              F1D09920
      ENDIF                                                          F1D09930
C                                                                    F1D09940
      DO 10 J=1,NN                                                   F1D09950
      IF(ITEM.LE.2)THEN                                              F1D09960
          ELF(J) = 0.0                                               F1D09970
      ENDIF                                                          F1D09980
      DO 10 I=1,NN                                                   F1D09990
      IF(ITEM.GT.0)THEN                                              F1D10000
          ELM(I,J)=0.0                                               F1D10010
      ENDIF                                                          F1D10020
   10 ELK(I,J)=0.0                                                   F1D10030
C                                                                    F1D10040
      IF(MODEL.NE.2)THEN                                             F1D10050
C                                                                    F1D10060
C     DO-LOOP on number of Gauss points begins here                  F1D10070
C                                                                    F1D10080
          DO 100 NI=1,NGP                                            F1D10090
          XI = GAUSPT(NI,NGP)                                        F1D10100
C                                                                    F1D10110
C     Call subroutine SHP1D to evaluate the interpolation functions  F1D10120
C          and their global derivatives at the Gauss point XI        F1D10130
C                                                                    F1D10140
          CALL SHP1D(H,IELEM,NPE,XI)                                 F1D10150
C                                                                    F1D10160
          CONST = GJ*GAUSWT(NI,NGP)                                  F1D10170
          IF(IELEM.EQ.0)THEN                                         F1D10180
              X = ELX(1) + 0.5*H*(1.0+XI)                            F1D10190
          ELSE                                                       F1D10200
              X = 0.0                                                F1D10210
              DO 30 J=1,NPE                                          F1D10220
   30         X = X + SF(J)*ELX(J)                                   F1D10230
          ENDIF                                                      F1D10240
C                                                                    F1D10250
C     Compute coefficient matrices and vectors for vaious model problemsF1D10260
C         governed by single second-order and fourth-order equations  F1D10270
C                   (MODEL = 1 or 3; NTYPE = 0 or 1)                 F1D10280
C                                                                    F1D10290
          CX=CX0+CX1*X                                               F1D10300
          IF (ITEM.NE.3) THEN                                        F1D10310
              FX=FX0+FX1*X+FX2*X*X                                   F1D10320
          ENDIF                                                      F1D10330
          IF (ITEM.GT.0)THEN                                         F1D10340
              CT=CT0+CT1*X                                           F1D10350
          ENDIF                                                      F1D10360
          IF (MODEL.EQ.1)THEN                                        F1D10370
C                                                                    F1D10380
C         Coefficients for ALL SINGLE-VARIABLE PROBLEMS (MODEL=1)    F1D10390
C                                                                    F1D10400
              IF (NTYPE.EQ.0)THEN                                    F1D10410
C                                                                    F1D10420
C             All problems governed by MODEL EQUATION (3.1) (NTYPE=0)  F1D10430
C                                                                    F1D10440
                  AX=AX0+AX1*X                                       F1D10450
                  DO 50 J = 1,NN                                     F1D10460
                  IF (ITEM.LE.2)THEN                                 F1D10470
                      ELF(J) = ELF(J) + CONST*SF(J)*FX               F1D10480
                  ENDIF                                              F1D10490
                  DO 50 I = 1,NN                                     F1D10500
                  IF (ITEM.NE.0)THEN                                 F1D10510
                      ELM(I,J) = ELM(I,J) + CONST*SF(I)*SF(J)*CT     F1D10520
                  ENDIF                                              F1D10530
                  AIJ = CONST*GDSF(I)*GDSF(J)                        F1D10540
                  CIJ = CONST*SF(I)*SF(J)                            F1D10550
   50             ELK(I,J)=ELK(I,J) + AX*AIJ + CX*CIJ                F1D10560
              ELSE                                                   F1D10570
C                                                                    F1D10580
C             RADIALLY SYMMETRIC ELASTICITY problems (MODEL=1, NTYPE>0) F1D10590
C             AX0=E1, AX1=E2, BX0=NU12, BX1=H, thickness             F1D10600
C                                                                    F1D10610
                  ANU21=BX0*AX0/AX1                                  F1D10620
                  IF (NTYPE.EQ.1)THEN                                F1D10630
```

```
                    C11=BX1*AX0/(1.0-BX0*ANU21)                      F1D10640
                    C22=C11*(AX1/AX0)                                F1D10650
                    C12=BX0*C22                                      F1D10660
                ELSE                                                 F1D10670
                    DENOM=1.0-BX0-ANU21                              F1D10680
                    C11=BX1*AX0*(1.0-BX0)/(1.0+BX0)/DENOM            F1D10690
                    C22=BX1*AX1*(1.0-ANU21)/(1.0+ANU21)/DENOM        F1D10700
                    C12=BX0*C22                                      F1D10710
                ENDIF                                                F1D10720
                DO 60 J=1,NN                                         F1D10730
                IF (ITEM.LE.2)THEN                                   F1D10740
                    ELF(J) = ELF(J) + CONST*SF(J)*FX*X               F1D10750
                ENDIF                                                F1D10760
                DO 60 I=1,NN                                         F1D10770
                IF (ITEM.NE.0)THEN                                   F1D10780
                    ELM(I,J) = ELM(I,J) + CONST*SF(I)*SF(J)*CT*X     F1D10790
                ENDIF                                                F1D10800
                AIJ = CONST*GDSF(I)*GDSF(J)*C11*X                    F1D10810
                CIJ = CONST*SF(I)*SF(J)*CX*X                         F1D10820
                DIJ = CONST*(GDSF(I)*SF(J)+SF(I)*GDSF(J))*C12        F1D10830
                EIJ = CONST*SF(I)*SF(J)*C22/X                        F1D10840
   60           ELK(I,J)=ELK(I,J) + AIJ + CIJ + DIJ + EIJ            F1D10850
            ENDIF                                                    F1D10860
        ELSE                                                         F1D10870
C                                                                    F1D10880
C       Coefficients for the EULER-BERNOULLI theory (MODEL=2)        F1D10890
C                                                                    F1D10900
            IF (NTYPE.EQ.0)THEN                                      F1D10910
C                                                                    F1D10920
C           The Euler-Bernoulli BEAM element (MODEL=1 and NTYPE=0)   F1D10930
C                                                                    F1D10940
                BX=BX0+BX1*X                                         F1D10950
                CX=CX0+CX1*X                                         F1D10960
                DO 70 J = 1,NN                                       F1D10970
                IF (ITEM.LE.2)THEN                                   F1D10980
                    ELF(J) = ELF(J) + CONST*SF(J)*FX                 F1D10990
                ENDIF                                                F1D11000
                DO 70 I = 1,NN                                       F1D11010
                IF (ITEM.NE.0)THEN                                   F1D11020
                    ELM(I,J) = ELM(I,J) + CONST*SF(I)*SF(J)*CT       F1D11030
                ENDIF                                                F1D11040
                BIJ = CONST*GDDSF(I)*GDDSF(J)                        F1D11050
                CIJ = CONST*SF(I)*SF(J)                              F1D11060
   70           ELK(I,J)=ELK(I,J) + BX*BIJ + CX*CIJ                  F1D11070
            ELSE                                                     F1D11080
C                                                                    F1D11090
C           The E-B CIRCULAR PLATE element (MODEL=1 and NTYPE>0)     F1D11100
C                                                                    F1D11110
                ANU21=BX0*AX0/AX1                                    F1D11120
                DI=(BX1**3)/12.0                                     F1D11130
                D11=DI*AX0/(1.0-BX0*ANU21)                           F1D11140
                D22=D11*(AX1/AX0)                                    F1D11150
                D12=BX0*D22                                          F1D11160
                DO 80 J=1,NN                                         F1D11170
                IF (ITEM.LE.2)THEN                                   F1D11180
                    ELF(J) = ELF(J) + CONST*SF(J)*FX*X               F1D11190
                ENDIF                                                F1D11200
                DO 80 I=1,NN                                         F1D11210
                BIJ = CONST*GDDSF(I)*GDDSF(J)*D11*X                  F1D11220
                CIJ = CONST*SF(I)*SF(J)*CX*X                         F1D11230
                DIJ = CONST*(GDDSF(I)*GDSF(J)+GDSF(I)*GDDSF(J))*D12  F1D11240
                EIJ = CONST*GDSF(I)*GDSF(J)*D22/X                    F1D11250
   80           ELK(I,J)=ELK(I,J) + BIJ + CIJ + DIJ + EIJ            F1D11260
            ENDIF                                                    F1D11270
        ENDIF                                                        F1D11280
  100   CONTINUE                                                     F1D11290
      ELSE                                                           F1D11300
C                                                                    F1D11310
C     Coefficients for the TIMOSHENKO beam and circular plate (MODEL=2) F1D11320
C         Full integration for bending coefficients                  F1D11330
C                                                                    F1D11340
        DO 160 NI=1,NGP                                              F1D11350
        XI=GAUSPT(NI,NGP)                                            F1D11360
C                                                                    F1D11370
        CALL SHP1D(H,IELEM,NPE,XI)                                   F1D11380
```

```
C                                                                       F1D11390
         CONST=GJ*GAUSWT(NI,NGP)                                        F1D11400
         X = 0.0                                                        F1D11410
         DO 110 J=1,NPE                                                 F1D11420
  110    X = X + SF(J)*ELX(J)                                           F1D11430
         IF(NTYPE.EQ.0 .OR. NTYPE.EQ.2)THEN                             F1D11440
C                                                                       F1D11450
C        The Timoshenko BEAM element (MODEL=2 and NTYPE=0 OR 2)         F1D11460
C                                                                       F1D11470
             BX=BX0+BX1*X                                               F1D11480
             CX=CX0+CX1*X                                               F1D11490
             FX=FX0+FX1*X+FX2*X*X                                       F1D11500
             JJ=1                                                       F1D11510
             DO 130 J=1,NPE                                             F1D11520
             IF(ITEM.LE.2)THEN                                          F1D11530
                 ELF(JJ)=ELF(JJ)+FX*SF(J)*CONST                         F1D11540
             ENDIF                                                      F1D11550
             II=1                                                       F1D11560
             DO 120 I=1,NPE                                             F1D11570
             CIJ=SF(I)*SF(J)*CONST                                      F1D11580
             BIJ=GDSF(I)*GDSF(J)*CONST                                  F1D11590
             ELK(II,JJ)     =ELK(II,JJ)     +CX*CIJ                     F1D11600
             ELK(II+1,JJ+1)=ELK(II+1,JJ+1)+BX*BIJ                       F1D11610
             IF(ITEM.NE.0)THEN                                          F1D11620
                 ELM(II,JJ)     =ELM(II,JJ)     +CT0*CIJ                F1D11630
                 ELM(II+1,JJ+1)=ELM(II+1,JJ+1)+CT1*CIJ                  F1D11640
             ENDIF                                                      F1D11650
  120        II=NDF*I+1                                                 F1D11660
  130        JJ=NDF*J+1                                                 F1D11670
         ELSE                                                           F1D11680
C                                                                       F1D11690
C        Timoshenko CIRCULAR PLATE element (MODEL=2 and NTYPE=1 or 3)   F1D11700
C             AX0=E1, AX1=E2, BX0=ANU12, BX1=H, FX2=G13*AREA*AK         F1D11710
C                                                                       F1D11720
             ANU21=BX0*AX0/AX1                                          F1D11730
             CX=CX0+CX1*X                                               F1D11740
             FX=FX0+FX1*X                                               F1D11750
             DI=(BX1**3)/12.0                                           F1D11760
             D11=DI*AX0/(1.0-BX0*ANU21)                                 F1D11770
             D22=D11*(AX1/AX0)                                          F1D11780
             D12=BX0*D22                                                F1D11790
             JJ=1                                                       F1D11800
             DO 150 J=1,NPE                                             F1D11810
             IF(ITEM.LE.2)THEN                                          F1D11820
                 ELF(JJ)=ELF(JJ)+FX*SF(J)*CONST*X                       F1D11830
             ENDIF                                                      F1D11840
             II=1                                                       F1D11850
             DO 140 I=1,NPE                                             F1D11860
             BIJ = CONST*GDSF(I)*GDSF(J)*D11*X                          F1D11870
             CIJ = CONST*SF(I)*SF(J)*X                                  F1D11880
             DIJ = CONST*(GDSF(I)*SF(J)+SF(I)*GDSF(J))*D12              F1D11890
             EIJ = CONST*SF(I)*SF(J)*D22/X                              F1D11900
             ELK(II,JJ)     =ELK(II,JJ)      + CX*CIJ                   F1D11910
             ELK(II+1,JJ+1)=ELK(II+1,JJ+1) + BIJ + DIJ + EIJ            F1D11920
             IF(ITEM.NE.0)THEN                                          F1D11930
                 ELM(II,JJ)     =ELM(II,JJ)     +CT0*CIJ                F1D11940
                 ELM(II+1,JJ+1)=ELM(II+1,JJ+1)+CT1*CIJ                  F1D11950
             ENDIF                                                      F1D11960
  140        II=NDF*I+1                                                 F1D11970
  150        JJ=NDF*J+1                                                 F1D11980
         ENDIF                                                          F1D11990
  160    CONTINUE                                                       F1D12000
C                                                                       F1D12010
C     Reduced integration is used to evaluate the transverse shear termsF1D12020
C                                                                       F1D12030
         LGP=NGP-1                                                      F1D12040
         DO 230 NI=1,LGP                                                F1D12050
         XI=GAUSPT(NI,LGP)                                              F1D12060
C                                                                       F1D12070
         CALL SHP1D(H,IELEM,NPE,XI)                                     F1D12080
         CONST=GJ*GAUSWT(NI,LGP)                                        F1D12090
C                                                                       F1D12100
         X = 0.0                                                        F1D12110
         DO 170 J=1,NPE                                                 F1D12120
  170    X = X + SF(J)*ELX(J)                                           F1D12130
         IF(NTYPE.EQ.0 .OR. NTYPE.EQ.2)THEN                             F1D12140
```

```
C                                                                       F1D12150
C        The Timoshenko BEAM element (MODEL=2 and NTYPE=0 or 2)        F1D12160
C        AX = GAK = AX0 + AX1*X  (reduced integration)                 F1D12170
C                                                                       F1D12180
            AX=AX0+AX1*X                                                F1D12190
            JJ=1                                                        F1D12200
            DO 190 J=1,NPE                                              F1D12210
            II=1                                                        F1D12220
            DO 180 I=1,NPE                                              F1D12230
            B11=GDSF(I)*GDSF(J)*CONST                                   F1D12240
            B01=SF(I)*GDSF(J)*CONST                                     F1D12250
            B10=GDSF(I)*SF(J)*CONST                                     F1D12260
            B00=SF(I)*SF(J)*CONST                                       F1D12270
            ELK(II,JJ)      =ELK(II,JJ)      +AX*B11                    F1D12280
            ELK(II,JJ+1)    =ELK(II,JJ+1)    +AX*B10                    F1D12290
            ELK(II+1,JJ)    =ELK(II+1,JJ)    +AX*B01                    F1D12300
            ELK(II+1,JJ+1)=ELK(II+1,JJ+1)+AX*B00                        F1D12310
  180       II=I*NDF+1                                                  F1D12320
  190       JJ=J*NDF+1                                                  F1D12330
        ELSE                                                            F1D12340
C                                                                       F1D12350
C        Timoshenko CIRCULAR PLATE element (MODEL=2 and NTYPE=1 or 3)  F1D12360
C                    BX1=H, FX2=G13*AREA*K (reduced integration)       F1D12370
C                                                                       F1D12380
            A33=BX1*FX2                                                 F1D12390
            JJ=1                                                        F1D12400
            DO 210 J=1,NPE                                              F1D12410
            II=1                                                        F1D12420
            DO 200 I=1,NPE                                              F1D12430
            BIJ = CONST*GDSF(I)*GDSF(J)*X                               F1D12440
            CIJ = CONST*SF(I)*SF(J)*X                                   F1D12450
            DIJ = CONST*GDSF(I)*SF(J)*X                                 F1D12460
            DJI = CONST*SF(I)*GDSF(J)*X                                 F1D12470
            ELK(II,JJ)      =ELK(II,JJ)       + A33*BIJ                 F1D12480
            ELK(II,JJ+1)    =ELK(II,JJ+1)     + A33*DIJ                 F1D12490
            ELK(II+1,JJ)    =ELK(II+1,JJ)     + A33*DJI                 F1D12500
            ELK(II+1,JJ+1)=ELK(II+1,JJ+1)  + A33*CIJ                    F1D12510
  200       II=NDF*I+1                                                  F1D12520
  210       JJ=NDF*J+1                                                  F1D12530
        ENDIF                                                           F1D12540
  230   CONTINUE                                                        F1D12550
        IF(ITEM.EQ.0 .AND. NTYPE.GT.1)THEN                              F1D12560
            CALL TMSFRC(ELF,ELX,FX0,FX1,FX2,H,NTYPE,NE,F3,MXELM)        F1D12570
        ENDIF                                                           F1D12580
      ENDIF                                                             F1D12590
        IF(ITEM.GT.2)RETURN                                             F1D12600
      IF(ITEM.EQ.1 .OR. ITEM.EQ.2)THEN                                  F1D12610
C                                                                       F1D12620
C     Equivalent coefficient matrices for TIME-DEPENDENT problems       F1D12630
C                                                                       F1D12640
        IF(ITEM .EQ. 1)THEN                                             F1D12650
C                                                                       F1D12660
C        Alfa-family of time approximation for PARABOLIC equations     F1D12670
C                                                                       F1D12680
            DO 250 J=1,NN                                               F1D12690
            SUM=0.0                                                     F1D12700
            DO 240 I=1,NN                                               F1D12710
            SUM=SUM+(ELM(I,J)-A2*ELK(I,J))*ELU(I)                       F1D12720
  240       ELK(I,J)=ELM(I,J)+A1*ELK(I,J)                               F1D12730
  250       ELF(J)=(A1+A2)*ELF(J)+SUM                                   F1D12740
        ELSE                                                            F1D12750
C                                                                       F1D12760
C        Newmark-family of approximation for HYPERBOLIC equations      F1D12770
C                                                                       F1D12780
            IF(TIME.EQ.0.0)THEN                                         F1D12790
                DO 260 J=1,NN                                           F1D12800
                DO 260 I=1,NN                                           F1D12810
                ELF(J)=ELF(J)-ELK(I,J)*ELU(I)                           F1D12820
  260           ELK(I,J)=ELM(I,J)                                       F1D12830
            ELSE                                                        F1D12840
                DO 270 J=1,NN                                           F1D12850
                DO 270 I=1,NN                                           F1D12860
                ELF(J)=ELF(J)+ELM(I,J)*(A3*ELU(I)+A4*ELV(I)+A5*ELA(I))F1D12870
  270           ELK(I,J)=ELK(I,J)+A3*ELM(I,J)                           F1D12880
            ENDIF                                                       F1D12890
```

```
         ENDIF                                                          F1D12900
      ENDIF                                                             F1D12910
      RETURN                                                            F1D12920
      END                                                               F1D12930

      SUBROUTINE ECHO(IN,IT)                                            F1D12940
      IMPLICIT REAL*8(A-H,O-Z)                                          F1D12950
C                                                                       F1D12960
      DIMENSION AA(20)                                                  F1D12970
      WRITE(IT,40)                                                      F1D12980
   10 CONTINUE                                                          F1D12990
      READ(IN,30,END=20) AA                                             F1D13000
      WRITE(IT,60) AA                                                   F1D13010
      GO TO 10                                                          F1D13020
   20 CONTINUE                                                          F1D13030
      REWIND(IN)                                                        F1D13040
      WRITE(IT,50)                                                      F1D13050
      RETURN                                                            F1D13060
   30 FORMAT(20A4)                                                      F1D13070
   40 FORMAT(5X,'*** ECHO OF THE INPUT DATA STARTS ***',/)              F1D13080
   50 FORMAT(5X,'**** ECHO OF THE INPUT DATA ENDS ****',/)              F1D13090
   60 FORMAT(1X,20A4)                                                   F1D13100
      END                                                               F1D13110

      SUBROUTINE JACOBI (N,Q,JVEC,M,V,X,IH,MXNEQ)                       F1D13120
C     __________________________________________________________________F1D13130
C                                                                       F1D13140
C        PURPOSE: To diagonalize matrix [Q] by successive rotations     F1D13150
C                                                                       F1D13160
C        DESCRIPTION OF THE VARIABLES:                                  F1D13170
C                                                                       F1D13180
C          N    .... Order of the real, symmetric matrix [Q] (N > 2)    F1D13190
C         [Q]   .... The matrix to be diagonalized (destroyed)          F1D13200
C         JVEC  .... 0, when only eigenvalues alone have to be found    F1D13210
C         [V]   .... Matrix of eigenvectors                             F1D13220
C          M    .... Number of rotations performed                      F1D13230
C                                                                       F1D13240
C     __________________________________________________________________F1D13250
      IMPLICIT REAL*8 (A-H,O-Z)                                         F1D13260
      DIMENSION Q(MXNEQ,MXNEQ),V(MXNEQ,MXNEQ),X(MXNEQ),IH(MXNEQ)        F1D13270
      EPSI=1.0D-08                                                      F1D13280
C                                                                       F1D13290
      IF(JVEC)10,50,10                                                  F1D13300
   10 DO 40 I=1,N                                                       F1D13310
      DO 40 J=1,N                                                       F1D13320
      IF(I-J)30,20,30                                                   F1D13330
   20 V(I,J)=1.0                                                        F1D13340
      GO TO 40                                                          F1D13350
   30 V(I,J)=0.0                                                        F1D13360
   40 CONTINUE                                                          F1D13370
   50 M=0                                                               F1D13380
      MI=N-1                                                            F1D13390
      DO 70 I=1,MI                                                      F1D13400
      X(I)=0.0                                                          F1D13410
      MJ=I+1                                                            F1D13420
      DO 70 J=MJ,N                                                      F1D13430
      IF(X(I)-DABS(Q(I,J)))60,60,70                                     F1D13440
   60 X(I)=DABS(Q(I,J))                                                 F1D13450
      IH(I)=J                                                           F1D13460
   70 CONTINUE                                                          F1D13470
   75 DO 100 I=1,MI                                                     F1D13480
      IF(I-1)90,90,80                                                   F1D13490
   80 IF(XMAX-X(I))90,100,100                                           F1D13500
   90 XMAX=X(I)                                                         F1D13510
      IP=I                                                              F1D13520
      JP=IH(I)                                                          F1D13530
  100 CONTINUE                                                          F1D13540
      IF(XMAX-EPSI)500,500,110                                          F1D13550
  110 M=M+1                                                             F1D13560
      IF(Q(IP,IP)-Q(JP,JP))120,130,130                                  F1D13570
  120 TANG=-2.0*Q(IP,JP)/(DABS(Q(IP,IP)-Q(JP,JP))+DSQRT((Q(IP,IP)       F1D13580
     1     -Q(JP,JP))**2+4.0*Q(IP,JP)**2))                              F1D13590
      GO TO 140                                                         F1D13600
```

```
130 TANG= 2.0*Q(IP,JP)/(DABS(Q(IP,IP)-Q(JP,JP))+DSQRT((Q(IP,IP)      F1D13610
  1      -Q(JP,JP))**2+4.0*Q(IP,JP)**2))                             F1D13620
140 COSN=1.0/DSQRT(1.0+TANG**2)                                       F1D13630
    SINE=TANG*COSN                                                    F1D13640
    QII=Q(IP,IP)                                                      F1D13650
    Q(IP,IP)=COSN**2*(QII+TANG*(2.*Q(IP,JP)+TANG*Q(JP,JP)))           F1D13660
    Q(JP,JP)=COSN**2*(Q(JP,JP)-TANG*(2.*Q(IP,JP)-TANG*QII))           F1D13670
    Q(IP,JP)=0.0                                                      F1D13680
    IF (Q(IP,IP)-Q(JP,JP)) 150,190,190                                F1D13690
150 TEMP=Q(IP,IP)                                                     F1D13700
    Q(IP,IP)=Q(JP,JP)                                                 F1D13710
    Q(JP,JP)=TEMP                                                     F1D13720
    IF(SINE) 160,170,170                                              F1D13730
160 TEMP=COSN                                                         F1D13740
    GOTO 180                                                          F1D13750
170 TEMP=-COSN                                                        F1D13760
180 COSN=DABS(SINE)                                                   F1D13770
    SINE=TEMP                                                         F1D13780
190 DO 260 I=1,MI                                                     F1D13790
    IF (I-IP) 210,260,200                                             F1D13800
200 IF (I-JP) 210,260,210                                             F1D13810
210 IF (IH(I)-IP) 220,230,220                                         F1D13820
220 IF (IH(I)-JP) 260,230,260                                         F1D13830
230 K=IH(I)                                                           F1D13840
    TEMP=Q(I,K)                                                       F1D13850
    Q(I,K)=0.0                                                        F1D13860
    MJ=I+1                                                            F1D13870
    X(I)=0.0                                                          F1D13880
    DO 250 J=MJ,N                                                     F1D13890
    IF (X(I)-DABS(Q(I,J))) 240,240,250                                F1D13900
240 X(I)=DABS(Q(I,J))                                                 F1D13910
    IH(I)=J                                                           F1D13920
250 CONTINUE                                                          F1D13930
    Q(I,K)=TEMP                                                       F1D13940
260 CONTINUE                                                          F1D13950
    X(IP)=0.0                                                         F1D13960
    X(JP)=0.0                                                         F1D13970
    DO 430 I=1,N                                                      F1D13980
    IF(I-IP) 270,430,320                                              F1D13990
270 TEMP=Q(I,IP)                                                      F1D14000
    Q(I,IP)=COSN*TEMP+SINE*Q(I,JP)                                    F1D14010
    IF (X(I)-DABS(Q(I,IP))) 280,290,290                               F1D14020
280 X(I)=DABS(Q(I,IP))                                                F1D14030
    IH(I)=IP                                                          F1D14040
290 Q(I,JP)=-SINE*TEMP+COSN*Q(I,JP)                                   F1D14050
    IF (X(I)-DABS(Q(I,JP))) 300,430,430                               F1D14060
300 X(I)=DABS(Q(I,JP))                                                F1D14070
    IH(I)=JP                                                          F1D14080
    GO TO 430                                                         F1D14090
320 IF(I-JP) 330,430,380                                              F1D14100
330 TEMP=Q(IP,I)                                                      F1D14110
    Q(IP,I)=COSN*TEMP+SINE*Q(I,JP)                                    F1D14120
    IF(X(IP)-DABS(Q(IP,I)))340,350,350                                F1D14130
340 X(IP)=DABS(Q(IP,I))                                               F1D14140
    IH(IP)=I                                                          F1D14150
350 Q(I,JP)=-SINE*TEMP+COSN*Q(I,JP)                                   F1D14160
    IF (X(I)-DABS(Q(I,JP))) 300,430,430                               F1D14170
380 TEMP=Q(IP,I)                                                      F1D14180
    Q(IP,I)=COSN*TEMP+SINE*Q(JP,I)                                    F1D14190
    IF(X(IP)-DABS(Q(IP,I)))390,400,400                                F1D14200
390 X(IP)=DABS(Q(IP,I))                                               F1D14210
    IH(IP)=I                                                          F1D14220
400 Q(JP,I)=-SINE*TEMP+COSN*Q(JP,I)                                   F1D14230
    IF(X(JP)-DABS(Q(JP,I)))410,430,430                                F1D14240
410 X(JP)=DABS(Q(JP,I))                                               F1D14250
    IH(JP)=I                                                          F1D14260
430 CONTINUE                                                          F1D14270
    IF(JVEC)440,75,440                                                F1D14280
440 DO 450 I=1,N                                                      F1D14290
    TEMP=V(I,IP)                                                      F1D14300
    V(I,IP)=COSN*TEMP+SINE*V(I,JP)                                    F1D14310
450 V(I,JP)=-SINE*TEMP+COSN*V(I,JP)                                   F1D14320
    GOTO 75                                                           F1D14330
500 RETURN                                                            F1D14340
    END                                                               F1D14350
```

```
      SUBROUTINE MESH1D(NEM,NNM,NPE,NOD,MXELM,MXNOD,DX,GLX)             F1D14360
C     __________________________________________________________________F1D14370
C                                                                       F1D14380
C          The subroutine computes the arrays {GLX} and [NOD]           F1D14390
C                                                                       F1D14400
C        {GLX}.... Vector of global coordinates                         F1D14410
C        {DX}..... Vector of element lengths [DX(1) = node 1 coord.]    F1D14420
C        [NOD].... Connectivity matrix                                  F1D14430
C     __________________________________________________________________F1D14440
C                                                                       F1D14450
      IMPLICIT REAL*8 (A-H,O-Z)                                         F1D14460
      DIMENSION GLX(MXNOD),DX(MXNOD),NOD(MXELM,4)                       F1D14470
C                                                                       F1D14480
C     Generate the elements of the connectivity matrix                  F1D14490
C                                                                       F1D14500
      DO 10 I=1,NPE                                                     F1D14510
   10 NOD(1,I)=I                                                        F1D14520
      DO 20 N=2,NEM                                                     F1D14530
      DO 20 I=1,NPE                                                     F1D14540
   20 NOD(N,I) = NOD(N-1,I)+NPE-1                                       F1D14550
C                                                                       F1D14560
C     Generate global coordinates of the global nodes                   F1D14570
C                                                                       F1D14580
      GLX(1)=DX(1)                                                      F1D14590
      IF(NPE.EQ.2)THEN                                                  F1D14600
          DO 30 I=1,NEM                                                 F1D14610
   30     GLX(I+1) = GLX(I) + DX(I+1)                                   F1D14620
      ELSE                                                              F1D14630
          DO 40 I=1,NEM                                                 F1D14640
          II=2*I                                                        F1D14650
          GLX(II) = GLX(II-1) + 0.5*DX(I+1)                             F1D14660
   40     GLX(II+1)=GLX(II-1) + DX(I+1)                                 F1D14670
      ENDIF                                                             F1D14680
      RETURN                                                            F1D14690
      END                                                               F1D14700

      SUBROUTINE MTRXML(A,N,M,B,L,C)                                    F1D14710
C     ________________________________________________________________  F1D14720
C                                                                       F1D14730
C         COMPUTE THE PRODUCT OF MATRICES [A] AND [B]: [C]=[A][B]       F1D14740
C                                                                       F1D14750
C     ________________________________________________________________  F1D14760
      IMPLICIT REAL*8 (A-H,O-Z)                                         F1D14770
      DIMENSION A(50,50),B(50,50),C(50,50)                              F1D14780
      DO 10 I=1,N                                                       F1D14790
      DO 10 J=1,L                                                       F1D14800
      C(I,J)=0.0                                                        F1D14810
      DO 10 K=1,M                                                       F1D14820
   10 C(I,J)=C(I,J)+A(I,K)*B(K,J)                                       F1D14830
      RETURN                                                            F1D14840
      END                                                               F1D14850
```

```
      SUBROUTINE PSTPRC(F3,GLF,GLX,NOD,GA,IELEM,NPE,MODEL,NTYPE,ITEM,    F1D14860
     *                  MXELM,MXNEQ,MXNOD,NEM,NDF)                       F1D14870
C     ________________________________________________________________   F1D14880
C                                                                        F1D14890
C     The subroutine computes the solution and its derivatives at five  F1D14900
C     points, including the nodes of the element. Note that the  post-  F1D14910
C     computed bending moment (BM) and shear force (VF) are as per the  F1D14920
C     sign convention shown in Figure 3.33 of the text; they differ in  F1D14930
C     sign from the element degrees of freedom.                         F1D14940
C                                                                        F1D14950
C        X........ Global (i.e., problem) coordinate                     F1D14960
C        XI ...... Local (i.e., element) coordinate                      F1D14970
C        SF....... Element interpolation (or shape) functions            F1D14980
C        GDSF..... First derivative of SF w.r.t. global coordinate       F1D14990
C        GDDSF.... Second derivative of SF w.r.t. global coordinate      F1D15000
C        ELU...... Element solution vector                               F1D15010
C        U........ Interpolated solution                                 F1D15020
C        DU....... Interpolated derivative of the solution               F1D15030
C        W........ Interpolated transverse deflection                    F1D15040
C        S........ Interpolated rotation function                        F1D15050
C        DS....... Interpolated derivative of the rotation               F1D15060
C        DW....... Interpolated derivative of the transverse deflection F1D15070
C        DDW...... Interpolated second derivative of transverse defl.    F1D15080
C        DDDW..... Interpolated third derivative of transverse defl.     F1D15090
C     ________________________________________________________________   F1D15100
C                                                                        F1D15110
      IMPLICIT REAL*8 (A-H,O-Z)                                          F1D15120
      DIMENSION F3(MXELM),GLF(MXNEQ),GLX(MXNOD),NOD(MXELM,4)             F1D15130
      DIMENSION XP(9),ELX(4),ELU(9)                                      F1D15140
      COMMON/IO/IN,IT                                                    F1D15150
      COMMON/SHP/SF(4),GDSF(4),GDDSF(4),GJ                               F1D15160
      COMMON/STF2/A1,A2,A3,A4,A5,AX0,AX1,BX0,BX1,CX0,CX1,CT0,CT1,FX0,    F1D15170
     *            FX1,FX2                                                F1D15180
      DATA XP/-1.0D0, -0.750D0, -0.50D0, -0.250D0, 0.0D0, 0.250D0,       F1D15190
     *                          0.50D0,  0.750D0, 1.0D0/                 F1D15200
C                                                                        F1D15210
      NPTS=9                                                             F1D15220
      DO 80 NE = 1, NEM                                                  F1D15230
      L=0                                                                F1D15240
      DO 10 I=1,NPE                                                      F1D15250
      NI=NOD(NE,I)                                                       F1D15260
      ELX(I)=GLX(NI)                                                     F1D15270
      LI=(NI-1)*NDF                                                      F1D15280
      DO 10 J=1,NDF                                                      F1D15290
      LI=LI+1                                                            F1D15300
      L=L+1                                                              F1D15310
   10 ELU(L)=GLF(LI)                                                     F1D15320
      H = ELX(NPE) - ELX(1)                                              F1D15330
C                                                                        F1D15340
      DO 70 NI=1,NPTS                                                    F1D15350
      XI = XP(NI)                                                        F1D15360
      CALL SHP1D(H,IELEM,NPE,XI)                                         F1D15370
      IF(MODEL.EQ.3)THEN                                                 F1D15380
          W=0.0                                                          F1D15390
          DW=0.0                                                         F1D15400
          DDW=0.0                                                        F1D15410
          XC=ELX(1)+0.5*H*(1.0+XI)                                       F1D15420
          DO 20 I=1,4                                                    F1D15430
          W  =W  + SF(I)*ELU(I)                                          F1D15440
          DW =DW + GDSF(I)*ELU(I)                                        F1D15450
   20     DDW=DDW+ GDDSF(I)*ELU(I)                                       F1D15460
          DDDW=((ELU(1)-ELU(3))*2.0/H-(ELU(4)+ELU(2)))*6.0/(H*H)         F1D15470
          ROT=-DW                                                        F1D15480
          IF(NTYPE.EQ.0)THEN                                             F1D15490
              BM=(BX0+XC*BX1)*DDW                                        F1D15500
              VF=(BX0+XC*BX1)*DDDW + BX1*DDW                             F1D15510
              WRITE(IT,90)XC,W,ROT,BM,VF                                 F1D15520
          ELSE                                                           F1D15530
              ANU21=BX0*AX0/AX1                                          F1D15540
              DI=(BX1**3)/12.0                                           F1D15550
              D11=DI*AX0/(1.0-BX0*ANU21)                                 F1D15560
              D22=D11*(AX1/AX0)                                          F1D15570
              D12=BX0*D22                                                F1D15580
              BMR=D11*DDW*XC+D12*DW                                      F1D15590
              BMT=D12*DDW*XC+D22*DW                                      F1D15600
```

```
            IF(XC.NE.0.0)THEN                                            F1D15610
                SFV=D11*(XC*DDDW+DDW)-D22*DW/XC                          F1D15620
                WRITE(IT,90)XC,W,ROT,BMR,BMT,SFV                         F1D15630
            ELSE                                                         F1D15640
                WRITE(IT,90)XC,W,ROT,BMR,BMT                             F1D15650
            ENDIF                                                        F1D15660
        ENDIF                                                            F1D15670
    ELSE                                                                 F1D15680
        XC=0.0                                                           F1D15690
        DO 30 I=1,NPE                                                    F1D15700
 30     XC=XC+SF(I)*ELX(I)                                               F1D15710
        IF(MODEL.EQ.1)THEN                                               F1D15720
            U=0.0                                                        F1D15730
            DU=0.0                                                       F1D15740
            DO 40 I=1,NPE                                                F1D15750
            U=U+SF(I)*ELU(I)                                             F1D15760
 40         DU=DU+GDSF(I)*ELU(I)                                         F1D15770
            IF(NTYPE.EQ.0)THEN                                           F1D15780
                SV=(AX0+AX1*XC)*DU                                       F1D15790
                WRITE(IT,90)XC,U,SV                                      F1D15800
            ELSE                                                         F1D15810
                ANU21=BX0*AX0/AX1                                        F1D15820
                IF(NTYPE.EQ.1)THEN                                       F1D15830
                    C11=BX1*AX0/(1.0-BX0*ANU21)                          F1D15840
                    C22=C11*(AX1/AX0)                                    F1D15850
                    C12=BX0*C22                                          F1D15860
                ELSE                                                     F1D15870
                    DENOM=1.0-BX0-ANU21                                  F1D15880
                    C11=BX1*AX0*(1.0-BX0)/(1.0+BX0)/DENOM                F1D15890
                    C22=BX1*AX1*(1.0-ANU21)/(1.0+ANU21)/DENOM            F1D15900
                    C12=BX0*C22                                          F1D15910
                ENDIF                                                    F1D15920
                IF(XC.NE.0.0)THEN                                        F1D15930
                    SR=C11*DU+C12*U/XC                                   F1D15940
                    ST=C12*DU+C22*U/XC                                   F1D15950
                    WRITE(IT,90)XC,U,SR,ST                               F1D15960
                ELSE                                                     F1D15970
                    WRITE(IT,90)XC,U,DU                                  F1D15980
                ENDIF                                                    F1D15990
            ENDIF                                                        F1D16000
        ELSE                                                             F1D16010
C       MODEL.EQ.2  Calculations                                         F1D16020
            IF(ITEM.EQ.0 .AND. NTYPE.GT.1)THEN                           F1D16030
                H=ELX(NPE)-ELX(1)                                        F1D16040
                CALL TMSTRS(GA,ELU,XI,W,DW,S,DS,NE,F3,H,MXELM)           F1D16050
            ELSE                                                         F1D16060
                W=0.0                                                    F1D16070
                DW=0.0                                                   F1D16080
                S = 0.0                                                  F1D16090
                DS=0.0                                                   F1D16100
                DDS=0.0                                                  F1D16110
                DO 50 I=1,NPE                                            F1D16120
                L=2*I-1                                                  F1D16130
                W = W + SF(I)*ELU(L)                                     F1D16140
                DW=DW+GDSF(I)*ELU(L)                                     F1D16150
                S = S + SF(I)*ELU(L+1)                                   F1D16160
                DS=DS+GDSF(I)*ELU(L+1)                                   F1D16170
 50             DDS=DDS+GDDSF(I)*ELU(L+1)                                F1D16180
            ENDIF                                                        F1D16190
                IF(NTYPE.EQ.0 .OR. NTYPE.EQ.2)THEN                       F1D16200
                    BM=-(BX0+BX1*XC)*DS                                  F1D16210
                    VF=-(AX0+AX1*XC)*(DW+S)                              F1D16220
                    WRITE(IT,90)XC,W,S,BM,VF                             F1D16230
                ELSE                                                     F1D16240
                    ANU21=BX0*AX0/AX1                                    F1D16250
                    DI=(BX1**3)/12.0                                     F1D16260
                    D11=DI*AX0/(1.0-BX0*ANU21)                           F1D16270
                    D22=D11*(AX1/AX0)                                    F1D16280
                    D12=BX0*D22                                          F1D16290
                    BMR=-(D11*DS*XC+D12*S)                               F1D16300
                    BMT=-(D12*DS*XC+D22*S)                               F1D16310
                    SFV=-FX2*(DW+S)*XC                                   F1D16320
                    WRITE(IT,90)XC,W,S,BMR,BMT,SFV                       F1D16330
                ENDIF                                                    F1D16340
        ENDIF                                                            F1D16350
```

```
      ENDIF                                                             F1D16360
   70 CONTINUE                                                          F1D16370
   80 CONTINUE                                                          F1D16380
      RETURN                                                            F1D16390
   90 FORMAT (2X,6E13.5)                                                F1D16400
      END                                                               F1D16410

      SUBROUTINE REACTN(MXELM,MXNEQ,NDF,NEM,NOD,NPE,NTYPE,PR,GLF,       F1D16420
     *                  SE,SL,SA,SI,CS,SN,CNT,SNT,HF,VF,PF,XB)          F1D16430
C                                                                       F1D16440
C     __________________________________________________________________F1D16450
C                                                                       F1D16460
C     The subroutine computes the generalized reaction forces in each   F1D16470
C         element of the truss (NDF=2) or frame (NDF=3) structure       F1D16480
C                                                                       F1D16490
C     __________________________________________________________________F1D16500
C                                                                       F1D16510
      IMPLICIT REAL*8(A-H,O-Z)                                          F1D16520
      DIMENSION  PR(MXELM),SE(MXELM),SL(MXELM),SA(MXELM),SI(MXELM)      F1D16530
      DIMENSION  CS(MXELM),SN(MXELM),CNT(MXELM),SNT(MXELM)              F1D16540
      DIMENSION  HF(MXELM),VF(MXELM),PF(MXELM),XB(MXELM)                F1D16550
      DIMENSION  NOD(MXELM,4),GLF(MXNEQ)                                F1D16560
      DIMENSION  ELR(6),TRM(6,6),TMPK(6,6)                              F1D16570
      COMMON/STF1/ELK(9,9),ELM(9,9),ELF(9),ELX(4),ELU(9),ELV(9),ELA(9)  F1D16580
C                                                                       F1D16590
      NN=NPE*NDF                                                        F1D16600
      DO 140 N=1,NEM                                                    F1D16610
      CN1=CS(N)                                                         F1D16620
      SN1=SN(N)                                                         F1D16630
C                                                                       F1D16640
C     Call TRSFRM to compute element stiffness matrix and force vector  F1D16650
C                                                                       F1D16660
      L=0                                                               F1D16670
      DO 100 I=1,NPE                                                    F1D16680
      NI=NOD(N,I)                                                       F1D16690
      LI=(NI-1)*NDF                                                     F1D16700
      DO 100 J=1,NDF                                                    F1D16710
      LI=LI+1                                                           F1D16720
      L=L+1                                                             F1D16730
  100 ELU(L)=GLF(LI)                                                    F1D16740
      CALL TRSFRM (MXELM,MXNEQ,N,NTYPE,PR,SE,SL,SA,SI,CS,SN,            F1D16750
     *             CNT,SNT,HF,VF,PF,XB)                                 F1D16760
C                                                                       F1D16770
C     Compute the FORCE and MOMENT RESULTANTS                           F1D16780
C                                                                       F1D16790
      DO 120 I=1,NN                                                     F1D16800
      ELR(I) = 0.0                                                      F1D16810
      DO 110 J=1,NN                                                     F1D16820
  110 ELR(I) =  ELR(I) + ELK(I,J)*ELU(J)                                F1D16830
  120 ELR(I) =  ELR(I) - ELF(I)                                         F1D16840
      ELF(1) =  ELR(1)*CN1+ELR(2)*SN1                                   F1D16850
      ELF(2) = -ELR(1)*SN1+ELR(2)*CN1                                   F1D16860
      IF(NTYPE.NE.0) THEN                                               F1D16870
          ELF(3) =  ELR(3)                                              F1D16880
          ELF(4) =  ELR(4)*CN1+ELR(5)*SN1                               F1D16890
          ELF(5) = -ELR(4)*SN1+ELR(5)*CN1                               F1D16900
          ELF(6) =  ELR(6)                                              F1D16910
      ELSE                                                              F1D16920
          ELF(3) =  ELR(3)*CN1+ELR(4)*SN1                               F1D16930
          ELF(4) = -ELR(3)*SN1+ELR(4)*CN1                               F1D16940
      ENDIF                                                             F1D16950
      WRITE(6,150)N, (ELF(I),I=1,NN)                                    F1D16960
      WRITE(6,160)   (ELR(I),I=1,NN)                                    F1D16970
  140 CONTINUE                                                          F1D16980
      RETURN                                                            F1D16990
  150 FORMAT (3X,I2,6E12.4)                                             F1D17000
  160 FORMAT (5X,6E12.4,/)                                              F1D17010
      END                                                               F1D17020
```

```
      SUBROUTINE SHP1D(H,IELEM,NPE,XI)                                  F1D17030
C     __________________________________________________________________F1D17040
C                                                                       F1D17050
C     The subroutine computes shape functions and their derivatives for F1D17060
C      Hermite cubic and Lagrange linear, quadratic and cubic elements  F1D17070
C                                                                       F1D17080
C        X.........  Global (i.e., problem) coordinate                   F1D17090
C        XI .......  Local (i.e., element) coordinate                   F1D17100
C        H.........  Element length                                     F1D17110
C        {SF}......  Interpolation (or shape) functions                 F1D17120
C        {DSF}.....  First derivative of SF w.r.t. XI                   F1D17130
C        {DDSF}....  Second derivative of SFH w.r.t. XI                 F1D17140
C        {GDSF}....  First derivative of SF w.r.t. X                    F1D17150
C        {GDDSF}...  Second derivative of SFH w.r.t. X                  F1D17160
C        GJ........  Jacobian of the transformation                     F1D17170
C     __________________________________________________________________F1D17180
C                                                                       F1D17190
      IMPLICIT REAL*8 (A-H,O-Z)                                         F1D17200
      COMMON/SHP/SF(4),GDSF(4),GDDSF(4),GJ                              F1D17210
      DIMENSION DSF(4),DDSF(4)                                          F1D17220
      IF(IELEM.EQ.0)THEN                                                F1D17230
C                                                                       F1D17240
C     HERMITE interpolation functions (for the Euler-Bernoulli theory)  F1D17250
C                                                                       F1D17260
          NET=4                                                         F1D17270
          SF(1) = 0.25*(2.0-3.0*XI+XI**3)                               F1D17280
          SF(2) =  -H*(1.0-XI)*(1.0-XI*XI)/8.0                          F1D17290
          SF(3) = 0.25*(2.0+3.0*XI-XI**3)                               F1D17300
          SF(4) =   H*(1.0+XI)*(1.0-XI*XI)/8.0                          F1D17310
          DSF(1) = -0.75*(1.0-XI*XI)                                    F1D17320
          DSF(2) =  H*(1.0+2.0*XI-3.0*XI*XI)/8.0                        F1D17330
          DSF(3) =  0.75*(1.0-XI*XI)                                    F1D17340
          DSF(4) =  H*(1.0-2.0*XI-3.0*XI*XI)/8.0                        F1D17350
          DDSF(1) = 1.5*XI                                              F1D17360
          DDSF(2) = 0.25*H*(1.0-3.0*XI)                                 F1D17370
          DDSF(3) = -1.5*XI                                             F1D17380
          DDSF(4) =-0.25*(1.0+3.0*XI)*H                                 F1D17390
      ELSE                                                              F1D17400
          NET=NPE                                                       F1D17410
          IF(IELEM.EQ.1)THEN                                            F1D17420
C                                                                       F1D17430
C     LAGRANGE interpolation functions used for linear, quadratic and   F1D17440
C             cubic approximation of second-order equations             F1D17450
C                                                                       F1D17460
C     LINEAR interpolation functions                                    F1D17470
C                                                                       F1D17480
              SF(1) = 0.5*(1.0-XI)                                      F1D17490
              SF(2) = 0.5*(1.0+XI)                                      F1D17500
              DSF(1) = -0.5                                             F1D17510
              DSF(2) = 0.5                                              F1D17520
              DDSF(1) = 0.0                                             F1D17530
              DDSF(2) = 0.0                                             F1D17540
          ELSE                                                          F1D17550
              IF(IELEM.EQ.2)THEN                                        F1D17560
C                                                                       F1D17570
C     QUADRATIC interpolation functions                                 F1D17580
C                                                                       F1D17590
                  SF(1) = -0.5*XI*(1.0-XI)                              F1D17600
                  SF(2) = 1.0-XI*XI                                     F1D17610
                  SF(3) = 0.5*XI*(1.0+XI)                               F1D17620
                  DSF(1) = -0.5*(1.0-2.0*XI)                            F1D17630
                  DSF(2) = -2.0*XI                                      F1D17640
                  DSF(3) = 0.5*(1.0+2.0*XI)                             F1D17650
                  DDSF(1) = 1.0                                         F1D17660
                  DDSF(2) = -2.0                                        F1D17670
                  DDSF(3) = 1.0                                         F1D17680
              ELSE                                                      F1D17690
C                                                                       F1D17700
C     CUBIC interpolation functions                                     F1D17710
C                                                                       F1D17720
                  SF(1) = 0.0625*(1.0-XI)*(9.0*XI*XI-1.)                F1D17730
                  SF(2) = 0.5625*(1.0-XI*XI)*(1.0-3.0*XI)               F1D17740
                  SF(3) = 0.5625*(1.0-XI*XI)*(1.0+3.0*XI)               F1D17750
                  SF(4) = 0.0625*(9.0*XI*XI-1.0)*(1.0+XI)               F1D17760
                  DSF(1) = 0.0625*(1.0+18.0*XI-27.0*XI*XI)              F1D17770
```

```
                  DSF(2) = 0.5625*(-3.0-2.0*XI+9.0*XI*XI)              F1D17780
                  DSF(3) = 0.5625*(3.0-2.0*XI-9.0*XI*XI)               F1D17790
                  DSF(4) = 0.0625*(18.0*XI+27.0*XI*XI-1.0)             F1D17800
                  DDSF(1) = 0.0625*(18.0-54.0*XI)                      F1D17810
                  DDSF(2) = 0.5625*(-2.0+18.0*XI)                      F1D17820
                  DDSF(3) = 0.5625*(-2.0-18.0*XI)                      F1D17830
                  DDSF(4) = 0.0625*(18.0+54.0*XI)                      F1D17840
              ENDIF                                                    F1D17850
          ENDIF                                                        F1D17860
C                                                                      F1D17870
C     Compute derivatives of the interpolation functions w.r.t. X      F1D17880
C                                                                      F1D17890
      ENDIF                                                            F1D17900
   80 GJ =  H*0.5                                                      F1D17910
      DO 90 I = 1,NET                                                  F1D17920
      GDSF(I) = DSF(I)/GJ                                              F1D17930
   90 GDDSF(I) = DDSF(I)/GJ/GJ                                         F1D17940
      RETURN                                                           F1D17950
      END                                                              F1D17960

      SUBROUTINE SOLVER(NRM,NCM,NEQNS,NBW,BAND,RHS,IRES)               F1D17970
C     ________________________________________________________________ F1D17980
C                                                                      F1D17990
C     The subroutine solves a  banded,  symmetric,  system of algebraic F1D18000
C     equations using the Gauss elimination method:  [BAND]{U} = {RHS}. F1D18010
C     The coefficient matrix is input as BAND(NEQNS,NBW) and the column F1D18020
C     vector is input as  RHS(NEQNS),  where NEQNS is the actual number F1D18030
C     of equations and NBW is the half band width.  The true dimensions F1D18040
C     of the matrix [BAND] in the calling program, are NRM by NCM. When F1D18050
C     IRES is greater than zero, the right hand elimination is skipped. F1D18060
C     ________________________________________________________________ F1D18070
C                                                                      F1D18080
      IMPLICIT REAL*8(A-H,O-Z)                                         F1D18090
      DIMENSION BAND(NRM,NCM),RHS(NRM)                                 F1D18100
C                                                                      F1D18110
      MEQNS=NEQNS-1                                                    F1D18120
      IF(IRES.LE.0) THEN                                               F1D18130
          DO 30 NPIV=1,MEQNS                                           F1D18140
          NPIVOT=NPIV+1                                                F1D18150
          LSTSUB=NPIV+NBW-1                                            F1D18160
          IF(LSTSUB.GT.NEQNS) THEN                                     F1D18170
              LSTSUB=NEQNS                                             F1D18180
          ENDIF                                                        F1D18190
C                                                                      F1D18200
          DO 20 NROW=NPIVOT,LSTSUB                                     F1D18210
          NCOL=NROW-NPIV+1                                             F1D18220
          FACTOR=BAND(NPIV,NCOL)/BAND(NPIV,1)                          F1D18230
          DO 10 NCOL=NROW,LSTSUB                                       F1D18240
          ICOL=NCOL-NROW+1                                             F1D18250
          JCOL=NCOL-NPIV+1                                             F1D18260
   10     BAND(NROW,ICOL)=BAND(NROW,ICOL)-FACTOR*BAND(NPIV,JCOL)       F1D18270
   20     RHS(NROW)=RHS(NROW)-FACTOR*RHS(NPIV)                         F1D18280
   30     CONTINUE                                                     F1D18290
      ELSE                                                             F1D18300
   40     DO 60 NPIV=1,MEQNS                                           F1D18310
          NPIVOT=NPIV+1                                                F1D18320
          LSTSUB=NPIV+NBW-1                                            F1D18330
          IF(LSTSUB.GT.NEQNS) THEN                                     F1D18340
              LSTSUB=NEQNS                                             F1D18350
          ENDIF                                                        F1D18360
          DO 50 NROW=NPIVOT,LSTSUB                                     F1D18370
          NCOL=NROW-NPIV+1                                             F1D18380
          FACTOR=BAND(NPIV,NCOL)/BAND(NPIV,1)                          F1D18390
   50     RHS(NROW)=RHS(NROW)-FACTOR*RHS(NPIV)                         F1D18400
   60     CONTINUE                                                     F1D18410
      ENDIF                                                            F1D18420
C                                                                      F1D18430
C     Back substitution                                                F1D18440
C                                                                      F1D18450
      DO 90 IJK=2,NEQNS                                                F1D18460
      NPIV=NEQNS-IJK+2                                                 F1D18470
      RHS(NPIV)=RHS(NPIV)/BAND(NPIV,1)                                 F1D18480
      LSTSUB=NPIV-NBW+1                                                F1D18490
      IF(LSTSUB.LT.1) THEN                                             F1D18500
          LSTSUB=1                                                     F1D18510
```

```
      ENDIF                                                             F1D18520
      NPIVOT=NPIV-1                                                     F1D18530
      DO 80 JKI=LSTSUB,NPIVOT                                           F1D18540
      NROW=NPIVOT-JKI+LSTSUB                                            F1D18550
      NCOL=NPIV-NROW+1                                                  F1D18560
      FACTOR=BAND(NROW,NCOL)                                            F1D18570
   80 RHS(NROW)=RHS(NROW)-FACTOR*RHS(NPIV)                              F1D18580
   90 CONTINUE                                                          F1D18590
      RHS(1)=RHS(1)/BAND(1,1)                                           F1D18600
      RETURN                                                            F1D18610
      END                                                               F1D18620


      SUBROUTINE TMSFRC(ELF,ELX,FX0,FX1,FX2,H,NTYPE,NE,F3,MXELM)        F1D18630
C     __________________________________________________________________F1D18640
C                                                                       F1D18650
C       The subroutine computes element force vector for the consistent F1D18660
C                  interpolation Timoshenko element (CIE)               F1D18670
C     __________________________________________________________________F1D18680
C                                                                       F1D18690
      IMPLICIT REAL*8(A-H,O-Z)                                          F1D18700
      COMMON/SHP/SF(4),GDSF(4),GDDSF(4),GJ                              F1D18710
      DIMENSION GAUSPT(5,5),GAUSWT(5,5),ELF(9),ELX(4),EX(3),F3(MXELM)   F1D18720
C                                                                       F1D18730
      DATA GAUSPT/5*0.0D0,-.57735027D0,.57735027D0,3*0.0D0,-.77459667D0,F1D18740
     * 0.0D0,.77459667D0,2*0.0D0,-.86113631D0,-.33998104D0,.33998104D0, F1D18750
     *.86113631D0,0.0D0,-.906180D0,-.538469D0,0.0D0,.538469D0,.906180D0/F1D18760
C                                                                       F1D18770
      DATA GAUSWT/2.0D0,4*0.0D0,2*1.0D0,3*0.0D0,.55555555D0,.88888888D0,F1D18780
     * 0.55555555D0,2*0.0D0,.34785485D0,2*.65214515D0,.34785485D0,0.0D0,F1D18790
     * 0.236927D0,.478629D0,.568889D0,.478629D0,.236927D0/              F1D18800
C                                                                       F1D18810
      NPE=3                                                             F1D18820
      IEL=2                                                             F1D18830
      NDF=2                                                             F1D18840
      NGP=IEL+1                                                         F1D18850
      DO 10 I=1,6                                                       F1D18860
10    ELF(I)=0.0                                                        F1D18870
C                                                                       F1D18880
      EX(1)=ELX(1)                                                      F1D18890
      EX(2)=ELX(1)+0.5*H                                                F1D18900
      EX(3)=ELX(2)                                                      F1D18910
C                                                                       F1D18920
      DO 50 NI=1,NGP                                                    F1D18930
      XI=GAUSPT(NI,NGP)                                                 F1D18940
      CALL SHP1D(H,IEL,NPE,XI)                                          F1D18950
      CONST=GJ*GAUSWT(NI,NGP)                                           F1D18960
      X = 0.0                                                           F1D18970
      DO 20 J=1,NPE                                                     F1D18980
20    X = X + SF(J)*EX(J)                                               F1D18990
C                                                                       F1D19000
C     Compute the polynomial variation of FX                            F1D19010
C                                                                       F1D19020
      IF(NTYPE.EQ.2)THEN                                                F1D19030
          FX=FX0+(FX1+FX2*X)*X                                          F1D19040
      ELSE                                                              F1D19050
          FX=(FX0+FX1*X)*X                                              F1D19060
      ENDIF                                                             F1D19070
C                                                                       F1D19080
C     Element force vector for the consistent interpolation beam elementF1D19090
C                                                                       F1D19100
25    II=1                                                              F1D19110
      DO 40 I=1,NPE                                                     F1D19120
      ELF(II)=ELF(II)+FX*SF(I)*CONST                                    F1D19130
40    II=NDF*I+1                                                        F1D19140
50    CONTINUE                                                          F1D19150
C                                                                       F1D19160
C     Rearrange the element coefficients                                F1D19170
C                                                                       F1D19180
      F3(NE)=ELF(3)                                                     F1D19190
      ELF(1)=ELF(1)+0.5*F3(NE)                                          F1D19200
      ELF(2)=-0.125*F3(NE)*H                                            F1D19210
      ELF(3)=ELF(5)+0.5*F3(NE)                                          F1D19220
      ELF(4)= 0.125*F3(NE)*H                                            F1D19230
      RETURN                                                            F1D19240
      END                                                               F1D19250
```

```
      SUBROUTINE TMSTRS(GA,ELU,XI,W,DW,S,DS,NE,F3,H,MXELM)              F1D19260
C                                                                       F1D19270
C     __________________________________________________________________F1D19280
C                                                                       F1D19280
C     The subroutine computes the solution and its global derivatives atF1D19290
C      the nine points (including the nodes) of the Timoshenko element  F1D19300
C                                                                       F1D19310
C        XC........ Global (i.e., problem) coordinate                   F1D19320
C        XI ....... ocal (i.e. element) coordinate                      F1D19330
C        SFL, SFQ.. Lagrange linear and quadratic shape functions       F1D19340
C        DSFL,DSFQ: First derivative of SF w.r.t. global coordinate     F1D19350
C        ELU....... Column vector of generalized displacements          F1D19360
C        W, DW..... Transverse deflection and its derivative            F1D19370
C        S, DS..... Rotation and its derivative                         F1D19380
C     __________________________________________________________________F1D19390
C                                                                       F1D19400
      IMPLICIT REAL*8 (A-H,O-Z)                                         F1D19410
      COMMON/IO/IN,IT                                                   F1D19420
      DIMENSION ELU(9),SFL(2),SFQ(3),DSFL(2),DSFQ(3),F3(MXELM)          F1D19430
C                                                                       F1D19440
      GJ =  H*0.5                                                       F1D19450
C                                                                       F1D19460
C     Interpolation functions for the Lagrange LINEAR element           F1D19470
C                                                                       F1D19480
      SFL(1) = 0.5*(1.0-XI)                                             F1D19490
      SFL(2) = 0.5*(1.0+XI)                                             F1D19500
      DSFL(1) = -0.5/GJ                                                 F1D19510
      DSFL(2) = 0.5/GJ                                                  F1D19520
C                                                                       F1D19530
C     Interpolation functions for the Lagrange QUADRATIC element        F1D19540
C                                                                       F1D19550
      SFQ(1) = -0.5*XI*(1.0-XI)                                         F1D19560
      SFQ(2) = 1.0-XI*XI                                                F1D19570
      SFQ(3) = 0.5*XI*(1.0+XI)                                          F1D19580
      DSFQ(1) = -0.5*(1.0-2.0*XI)/GJ                                    F1D19590
      DSFQ(2) = -2.0*XI/GJ                                              F1D19600
      DSFQ(3) = 0.5*(1.0+2.0*XI)/GJ                                     F1D19610
C                                                                       F1D19620
      W3=(3.0*H*F3(NE)/GA + 8.0*(ELU(1)+ELU(3))                         F1D19630
     *                    + 2.0*(ELU(4)-ELU(2))*H)/16.0                 F1D19640
      W =  SFQ(1)*ELU(1) + SFQ(2)*W3 + SFQ(3)*ELU(3)                    F1D19650
      DW= DSFQ(1)*ELU(1) +DSFQ(2)*W3 +DSFQ(3)*ELU(3)                    F1D19660
      S =  SFL(1)*ELU(2) + SFL(2)*ELU(4)                                F1D19670
      DS= DSFL(1)*ELU(2) +DSFL(2)*ELU(4)                                F1D19680
C                                                                       F1D19690
      RETURN                                                            F1D19700
      END                                                               F1D19710

      SUBROUTINE TRSFRM (MXELM,MXNEQ,N,NTYPE,PR,SE,SL,SA,SI,CS,SN,      F1D19720
     *                   CNT,SNT,HF,VF,PF,XB)                           F1D19730
C     __________________________________________________________________F1D19740
C                                                                       F1D19750
C     The subroutine computes stiffness matrix and force vector for the F1D19760
C              truss (NDF=2) and frame (NDF=3) elements                 F1D19770
C                                                                       F1D19780
C     SE......Young's modulus                                           F1D19790
C     SL......Element length                                            F1D19800
C     SA......Cross-sectional area                                      F1D19810
C     SI......Moment of inertia                                         F1D19820
C     CS......Cosine of the angle of orientation                        F1D19830
C     SN......Sine of the angle of orientation                          F1D19840
C     HF......Distributed force along the length of the element         F1D19850
C     VF......Distributed force transverse to the element               F1D19860
C     PF......Point force at point other than nodes                     F1D19870
C     XB......Distance along the length from node 1 of the element      F1D19880
C             of the location of the point force, PF                    F1D19890
C     CNT,SNT:Direction cosines of the point force's line of applicationF1D19900
C     __________________________________________________________________F1D19910
C                                                                       F1D19920
      IMPLICIT REAL*8(A-H,O-Z)                                          F1D19930
      DIMENSION  PR(MXELM),SE(MXELM),SL(MXELM),SA(MXELM),SI(MXELM)      F1D19940
      DIMENSION  CS(MXELM),SN(MXELM),CNT(MXELM),SNT(MXELM)              F1D19950
      DIMENSION  HF(MXELM),VF(MXELM),PF(MXELM),XB(MXELM)                F1D19960
      DIMENSION  ELR(6),TRM(6,6),TMPK(6,6)                              F1D19970
      COMMON/STF1/ELK(9,9),ELM(9,9),ELF(9),ELX(4),ELU(9),ELV(9),ELA(9)  F1D19980
```

```
C                                                                        F1D19990
      CN1=CS(N)                                                          F1D20000
      SN1=SN(N)                                                          F1D20010
      CN2=CN1*CN1                                                        F1D20020
      SN2=SN1*SN1                                                        F1D20030
      CSN=CN1*SN1                                                        F1D20040
C                                                                        F1D20050
C     Element coefficients                                               F1D20060
C                                                                        F1D20070
      IF(NTYPE.EQ.0) THEN                                                F1D20080
C                                                                        F1D20090
C         The plane TRUSS element                                        F1D20100
C                                                                        F1D20110
          NN=4                                                           F1D20120
          C1=SA(N)*SE(N)/SL(N)                                           F1D20130
          ELK(1,1) = C1*CN2                                              F1D20140
          ELK(2,1) = C1*CSN                                              F1D20150
          ELK(2,2) = C1*SN2                                              F1D20160
          ELK(3,1) = -ELK(1,1)                                           F1D20170
          ELK(3,2) = -ELK(2,1)                                           F1D20180
          ELK(3,3) =  ELK(1,1)                                           F1D20190
          ELK(4,1) = -ELK(2,1)                                           F1D20200
          ELK(4,2) = -ELK(2,2)                                           F1D20210
          ELK(4,3) = -ELK(3,2)                                           F1D20220
          ELK(4,4) =  ELK(2,2)                                           F1D20230
C                                                                        F1D20240
          DO 10 I=1,NN                                                   F1D20250
          DO 10 J=I,NN                                                   F1D20260
   10     ELK(I,J) = ELK(J,I)                                            F1D20270
C                                                                        F1D20280
C         Contribution of the point force to nodal forces                F1D20290
C                                                                        F1D20300
          XI=XB(N)/SL(N)                                                 F1D20310
          SFL1 = 1.0-XI                                                  F1D20320
          SFL2 = XI                                                      F1D20330
C                                                                        F1D20340
          F1=0.5*HF(N)*SL(N)                                             F1D20350
          F3=0.5*HF(N)*SL(N)                                             F1D20360
          ELF(1) = F1*CN1                                                F1D20370
          ELF(2) = F1*SN1                                                F1D20380
          ELF(3) = F3*CN1                                                F1D20390
          ELF(4) = F3*SN1                                                F1D20400
      ELSE                                                               F1D20410
          NN=6                                                           F1D20420
          IF(NTYPE.EQ.1)THEN                                             F1D20430
C                                                                        F1D20440
C         The EULER-BERNOULLI FRAME element                              F1D20450
C                                                                        F1D20460
              AMU=0.5*SA(N)*SL(N)*SL(N)/SI(N)                            F1D20470
              C1=2.0*SE(N)*SI(N)/(SL(N)**3)                              F1D20480
              C2=6.0*SE(N)*SI(N)/(SL(N)*SL(N))                           F1D20490
              C3=C1*(AMU*CN2+6.0*SN2)                                    F1D20500
              C4=C1*(AMU-6.0)*CSN                                        F1D20510
              C5=C1*(AMU*SN2+6.0*CN2)                                    F1D20520
              C6=4.0*SE(N)*SI(N)/SL(N)                                   F1D20530
C                                                                        F1D20540
              ELK(1,1) = C3                                              F1D20550
              ELK(2,1) = C4                                              F1D20560
              ELK(2,2) = C5                                              F1D20570
              ELK(3,1) = C2*SN1                                          F1D20580
              ELK(3,2) =-C2*CN1                                          F1D20590
              ELK(3,3) = C6                                              F1D20600
              ELK(4,1) =-C3                                              F1D20610
              ELK(4,2) =-C4                                              F1D20620
              ELK(4,3) =-C2*SN1                                          F1D20630
              ELK(4,4) = C3                                              F1D20640
              ELK(5,1) =-C4                                              F1D20650
              ELK(5,2) =-C5                                              F1D20660
              ELK(5,3) = C2*CN1                                          F1D20670
              ELK(5,4) = C4                                              F1D20680
              ELK(5,5) = C5                                              F1D20690
              ELK(6,1) = C2*SN1                                          F1D20700
              ELK(6,2) =-C2*CN1                                          F1D20710
              ELK(6,3) = 0.5*C6                                          F1D20720
              ELK(6,4) =-C2*SN1                                          F1D20730
```

```
          ELK(6,5) = C2*CN1                                                F1D20740
          ELK(6,6) = C6                                                    F1D20750
C                                                                          F1D20760
          DO 20 I=1,NN                                                     F1D20770
          DO 20 J=I,NN                                                     F1D20780
   20     ELK(I,J) = ELK(J,I)                                              F1D20790
C                                                                          F1D20800
C       Contribution of the point force to nodal generalized forces        F1D20810
C                                                                          F1D20820
          XI=XB(N)/SL(N)                                                   F1D20830
          TF=PF(N)*SNT(N)                                                  F1D20840
          AF=PF(N)*CNT(N)                                                  F1D20850
          SFL1 = 1.0-XI                                                    F1D20860
          SFL2 = XI                                                        F1D20870
          SFH1 = 1.0 - 3.0*XI*XI + 2.0*(XI**3)                             F1D20880
          SFH2 = -XI*(1.0+XI*XI-2.0*XI)*SL(N)                              F1D20890
          SFH3 = 3.0*XI*XI - 2.0*(XI**3)                                   F1D20900
          SFH4 = -XI*(XI*XI - XI)*SL(N)                                    F1D20910
C                                                                          F1D20920
          F1=0.5*HF(N)*SL(N)           + SFL1*AF                           F1D20930
          F2=0.5*VF(N)*SL(N)           + SFH1*TF                           F1D20940
          F3=-VF(N)*SL(N)*SL(N)/12.0 + SFH2*TF                             F1D20950
          F4=0.5*HF(N)*SL(N)           + SFL2*AF                           F1D20960
          F5=0.5*VF(N)*SL(N)           + SFH3*TF                           F1D20970
          F6=VF(N)*SL(N)*SL(N)/12.0    + SFH4*TF                           F1D20980
          ELF(1) = F1*CN1-F2*SN1                                           F1D20990
          ELF(2) = F1*SN1+F2*CN1                                           F1D21000
          ELF(3) = F3                                                      F1D21010
          ELF(4) = F4*CN1-F5*SN1                                           F1D21020
          ELF(5) = F4*SN1+F5*CN1                                           F1D21030
          ELF(6) = F6                                                      F1D21040
      ELSE                                                                 F1D21050
C                                                                          F1D21060
C       The TIMOSHENKO FRAME element (shear coefficient=5/6)               F1D21070
C                                                                          F1D21080
          SG=5.0*SE(N)/(1.0+PR(N))/12.0                                    F1D21090
          C1=SA(N)*SE(N)/SL(N)                                             F1D21100
          C2=SG*SA(N)/SL(N)                                                F1D21110
          C3=0.5*SG*SA(N)                                                  F1D21120
          C4=0.25*SG*SA(N)*SL(N)                                           F1D21130
          C5=SE(N)*SI(N)/SL(N)                                             F1D21140
          ELK(1,1)=C1                                                      F1D21150
          ELK(2,1)=0.0                                                     F1D21160
          ELK(2,2)=C2                                                      F1D21170
          ELK(3,1)=0.0                                                     F1D21180
          ELK(3,2)=-C3                                                     F1D21190
          ELK(3,3)=C4+C5                                                   F1D21200
          ELK(4,1)=-C1                                                     F1D21210
          ELK(4,2)=0.0                                                     F1D21220
          ELK(4,3)=0.0                                                     F1D21230
          ELK(4,4)=C1                                                      F1D21240
          ELK(5,1)=0.0                                                     F1D21250
          ELK(5,2)=-C2                                                     F1D21260
          ELK(5,3)=C3                                                      F1D21270
          ELK(5,4)=0.0                                                     F1D21280
          ELK(5,5)=C2                                                      F1D21290
          ELK(6,1)=0.0                                                     F1D21300
          ELK(6,2)=-C3                                                     F1D21310
          ELK(6,3)=C4-C5                                                   F1D21320
          ELK(6,4)=0.0                                                     F1D21330
          ELK(6,5)=C3                                                      F1D21340
          ELK(6,6)=C4+C5                                                   F1D21350
C                                                                          F1D21360
          DO 25 I=1,NN                                                     F1D21370
          DO 25 J=1,NN                                                     F1D21380
   25     TRM(J,I)=0.0                                                     F1D21390
C                                                                          F1D21400
          TRM(1,1)=CN1                                                     F1D21410
          TRM(1,2)=SN1                                                     F1D21420
          TRM(2,1)=-SN1                                                    F1D21430
          TRM(2,2)=CN1                                                     F1D21440
          TRM(3,3)=1.0                                                     F1D21450
          TRM(4,4)=CN1                                                     F1D21460
          TRM(4,5)=SN1                                                     F1D21470
          TRM(5,4)=-SN1                                                    F1D21480
```

```
            TRM(5,5)=CN1                                                F1D21490
            TRM(6,6)=1.0                                                F1D21500
C                                                                       F1D21510
            DO 30 I=1,NN                                                F1D21520
            DO 30 J=I,NN                                                F1D21530
   30       ELK(I,J) = ELK(J,I)                                         F1D21540
C                                                                       F1D21550
            DO 40 I=1,NN                                                F1D21560
            DO 40 J=1,NN                                                F1D21570
            TMPK(I,J)=0.0                                               F1D21580
            DO 40 K=1,NN                                                F1D21590
   40       TMPK(I,J)=TMPK(I,J)+TRM(K,I)*ELK(K,J)                       F1D21600
C                                                                       F1D21610
            DO 50 I=1,NN                                                F1D21620
            DO 50 J=1,NN                                                F1D21630
            ELK(I,J)=0.0                                                F1D21640
            DO 50 K=1,NN                                                F1D21650
   50       ELK(I,J)=ELK(I,J)+TMPK(I,K)*TRM(K,J)                        F1D21660
C                                                                       F1D21670
C        Contribution of the point force to nodal generalized forces   F1D21680
C                                                                       F1D21690
            XI=XB(N)/SL(N)                                              F1D21700
            TF=PF(N)*SNT(N)                                             F1D21710
            AF=PF(N)*CNT(N)                                             F1D21720
            SFL1 = 1.0-XI                                               F1D21730
            SFL2 = XI                                                   F1D21740
            SFQ1 = (1.0-XI)*(1.0-2.0*XI)                                F1D21750
            SFQ2 = -XI*(1.0-2.0*XI)                                     F1D21760
            SFQ3 = 4.0*XI*(1.0-XI)                                      F1D21770
C                                                                       F1D21780
            F1=0.5*HF(N)*SL(N)          + SFL1*AF                       F1D21790
            F2=0.5*VF(N)*SL(N)          + (SFQ1+0.5*SFQ3)*TF            F1D21800
            F3=-VF(N)*SL(N)*SL(N)/12.0 - 0.125*SFQ3*SL(N)*TF            F1D21810
            F4=0.5*HF(N)*SL(N)          + SFL2*AF                       F1D21820
            F5=0.5*VF(N)*SL(N)          + (SFQ2+0.5*SFQ3)*TF            F1D21830
            F6=VF(N)*SL(N)*SL(N)/12.0  + 0.125*SFQ3*SL(N)*TF            F1D21840
            ELF(1) = F1*CN1-F2*SN1                                      F1D21850
            ELF(2) = F1*SN1+F2*CN1                                      F1D21860
            ELF(3) = F3                                                 F1D21870
            ELF(4) = F4*CN1-F5*SN1                                      F1D21880
            ELF(5) = F4*SN1+F5*CN1                                      F1D21890
            ELF(6) = F6                                                 F1D21900
         ENDIF                                                          F1D21910
      ENDIF                                                             F1D21920
      RETURN                                                            F1D21930
      END                                                               F1D21940
```

APPENDIX 2

COMPUTER PROGRAM FEM2DV2

```
C      Program name: FEM2DV2                         Length: 2740 lines    F2D00010
C     * * * * * * * * * * * * * * * * * * * * * * * * * * * * * * * * *    F2D00020
C     *                      Program  FEM2DV2                          *    F2D00030
C     *   (AN IN-CORE FINITE ELEMENT ANALYSIS COMPUTER PROGRAM)        *    F2D00040
C     * * * * * * * * * * * * * * * * * * * * * * * * * * * * * * * * *    F2D00050
C     . . . . . . . . . . . . . . . . . . . . . . . . . . . . . . . . .    F2D00060
C     .                                                                .    F2D00070
C     .     This  is  a  finite  element  computer  program  for  the  .    F2D00080
C     . analysis of two-dimensional problems governed by second-order .    F2D00090
C     . partial  differential  equations  arising  in:  heat transfer, .    F2D00100
C     . electrical engineering, fluid dynamics, and solid mechanics.   .    F2D00110
C     .                                                                .    F2D00120
C     .     The  program  uses  linear and quadratic,  triangular and .    F2D00130
C     . rectangular, elements with isoparametric formulations. Meshes .    F2D00140
C     . of only one type of element are allowed for a problem  (i.e., .    F2D00150
C     . two different types of elements cannot be used in a problem). .    F2D00160
C     .                                                                .    F2D00170
C     .     Many  field  problems  of engineering and applied science .    F2D00180
C     . can be analyzed using this program.  In  particular,  FEM2DV2 .    F2D00190
C     . can be used in the finite element analysis of problems in the .    F2D00200
C     . following fields:                                              .    F2D00210
C     .                                                                .    F2D00220
C     .     1.  Heat conduction and convection                         .    F2D00230
C     .     2.  Flows of viscous incompressible fluids (by penalty     .    F2D00240
C     .         function formulation)                                  .    F2D00250
C     .     3.  Plane elasticity problems                              .    F2D00260
C     .     4.  Plate bending problems using rectangular elements      .    F2D00270
C     .         based on the classical and first-order (or Mindlin)    .    F2D00280
C     .         plate theory.                                          .    F2D00290
C     .                                                                .    F2D00300
C     .     The main objective of this  program  is to illustrate how .    F2D00310
C     . finite element formulations developed in  Chapters  8 thru 12 .    F2D00320
C     . can be implemented on a computer  and used in the analysis of .    F2D00330
C     . engineering problems.  Modeling of large and complex problems .    F2D00340
C     . was not an objective of the program.  The program or parts of .    F2D00350
C     . it can be modified to analyze field problems not discussed in .    F2D00360
C     . the book.                                                      .    F2D00370
C     . . . . . . . . . . . . . . . . . . . . . . . . . . . . . . . . .    F2D00380
C                                                                           F2D00390
```

```
C            DESCRIPTION OF SOME KEY VARIABLES USED IN THE PROGRAM      F2D00400
C          (See Table 13.1 for a description of other variables)       F2D00410
C                                                                      F2D00420
C     [CMAT]    Matrix of stiffnesses in elasticity and plate bending  F2D00430
C               problems (computed  in  the  program from engineering  F2D00440
C               constants, E1, E2, G12, ANU12, etc. and thickness)     F2D00450
C                                                                      F2D00460
C     {ELA}     Vector of elemental nodal accelerations                F2D00470
C     {ELF}     Vector of element nodal source (or force) vector       F2D00480
C     [ELK]     Element coefficient (or stiffness) matrix              F2D00490
C     {ELU}     Vector of element nodal values of primary variables    F2D00500
C     {ELV}     Vector of elemental nodal velocities                   F2D00510
C     {ELXY}    Vector of elemental global coordinates:                F2D00520
C               ELXY(I,1)=x-coordinate;  ELXY(I,2)=y-coordinate        F2D00530
C     {GLA}     Vector of global nodal accelerations                   F2D00540
C     {GLF}     Vector of global nodal source (or force) vector        F2D00550
C     [GLK]     Global coefficient (or stiffness) matrix               F2D00560
C     {GLU}     Vector of global nodal values of primary variables     F2D00570
C     {GLV}     Vector of global nodal velocities                      F2D00580
C                                                                      F2D00590
C     NDF       Number of degrees of freedom per node:                 F2D00600
C               NDF=1, For SINGLE VARIABLE problems                    F2D00610
C               NDF=2, For ELASTICITY and VISCOUS FLUID FLOW           F2D00620
C               NDF=3, For PLATE BENDING when FSDT or CST(N)           F2D00630
C                      elements are used                               F2D00640
C               NDF=4, For PLATE BENDING when CST(C) element           F2D00650
C                      is used                                         F2D00660
C                                                                      F2D00670
C     NEQ       Total number of equations in the problem (=NNM*NDF)    F2D00680
C     NHBW      Half band width of the global coefficient matrix, GLK  F2D00690
C     NN        Total number of degrees of freedom per element         F2D00700
C     -----------------------------------------------------------------F2D00710
C                                                                      F2D00720
C          DESCRIPTION OF PARAMETERS USED TO DIMENSION THE ARRAYS      F2D00730
C                                                                      F2D00740
C     MAXCNV... Maximum number of elements with convection B.C.        F2D00750
C     MAXELM... Maximum number of elements allowed in the program      F2D00760
C     MAXNOD... Maximum number of nodes allowed in the program         F2D00770
C     MAXNX.... Maximum number of allowed subdivisions DX(I) along x   F2D00780
C     MAXNY.... Maximum number of allowed subdivisions DY(I) along y   F2D00790
C     MAXSPV... Maximum number of specified primary variables          F2D00800
C     MAXSSV... Maximum number of specified secondary variables        F2D00810
C     NCMAX.... Actual column dimension of: [GLK],[GLM],{GLU},{GLV},   F2D00820
C               {GLA}, and {GLF}                                       F2D00830
C                                                                      F2D00840
C               The actual row dimension of the assembled coefficient  F2D00850
C               matrix  should be greater than or equal to  the total  F2D00860
C               number of algebraic equations in the FE model.         F2D00870
C                                                                      F2D00880
C     NRMAX.... Actual row dimension of: [GLK] and [GLM]               F2D00890
C                                                                      F2D00900
C               The actual column-dimension of  assembled coefficient  F2D00910
C               matrix  should be greater than or equal to  the  half  F2D00920
C               bandwidth for static analysis or the total  number of  F2D00930
C               equations for the dynamic analysis.                    F2D00940
C                                                                      F2D00950
C     NOTE:_____The  values  of NRMAX, NCMAX, MAXELM, MAXNOD, MAXCNV,  F2D00960
C               MAXSSV and MAXSPV in the 'PARAMETER' statement should  F2D00970
C               modified as required by the size of the problem.       F2D00980
C     -----------------------------------------------------------------F2D00990
C                        SUBROUTINES USED IN THE PROGRAM               F2D01000
C                                                                      F2D01010
C      BOUNRY, CNCTVT, EGNBOU, INVRSE, ECHO*, ELKMFR, ELKMFT, MSH2DG,  F2D01020
C           MSH2DR, PSTPRC, QUADRT, SHPRCT, SHPTRI, SOLVER*, TIMER     F2D01030
C                                                                      F2D01040
C     Subroutines IWKIN and DGVCSP are the system subroutines used to  F2D01050
C     calculate  eigenvalues and eigenvectors.  Any other subroutines  F2D01060
C     (e.g.,  AXLBX  and  JACOBI  from FEM1DV2) can be used in place.  F2D01070
C ____________________________________________________________________ F2D01080
C                                                                      F2D01090
      IMPLICIT REAL*8(A-H,O-Z)                                         F2D01100
      PARAMETER (NRMAX=310, NCMAX=310, MAXELM=250, MAXNOD=200,         F2D01110
     1           MAXSPV=200, MAXSSV=50, MAXCNV=20, MAXNX=25, MAXNY=25) F2D01120
C                                                                      F2D01130
      DIMENSION ISPV(MAXSPV,2),VSPV(MAXSPV),ISSV(MAXSSV,2),VSSV(MAXSSV) F2D01140
```

```
      DIMENSION IBN(MAXCNV),INOD(MAXCNV,3),BETA(MAXSPV),TINF(MAXSSV)       F2D01150
      DIMENSION GLF(NRMAX),TITLE(20),IBS(3),IBP(3),GLM(NRMAX,NRMAX)         F2D01160
      DIMENSION GLK(NRMAX,NCMAX),GLU(NRMAX),GLV(NRMAX),GLA(NRMAX)           F2D01170
      DIMENSION NOD(MAXELM,9),GLXY(MAXNOD,2),DX(MAXNX),DY(MAXNY)            F2D01180
      DIMENSION EGNVAL(NRMAX),EGNVEC(NRMAX,NRMAX),IBDY(MAXSPV)              F2D01190
C                                                                           F2D01200
      COMMON/STF/ELF(27),ELK(27,27),ELM(27,27),ELXY(9,2),ELU(27),           F2D01210
     1           ELV(27),ELA(27),A1,A2,A3,A4,A5                             F2D01220
      COMMON/PST/A10,A1X,A1Y,A20,A2X,A2Y,A00,C0,CX,CY,F0,FX,FY,             F2D01230
     1           C44,C55,AMU,PENLTY,CMAT(3,3)                               F2D01240
      COMMON/PNT/IPDF,IPDR,NIPF,NIPR                                        F2D01250
      COMMON/IO/IN,ITT                                                      F2D01260
      COMMON/WORKSP/RWKSP                                                   F2D01270
      REAL RWKSP(228000)                                                    F2D01280
      CALL IWKIN(228000)                                                    F2D01290
C                                                                           F2D01300
C     * * * * * * * * * * * * * * * * * * * * * * * * * * * * * * * * *     F2D01310
C     *                                                               *     F2D01320
C     *               P R E P R O C E S S O R   U N I T               *     F2D01330
C     *                                                               *     F2D01340
C     * * * * * * * * * * * * * * * * * * * * * * * * * * * * * * * * *     F2D01350
C                                                                           F2D01360
      IN=5                                                                  F2D01370
      ITT=6                                                                 F2D01380
      CALL ECHO(IN,ITT)                                                     F2D01390
      ICONV=0                                                               F2D01400
      INTIAL=0                                                              F2D01410
      NSSV=0                                                                F2D01420
      NFLAG=1                                                               F2D01430
C                                                                           F2D01440
C     R E A D I N   T H E   I N P U T   D A T A   H E R E                   F2D01450
C                                                                           F2D01460
      READ(IN,400) TITLE                                                    F2D01470
C                                                                           F2D01480
C     Read problem and analysis type                                        F2D01490
C                                                                           F2D01500
      READ(IN,*)ITYPE,IGRAD,ITEM,NEIGN                                      F2D01510
      IF(ITEM.EQ.0) NEIGN=0                                                 F2D01520
      IF(NEIGN.NE.0) THEN                                                   F2D01530
          IF(ITYPE.LE.3 .AND. NEIGN.GT.1) THEN                              F2D01540
              WRITE(ITT,991)                                                F2D01550
              STOP                                                          F2D01560
          ELSE                                                              F2D01570
              READ(IN,*) NVALU,NVCTR                                        F2D01580
          ENDIF                                                             F2D01590
      ENDIF                                                                 F2D01600
C                                                                           F2D01610
C     Read finite element mesh information                                  F2D01620
C                                                                           F2D01630
      READ(IN,*) IELTYP,NPE,MESH,NPRNT                                      F2D01640
      IF(ITYPE.GE.3 .AND. IELTYP.EQ.0) THEN                                 F2D01650
          WRITE(ITT,990)                                                    F2D01660
          STOP                                                              F2D01670
      ENDIF                                                                 F2D01680
      IF(NPE.LE.4) THEN                                                     F2D01690
          IEL=1                                                             F2D01700
      ELSE                                                                  F2D01710
          IEL=2                                                             F2D01720
      ENDIF                                                                 F2D01730
      IF(MESH.NE.1) THEN                                                    F2D01740
          READ(IN,*) NEM,NNM                                                F2D01750
          IF(MESH.EQ.0)THEN                                                 F2D01760
C                                                                           F2D01770
C     If mesh CANNOT be generated by the program, read the mesh data in     F2D01780
C     the next three statements                                             F2D01790
C                                                                           F2D01800
              DO 10 N=1,NEM                                                 F2D01810
   10         READ(IN,*) (NOD(N,I),I=1,NPE)                                 F2D01820
              READ(IN,*) ((GLXY(I,J),J=1,2),I=1,NNM)                        F2D01830
          ELSE                                                              F2D01840
C                                                                           F2D01850
C     When mesh is to be generated by the program for more complicated      F2D01860
C     geometries, call MSH2DGeneral (which reads pertinent data there)      F2D01870
C                                                                           F2D01880
              CALL MSH2DG(NEM,NNM,NOD,MAXELM,MAXNOD,GLXY)                   F2D01890
```

```
          ENDIF                                                          F2D01900
      ELSE                                                               F2D01910
C                                                                        F2D01920
C     When mesh is to be generated for rectangular domains, call programF2D01930
C     MSH2DRectangular, which requires the following data:               F2D01940
C                                                                        F2D01950
          READ(IN,*) NX,NY                                               F2D01960
          READ(IN,*) X0,(DX(I),I=1,NX)                                   F2D01970
          READ(IN,*) Y0,(DY(I),I=1,NY)                                   F2D01980
          CALL MSH2DR (IEL,IELTYP,NX,NY,NPE,NNM,NEM,NOD,DX,DY,X0,Y0,     F2D01990
     *                GLXY,MAXELM,MAXNOD,MAXNX,MAXNY)                    F2D02000
      ENDIF                                                              F2D02010
C                                                                        F2D02020
      IF(ITYPE.EQ.0) THEN                                                F2D02030
          NDF = 1                                                        F2D02040
      ELSE                                                               F2D02050
          IF(ITYPE.GE.3) THEN                                            F2D02060
              NDF = 3                                                    F2D02070
          ELSE                                                           F2D02080
              NDF = 2                                                    F2D02090
          ENDIF                                                          F2D02100
      ENDIF                                                              F2D02110
      IF(ITYPE.EQ.5) NDF=4                                               F2D02120
C                                                                        F2D02130
      NEQ=NNM*NDF                                                        F2D02140
      NN=NPE*NDF                                                         F2D02150
      IF(NEIGN.EQ.0) THEN                                                F2D02160
C                                                                        F2D02170
C     Compute the half bandwidth of the global coefficient matrix        F2D02180
C                                                                        F2D02190
          NHBW=0                                                         F2D02200
          DO 20 N=1,NEM                                                  F2D02210
          DO 20 I=1,NPE                                                  F2D02220
          DO 20 J=1,NPE                                                  F2D02230
          NW=(IABS(NOD(N,I)-NOD(N,J))+1)*NDF                             F2D02240
   20     IF (NHBW.LT.NW) NHBW=NW                                        F2D02250
      ELSE                                                               F2D02260
          NHBW=NEQ                                                       F2D02270
      ENDIF                                                              F2D02280
C                                                                        F2D02290
C     Read specified primary and secondary degrees of freedom: node      F2D02300
C     number, local degree of freedom number, and specified value.       F2D02310
C                                                                        F2D02320
      READ(IN,*) NSPV                                                    F2D02330
      IF(NSPV.NE.0) THEN                                                 F2D02340
          READ(IN,*) ((ISPV(I,J),J=1,2),I=1,NSPV)                        F2D02350
          IF(NEIGN.EQ.0) THEN                                            F2D02360
              READ(IN,*) (VSPV(I),I=1,NSPV)                              F2D02370
          ENDIF                                                          F2D02380
      ENDIF                                                              F2D02390
      IF(NEIGN.EQ.0) THEN                                                F2D02400
          READ(IN,*) NSSV                                                F2D02410
          IF(NSSV.NE.0) THEN                                             F2D02420
              READ(IN,*) ((ISSV(I,J),J=1,2),I=1,NSSV)                    F2D02430
              READ(IN,*) (VSSV(I),I=1,NSSV)                              F2D02440
          ENDIF                                                          F2D02450
      ENDIF                                                              F2D02460
C                                                                        F2D02470
      WRITE(ITT,400) TITLE                                               F2D02480
      WRITE(ITT,910)                                                     F2D02490
      WRITE(ITT,890)                                                     F2D02500
      WRITE(ITT,910)                                                     F2D02510
      IF(ITYPE.EQ.0) THEN                                                F2D02520
C                                                                        F2D02530
C     Heat transfer and like problems:__________________________________F2D02540
C                                                                        F2D02550
          WRITE(ITT,410)                                                 F2D02560
C                                                                        F2D02570
C     Read the coefficients of the differential equation modeled         F2D02580
C                                                                        F2D02590
C     A11 = A10 + A1X*X + A1Y*Y; A22 = A20 + A2X*X + A2Y*Y; A00=CONST.   F2D02600
C                                                                        F2D02610
          READ(IN,*)A10,A1X,A1Y                                          F2D02620
          READ(IN,*)A20,A2X,A2Y                                          F2D02630
          READ(IN,*)A00                                                  F2D02640
```

```
      WRITE(ITT,420) A10,A1X,A1Y,A20,A2X,A2Y,A00                        F2D02650
      READ(IN,*)ICONV                                                   F2D02660
      IF(ICONV.NE.0) THEN                                               F2D02670
          READ(IN,*) NBE                                                F2D02680
          READ(IN,*) (IBN(I),BETA(I),TINF(I),I=1,NBE)                   F2D02690
          READ(IN,*) ((INOD(I,J),J=1,2),I=1,NBE)                        F2D02700
          WRITE(ITT,440) NBE                                            F2D02710
          DO 30 I=1,NBE                                                 F2D02720
   30     WRITE(ITT,860) IBN(I),(INOD(I,J),J=1,2),BETA(I),TINF(I)       F2D02730
      ENDIF                                                             F2D02740
   ELSE                                                                 F2D02750
      IF(ITYPE.EQ.1) THEN                                               F2D02760
C                                                                       F2D02770
C     Viscous incompressible flows:_____________________________________F2D02780
C                                                                       F2D02790
          WRITE(ITT,450)                                                F2D02800
          READ(IN,*)AMU,PENLTY                                          F2D02810
          WRITE(ITT,460) AMU,PENLTY                                     F2D02820
      ELSE                                                              F2D02830
          IF(ITYPE.EQ.2) THEN                                           F2D02840
C                                                                       F2D02850
C     Plane elasticity problems:________________________________________F2D02860
C                                                                       F2D02870
              READ(IN,*) LNSTRS                                         F2D02880
              WRITE(ITT,470)                                            F2D02890
              READ(IN,*) E1,E2,ANU12,G12,THKNS                          F2D02900
              WRITE(ITT,520) THKNS,E1,E2,ANU12,G12                      F2D02910
C                                                                       F2D02920
C     Compute the material coefficient matrix, CMAT(I,J), I,J=1,2,3.    F2D02930
C                                                                       F2D02940
              ANU21=ANU12*E2/E1                                         F2D02950
              DENOM=1.0-ANU12*ANU21                                     F2D02960
              CMAT(3,3)=G12*THKNS                                       F2D02970
              IF(LNSTRS.EQ.0) THEN                                      F2D02980
C                                                                       F2D02990
C     Plane strain  (ANU23 = ANU12)                                     F2D03000
C                                                                       F2D03010
                  WRITE(ITT,490)                                        F2D03020
                  S0=(1.0-ANU12-2.0*ANU12*ANU21)                        F2D03030
                  CMAT(1,1)=THKNS*E1*(1.0-ANU12)/S0                     F2D03040
                  CMAT(1,2)=THKNS*E1*ANU21/S0                           F2D03050
                  CMAT(2,2)=THKNS*E2*DENOM/S0/(1.0+ANU12)               F2D03060
              ELSE                                                      F2D03070
C                                                                       F2D03080
C     Plane stress                                                      F2D03090
C                                                                       F2D03100
                  WRITE(ITT,510)                                        F2D03110
                  CMAT(1,1)=THKNS*E1/DENOM                              F2D03120
                  CMAT(1,2)=ANU21*CMAT(1,1)                             F2D03130
                  CMAT(2,2)=E2*CMAT(1,1)/E1                             F2D03140
              ENDIF                                                     F2D03150
C                                                                       F2D03160
          ELSE                                                          F2D03170
C                                                                       F2D03180
C     Plate bending problems:___________________________________________F2D03190
C                                                                       F2D03200
              WRITE(ITT,500)                                            F2D03210
              IF(ITYPE.EQ.3) THEN                                       F2D03220
                  WRITE(ITT,505)                                        F2D03230
              ELSE                                                      F2D03240
                  WRITE(ITT,506)                                        F2D03250
              ENDIF                                                     F2D03260
              READ(IN,*) E1,E2,ANU12,G12,G13,G23,THKNS                  F2D03270
              WRITE(ITT,520) THKNS,E1,E2,ANU12,G12                      F2D03280
              WRITE(ITT,530) G13,G23                                    F2D03290
              ANU21=ANU12*E2/E1                                         F2D03300
              DENOM=1.0-ANU12*ANU21                                     F2D03310
              CMAT(1,1)=(THKNS**3)*E1/DENOM/12.0D0                      F2D03320
              CMAT(1,2)=ANU21*CMAT(1,1)                                 F2D03330
              CMAT(2,2)=E2*CMAT(1,1)/E1                                 F2D03340
              CMAT(3,3)=G12*(THKNS**3)/12.0D0                           F2D03350
              SCF=5.0D0/6.0D0                                           F2D03360
              C44=SCF*G23*THKNS                                         F2D03370
              C55=SCF*G13*THKNS                                         F2D03380
          ENDIF                                                         F2D03390
```

```
C                                                                  F2D03400
            CMAT(1,3)=0.0                                          F2D03410
            CMAT(2,3)=0.0                                          F2D03420
            CMAT(2,1)=CMAT(1,2)                                    F2D03430
            CMAT(3,1)=CMAT(1,3)                                    F2D03440
            CMAT(3,2)=CMAT(2,3)                                    F2D03450
C                                                                  F2D03460
        ENDIF                                                      F2D03470
      ENDIF                                                        F2D03480
      IF(NEIGN.EQ.0) THEN                                          F2D03490
        READ(IN,*)F0,FX,FY                                         F2D03500
        WRITE(ITT,430) F0,FX,FY                                    F2D03510
      ENDIF                                                        F2D03520
C                                                                  F2D03530
      IF(ITEM.NE.0) THEN                                           F2D03540
        READ(IN,*) C0,CX,CY                                        F2D03550
        IF(ITYPE.GT.1) THEN                                        F2D03560
            IF(ITYPE.EQ.2)THEN                                     F2D03570
                C0=THKNS*C0                                        F2D03580
                CX=THKNS*CX                                        F2D03590
                CY=THKNS*CY                                        F2D03600
            ELSE                                                   F2D03610
                IF(NEIGN.LE.1) THEN                                F2D03620
                    C0=THKNS*C0                                    F2D03630
                    CX=(THKNS**2)*C0/12.0D0                        F2D03640
                    CY=CX                                          F2D03650
                ENDIF                                              F2D03660
            ENDIF                                                  F2D03670
        ENDIF                                                      F2D03680
C                                                                  F2D03690
        IF(NEIGN.NE.0) THEN                                        F2D03700
            WRITE(ITT,810)                                         F2D03710
            WRITE(ITT,540) C0,CX,CY                                F2D03720
        ELSE                                                       F2D03730
            WRITE(ITT,820)                                         F2D03740
            WRITE(ITT,540) C0,CX,CY                                F2D03750
C                                                                  F2D03760
C     Read the necessary data for time-dependent problems          F2D03770
C                                                                  F2D03780
            READ(IN,*) NTIME,NSTP,INTVL,INTIAL                     F2D03790
            IF(INTVL.LE.0)INTVL=1                                  F2D03800
            READ(IN,*) DT,ALFA,GAMA,EPSLN                          F2D03810
            A1=ALFA*DT                                             F2D03820
            A2=(1.0-ALFA)*DT                                       F2D03830
            WRITE(ITT,550) DT,ALFA,GAMA,NTIME,NSTP,INTVL           F2D03840
            IF(ITEM.EQ.1) THEN                                     F2D03850
                IF(NSSV.NE.0) THEN                                 F2D03860
                    DO 40 I=1,NSSV                                 F2D03870
   40               VSSV(I)=VSSV(I)*DT                             F2D03880
                ENDIF                                              F2D03890
                IF(INTIAL.NE.0) THEN                               F2D03900
                    READ(IN,*) (GLU(I),I=1,NEQ)                    F2D03910
                ELSE                                               F2D03920
                    DO 50 I=1,NEQ                                  F2D03930
   50               GLU(I)=0.0                                     F2D03940
                ENDIF                                              F2D03950
            ELSE                                                   F2D03960
                DT2=DT*DT                                          F2D03970
                A3=2.0/GAMA/DT2                                    F2D03980
                A4=A3*DT                                           F2D03990
                A5=1.0/GAMA-1.0                                    F2D04000
                IF(INTIAL.NE.0) THEN                               F2D04010
                    READ(IN,*) (GLU(I),I=1,NEQ)                    F2D04020
                    READ(IN,*) (GLV(I),I=1,NEQ)                    F2D04030
                    DO 60 I=1,NEQ                                  F2D04040
   60               GLA(I)=0.0                                     F2D04050
                ELSE                                               F2D04060
                    DO 70 I=1,NEQ                                  F2D04070
                    GLU(I)=0.0                                     F2D04080
                    GLV(I)=0.0                                     F2D04090
   70               GLA(I)=0.0                                     F2D04100
                ENDIF                                              F2D04110
            ENDIF                                                  F2D04120
        ENDIF                                                      F2D04130
```

```
      ELSE                                                              F2D04140
         WRITE(ITT,830)                                                 F2D04150
      ENDIF                                                             F2D04160
C                                                                       F2D04170
C     *****     E N D   O F   T H E   D A T A   I N P U T       *****   F2D04180
C                                                                       F2D04190
      IF(IELTYP.EQ.0) THEN                                              F2D04200
         WRITE(ITT,790)                                                 F2D04210
      ELSE                                                              F2D04220
         WRITE(ITT,800)                                                 F2D04230
      ENDIF                                                             F2D04240
C                                                                       F2D04250
      WRITE(ITT,560) IELTYP,NPE,NDF,NEM,NNM,NEQ,NHBW                    F2D04260
      IF(MESH.EQ.1) WRITE(ITT,570) NX,NY                                F2D04270
      WRITE(ITT,710)NSPV                                                F2D04280
      IF(NSSV.NE.0) THEN                                                F2D04290
         WRITE(ITT,715)NSSV                                             F2D04300
         WRITE(ITT,720)                                                 F2D04310
         DO 80 IB=1,NSSV                                                F2D04320
   80    WRITE(ITT,960)(ISSV(IB,JB),JB=1,2),VSSV(IB)                    F2D04330
      ENDIF                                                             F2D04340
C                                                                       F2D04350
      IF(NPRNT.EQ.1) THEN                                               F2D04360
         WRITE(ITT,700)                                                 F2D04370
         DO 100 I=1,NEM                                                 F2D04380
  100    WRITE(ITT,900) I,(NOD(I,J),J=1,NPE)                            F2D04390
      ENDIF                                                             F2D04400
C                                                                       F2D04410
      WRITE(ITT,910)                                                    F2D04420
      WRITE(ITT,580)                                                    F2D04430
      WRITE(ITT,910)                                                    F2D04440
      DO 150 IM=1,NNM                                                   F2D04450
      DO 110 K=1,NDF                                                    F2D04460
      IBP(K)=0                                                          F2D04470
  110 IBS(K)=0                                                          F2D04480
      IF(NSPV.NE.0) THEN                                                F2D04490
         DO 120 JP=1,NSPV                                               F2D04500
         NODE=ISPV(JP,1)                                                F2D04510
         NDOF=ISPV(JP,2)                                                F2D04520
         IF(NODE.EQ.IM) THEN                                            F2D04530
            IBP(NDOF)=NDOF                                              F2D04540
         ENDIF                                                          F2D04550
  120    CONTINUE                                                       F2D04560
      ENDIF                                                             F2D04570
C                                                                       F2D04580
      IF(NSSV.NE.0) THEN                                                F2D04590
         DO 140 JS=1,NSSV                                               F2D04600
         NODE=ISSV(JS,1)                                                F2D04610
         NDOF=ISSV(JS,2)                                                F2D04620
         IF(NODE.EQ.IM) THEN                                            F2D04630
            IBS(NDOF)=NDOF                                              F2D04640
         ENDIF                                                          F2D04650
  140    CONTINUE                                                       F2D04660
      ENDIF                                                             F2D04670
C                                                                       F2D04680
      IF(NDF.EQ.1) THEN                                                 F2D04690
         WRITE(ITT,870)IM,(GLXY(IM,J),J=1,2),(IBP(K),K=1,NDF),          F2D04700
     *                (IBS(K),K=1,NDF)                                  F2D04710
      ELSE                                                              F2D04720
         IF(NDF.EQ.2) THEN                                              F2D04730
            WRITE(ITT,920)IM,(GLXY(IM,J),J=1,2),(IBP(K),K=1,NDF),       F2D04740
     *                   (IBS(K),K=1,NDF)                               F2D04750
         ELSE                                                           F2D04760
            IF(NDF.EQ.3) THEN                                           F2D04770
               WRITE(ITT,880)IM,(GLXY(IM,J),J=1,2),(IBP(K),K=1,NDF),    F2D04780
     *                      (IBS(K),K=1,NDF)                            F2D04790
            ELSE                                                        F2D04800
               WRITE(ITT,885)IM,(GLXY(IM,J),J=1,2),(IBP(K),K=1,NDF),    F2D04810
     *                      (IBS(K),K=1,NDF)                            F2D04820
            ENDIF                                                       F2D04830
         ENDIF                                                          F2D04840
      ENDIF                                                             F2D04850
  150 CONTINUE                                                          F2D04860
      WRITE(ITT,910)                                                    F2D04870
C                                                                       F2D04880
```

```
C     Define the polynomial degree and number of integration points      F2D04890
C     (based on the assumed variation of the coefficients AX, BX, etc.)  F2D04900
C                                                                        F2D04910
      IPDR = IEL                                                         F2D04920
      NIPR = IPDR+IEL-1                                                  F2D04930
      IF(IELTYP.EQ.0) THEN                                               F2D04940
          IF(ITYPE.EQ.0) THEN                                            F2D04950
              IPDF = 2*IEL+1                                             F2D04960
              NIPF = IPDF+IEL                                            F2D04970
          ELSE                                                           F2D04980
              IF(ITEM.NE.0) THEN                                         F2D04990
                  IPDF = 2*IEL+1                                         F2D05000
                  NIPF = IPDF+IEL                                        F2D05010
              ELSE                                                       F2D05020
                  IPDF = IEL+1                                           F2D05030
                  NIPF = IPDF+1                                          F2D05040
              ENDIF                                                      F2D05050
          ENDIF                                                          F2D05060
          ISTR = 1                                                       F2D05070
          NSTR = 1                                                       F2D05080
          WRITE(ITT,480) IPDF,NIPF,IPDR,NIPR,ISTR,NSTR                   F2D05090
      ELSE                                                               F2D05100
          IF(ITYPE.GE.4) THEN                                            F2D05110
              IPDF = 4                                                   F2D05120
              ISTR = 2                                                   F2D05130
          ELSE                                                           F2D05140
              IPDF = IEL+1                                               F2D05150
              ISTR = IEL                                                 F2D05160
          ENDIF                                                          F2D05170
          WRITE(ITT,485) IPDF,IPDR,ISTR                                  F2D05180
      ENDIF                                                              F2D05190
C                                                                        F2D05200
C     * * * * * * * * * * * * * * * * * * * * * * * * * * * * * * * *    F2D05210
C     *                                                             *    F2D05220
C     *                   P R O C E S S O R   U N I T               *    F2D05230
C     *                                                             *    F2D05240
C     * * * * * * * * * * * * * * * * * * * * * * * * * * * * * * * *    F2D05250
C                                                                        F2D05260
      IF(ITEM.NE.0) THEN                                                 F2D05270
          TIME=0.0                                                       F2D05280
      ENDIF                                                              F2D05290
C                                                                        F2D05300
C     Counter on number of TIME steps begins here                        F2D05310
C                                                                        F2D05320
      NT = 0                                                             F2D05330
      NCOUNT=0                                                           F2D05340
  170 NCOUNT=NCOUNT+1                                                    F2D05350
      IF(ITEM.NE.0 .AND. NEIGN.EQ.0) THEN                                F2D05360
          IF(NCOUNT.GE.NSTP) THEN                                        F2D05370
              F0=0.0                                                     F2D05380
              FX=0.0                                                     F2D05390
              FY=0.0                                                     F2D05400
          ENDIF                                                          F2D05410
      ENDIF                                                              F2D05420
C                                                                        F2D05430
C     Initialize the global coefficient matrices and vectors             F2D05440
C                                                                        F2D05450
      DO 180 I=1,NEQ                                                     F2D05460
      GLF(I)=0.0                                                         F2D05470
      DO 180 J=1,NHBW                                                    F2D05480
      IF(NEIGN.NE.0) GLM(I,J)=0.0                                        F2D05490
  180 GLK(I,J)=0.0                                                       F2D05500
C                                                                        F2D05510
C     Do-loop on the number of ELEMENTS to compute element matrices      F2D05520
C     and their assembly begins here                                     F2D05530
C                                                                        F2D05540
      DO 250 N=1,NEM                                                     F2D05550
      DO 200 I=1,NPE                                                     F2D05560
      NI=NOD(N,I)                                                        F2D05570
      ELXY(I,1)=GLXY(NI,1)                                               F2D05580
      ELXY(I,2)=GLXY(NI,2)                                               F2D05590
      IF(NEIGN.EQ.0) THEN                                                F2D05600
          IF(ITEM.NE.0) THEN                                             F2D05610
              LI=(NI-1)*NDF                                              F2D05620
              L = (I-1)*NDF                                              F2D05630
```

```
          DO 190 J=1,NDF                                               F2D05640
          LI=LI+1                                                      F2D05650
          L=L+1                                                        F2D05660
          ELU(L)=GLU(LI)                                               F2D05670
          IF(ITEM.EQ.2) THEN                                           F2D05680
              ELV(L)=GLV(LI)                                           F2D05690
              ELA(L)=GLA(LI)                                           F2D05700
          ENDIF                                                        F2D05710
  190     CONTINUE                                                     F2D05720
      ENDIF                                                            F2D05730
    ENDIF                                                              F2D05740
  200 CONTINUE                                                         F2D05750
C                                                                      F2D05760
C     Call subroutine ELKMFT (for Triangular elements) or ELKMFR (for  F2D05770
C     Rectangular elements) to compute the ELement [K], [M] and {F}.   F2D05780
C                                                                      F2D05790
      IF(IELTYP.EQ.0) THEN                                             F2D05800
          CALL ELKMFT (NCOUNT,NEIGN,NPE,NN,ITYPE,ITEM)                 F2D05810
      ELSE                                                             F2D05820
          CALL ELKMFR (NCOUNT,NEIGN,NPE,NN,ITYPE,ITEM)                 F2D05830
      ENDIF                                                            F2D05840
C                                                                      F2D05850
      IF(ICONV.NE.0) THEN                                              F2D05860
C                                                                      F2D05870
C     Add the convective terms for CONVECTION type boundary conditions F2D05880
C     (exact for straight sided elements; otherwise approximate values) F2D05890
C                                                                      F2D05900
          DO 210 M = 1,NBE                                             F2D05910
          IF(IBN(M).EQ.N) THEN                                         F2D05920
              M1  = INOD(M,1)                                          F2D05930
              M2  = INOD(M,2)                                          F2D05940
              NM1 = NOD(N,M1)                                          F2D05950
              NM2 = NOD(N,M2)                                          F2D05960
              DL  = DSQRT((GLXY(NM2,1)-GLXY(NM1,1))**2                 F2D05970
     *                   +(GLXY(NM2,2)-GLXY(NM1,2))**2)                F2D05980
              BL  = BETA(M)*DL                                         F2D05990
              TF  = TINF(M)*BL                                         F2D06000
              IF(IEL.EQ.1)THEN                                         F2D06010
                  ELK(M1,M1)=ELK(M1,M1)+BL/3.0                         F2D06020
                  ELK(M1,M2)=ELK(M1,M2)+BL/6.0                         F2D06030
                  ELK(M2,M1)=ELK(M2,M1)+BL/6.0                         F2D06040
                  ELK(M2,M2)=ELK(M2,M2)+BL/3.0                         F2D06050
                  ELF(M1)=ELF(M1)+0.5*TF                               F2D06060
                  ELF(M2)=ELF(M2)+0.5*TF                               F2D06070
              ELSE                                                     F2D06080
                  IF(NPE.GE.8) THEN                                    F2D06090
                      NPEL=4                                           F2D06100
                  ELSE                                                 F2D06110
                      NPEL=3                                           F2D06120
                  ENDIF                                                F2D06130
                  M3=M1+NPEL                                           F2D06140
                  ELK(M1,M1)=ELK(M1,M1)+4.0*BL/30.0                    F2D06150
                  ELK(M1,M3)=ELK(M1,M3)+2.0*BL/30.0                    F2D06160
                  ELK(M1,M2)=ELK(M1,M2)-BL/30.0                        F2D06170
                  ELK(M3,M1)=ELK(M3,M1)+2.0*BL/30.0                    F2D06180
                  ELK(M3,M3)=ELK(M3,M3)+16.0*BL/30.0                   F2D06190
                  ELK(M2,M3)=ELK(M2,M3)+2.0*BL/30.0                    F2D06200
                  ELK(M2,M1)=ELK(M2,M1)-BL/30.0                        F2D06210
                  ELK(M3,M2)=ELK(M3,M2)+2.0*BL/30.0                    F2D06220
                  ELK(M2,M2)=ELK(M2,M2)+4.0*BL/30.0                    F2D06230
                  ELF(M1)=ELF(M1)+TF/6.0                               F2D06240
                  ELF(M3)=ELF(M3)+4.0*TF/6.0                           F2D06250
                  ELF(M2)=ELF(M2)+TF/6.0                               F2D06260
              ENDIF                                                    F2D06270
          ENDIF                                                        F2D06280
  210     CONTINUE                                                     F2D06290
      ENDIF                                                            F2D06300
C                                                                      F2D06310
      IF(NCOUNT.EQ.1) THEN                                             F2D06320
          IF(NPRNT.EQ.1 .OR. NPRNT.EQ.3) THEN                          F2D06330
              IF(N.EQ.1) THEN                                          F2D06340
C                                                                      F2D06350
C     Print element matrices and vectors (only when NPRNT=1 or NPRNT=3) F2D06360
C                                                                      F2D06370
                  WRITE(ITT,610)                                       F2D06380
```

```
                DO 220 I=1,NN                                           F2D06390
  220           WRITE(ITT,930) (ELK(I,J),J=1,NN)                        F2D06400
                IF(NEIGN.EQ.0) THEN                                     F2D06410
                    WRITE(ITT,630)                                      F2D06420
                    WRITE(ITT,930) (ELF(I),I=1,NN)                      F2D06430
                ELSE                                                    F2D06440
                    WRITE(ITT,620)                                      F2D06450
                    DO 230 I=1,NN                                       F2D06460
  230               WRITE(ITT,930) (ELM(I,J),J=1,NN)                    F2D06470
                ENDIF                                                   F2D06480
            ENDIF                                                       F2D06490
        ENDIF                                                           F2D06500
    ENDIF                                                               F2D06510
C                                                                       F2D06520
    IF(NEIGN.EQ.0) THEN                                                 F2D06530
        IF(ITEM.NE.0) THEN                                              F2D06540
C                                                                       F2D06550
C   Compute the element coefficient matrices [K-hat] and {F-hat}        F2D06560
C   (i.e., after time approximation) in the transient analysis:_______F2D06570
C                                                                       F2D06580
            CALL TIMER(NCOUNT,INTIAL,ITEM,NN)                           F2D06590
        ENDIF                                                           F2D06600
    ENDIF                                                               F2D06610
C                                                                       F2D06620
C   ASSEMBLE element matrices to obtain global matrices:______________F2D06630
C                                                                       F2D06640
    DO 240 I=1,NPE                                                      F2D06650
    NR=(NOD(N,I)-1)*NDF                                                 F2D06660
    DO 240 II=1,NDF                                                     F2D06670
    NR=NR+1                                                             F2D06680
    L=(I-1)*NDF+II                                                      F2D06690
    IF(NEIGN.EQ.0) THEN                                                 F2D06700
        GLF(NR)=GLF(NR)+ELF(L)                                          F2D06710
    ENDIF                                                               F2D06720
    DO 240 J=1,NPE                                                      F2D06730
    IF(NEIGN.EQ.0) THEN                                                 F2D06740
        NCL=(NOD(N,J)-1)*NDF                                            F2D06750
    ELSE                                                                F2D06760
        NC=(NOD(N,J)-1)*NDF                                             F2D06770
    ENDIF                                                               F2D06780
    DO 240 JJ=1,NDF                                                     F2D06790
    M=(J-1)*NDF+JJ                                                      F2D06800
    IF(NEIGN.EQ.0) THEN                                                 F2D06810
        NC=NCL+JJ+1-NR                                                  F2D06820
        IF(NC.GT.0) THEN                                                F2D06830
            GLK(NR,NC)=GLK(NR,NC)+ELK(L,M)                              F2D06840
        ENDIF                                                           F2D06850
    ELSE                                                                F2D06860
        NC=NC+1                                                         F2D06870
        GLK(NR,NC)=GLK(NR,NC)+ELK(L,M)                                  F2D06880
        GLM(NR,NC)=GLM(NR,NC)+ELM(L,M)                                  F2D06890
    ENDIF                                                               F2D06900
  240 CONTINUE                                                          F2D06910
  250 CONTINUE                                                          F2D06920
C                                                                       F2D06930
C   Print global matrices when NPRNT > 2                                F2D06940
C                                                                       F2D06950
    IF(NCOUNT.LE.1) THEN                                                F2D06960
        IF(NPRNT.GE.2) THEN                                             F2D06970
            WRITE(ITT,640)                                              F2D06980
            DO 260 I=1,NEQ                                              F2D06990
  260       WRITE(ITT,930) (GLK(I,J),J=1,NHBW)                          F2D07000
            IF(NEIGN.EQ.0) THEN                                         F2D07010
                WRITE(ITT,650)                                          F2D07020
                WRITE(ITT,930) (GLF(I),I=1,NEQ)                         F2D07030
            ENDIF                                                       F2D07040
        ENDIF                                                           F2D07050
    ENDIF                                                               F2D07060
C                                                                       F2D07070
C   Impose BOUNDARY CONDITIONS on primary and secondary variables       F2D07080
C                                                                       F2D07090
    IF(NEIGN.NE.0) THEN                                                 F2D07100
        CALL EGNBOU(GLK,GLM,IBDY,ISPV,MAXSPV,NDF,NEQ,NEQR,NSPV,NRMAX)   F2D07110
C                                                                       F2D07120
C   Call SYSTEM subroutine 'DGVCSP' to solve the EIGENVALUE PROBLEM     F2D07130
```

```
C     to compute the eigenvalues and eigenvectors, and print them:      F2D07140
C                                                                       F2D07150
C      EGNVAL(I)   = Ith eigenvalue                                     F2D07160
C      EGNVEC(I,J) = Jth component of the Ith eigenvector               F2D07170
C                                                                       F2D07180
         CALL DGVCSP(NEQR,GLK,NRMAX,GLM,NRMAX,EGNVAL,EGNVEC,NRMAX)      F2D07190
C                                                                       F2D07200
         WRITE(ITT,660)                                                 F2D07210
         IF(NVALU.GT.NEQR)NVALU=NEQR                                    F2D07220
         DO 270 I=1,NVALU                                               F2D07230
         IF(ITEM.GE.2 .AND. NEIGN.EQ.1) THEN                            F2D07240
             VALUE = DSQRT(EGNVAL(I))                                   F2D07250
             WRITE(ITT,840)I,EGNVAL(I),VALUE                            F2D07260
         ELSE                                                           F2D07270
             WRITE(ITT,845)I,EGNVAL(I)                                  F2D07280
         ENDIF                                                          F2D07290
         IF(NVCTR.NE.0) THEN                                            F2D07300
             WRITE(ITT,850)                                             F2D07310
             WRITE(ITT,930)(EGNVEC(J,I),J=1,NEQR)                       F2D07320
         ENDIF                                                          F2D07330
  270    CONTINUE                                                       F2D07340
         STOP                                                           F2D07350
     ELSE                                                               F2D07360
         CALL BOUNRY(ISPV,ISSV,MAXSPV,MAXSSV,NDF,NCMAX,NRMAX,NEQ,NHBW,  F2D07370
    *                NSPV,NSSV,GLK,GLF,VSPV,VSSV,NCOUNT,INTIAL)         F2D07380
C                                                                       F2D07390
C     Call subroutine  SOLVE  to solve the system of algebraic equationsF2D07400
C     The solution is returned in the array GLF                         F2D07410
C                                                                       F2D07420
         IRES=0                                                         F2D07430
         CALL SOLVER(NRMAX,NCMAX,NEQ,NHBW,GLK,GLF,IRES)                 F2D07440
C                                                                       F2D07450
         IF(ITEM.NE.0) THEN                                             F2D07460
C                                                                       F2D07470
C     For nonzero initial conditions, GLF in the very first solution    F2D07480
C     is the acceleration, {A}=[MINV]({F}-[K]{U})                       F2D07490
C                                                                       F2D07500
             IF(NCOUNT.EQ.1 .AND. INTIAL.NE.0) THEN                     F2D07510
               IF(ITEM.EQ.2) THEN                                       F2D07520
                 DO 280 I=1,NEQ                                         F2D07530
  280            GLA(I)=GLF(I)                                          F2D07540
                 WRITE(ITT,600) TIME                                    F2D07550
                 WRITE(ITT,930) (GLA(I),I=1,NEQ)                        F2D07560
                 GOTO 170                                               F2D07570
               ENDIF                                                    F2D07580
             ELSE                                                       F2D07590
                 NT = NT + 1                                            F2D07600
                 TIME=TIME+DT                                           F2D07610
             ENDIF                                                      F2D07620
         ENDIF                                                          F2D07630
C                                                                       F2D07640
C     Compute the difference between solutions at two consecutive times,F2D07650
C     and calculate new velocities and accelerations                    F2D07660
C                                                                       F2D07670
         DIFF=0.0                                                       F2D07680
         SOLN=0.0                                                       F2D07690
         DO 290 I=1,NEQ                                                 F2D07700
         IF(ITEM.NE.0) THEN                                             F2D07710
             SOLN=SOLN+GLF(I)*GLF(I)                                    F2D07720
             DIFF=DIFF+(GLF(I)-GLU(I))*(GLF(I)-GLU(I))                  F2D07730
         ENDIF                                                          F2D07740
         IF(ITEM.EQ.2) THEN                                             F2D07750
             GLU(I)=A3*(GLF(I)-GLU(I))-A4*GLV(I)-A5*GLA(I)              F2D07760
             GLV(I)=GLV(I)+A1*GLU(I)+A2*GLA(I)                          F2D07770
             GLA(I)=GLU(I)                                              F2D07780
         ENDIF                                                          F2D07790
  290    GLU(I)=GLF(I)                                                  F2D07800
         IF(ITEM.NE.0 .AND. NT.GT.1) THEN                               F2D07810
             NFLAG=0                                                    F2D07820
             PERCNT=DSQRT(DIFF/SOLN)                                    F2D07830
             IF(PERCNT.LE.EPSLN) THEN                                   F2D07840
                 WRITE(ITT,980)                                         F2D07850
                 STOP                                                   F2D07860
             ELSE                                                       F2D07870
                 INTGR=(NT/INTVL)*INTVL                                 F2D07880
```

```
               IF(INTGR.EQ.NT) NFLAG=1                                   F2D07890
            ENDIF                                                        F2D07900
         ENDIF                                                           F2D07910
         IF(NFLAG.NE.0) THEN                                             F2D07920
C                                                                        F2D07930
C     Print the solution (i.e., nodal values of the primary variables)  F2D07940
C                                                                        F2D07950
            IF(ITEM.NE.0) THEN                                           F2D07960
               WRITE(ITT,590) TIME,NT                                    F2D07970
            ENDIF                                                        F2D07980
            WRITE(ITT,660)                                               F2D07990
            IF(NDF.LE.3) THEN                                            F2D08000
               MDF=NDF                                                   F2D08010
            ELSE                                                         F2D08020
               MDF=3                                                     F2D08030
               WRITE(ITT,666)                                            F2D08040
               WRITE(ITT,930)(GLU(J),J=NDF,NEQ,NDF)                      F2D08050
            ENDIF                                                        F2D08060
            IF(ITYPE.EQ.0) THEN                                          F2D08070
               WRITE(ITT,940)                                            F2D08080
            ELSE                                                         F2D08090
               WRITE(ITT,970)                                            F2D08100
            ENDIF                                                        F2D08110
            IF(NDF.EQ.1)WRITE(ITT,670)                                   F2D08120
            IF(NDF.EQ.2)WRITE(ITT,680)                                   F2D08130
            IF(NDF.GE.3)WRITE(ITT,690)                                   F2D08140
            IF(ITYPE.EQ.0) THEN                                          F2D08150
               WRITE(ITT,940)                                            F2D08160
            ELSE                                                         F2D08170
               WRITE(ITT,970)                                            F2D08180
            ENDIF                                                        F2D08190
            DO 300 I=1,NNM                                               F2D08200
            II=NDF*(I-1)+1                                               F2D08210
            JJ=II+MDF-1                                                  F2D08220
  300       WRITE(ITT,950)I,(GLXY(I,J),J=1,2),(GLU(J),J=II,JJ)           F2D08230
            WRITE(ITT,970)                                               F2D08240
         ENDIF                                                           F2D08250
C                                                                        F2D08260
         IF(IGRAD.NE.0) THEN                                             F2D08270
            IF(NFLAG.EQ.1) THEN                                          F2D08280
C                                                                        F2D08290
C     * * * * * * * * * * * * * * * * * * * * * * * * * * * * * * *     F2D08300
C     *                                                           *     F2D08310
C     *           P O S T P R O C E S S O R   U N I T             *     F2D08320
C     *                                                           *     F2D08330
C     * * * * * * * * * * * * * * * * * * * * * * * * * * * * * * *     F2D08340
C                                                                        F2D08350
               IF(ITYPE.LE.1) THEN                                       F2D08360
                  WRITE(ITT,970)                                         F2D08370
               ELSE                                                      F2D08380
                  WRITE(ITT,940)                                         F2D08390
               ENDIF                                                     F2D08400
               IF(ITYPE.LE.0) THEN                                       F2D08410
                  WRITE(ITT,730)                                         F2D08420
                  IF(IGRAD.EQ.1) THEN                                    F2D08430
                     WRITE(6,740)                                        F2D08440
                  ELSE                                                   F2D08450
                     WRITE(6,750)                                        F2D08460
                  ENDIF                                                  F2D08470
               ELSE                                                      F2D08480
                  IF(ITYPE.EQ.1)WRITE(ITT,760)                           F2D08490
                  IF(ITYPE.GE.2)WRITE(ITT,770)                           F2D08500
                  IF(ITYPE.EQ.3)WRITE(ITT,780)                           F2D08510
               ENDIF                                                     F2D08520
               IF(ITYPE.LE.1) THEN                                       F2D08530
                  WRITE(ITT,970)                                         F2D08540
               ELSE                                                      F2D08550
                  WRITE(ITT,940)                                         F2D08560
               ENDIF                                                     F2D08570
C                                                                        F2D08580
C     Compute the GRADIENT of the solution for single-variable problems F2D08590
C     or STRESSES for viscous flows, plane elasticity and plate bending F2D08600
C                                                                        F2D08610
               DO 320 N=1,NEM                                            F2D08620
               DO 310 I=1,NPE                                            F2D08630
```

```
                NI=NOD(N,I)                                       F2D08640
                ELXY(I,1)=GLXY(NI,1)                              F2D08650
                ELXY(I,2)=GLXY(NI,2)                              F2D08660
                LI=(NI-1)*NDF                                     F2D08670
                L=(I-1)*NDF                                       F2D08680
                DO 310 J=1,NDF                                    F2D08690
                LI=LI+1                                           F2D08700
                L=L+1                                             F2D08710
                ELU(L)=GLU(LI)                                    F2D08720
  310           CONTINUE                                          F2D08730
  320           CALL PSTPRC(ELXY,ITYPE,IELTYP,IGRAD,NDF,NPE,THKNS,  F2D08740
     *                      ELU,ISTR,NSTR)                        F2D08750
                IF(ITYPE.LE.1) THEN                               F2D08760
                   WRITE(ITT,970)                                 F2D08770
                ELSE                                              F2D08780
                   WRITE(ITT,940)                                 F2D08790
                ENDIF                                             F2D08800
             ENDIF                                                F2D08810
          ENDIF                                                   F2D08820
C                                                                 F2D08830
          IF(ITEM.NE.0) THEN                                      F2D08840
             IF(NT.GE.NTIME) THEN                                 F2D08850
                STOP                                              F2D08860
             ELSE                                                 F2D08870
                GOTO 170                                          F2D08880
             ENDIF                                                F2D08890
          ENDIF                                                   F2D08900
       ENDIF                                                      F2D08910
       STOP                                                       F2D08920
C                                                                 F2D08930
C                       F  O  R  M  A  T  S                       F2D08940
C                                                                 F2D08950
  400 FORMAT(20A4)                                                F2D08960
  410 FORMAT (/,16X,'ANALYSIS  OF   A  POISSON/LAPLACE  EQUATION')  F2D08970
  420 FORMAT (/,5X,'COEFFICIENTS OF THE DIFFERENTIAL EQUATION:',//,  F2D08980
     *        8X,'Coefficient, A10 .......................=',E12.4,/,  F2D08990
     *        8X,'Coefficient, A1X .......................=',E12.4,/,  F2D09000
     *        8X,'Coefficient, A1Y .......................=',E12.4,/,  F2D09010
     *        8X,'Coefficient, A20 .......................=',E12.4,/,  F2D09020
     *        8X,'Coefficient, A2X .......................=',E12.4,/,  F2D09030
     *        8X,'Coefficient, A2Y .......................=',E12.4,/,  F2D09040
     *        8X,'Coefficient, A00 .......................=',E12.4,/)  F2D09050
  430 FORMAT (/,5X,'CONTINUOUS SOURCE COEFFICIENTS:',//,          F2D09060
     *        8X,'Coefficient, F0  .......................=',E12.4,/,  F2D09070
     *        8X,'Coefficient, FX  .......................=',E12.4,/,  F2D09080
     *        8X,'Coefficient, FY  .......................=',E12.4,/)  F2D09090
  440 FORMAT (/,5X,'CONVECTIVE HEAT TRANSFER DATA:',//,           F2D09100
     *        8X,'Number of elements with convection, NBE .=',I4,/,  F2D09110
     *        8X,'Elements, their LOCAL nodes and convective',/,  F2D09120
     *        8X,'heat transfer data:',/,                         F2D09130
     *        8X,'Ele. No.',4X,'End Nodes',8X,'Film Coeff.',6X,   F2D09140
     *        'T-Infinity',/)                                     F2D09150
  450 FORMAT (/,16X,'A  VISCOUS INCOMPRESSIBLE FLOW IS ANALYZED')  F2D09160
  460 FORMAT (/,5X,'PARAMETERS OF THE FLUID FLOW PROBLEM:',//,    F2D09170
     *        8X,'Viscosity of the fluid, AMU ............=',E12.4,/,  F2D09180
     *        8X,'Penalty parameter, PENLTY ..............=',E12.4,/)  F2D09190
  470 FORMAT (/,16X,'A  2-D  ELASTICITY   PROBLEM  IS  ANALYZED')  F2D09200
  480 FORMAT (/,5X,'NUMERICAL INTEGRATION DATA:',//,              F2D09210
     *        8X,'Full Integration polynomial degree, IPDF =',I4,/,  F2D09220
     *        8X,'Number of full integration points,  NIPF =',I4,/,  F2D09230
     *        8X,'Reduced Integration polynomial deg.,IPDR =',I4,/,  F2D09240
     *        8X,'No.  of reduced integration points, NIPR =',I4,/,  F2D09250
     *        8X,'Integ. poly. deg. for stress comp., ISTR =',I4,/,  F2D09260
     *        8X,'No. of integ. pts. for stress comp.,NSTR =',I4,/)  F2D09270
  485 FORMAT (/,5X,'NUMERICAL INTEGRATION DATA:',//,              F2D09280
     *        8X,'Full quadrature (IPDF x IPDF) rule, IPDF =',I4,/,  F2D09290
     *        8X,'Reduced quadrature (IPDR x IPDR),   IPDR =',I4,/,  F2D09300
     *        8X,'Quadrature rule used in postproc.,  ISTR =',I4,/)  F2D09310
  490 FORMAT (9X,'**PLANE STRAIN assumption is selected by user**',/)  F2D09320
  500 FORMAT (/,16X,'A  PLATE   BENDING   PROBLEM  IS  ANALYZED')  F2D09330
  505 FORMAT (16X,  '*** using the shear deformation theory ***')  F2D09340
  506 FORMAT (16X,  '**** using the classical plate theory ****')  F2D09350
  510 FORMAT (/,8X,'***PLANE STRESS assumption is selected by user**',/)F2D09360
  520 FORMAT (/,5X,'MATERIAL PROPERTIES OF THE SOLID ANALYZED:',//,  F2D09370
     *        8X,'Thickness of the body, THKNS ...........=',E12.4,/,  F2D09380
```

```
     *        8X,'Modulus of elasticity, E1 ..............=',E12.4,/,  F2D09390
     *        8X,'Modulus of elasticity, E2 ..............=',E12.4,/,  F2D09400
     *        8X,'Poisson s ratio, ANU12 .................=',E12.4,/,  F2D09410
     *        8X,'Shear modulus, G12 .....................=',E12.4)    F2D09420
 530 FORMAT (8X,'Shear modulus, G13 .....................=',E12.4,/,  F2D09430
     *        8X,'Shear modulus, G23 .....................=',E12.4,/)  F2D09440
 540 FORMAT (/,5X,'PARAMETERS OF THE DYNAMIC ANALYSIS:',//,            F2D09450
     *        8X,'Coefficient, C0 ........................=',E12.4,/,  F2D09460
     *        8X,'Coefficient, CX ........................=',E12.4,/,  F2D09470
     *        8X,'Coefficient, CY ........................=',E12.4)    F2D09480
 550 FORMAT (8X,'Time increment used, DT ................=',E12.4,/,  F2D09490
     *        8X,'Parameter, ALFA ........................=',E12.4,/,  F2D09500
     *        8X,'Parameter, GAMA ........................=',E12.4,/,  F2D09510
     *        8X,'Number of time steps used, NTIME .......=',I4,/,     F2D09520
     *        8X,'Time step at which load is removed, NSTP.=',I4,/,    F2D09530
     *        8X,'Time interval at which soln. is printed..=',I4,/)    F2D09540
 560 FORMAT (/,5X,'FINITE ELEMENT MESH INFORMATION:',//,               F2D09550
     *        8X,'Element type: 0 = Triangle; >0 = Quad.)..=',I4,/,    F2D09560
     *        8X,'Number of nodes per element, NPE .......=',I4,/,     F2D09570
     *        8X,'No. of primary deg. of freedom/node, NDF =',I4,/,    F2D09580
     *        8X,'Number of elements in the mesh, NEM .....=',I4,/,    F2D09590
     *        8X,'Number of nodes in the mesh, NNM .......=',I4,/,     F2D09600
     *        8X,'Number of equations to be solved, NEQ ...=',I4,/,    F2D09610
     *        8X,'Half bandwidth of the matrix GLK, NHBW ..=',I4)      F2D09620
 570 FORMAT (8X,'Mesh subdivisions, NX and NY ...........=',2I4,/)     F2D09630
 580 FORMAT (5X,'Node   x-coord.   y-coord.    Speci. primary & secondaF2D09640
     *ry variables',/,38X,'(0, unspecified; >0, specified)',           F2D09650
     *            /,41X,'Primary DOF  Secondary DOF')                  F2D09660
 590 FORMAT (/,5X,'*TIME* =',E12.5,5X,'Time Step Number =',I3)         F2D09670
 600 FORMAT (/,5X,'*TIME* =',E12.5,' (Initial acceleration vector:)',/)F2D09680
 610 FORMAT (/,5X,'Element coefficient matrix: ',/)                    F2D09690
 620 FORMAT (/,5X,'Element mass matrix: ',/)                           F2D09700
 630 FORMAT (/,5X,'Element source vector:',/)                          F2D09710
 640 FORMAT (/,5X,'Global coefficient matrix:',/)                      F2D09720
 650 FORMAT (/,5X,'Global source vector:',/)                           F2D09730
 660 FORMAT (/,5X,'S O L U T I O N :',/)                               F2D09740
 666 FORMAT (5X,'Nodal values of W,xy for conforming plate element:',/)F2D09750
 670 FORMAT (5X,'Node    x-coord.      y-coord.    Primary DOF')       F2D09760
 680 FORMAT (5X,'Node    x-coord.      y-coord.     Value of u',       F2D09770
     *        '    Value of v')                                        F2D09780
 690 FORMAT (5X,'Node    x-coord.      y-coord.     deflec. w',        F2D09790
     *        '     x-rotation     y-rotation')                        F2D09800
 700 FORMAT (/,5X,'Connectivity Matrix, [NOD]',/)                      F2D09810
 710 FORMAT (8X,'No. of specified PRIMARY variables, NSPV =',I4)       F2D09820
 715 FORMAT (8X,'No. of speci.  SECONDARY variables, NSSV =',I4,/)     F2D09830
 720 FORMAT (6X,'Node  DOF     Value',/)                               F2D09840
 730 FORMAT (4X,'The orientation of  gradient vector is measured from  F2D09850
    1the positive x-axis',/)                                           F2D09860
 740 FORMAT (4X,'x-coord.      y-coord.    -a11(du/dx)   -a22(du/dy)', F2D09870
    1        3X,'Flux Mgntd  Orientation')                             F2D09880
 750 FORMAT (4X,'x-coord.      y-coord.     a22(du/dy)   -a11(du/dx)', F2D09890
    1        3X,'Flux Mgntd  Orientation')                             F2D09900
 760 FORMAT (5X,'x-coord.      y-coord.      sigma-x       sigma-y',   F2D09910
     *'     sigma-xy     pressure')                                    F2D09920
 770 FORMAT (5X,'x-coord.      y-coord.      sigma-x       sigma-y',   F2D09930
     *'     sigma-xy')                                                 F2D09940
 780 FORMAT (5X,'                            sigma-xz      sigma-yz')  F2D09950
 790 FORMAT (/,8X,'*** A mesh of   TRIANGLES     is chosen by user ***')F2D09960
 800 FORMAT (/,8X,'*** A mesh of QUADRILATERALS is chosen by user ***')F2D09970
 810 FORMAT (/,8X,'******* An EIGENVALUE PROBLEM   is analyzed *******')F2D09980
 820 FORMAT (/,8X,'******** A TRANSIENT PROBLEM   is analyzed ********')F2D09990
 830 FORMAT (/,8X,'******* A STEADY-STATE PROBLEM is analyzed *******')F2D10000
 840 FORMAT(/,3X,'Eigenvalue(',I3,') =',E15.6,3X,'Frequency =',E13.5)  F2D10010
 845 FORMAT(5X,'E I G E N V A L U E (',I3,') =',E15.6)                 F2D10020
 850 FORMAT(/,5X,'E I G E N V E C T O R :',/)                          F2D10030
 860 FORMAT (8X,I5,5X,2I5,6X,E13.5,5X,E13.5)                           F2D10040
 870 FORMAT (5X,I3,2E12.4,8X,I9,9X,I5)                                 F2D10050
 880 FORMAT (5X,I3,2E12.4,7X,3I4,2X,3I4)                               F2D10060
 885 FORMAT (5X,I3,2E12.4,5X,4I4,2X,4I4)                               F2D10070
 890 FORMAT (12X,'OUTPUT  FROM  PROGRAM  *FEM2DV2*   BY  J. N. REDDY') F2D10080
 900 FORMAT (10X,10I5)                                                 F2D10090
 910 FORMAT (2X,70('_'),/)                                             F2D10100
 920 FORMAT (5X,I3,2E12.4,8X,2I5,4X,2I5)                               F2D10110
 930 FORMAT (8X,5E14.5)                                                F2D10120
 940 FORMAT (2X,65(' '),/)                                             F2D10130
```

```
  950 FORMAT (5X,I3,5E14.5)                                          F2D10140
  960 FORMAT (5X,I5,I4,E14.5)                                        F2D10150
  970 FORMAT (2X,77('_'),/)                                          F2D10160
  980 FORMAT (/,3X,'*** THE SOLUTION HAS REACHED  A  STEADY STATE ***') F2D10170
  990 FORMAT (/,3X,'**TRIANGULAR ELEMENTS ARE NOT ALLOWED FOR PLATES**')F2D10180
  991 FORMAT (/,3X,'*STABILITY ANALYSIS IS ONLY FOR BENDING OF PLATES*',F2D10190
     *        /,3X,'**** according to the  classical plate theory ****')F2D10200
      END                                                            F2D10210


      SUBROUTINE BOUNRY(ISPV,ISSV,MAXSPV,MAXSSV,NDF,NCMAX,NRMAX,NEQ,    F2D10220
     *                  NHBW,NSPV,NSSV,S,SL,VSPV,VSSV,NCOUNT,INTIAL)    F2D10230
C     ___________________________________________________________      F2D10240
C                                                                      F2D10250
C     The subroutine implements  specified  values of the primary and  F2D10260
C     secondary variables by modifying the coefficient matrix [S] and  F2D10270
C     (banded and symmetric) and the right-hand side vector {SL}.      F2D10280
C                                                                      F2D10290
C     ___________________________________________________________      F2D10300
      IMPLICIT REAL*8(A-H,O-Z)                                         F2D10310
      DIMENSION S(NRMAX,NCMAX),SL(NRMAX),ISPV(MAXSPV,2),VSPV(MAXSPV),  F2D10320
     *          ISSV(MAXSSV,2),VSSV(MAXSSV)                            F2D10330
      COMMON/IO/IN,ITT                                                 F2D10340
C                                                                      F2D10350
      IF(NSSV.NE.0) THEN                                               F2D10360
         IF(INTIAL.EQ.0 .OR. NCOUNT.NE. 1) THEN                        F2D10370
C                                                                      F2D10380
C     Implement specified values of the SECONDARY VARIABLES:___________F2D10390
C                                                                      F2D10400
             DO 10 I=1,NSSV                                            F2D10410
             II=(ISSV(I,1)-1)*NDF+ISSV(I,2)                            F2D10420
   10        SL(II)=SL(II)+VSSV(I)                                     F2D10430
         ENDIF                                                         F2D10440
      ENDIF                                                            F2D10450
C                                                                      F2D10460
C     Implement specified values of the PRIMARY VARIABLES:_____________F2D10470
C                                                                      F2D10480
      IF(NSPV.NE.0) THEN                                               F2D10490
         DO 50 NB=1,NSPV                                               F2D10500
         IE=(ISPV(NB,1)-1)*NDF+ISPV(NB,2)                              F2D10510
         VALUE=VSPV(NB)                                                F2D10520
         IT=NHBW-1                                                     F2D10530
         I=IE-NHBW                                                     F2D10540
         DO 30 II=1,IT                                                 F2D10550
         I=I+1                                                         F2D10560
         IF(I.GE.1) THEN                                               F2D10570
             J=IE-I+1                                                  F2D10580
             SL(I)=SL(I)-S(I,J)*VALUE                                  F2D10590
             S(I,J)=0.0                                                F2D10600
         ENDIF                                                         F2D10610
   30    CONTINUE                                                      F2D10620
         S(IE,1)=1.0                                                   F2D10630
         SL(IE)=VALUE                                                  F2D10640
         I=IE                                                          F2D10650
         DO 40 II=2,NHBW                                               F2D10660
         I=I+1                                                         F2D10670
         IF(I.LE.NEQ) THEN                                             F2D10680
            SL(I)=SL(I)-S(IE,II)*VALUE                                 F2D10690
            S(IE,II)=0.0                                               F2D10700
         ENDIF                                                         F2D10710
   40    CONTINUE                                                      F2D10720
   50    CONTINUE                                                      F2D10730
      ENDIF                                                            F2D10740
      RETURN                                                           F2D10750
      END                                                              F2D10760
```

```
      SUBROUTINE CNCTVT(NELEM,NNODE,NODES,MAXELM,MAXNOD,GLXY)          F2D10770
C     ________________________________________________________________ F2D10780
C                                                                      F2D10790
C     Generates nodal connectivity array for a specified type of mesh  F2D10800
C                                                                      F2D10810
C     NEL1    = First element in the row of elements                   F2D10820
C     NELL    = Last element in the row                                F2D10830
C     IELINC  = Increment from element to the next in the row          F2D10840
C     NODINC  = Node increment from one element to the next            F2D10850
C     NPE     = Number of nodes per element                            F2D10860
C     NODE(I) = Global node numbers corresponding to the local nodes   F2D10870
C               of the first element in the row                        F2D10880
C     ________________________________________________________________ F2D10890
C                                                                      F2D10900
      IMPLICIT REAL*8(A-H,O-Z)                                         F2D10910
      DIMENSION  NODES(MAXELM,9),GLXY(MAXNOD,2),NODE(9)                F2D10920
C                                                                      F2D10930
C     Read element data                                                F2D10940
C                                                                      F2D10950
      READ(5,*) NRECEL                                                 F2D10960
      DO 30 IREC=1,NRECEL                                              F2D10970
      READ(5,*) NEL1,NELL,IELINC,NODINC,NPE,(NODE(I),I=1,NPE)          F2D10980
      IF(IELINC.LE.0) IELINC=1                                         F2D10990
      IF(NODINC.LE.0) NODINC=1                                         F2D11000
      IF(NELL.LE.NEL1) NELL=NEL1                                       F2D11010
      IF(NELL.GT.NELEM) THEN                                           F2D11020
          WRITE(6,60)                                                  F2D11030
          STOP                                                         F2D11040
      ELSE                                                             F2D11050
          NINC=-1                                                      F2D11060
          DO 20 N=NEL1,NELL,IELINC                                     F2D11070
          NINC=NINC+1                                                  F2D11080
          DO 10 M=1,NPE                                                F2D11090
   10     NODES(N,M)=NODE(M)+NINC*NODINC                               F2D11100
   20     CONTINUE                                                     F2D11110
      ENDIF                                                            F2D11120
   30 CONTINUE                                                         F2D11130
C                                                                      F2D11140
      DO 50 N=1,NELEM                                                  F2D11150
      SUMX=0.0                                                         F2D11160
      SUMY=0.0                                                         F2D11170
      NEN=NPE                                                          F2D11180
      IF(NEN.NE.4) THEN                                                F2D11190
          DO 40 M=5,NEN                                                F2D11200
          MM=NODES(N,M)                                                F2D11210
          IF(M.NE.9 .OR. M.NE.6) THEN                                  F2D11220
              M4=NODES(N,M-4)                                          F2D11230
              M3=NODES(N,M-3)                                          F2D11240
              IF(M.EQ.8) M3=NODES(N,1)                                 F2D11250
              IF(GLXY(MM,1).EQ.1.E20)                                  F2D11260
     *           GLXY(MM,1)=0.5*(GLXY(M4,1)+GLXY(M3,1))                F2D11270
              IF(GLXY(MM,2).EQ.1.E20)                                  F2D11280
     *           GLXY(MM,2)=0.5*(GLXY(M4,2)+GLXY(M3,2))                F2D11290
              IF(NEN.NE.8) THEN                                        F2D11300
                  SUMX=SUMX+GLXY(M4,1)                                 F2D11310
                  SUMY=SUMY+GLXY(M4,2)                                 F2D11320
              ENDIF                                                    F2D11330
          ELSE                                                         F2D11340
              IF(GLXY(MM,1).EQ.1.E20) GLXY(MM,1)=0.25*SUMX             F2D11350
              IF(GLXY(MM,2).EQ.1.E20) GLXY(MM,2)=0.25*SUMY             F2D11360
          ENDIF                                                        F2D11370
   40 CONTINUE                                                         F2D11380
      ENDIF                                                            F2D11390
   50 CONTINUE                                                         F2D11400
   60 FORMAT(/,'MSG from CNCTVT: Element number exceeds maximum value') F2D11410
      RETURN                                                           F2D11420
      END                                                              F2D11430

      SUBROUTINE EGNBOU(A,D,IBDY,ISPV,MXPV,NDF,NEQ,NEQR,NSPV,NRM)      F2D11440
C     ________________________________________________________________ F2D11450
C                                                                      F2D11460
C     Imposes specified homogeneous boundary conditions on the primary F2D11470
C     variables  by  eliminating rows and columns corresponding to the F2D11480
C     specified degrees of freedom                                     F2D11490
C     ________________________________________________________________ F2D11500
```

```
C                                                                 F2D11510
      IMPLICIT REAL*8 (A-H,O-Z)                                   F2D11520
      DIMENSION A(NRM,NRM),D(NRM,NRM),ISPV(MXPV,2),IBDY(MXPV)     F2D11530
C                                                                 F2D11540
      DO 10 I=1,NSPV                                              F2D11550
   10 IBDY(I)=(ISPV(I,1)-1)*NDF+ISPV(I,2)                         F2D11560
      DO 30 I=1,NSPV                                              F2D11570
      IMAX=IBDY(I)                                                F2D11580
      DO 20 J=I,NSPV                                              F2D11590
      IF(IBDY(J).GE.IMAX) THEN                                    F2D11600
          IMAX=IBDY(J)                                            F2D11610
          IKEPT=J                                                 F2D11620
      ENDIF                                                       F2D11630
   20 CONTINUE                                                    F2D11640
      IBDY(IKEPT)=IBDY(I)                                         F2D11650
      IBDY(I)=IMAX                                                F2D11660
   30 CONTINUE                                                    F2D11670
      NEQR = NEQ                                                  F2D11680
      DO 80 I=1,NSPV                                              F2D11690
      IB=IBDY(I)                                                  F2D11700
      IF(IB .LT. NEQR) THEN                                       F2D11710
          NEQR1=NEQR-1                                            F2D11720
          DO 60 II=IB,NEQR1                                       F2D11730
          DO 40 JJ=1,NEQR                                         F2D11740
          D(II,JJ)=D(II+1,JJ)                                     F2D11750
   40     A(II,JJ)=A(II+1,JJ)                                     F2D11760
          DO 50 JJ=1,NEQR                                         F2D11770
          D(JJ,II)=D(JJ,II+1)                                     F2D11780
   50     A(JJ,II)=A(JJ,II+1)                                     F2D11790
   60     CONTINUE                                                F2D11800
      ENDIF                                                       F2D11810
      NEQR=NEQR-1                                                 F2D11820
   80 CONTINUE                                                    F2D11830
      RETURN                                                      F2D11840
      END                                                         F2D11850

      SUBROUTINE INVRSE(A,B)                                      F2D11860
      IMPLICIT REAL*8 (A-H,O-Z)                                   F2D11870
C     ___________________________________________________________ F2D11880
C                                                                 F2D11890
C     The subroutine computes the inverse of a 3x3 matrix,[A]; the F2D11900
C     inverse is stored in matrix [B]                             F2D11910
C     ___________________________________________________________ F2D11920
C                                                                 F2D11930
      DIMENSION  A(3,3), B(3,3)                                   F2D11940
C                                                                 F2D11950
      G(Z1,Z2,Z3,Z4) = Z1*Z2 - Z3*Z4                              F2D11960
      F(Z1,Z2,Z3,Z4) = G(Z1,Z2,Z3,Z4) / DET                       F2D11970
      C1  = G(A(2,2),A(3,3),A(2,3),A(3,2))                        F2D11980
      C2  = G(A(2,3),A(3,1),A(2,1),A(3,3))                        F2D11990
      C3  = G(A(2,1),A(3,2),A(2,2),A(3,1))                        F2D12000
      DET = A(1,1)*C1 + A(1,2)*C2 + A(1,3)*C3                     F2D12010
      B(1,1) =  F(A(2,2),A(3,3),A(3,2),A(2,3))                    F2D12020
      B(1,2) = -F(A(1,2),A(3,3),A(1,3),A(3,2))                    F2D12030
      B(1,3) =  F(A(1,2),A(2,3),A(1,3),A(2,2))                    F2D12040
      B(2,1) = -F(A(2,1),A(3,3),A(2,3),A(3,1))                    F2D12050
      B(2,2) =  F(A(1,1),A(3,3),A(3,1),A(1,3))                    F2D12060
      B(2,3) = -F(A(1,1),A(2,3),A(1,3),A(2,1))                    F2D12070
      B(3,1) =  F(A(2,1),A(3,2),A(3,1),A(2,2))                    F2D12080
      B(3,2) = -F(A(1,1),A(3,2),A(1,2),A(3,1))                    F2D12090
      B(3,3) =  F(A(1,1),A(2,2),A(2,1),A(1,2))                    F2D12100
      RETURN                                                      F2D12110
      END                                                         F2D12120

      SUBROUTINE ECHO(IN,IT)                                      F2D12130
C                                                                 F2D12140
      DIMENSION AA(20)                                            F2D12150
      WRITE(IT,40)                                                F2D12160
   10 CONTINUE                                                    F2D12170
      READ(IN,30,END=20) AA                                       F2D12180
      WRITE(IT,30) AA                                             F2D12190
      GO TO 10                                                    F2D12200
   20 CONTINUE                                                    F2D12210
```

```
      REWIND(IN)                                                        F2D12220
      WRITE(IT,50)                                                      F2D12230
      RETURN                                                            F2D12240
   30 FORMAT(20A4)                                                      F2D12250
   40 FORMAT(5X,'*** ECHO OF THE INPUT DATA STARTS ***',/)              F2D12260
   50 FORMAT(5X,'**** ECHO OF THE INPUT DATA ENDS ****',/)              F2D12270
      END                                                               F2D12280


      SUBROUTINE ELKMFR(NCOUNT,NEIGN,NPE,NN,ITYPE,ITEM)                 F2D12290
C     __________________________________________________________________F2D12300
C                                                                       F2D12310
C     Element  calculations based on  linear and quadratic rectangular  F2D12320
C     elements and  isoparametric  formulation are carried out for all  F2D12330
C     classes problems of the  book.  Reduced integrations are used on  F2D12340
C     certain terms of viscous flow and plate bending problems.         F2D12350
C     __________________________________________________________________F2D12360
C                                                                       F2D12370
      IMPLICIT REAL*8(A-H,O-Z)                                          F2D12380
      COMMON/STF/ELF(27),ELK(27,27),ELM(27,27),ELXY(9,2),ELU(27),       F2D12390
     1           ELV(27),ELA(27),A1,A2,A3,A4,A5                         F2D12400
      COMMON/PST/A10,A1X,A1Y,A20,A2X,A2Y,A00,C0,CX,CY,F0,FX,FY,         F2D12410
     1           C44,C55,AMU,PENLTY,CMAT(3,3)                           F2D12420
      COMMON/SHP/SF(9),GDSF(2,9),SFH(16),GDSFH(2,16),GDDSFH(3,16)       F2D12430
      COMMON/PNT/IPDF,IPDR,NIPF,NIPR                                    F2D12440
      DIMENSION  GAUSPT(5,5),GAUSWT(5,5)                                F2D12450
      COMMON/IO/IN,ITT                                                  F2D12460
C                                                                       F2D12470
      DATA GAUSPT/5*0.0D0, -0.57735027D0, 0.57735027D0, 3*0.0D0,        F2D12480
     2  -0.77459667D0, 0.0D0, 0.77459667D0, 2*0.0D0, -0.86113631D0,     F2D12490
     3  -0.33998104D0, 0.33998104D0, 0.86113631D0, 0.0D0, -0.90617984D0,F2D12500
     4  -0.53846931D0,0.0D0,0.53846931D0,0.90617984D0/                  F2D12510
C                                                                       F2D12520
      DATA GAUSWT/2.0D0, 4*0.0D0, 2*1.0D0, 3*0.0D0, 0.55555555D0,       F2D12530
     2    0.88888888D0, 0.55555555D0, 2*0.0D0, 0.34785485D0,            F2D12540
     3 2*0.65214515D0, 0.34785485D0, 0.0D0, 0.23692688D0,               F2D12550
     4    0.47862867D0, 0.56888888D0, 0.47862867D0, 0.23692688D0/       F2D12560
C                                                                       F2D12570
      NDF = NN/NPE                                                      F2D12580
      IF(ITYPE.LE.3) THEN                                               F2D12590
          NET=NPE                                                       F2D12600
      ELSE                                                              F2D12610
          NET=NN                                                        F2D12620
      ENDIF                                                             F2D12630
C                                                                       F2D12640
C     Initialize the arrays                                             F2D12650
C                                                                       F2D12660
      DO 120 I = 1,NN                                                   F2D12670
      IF(NEIGN.EQ.0) THEN                                               F2D12680
          ELF(I)  = 0.0                                                 F2D12690
      ENDIF                                                             F2D12700
      DO 120 J = 1,NN                                                   F2D12710
      IF(ITEM.NE.0) THEN                                                F2D12720
          ELM(I,J)= 0.0                                                 F2D12730
      ENDIF                                                             F2D12740
  120 ELK(I,J)= 0.0                                                     F2D12750
C                                                                       F2D12760
C     Do-loops on numerical (Gauss) integration begin here. Subroutine  F2D12770
C     SHPRCT (SHaPe functions for ReCTangular elements) is called here  F2D12780
C                                                                       F2D12790
      DO 200 NI = 1,IPDF                                                F2D12800
      DO 200 NJ = 1,IPDF                                                F2D12810
      XI  = GAUSPT(NI,IPDF)                                             F2D12820
      ETA = GAUSPT(NJ,IPDF)                                             F2D12830
      CALL SHPRCT (NPE,XI,ETA,DET,ELXY,NDF,ITYPE)                       F2D12840
      CNST = DET*GAUSWT(NI,IPDF)*GAUSWT(NJ,IPDF)                        F2D12850
      X=0.0                                                             F2D12860
      Y=0.0                                                             F2D12870
      DO 140 I=1,NPE                                                    F2D12880
      X=X+ELXY(I,1)*SF(I)                                               F2D12890
  140 Y=Y+ELXY(I,2)*SF(I)                                               F2D12900
C                                                                       F2D12910
      IF(NEIGN.EQ.0) THEN                                               F2D12920
          SOURCE=F0+FX*X+FY*Y                                           F2D12930
      ENDIF                                                             F2D12940
```

```
      IF(ITEM.NE.0) THEN                                               F2D12950
          IF(ITYPE.LE.2)THEN                                           F2D12960
              CT=C0+CX*X+CY*Y                                          F2D12970
          ENDIF                                                        F2D12980
      ENDIF                                                            F2D12990
      IF(ITYPE.LE.0) THEN                                              F2D13000
          A11=A10+A1X*X+A1Y*Y                                          F2D13010
          A22=A20+A2X*X+A2Y*Y                                          F2D13020
      ENDIF                                                            F2D13030
C                                                                      F2D13040
      II=1                                                             F2D13050
      DO 180 I=1,NET                                                   F2D13060
      JJ=1                                                             F2D13070
      DO 160 J=1,NET                                                   F2D13080
      IF(ITYPE.LE.3) THEN                                              F2D13090
          S00=SF(I)*SF(J)*CNST                                         F2D13100
          S11=GDSF(1,I)*GDSF(1,J)*CNST                                 F2D13110
          S22=GDSF(2,I)*GDSF(2,J)*CNST                                 F2D13120
          S12=GDSF(1,I)*GDSF(2,J)*CNST                                 F2D13130
          S21=GDSF(2,I)*GDSF(1,J)*CNST                                 F2D13140
      ENDIF                                                            F2D13150
      IF(ITYPE.EQ.0) THEN                                              F2D13160
C                                                                      F2D13170
C     Heat transfer and like problems (i.e. single DOF problems):_______F2D13180
C                                                                      F2D13190
          ELK(I,J) = ELK(I,J) + A11*S11 + A22*S22 + A00*S00            F2D13200
          IF(ITEM.NE.0) THEN                                           F2D13210
              ELM(I,J) = ELM(I,J) + CT*S00                             F2D13220
          ENDIF                                                        F2D13230
      ELSE                                                             F2D13240
          IF(ITYPE.EQ.1) THEN                                          F2D13250
C                                                                      F2D13260
C     Viscous incompressible fluids:____________________________________F2D13270
C     Compute coefficients associated with viscous terms (full integ.) F2D13280
C                                                                      F2D13290
              ELK(II,JJ)     = ELK(II,JJ)     + AMU*(2.0*S11 + S22)    F2D13300
              ELK(II+1,JJ)   = ELK(II+1,JJ)   + AMU*S12                F2D13310
              ELK(II,JJ+1)   = ELK(II,JJ+1)   + AMU*S21                F2D13320
              ELK(II+1,JJ+1)= ELK(II+1,JJ+1) + AMU*(S11 + 2.0*S22)     F2D13330
              IF(ITEM.NE.0) THEN                                       F2D13340
                  ELM(II,JJ)     = ELM(II,JJ)     + CT*S00             F2D13350
                  ELM(II+1,JJ+1)= ELM(II+1,JJ+1) + CT*S00              F2D13360
              ENDIF                                                    F2D13370
          ELSE                                                         F2D13380
              IF(ITYPE.EQ.2) THEN                                      F2D13390
C                                                                      F2D13400
C     Plane elasticity problems:________________________________________F2D13410
C                                                                      F2D13420
              ELK(II,JJ)    =ELK(II,JJ)    +CMAT(1,1)*S11+CMAT(3,3)*S22 F2D13430
              ELK(II,JJ+1)  =ELK(II,JJ+1)  +CMAT(1,2)*S12+CMAT(3,3)*S21 F2D13440
              ELK(II+1,JJ)  =ELK(II+1,JJ)  +CMAT(1,2)*S21+CMAT(3,3)*S12 F2D13450
              ELK(II+1,JJ+1)=ELK(II+1,JJ+1)+CMAT(3,3)*S11+CMAT(2,2)*S22 F2D13460
              IF(ITEM.NE.0) THEN                                       F2D13470
                  ELM(II,JJ)     = ELM(II,JJ)     + CT*S00             F2D13480
                  ELM(II+1,JJ+1)= ELM(II+1,JJ+1) + CT*S00              F2D13490
              ENDIF                                                    F2D13500
              ELSE                                                     F2D13510
                  IF(ITYPE.GE.4) THEN                                  F2D13520
C                                                                      F2D13530
C     Classical plate theory:___________________________________________F2D13540
C                                                                      F2D13550
                      BM1=CMAT(1,1)*GDDSFH(1,J)+CMAT(1,2)*GDDSFH(2,J)  F2D13560
                      BM2=CMAT(1,2)*GDDSFH(1,J)+CMAT(2,2)*GDDSFH(2,J)  F2D13570
                      BM6=2.0*CMAT(3,3)*GDDSFH(3,J)                    F2D13580
                      ELK(I,J)=ELK(I,J)+CNST*(GDDSFH(1,I)*BM1+         F2D13590
     *                     GDDSFH(2,I)*BM2+2.0*GDDSFH(3,I)*BM6)        F2D13600
                      IF(ITEM.NE.0) THEN                               F2D13610
                          S00=SFH(I)*SFH(J)*CNST                       F2D13620
                          SXX=GDSFH(1,I)*GDSFH(1,J)*CNST               F2D13630
                          SYY=GDSFH(2,I)*GDSFH(2,J)*CNST               F2D13640
                          IF(NEIGN.LE.1) THEN                          F2D13650
                              ELM(I,J)=ELM(I,J) + C0*S00+CX*SXX+CY*SYY F2D13660
                          ELSE                                         F2D13670
                              SXY=GDSFH(1,I)*GDSFH(2,J)*CNST           F2D13680
                              SYX=GDSFH(2,I)*GDSFH(1,J)*CNST           F2D13690
```

```
                          ELM(I,J)=ELM(I,J)  + C0*SXX + CX*SYY         F2D13700
     *                                       + CY*(SXY + SYX)          F2D13710
                       ENDIF                                           F2D13720
                    ENDIF                                              F2D13730
                 ELSE                                                  F2D13740
C                                                                      F2D13750
C      Shear deformable plate theory:_________________________________F2D13760
C                                                                      F2D13770
                    ELK(II+1,JJ+1)= ELK(II+1,JJ+1) +                   F2D13780
     *                              CMAT(1,1)*S11+CMAT(3,3)*S22        F2D13790
                    ELK(II+1,JJ+2)= ELK(II+1,JJ+2) +                   F2D13800
     *                              CMAT(1,2)*S12+CMAT(3,3)*S21        F2D13810
                    ELK(II+2,JJ+1)= ELK(II+2,JJ+1) +                   F2D13820
     *                              CMAT(3,3)*S12+CMAT(1,2)*S21        F2D13830
                    ELK(II+2,JJ+2)= ELK(II+2,JJ+2) +                   F2D13840
     *                              CMAT(3,3)*S11+CMAT(2,2)*S22        F2D13850
                    IF(ITEM.NE.0) THEN                                 F2D13860
                       IF(NEIGN.LE.1) THEN                             F2D13870
                          ELM(II,JJ)     = ELM(II,JJ)      + C0*S00    F2D13880
                          ELM(II+1,JJ+1)= ELM(II+1,JJ+1)  + CX*S00    F2D13890
                          ELM(II+2,JJ+2)= ELM(II+2,JJ+2)  + CY*S00    F2D13900
                       ELSE                                            F2D13910
                          ELM(II,JJ)     = ELM(II,JJ)+C0*S11+CX*S22   F2D13920
     *                                             +CY*(S12+S21)      F2D13930
                       ENDIF                                           F2D13940
                    ENDIF                                              F2D13950
                 ENDIF                                                 F2D13960
              ENDIF                                                    F2D13970
           ENDIF                                                       F2D13980
        ENDIF                                                          F2D13990
  160 JJ = NDF*J+1                                                     F2D14000
        IF(NEIGN.EQ.0) THEN                                            F2D14010
C                                                                      F2D14020
C       Source of the form fx = F0 + FX*X + FY*Y is assumed            F2D14030
C                                                                      F2D14040
           IF(ITYPE.LE.3) THEN                                         F2D14050
              L=(I-1)*NDF+1                                            F2D14060
              ELF(L) = ELF(L)+CNST*SF(I)*SOURCE                        F2D14070
           ELSE                                                        F2D14080
              ELF(I) = ELF(I)+CNST*SFH(I)*SOURCE                       F2D14090
           ENDIF                                                       F2D14100
        ENDIF                                                          F2D14110
  180 II = NDF*I+1                                                     F2D14120
  200 CONTINUE                                                         F2D14130
C                                                                      F2D14140
      IF(ITYPE.EQ.1 .OR. ITYPE.EQ.3) THEN                              F2D14150
C                                                                      F2D14160
C     Use reduced integration to evaluate coefficients associated with F2D14170
C     penalty terms for flows and transverse shear terms for plates.   F2D14180
C                                                                      F2D14190
         DO 280 NI=1,IPDR                                              F2D14200
         DO 280 NJ=1,IPDR                                              F2D14210
         XI  = GAUSPT(NI,IPDR)                                         F2D14220
         ETA = GAUSPT(NJ,IPDR)                                         F2D14230
         CALL SHPRCT (NPE,XI,ETA,DET,ELXY,NDF,ITYPE)                   F2D14240
         CNST=DET*GAUSWT(NI,IPDR)*GAUSWT(NJ,IPDR)                      F2D14250
C                                                                      F2D14260
         II=1                                                          F2D14270
         DO 260 I=1,NPE                                                F2D14280
         JJ = 1                                                        F2D14290
         DO 240 J=1,NPE                                                F2D14300
         S11=GDSF(1,I)*GDSF(1,J)*CNST                                  F2D14310
         S22=GDSF(2,I)*GDSF(2,J)*CNST                                  F2D14320
         S12=GDSF(1,I)*GDSF(2,J)*CNST                                  F2D14330
         S21=GDSF(2,I)*GDSF(1,J)*CNST                                  F2D14340
         IF(ITYPE.EQ.1) THEN                                           F2D14350
C                                                                      F2D14360
C     Viscous incompressible fluids (penalty terms):__________________F2D14370
C                                                                      F2D14380
            ELK(II,JJ)     = ELK(II,JJ)      + PENLTY*S11              F2D14390
            ELK(II+1,JJ)   = ELK(II+1,JJ)    + PENLTY*S21              F2D14400
            ELK(II,JJ+1)   = ELK(II,JJ+1)    + PENLTY*S12              F2D14410
            ELK(II+1,JJ+1)= ELK(II+1,JJ+1)  + PENLTY*S22              F2D14420
         ELSE                                                          F2D14430
C                                                                      F2D14440
```

```
C     Shear deformable plates (transverse shear terms):_______________F2D14450
C                                                                     F2D14460
              S00=SF(I)*SF(J)*CNST                                    F2D14470
              S10 = GDSF(1,I)*SF(J)*CNST                              F2D14480
              S01 = SF(I)*GDSF(1,J)*CNST                              F2D14490
              S20 = GDSF(2,I)*SF(J)*CNST                              F2D14500
              S02 = SF(I)*GDSF(2,J)*CNST                              F2D14510
              ELK(II,JJ)     = ELK(II,JJ)      + C55*S11+C44*S22      F2D14520
              ELK(II,JJ+1)   = ELK(II,JJ+1)    + C55*S10              F2D14530
              ELK(II+1,JJ)   = ELK(II+1,JJ)    + C55*S01              F2D14540
              ELK(II,JJ+2)   = ELK(II,JJ+2)    + C44*S20              F2D14550
              ELK(II+2,JJ)   = ELK(II+2,JJ)    + C44*S02              F2D14560
              ELK(II+1,JJ+1)= ELK(II+1,JJ+1)  + C55*S00              F2D14570
              ELK(II+2,JJ+2)= ELK(II+2,JJ+2)  + C44*S00              F2D14580
          ENDIF                                                       F2D14590
  240 JJ=NDF*J+1                                                      F2D14600
  260 II=NDF*I+1                                                      F2D14610
  280 CONTINUE                                                        F2D14620
      ENDIF                                                           F2D14630
      RETURN                                                          F2D14640
      END                                                             F2D14650

      SUBROUTINE ELKMFT(NCOUNT,NEIGN,NPE,NN,ITYPE,ITEM)               F2D14660
C     ________________________________________________________________F2D14670
C                                                                     F2D14680
C     Element  calculations based on linear and quadratic rectangular F2D14690
C     elements and  isoparametric formulation are carried out for all F2D14700
C     classes problems of the book.  Reduced integration  is  used on F2D14710
C     penalty terms of viscous flow and plate bending problems.       F2D14720
C     ________________________________________________________________F2D14730
C                                                                     F2D14740
      IMPLICIT REAL*8(A-H,O-Z)                                        F2D14750
      COMMON/STF/ELF(27),ELK(27,27),ELM(27,27),ELXY(9,2),ELU(27),     F2D14760
     1           ELV(27),ELA(27),A1,A2,A3,A4,A5                       F2D14770
      COMMON/PST/A10,A1X,A1Y,A20,A2X,A2Y,A00,C0,CX,CY,F0,FX,FY,       F2D14780
     1           C44,C55,AMU,PENLTY,CMAT(3,3)                         F2D14790
      COMMON/QUAD/AL1(7,5),AL2(7,5),AL3(7,5),ALWT(7,5)                F2D14800
      COMMON/PNT/IPDF,IPDR,NIPF,NIPR                                  F2D14810
      COMMON/SHP/SF(9),GDSF(2,9)                                      F2D14820
      COMMON/IO/IN,ITT                                                F2D14830
C                                                                     F2D14840
      NDF = NN/NPE                                                    F2D14850
C                                                                     F2D14860
C     Call subroutine QUADRaTure to compute arrays of integration     F2D14870
C     points and weights for the given NIPF and IPDF                  F2D14880
C                                                                     F2D14890
      CALL QUADRT (NIPF,IPDF)                                         F2D14900
C                                                                     F2D14910
C     Initialize the arrays                                           F2D14920
C                                                                     F2D14930
      DO 120 I = 1,NN                                                 F2D14940
      IF(NEIGN.EQ.0) THEN                                             F2D14950
          ELF(I)   = 0.0                                              F2D14960
      ENDIF                                                           F2D14970
      DO 120 J = 1,NN                                                 F2D14980
      IF(ITEM.NE.0) THEN                                              F2D14990
          ELM(I,J)= 0.0                                               F2D15000
      ENDIF                                                           F2D15010
  120 ELK(I,J)= 0.0                                                   F2D15020
C                                                                     F2D15030
C     Do-loop on the numerical integration begins here                F2D15040
C                                                                     F2D15050
      DO 200 NI = 1,NIPF                                              F2D15060
      AC1 = AL1(NI,IPDF)                                              F2D15070
      AC2 = AL2(NI,IPDF)                                              F2D15080
      AC3 = AL3(NI,IPDF)                                              F2D15090
      CALL SHPTRI(NPE,AC1,AC2,AC3,DET,ELXY)                           F2D15100
      CNST = 0.50D0*DET*ALWT(NI,IPDF)                                 F2D15110
      X=0.0                                                           F2D15120
      Y=0.0                                                           F2D15130
      DO 140 I=1,NPE                                                  F2D15140
      X=X+ELXY(I,1)*SF(I)                                             F2D15150
  140 Y=Y+ELXY(I,2)*SF(I)                                             F2D15160
```

```
C                                                                        F2D15170
      IF(NEIGN.EQ.0) THEN                                                F2D15180
          SOURCE=F0+FX*X+FY*Y                                            F2D15190
      ENDIF                                                              F2D15200
      IF(ITEM.NE.0) THEN                                                 F2D15210
          CT =C0+CX*X+CY*Y                                               F2D15220
      ENDIF                                                              F2D15230
      IF(ITYPE.LE.0) THEN                                                F2D15240
          A11=A10+A1X*X+A1Y*Y                                            F2D15250
          A22=A20+A2X*X+A2Y*Y                                            F2D15260
      ENDIF                                                              F2D15270
C                                                                        F2D15280
      II=1                                                               F2D15290
      DO 180 I=1,NPE                                                     F2D15300
      JJ=1                                                               F2D15310
      DO 160 J=1,NPE                                                     F2D15320
      S00=SF(I)*SF(J)*CNST                                               F2D15330
      S11=GDSF(1,I)*GDSF(1,J)*CNST                                       F2D15340
      S22=GDSF(2,I)*GDSF(2,J)*CNST                                       F2D15350
      S12=GDSF(1,I)*GDSF(2,J)*CNST                                       F2D15360
      S21=GDSF(2,I)*GDSF(1,J)*CNST                                       F2D15370
      IF(ITYPE.EQ.0) THEN                                                F2D15380
C                                                                        F2D15390
C     Heat transfer and like problems (i.e. single DOF problems):_______F2D15400
C                                                                        F2D15410
          ELK(I,J) = ELK(I,J) + A11*S11 + A22*S22 + A00*S00              F2D15420
          IF(ITEM.NE.0) THEN                                             F2D15430
              ELM(I,J) = ELM(I,J) + CT*S00                               F2D15440
          ENDIF                                                          F2D15450
      ELSE                                                               F2D15460
          IF(ITYPE.EQ.1) THEN                                            F2D15470
C                                                                        F2D15480
C     Viscous incompressible fluids:____________________________________F2D15490
C     Compute coefficients associated with viscous terms (full integ.)  F2D15500
C                                                                        F2D15510
              ELK(II,JJ)     = ELK(II,JJ)     + AMU*(2.0*S11 + S22)      F2D15520
              ELK(II+1,JJ)   = ELK(II+1,JJ)   + AMU*S12                  F2D15530
              ELK(II,JJ+1)   = ELK(II,JJ+1)   + AMU*S21                  F2D15540
              ELK(II+1,JJ+1)= ELK(II+1,JJ+1) + AMU*(S11 + 2.0*S22)       F2D15550
              IF(ITEM.NE.0) THEN                                         F2D15560
                  ELM(II,JJ)     = ELM(II,JJ)     + CT*S00               F2D15570
                  ELM(II+1,JJ+1)= ELM(II+1,JJ+1) + CT*S00                F2D15580
              ENDIF                                                      F2D15590
          ELSE                                                           F2D15600
C                                                                        F2D15610
C     Plane elasticity problems:________________________________________F2D15620
C                                                                        F2D15630
              ELK(II,JJ)     =ELK(II,JJ)     +CMAT(1,1)*S11+CMAT(3,3)*S22 F2D15640
              ELK(II,JJ+1)   =ELK(II,JJ+1)   +CMAT(1,2)*S12+CMAT(3,3)*S21 F2D15650
              ELK(II+1,JJ)   =ELK(II+1,JJ)   +CMAT(1,2)*S21+CMAT(3,3)*S12 F2D15660
              ELK(II+1,JJ+1)=ELK(II+1,JJ+1)+CMAT(3,3)*S11+CMAT(2,2)*S22 F2D15670
              IF(ITEM.NE.0) THEN                                         F2D15680
                  ELM(II,JJ)     = ELM(II,JJ)     + CT*S00               F2D15690
                  ELM(II+1,JJ+1)= ELM(II+1,JJ+1) + CT*S00                F2D15700
              ENDIF                                                      F2D15710
          ENDIF                                                          F2D15720
      ENDIF                                                              F2D15730
  160 JJ = NDF*J+1                                                       F2D15740
      IF(NEIGN.EQ.0) THEN                                                F2D15750
C                                                                        F2D15760
C     Source of the form fx = F0 + FX*X + FY*Y is assumed                F2D15770
C                                                                        F2D15780
          L=(I-1)*NDF+1                                                  F2D15790
          ELF(L) = ELF(L)+CNST*SF(I)*SOURCE                              F2D15800
      ENDIF                                                              F2D15810
  180 II = NDF*I+1                                                       F2D15820
  200 CONTINUE                                                           F2D15830
C                                                                        F2D15840
      IF(ITYPE.EQ.1 .OR. ITYPE.EQ.3) THEN                                F2D15850
C                                                                        F2D15860
C     Use reduced integration to evaluate coefficients associated with   F2D15870
C     penalty terms for flows and transverse shear terms for plates.     F2D15880
C                                                                        F2D15890
C     Call subroutine QUADRaTure to compute arrays of integration        F2D15900
C     points and weights for the given NIPR and IPDR                     F2D15910
```

```
C                                                                       F2D15920
          CALL QUADRT (NIPR,IPDR)                                       F2D15930
C                                                                       F2D15940
          DO 280 NI=1,NIPR                                              F2D15950
          AC1 = AL1(NI,IPDR)                                            F2D15960
          AC2 = AL2(NI,IPDR)                                            F2D15970
          AC3 = AL3(NI,IPDR)                                            F2D15980
          CALL SHPTRI(NPE,AC1,AC2,AC3,DET,ELXY)                         F2D15990
          CNST = 0.50D0*DET*ALWT(NI,IPDR)                               F2D16000
C                                                                       F2D16010
          II=1                                                          F2D16020
          DO 260 I=1,NPE                                                F2D16030
          JJ = 1                                                        F2D16040
          DO 240 J=1,NPE                                                F2D16050
          S11=GDSF(1,I)*GDSF(1,J)*CNST                                  F2D16060
          S22=GDSF(2,I)*GDSF(2,J)*CNST                                  F2D16070
          S12=GDSF(1,I)*GDSF(2,J)*CNST                                  F2D16080
          S21=GDSF(2,I)*GDSF(1,J)*CNST                                  F2D16090
          IF(ITYPE.EQ.1) THEN                                           F2D16100
C                                                                       F2D16110
C     Viscous incompressible fluids (penalty terms):____________________F2D16120
C                                                                       F2D16130
              ELK(II,JJ)     = ELK(II,JJ)      + PENLTY*S11             F2D16140
              ELK(II+1,JJ)   = ELK(II+1,JJ)    + PENLTY*S21             F2D16150
              ELK(II,JJ+1)   = ELK(II,JJ+1)    + PENLTY*S12             F2D16160
              ELK(II+1,JJ+1)= ELK(II+1,JJ+1)  + PENLTY*S22             F2D16170
          ELSE                                                          F2D16180
C                                                                       F2D16190
C     Shear deformable plates (transverse shear terms):_________________F2D16200
C                                                                       F2D16210
              S00=SF(I)*SF(J)*CNST                                      F2D16220
              S10 = GDSF(1,I)*SF(J)*CNST                                F2D16230
              S01 = SF(I)*GDSF(1,J)*CNST                                F2D16240
              S20 = GDSF(2,I)*SF(J)*CNST                                F2D16250
              S02 = SF(I)*GDSF(2,J)*CNST                                F2D16260
              ELK(II,JJ)     = ELK(II,JJ)      + C55*S11+C44*S22        F2D16270
              ELK(II,JJ+1)   = ELK(II,JJ+1)    + C55*S10                F2D16280
              ELK(II+1,JJ)   = ELK(II+1,JJ)    + C55*S01                F2D16290
              ELK(II,JJ+2)   = ELK(II,JJ+2)    + C44*S20                F2D16300
              ELK(II+2,JJ)   = ELK(II+2,JJ)    + C44*S02                F2D16310
              ELK(II+1,JJ+1)= ELK(II+1,JJ+1)  + C55*S00                F2D16320
              ELK(II+2,JJ+2)= ELK(II+2,JJ+2)  + C44*S00                F2D16330
          ENDIF                                                         F2D16340
  240 JJ=NDF*J+1                                                        F2D16350
  260 II=NDF*I+1                                                        F2D16360
  280 CONTINUE                                                          F2D16370
      ENDIF                                                             F2D16380
      RETURN                                                            F2D16390
      END                                                               F2D16400

      SUBROUTINE MSH2DG(NELEM,NNODE,NODES,MAXELM,MAXNOD,GLXY)           F2D16410
C     __________________________________________________________________F2D16420
C                                                                       F2D16430
C     Generates nodal point coordinates for specified type meshes       F2D16440
C                                                                       F2D16450
C     NOD1  = First node number in the line segment                     F2D16460
C     NODL  = Last node number in the line segment                      F2D16470
C     NODINC= Node increment from one node to the next along the line   F2D16480
C     X1,Y1 = Global coordinates of the first node on the line          F2D16490
C     XL,YL = Global coordinates of the last node on the line           F2D16500
C     RATIO = The ratio of the first element to the last element        F2D16510
C                                                                       F2D16520
C     __________________________________________________________________F2D16530
      IMPLICIT REAL*8(A-H,O-Z)                                          F2D16540
      DIMENSION  GLXY(MAXNOD,2)                                         F2D16550
C                                                                       F2D16560
      DO 10 I=1,NNODE                                                   F2D16570
      GLXY(I,1)=1.E20                                                   F2D16580
   10 GLXY(I,2)=1.E20                                                   F2D16590
C                                                                       F2D16600
C     Read number of the records (line segmments) and data in each line F2D16610
C                                                                       F2D16620
      READ(5,*)NRECL                                                    F2D16630
      DO 30 IREC=1,NRECL                                                F2D16640
```

```
      READ(5,*)NOD1,NODL,NODINC,X1,Y1,XL,YL,RATIO                       F2D16650
      IF(NODL.LT.NOD1) NODL = NOD1                                      F2D16660
      IF(NODL.NE.NOD1) THEN                                             F2D16670
          IF(NODINC.LE.0) NODINC = 1                                    F2D16680
          IF(RATIO.LE.0.0) RATIO=1.0                                    F2D16690
          NODIF = (NODL-NOD1)/NODINC                                    F2D16700
          XL1=XL-X1                                                     F2D16710
          YL1=YL-Y1                                                     F2D16720
          GLXY(NOD1,1)=X1                                               F2D16730
          GLXY(NOD1,2)=Y1                                               F2D16740
          ALNGTH=DSQRT(XL1*XL1+YL1*YL1)                                 F2D16750
          ALINC=(2.0*ALNGTH/NODIF)*RATIO/(RATIO+1.0)                    F2D16760
          ALRAT=ALINC/RATIO                                             F2D16770
          IF(NODIF.NE.1) DEL=(ALINC-ALRAT)/(NODIF-1)                    F2D16780
          IF(NODIF.EQ.1) DEL=0.0                                        F2D16790
          SUM=0.0                                                       F2D16800
          I=-1                                                          F2D16810
          DO 20 N=1,NODIF                                               F2D16820
          I=I+1                                                         F2D16830
          SUM=SUM+ALINC-I*DEL                                           F2D16840
          NI=NOD1+N*NODINC                                              F2D16850
          GLXY(NI,1)=X1+XL1*SUM/ALNGTH                                  F2D16860
          GLXY(NI,2)=Y1+YL1*SUM/ALNGTH                                  F2D16870
   20     CONTINUE                                                      F2D16880
      ENDIF                                                             F2D16890
   30 CONTINUE                                                          F2D16900
      CALL CNCTVT(NELEM,NNODE,NODES,MAXELM,MAXNOD,GLXY)                 F2D16910
      RETURN                                                            F2D16920
      END                                                               F2D16930


      SUBROUTINE MSH2DR(IEL,IELTYP,NX,NY,NPE,NNM,NEM,NOD,DX,DY,X0,Y0,   F2D16940
     1                  GLXY,MAXNEM,MAXNNM,MAXNX,MAXNY)                 F2D16950
C     __________________________________________________________________F2D16960
C                                                                       F2D16970
C     The subroutine generates arrays [NOD] and [GLXY] for rectangular  F2D16980
C     domains. The domain is divided  into  NX  subdivisions along the  F2D16990
C     x-direction and NY subdivisions in the y-direction. The subdivi-  F2D17000
C     sions define  rectangular elements  of the type required.  For a  F2D17010
C     triangular element mesh, the subdivision defines two linear ele-  F2D17020
C     ments with their common diagonal being inclined to the right.     F2D17030
C                                                                       F2D17040
C     __________________________________________________________________F2D17050
      IMPLICIT REAL*8 (A-H,O-Z)                                         F2D17060
      DIMENSION NOD(MAXNEM,9),GLXY(MAXNNM,2),DX(MAXNX),DY(MAXNY)        F2D17070
      COMMON/IO/IN,ITT                                                  F2D17080
C                                                                       F2D17090
      NEX1 = NX+1                                                       F2D17100
      NEY1 = NY+1                                                       F2D17110
      NXX  = IEL*NX                                                     F2D17120
      NYY  = IEL*NY                                                     F2D17130
      NXX1 = NXX + 1                                                    F2D17140
      NYY1 = NYY + 1                                                    F2D17150
      NEM  = NX*NY                                                      F2D17160
      IF(IELTYP.EQ.0)NEM=2*NX*NY                                        F2D17170
      NNM=NXX1*NYY1                                                     F2D17180
      IF(NPE.EQ.8)NNM = NXX1*NYY1 - NX*NY                               F2D17190
      IF(IELTYP.EQ.0) THEN                                              F2D17200
C                                                                       F2D17210
C     Generate the array [NOD]:_________________________________________F2D17220
C                                                                       F2D17230
C     TRIANGULAR ELEMENTS                                               F2D17240
C                                                                       F2D17250
          NX2=2*NX                                                      F2D17260
          NY2=2*NY                                                      F2D17270
          NOD(1,1) = 1                                                  F2D17280
          NOD(1,2) = IEL+1                                              F2D17290
          NOD(1,3) = IEL*NXX1+IEL+1                                     F2D17300
          IF(NPE .GT. 3) THEN                                           F2D17310
              NOD(1,4) = 2                                              F2D17320
              NOD(1,5) = NXX1 + 3                                       F2D17330
              NOD(1,6) = NXX1 + 2                                       F2D17340
          ENDIF                                                         F2D17350
C                                                                       F2D17360
          NOD(2,1) = 1                                                  F2D17370
```

```
            NOD(2,2) = NOD(1,3)                                              F2D17380
            NOD(2,3) = IEL*NXX1+1                                            F2D17390
            IF(NPE .GT. 3) THEN                                              F2D17400
                NOD(2,4) = NOD(1,6)                                          F2D17410
                NOD(2,5) = NOD(1,3) - 1                                      F2D17420
                NOD(2,6) = NOD(2,4) - 1                                      F2D17430
            ENDIF                                                            F2D17440
C                                                                            F2D17450
            K=3                                                              F2D17460
            DO 60 IY=1,NY                                                    F2D17470
            L=IY*NX2                                                         F2D17480
            M=(IY-1)*NX2                                                     F2D17490
            IF(NX.GT.1) THEN                                                 F2D17500
                DO 30 N=K,L,2                                                F2D17510
                DO 20 I=1,NPE                                                F2D17520
                NOD(N,I)   = NOD(N-2,I)+IEL                                  F2D17530
   20           NOD(N+1,I)= NOD(N-1,I)+IEL                                   F2D17540
   30           CONTINUE                                                     F2D17550
            ENDIF                                                            F2D17560
            IF(IY.LT.NY) THEN                                                F2D17570
                DO 40 I=1,NPE                                                F2D17580
                NOD(L+1,I)=NOD(M+1,I)+IEL*NXX1                               F2D17590
   40           NOD(L+2,I)=NOD(M+2,I)+IEL*NXX1                               F2D17600
            ENDIF                                                            F2D17610
   60       K=L+3                                                            F2D17620
        ELSE                                                                 F2D17630
C                                                                            F2D17640
C       RECTANGULAR ELEMENTS                                                 F2D17650
C                                                                            F2D17660
            K0 = 0                                                           F2D17670
            IF (NPE .EQ. 9) K0=1                                             F2D17680
C                                                                            F2D17690
            NOD(1,1) = 1                                                     F2D17700
            NOD(1,2) = IEL+1                                                 F2D17710
            NOD(1,3) = NXX1+(IEL-1)*NEX1+IEL+1                               F2D17720
            IF(NPE .EQ. 9) NOD(1,3)=4*NX+5                                   F2D17730
            NOD(1,4) = NOD(1,3) - IEL                                        F2D17740
            IF(NPE .GT. 4) THEN                                              F2D17750
                NOD(1,5) = 2                                                 F2D17760
                NOD(1,6) = NXX1 + (NPE-6)                                    F2D17770
                NOD(1,7) = NOD(1,3) - 1                                      F2D17780
                NOD(1,8) = NXX1+1                                            F2D17790
                IF(NPE .EQ. 9) THEN                                          F2D17800
                    NOD(1,9)=NXX1+2                                          F2D17810
                ENDIF                                                        F2D17820
            ENDIF                                                            F2D17830
C                                                                            F2D17840
            IF(NY .GT. 1) THEN                                               F2D17850
                M = 1                                                        F2D17860
                DO 110 N = 2,NY                                              F2D17870
                L = (N-1)*NX + 1                                             F2D17880
                DO 100 I = 1,NPE                                             F2D17890
  100           NOD(L,I) = NOD(M,I)+NXX1+(IEL-1)*NEX1+K0*NX                  F2D17900
  110           M=L                                                          F2D17910
            ENDIF                                                            F2D17920
C                                                                            F2D17930
            IF(NX .GT .1) THEN                                               F2D17940
                DO 140 NI = 2,NX                                             F2D17950
                DO 120  I = 1,NPE                                            F2D17960
                K1 = IEL                                                     F2D17970
                IF(I .EQ. 6 .OR. I .EQ. 8)K1=1+K0                            F2D17980
  120           NOD(NI,I) = NOD(NI-1,I)+K1                                   F2D17990
                M = NI                                                       F2D18000
                DO 140 NJ = 2,NY                                             F2D18010
                L = (NJ-1)*NX+NI                                             F2D18020
                DO 130 J = 1,NPE                                             F2D18030
  130           NOD(L,J) = NOD(M,J)+NXX1+(IEL-1)*NEX1+K0*NX                  F2D18040
  140           M = L                                                        F2D18050
            ENDIF                                                            F2D18060
        ENDIF                                                                F2D18070
C                                                                            F2D18080
C       Generate the global coordinates of the nodes, [GLXY]:______________F2D18090
C                                                                            F2D18100
        DX(NEX1)=0.0                                                         F2D18110
        DY(NEY1)=0.0                                                         F2D18120
```

```
      XC=X0                                                             F2D18130
      YC=Y0                                                             F2D18140
      IF(NPE .EQ. 8) THEN                                               F2D18150
          DO 180 NI = 1, NEY1                                           F2D18160
          I = (NXX1+NEX1)*(NI-1)+1                                      F2D18170
          J = 2*NI-1                                                    F2D18180
          GLXY(I,1) = XC                                                F2D18190
          GLXY(I,2) = YC                                                F2D18200
          DO 150 NJ = 1,NX                                              F2D18210
          DELX=0.5*DX(NJ)                                               F2D18220
          I=I+1                                                         F2D18230
          GLXY(I,1) = GLXY(I-1,1)+DELX                                  F2D18240
          GLXY(I,2) = YC                                                F2D18250
          I=I+1                                                         F2D18260
          GLXY(I,1) = GLXY(I-1,1)+DELX                                  F2D18270
          GLXY(I,2) = YC                                                F2D18280
  150     CONTINUE                                                      F2D18290
          IF(NI.LE.NY) THEN                                             F2D18300
              I = I+1                                                   F2D18310
              YC= YC+0.5*DY(NI)                                         F2D18320
              GLXY(I,1) = XC                                            F2D18330
              GLXY(I,2) = YC                                            F2D18340
              DO 160 II = 1, NX                                         F2D18350
              I = I+1                                                   F2D18360
              GLXY(I,1) = GLXY(I-1,1)+DX(II)                            F2D18370
  160         GLXY(I,2) = YC                                            F2D18380
          ENDIF                                                         F2D18390
  180     YC = YC+0.5*DY(NI)                                            F2D18400
C                                                                       F2D18410
      ELSE                                                              F2D18420
          YC=Y0                                                         F2D18430
          DO 200 NI = 1, NEY1                                           F2D18440
          XC = X0                                                       F2D18450
          I = NXX1*IEL*(NI-1)                                           F2D18460
          DO 190 NJ = 1, NEX1                                           F2D18470
          I=I+1                                                         F2D18480
          GLXY(I,1) = XC                                                F2D18490
          GLXY(I,2) = YC                                                F2D18500
          IF(NJ.LT.NEX1) THEN                                           F2D18510
              IF(IEL.EQ.2) THEN                                         F2D18520
                  I=I+1                                                 F2D18530
                  XC = XC + 0.5*DX(NJ)                                  F2D18540
                  GLXY(I,1) = XC                                        F2D18550
                  GLXY(I,2) = YC                                        F2D18560
              ENDIF                                                     F2D18570
          ENDIF                                                         F2D18580
  190     XC = XC + DX(NJ)/IEL                                          F2D18590
          XC = X0                                                       F2D18600
          IF(IEL.EQ.2) THEN                                             F2D18610
              YC = YC + 0.5*DY(NI)                                      F2D18620
              DO 195 NJ = 1, NEX1                                       F2D18630
              I=I+1                                                     F2D18640
              GLXY(I,1) = XC                                            F2D18650
              GLXY(I,2) = YC                                            F2D18660
              IF(NJ.LT.NEX1) THEN                                       F2D18670
                  I=I+1                                                 F2D18680
                  XC = XC + 0.5*DX(NJ)                                  F2D18690
                  GLXY(I,1) = XC                                        F2D18700
                  GLXY(I,2) = YC                                        F2D18710
              ENDIF                                                     F2D18720
  195     XC = XC + 0.5*DX(NJ)                                          F2D18730
          ENDIF                                                         F2D18740
  200     YC = YC + DY(NI)/IEL                                          F2D18750
      ENDIF                                                             F2D18760
      RETURN                                                            F2D18770
      END                                                               F2D18780

      SUBROUTINE PSTPRC(ELXY,ITYPE,IELTYP,IGRAD,NDF,NPE,THKNS,ELU,      F2D18790
     *                  ISTR,NSTR)                                      F2D18800
C     __________________________________________________________________F2D18810
C                                                                       F2D18820
C     Computes the derivatives of the solution for  heat transfer like  F2D18830
C     problems and stresses for fluid flow, plane elasticity and plate  F2D18840
C     bending problems.                                                 F2D18850
C     __________________________________________________________________F2D18860
```

```
C                                                                       F2D18870
      IMPLICIT REAL*8 (A-H,O-Z)                                         F2D18880
      DIMENSION  X(3),Y(3),GAMA(3),BETA(3),ELXY(9,2),ELU(27),GAUSPT(4,4)F2D18890
      COMMON/PST/A10,A1X,A1Y,A20,A2X,A2Y,A00,C0,CX,CY,F0,FX,FY,         F2D18900
     1          C44,C55,AMU,PENLTY,CMAT(3,3)                            F2D18910
      COMMON/SHP/SF(9),GDSF(2,9),SFH(16),GDSFH(2,16),GDDSFH(3,16)      F2D18920
      COMMON/QUAD/AL1(7,5),AL2(7,5),AL3(7,5),ALWT(7,5)                 F2D18930
      COMMON/IO/IN,ITT                                                  F2D18940
C                                                                       F2D18950
      DATA GAUSPT/4*0.0D0, -0.57735027D0, 0.57735027D0, 2*0.0D0,       F2D18960
     2     -0.77459667D0, 0.0D0, 0.77459667D0, 0.0D0, -0.86113631D0,   F2D18970
     3     -0.33998104D0, 0.33998104D0, 0.86113631D0/                   F2D18980
C                                                                       F2D18990
      PI=4.0D0*DATAN(1.0D0)                                             F2D19000
      CONST=180.0D0/PI                                                  F2D19010
      IF(IELTYP.EQ.0) THEN                                              F2D19020
C                                                                       F2D19030
C     Computation of the gradient/stresses at the reduced-integration   F2D19040
C     points of TRIANGULAR ELEMENTS:____________________________________F2D19050
C                                                                       F2D19060
          CALL QUADRT (NSTR,ISTR)                                       F2D19070
C                                                                       F2D19080
          DO 40 NI=1,NSTR                                               F2D19090
          AC1 = AL1(NI,ISTR)                                            F2D19100
          AC2 = AL2(NI,ISTR)                                            F2D19110
          AC3 = AL3(NI,ISTR)                                            F2D19120
          CALL SHPTRI(NPE,AC1,AC2,AC3,DET,ELXY)                         F2D19130
          XC  = 0.0                                                     F2D19140
          YC  = 0.0                                                     F2D19150
          DO 10 I=1,NPE                                                 F2D19160
          XC  = XC+SF(I)*ELXY(I,1)                                      F2D19170
   10     YC  = YC+SF(I)*ELXY(I,2)                                      F2D19180
          IF(ITYPE.LT.3) THEN                                           F2D19190
              UX  = 0.0                                                 F2D19200
              UY  = 0.0                                                 F2D19210
              VX  = 0.0                                                 F2D19220
              VY  = 0.0                                                 F2D19230
              DO 20 I=1,NPE                                             F2D19240
              J=NDF*I-1                                                 F2D19250
              IF(ITYPE.EQ.0)J=I                                         F2D19260
              UX  = UX + ELU(J)*GDSF(1,I)                               F2D19270
              UY  = UY + ELU(J)*GDSF(2,I)                               F2D19280
              IF(ITYPE.GE.1) THEN                                       F2D19290
                  K=J+1                                                 F2D19300
                  VX  = VX + ELU(K)*GDSF(1,I)                           F2D19310
                  VY  = VY + ELU(K)*GDSF(2,I)                           F2D19320
              ENDIF                                                     F2D19330
   20         CONTINUE                                                  F2D19340
C                                                                       F2D19350
              IF(ITYPE.EQ.0) THEN                                       F2D19360
C                                                                       F2D19370
C     Single-degree-of-freedom problems:--------------------------------F2D19380
C                                                                       F2D19390
                  SX   = -(A10+A1X*XC+A1Y*YC)*UX                        F2D19400
                  SY   = -(A20+A2X*XC+A2Y*YC)*UY                        F2D19410
                  VALUE= DSQRT(SX**2+SY**2)                             F2D19420
                  IF(IGRAD.EQ.1) THEN                                   F2D19430
                      QX=SX                                             F2D19440
                      QY=SY                                             F2D19450
                  ELSE                                                  F2D19460
                      QX=-SY                                            F2D19470
                      QY= SX                                            F2D19480
                  ENDIF                                                 F2D19490
                  IF(QX.EQ.0.0) THEN                                    F2D19500
                      IF(QY.LT.0.0) THEN                                F2D19510
                          ANGLE =-90.0                                  F2D19520
                      ELSE                                              F2D19530
                          ANGLE = 90.0                                  F2D19540
                      ENDIF                                             F2D19550
                  ELSE                                                  F2D19560
                      ANGLE=DATAN2(QY,QX)*CONST                         F2D19570
                  ENDIF                                                 F2D19580
                  WRITE(ITT,200) XC,YC,QX,QY,VALUE,ANGLE                F2D19590
              ELSE                                                      F2D19600
```

```
C                                                                       F2D19610
                  IF(ITYPE.EQ.1) THEN                                   F2D19620
C                                                                       F2D19630
C     Viscous incompressible flows (penalty model):---------------------F2D19640
C                                                                       F2D19650
                     PRESSR = -PENLTY*(UX+VY)                           F2D19660
                     STRESX = 2.0*AMU*UX-PRESSR                         F2D19670
                     STRESY = 2.0*AMU*VY-PRESSR                         F2D19680
                     STRSXY = AMU*(UY+VX)                               F2D19690
                     WRITE(ITT,300) XC,YC,STRESX,STRESY,STRSXY,PRESSR   F2D19700
                  ELSE                                                  F2D19710
C                                                                       F2D19720
C     Plane elasticity problems:----------------------------------------F2D19730
C                                                                       F2D19740
                     STRESX = (CMAT(1,1)*UX+CMAT(1,2)*VY)/THKNS         F2D19750
                     STRESY = (CMAT(1,2)*UX+CMAT(2,2)*VY)/THKNS         F2D19760
                     STRSXY = CMAT(3,3)*(UY+VX)/THKNS                   F2D19770
                     WRITE(ITT,300) XC,YC,STRESX,STRESY,STRSXY          F2D19780
                  ENDIF                                                 F2D19790
               ENDIF                                                    F2D19800
            ENDIF                                                       F2D19810
   40       CONTINUE                                                    F2D19820
      ELSE                                                              F2D19830
C                                                                       F2D19840
C     Calculation of the gradient/stresses at the reduced integration   F2D19850
C     gauss points of RECTANGULAR ELEMENTS:_____________________________F2D19860
C                                                                       F2D19870
         DO 100 NI=1,ISTR                                               F2D19880
         DO 100 NJ=1,ISTR                                               F2D19890
         XI  = GAUSPT(NI,ISTR)                                          F2D19900
         ETA = GAUSPT(NJ,ISTR)                                          F2D19910
         CALL SHPRCT (NPE,XI,ETA,DET,ELXY,NDF,ITYPE)                    F2D19920
         XC  = 0.0                                                      F2D19930
         YC  = 0.0                                                      F2D19940
         DO 50 I=1,NPE                                                  F2D19950
         XC  = XC+SF(I)*ELXY(I,1)                                       F2D19960
   50    YC  = YC+SF(I)*ELXY(I,2)                                       F2D19970
         IF(ITYPE.LT.3) THEN                                            F2D19980
            UX  = 0.0                                                   F2D19990
            UY  = 0.0                                                   F2D20000
            VX  = 0.0                                                   F2D20010
            VY  = 0.0                                                   F2D20020
            DO 60 I=1,NPE                                               F2D20030
            J=NDF*I-1                                                   F2D20040
            IF(ITYPE.EQ.0)J=I                                           F2D20050
            UX  = UX + ELU(J)*GDSF(1,I)                                 F2D20060
            UY  = UY + ELU(J)*GDSF(2,I)                                 F2D20070
            IF(ITYPE.GE.1) THEN                                         F2D20080
               K=J+1                                                    F2D20090
               VX  = VX + ELU(K)*GDSF(1,I)                              F2D20100
               VY  = VY + ELU(K)*GDSF(2,I)                              F2D20110
            ENDIF                                                       F2D20120
   60 CONTINUE                                                          F2D20130
C                                                                       F2D20140
            IF(ITYPE.EQ.0) THEN                                         F2D20150
C                                                                       F2D20160
C     Single-degree-of-freedom problems:--------------------------------F2D20170
C                                                                       F2D20180
               SX    = -(A10+A1X*XC+A1Y*YC)*UX                          F2D20190
               SY    = -(A20+A2X*XC+A2Y*YC)*UY                          F2D20200
               VALUE= DSQRT(SX**2+SY**2)                                F2D20210
               IF(IGRAD.EQ.1) THEN                                      F2D20220
                  QX=SX                                                 F2D20230
                  QY=SY                                                 F2D20240
               ELSE                                                     F2D20250
                  QX=-SY                                                F2D20260
                  QY= SX                                                F2D20270
               ENDIF                                                    F2D20280
               IF(QX.EQ.0.0) THEN                                       F2D20290
                  IF(QY.LT.0.0) THEN                                    F2D20300
                     ANGLE =-90.0                                       F2D20310
                  ELSE                                                  F2D20320
                     ANGLE = 90.0                                       F2D20330
                  ENDIF                                                 F2D20340
```

```
                ELSE                                                    F2D20350
                    ANGLE=DATAN2(QY,QX)*CONST                           F2D20360
                ENDIF                                                   F2D20370
                WRITE(ITT,200) XC,YC,QX,QY,VALUE,ANGLE                  F2D20380
            ELSE                                                        F2D20390
C                                                                       F2D20400
                IF(ITYPE.EQ.1) THEN                                     F2D20410
C                                                                       F2D20420
C     Viscous incompressible flows (penalty model):---------------------F2D20430
C                                                                       F2D20440
                    PRESSR = -PENLTY*(UX+VY)                            F2D20450
                    STRESX = 2.0*AMU*UX-PRESSR                          F2D20460
                    STRESY = 2.0*AMU*VY-PRESSR                          F2D20470
                    STRSXY = AMU*(UY+VX)                                F2D20480
                    WRITE(ITT,300) XC,YC,STRESX,STRESY,STRSXY,PRESSR    F2D20490
                ELSE                                                    F2D20500
C                                                                       F2D20510
C     Plane elasticity problems:----------------------------------------F2D20520
C                                                                       F2D20530
                    STRESX = (CMAT(1,1)*UX+CMAT(1,2)*VY)/THKNS          F2D20540
                    STRESY = (CMAT(1,2)*UX+CMAT(2,2)*VY)/THKNS          F2D20550
                    STRSXY = CMAT(3,3)*(UY+VX)/THKNS                    F2D20560
                    WRITE(ITT,300) XC,YC,STRESX,STRESY,STRSXY           F2D20570
                ENDIF                                                   F2D20580
            ENDIF                                                       F2D20590
        ELSE                                                            F2D20600
C                                                                       F2D20610
C     Plate bending problems:-------------------------------------------F2D20620
C     Stresses SGMAX, SGMAY and SGMXY are computed at the top/bottom of F2D20630
C     the plate (and SGMXZ and SGMYZ are constant through thickness)    F2D20640
C                                                                       F2D20650
            PLTD=(THKNS*THKNS)/6.0D0                                    F2D20660
            SIX  = 0.0                                                  F2D20670
            SIY  = 0.0                                                  F2D20680
            DWX  = 0.0                                                  F2D20690
            DWY  = 0.0                                                  F2D20700
            DSXY = 0.0                                                  F2D20710
            DSYX = 0.0                                                  F2D20720
            DSXX = 0.0                                                  F2D20730
            DSYY = 0.0                                                  F2D20740
            IF(ITYPE.EQ.3) THEN                                         F2D20750
C                                                                       F2D20760
C     First-order shear deformation theory of plates:-------------------F2D20770
C                                                                       F2D20780
                DO 80 I=1,NPE                                           F2D20790
                J=NDF*(I-1)+1                                           F2D20800
                K=J+1                                                   F2D20810
                L=K+1                                                   F2D20820
                DWX  = DWX+GDSF(1,I)*ELU(J)                             F2D20830
                DWY  = DWY+GDSF(2,I)*ELU(J)                             F2D20840
                SIX  = SIX+SF(I)*ELU(K)                                 F2D20850
                SIY  = SIY+SF(I)*ELU(L)                                 F2D20860
                DSXX = DSXX+GDSF(1,I)*ELU(K)                            F2D20870
                DSXY = DSXY+GDSF(2,I)*ELU(K)                            F2D20880
                DSYX = DSYX+GDSF(1,I)*ELU(L)                            F2D20890
   80           DSYY = DSYY+GDSF(2,I)*ELU(L)                            F2D20900
                SGMAX = (CMAT(1,1)*DSXX+CMAT(1,2)*DSYY)/PLTD            F2D20910
                SGMAY = (CMAT(1,2)*DSXX+CMAT(2,2)*DSYY)/PLTD            F2D20920
                SGMXY = CMAT(3,3)*(DSXY+DSYX)/PLTD                      F2D20930
                SGMXZ = 1.2*C55*(DWX+SIX)/THKNS                         F2D20940
                SGMYZ = 1.2*C44*(DWY+SIY)/THKNS                         F2D20950
                WRITE(ITT,300) XC,YC,SGMAX,SGMAY,SGMXY                  F2D20960
                WRITE(ITT,400) SGMXZ,SGMYZ                              F2D20970
            ELSE                                                        F2D20980
C                                                                       F2D20990
C     Classical theory of plates:---------------------------------------F2D21000
C                                                                       F2D21010
                NN=NPE*NDF                                              F2D21020
                DO 90 I=1,NN                                            F2D21030
                DSXX = DSXX+GDDSFH(1,I)*ELU(I)                          F2D21040
                DSYY = DSYY+GDDSFH(2,I)*ELU(I)                          F2D21050
   90           DSXY = DSXY+GDDSFH(3,I)*ELU(I)                          F2D21060
C                                                                       F2D21070
                SGMAX =-(CMAT(1,1)*DSXX+CMAT(1,2)*DSYY)/PLTD            F2D21080
                SGMAY =-(CMAT(1,2)*DSXX+CMAT(2,2)*DSYY)/PLTD            F2D21090
```

```
                  SGMXY =-4.0*CMAT(3,3)*DSXY/PLTD                     F2D21100
                  WRITE(ITT,300) XC,YC,SGMAX,SGMAY,SGMXY              F2D21110
              ENDIF                                                   F2D21120
          ENDIF                                                       F2D21130
  100     CONTINUE                                                    F2D21140
      ENDIF                                                           F2D21150
  200 FORMAT(5E13.4,3X,F7.2)                                          F2D21160
  300 FORMAT(6E13.4)                                                  F2D21170
  400 FORMAT(26X,2E13.4)                                              F2D21180
      RETURN                                                          F2D21190
      END                                                             F2D21200


      SUBROUTINE QUADRT(NIP,IPD)                                      F2D21210
C     ________________________________________________________________F2D21220
C                                                                     F2D21230
C     Quadrature points and weights for triangular elements           F2D21240
C       IPD = Integrand Polynomial Degree                             F2D21250
C       NIP = Number of Integration Points                            F2D21260
C     ________________________________________________________________F2D21270
C                                                                     F2D21280
      IMPLICIT REAL*8(A-H,O-Z)                                        F2D21290
      COMMON/QUAD/AL1(7,5),AL2(7,5),AL3(7,5),ALWT(7,5)               F2D21300
C                                                                     F2D21310
C     Initialize arrays                                               F2D21320
C                                                                     F2D21330
      DO 20 I = 1, NIP                                                F2D21340
      DO 10 J = 1, IPD                                                F2D21350
      AL1(I,J)   =  0.000000000000000                                 F2D21360
      AL2(I,J)   =  0.000000000000000                                 F2D21370
      AL3(I,J)   =  0.000000000000000                                 F2D21380
      ALWT(I,J)  =  0.000000000000000                                 F2D21390
   10 CONTINUE                                                        F2D21400
   20 CONTINUE                                                        F2D21410
C                                                                     F2D21420
C     One-point quadrature (for polynomials of order 1):______________F2D21430
C                                                                     F2D21440
      AL1(1,1)   =  0.333333333333333                                 F2D21450
      AL2(1,1)   =  0.333333333333333                                 F2D21460
      AL3(1,1)   =  0.333333333333333                                 F2D21470
      ALWT(1,1)  =  1.000000000000000                                 F2D21480
C                                                                     F2D21490
C     Three-point quadrature (for polynomials of order 2):____________F2D21500
C                                                                     F2D21510
      AL1(1,2)   =  0.000000000000000                                 F2D21520
      AL2(1,2)   =  0.500000000000000                                 F2D21530
      AL3(1,2)   =  0.500000000000000                                 F2D21540
      AL1(2,2)   =  0.500000000000000                                 F2D21550
      AL2(2,2)   =  0.000000000000000                                 F2D21560
      AL3(2,2)   =  0.500000000000000                                 F2D21570
      AL1(3,2)   =  0.500000000000000                                 F2D21580
      AL2(3,2)   =  0.500000000000000                                 F2D21590
      AL3(3,2)   =  0.000000000000000                                 F2D21600
      ALWT(1,2)  =  0.333333333333333                                 F2D21610
      ALWT(2,2)  =  0.333333333333333                                 F2D21620
      ALWT(3,2)  =  0.333333333333333                                 F2D21630
C                                                                     F2D21640
C     Four-point quadrature (for polynomials of order 3):_____________F2D21650
C                                                                     F2D21660
      AL1(1,3)   =  0.333333333333333                                 F2D21670
      AL2(1,3)   =  0.333333333333333                                 F2D21680
      AL3(1,3)   =  0.333333333333333                                 F2D21690
      AL1(2,3)   =  0.600000000000000                                 F2D21700
      AL2(2,3)   =  0.200000000000000                                 F2D21710
      AL3(2,3)   =  0.200000000000000                                 F2D21720
      AL1(3,3)   =  0.200000000000000                                 F2D21730
      AL2(3,3)   =  0.600000000000000                                 F2D21740
      AL3(3,3)   =  0.200000000000000                                 F2D21750
      AL1(4,3)   =  0.200000000000000                                 F2D21760
      AL2(4,3)   =  0.200000000000000                                 F2D21770
      AL3(4,3)   =  0.600000000000000                                 F2D21780
      ALWT(1,3)  = -0.562500000000000                                 F2D21790
      ALWT(2,3)  =  0.520833333333333                                 F2D21800
      ALWT(3,3)  =  0.520833333333333                                 F2D21810
      ALWT(4,3)  =  0.520833333333333                                 F2D21820
```

```
C                                                                       F2D21830
C     Six-point quadrature (for polynomials of order 4):________________F2D21840
C                                                                       F2D21850
      AL1(1,4)   =  0.816847572980459                                   F2D21860
      AL2(1,4)   =  0.091576213509771                                   F2D21870
      AL3(1,4)   =  0.091576213509771                                   F2D21880
      AL1(2,4)   =  0.091576213509771                                   F2D21890
      AL2(2,4)   =  0.816847572980459                                   F2D21900
      AL3(2,4)   =  0.091576213509771                                   F2D21910
      AL1(3,4)   =  0.091576213509771                                   F2D21920
      AL2(3,4)   =  0.091576213509771                                   F2D21930
      AL3(3,4)   =  0.816847572980459                                   F2D21940
      AL1(4,4)   =  0.108103018168070                                   F2D21950
      AL2(4,4)   =  0.445948490915965                                   F2D21960
      AL3(4,4)   =  0.445948490915965                                   F2D21970
      AL1(5,4)   =  0.445948490915965                                   F2D21980
      AL2(5,4)   =  0.108103018168070                                   F2D21990
      AL3(5,4)   =  0.445948490915965                                   F2D22000
      AL1(6,4)   =  0.445948490915965                                   F2D22010
      AL2(6,4)   =  0.445948490915965                                   F2D22020
      AL3(6,4)   =  0.108103018168070                                   F2D22030
      ALWT(1,4)  =  0.109951743655322                                   F2D22040
      ALWT(2,4)  =  0.109951743655322                                   F2D22050
      ALWT(3,4)  =  0.109951743655322                                   F2D22060
      ALWT(4,4)  =  0.223381589678011                                   F2D22070
      ALWT(5,4)  =  0.223381589678011                                   F2D22080
      ALWT(6,4)  =  0.223381589678011                                   F2D22090
C                                                                       F2D22100
C     Seven-point quadrature (for polynomials of order 5):______________F2D22110
C                                                                       F2D22120
      AL1(1,5)   =  0.333333333333333                                   F2D22130
      AL2(1,5)   =  0.333333333333333                                   F2D22140
      AL3(1,5)   =  0.333333333333333                                   F2D22150
      AL1(2,5)   =  0.797426985353087                                   F2D22160
      AL2(2,5)   =  0.101286507323456                                   F2D22170
      AL3(2,5)   =  0.101286507323456                                   F2D22180
      AL1(3,5)   =  0.101286507323456                                   F2D22190
      AL2(3,5)   =  0.797426985353087                                   F2D22200
      AL3(3,5)   =  0.101286507323456                                   F2D22210
      AL1(4,5)   =  0.101286507323456                                   F2D22220
      AL2(4,5)   =  0.101286507323456                                   F2D22230
      AL3(4,5)   =  0.797426985353087                                   F2D22240
      AL1(5,5)   =  0.059715871789770                                   F2D22250
      AL2(5,5)   =  0.470142064105115                                   F2D22260
      AL3(5,5)   =  0.470142064105115                                   F2D22270
      AL1(6,5)   =  0.470142064105115                                   F2D22280
      AL2(6,5)   =  0.059715871789770                                   F2D22290
      AL3(6,5)   =  0.470142064105115                                   F2D22300
      AL1(7,5)   =  0.470142064105115                                   F2D22310
      AL2(7,5)   =  0.470142064105115                                   F2D22320
      AL3(7,5)   =  0.059715871789770                                   F2D22330
      ALWT(1,5)  =  0.225000000000000                                   F2D22340
      ALWT(2,5)  =  0.125939180544827                                   F2D22350
      ALWT(3,5)  =  0.125939180544827                                   F2D22360
      ALWT(4,5)  =  0.125939180544827                                   F2D22370
      ALWT(5,5)  =  0.132394152788506                                   F2D22380
      ALWT(6,5)  =  0.132394152788506                                   F2D22390
      ALWT(7,5)  =  0.132394152788506                                   F2D22400
C                                                                       F2D22410
      RETURN                                                            F2D22420
      END                                                               F2D22430

      SUBROUTINE SHPRCT(NPE,XI,ETA,DET,ELXY,NDF,ITYPE)                  F2D22440
C     __________________________________________________________________F2D22450
C                                                                       F2D22460
C     The subroutine evaluates the interpolation functions (SF(I)) and  F2D22470
C     their derivatives with respect to global coordinates (GDSF(I,J))  F2D22480
C     for Lagrange linear & quadratic rectangular elements, using  the  F2D22490
C     isoparametric formulation. The subroutine also evaluates Hermite  F2D22500
C     interpolation functions and their  global derivatives  using the  F2D22510
C     subparametric formulation.                                        F2D22520
C                                                                       F2D22530
C     SF(I).......Interpolation function for node I of the element      F2D22540
C     DSF(J,I).....Derivative of SF(I) with respect to XI if J=1 and    F2D22550
C                 and ETA if J=2                                        F2D22560
```

```
C     GDSF(J,I)....Derivative of SF(I) with respect to  X if J=1 and      F2D22570
C                  and  Y  if J=2                                         F2D22580
C     XNODE(I,J)...J-TH (J=1,2) Coordinate of node I of the element       F2D22590
C     NP(I)........Array of element nodes (used to define SF and DSF)     F2D22600
C     GJ(I,J)......Jacobian matrix                                        F2D22610
C     GJINV(I,J)...Inverse of the jacobian matrix                         F2D22620
C     ____________________________________________________________________ F2D22630
C                                                                         F2D22640
      IMPLICIT REAL*8 (A-H,O-Z)                                           F2D22650
      DIMENSION ELXY(9,2),XNODE(9,2),NP(9),DSF(2,9),GJ(2,2),GJINV(2,2)    F2D22660
      DIMENSION GGJ(3,3),GGINV(3,3),DDSJ(3,16),DDSF(3,4),DJCB(3,2),       F2D22670
     *          DSFH(3,16),DDSFH(3,16)                                    F2D22680
      COMMON/SHP/SF(9),GDSF(2,9),SFH(16),GDSFH(2,16),GDDSFH(3,16)         F2D22690
      COMMON/IO/IN,ITT                                                    F2D22700
      DATA XNODE/-1.0D0, 2*1.0D0, -1.0D0, 0.0D0, 1.0D0, 0.0D0, -1.0D0,    F2D22710
     *        0.0D0, 2*-1.0D0, 2*1.0D0, -1.0D0, 0.0D0, 1.0D0, 2*0.0D0/    F2D22720
      DATA NP/1,2,3,4,5,7,6,8,9/                                          F2D22730
C                                                                         F2D22740
      FNC(A,B) = A*B                                                      F2D22750
      IF(NPE.EQ.4) THEN                                                   F2D22760
C                                                                         F2D22770
C     LINEAR Lagrange interpolation functions for FOUR-NODE element       F2D22780
C                                                                         F2D22790
         DO 10 I = 1, NPE                                                 F2D22800
         XP  = XNODE(I,1)                                                 F2D22810
         YP  = XNODE(I,2)                                                 F2D22820
         XI0 = 1.0+XI*XP                                                  F2D22830
         ETA0=1.0+ETA*YP                                                  F2D22840
         SF(I)   = 0.25*FNC(XI0,ETA0)                                     F2D22850
         DSF(1,I)= 0.25*FNC(XP,ETA0)                                      F2D22860
   10    DSF(2,I)= 0.25*FNC(YP,XI0)                                       F2D22870
      ELSE                                                                F2D22880
         IF(NPE.EQ.8) THEN                                                F2D22890
C                                                                         F2D22900
C     QUADRATIC Lagrange interpolation functions for EIGHT-NODE element F2D22910
C                                                                         F2D22920
            DO 20 I = 1, NPE                                              F2D22930
            NI   = NP(I)                                                  F2D22940
            XP   = XNODE(NI,1)                                            F2D22950
            YP   = XNODE(NI,2)                                            F2D22960
            XI0  = 1.0+XI*XP                                              F2D22970
            ETA0 = 1.0+ETA*YP                                             F2D22980
            XI1  = 1.0-XI*XI                                              F2D22990
            ETA1 = 1.0-ETA*ETA                                            F2D23000
            IF(I.LE.4) THEN                                               F2D23010
               SF(NI)     = 0.25*FNC(XI0,ETA0)*(XI*XP+ETA*YP-1.0)         F2D23020
               DSF(1,NI)  = 0.25*FNC(ETA0,XP)*(2.0*XI*XP+ETA*YP)          F2D23030
               DSF(2,NI)  = 0.25*FNC(XI0,YP)*(2.0*ETA*YP+XI*XP)           F2D23040
            ELSE                                                          F2D23050
               IF(I.LE.6) THEN                                            F2D23060
                  SF(NI)     = 0.5*FNC(XI1,ETA0)                          F2D23070
                  DSF(1,NI)  = -FNC(XI,ETA0)                              F2D23080
                  DSF(2,NI)  = 0.5*FNC(YP,XI1)                            F2D23090
               ELSE                                                       F2D23100
                  SF(NI)     = 0.5*FNC(ETA1,XI0)                          F2D23110
                  DSF(1,NI)  = 0.5*FNC(XP,ETA1)                           F2D23120
                  DSF(2,NI)  = -FNC(ETA,XI0)                              F2D23130
               ENDIF                                                      F2D23140
            ENDIF                                                         F2D23150
   20       CONTINUE                                                      F2D23160
         ELSE                                                             F2D23170
C                                                                         F2D23180
C     QUADRATIC Lagrange interpolation functions for NINE-NODE element  F2D23190
C                                                                         F2D23200
            DO 30 I=1,NPE                                                 F2D23210
            NI   = NP(I)                                                  F2D23220
            XP   = XNODE(NI,1)                                            F2D23230
            YP   = XNODE(NI,2)                                            F2D23240
            XI0  = 1.0+XI*XP                                              F2D23250
            ETA0 = 1.0+ETA*YP                                             F2D23260
            XI1  = 1.0-XI*XI                                              F2D23270
            ETA1 = 1.0-ETA*ETA                                            F2D23280
            XI2  = XP*XI                                                  F2D23290
            ETA2 = YP*ETA                                                 F2D23300
            IF(I .LE. 4) THEN                                             F2D23310
```

```
                  SF(NI)    = 0.25*FNC(XI0,ETA0)*XI2*ETA2                    F2D23320
                  DSF(1,NI)= 0.25*XP*FNC(ETA2,ETA0)*(1.0+2.0*XI2)            F2D23330
                  DSF(2,NI)= 0.25*YP*FNC(XI2,XI0)*(1.0+2.0*ETA2)             F2D23340
              ELSE                                                           F2D23350
                  IF(I .LE. 6) THEN                                          F2D23360
                      SF(NI)    = 0.5*FNC(XI1,ETA0)*ETA2                     F2D23370
                      DSF(1,NI) = -XI*FNC(ETA2,ETA0)                         F2D23380
                      DSF(2,NI) = 0.5*FNC(XI1,YP)*(1.0+2.0*ETA2)             F2D23390
                  ELSE                                                       F2D23400
                      IF(I .LE. 8) THEN                                      F2D23410
                          SF(NI)    = 0.5*FNC(ETA1,XI0)*XI2                  F2D23420
                          DSF(2,NI) = -ETA*FNC(XI2,XI0)                      F2D23430
                          DSF(1,NI) = 0.5*FNC(ETA1,XP)*(1.0+2.0*XI2)         F2D23440
                      ELSE                                                   F2D23450
                          SF(NI)    = FNC(XI1,ETA1)                          F2D23460
                          DSF(1,NI) = -2.0*XI*ETA1                           F2D23470
                          DSF(2,NI) = -2.0*ETA*XI1                           F2D23480
                      ENDIF                                                  F2D23490
                  ENDIF                                                      F2D23500
              ENDIF                                                          F2D23510
   30         CONTINUE                                                       F2D23520
          ENDIF                                                              F2D23530
      ENDIF                                                                  F2D23540
C                                                                            F2D23550
C     Compute the Jacobian matrix [GJ] and its inverse [GJINV]               F2D23560
C                                                                            F2D23570
      DO 40 I = 1,2                                                          F2D23580
      DO 40 J = 1,2                                                          F2D23590
      GJ(I,J)  = 0.0                                                         F2D23600
      DO 40 K = 1,NPE                                                        F2D23610
   40 GJ(I,J)  = GJ(I,J) + DSF(I,K)*ELXY(K,J)                                F2D23620
C                                                                            F2D23630
      DET = GJ(1,1)*GJ(2,2)-GJ(1,2)*GJ(2,1)                                  F2D23640
      GJINV(1,1) = GJ(2,2)/DET                                               F2D23650
      GJINV(2,2) = GJ(1,1)/DET                                               F2D23660
      GJINV(1,2) = -GJ(1,2)/DET                                              F2D23670
      GJINV(2,1) = -GJ(2,1)/DET                                              F2D23680
C                                                                            F2D23690
      IF(ITYPE.LE.3) THEN                                                    F2D23700
C                                                                            F2D23710
C     Compute the derivatives of the interpolation functions with            F2D23720
C     respect to the global coordinates (x,y): [GDSF]                        F2D23730
C                                                                            F2D23740
          DO 50 I  = 1,2                                                     F2D23750
          DO 50 J  = 1,NPE                                                   F2D23760
          GDSF(I,J) = 0.0                                                    F2D23770
          DO 50 K  = 1, 2                                                    F2D23780
   50     GDSF(I,J) = GDSF(I,J) + GJINV(I,K)*DSF(K,J)                        F2D23790
      ELSE                                                                   F2D23800
C                                                                            F2D23810
C     Conforming Hermite interpolation functions (four-node element)         F2D23820
C                                                                            F2D23830
          IF(NDF.EQ.4) THEN                                                  F2D23840
              II = 1                                                         F2D23850
              DO 60 I = 1, NPE                                               F2D23860
              XP   = XNODE(I,1)                                              F2D23870
              YP   = XNODE(I,2)                                              F2D23880
              XI1  = XI*XP-1.0                                               F2D23890
              XI2  = XI1-1.0                                                 F2D23900
              ETA1 = ETA*YP-1.0                                              F2D23910
              ETA2 = ETA1-1.0                                                F2D23920
              XI0  = (XI+XP)**2                                              F2D23930
              ETA0 = (ETA+YP)**2                                             F2D23940
              XIP0 = XI+XP                                                   F2D23950
              XIP1 = 3.0*XI*XP+XP*XP                                         F2D23960
              XIP2 = 3.0*XI*XP+2.0*XP*XP                                     F2D23970
              YIP0 = ETA+YP                                                  F2D23980
              YIP1 = 3.0*ETA*YP+YP*YP                                        F2D23990
              YIP2 = 3.0*ETA*YP+2.0*YP*YP                                    F2D24000
C                                                                            F2D24010
              SFH(II)       = 0.0625*FNC(ETA0,ETA2)*FNC(XI0,XI2)             F2D24020
              DSFH(1,II)    = 0.0625*FNC(ETA0,ETA2)*XIP0*(XIP1-4.0)          F2D24030
              DSFH(2,II)    = 0.0625*FNC(XI0,XI2)*YIP0*(YIP1-4.0)            F2D24040
              DDSFH(1,II)   = 0.125*FNC(ETA0,ETA2)*(XIP2-2.0)                F2D24050
              DDSFH(2,II)   = 0.125*FNC(XI0,XI2)*(YIP2-2.0)                  F2D24060
              DDSFH(3,II)   = 0.0625*(XIP1-4.0)*(YIP1-4.0)*XIP0*YIP0         F2D24070
```

```
C                                                                       F2D24080
            SFH(II+1)     = -0.0625*XP*FNC(XI0,XI1)*FNC(ETA0,ETA2)      F2D24090
            DSFH(1,II+1)  = -0.0625*FNC(ETA0,ETA2)*XP*XIP0*(XIP1-2.0)   F2D24100
            DSFH(2,II+1)  = -0.0625*FNC(XI0,XI1)*XP*YIP0*(YIP1-4.)      F2D24110
            DDSFH(1,II+1) = -0.125*FNC(ETA0,ETA2)*XP*(XIP2-1.0)         F2D24120
            DDSFH(2,II+1) = -0.125*FNC(XI0,XI1)*(YIP2-2.0)*XP           F2D24130
            DDSFH(3,II+1) = -0.0625*XP*XIP0*(XIP1-2.)*(YIP1-4.)*YIP0    F2D24140
C                                                                       F2D24150
            SFH(II+2)     = -0.0625*YP*FNC(XI0,XI2)*FNC(ETA0,ETA1)      F2D24160
            DSFH(1,II+2)  = -0.0625*FNC(ETA0,ETA1)*YP*XIP0*(XIP1-4.)    F2D24170
            DSFH(2,II+2)  = -0.0625*FNC(XI0,XI2)*YP*YIP0*(YIP1-2.)      F2D24180
            DDSFH(1,II+2) = -0.125*FNC(ETA0,ETA1)*YP*(XIP2-2.)          F2D24190
            DDSFH(2,II+2) = -0.125*FNC(XI0,XI2)*YP*(YIP2-1.0)           F2D24200
            DDSFH(3,II+2) = -0.0625*YP*YIP0*(YIP1-2.)*(XIP1-4.0)*XIP0   F2D24210
C                                                                       F2D24220
            SFH(II+3)     = 0.0625*XP*YP*FNC(XI0,XI1)*FNC(ETA0,ETA1)    F2D24230
            DSFH(1,II+3)  = 0.0625*FNC(ETA0,ETA1)*XP*YP*(XIP1-2.)*XIP0F2D24240
            DSFH(2,II+3)  = 0.0625*FNC(XI0,XI1)*XP*YP*(YIP1-2.)*YIP0    F2D24250
            DDSFH(1,II+3) = 0.125*FNC(ETA0,ETA1)*XP*YP*(XIP2-1.)        F2D24260
            DDSFH(2,II+3) = 0.125*FNC(XI0,XI1)*XP*YP*(YIP2-1.0)         F2D24270
            DDSFH(3,II+3) = 0.0625*XP*YP*YIP0*XIP0*(YIP1-2.)*(XIP1-2.)F2D24280
            II = I*NDF + 1                                              F2D24290
   60       CONTINUE                                                    F2D24300
        ELSE                                                            F2D24310
C                                                                       F2D24320
C     Non-conforming Hermite interpolation functions (Four-node element)F2D24330
C                                                                       F2D24340
            II = 1                                                      F2D24350
            DO 80 I = 1, NPE                                            F2D24360
            XP     = XNODE(I,1)                                         F2D24370
            YP     = XNODE(I,2)                                         F2D24380
            XI0    = XI*XP                                              F2D24390
            ETA0   = ETA*YP                                             F2D24400
            XIP1   = XI0+1                                              F2D24410
            ETAP1  = ETA0+1                                             F2D24420
            XIM1   = XI0-1                                              F2D24430
            ETAM1  = ETA0-1                                             F2D24440
            XID    = 3.0+2.0*XI0+ETA0-3.0*XI*XI-ETA*ETA-2.0*XI/XP       F2D24450
            ETAD   = 3.0+XI0+2.0*ETA0-XI*XI-3.0*ETA*ETA-2.0*ETA/YP      F2D24460
            ETAXI  = 4.0+2.0*(XI0+ETA0)-3.0*(XI*XI+ETA*ETA)             F2D24470
     *                                   -2.0*(ETA/YP+XI/XP)            F2D24480
C                                                                       F2D24490
            SFH(II) = 0.125*XIP1*ETAP1*(2.0+XI0+ETA0-XI*XI-ETA*ETA)     F2D24500
            DSFH(1,II)    = 0.125*XP*ETAP1*XID                          F2D24510
            DSFH(2,II)    = 0.125*YP*XIP1*ETAD                          F2D24520
            DDSFH(1,II)   = 0.250*XP*ETAP1*(XP-3.0*XI-1.0/XP)           F2D24530
            DDSFH(2,II)   = 0.250*YP*XIP1*(YP-3.0*ETA-1.0/YP)           F2D24540
            DDSFH(3,II)   = 0.125*XP*YP*ETAXI                           F2D24550
C                                                                       F2D24560
            SFH(II+1)     = 0.125*XP*XIP1*XIP1*XIM1*ETAP1               F2D24570
            DSFH(1,II+1)  = 0.125*XP*XP*ETAP1*(3.0*XI0-1.0)*XIP1        F2D24580
            DSFH(2,II+1)  = 0.125*XP*YP*XIP1*XIP1*XIM1                  F2D24590
            DDSFH(1,II+1) = 0.250*XP*XP*XP*ETAP1*(3.0*XI0+1.0)          F2D24600
            DDSFH(2,II+1) = 0.0                                         F2D24610
            DDSFH(3,II+1) = 0.125*XP*XP*YP*(3.0*XI0-1.0)*XIP1           F2D24620
C                                                                       F2D24630
            SFH(II+2)     = 0.125*YP*XIP1*ETAP1*ETAP1*ETAM1             F2D24640
            DSFH(1,II+2)  = 0.125*XP*YP*ETAP1*ETAP1*ETAM1               F2D24650
            DSFH(2,II+2)  = 0.125*YP*YP*XIP1*(3.0*ETA0-1.0)*ETAP1       F2D24660
            DDSFH(1,II+2) = 0.0                                         F2D24670
            DDSFH(2,II+2) = 0.250*YP*YP*YP*XIP1*(3.0*ETA0+1.0)          F2D24680
            DDSFH(3,II+2) = 0.125*XP*YP*YP*(3.0*ETA0-1.0)*ETAP1         F2D24690
            II = I*NDF + 1                                              F2D24700
   80       CONTINUE                                                    F2D24710
        ENDIF                                                           F2D24720
C                                                                       F2D24730
C     Compute the global first and second derivatives of the Hermite    F2D24740
C     interpolation functions.  The geometry is approximated using the  F2D24750
C     linear Lagrane interpolation functions (Subparametric formulation)F2D24760
C                                                                       F2D24770
        DDSF(1,1) =   0.0D0                                             F2D24780
        DDSF(2,1) =   0.0D0                                             F2D24790
        DDSF(3,1) =   0.250D0                                           F2D24800
        DDSF(1,2) =   0.0D0                                             F2D24810
        DDSF(2,2) =   0.0D0                                             F2D24820
        DDSF(3,2) = - 0.250D0                                           F2D24830
```

```
          DDSF(1,3) =   0.0D0                                            F2D24840
          DDSF(2,3) =   0.0D0                                            F2D24850
          DDSF(3,3) =   0.250D0                                          F2D24860
          DDSF(1,4) =   0.0D0                                            F2D24870
          DDSF(2,4) =   0.0D0                                            F2D24880
          DDSF(3,4) = - 0.250D0                                          F2D24890
C                                                                        F2D24900
C     Compute global first derivatives of Hermite functions              F2D24910
C                                                                        F2D24920
          NN=NDF*NPE                                                     F2D24930
          DO 110 I = 1, 2                                                F2D24940
             DO 100 J = 1, NN                                            F2D24950
             SUM = 0.0D0                                                 F2D24960
             DO 90 K = 1, 2                                              F2D24970
                SUM = SUM + GJINV(I,K)*DSFH(K,J)                         F2D24980
   90        CONTINUE                                                    F2D24990
             GDSFH(I,J) = SUM                                            F2D25000
  100        CONTINUE                                                    F2D25010
  110     CONTINUE                                                       F2D25020
C                                                                        F2D25030
C     Compute global second derivatives of Hermite functions             F2D25040
C                                                                        F2D25050
          DO 140 I = 1, 3                                                F2D25060
             DO 130 J = 1, 2                                             F2D25070
             SUM = 0.0D0                                                 F2D25080
             DO 120 K = 1, NPE                                           F2D25090
                SUM = SUM + DDSF(I,K)*ELXY(K,J)                          F2D25100
  120        CONTINUE                                                    F2D25110
             DJCB(I,J) = SUM                                             F2D25120
  130        CONTINUE                                                    F2D25130
  140     CONTINUE                                                       F2D25140
C                                                                        F2D25150
          DO 170 K = 1, 3                                                F2D25160
             DO 160 J = 1, NN                                            F2D25170
                SUM = 0.0D0                                              F2D25180
                DO 150 L = 1, 2                                          F2D25190
                   SUM = SUM + DJCB(K,L)*GDSFH(L,J)                      F2D25200
  150           CONTINUE                                                 F2D25210
                DDSJ(K,J) = SUM                                          F2D25220
  160        CONTINUE                                                    F2D25230
  170     CONTINUE                                                       F2D25240
C                                                                        F2D25250
C     Compute the jacobian of the transformation                         F2D25260
C                                                                        F2D25270
          GGJ(1,1)=GJ(1,1)*GJ(1,1)                                       F2D25280
          GGJ(1,2)=GJ(1,2)*GJ(1,2)                                       F2D25290
          GGJ(1,3)=2.0*GJ(1,1)*GJ(1,2)                                   F2D25300
          GGJ(2,1)=GJ(2,1)*GJ(2,1)                                       F2D25310
          GGJ(2,2)=GJ(2,2)*GJ(2,2)                                       F2D25320
          GGJ(2,3)=2.0*GJ(2,1)*GJ(2,2)                                   F2D25330
          GGJ(3,1)=GJ(2,1)*GJ(1,1)                                       F2D25340
          GGJ(3,2)=GJ(2,2)*GJ(1,2)                                       F2D25350
          GGJ(3,3)=GJ(2,1)*GJ(1,2)+GJ(1,1)*GJ(2,2)                       F2D25360
          CALL INVRSE(GGJ,GGINV)                                         F2D25370
C                                                                        F2D25380
          DO 200 I = 1, 3                                                F2D25390
             DO 190 J = 1, NN                                            F2D25400
                SUM = 0.0D0                                              F2D25410
                DO 180 K = 1, 3                                          F2D25420
                   SUM = SUM + GGINV(I,K)*(DDSFH(K,J)-DDSJ(K,J))         F2D25430
  180           CONTINUE                                                 F2D25440
                GDDSFH(I,J) = SUM                                        F2D25450
  190        CONTINUE                                                    F2D25460
  200     CONTINUE                                                       F2D25470
      ENDIF                                                              F2D25480
      RETURN                                                             F2D25490
      END                                                                F2D25500

      SUBROUTINE SHPTRI(NPE,AL1,AL2,AL3,DET,ELXY)                        F2D25510
C                                                                        F2D25520
C     ________________________________________________________________   F2D25530
C     The subroutine computes the  Lagrangian  interpolation functions   F2D25540
C     and their global derivatives at quadrature points for the linear   F2D25550
C     and quadratic (i.e. three-node and six-node) triangular elements   F2D25560
```

```
C     ----------------------------------------------------------------------- F2D25570
C                                                                             F2D25580
      IMPLICIT REAL*8(A-H,O-Z)                                                F2D25590
      COMMON/SHP/SF(9),GDSF(2,9)                                              F2D25600
      DIMENSION DSF(3,9),ELXY(9,2),GJ(2,2),GJINV(2,2)                         F2D25610
C                                                                             F2D25620
C     Initialize the arrays                                                   F2D25630
C                                                                             F2D25640
      DO 10 I = 1, NPE                                                        F2D25650
         DSF(1,I) = 0.0D0                                                     F2D25660
         DSF(2,I) = 0.0D0                                                     F2D25670
         DSF(3,I) = 0.0D0                                                     F2D25680
   10 CONTINUE                                                                F2D25690
C                                                                             F2D25700
      IF(NPE.EQ.3) THEN                                                       F2D25710
C                                                                             F2D25720
C     Linear Lagrane interpolation for three-node element                     F2D25730
C                                                                             F2D25740
   20     SF(1) =  AL1                                                        F2D25750
          SF(2) =  AL2                                                        F2D25760
          SF(3) =  AL3                                                        F2D25770
          DSF(1,1) = 1.0D0                                                    F2D25780
          DSF(2,2) = 1.0D0                                                    F2D25790
          DSF(3,3) = 1.0D0                                                    F2D25800
      ELSE                                                                    F2D25810
C                                                                             F2D25820
C     Quadratic Lagrange interpolation functions for six-nde element          F2D25830
C                                                                             F2D25840
          SF(1) = AL1 * (2.0D0 * AL1 - 1)                                     F2D25850
          SF(2) = AL2 * (2.0D0 * AL2 - 1)                                     F2D25860
          SF(3) = AL3 * (2.0D0 * AL3 - 1)                                     F2D25870
          SF(4) = 4.0D0 * AL1 * AL2                                           F2D25880
          SF(5) = 4.0D0 * AL2 * AL3                                           F2D25890
          SF(6) = 4.0D0 * AL3 * AL1                                           F2D25900
          DSF(1,1) = 4.0D0 * AL1 - 1                                          F2D25910
          DSF(2,2) = 4.0D0 * AL2 - 1                                          F2D25920
          DSF(3,3) = 4.0D0 * AL3 - 1                                          F2D25930
          DSF(1,4) = 4.0D0 * AL2                                              F2D25940
          DSF(2,4) = 4.0D0 * AL1                                              F2D25950
          DSF(2,5) = 4.0D0 * AL3                                              F2D25960
          DSF(3,5) = 4.0D0 * AL2                                              F2D25970
          DSF(1,6) = 4.0D0 * AL3                                              F2D25980
          DSF(3,6) = 4.0D0 * AL1                                              F2D25990
      ENDIF                                                                   F2D26000
C                                                                             F2D26010
C     Compute the global derivatives of SF(I).  Note that the special         F2D26020
C     form of the jacobian for area coordinates,  AL3 = 1-AL1-AL2  is         F2D26030
C     substituted                                                             F2D26040
C                                                                             F2D26050
      DO 60 I = 1,2                                                           F2D26060
      DO 50 J = 1,2                                                           F2D26070
      SUM = 0.0D0                                                             F2D26080
      DO 40 K = 1, NPE                                                        F2D26090
      SUM = SUM + (DSF(I,K) - DSF(3,K))*ELXY(K,J)                             F2D26100
   40 CONTINUE                                                                F2D26110
      GJ(I,J) = SUM                                                           F2D26120
   50 CONTINUE                                                                F2D26130
   60 CONTINUE                                                                F2D26140
C                                                                             F2D26150
      DET = GJ(1,1)*GJ(2,2) - GJ(1,2)*GJ(2,1)                                 F2D26160
      GJINV(1,1) = GJ(2,2)/DET                                                F2D26170
      GJINV(2,2) = GJ(1,1)/DET                                                F2D26180
      GJINV(1,2) = -GJ(1,2)/DET                                               F2D26190
      GJINV(2,1) = -GJ(2,1)/DET                                               F2D26200
      DO 100 I = 1, 2                                                         F2D26210
      DO 90 J = 1, NPE                                                        F2D26220
      SUM = 0.0D0                                                             F2D26230
      DO 80 K = 1, 2                                                          F2D26240
      SUM = SUM + GJINV(I,K) * (DSF(K,J) - DSF(3,J))                          F2D26250
   80 CONTINUE                                                                F2D26260
      GDSF(I,J) = SUM                                                         F2D26270
   90 CONTINUE                                                                F2D26280
  100 CONTINUE                                                                F2D26290
      RETURN                                                                  F2D26300
      END                                                                     F2D26310
```

```
      SUBROUTINE SOLVER(NRM,NCM,NEQNS,NBW,BAND,RHS,IRES)                F2D26320
C     ____________________________________________________________      F2D26330
C                                                                       F2D26340
C     The subroutine solves a  banded,  symmetric,  system of algebraic F2D26350
C     equations using the Gauss elimination method:  [BAND]{U} = {RHS}. F2D26360
C     The coefficient matrix is input as BAND(NEQNS,NBW) and the column F2D26370
C     vector is input as  RHS(NEQNS),  where NEQNS is the actual number F2D26380
C     of equations and NBW is the half band width.  The true dimensions F2D26390
C     of the matrix [BAND] in the calling program, are NRM by NCM. When F2D26400
C     IRES is greater than zero, the right hand elimination is skipped. F2D26410
C     ____________________________________________________________      F2D26420
C                                                                       F2D26430
      IMPLICIT REAL*8(A-H,O-Z)                                          F2D26440
      DIMENSION BAND(NRM,NCM),RHS(NRM)                                  F2D26450
C                                                                       F2D26460
      MEQNS=NEQNS-1                                                     F2D26470
      IF(IRES.LE.0) THEN                                                F2D26480
          DO 30 NPIV=1,MEQNS                                            F2D26490
          NPIVOT=NPIV+1                                                 F2D26500
          LSTSUB=NPIV+NBW-1                                             F2D26510
          IF(LSTSUB.GT.NEQNS) THEN                                      F2D26520
              LSTSUB=NEQNS                                              F2D26530
          ENDIF                                                         F2D26540
C                                                                       F2D26550
          DO 20 NROW=NPIVOT,LSTSUB                                      F2D26560
          NCOL=NROW-NPIV+1                                              F2D26570
          FACTOR=BAND(NPIV,NCOL)/BAND(NPIV,1)                           F2D26580
          DO 10 NCOL=NROW,LSTSUB                                        F2D26590
          ICOL=NCOL-NROW+1                                              F2D26600
          JCOL=NCOL-NPIV+1                                              F2D26610
   10     BAND(NROW,ICOL)=BAND(NROW,ICOL)-FACTOR*BAND(NPIV,JCOL)        F2D26620
   20     RHS(NROW)=RHS(NROW)-FACTOR*RHS(NPIV)                          F2D26630
   30     CONTINUE                                                      F2D26640
      ELSE                                                              F2D26650
   40     DO 60 NPIV=1,MEQNS                                            F2D26660
          NPIVOT=NPIV+1                                                 F2D26670
          LSTSUB=NPIV+NBW-1                                             F2D26680
          IF(LSTSUB.GT.NEQNS) THEN                                      F2D26690
              LSTSUB=NEQNS                                              F2D26700
          ENDIF                                                         F2D26710
          DO 50 NROW=NPIVOT,LSTSUB                                      F2D26720
          NCOL=NROW-NPIV+1                                              F2D26730
          FACTOR=BAND(NPIV,NCOL)/BAND(NPIV,1)                           F2D26740
   50     RHS(NROW)=RHS(NROW)-FACTOR*RHS(NPIV)                          F2D26750
   60     CONTINUE                                                      F2D26760
      ENDIF                                                             F2D26770
C                                                                       F2D26780
C     Back substitution                                                 F2D26790
C                                                                       F2D26800
      DO 90 IJK=2,NEQNS                                                 F2D26810
      NPIV=NEQNS-IJK+2                                                  F2D26820
      RHS(NPIV)=RHS(NPIV)/BAND(NPIV,1)                                  F2D26830
      LSTSUB=NPIV-NBW+1                                                 F2D26840
      IF(LSTSUB.LT.1) THEN                                              F2D26850
          LSTSUB=1                                                      F2D26860
      ENDIF                                                             F2D26870
      NPIVOT=NPIV-1                                                     F2D26880
      DO 80 JKI=LSTSUB,NPIVOT                                           F2D26890
      NROW=NPIVOT-JKI+LSTSUB                                            F2D26900
      NCOL=NPIV-NROW+1                                                  F2D26910
      FACTOR=BAND(NROW,NCOL)                                            F2D26920
   80 RHS(NROW)=RHS(NROW)-FACTOR*RHS(NPIV)                              F2D26930
   90 CONTINUE                                                          F2D26940
      RHS(1)=RHS(1)/BAND(1,1)                                           F2D26950
      RETURN                                                            F2D26960
      END                                                               F2D26970

      SUBROUTINE TIMER(NCOUNT,INTIAL,ITEM,NN)                           F2D26980
C     ____________________________________________________________      F2D26990
C                                                                       F2D27000
C    The  subroutine computes  the algebraic equations associated with  F2D27010
C    the  parabolic and hyperbolic differential equations bu using the  F2D27020
C    alfa-family and Newmark family of approximations, respectively.    F2D27030
C     ____________________________________________________________      F2D27040
```

```
C                                                                       F2D27050
      IMPLICIT REAL*8(A-H,O-Z)                                          F2D27060
      COMMON/STF/ELF(27),ELK(27,27),ELM(27,27),ELXY(9,2),ELU(27),       F2D27070
     1           ELV(27),ELA(27),A1,A2,A3,A4,A5                         F2D27080
C                                                                       F2D27090
      IF(ITEM.EQ.1) THEN                                                F2D27100
C                                                                       F2D27110
C     The alfa-family of time approximation for parabolic equations     F2D27120
C                                                                       F2D27130
          DO 20 I=1,NN                                                  F2D27140
          SUM=0.0                                                       F2D27150
          DO 10 J=1,NN                                                  F2D27160
          SUM=SUM+(ELM(I,J)-A2*ELK(I,J))*ELU(J)                         F2D27170
   10     ELK(I,J)=ELM(I,J)+A1*ELK(I,J)                                 F2D27180
   20     ELF(I)=(A1+A2)*ELF(I)+SUM                                     F2D27190
      ELSE                                                              F2D27200
C                                                                       F2D27210
C     The Newmark intergration scheme for hyperbolic equations          F2D27220
C                                                                       F2D27230
          IF(NCOUNT.EQ.1 .AND. INTIAL.NE.0) THEN                        F2D27240
              DO 40 I = 1,NN                                            F2D27250
              ELF(I)  = 0.0                                             F2D27260
              DO 40 J = 1,NN                                            F2D27270
              ELF(I)  = ELF(I)-ELK(I,J)*ELU(J)                          F2D27280
   40         ELK(I,J)= ELM(I,J)                                        F2D27290
          ELSE                                                          F2D27300
              DO 70 I = 1,NN                                            F2D27310
              SUM     = 0.0                                             F2D27320
              DO 60 J = 1,NN                                            F2D27330
              SUM     = SUM+ELM(I,J)*(A3*ELU(J)+A4*ELV(J)+A5*ELA(J))    F2D27340
   60         ELK(I,J)= ELK(I,J)+A3*ELM(I,J)                            F2D27350
   70         ELF(I)  = ELF(I)+SUM                                      F2D27360
          ENDIF                                                         F2D27370
      ENDIF                                                             F2D27380
      RETURN                                                            F2D27390
      END                                                               F2D27400
```

INDEX